Lecture Notes on Data Engineering and Communications Technologies

Volume 132

Series Editor

The aim of the book series is to present cutting edge engineering approaches to data technologies and communications. It will publish latest advances on the engineering task of building and deploying distributed, scalable and reliable data infrastructures and communication systems.

The series will have a prominent applied focus on data technologies and communications with aim to promote the bridging from fundamental research on data science and networking to data engineering and communications that lead to industry products, business knowledge and standardisation.

Indexed by SCOPUS, INSPEC, EI Compendex.

All books published in the series are submitted for consideration in Web of Science.

More information about this series at https://link.springer.com/bookseries/15362

Vaclav Skala · T. P. Singh · Tanupriya Choudhury ·
Ravi Tomar · Md. Abul Bashar
Editors

Machine Intelligence and Data Science Applications

Proceedings of MIDAS 2021

Springer

Editors
Vaclav Skala
Department of Computer Science
and Engineering
Faculty of Applied Sciences
University of West Bohemia
Plzeň, Czech Republic

Tanupriya Choudhury
Department of Informatics
University of Petroleum and Energy Studies
Dehradun, India

Md. Abul Bashar
Department of Computer Science
and Engineering
Comilla University
Kotbari, Bangladesh

T. P. Singh
Department of Informatics
University of Petroleum and Energy Studies
Dehradun, India

Ravi Tomar
Department of Informatics
University of Petroleum and Energy Studies
Dehradun, India

ISSN 2367-4512 ISSN 2367-4520 (electronic)
Lecture Notes on Data Engineering and Communications Technologies
ISBN 978-981-19-2346-3 ISBN 978-981-19-2347-0 (eBook)
https://doi.org/10.1007/978-981-19-2347-0

This Springer imprint is published by the registered company Springer Nature Singapore Pte Ltd.
The registered company address is: 152 Beach Road, #21-01/04 Gateway East, Singapore 189721, Singapore

Preface

MIDAS 2021 aims to promote and provide a platform for researchers, academia, and practitioners to meet and exchange ideas on recent theoretical and applied machine and artificial intelligence and data sciences research. The conference provides a platform for the computer science, computer engineering, and information technology professionals, scientists, researchers, engineers, educators, and students from all over the world to share their scientific contributions, ideas, and views. We feel glad to welcome all readers to the International Conference MIDAS 2021 at Comilla University, Bangladesh, being organized in association with Nikhil Bharat Siksha Parisad (NBSP-India), South Asian Chamber of Scientific, Research and Development, International Association of Professional and Fellow Engineers, USA. This time, due to COVID-19 pandemic conditions, the conference is taking place in virtual space at Comilla University, Bangladesh, but interestingly with a larger participation than ever. The theme of the conference was very thoughtfully decided as machine intelligence and data sciences. With increasing interest in building more and more sophisticated systems on the wheels of tremendous data generation and its processing capabilities, the topic appears to be most suitable to the current research inclination of the academic fraternity. The type of research submissions received further strengthens the suitability of the theme of the conference. Again a great pleasure to share that the researchers from 17 different countries around the continents submitted their research contributions in the form of articles and thus making the conference a wonderful platform for exchange of ideas and further strengthening the collaborations among the interested people in the research fraternity. This volume is a compilation of the chapters of various presentations presented in conference with the aim to be a memoir to the event. The theme of the International Conference is Machine Intelligence and Applications and it comprises of four tracks. Track 1 addresses the algorithmic aspect of machine intelligence, while Track 2 includes the framework and optimization of various algorithms. Track 3 includes all the papers related to wide applications in various fields, and the book volume may end with

Track 4 which will include interdisciplinary applications. We truly believe that the book will fit as a good read for those looking forward to exploring areas of machine learning and its applications.

Plzeň, Czech Republic	Vaclav Skala
Dehradun, India	T. P. Singh
Dehradun, India	Tanupriya Choudhury
Dehradun, India	Ravi Tomar
Kotbari, Bangladesh	Md. Abul Bashar

Contents

Rise of Blockchain-Based Non-fungible Tokens (NFTs): Overview, Trends, and Future Prospects 1
Harsh Vardhan Singh Rawat, Diksha Bisht, Sandeep Kumar, and Sarishma Dangi

An Efficient Data Preparation Strategy for Sentiment Analysis with Associative Database 11
Dipto Biswas, Md. Samsuddoha, and Partha Chakraborty

A Comparison of Traditional and Ensemble Machine Learning Approaches for Parkinson's Disease Classification 25
Kevin Sabu, Maddula Ramnath, Ankur Choudhary, Gaurav Raj, and Arun Prakash Agrawal

Reducing Error Rate for Eye-Tracking System by Applying SVM 35
Nafiz Ishtiaque Ahmed and Fatema Nasrin

Eye-Gaze-Controlled Wheelchair System with Virtual Keyboard for Disabled Person Using Raspberry Pi 49
Partha Chakraborty, Md. Mofizul Alam Mozumder, and Md. Saif Hasan

SATLabel: A Framework for Sentiment and Aspect Terms Based Automatic Topic Labelling 63
Khandaker Tayef Shahriar, Mohammad Ali Moni, Mohammed Moshiul Hoque, Muhammad Nazrul Islam, and Iqbal H. Sarker

Deriving Soft Biometric Feature from Facial Images 77
Mazida A. Ahmed, Ridip D. Choudhury, Vaskar Deka, Manash P. Bhuyan, and Parvez A. Boruah

Retinal Disease Classification from Retinal-OCT Images Using Deep Learning Methods 95
Archana Naik, B. S. Pavana, and Kavitha Sooda

Stock Recommendation BOT for Swing Trading and Long-Term Investments in Indian Stock Markets 105
Samarth Patgaonkar, Sneha Dharamsi, Ayush Jain, and Nimesh Marfatia

Issues in Machine Translation—A Case Study of the Kashmiri Language 117
Nawaz Ali Lone, Kaiser J. Giri, and Rumaan Bashir

Bangladeshi Land Cover Change Detection with Satelite Image Using GIS Techniques 125
Kazi Atai Kibria, Al Akram Chowdhury, Abu Saleh Musa Miah, Md. Ragib Shahriar, Shahanaz Pervin, Jungpil Shin, Md. Mamunur Rashid, and Atiquer Rahman Sarkar

Genetic Algorithm-Based Optimal Deep Neural Network for Detecting Network Intrusions 145
Sourav Adhikary, Md. Musfique Anwar, Mohammad Jabed Morshed Chowdhury, and Iqbal H. Sarker

Deep Convolutional Neural Network-Based Bangla Sign Language Detection on a Novel Dataset 157
Md. Jahid Hasan, S. K. Nahid Hasan, and Kazi Saeed Alam

Avyanna: A Website for Women's Safety 169
Riya Sil, Avijoy Das Bhowmick, Bharat Bhushan, and Ayan Sikdar

Web Page Classification Based on Novel Black Widow Meta-Heuristic Optimization with Deep Learning Technique 177
V. Gokula Krishnan, J. Deepa, Pinagadi Venkateswara Rao, and V. Divya

A Systematic Study on Network Attacks and Intrusion Detection System 195
Milan Samantaray, Suneeta Satapathy, and Arundhati Lenka

Achieving Data Privacy Using Extended NMF 211
Neetika Bhandari and Payal Pahwa

Challenges of Robotics: A Quest for an Integrated Action 227
Md. Toriqul Islam, Ridoan Karim, and Sonali Vyas

Developing an Integrated Hybrid App to Reduce Overproduction and Waiting Time Using Kanban Board 239
Akansha Yadav and Girija Jha

Implementation of Gamified Navigation and Location Mapping Using Augmented Reality 265
R. Janarthanan, A. Annapoorani, S. Abhilash, and P. Dinesh

A Survey on Autonomous Vehicles 277
Md. Hashim and Pooja Dehraj

Impact of Innovative Technology on Quality Education: The Resourceful Intelligence for Smart Society Development 293
Suplab Kanti Podder, Benny Thomas, and Debabrata Samanta

A Review on Virtual Reality and Applications for Next Generation Systems and Society 5.0 303
Dishant Khosla, Sarwan Singh, Manvinder Sharma, Ayush Sharma, Gaurav Bharti, and Geetendra Rajput

Facial Recognition with Computer Vision 313
Vishesh Jindal, Shailendra Narayan Singh, and Soumya Suvra Khan

Community Strength Analysis in Social Network Using Cloud Computing 331
A. Rohini, T. Sudalai Muthu, Tanupriya Choudhury, and S. Visalaxi

Object Detection on Dental X-ray Images Using Region-Based Convolutional Neural Networks 341
Rakib Hossen, Minhazul Arefin, and Mohammed Nasir Uddin

Real-Time and Context-Based Approach to Provide Users with Recommendations for Scaling of Cloud Resources Using Machine Learning and Recommendation Systems 355
Atishay Jain

Traffic Density Estimation Using Transfer Learning with Pre-trained InceptionResNetV2 Network 363
Md. Nafis Tahmid Akhand, Sunanda Das, and Mahmudul Hasan

A Mobile Application for Sales Representatives: A Case Study of a Liquor Brand 377
Sharon Xavier, Saroj Kumar Panigrahy, and Asish Kumar Dalai

Wireless Sensor Networks (WSNs): Toward an Energy-Efficient Routing Protocol Design 389
Nitish Pathak, Neelam Sharma, and Harshita Chadha

Empirical Analysis of Machine Learning and Deep Learning Techniques for COVID-19 Detection Using Chest X-rays 399
Vittesha Gupta and Arunima Jaiswal

Comparative Analysis of Machine Learning and Deep Learning Algorithms for Skin Cancer Detection 409
Nikita Thakur and Arunima Jaiswal

Investigating Efficacy of Transfer Learning for Fruit Classification 419
Vikas Khullar, Raj Gaurang Tiwari, Ambuj Kumar Agarwal, and Alok Misra

BERT-Based Secure and Smart Management System for Processing Software Development Requirements from Security Perspective 427
Raghavendra Rao Althar and Debabrata Samanta

Hyper Parameter Optimization Technique for Network Intrusion Detection System Using Machine Learning Algorithms 441
M. Swarnamalya, C. K. Raghavendra, and M. Seshamalini

Energy Efficient VM Consolidation Technique in Cloud Computing Using Cat Swarm Optimization 457
Sudheer Mangalampalli, Kiran Sree Pokkuluri, Pothuraju Raju, P. J. R. Shalem Raju, S. S. S. N. Usha Devi N, and Vamsi Krishna Mangalampalli

A Forecast of Geohazard and Factors Influencing Geohazard Using Transfer Learning 469
S. Visalaxi, T. Sudalaimuthu, Tanupriya Choudhury, and A. Rohini

Application and Uses of Big Data Analytics in Different Domain 481
Abhineet Anand, Naresh Kumar Trivedi, Md Abdul Wassay, Yousef AlSaud, and Shikha Maheshwari

Pedestrian Detection with Anchor-Free and FPN Enhanced Deep Learning Approach 501
J. Sangeetha, P. Rajendiran, and Hariraj Venkatesan

Efficient Machine Learning Approaches to Detect Fake News of Covid-19 513
Shagoto Rahman, M. Raihan, Kamrul Hasan Talukder, Sabia Khatun Mithila, Md. Mehedi Hassan, Laboni Akter, and Md. Mohsin Sarker Raihan

Application of Digital Technology in Accounting Profession for Achieving Business Goals and Sustainable Development 527
V. Sukthankar Sitaram, Sukanta Kumar Baral, Ramesh Chandra Rath, and Richa Goel

Application of Robotics in the Healthcare Industry 539
Vishesh Jindal, Shailendra Narayan Singh, and Soumya Suvra Khan

Context-Driven Method for Smarter and Connected Traffic Lights Using Machine Learning with the Edge Servers 551
Parminder Singh Sethi, Atishay Jain, Sandeep Kumar, and Ravi Tomar

Forecasting of COVID-19 Trends in Bangladesh Using Machine Learning Approaches 561
Chayti Saha, Fozilatunnesa Masuma, Nayan Banik, and Partha Chakraborty

A Comparative Study of Machine Learning Algorithms to Detect Cardiovascular Disease with Feature Selection Method 573
Md. Jubier Ali, Badhan Chandra Das, Suman Saha, Al Amin Biswas, and Partha Chakraborty

A Pilot Study for Devanagari Script Character Recognition Using Deep Learning Models 587
Ayush Sharma, Mahendra Soni, Chandra Prakash, Gaurav Raj, Ankur Choudhary, and Arun Prakash Agrawal

COVID-19 in Bangladesh: An Exploratory Data Analysis and Prediction of Neurological Syndrome Using Machine Learning Algorithms Based on Comorbidity 595
Shuvo Chandra Das, Aditi Sarker, Sourav Saha, and Partha Chakraborty

Ransomware Family Classification with Ensemble Model Based on Behavior Analysis 609
Nowshin Tasnim, Khandaker Tayef Shahriar, Hamed Alqahtani, and Iqbal H. Sarker

UCSP: A Framework to Tackle the Challenge of Dependency Chain in Cloud Forensics 621
Prajwal Bhardwaj, Kaustubh Lohani, Navtej Singh, Vivudh Fore, and Ravi Tomar

DeshiFoodBD: Development of a Bangladeshi Traditional Food Image Dataset and Recognition Model Using *Inception V3* 639
Samrat Kumar Dey, Lubana Akter, Dola Saha, Mshura Akter, and Md. Mahbubur Rahman

Sentiment Analysis of E-commerce Consumer Based on Product Delivery Time Using Machine Learning 649
Hasnur Jahan, Abu Kowshir Bitto, Md. Shohel Arman, Imran Mahmud, Shah Fahad Hossain, Rakhi Moni Saha, and Md. Mahfuj Hasan Shohug

Applications of Artificial Intelligence in IT Disaster Recovery 663
Kaustubh Lohani, Prajwal Bhardwaj, Aryaman Atrey, Sandeep Kumar, and Ravi Tomar

Mushroom Classification Using MI, PCA, and MIPCA Techniques 679
Sunil Kaushik and Tanupriya Choudhury

Factors Influencing University Students' E-Learning Adoption in Bangladesh During COVID-19: An Empirical Study with Machine Learning 695
Rakib Ahmed Saleh, Md. Tariqul Islam, and Rozi Nor Haizan Nor

Nano Rover: A Multi-sensory Full-Functional Surveillance Robot with Modified Inception-Net ... 707
Sheekar Banerjee, Aminun Nahar Jhumur, and Md. Ezharul Islam

Bangla Handwritten Character Recognition Using Convolutional Neural Network ... 721
Partha Chakraborty, Afroza Islam, Mohammad Abu Yousuf, Ritu Agarwal, and Tanupriya Choudhury

Music Genre Classification Using Light Gradient Boosting Machine: A Pilot Study ... 733
Akhil Sibi, Rahul Singh, Kumar Anurag, Ankur Choudhary, Arun Prakash Agrawal, and Gaurav Raj

HRPro: A Machine Learning Approach for Recruitment Process Automation ... 749
Atri Biswas, Shourjya Chakraborty, Debarshi Deb, and Rajdeep Chatterjee

A Deep Convolutional Generative Adversarial Network-Based Model to Analyze Histopathological Breast Cancer Images ... 761
Tanzina Akter Tani, Mir Moynuddin Ahmed Shibly, and Shamim Ripon

What Drives Adoption of Cloud-Based Online Games in an Emerging Market? An Investigation Using Flow Theory ... 775
Ashok S. Malhi, Raj K. Kovid, Abhisek Dutta, and Rajeev Sijariya

Method to Recommend the Investment Portfolio with the Mix of Various Schemes and Options Using Reinforcement Learning ... 789
Parminder Singh Sethi, Vasanth Sathyanaryanan, Nalam Lakshmi, Ravi Tomar, and Vivudh Fore

Deep Learning Approach for Electricity Load Forecasting Using Multivariate Time Series Data ... 805
Shishir Zaman, Md. Nayeem, Rifah Tatrapi, and Shamim Ripon

Social Media Role in Pricing Value of Cryptocurrency ... 819
Anju Mishra, Vedansh Gupta, Sandeep Srivastava, Arvind K. Pandey, Lalan Kumar, and Tanupriya Choudhury

Twitter Spatio-temporal Topic Dynamics and Sentiment Analysis During the First COVID-19 Lockdown in India ... 831
Arunkumar Dhandapani, Anandkumar Balasubramaniam, Thirunavukarasu Balasubramaniam, and Anand Paul

Natural Language Interface in Dogri to Database ... 843
Shubhnandan S. Jamwal and Vijay Singh Sen

The Mutation Study of Social Media in Student's Life ... 853
Anju Mishra, Nikhil Kumar Varun, Sandeep Srivastava, Arvind K. Pandey, Lalan Kumar, and Tanupriya Choudhury

GPS-Based Route Choice Model for Smart Transportation System: Bringing Intelligence into Vehicular Cloud ... 865
Sirisha Potluri, Sachi Nandan Mohanty, Katta Subba Rao, and Tanupriya Choudhury

Social Listening on Budget—A Study of Sentimental Analysis and Prediction of Sentiments Using Text Analytics & Predictive Algorithms ... 879
A. Mansurali, P. Mary Jayanthi, R. Swamynathan, and Tanupriya Choudhury

An Application of Internet of Things for Cybersecurity and Control: Emerging Needs and Challenges ... 893
Sukanta Kumar Baral, Ramesh Chandra Rath, Richa Goel, and Tilottama Singh

Comparative Performance Evaluation of Supervised Classification Models on Large Static Malware Dataset ... 905
Kaniz Tasmim, Tamanna Akter, Nayan Banik, and Partha Chakraborty

Author Index ... 919

About the Editors

Prof. Vaclav Skala is Full professor of Computer Science at the University of West Bohemia, Plzen, and VSB-Technical University Ostrava, Czech Republic. He received his ING. (equivalent of M.Sc.) degree in 1975 from the Institute of Technology in Plzen and C.Sc. (equivalent of Ph.D.) degree from the Czech Technical University in Prague in 1981. In 1996, he became Full Professor in Computer Science. In 1997, the Centre of Computer Graphics and Visualization (CCGV) was formally established, and since then, he is the head of the CCGV in Plzen. Professor Vaclav Skala is Associate Editor of *The Visual Computer* (Springer), *Computers and Graphics* (Elsevier), Member of the Editorial Board of *Machine Graphics and Vision* (Polish Academy of Sciences) and Editor in Chief of the *Journal of WSCG*. He is Member of international programme committees of prestigious conferences and workshops. He is Member of ACM SIGGRAPH, IEEE and Eurographics Association. Professor Vaclav Skala has published over 200 research papers at conferences and research journals. His current research interests are computer graphics and visualization, mathematics, especially geometrical algebra, algorithms and data structures.

Dr. T. P. Singh is currently positioned as Professor and Head of Department of Department at School of Computer Science, University of Petroleum & Energy Studies, Dehradun, UK, India. He holds Doctorate in Computer Science from Jamia Millia Islamia University, New Delhi. He carries 25 years of rich experience with him. He has been associated with Tata Group and Sharda University, Greater Noida, NCR, India. His research interests include machine intelligence, pattern recognition and development of hybrid intelligent systems. He has guided 15 Master's theses. Currently, five research scholars are working towards their Doctoral degree under him. There are more than dozens of publications to his credit in various national and international journals. Dr. Singh is Senior Member of various professional bodies including IEEE, ISTE, IAENG, ACM, etc., and also on the editorial/reviewer panel of different journals.

Dr. Tanupriya Choudhury received his Bachelor's degree in CSE from West Bengal University of Technology, Kolkata, India, and Master's degree in CSE from Dr. M. G. R. University, Chennai, India. He has received his Ph.D. degree in the year 2016. He has ten years of experience in teaching as well as in research. Currently, he is working as Associate Professor in dept. of CSE at UPES, Dehradun. Recently, he has received Global Outreach Education Award for Excellence in Best Young Researcher Award in GOECA 2018. His areas of interests include human computing, soft computing, cloud computing, data mining, etc. He has filed 14 patents till date and received 16 copyrights from MHRD for his own software. He has been associated with many conferences in India and abroad. He has authored more than 230 research papers till date. He has delivered invited talk and guest lecture in Jamia Millia Islamia University, Maharaja Agersen College of Delhi University, Duy Tan University, Vietnam, etc. He has been associated with many conferences throughout India as TPC Member and Session Chair, etc. He is Lifetime Member of IETA, Senior Member of IEEE and Member of IET (UK) and other renowned technical societies. He is associated with corporate, and he is Technical Adviser of Deetya Soft Pvt. Ltd. Noida, IVRGURU and Mydigital360, etc. He is holding the post of Secretary in Indian Engineering Teacher's Association-India (IETA), and he is also holding Advisor position in INDO-UK Confederation of Science, Technology and Research Ltd., London, UK, and International Association of Professional and Fellow Engineers-Delaware-USA.

Dr. Ravi Tomar is currently working as Associate Professor in the School of Computer Science at the University of Petroleum and Energy Studies, Dehradun, India. He is an experienced academician with a demonstrated history of working in the higher education industry. Skilled in Programming, Computer Networking, Stream processing, Python, Oracle Database, C++, Core Java, J2EE, RPA and CorDApp. His research interests include Wireless Sensor Networks, Image Processing, Data Mining and Warehousing, Computer Networks, big data technologies and VANET. He has authored 51+ papers in different research areas, filled four Indian patent, edited five books and have authored four books. He has delivered Training to corporates nationally and internationally on Confluent Apache Kafka, Stream Processing, RPA, CordaApp, J2EE and IoT to clients like KeyBank, Accenture, Union Bank of Philippines, Ernst and Young and Deloitte. Dr. Tomar is officially recognized as Instructor for Confluent and CordApp. He has conducted various International conferences in India, France and Nepal. He has been awarded a young researcher in Computer Science and Engineering by RedInno, India in 2018, Academic Excellence and Research Excellence Award by UPES in 2021 and Young Scientist Award by UCOST, Dehradun.

Dr. Md. Abul Bashar is Expert in data mining, machine learning, deep learning and artificial intelligence. He has more than eight years of research and development experiences in these areas. He is currently working as Data Science Postdoctoral Research Fellow at the School of Computer Science and Associate Investigator at Centre for Data Science, QUT. He worked on projects that conducted impactful theoretical research and real-life applications. He worked on several projects funded by the Australian Research Council, government and private organizations. These projects made a practical impact in the data science community and industry. Some outcomes of these projects were published in high-quality journals and conferences, recognized by many national and international mainstream news media and others benefited industry partners. Dr. Bashar received his Ph.D. degree in Data Science from QUT and B.Sc. (Engg.) degree from Shah Jalal University of Science and Technology (SUST), Sylhet, Bangladesh.

Rise of Blockchain-Based Non-fungible Tokens (NFTs): Overview, Trends, and Future Prospects

Harsh Vardhan Singh Rawat, Diksha Bisht, Sandeep Kumar, and Sarishma Dangi

Abstract Non-fungible tokens (NFTs) are blockchain-based unique, transferrable digital assets. NFTs have recently gained popularity due to their inherent ability of being traded and re-traded over the public blockchain. The financial market witnessed a huge spike in NFT trading in 2021. However, the development and research work in the area of NFTs is still in its early nascent stages. Therefore, in this work, we aim to discuss the working of NFT including its technical components, protocols, security, etc. We then discuss the key benefits as well as challenges facing the NFTs. Lastly, we provide a critical trend analysis to predict where this industry is headed and how it will bloom in upcoming years.

Keywords Blockchain · Non-fungible tokens

1 Introduction

Blockchain is a decentralized and distributed network of ledger that is shared among the other nodes of that network. Blockchain ensures the user with unmatched security of the recorded data due to its fault-tolerance which comes from it being democratic unlike the issues faced in centralized cloud services [1]. It can authenticate and perform transactions between peers securely without the need for a third party. Blockchain is either public or private. Private blockchain has been used for creating solutions for specific organizations and their issue. Public blockchain has a crucial role in the development of cryptocurrency and assets in real-time, that can be evaluated and traded, giving a huge rise to finance technologies as well. Due to its unique properties such as immutability, decentralization, distributedness, and security, blockchain is a technology that provides the foundation for the development of

H. Vardhan Singh Rawat · D. Bisht · S. Dangi (✉)
Graphic Era Deemed to be University, Dehradun, Uttarakhand, India
e-mail: sarishmasingh@gmail.com; sarishma.cse@geu.ac.in

S. Kumar
IIMT College of Engineering, Noida, Uttar Pradesh, India

V. Skala et al. (eds.), *Machine Intelligence and Data Science Applications*, Lecture Notes on Data Engineering and Communications Technologies 132,
https://doi.org/10.1007/978-981-19-2347-0_1

numerous real-time solutions in diverse industries such as healthcare [2], business [3], IoT [4], financial services [5], forensic investigation [6], and many others.

An NFT is a blockchain-based data unit that confirms a digital asset's uniqueness and thus non-interchangeability, as well as providing a unique digital certificate of ownership for the NFT. A non-fungible token (NFTs) is nothing more than a one-of-a-kind digital asset. Bitcoins are fungible, which means that they are all the same and may be used interchangeably. A work of art is an example of a non-fungible token which means one can have two identical pieces of digital art, yet each one is completely different. In 2012, Colored Coins could be said to be the first NFTs ever created. Colored Coins are made up of minuscule bitcoin denominations, sometimes as small as a single satoshi, the smallest unit of bitcoin. Colored Coins can be used to represent a variety of assets and have a variety of applications, including property, coupons, the ability to create your own cryptocurrency, corporate shares, subscriptions, access tokens, and digital collectibles [7]. Colored Coins represented a significant advancement in Bitcoin's capabilities, but their drawback was that they could only signify specific values if everyone agreed on their value.

A quite noticeable amount of individuals have come to the realization of the enormous potential for issuing the assets onto public blockchain after the emergence of Colored Coins. Counterparty, a peer-to-peer financial platform and distributed, open-source Internet protocol was built on the top of the Bitcoin blockchain, was discovered by Robert Dermody, Adam Krellenstein, and Evan Wagner in 2014. Counterparty granted the permission for asset minting and even had a crypto coin with the ticker XCP. Spells of Genesis raised funds for development by establishing BitCrystals, a cryptocurrency that served as the in-game currency. As Ethereum became more popular, two" creative technologists" decided to launch their own NFT project. John Watkinson and Matt Hall used the Ethereum blockchain to produce unique characters. The number of characters would be limited to 10,000 and no two numbers would be identical. Their initiative was dubbed Cryptopunks in honor of the Cypherpunks who experimented with Bitcoin predecessors in the 1990s.

The NFT ecosystem has witnessed significant expansion in 2018 and 2019. There are already over a hundred projects in the space, with more in the works. NFT marketplaces are prospering, with OpenSea leading the way and SuperRare gaining traction. Although the transaction volumes are minor compared to other crypto marketplaces, they are rapidly rising and have made great strides. As more people and businesses discover the impact that NFTs can have and apply them; the growth of the NFT ecosystem thereby increases and multiply in the future. In this work, we aim to cover the working of NFT and their key benefits as well challenges.

The key contributions of this work are outlined as follows:

- To provide an overview of Non-Fungible Tokens and their work in the digital asset management.
- To discuss the rise of NFTs during the Covid-19 pandemic era.
- To identify the benefits and challenges of adopting NFTs as a form of token in digital asset management.
- To identify the key factors which have boosted the rise of NFTs.

- To provide a discussion on future prospects of NFTs.

The rest of this paper is categorized as: Sect. 2 outlines the related work in this area. Section 3 provides an overview of Non-Fungible Tokens and their working. Section 4 gives the comprehensive review analyzing trends in this field. Section 5 discusses the various benefits which are provided by NFTs. Section 6 includes various challenges faced by NFTs. Section 7 provides the future prospects of the NFTs and outlines research directions in the near future. Lastly, we conclude our work.

2 Related Work

The history of non-fungible tokens is much longer, however, NFTs have emerged over the past few years and not much research work has been done in this field. In 2021, Foteini Valeontin et al. [8] highlighted the current state of affairs with respect to NFTs and the cultural heritage sector, emphasizing obstacles while highlighting revenue-generating prospects, in order to assist galleries and museums in addressing their ever-increasing financial challenges. In the same year, Qin Wang et al. [9] examined the NFT ecosystems from a variety of perspectives, giving an overview of current NFT solutions, as well as their technological components, protocols, standards, and desirable properties. It also discussed the evolution of security, including the viewpoints of their design models, as well as opportunities and difficulties. With the use of a blockchain-based image matching game, Akash arora et al. [10] examined the technical underpinnings of the blockchain and cryptocurrencies, notably non-fungible tokens or"crypto-collectibles". Ramakrishnan Raman and Benson Edwin Raj [11] discussed how NFTs will have a greater impact on digital transactions in the future. The technical elements, security impacts, and successful implementations of NFTs in many sectors are also covered in this chapter. The prerequisites for presenting intellectual property assets, notably patents, as NFTs were investigated by Seyed Mojtaba Hosseini Bamakan [12]. It provided a layered conceptual NFT-based patent structure with detailed explanations of each layer, including storage, decentralized authentication, decentralized verification, blockchain, and application layer. Raeesah Chohana and Jeannette Paschenb [13] use a modified AIDA (awareness, desire, action, and repeated action) hierarchy to explain NFTs in basic terms and analyze the marketing ramifications. These implications can provide marketing managers and executives with guidance on how to encourage consumers to buy NFTs based on their unique properties including scarcity, non-fungibility, confirmed authenticity, evidence of ownership, royalties, and direct distribution infrastructure. Most of the research work has been done in the recent times only and this area will receive increased attention in the near future.

3 Non-fungible Tokens

Non-Fungible Tokens are pieces of distinctive data and metadata that stays unalterable. They can represent digital as well real-world assets. NFTs are uniquely identical and cannot be at parity with other NFTs, unlike cryptocurrency. They are authenticated by the ledger that is maintained on the blockchain network through its unique code. Some of the properties of NFTs are discussed as follows:

- Verifiability: All the processes including buying, selling, ownership, etc. can be verified publicly using the public blockchain. Ease of verification provides trust and validity to the transactions and the concerned parties.
- Availability: NFTs can be traded at any place and time, it does not require any particular place or time compared to real-world assets.
- Tamper-proof: It is not possible to tamper with the data of an NFT as it is based on the underlying blockchain technology which is inherently tamper-proof by its nature.
- Usability: NFT is instantly updated which states that it will always provide the latest information of its ownership. Thus, it provides an ease of conducting business and transactions in minimum possible time.
- Atomicity: NFTs transfer is done in one atomic, consistent, isolated, and durable (ACID) transaction.
- Tradability: Every NFT can be easily traded, irrespective of the value of cryptocurrency.

NFTs can be created on many blockchain networks but currently, ethereum's network has the best community support. Minting of an NFT costs cryptocurrency thus requires a crypto wallet that supports the cryptocurrency of the network it is going to be minted upon. Today, we have various platforms to create NFTs such as OpenSea and Rarible where creating an NFT can even cost you free. After minting the NFTs can be preserved or listed on any specific marketplace for auction or selling.

Before minting the attributes need to be specified along with assets, minting is done by execution of code in smart contracts made in solidity for etherium that verifies the ERC-721 standards which define the NFT standards, a smart contract allots unique ID to the NFT. NFTs can modify information such as the metadata, lock-time, date, supply, or other different kinds of attributes. After the creation and validation, a transaction ID is produced when the information is finally registered on the blockchain network, transaction is performed successfully.

The holder of the unique piece of hex values signed by the creator can only claim over NFT-based intellectual property. To transfer NFTs, the owner must prove in possession of the corresponding private key and send the assets to another address (es) with a correct digital signature. This simple operation is usually performed using a cryptocurrency wallet and is represented as sending a transaction to involve smart contracts in the ERC-777 standard.

Today, The NFT and its understanding are in their starting phase but coming to recognition for its efficiency, uniqueness, and proof of ownership. Business models are creating NFTs for better asset allocation, Govt. for land registry systems, Artists for tokenizing their art, etc. On the other hand, there is a rising marketplace for various types of NFTs such as characters, music, images, etc. created by artists of all kinds playing a key role in growth of the blockchain community.

4 Analyzing Trends

The marketplace has been welcoming for the NFTs from the recreational industry to the administrative authorities. The awareness is rising about the NFTs in the market really fast. Billions of dollars have been spent on NFTs already, looking at the market trends no particular region has any kind of monopoly.

NFT Market is growing much exponentially, in 2021 average NFT sales were between 10 and 20 Million $ per week, with abnormal hikes up to 200 Million $ per week whereas in years before it could not even touch that mark.

Currently, NFT market is flooded with a lot of gaming, metaverse, sports, collectible, and art content in major where the major providers are: NBA Top Shot, Cryptopunks, Hashmasks, Sorare, CryptoKitties, Decentraland, Rarible, OpenSea, SuperRare, etc.

Figure 1 represents the search interest of consumers in the NFT in different countries around the world. The interest values are based on the comparative percentage values. Figure 2 represents the distribution of the NFT market among the different fields of interest and their market presence according to the estimated value. Table 1

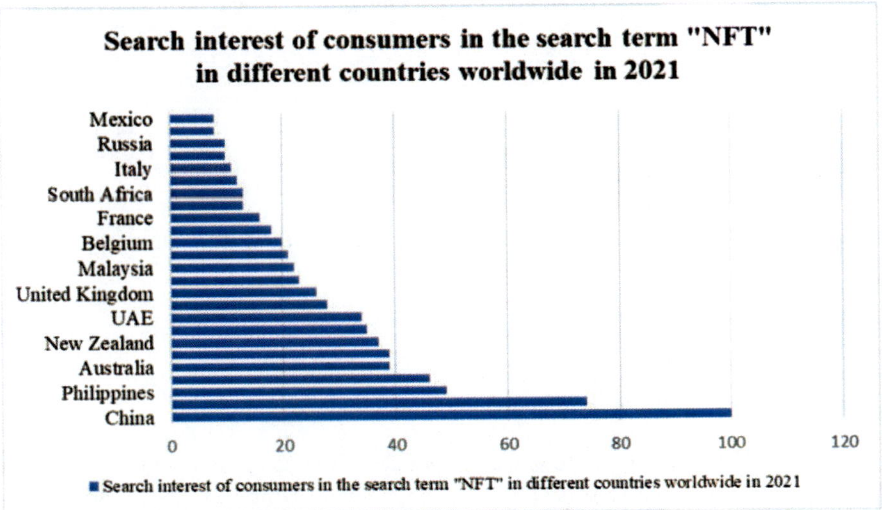

Fig. 1 Search interest of consumers in the search term"NFT" in different countries worldwide in 2021

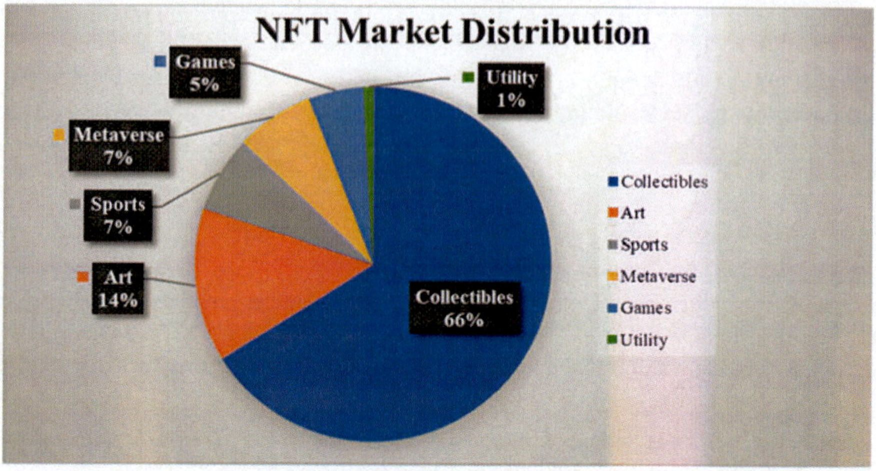

Fig. 2 NFT market distribution

Table 1 Top ten markets for NFTs

Market	Average price	Traders	Volume
OpenSea	$938.99	1,387,357	$14.68B
Axie infinity	$216.51	1,624,169	$3.94B
CryptoPunks	$123.69 k	5600	$2.4B
NBA top shot	$63.63	492,039	$776.49 M
Solanart	$1.1 k	170,703	$593.46 M
Mobox	$790.81	59,178	$530.85 M
Magic eden	$316.46	226,454	$380.62 M
AtomicMarket	$24.98	895,199	$323.09 M
Rarible	$990.79	92,115	$277.9 M
SuperRare.co	$7.94 k	5,493	$212.3 M

represents the top ten markets for NFTs creation/buying/selling with their Average Price, Traders, and Volume [14].

5 Benefits

The market for NFTs in public and even private domains is tremendously evolving and the key reasons why NFTs are so trustable and growing fast are following:

- Proof of authority: NFTs at any point of time is live and transparent about the most recent activity of trading conducted, stating their current ownership status.

The ownership of an NFT cannot be mimicked as the NFT to hold its own "proof of authority" at any interval of time, NFT can be verified easily.

- Truly unique: NFTs have a specific ID that identifies them at any point of interval, each NFT is isolated from others and can easily be differently identified. The NFTs due to their unique identification ensure authenticity as no other NFTs can point to the same asset. Neither the ID can be duplicated nor tampered with.
- Indivisible: Any Non-Fungible Token is indivisible and can only stay in existence as an individual token, unlike the cryptocurrencies which act as a fungible tokens and are divisible. This states that the NFTs cannot be used for transaction at parity and are independent of the cryptocurrency market, unlike fungible tokens.
- Unshareable: NFTs are individual assets and cannot have multiple owners at any instance of time, unlike fungible tokens which are traded at parity and used for digital transactions just the way fiat is used in the physical world.
- Unalterable: NFTs records of ownership cannot be modified as the information is maintained at all times in the blockchain ledger. All transactions or changes that occur with NFTs ownership are maintained at multiple decentralized nodes proving it to be fault-tolerant, fake NFT records can neither be generated nor be tampered with.
- Duplicity resistant: The NFTs cannot be imitated as they have a uniquely identifiable ID that cannot be repeated. At any instance there will always be only one authentic NFT that will represent any particular authentic asset, there can exist no other NFT at the same instance that can identify the same asset.
- Transferable: NFTs are easily transferable and can be bought and sold multiple times. NFTs are quite easy to transfer from one owner to another with the help of smart contracts that are executed in the background. The owner is verified through its particular authentication ID and then can transfer it easily.
- Copyright: NFTs cannot be modified after the creation, no one can make changes within that information. This permits the creators to keep the copyright even after the NFTs are traded by defining personal digital signatures, ensuring the betterment of the creator community of all sorts and kinds.
- Secure: NFTs utilize blockchain technology which is fault-tolerant since there is always an identical record of the database on other nodes as well. All nodes keep an immutable ledger and it is maintained at every location simultaneously making the detection, prevention, and auditing of the information easier. The technology itself works on the principle of democracy of decentralized and distributed nodes, stating that if a node is compromised it will not affect the whole system.
- Efficient: Trading any asset through NFT is easier as it removes the human interaction and physical interference that can be faced by any individual. Trading NFTs does not demand for any topological requirements and saves the act of physical trading, paperwork, and unnecessary hustle of real physical life.
- Robust—NFTs carry the information of the asset assigned at all times. NFT can be verified even without the presence of the owner as the data of previous and current ownership can be found and identified uniquely. Even if the NFTs access is lost, the NFT will still point to the right owner of that particular NFT.

- Investment: NFTs are the assets of tomorrow and the right assets associated with their real-world uniqueness and authenticity can prove to give greater monetary benefits in future. Investing in the right NFTs can bring good returns in future.

6 Challenges

NFTs have been distributed well but still, they are a new concept for the general market and since it is new and in trend, many underlying challenges can be invisible to the common public. There are these following challenges concerning NFTs and their market:

- Value: The NFTs are costly and not all NFTs age well. The market of NFTs is growing at exponential pace but in the hype of NFTs, not all NFTs will ensure higher value returns as the digital assets generally rely on demands and not needs.
- Control of asset: NFTs do verify the ownership and provide a better authentication system but cannot resist that it can be copied and pasted. The data of NFT can be claimed by the owner at any instance if used for unlawful practices without permission but it can be circulated over the internet just as easily.
- Risky market: Currently the NFTs count is growing in the market, even though every NFT is authentic in itself but does not guarantee value in the long run. The market of digital NFTs is associated with the trends and the point that they can act as collectibles associated with any demanded field.
- Environmental impact: Since NFTs are traded through blockchain-based cryptocurrency, it is not sustainable for the environment. Installing a cryptocurrency mining plant is very expensive and not good for environment. Mining of blockchain puts great impact on the environment as it requires huge computation power, electricity and causes heat pollution.
- Intangible: Digital NFTs hold sentimental value and since they are not tangible, they have no practical use. Currently, the NFTs are non-tangible even if we look at metaverse coming up with such technologies that can provide us a haptic experience, it is very limited and in its initial stages.
- Uncertain: The NFTs are not dependent upon any sort of currency or item thus making them purely dependent upon hype, emotional demands, and capability of a person. Real-world assets have a certain value to them at any stage of life due to their usability and material value but since NFTs are not tangible they do not provide any particular usability for most fields and have no materialistic value.
- Theft vulnerability: Even though the NFTs are secure, few trading platforms are not. NFTs can be stolen if the marketplace where NFTs are traded is unsafe. Whereas the NFTs cannot be tampered with but still using a wrong place for performing transactions may lead to theft of NFT.
- Unrecoverable: NFTs can prove their ownership but if the private key is compromised then NFTs cannot be accessed. If the access to NFT is lost once, the NFT cannot be transferred at all. The access cannot be recovered if lost once.

- Acceptance: The NFTs are spreading well in the whole world but still are not very accepted by governments and the population on a general level. Neither the whole world has the access to be able to buy an NFT nor it is accepted as it is not traded with fiat in the digital world.

7 Future Prospects

There is a huge spike in the NFT market but it is just an initial phase in the evolution of NFTs and their market. NFTs are yet to be at their final stage as they carry a lot of potentials. Following are future prospects and implementations that can appear in NFTs:

- Haptic virtual reality—There is an obvious claim that a haptic virtual environment that can interact with NFTs will surely arrive creating a higher sense of exclusive experience. Recently this technology has been used by metaverse to create such an experience for their users, but the full-blown implementation of this technology can bring huge change in how humans experience. NFTs have the potential along with this technology to create a literal virtual world that can not only entertain but help in various medical conditions.
- Refined registry—Since physical assets can be represented through NFTs, most electronic devices, passes, and IDs can be represented by NFTs, verifying the ownership and other attributes created while purchase.
- Identity management—NFTs can be used to represent the identity of a person within a network and because of their uniqueness they cannot be mimicked.
- One portable market—NFTs are stored in crypto wallets thus can be carried along at any time which makes real-time transactions authenticable and efficient. When the NFTs will take over the major industries, the similar properties will create one portable market.
- Forgery free environment: As the NFTs evolve, soon it will become impossible to forge any authentic item from a branded shoe to human identity.
- ML & AI with NFTs: ML-trained digital objects can become NFTs providing a human-like interaction with help of haptic virtual reality where ML-trained models [15] can be used to enhance the human capabilities [16] of artificially intelligent NFTs.
- Exclusivity: Collectibles and various recreational digital assets such as songs, poems, videos, photos, etc. can be created into limited NFTs and distributed to a constrained public to maintain exclusivity.

8 Conclusion

Non-fungible tokens are now these days used in trading art, GIFs, games, virtual assets, etc. The market has witnessed a boom in the industry with a projection of higher involvement in future in the process of trading assets. NFTs are digital assets traded over public blockchain with an embedded smart contract to define their behavior. In this work, we provide the working of blockchain-based NFTs and discuss the associated benefits and challenges of adopting NFTs. We also analyze the market trends and present future prospects for NFTs.

References

1. Anuj Kumar Y, Tomar R, Kumar D, Gupta H (2012) Security and privacy concerns in cloud computing. Int J Adv Res Comput Sci Softw Eng 2
2. Sharma A, Sarishma TR, Chilamkurti N, Kim BG (2020) Blockchain based smart contracts for internet of medical things in e-healthcare. Electron 9:1–14. https://doi.org/10.3390/electronics9101609
3. Ali Syed T, Alzahrani A, Jan S, Siddiqui MS, Nadeem A, Alghamdi T (2019) A comparative analysis of blockchain architecture and its applications: problems and recommendations. IEEE Access. 7:176838–176869. https://doi.org/10.1109/ACCESS.2019.2957660
4. Abbas QE, Sung-Bong J (2019) A survey of blockchain and its applications. In: 2019 international conference on artificial intelligence in information and communication (ICAIIC)
5. Tama BA (2017) A critical review of blockchain and its current applications. pp 109–113
6. Sarishma GA, Mishra P (2021) Blockchain based framework to maintain chain of custody (CoC) in a forensic investigation. In: Communications in computer and information science. pp 37–46
7. Bansal P, Aggarwal B, Tomar R (2019) Low-voltage multi-input high trans-conductance amplifier using flipped voltage follower and its application in high pass filter. In: 2019 international conference automation computing technology management ICACTM 2019. vol 2. pp 525–529. https://doi.org/10.1109/ICACTM.2019.8776789
8. Valeonti F, Bikakis A, Terras M, Speed C, Hudson-Smith A, Chalkias K (2021) Crypto collectibles, museum funding and openGLAM: challenges, opportunities and the potential of non-fungible tokens (NFTs). Appl Sci 11. https://doi.org/10.3390/app11219931
9. Wang Q, Li R, Wang Q, Chen S (2021) Non-fungible token (NFT): overview, evaluation, opportunities and challenges
10. Arora A, Kanisk Kumar S (2022) Smart contracts and NFTs: non-fungible tokens as a core component of blockchain to be used as collectibles. In: Lecture notes on data engineering and communications technologies. pp 401–422
11. Raman R, Edwin Raj B (2021) The world of NFTs (Non-Fungible Tokens). In: Enabling blockchain technology for secure networking and communications. pp 89–108
12. Mojtaba S, Bamakan H, Nezhadsistani N, Bodaghi O, Qu Q (2021) A decentralized framework for patents and intellectual property as NFT in blockchain networks. 0–11
13. Chohan R, Paschen J (2021) What marketers need to know about non-fungible tokens (NFTs). Bus Horiz
14. https://dappradar.com/nft/marketplaces
15. Jayaraman S, Tanupriya C, Kumar P (2017) Analysis of classification models based on cuisine prediction using machine learning. In: 2017 international conference on smart technologies for smart nation (SmartTechCon). pp 1485–1490
16. Narendra JK, Vikrant K, Praveen K, Choudhury T (2018) Movie recommendation system: hybrid information filtering system. In: Intelligent computing and information and communication

An Efficient Data Preparation Strategy for Sentiment Analysis with Associative Database

Dipto Biswas, Md. Samsuddoha, and Partha Chakraborty

Abstract Sentiment analysis is a process of categorizing and determining the expressed sentiments. It provides an explicit overview of extensive mass sentiments about particular subjects. Sentiment analysis involves various challenges because expressed opinions and sentiments contain an immense number of anomalies. Data preparation is a prerequisite assignment that can deal with those anomalies for sentiment analysis. This paper represents an efficient data preparation strategy for sentiment analysis using the associative database model. The efficient data preparation strategy involves three subtasks, such as eliminating non-sentimental sentences, eradicating unnecessary tokens, in addition to extracting vocabulary, and arranging that vocabulary uniquely through the associative database model. The experimental results show that the performance of the proposed data preparation approach is comparatively efficient. A comparison with some existing sentiment analysis approaches demonstrates that the accuracy of sentiment analysis has been comparatively improved and enhanced by integrating the proposed efficient data preparation strategy into it.

Keywords NLP · RFC · ADBM · MBOW · Sentiment analysis

D. Biswas · Md. Samsuddoha (✉)
Department of Computer Science and Engineering, University of Barishal, Barishal 8200, Bangladesh
e-mail: sams.csebu@gmail.com

D. Biswas
e-mail: diptobiswas.cse.bu@gmail.com

P. Chakraborty
Department of Computer Science and Engineering, Comilla University, Cumilla 3506, Bangladesh
e-mail: partha.chak@cou.ac.bd

V. Skala et al. (eds.), *Machine Intelligence and Data Science Applications*, Lecture Notes on Data Engineering and Communications Technologies 132,
https://doi.org/10.1007/978-981-19-2347-0_2

1 Introduction

Sentiment Analysis (SA) is a computational strategy that determines whether an expressed sentiment is negative, neutral, or positive [1]. SA demonstrates a general overview of extensive mass opinions behind particular entities [2]. There are endless applications of SA, and this classification strategy helps to identify human expressions accurately [3]. However, SA is an incredibly complicated task because expressed opinions and sentiments contain an immense number of anomalies [4]. Human expressed opinions exist in a mixture of sentimental sentences, sarcastic sentences, non-sentimental sentences, and other unreasonable issues [5]. Moreover, human comments may have no typos, full of spelling mistakes, enormous punctuation, hyphenated words, numbers, section markers, stopwords, linking words, and other unnecessary tokens [6]. In addition, sentiments belong to vocabulary contained in annotations of opinions [7]. Those non-sentimental sentences and unnecessary tokens conceal the exact meanings of those vocabulary based on their utilization [8]. Those challenges create a vigorous hindrance to analyzing sentiments appropriately [10]. Data preparation (DP) is a preliminary assignment that can deal with all of those challenges and eradicate anomalies for sentiment analysis [11]. This research proposed an efficient data preparation strategy with an associative database model for sentiment analysis. This efficient data preparation strategy consists of three individual approaches, such as eliminating irrelevant non-sentimental sentences with features extraction methods [8] and modified Random Forest Classifier (RFC) [17], eradicating unnecessary tokens with a new data cleaning procedure that consists of various existing data cleaning methodologies [15], and extracting desirable vocabulary related to sentiments and representing that vocabulary uniquely through an associative database model [18]. This research has provided three contributions to the proposed efficient data preparation strategy for SA.

- A modified tree building procedure has been developed for RFC to detect irrelevant non-sentimental sentences for elimination.
- A precise data cleaning approach has been developed for eradicating unnecessary tokens that consist of various existing data cleaning methods.
- An associative database model has been integrated to represent desirable vocabulary uniquely to find out the grammatical correlations and exact meanings of those vocabulary based on their utilization.

All individual approaches involved in the proposed data preparation strategy have been implemented and applied to two distinct datasets. The experimental results show that the proposed DP strategy is comparatively efficient. Moreover, the accuracy of sentiment analysis has been comparatively improved and enhanced by integrating the proposed DP strategy into it. The rest of the paper has been consecutively organized as Literature Review in the Sect. 2, Proposed Methodology in the Sect. 3, Result and Discussion in the Sect. 4, and Conclusion in the Sect. 5.

2 Literature Review

Sentiment analysis is undoubtedly a hot issue in Natural Language Processing (NLP) background. SA refers to extracting appropriate sentiments from the expressed opinions of users, customers, and consumers about any particular entity. It is really a complex task because expressed opinions are not in a well-structured format. Many researchers have implemented sentiment analysis in various ways and also introduced various anomalies and challenges to sentiment analysis related to data preparation. Deng et al. [7] addressed some challenges to SA, such as human opinions as sentiments are not well-structured, and full of unnecessary sarcastic statements. SA has been implemented by a weighting scheme with SVM and feature selection methods. The average accuracy became 88.00–88.70% but the researchers did not mention how they had prepared data for analyzing sentiment [7]. Agarwal et al. [8] also introduced a few challenges related to SA, such as human opinions containing negations at a higher frequency and unnecessary statements. SA has been implemented by SVM and five different feature extraction combinations. The average accuracy is 60.83–75.39% but the researchers did not express anything about how the negations and unnecessary statements are handled [8]. Turney et al. [9] addressed a problem with SA that is handling negations. Even this study stated that negations in sentiments conceal the exact meanings of expressed opinions. SA has been implemented by different patterns of tags and PMI—information retrieval features with an average accuracy of about 65.83–84.00% [19]. Mullen et al. [10] implemented SA with osgodian semantic differentiation with WordNet, hybrid SVM, and PMI—information retrieval features with an average accuracy of about 87.00–89.00%. This research introduced a few challenges related to SA, such as dealing with negations, and missing and misspelled words [20]. Nasukawa et al. [11] implemented SA very crucially with syntactic dependencies among phrases and subject term modifiers with a great average accuracy of about 86.00–88.00%. This research introduced unnecessary words and sentences to express opinions and conceal the exact meanings of sentimental words [23]. Kanayama et al. [12] also introduced several obstacles to SA, such as spam detection, errors, and negations handling. SA has been implemented by considering two contextual coherencies, such as Intra-sentential and Inter-sentential, with 88.00–90.00% average accuracy [24].

Above, SA studies introduced various challenges to sentiment analysis. Dealing with negations, detecting spam and fake sentiments, identifying non-sentimental sentences, recognizing sarcastic statements, and handling missing and misspelled words are the most common challenges that create anomalies in data. This research has provided an effort to propose an efficient data preparation strategy for SA. The efficient DP strategy has been able to deal with all of those identified challenges. The experimental results show that the proposed DP strategy is comparatively efficient. The proposed DP strategy has been integrated into a sentiment analysis implementation called Newish Sentiment Analysis (NSA). A comparison demonstrates that the accuracy has been comparatively improved by integrating the proposed efficient data preparation strategy into sentiment analysis implementations.

3 Proposed Methodology

An efficient data preparation strategy has been described in this proposed methodology section. The data preparation approach consists of three individual approaches, such as eliminating non-sentimental sentences, eradicating unnecessary tokens, in addition to extracting vocabulary, and arranging that vocabulary uniquely through the associative database model depicted in Fig. 1.

3.1 Eliminating Non-sentimental Sentences or Reviews

The elimination process of non-sentimental sentences consists of four activities, such as preprocessing, sentimental and non-sentimental features extraction, detection of non-sentimental sentences and testing algorithmic results with collected datasets, and finally eliminating non-sentimental sentences. The preprocessing phase involves splitting sentences and words individually. The sentimental and non-sentimental features. The extraction phase consists of the unigram method and four level feature extractions named punctuation-related, top-word, sentiment-related, and lastly, lexical and syntactic features. The Unigram A method is used to split each sentence into unique words for extracting features. Afterward, sentiment-related features have been applied to count the number of positive and negative reviews, the number of positive and negative hashtags, and word contrast [13].

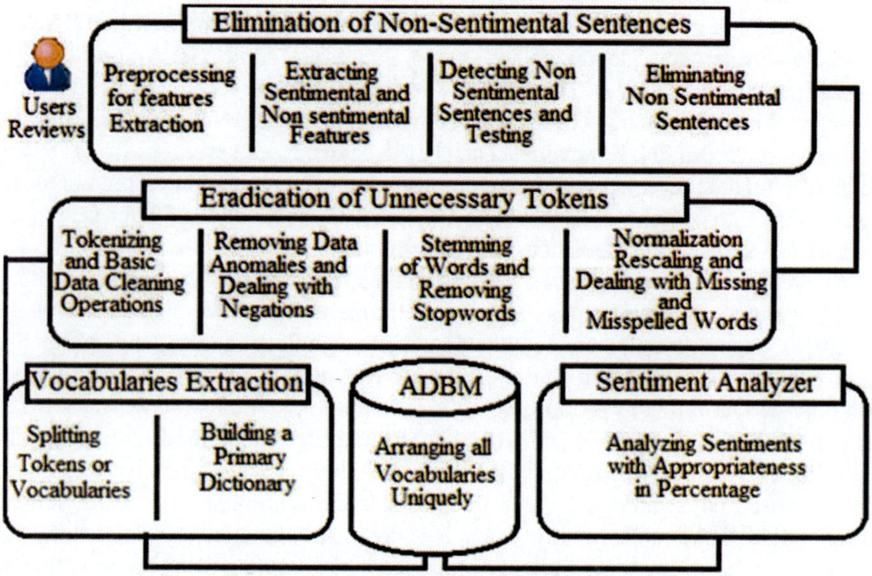

Fig. 1 Proposed efficient data preparation strategy for sentiment analysis

$$P(r) = \frac{(\phi \cdot \mathrm{PSW} + \mathrm{psw}) - (\phi \cdot \mathrm{NGW} + \mathrm{ngw})}{(\phi \cdot \mathrm{PSW} + \mathrm{psw}) + (\phi \cdot \mathrm{NGW} + \mathrm{ngw})} \quad (1)$$

A special weight $P(r)$ has been added and described in Eq. 1. Equation 1 consists of several parameters. Here, r mentioned in the reviews, ϕ is a constant considered 5 as its value at all times. ϕ is utilized as the weight of highly emotional words. The psw defines positive words and the PSW defines highly emotional positive words. Similarly, the ngw defines negative words and the NGW defines highly emotional negative words. Through punctuation-related features capital letters, questions, exclamations, and quotations have been identified to recognize non-sentimental sentences. The lexical and syntactic features perform to recognize those sentences that hide the original and exact sentiment-related sentences. By utilizing top word features, all the numerical values are identified such as 100, 50%, etc. After that non-sentimental sentences have been detected and tested through a modified Random Forest Classifier (RFC).

Algorithm 1 Algorithm for Random Forest Classifier (RFC)

1. *Begin*
2. *i := 0*
3. Perform ***bootstrap selection*** on training dataset.
4. Put bootstrap data to ***SubsetData[i].***
5. ***Build tree[i]*** by ***SubsetData[i]*** depicted in Algorithm 2.
6. Add ***tree[i]*** to ***NumberTree.***
7. ***i := i + 1***
8. ***If TreeCount*** is greater than ***i do***
9.   ***Go to step 3*** and perform the sequential instructions.
10.   Continue until ***TreeCount*** becomes less than ***i***.
11. ***End if***
12. ***Else do***
13.   Print the ***NumberTree*** as the output.
14. ***End else***
15. ***End***

Algorithm 2 Algorithm for Building Tree

```
1.Begin
2.   Sorting index SubsetData[j].
3.   CurrentNode := SubsetData[j]
4.   Put CurrentNode into Stack.
5.   If Stack is not empty do
6.       Pop Stack and take CurrentNode.
7.       Calculate gini index based on equation 2.
8.       Choose the best feature from CurrentNode.
9.       If CurrentNode != impure based on best feature do
10.          Go to step 5.
11.      End if
12.      Else
13.          Split CurrentNode to Right and Left Nodes.
14.          Add Right Node to ObjectTree and to Stack.
15.          Add Left Node to ObjecTree, and Stack
16.          Go to step 5.
17.      End Else
18.  End if
19.  Else
20.      Hold and Print ObjecTree as Output.
21.  End Else
22.End
```

The execution of modified Random Forest Classifier (RFC) and its tree building approach have been expressed in Algorithm 1 and Algorithm 2, respectively. With a view to choosing the best feature from the CurrentNodes, the gini index [13] has been utilized as a momentous component described in Eq. 2.

$$\text{gini}\ (R) = 1 - \sum_{k=1}^{N} P_k^2 \tag{2}$$

Here, R refers to the number of data sets. P_k refers to the relative frequency of class k, and N mentions the number of classes P_k. Presence of heterogeneous classes creates a state or condition called impurity [13] for the current node. Considering the best feature, impurity for CurrentNode is measured and CurrentNodes are split into left and right nodes. After that, the right and left nodes are assigned to ObjecTree and Stack, respectively.

3.2 Eradicating Unnecessary Tokens from Sentences

All words and tokens in sentences are not related to sentiments [14]. Irrelevant tokens need to be removed to analyze sentiment with more accuracy. The eradication process of unnecessary tokens consists of few phases depicted in Fig. 2.

Splitting sentences and words involves separating each sentence and word from review documents. Basic data cleaning operations generally involve removing punctuation with apostrophes, eradicating numbers like (20/20), removing single characters and non-alphabetical tokens, and eliminating less important words in sentences. Removing data anomalies involves managing coverage anomalies, semantic anomalies, and syntactic anomalies [15]. Negative words can express positivity or negativity [16] based on their utilization, such as "not good", "not bad", etc. Dealing with negations involves converting apostrophes into meaningful words, and extracting the exact meanings depending on their use in sentences. Normalization and rescaling have been included to convert the tokens into another compatible form that is comparatively suitable for understanding. Stemming of words is used to convert several words having different parts of speech into a staple word for better analysis, and it eradicates entropy [17]. Removing stop words means removing pronouns, identifiers, and other linking words such as (a, an, the, it, this, etc.) [17]. To enhance the performance, it is highly required to make corrections to missing and misspelled words. The section includes PyEnchant [17] which acts as a dictionary and provides an aid to identify misspelled and missing words for correction. After eradicating unnecessary tokens, all documents contain valuable vocabulary. Those vocabulary is in a standard format that has been further utilized in sentiment analysis. Eventually, desirable vocabulary has been extracted and those vocabulary has been arranged uniquely through the Associative Database Model (ADBM) to capture the appropriate sentiments.

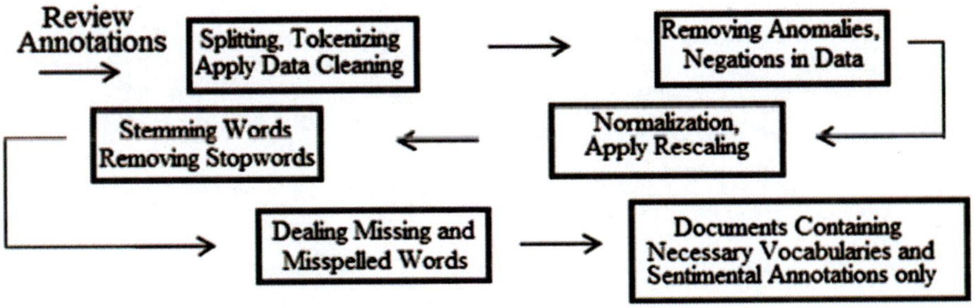

Fig. 2 Eradication process or pipeline for unnecessary tokens and vocabularies

3.3 Extracting Vocabularies and Arranging Uniquely

Sentiment belongs to words, more precisely in vocabulary [16]. To extract vocabulary, the split function has been used. After extraction, the vocabulary has been organized as a dictionary primarily. The dictionary can be considered as a storehouse of vocabulary and those vocabulary has been arranged uniquely for appropriate analysis, through the associative database model. The integrated ADBM in the proposed DP technique is depicted in Fig. 3. With examples such as "John likes to watch romantic movies. Mary likes romantic movies too", and "Mary also likes to watch football games".

Figure 3 demonstrates the unique representation of vocabularies using an associative database model. Figure 3 represents that vocabulary builds relationships among them as associations. The ADBM consists of two data structures, such as The terminals and directions are described in Fig. 3. The terminals have specific names, items, unique types, vocabularies, etc. On the other hand, directions are a collection of lines or links. Terminals such as tokens or Vocabularies are depicted as nodes and relationships among tokens or vocabularies are represented as lines called directions. Such unique representation of vocabulary as terminals and directions through ADBM has increased the efficiency of data preparation and sentiment analysis. According to the proposed DP strategy, non-sentimental sentences have been eliminated, unnecessary tokens have been eradicated, and vocabulary has been extracted and arranged. uniquely, Now, the data that means review documents are prepared and processed for utilization in the context of sentiment analysis. In order to evaluate the effectiveness of the proposed DP approach has been applied over an existing Movie Reviews data set and an own created Restaurant Reviews data set. As well as, the accuracy of SA has been comparatively improved as depicted in the Results and Discussion section.

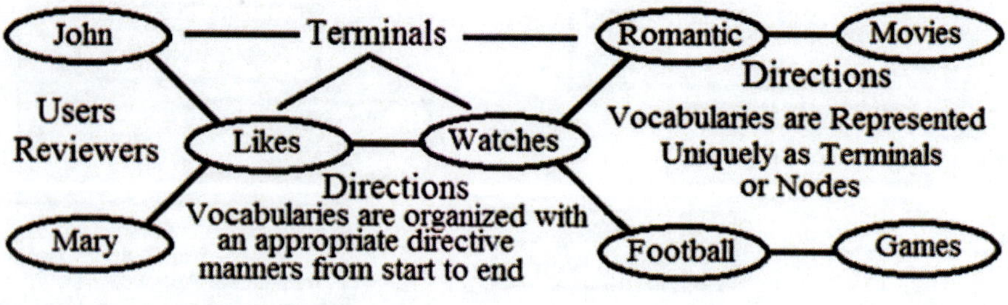

Fig. 3 Unique representation of vocabularies as terminals with the ADBM

4 Result and Discussion

4.1 Datasets Collection

The proposed efficient data preparation strategy has been applied to two distinct datasets for preparing data in a suitable form during analyzing sentiments. Sentiment analysis has been implemented on an existing movie review dataset and an own developed restaurant review dataset. The movie review dataset has 1000 positive reviews and 1000 negative reviews [21]. On the contrary, the restaurant review dataset has 100 positive reviews and 100 negative reviews [22].

4.2 Measurement Matrices

The evaluation has been implemented by utilizing measurement matrices. The measurement matrices are classified into two classes, such as the positive class and the negative class. For positive reviews, they are classified as true positive reviews (TPR) and false-negative reviews (FNR). Similarly, negative reviews are classified as false positive reviews (FPR) and true negative reviews (TNR). The performance of the proposed data preparation technique and the efficiency of sentiment analysis have been measured through three standard parameters, such as precision, recall, and accuracy. The precision parameter has defined a rate of prediction that helps to recognize the intimacy of the estimated values with other positive or negative reviews. The recall parameter has expressed a rate of prediction by which negative and positive reviews are predicted to be negative or positive reviews. The accuracy parameter has calculated a value through which the appropriate results are predicted as an evaluation of the performance of the proposed data preparation approach and sentiment analysis.

4.3 Model Development

The proposed efficient data preparation strategy consists of three crucial and vital tasks such as Task 1: elimination of non-sentimental sentences, Task 2: eradication of unnecessary tokens and words, Task 3: extraction of desirable vocabularies and unique arrangement of those vocabularies. The proper combinations of those three tasks have provided a significant benefit for preparing data in terms of sentiment analysis. In order to build combinations among tasks, an efficient binary representation for three bits has been followed. Here three bits represent three tasks, respectively. Since there are three tasks, the number of combinations has been considered as $2^3 = 8$. The combinations of tasks utilized for evaluations are C1(0,0,0), C2(0,0,1), C3(0,1,0), C4(0,1,1), C5(1,0,0), C6(1,0,1), C7(1,1,0), and C8(1,1,1). Here, 0 represents the

Table 1 Experimental results of sentiment analysis in which the proposed efficient data preparation strategy has been integrated for two distinct datasets

Binary combination	Movie reviews [21]			Restaurant reviews [22]		
	Accuracy (AVG)	Precision (AVG)	Recall (AVG)	Accuracy (AVG)	Precision (AVG)	Recall (AVG)
C1 (0,0,1)	0.211	0.201	0.203	0.398	0.311	0.352
C2 (0,0,1)	0.244	0.291	0.222	0.401	0.313	0.359
C3 (0,1,0)	0.509	0.498	0.501	0.687	0.615	0.672
C4 (0,1,1)	0.704	0.684	0.674	0.747	0.719	0.724
C5 (1,0,0)	0.666	0.611	0.651	0.696	0.629	0.677
C6 (1,0,1)	0.783	0.753	0.747	0.798	0.724	0.774
C7 (1,1,0)	0.821	0.834	0.798	0.844	0.833	0.811
C8 (1,1,1)	**0.947**	**0.843**	**0.861**	**0.956**	**0.872**	**0.927**

absence of any task and 1 represents the presence of any task in any combination. For example, C7 (1,1,0) means the evaluation has been implemented by a combination that consists of eliminating non-sentimental sentences, eradicating the unnecessary tokens but without the unique representation of desirable vocabularies. Eventually, the proposed combination C8(1,1,1) means the evaluation has been implemented by a combination that consists of eliminating non-sentimental sentences, eradicating the unnecessary tokens, and arranging desirable vocabularies uniquely.

A new lexicon-based sentiment analysis approach called newish sentiment analysis has been implemented with a modified Bag-of-Words (MBOW) model in which the proposed efficient DP strategy has been integrated during analyzing sentiments. The proficiency of sentiment analysis has been comparatively improved by integrating the proposed data preparation technique and experimental results are recorded in Table 1. All rows of Table 1 have defined the numerical values of standard parameters such as Accuracy (AVG), Precision (AVG), and Recall (AVG), respectively. The values were achieved through different combinations of tasks. The last row of Table 1 represents the best numerical values of standard parameters. These outstanding results have been achieved when the sentiment analysis has been implemented with the proposed combination of C8(1,1,1). The accuracy of SA for the movie review dataset and restaurant review dataset are 94.7 and 95.6%, respectively.

4.4 Discussion and Evaluation

The Discussion and Evaluation section describes some important issues such as complications encountered during analyzing sentiments, comparisons between NSA and other existing popular sentiment analysis implementations, and limitations. Various considerable complexities, such as overfitting for large data, stabilization of large amounts of data, preserving orders in vocabulary, and relations among

Table 2 A comparison of various existing lexicon-based sentiment analysis approaches with newish sentiment analysis (NSA) approach

Implementations	Techniques	Accuracy
[7]	Weighting scheme with SVM + feature selection methods	88.00–88.70%
[8]	SVM + Five different features extractions combinations	60.83–75.39%
[9]	Different patterns of tags + PMI-IR (Information retrieval)	65.83–84.00%
[10]	Osgodian semantic differentiation with wordNet + Hybrid SVM + PMI-IR	87.00–89.00%
[11]	Syntactic dependencies among the phrases and subject term modifiers	86.00–88.00%
[12]	Two contextual coherency such as intra-sentential and inter-sentential	88.00–90.00%
Newish sentiment analysis (NSA)	Modified bag-of-words (MBOW) model + Proposed efficient data preparation technique	94.70–95.60%

vocabulary with content, have been dealt with by NSA during its implementation for analyzing sentiments. The NSA has utilized an auxiliary dictionary for handling overfitting and stabilization complications in data. The ADBM has provided assistance to preserve the orders and maintain the exact correlation among vocabulary based on their utilization through unique representation of vocabulary. To evaluate the significance of the proposed efficient data preparation technique, a comparison has been depicted in Table 2. The comparison was based on accuracy with six different popular lexicon-based sentiment analysis implementations.

Table 2 contains the names of implementations, methodologies, and accuracy of those implementations, respectively. The Accuracy column holds two values as a range that defines the accuracy that varies between those two values on average for each implementation. For example, the last row of Table 2 defines that the NSA approach has been implemented by using a modified Bag-of-Word model and integrating the proposed efficient data preparation technique.

The NSA has provided accuracy of about 94.7 and 95.6% on average for distinct datasets. Moreover, the accuracy of SA has been comparatively improved by integrating the proposed DP approach into it. Several limitations belong to this implementation. This implementation is unable to perform well on superfluous labeled data and isolated domains. In addition, this implementation is not proficient at dealing with ambiguous meanings of vocabulary and complex sentimental sentences.

5 Conclusion

Data preparation can be considered an elementary task for sentiment analysis. However, there are a few works that are directly related to data preparation regarding sentiment analysis. An efficient data preparation technique has been proposed for sentiment analysis that is suitable for preparing data appropriately. Three crucial tasks have been integrated into the proposed data preparation approach, such as elimination of non-sentimental sentences, eradication of unnecessary tokens, and arranging desirable vocabulary uniquely. The experimental results demonstrate that the proposed data preparation technique is comparatively efficient. Moreover, the proposed technique has also enhanced the efficiency of sentiment analysis.

Our future work is to develop a sentiment analysis approach that will be able to perform brilliantly on an enormous amount of labeled data and different or isolated domains. In addition, the upcoming sentiment analysis approach will be capable of eradicating ambiguous meanings of vocabulary based on their utilization in complex sentences. Identifying appropriate word senses of vocabularies can enhance the performance of sentiment analysis with more accuracy.

References

1. Li D, Liu Y (2018) In: Deep learning in natural language processing. 2nd edn. Springer
2. Matthew J, Rosamond T (2020) Sentiment analysis.https://doi.org/10.1007/978-3-030-39643-5_14
3. Michael B, Christian B, Frank H, Frank K, Silipo R (2020) Data preparation. https://doi.org/10.1007/978-3-030-45574-3_6
4. Rahul Vasundhara R, Monika (2019) Sentiment analysis on product reviews. pp 5–9. https://doi.org/10.1109/ICCCIS48478.2019.8974527
5. Subhabrata M, Bhattacharyya P (2012) Feature specific sentiment analysis for productreviews. In: International conference on intelligent text processing and computational linguistics, Springer, Berlin, Heidelberg
6. Liu B, Zhang L (2013) A survey of opinion mining and sentiment analysis. Mining TextData, pp 415–463. https://doi.org/10.1007/978-1-4614-3223-4_13
7. Zhi-Hong D, Kun-Hu L, Hongliang Y (2014) A study of supervised term weightingscheme for sentiment analysis. Expert Syst Appl: An Int J 41:3506–3513. https://doi.org/10.1016/j.eswa.2013.10.056
8. Agarwal A, Boyi X, Ilia V, Owen R, Rebecca P (2011) Sentiment analysis of twitterdata. In: Proceedings of the workshop on languages in social media, association for computational linguistics, pp 30–38
9. Turney P (2002) Thumbs up or thumbs down? Semantic orientation applied to un-supervised classification of reviews. In: Proceedings of the 40th annual meeting on association for computational linguistics ACL'02, pp 417–424. https://doi.org/10.3115/1073083.1073153
10. Mullen T, Nigel C (2004) Sentiment analysis using support vector machines with diverse information sources. In: EMNLP vol 4. pp 412–418.
11. Nasukawa T, Yi J (2003) Sentiment analysis: capturing favorability using naturallanguage processing. pp 70–77. https://doi.org/10.1145/945645.945658
12. Kanayama H, Nasukawa T (2006) Fully automatic lexicon expansion for domain-oriented sentiment analysis. In: Proceedings of EMNLP, pp 355–363

13. Yessi Y, Aina M, Anny S (2019) SARCASM detection for sentiment analysis in indonesian language tweets. Indonesian J Comput Cybernet Syst 13:53–62. https://doi.org/10.22146/ijccs.41136
14. Diana M, John C (2003) Disambiguating nouns, verbs, and adjectives using auto-matically acquired selectional preferences. Comput Linguist 29:639–654. https://doi.org/10.1162/089120103322753365
15. Abdallah Z-S, Du L, Webb G-I (2017) Data preparation. Springer, Boston, MA. https://doi.org/10.1007/978-1-4899-7687-1_62
16. Mohey D, Din E (2016) Enhancement bag-of-words model for solving the challenges of sentiment analysis. Int J Adv Comput Sci Appl 7(1). https://doi.org/10.14569/IJACSA.2016.070134
17. Federico M, Tomaso F, Paolo F, Stefano M, Eleonora I (2016) A comparison be-tween preprocessing techniques for sentiment analysis in twitter
18. Joseph H, Kovacs P (2009) A comparison of the relational database model and the associative database model. pp 208–213
19. Partha C, Sabbir A, Mohammad Y, Azad A-K-M, Salem A, Moni M-A (2021) A human-robot interaction system calculating visual focus of human's attention level. IEEE Access 9:93409–93421.https://doi.org/10.1109/ACCESS.2021.3091642
20. Partha C, Mohammad Y, Saifur R (2020) Predicting level of visual fo-cus of human's attention using machine learning approaches. https://doi.org/10.1007/978-981-33-4673-4_56
21. Dataset. http://www.cs.cornell.edu/people/pabo/movie-review-data/reviewpolarity.tar.gz. Last accessed 29 Jun 2021
22. Datasets. https://github.com/diptobiswas2020/Data-Preparation-Dataset/tree/positive-reviews https://github.com/diptobiswas2020/Data-Preparation-Dataset/tree/negative-reviews. Last Accessed 1 Jul 2021
23. Partha C, Zahidur Z, Saifur R (2019) Movie success prediction using historical andcurrent data mining. Int J Comput Appl 178:1–5. https://doi.org/10.5120/ijca2019919415
24. Chen L (2021) Data preparation for deep learning. https://doi.org/10.1007/978-981-16-2233-5_14

A Comparison of Traditional and Ensemble Machine Learning Approaches for Parkinson's Disease Classification

Kevin Sabu, Maddula Ramnath, Ankur Choudhary, Gaurav Raj, and Arun Prakash Agrawal

Abstract Parkinson's disease (PD) is a movement-related disorder that negatively impacts the central nervous system. It is progressive, which means that patients' condition worsens over time. According to literature, it is more common among men as compared to women. It affects a person in various ways such as in its initial stages, it may cause a few tremors in certain body parts and the consequences worsen throughout its progression—difficulties in speech, writing, and movement all develop eventually. This paper performs an effective comparison of an individual as well as ensemble machine learning methods on quantitative acoustic measure data for PD classification. In individual models, category Support vector machine (SVM), Naive Bayes, Decision Tree, K-Nearest Neighbor (KNN), and Logistic Regression approaches have been adopted. In the category of ensemble models Random Forest, Gradient boosting, Adaptive Boosting, and XGBoost. The models were evaluated and compared using performance metrics such as Recall, F1 score, Precision, and Accuracy. Results indicate that XGBoost performs well than others.

Keywords Parkinson's disease · Machine learning · Classification methods · Ensemble learning · Gradient boosting · XGBoost

1 Introduction

Parkinson's disease was identified as a Neuro System disorder by James Parkinson as far back as the year 1817 [1]. It occurs when the neurons release dopamine in the brain get damaged. This leads to a deficiency of dopamine, which in turn hampers the generic movement of the affected individual. According to literature, after Alzheimer's disease, PD is the most recognized degenerative neurological disorder. In general, its onset is at about 40 years of age and as the person crosses the 60-year mark, its severity increases. PD is twice as prevalent among men compared to

K. Sabu · M. Ramnath · A. Choudhary · G. Raj (✉) · A. Prakash Agrawal
Department of Computer Science and Engineering, School of Engineering and Technology, Sharda University, Greater Noida, India
e-mail: er.gaurav.raj@gmail.com

V. Skala et al. (eds.), *Machine Intelligence and Data Science Applications*, Lecture Notes on Data Engineering and Communications Technologies 132,
https://doi.org/10.1007/978-981-19-2347-0_3

women. Global data reveals that 1% of all people above 65 years of age are suffering from PD [2]. It is a progressive disorder, which means that the individual's condition worsens over time. If it persists without diagnosis or medication for long enough, non-motor problems such as sleep deprivation, fatigue, difficulty in speaking, hallucinations, and depression are introduced in addition to the already existing difficulty in movement. Although there is no reliable cure for it, the right medication can help alleviate the symptoms as it uses methods to counter the deficiency of dopamine. That is why early diagnosis of PD is of utmost importance. It should be noted that manual PD diagnosis is not that intuitive. Symptoms can vary from individual to individual. PD patients' doctors usually need to perform multiple diagnostic tests, as well as a thorough review of their patients' medical histories. Based on the symptoms, they may need to perform tests for other disorders too. Weekly appointments are usually required to observe how the condition progresses. This makes manual PD diagnosis a time-taking process, and also the reason that better approaches that use ML and deep learning have been used over the past few years, as they prove to be the more effective and efficient classification for PD classification. To perform the PD classification using machine learning or deep learning approaches, different kinds of data have been explored. Data such as vocal frequency measures, gait patterns, the demographic and genetic information of healthy people as well as people affected by PD can all be utilized to construct algorithms that perform this task well.

According to literature the most common approach used for the task of PD classification is Machine Learning. Various studies have used many different machine learning methods, which include Random Forest(RF), SVM, Linear/Logistic regression, KNN, K Star, Naive Bayes, etc. [3–7]. Another relatively common approach used for PD classification is the usage of deep learning algorithms, some of which include artificial neural networks and convolutional neural networks [3, 8–9]. In some cases, hybrid approaches were used for PD classification, which is the combination of algorithms to introduce a new system [10–11]. Such combined approaches are found to be very effective and produce good results overall. Implementations of these approaches are done on different well-known datasets, which include the Oxford Parkinson's disease voice dataset, Gait pattern dataset, etc. Studies have also tried to identify the clinical biomarkers that may help in PD diagnosis [12–15].

This paper utilizes the Oxford PD detection voice dataset [16] and performs a comparison of how the ensemble machine learning techniques, such as Gradient Boosting (GBM), XGBoost, AdaBoost, etc. perform against the more established and explored algorithms, such as SVM, RF, Decision trees, etc. for the task of PD classification. The comparison has been done based on different performance measures like accuracy, precision, recall, etc.

2 Literature Review

PD is a disorder that degrades the nervous system, negatively affects motion and is highly prevalent in older people. Although it is not curable yet, early identification may help in suppressing the symptoms through medication before the condition worsens. Over the last two decades, a handful of studies have been done regarding Parkinson's disease classification as well as the identification of its biomarkers. Some of them have been explored and the findings are as follows.

Das et al. [3] compared four different classification algorithms for PD classification—Regression, Decision Tree, DMneural, and Neural networks and that claimed Neural networks performed the best with 92.9% accuracy as a comparison to selected classification algorithms. Sriram et al. [4] proposed an intelligent system that uses tools such as Orange and Weka (Waikato Environment Knowledge Analysis) for data visualization and modeling, respectively. Various supervised machine learning algorithms were implemented, such as Naive Bayes, KStar, RF, Logistic regression, etc. out of which RF performed the best with an accuracy of 90.26%. Bind et al. [5] presented an in-depth examination of the numerous ML techniques that have previously been used to predict PD over years. They included the results achieved by numerous researchers from 2008–2014. Sujatha et al. [6] compared 16–17 different data mining classification algorithms used for PD prediction. The comparison depends upon various evaluation metrics like Recall, F-score, Precision, etc. A KNN classifier known as IBk gave the best results. Shamrat et al. [7] implemented three supervised classification algorithms for PD classification. These were SVM, Logistic regression(LR), and KNN. The results proved that the SVM classifier performed the best while the KNN classifier performed the worst based on measures: Precision, Recall, and F1-score. Suganya et al. [17] performed a comparative study on five feature selection techniques, which are Fuzzy C-means, Particle Swarm Optimization, Artificial Bear Optimization (ABO), Ant Colony Optimization (ACO), and SCFW-KELM. These algorithms were compared based on performance measures like sensitivity, specificity, and accuracy. ABO algorithm performed the best, giving an accuracy of 97%. Little et al. [16] proposed two new approaches to differentiate between normal and disordered voice, namely recurrence and scale analysis. These approaches were better at utilizing the randomness and non-linearity of cluttered voice signals than the previous ones. Wenning et al. [12] identified the clinical features that differentiate Multiple System Atrophy (MSA) from PD, which include response to Levodopa, axial functions, etc. They concluded that these features yielded better results when they were available until death, thus not being too useful for early diagnosis. Hall et al. [13] studied the effects of indicators such as family history and gene information on PD prediction. It was seen that the models using these additional indicators performed better than models only using the standard demographic data. Darweesh et al. [14] studied the impact of a non-motor risk score, known as PREDICT-PD on PD occurrence. 6492 people were observed over 20 years to identify the impact. Risk regression models were implemented which include additional demographic data such as age and sex. They concluded that the addition of PREDICT-PD as a risk factor just marginally improved PD classification results. Lin et al. [15] determined whether the Plasma Neurofilament light chain (P-NfL) has an impact on PD progression or not. Statistical and Cox regression analysis along with a few other measures were used for the task. The results showed that higher levels of

P-NfL correlated to decline in motor aspects, which confirms that P-NfL is a valid biomarker to assess PD intensity and development.

Ibrahim et al. [11] proposed a system that uses Support vector regression, an Adaptive neuro-fuzzy inference system, and SVM for predicting PD progression. The results showed that the best performance was delivered by the combination of SVM-EM-PCA, with an AUC score of 0.9972. Campbell et al. [18] implemented Linear class analysis (LCA) to distinguish three PD subtypes—"motor only", "cognitive and motor", and "psychiatric and motor". Following this, Discriminant analysis and Cox regression were performed to identify unique features and observe the differences among PD subtypes based on Deep brain stimulation (DBS), mortality, and dementia. Karan et al. [9] proposed a deep learning approach using a Stacked Auto-Encoder (SAE) for PD voice signal classification. This has been done through first converting input speech signals into time-frequency representation using Short-time Fourier Transform and Continuous Wavelet Transform, which is then fed to a Stacked auto-encoder for feature extraction. These newly-obtained features get fed to SVM and Softmax classifier (SC). It is found that the SC gives better results. Mozhdeh Farahbakhsh et al. [8] proposed a Magnetic Resonance Imaging (MRI)-based Convolutional Neural Network (CNN) based model to differentiate between the PD stages. The Parksinson's progression Markers Initiative (PPMI) image dataset was used and preprocessed, following which the CNN system, which consists of nine layers, wherein the output layer used the softmax function. K-fold cross-validation was used for testing and validation of results. The results showed that the proposed approach got an accuracy of 94%. Warden et al. [10] presented additional medication data to improve on their previously developed penalized logistic regression (PLR) model for identifying prodromal PD. They developed models both including and excluding the additional medication data (Part D data). For comparison, they used a Random Forest classifier as well as a combination of the two approaches, by using RF in the form of a predictor in the PLR model. The combination of the two approaches performed the best, but all the aforementioned models performed relatively well too.

3 Methodology

3.1 Dataset and Feature Information

The adopted dataset consists of 24 columns and 195 rows. In this dataset quantitative acoustic measurements of 31 people were taken, out of which 23 were positive for PD. These measurements are documented at the national center for speech and voice, Denver, in collaboration with Max little of Oxford University [16].

These are the quantitative acoustic measurements used in this research:

- MDVP: Fo
- MDVP:Jitter(Abs)
- MDVP:RAP

- MDVP:PPQ
- Jitter:DDP
- MDVP:Shimmer
- Shimmer:APQ3
- Shimmer:APQ5
- MDVP:APQ
- Status:- (1) PD positive, (0) PD negative.

3.2 Classifiers Adopted in This Research

Individual classifiers

(a) ***Naive Bayes***: It is a probabilistic classification method based on the Bayes theorem as an underlying principle. It works on the assumption that all features are independent of each other.

(b) ***K-Nearest Neighbor***: It is a classification technique where values assigned to the recent data points depend on their similarity with existing data points. The similarity measure used is the distance between the points.

(c) ***Decision Tree***: One of the supervised machine learning techniques, in decision trees the data is constantly splitting according to specific parameters. Leaves are decisions of the final outcome and decision nodes are where data is split.

(d) ***Support Vector Machine***: SVM aims to generate a line that divides the data points into various classes. Both classification and regression problems can get solved using SVM.

(e) ***Logistic Regression:*** This technique is used to distinguish data into binary or multiple classes. The data being distinguished is categorical.

Ensemble classifiers

Ensemble classifiers can be categorized into bagging and boosting.

(a) ***Bagging***

(i) ***Bagging estimator***—It is used to remove variance in a noisy dataset. Bagging is also known as bootstrap aggregation, in bagging independent data points can be chosen more than once.

(ii) ***Random Forest***—It is a modification of the bagging estimator technique and consists of multiple decision trees. Features are selected at random and govern the ideal decision at each node of the decision tree.

(b) **Boosting**

(i) ***Gradient boosting***—Both classification and regression can be done using gradient boosting such as generally used to reduce bias error. In this technique, the base eliminator is always constant such that it is fixed.

(ii) ***AdaBoost***—It belongs to the boosting technique family. It creates strong classifiers through the use of multiple weak classifiers. Reassignment of weights is done for all data instances according to the correctness of the instances.

(iii) ***XGBoost***—XGboost combines the predictive power of multiple methods, resulting in a uni model that gives a combined output from various models. Due to this property, XGBoost has become one of the most powerful techniques.

Performance metrics

There exist various metrics available that may help to measure the results of classifiers for a classification task. These metrics use some common terms which are:

True positive (TP)—The situation wherein the values for prediction and the outcome are both positive.

True negative (TN)—The situation wherein the prediction and the outcome are both negative.

False-positive (FP)—The situation wherein the outcome is negative but the prediction by the classifier is positive.

False-negative (FN)—The situation wherein the outcome is positive but the prediction by the classifier is negative.

Precision—It is defined as the number of the correct true class observations per total number of true class observations made by the classifier. Given through the following:

$$\text{Precision} : \text{TP}/\text{TP} + \text{FP}$$

Recall—It is defined as the number of correct true class observations per the total number of observations. Given through the following:

$$\text{Recall} : \text{TP}/\text{TP} + \text{FN}$$

F1-score—It is obtained by computing the harmonic mean of two other metrics—Recall and Precision. Given though the following:

$$F1 - \text{score} : 2^*\left(\text{Recall}^*\text{Precision}\right)/(\text{Recall} + \text{Precision}).$$

Accuracy—It is defined as the number of correct observations per the total number of observations computed through the classifier. Given through the following:

$$\text{Accuracy} : \text{TP} + \text{TN}/(\text{TP} + \text{TN} + \text{FP} + \text{FN})$$

4 Results and Discussion

The data was firstly imported and checked for any missing values. No missing values were found in any of the columns. Now the dataset was split into two parts. One part held the features that are potentially responsible for PD classification while the other held the "status" column, which defines through binary values whether a person is PD positive or not. Now that the data was fit to be preprocessed, the data values were scaled and split into two sets—training data and testing data. The split resulted in 80% of the data being used for training while 20% was to be used for testing. Data modeling was the final step wherein five individual classifiers and five ensemble classifiers were implemented and their performance was evaluated using the previously mentioned performance metrics. The results of the individual and the ensemble classifiers are given in Tables 1 and 2, respectively.

It can be observed that from the list of individual classifiers, the k-Nearest Neighbor has performed the best with an accuracy score of 94.9%, while from the list of ensemble classifiers, XGBoost has the highest accuracy score at 94.9%. Although the best performing classifiers from both lists have the same accuracy, the ensemble classifiers have a better mean accuracy (92%) compared to the individual classifiers (87%).

It is also of utmost importance to identify what factors contribute the most toward identifying whether a person is suffering from PD or not. This is because identification of the correct factors helps in keeping a tab on them and consequently helps in the early diagnosis of PD. For this purpose, we have used one of the better performing classifiers (Random forest) as a base and calculated feature importance. We have computed the importance of 22 features that are present in the dataset and relevant to the classification of whether a person is suffering from PD or not. The feature importance is given in Fig. 1.

It can be seen that certain values have been assigned to features that signify their importance. The voice attribute that holds the most importance for PD classification according to this bar chart is "spread1" followed by "MDVP:Fo", "MDVP:Flo", etc.

Table 1 Performance of Individual classifiers

Individual classifiers	Precision (%)	Recall (%)	F1-score (%)	Accuracy (%)
Naive Bayes	92.3	75	82.8	74.4
Support vector machine	86.5	100	92.8	87.2
Decision tree	96.1	84.4	90	84.6
Logistic regression	91.4	100	95.5	92.3
K-nearest neighbor (for n = 7)	94.1	100	97	94.9

Table 2 Performance of ensemble classifiers

Ensemble classifiers	Precision (%)	Recall (%)	F1-score (%)	Accuracy (%)
Bagging meta-estimator	93.8	93.8	93.8	89.4
Random forest	93.9	96.9	95.4	92.3
Gradient boosting	93.9	96.9	95.4	92.3
Adaptive boosting	96.8	93.8	95.3	92.3
XGBoost	96.9	96.9	96.9	94.9

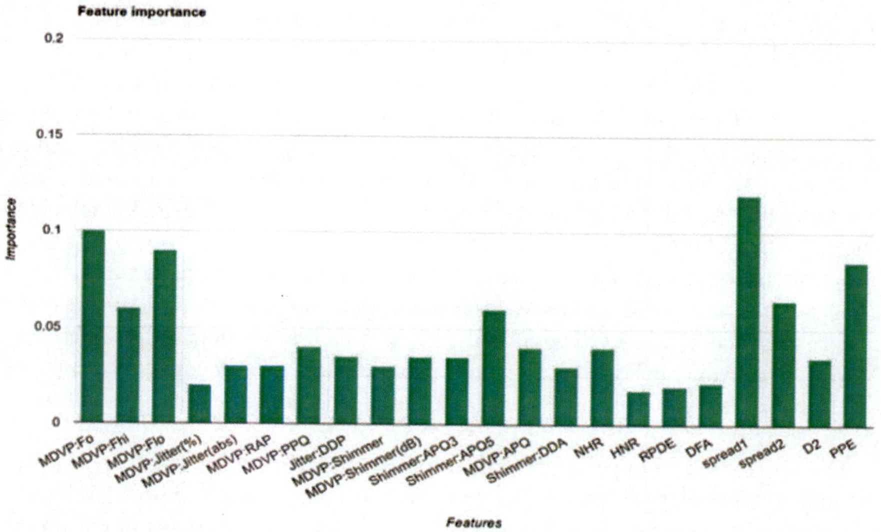

Fig. 1 Feature importance

5 Conclusion and Future Work

This research inspects a handful of ML algorithms to achieve our objective, which was to classify healthy people and people suffering from PD. The two sets of classifiers—Individual and Ensemble were utilized. Five classifiers were implemented from each set. Among the individual classifiers, the k-nearest neighbor obtained the highest accuracy (94.9%) when the value of n was 7 and among the ensemble classifiers, XGBoost obtained the highest accuracy (94.9%). Although the accuracy measure is the same for both of these classifiers, KNN has a higher recall and F1 score. It was also important to identify which quantitative acoustic measures accounted for the most for correctly classifying PD. We concluded that "spread1" was the most important measure among the 22 measures that were available in the dataset.

In the future, we intend to work on image and gait data for PD classification. We would also like to explore more algorithms to know how they perform on such data.

References

1. Goetz CG (2011) "The history of Parkinson's disease: early clinical descriptions and neurological therapies. Cold Spring Harb Perspect Med 1(1). https://doi.org/10.1101/cshperspect.a008862
2. Radhakrishnan DM, Goyal V (2018) Parkinson's disease: a review. Neurol India 66(7):S26–S35. https://doi.org/10.4103/0028-3886.226451
3. Das R (2010) A comparison of multiple classification methods for diagnosis of Parkinson disease. Expert Syst Appl 37(2):1568–1572. https://doi.org/10.1016/j.eswa.2009.06.040

4. Sriram TV, Rao MV, Narayana GVS, Kaladhar D, Vital TPR (2013) Intelligent Parkinson disease prediction using machine learning algorithms. Int J Eng Innov Technol 212–215. [Online]. Available: http://www.ijeit.com/Vol3/Issue 3/IJEIT1412201309_33.pdf
5. Bind S, Tiwari AK, Sahani AK (2015) A survey of machine learning based approaches for Parkinson disease prediction. Int J Comput Sci Inf Technol 6(2):1648–1655. [Online]. Available: http://www.ijcsit.com/docs/Volume6/vol6issue02/ijcsit20150602163.pdf
6. Sujatha J, Rajagopalan SP (2017) Performance evaluation of machine learning algorithms in the classification of parkinson disease using voice attributes. Int J Appl Eng Res 12(21):10669–10675
7. Javed Mehedi Shamrat FM, Asaduzzaman M, Rahman AKMS, Tusher RTH, Tasnim Z (2019) A comparative analysis of parkinson disease prediction using machine learning approaches. Int J Sci Technol Res 8(11):2576–2580
8. Mozhdehfarahbakhsh A, Chitsazian S, Chakrabarti S, Chakrabarti T, Kateb B, Nami M (2021) An MRI-based deep learning model to predict Parkinson's disease stages. medRxiv, pp 2021.02.19.21252081, 2021, [Online]. Available https://doi.org/10.1101/2021.02.19.21252081
9. Karan B, Sahu SS, Mahto K (2020) Stacked auto-encoder based time-frequency features of Speech signal for Parkinson disease prediction. In: 2020 international conference artificial intelligence signal processing AISP 2020, January, pp 1–5. https://doi.org/10.1109/AISP48273.2020.9073595
10. Warden MN, Searles Nielsen S, Camacho-Soto A, Garnett R, Racette BA (2021) A comparison of prediction approaches for identifying prodromal Parkinson disease. PLoS One 16(8):e0256592. https://doi.org/10.1371/journal.pone.0256592
11. Nilashi M, Ibrahim O, Ahani A (2016) Accuracy improvement for predicting parkinson's disease progression. Sci Rep 6:1–18. https://doi.org/10.1038/srep34181
12. Wenning GK, Ben-Shlomo Y, Hughes A, Daniel SE, Lees A, Quinn NP (2000) What clinical features are most useful to distinguish definite multiple system atrophy from Parkinson's disease? J Neurol Neurosurg Psychiatry 68(4):434–440. https://doi.org/10.1136/jnnp.68.4.434
13. Hall TO et al (2013) Risk prediction for complex diseases: application to Parkinson disease. Genet Med 15(5):361–367. https://doi.org/10.1038/gim.2012.109
14. Darweesh SKL, Koudstaal PJ, Stricker BH, Hofman A, Steyerberg EW, Ikram MA (2016) Predicting Parkinson disease in the community using a nonmotor risk score. Eur J Epidemiol 31(7):679–684. https://doi.org/10.1007/s10654-016-0130-1
15. Lin CH et al (2019) Blood NfL: a biomarker for disease severity and progression in Parkinson disease. Neurology 93(11):e1104–e1111. https://doi.org/10.1212/WNL.0000000000008088
16. Little MA, McSharry PE, Roberts SJ, Costello DAE, Moroz IM (2007) Exploiting nonlinear recurrence and fractal scaling properties for voice disorder detection. Biomed Eng Online 6. https://doi.org/10.1186/1475-925X-6-23
17. Suganya P, Sumathi CP (2015) A novel metaheuristic data mining algorithm for the detection and classification of Parkinson disease. Indian J Sci Technol 8(14). https://doi.org/10.17485/ijst/2015/v8i14/72685
18. Campbell MC et al (2020) Parkinson disease clinical subtypes: key features and clinical milestones. Ann Clin Transl Neurol 7(8):1272–1283. https://doi.org/10.1002/acn3.51102

Reducing Error Rate for Eye-Tracking System by Applying SVM

Nafiz Ishtiaque Ahmed and Fatema Nasrin

Abstract Electrooculography (EOG) is widely considered the most effective signal-processing technique for identifying distinct eye movements. The EOG signal was used to extract functionality to provide dependable assistance to visually impaired patients. In EOG studies, the extraction of new features is an adequate and reasonable phenomenon. The EOG system is less expensive than any other signal-processing system. Still, it has significant drawbacks, such as a high error rate. In our study, we measured the Euclidean distance error. We found that it is 3.95 cm, which is significantly less than the standard error rate. The main objective of our study is to investigate an EOG analysis with the least possible error rate. EOG is substantially less expensive than other eye-tracking systems, and the proposed method can be used to provide a consistent user experience for visually impaired patients at a low cost with a minimum error rate. Moreover, this method can be applied in drone controllers, mouse controllers, and wheelchair controllers.

Keywords Electrooculography (EOG) · Eye tracking · Arduino · BCI · Signal processing

1 Introduction

Detecting eye movement has undoubtedly become a common research subject in recent years. The movement of the eyeball is a vital sign for certain neurobiological disorders since it represents dysfunction. Electrooculography (EOG) signals are used to detect these types of disorders. Electrooculography (EOG) is a technique used to detect eye movements, gestures, and motions. EOG can be used to measure the potential difference between the cornea and the retina [1]. They have been used

N. Ishtiaque Ahmed
University of Ulsan, Ulsan, South Korea
e-mail: cse.ishtiaque@gmail.com

F. Nasrin (✉)
Jahangirnagar University, Dhaka, Bangladesh
e-mail: fatemanasrin1234@gmail.com

V. Skala et al. (eds.), *Machine Intelligence and Data Science Applications*, Lecture Notes on Data Engineering and Communications Technologies 132,
https://doi.org/10.1007/978-981-19-2347-0_4

for various purposes, including rehabilitation, robot control, wheelchair control, and desktop control [2–4].

Brain–computer interfaces (BCI) enable users to generate control commands using basic thoughts generated throughout their brains. In our study, we have used an EOG-based non-evasive BCI. Previous research has demonstrated the efficacy of EOG-based multimodal interfaces combined with the BCI. Different Euclidean distance error rates are found here at different levels at multiple stages [5]. With EOG, only feature extraction is performed, and calculating the error rate is secondary using multiple support vector machine regression [6, 7]. Previously, an EOG-based human–computer interface with an assistance feature for physically disabled people was demonstrated. Different eye movements showed the response of disabled people. The error handling was not seen because only the features were extracted [8]. Most of the cases have used SVM classifiers for feature extraction in different eye movements in the state of sleep and wakefulness [9]. In our eye-tracking tests, we used SVM and got better results. Deep learning-based classifiers have been used in the sleep stages of EOG [10–12].

EOG signal was differentiated by various eye movements of different subjects and extracted multiple features using multiple classifiers. The classifier has been prominent for detecting peak detection, blinking detection and rejection, and pattern recognition [13]. In addition, the classification system provides an accurate way of distinguishing eye movements (up, down, left, right, up, down, up, down, and down) [14]. The classification for the output assessment over EOG was used as an eye-writing recognition system [15].

In our study, we have used to identify EOG eye movements using a classification method. We have minimized the error rate for the eye-tracking system. As a result, reducing the error rate will increase the efficiency and accuracy of our study. The minimum error rate has been shown, and that will yield effective results over other systems.

2 Literature Review

The study's goal is to reduce the error rate for EOG signal analysis. The detection of eye movements using the Euclidean distance is prominent in EOG analysis. Several studies have used Euclidean distance to detect eye movements based on EOG. New features are presented in different eye movements, such as left and right, up and down, and clockwise and anticlockwise [16]. Eye-tracking based on multiple behaviors has been gradually developed [17]. Previous studies have used the forehead as a source of EOG-extracted features [18]. Other studies, such as driving protection, only evaluated Euclidean distance in different inputs [19, 20]. Most studies have not provided the Euclidean distance error. However, previous research has found that the average Euclidean error is 12.3 cm, a significantly high error rate [21].

Machine learning approaches provide better accuracy in several studies, such as auditory attention [22], natural language processing [23], anomaly detection [24], and missing data [25]. Subject-to-subject EOG signal and fast Fourier transform time signal have been implemented in our proposed process. We used the SVM as well and evaluated the results using single trials [26]. Most studies have discovered eye motions such as right, left, upright, downright, up, and down. In our proposed system, we have minimized the Euclidean error rate. Traditional research has emphasized the new function extraction in various ways without addressing error handling. We have shown the distance error rate for Euclidean distance. We discovered a Euclidean error of 3.95 cm in our study, which is lower than other studies. A low error rate will make it easier to extract new features and improve the work's accuracy. Furthermore, we have used a classification method to detect eye movement using EOG.

3 Materials and Methods

This research was conducted from three healthy right-handed subjects who self-reported normal or corrected-to-normal vision. The subject was at ease during the study. The electrode placement, data acquisition, and room setup were preprocessed, as shown in Fig. 1.

Stimuli were displayed on a 1920 × 1080 LCD color monitor with a refresh rate of 60 Hz. The viewing distance was 20 cm. The PhysioLab PSL-iEOG-2 V-100 electrooculography device was used to record eye movement, and the ADC port of the Arduino Uno microcontroller was used to convert the analog signal to digital signal at a sampling rate of 1000 Hz. We used MATLAB and Python to create the graphical user interface. We used an EOG signal-processing system to detect eye movements.

A 50° × 50° visual angel square region with a dark background was set up for the experiment. A green dot target (size ~ 0.5°) appears in the center of the screen and shifts its direction toward each of the five calibration points in each direction simultaneously. Several EOG-based eye-tracking studies used a similar visual stimuli paradigm [6]. The stimuli were pointed directly right, left, up, down, upright, upleft, downright, and downleft. We used one subject and 30 individual trials in our experiment. We used EOG to record the subjects' eye movements. The eye movements were calculated in degrees. The EOG electrode signals will be in volts, and the subject will be seated 20 cm away from the display.

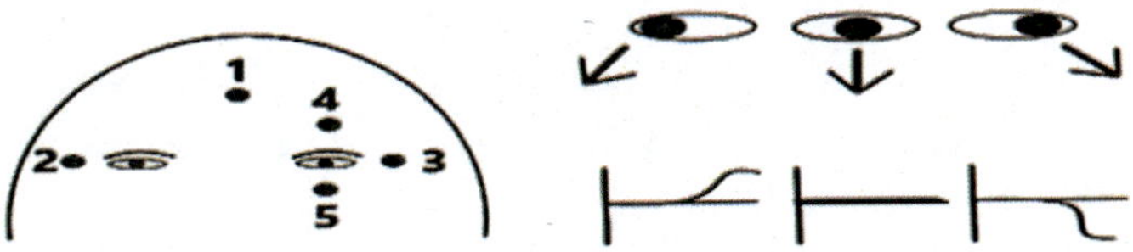

Fig. 1 EOG gaze tracking. **a** EOG electrode positions. 1: Ground electrode; 2 and 3: horizontal eye movement; 4 and 5: vertical eye movement. **b** Ocular dipole model shows variation by horizontal eye movement

Four feature runs with five-target stimuli in each direction were completed. Throughout the four runs, each subject focuses their eyes on a total of 20-target stimuli. Each subject's directions are presented in a clockwise rotation. Totally, five green dot goals appeared one by one in a 5° visual angle displacement in each direction state.

The participants were instructed to focus their eyes on the target and follow it when it appeared in the next stimulus position (Fig. 2). Five potential stimulation locations are configured at 5°, 10°, 15°, 20°, and 25° in each direction (up, down, left, right). A green dot goal (size ~ 0.5°) appears in the center of the screen for 2 s at the start of each trial. After 2 s, the green dot shifts to the next stimulus spot, 5 degrees to the left. The green dot returns to the initial stimulation spot at the center point after 1 s and stays for 2 s. Then, it starts to travel at 10°, 15°, 20°, and 25° visual angles, moving in each direction sequentially. Before each functional directional run, a five-point calibration point was applied to map eye position to screen coordinates. A visual angular error of less than 5° was needed for effective calibration. Before each trial run, fixation point stimuli were shown in the center of the screen. The first goal in each directional run appears at that spot. The Arduino Uno microcontroller was used to monitor eye movements to maintain a standard baseline continuously. To reduce contamination from 60 Hz power line noise, the analog EOG signal from 5 electrodes at each recording site was de-noised using an adaptive notch filtering technique (Fig. 3).

The analog signal was then converted to digital data at a sampling rate of 1000 Hz by connecting the EOG system to the Arduino Uno microcontroller's ADC port. Each trial lasted 1000 ms and consisted of five visual angels. The mean value of the pre-stimulus reference period (200 ms prior to stimulus onset) was subtracted from the data baseline across trials, as shown in Fig. 4. This study extracts three features from the EOG, including (1) power spectral density (PSD); in PSD, the delta band shows higher power in higher gaze angle, afterword focuses on the next appeared ball at the center, and it will continue for 5-, 10-, 15-, 25-, and 30-degree angles; (2) discrete wavelet transform; and (3) wavelet coefficients.

Fast Fourier transform is an algorithm, which computes the discrete Fourier transform of a sequence and inverse. Fast Fourier transform converts the signal from its

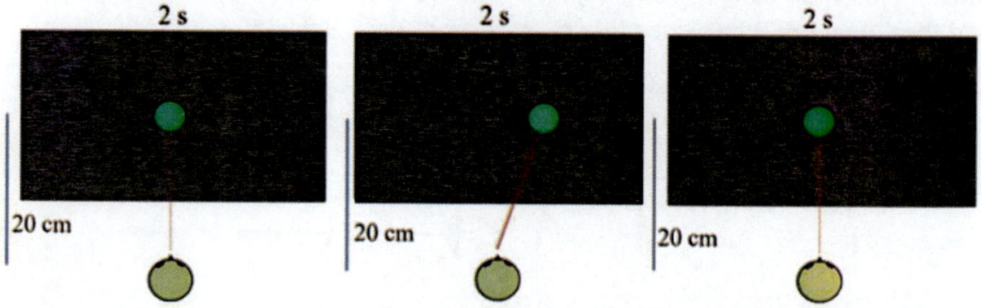

Fig. 2 Eye movement paradigm for the experimental trial shows the initial target condition for 2 s. and then, the target dot moves to next visual position and then back to its initial position

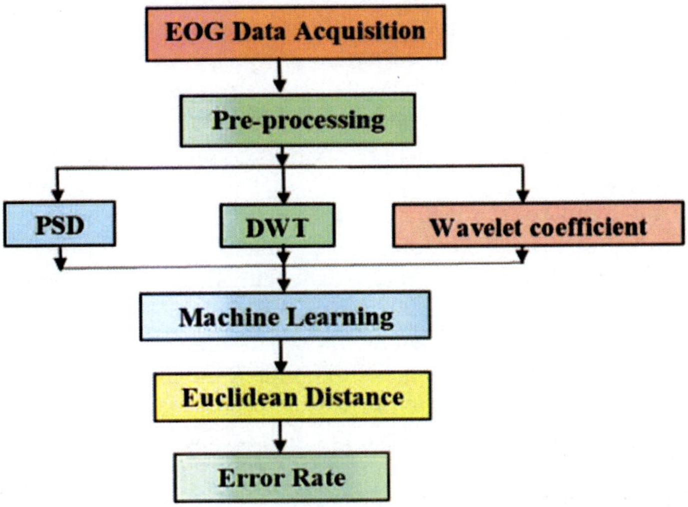

Fig. 3 EOG-based eye-tracking task classification method. This framework have applied for the SVM model

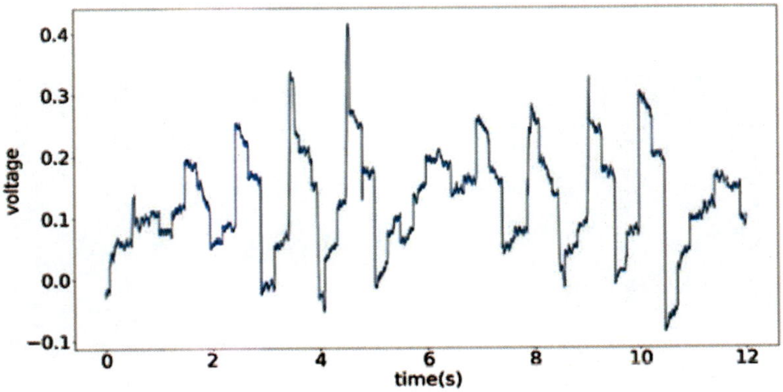

Fig. 4 Sample EOGH signal recorded, where in each second the target dot moves toward far distance and its corresponding EOG signal also shows higher voltage in higher angel

original domain to the frequency domain. Fast Fourier transform algorithm is beneficial for actual and symmetrical data analysis. Applying the power spectra in fast Fourier transform provides a plot of the portion of signal's power falling within given frequency bins. We compute the power spectra using fast Fourier transform (FFT) using the following equation,

$$P = \frac{\sum_{fb=fb_1}^{fb=fb_2} P(f)}{\sum_{fb=fb_{lo}}^{fb=fb_{hi}} P(f)} \tag{1}$$

Here, f_{bhi}, f_{blo}, f_{b1}, and f_{b2} represent the frequency sub-band. Next, the band-power feature vectors are extracted from the power spectra in each of the following frequency bands: delta (1–4 Hz), theta (4–7 Hz), alpha (8–13 Hz), beta (14–30 Hz), low gamma (30–50 Hz), high gamma (70–200 Hz), and all frequency components (1–200 Hz). As shown in Fig. 5, the low-frequency component, especially the delta band, is clearly responding according to the visual angels; we give a special focus on the delta band in this study. Finally, for normalizing the power spectra, we divided all trials into testing trials and training trials and then calculated z-scores from each power spectra across the training trials for the individual channel. We calculated z-scores from each power spectra across the testing to normalize testing trials without using those trials. We used a discrete wavelet in our research. Signal and time–frequency analysis are two applications of continuous wavelet tools. This analysis used discrete wavelet transforms to process the primary signal at 10-, 15-, and 20-degree gaze angles. Wavelet transform is a helpful method for feature extraction and image processing because it can process signals sparsely. It will also provide an accurate representation of signal processing. For analyzing EOG signals, the wavelet transform is a valuable method.

A wavelet is a wave-like oscillation with an amplitude that starts at zero, increases, and returns to zero. When a wavelet transform is applied to a function, it considers it in terms of localized oscillations in both time and frequency. The estimation or

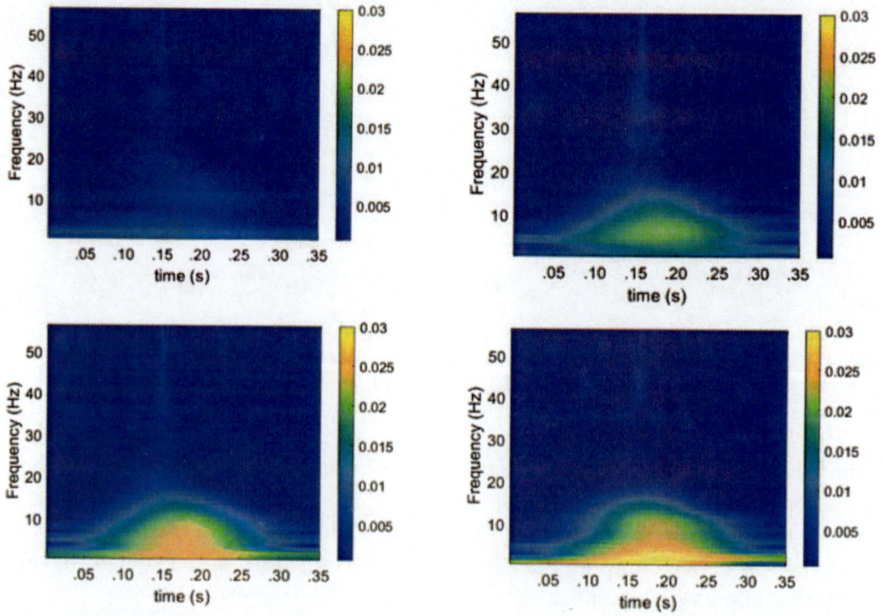

Fig. 5 Power spectral density shows strong power in low-frequency band component from 0°,5°,10°, and 15° in the figure. This feature reveals that low-band component from power spectra can make better classification

scaling coefficients are the signal's low pass presentation, and the details are the wavelet coefficients.

As a result, the importance of wavelength coefficients in EOG is for better signal representation. SVM classifiers are effective in high-dimensional spaces and use little memory. The LIBSVM toolbox was used for building the SVM classifier. The SVM classifier was trained by using the z-score of the band power from the power spectra and the DWT and wavelet coefficients.

4 Results

4.1 Simulation Platform

The user was seated comfortably in front of the monitor, and he completed his task as per the instruction given. Five wet electrodes were used to measure the EOG signals, and there were five channels to record the EOG signals. The eye movements experiment consisted of five parts. The first was a pretest used to adjust the classification parameters and the measurement system according to the user. Second, the aim of identifying the user's eye movements features and then train the subject on the EOG measurement system. We checked each of the directions fifteen times. In the individual trials, the green dot was in the center of the screen, and the center point was the starting point of the experiment. After two seconds, the user was instructed to look toward the green dot that was presented on the screen. After that, another two seconds the green dot vanished, and a new trial was started. The user should look at the angle that was instructed for the measurements. We collected all the eye movements' measurements through the EOG signal by following this instructed experimental procedure. In the experiment, a total of 20-target stimuli and 30 trials have been used. In our study, we calculated the distance and measured the distance error. The experimental data analysis of this study was implemented in MATLAB. In this research, we minimize the distance error.

4.2 Applied Classifier

Differentiation and peck detection play a significant role in the classification algorithm. In our study, we have used an SVM classifier. Using a classifier is prominent for eye movements' experimental results. We applied a classifier in every angle of the eye movements and calculated the distance error rate for individual angles.

4.3 EOG Analysis

Detecting eye movements through EOG is efficient. In our study, we have used the higher positive potential for a higher positive gaze angle, and a higher negative potential is also recorded for a higher negative gaze angle. Here, EOGH represents a linear relationship between potential and gaze angle. For measuring the degree of the angle, we have used a total of 5 stimuli and, over the four runs, give their gaze attention to the total of 20-target stimuli. Within each direction (up, down, left, right), 5 locations for the stimulus have been used, and the degree angle was 5°,10°,15°, 20°, and 25°. After 1 s, the green dot returns to the initial stimulus position at the center point for 2 s. After 2 s, the green dot moves to the next stimulus position, suppose in the left direction at 5°. Then, it continues to move at 10°, 15°, 20°, and 25° visual angel and sequentially continues in each direction. In every two points, we were calculating the distance and detecting the error rate of that distance. We successfully made our experimental result, and we found a minimum error rate by calculating Euclidean distance.

4.4 Performance Evaluation

In our study, we have calculated individual distances between the angles. When the eye movements were detected, it counted the time of each eye movement. After that, we applied the SVM classifier. Figure 6 shows the confusion matrix of SVM classification for four directions using delta band of PSD feature. The performance of the classification was evaluated by using the following performance measures:

$$\text{Accuracy} = \frac{\text{TP} + \text{TN}}{\text{TP} + \text{FP} + \text{TN} + \text{FN}} \times 100 \quad (2)$$

TP represents the true positive, TN represents the true negative, FP represents the false positive, and FN represents the false negative. The model has achieved 75% accuracy to detect the horizontal positive visual angle of four-target stimuli. For horizontal negative angle, the vertical positive and negative visual angle is the model that makes 60, 35, and 45% accuracy. The random chance of label is 25%. Figure 7 represents all applied features' accurateness for detecting horizontal positive visual angles.

4.5 Euclidean Distance

Euclidean distance calculates the distance between two points. In our study, we have used Euclidean distance for calculating the distance of eye movements. In this

A. Horizontal positive eye movement detection

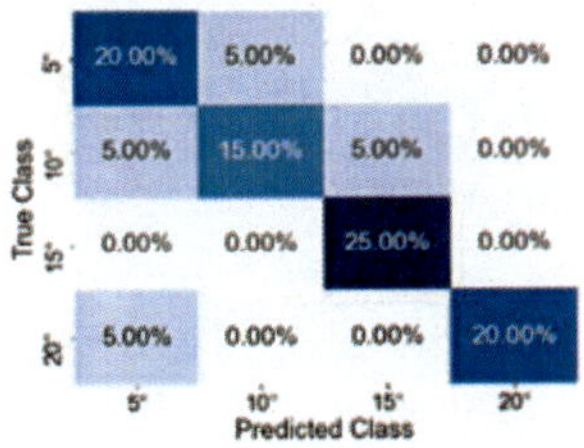

B. Horizontal negative eye movement detection

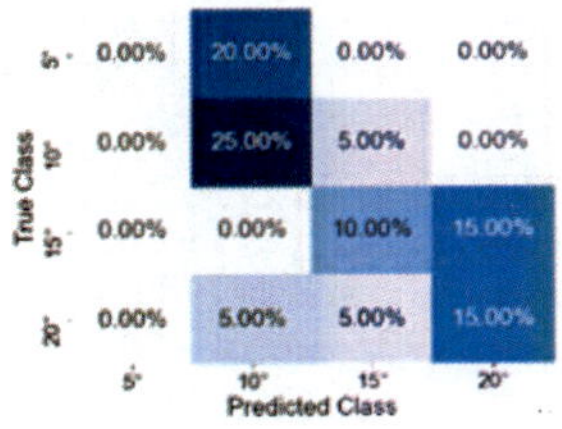

C. Vertical positive eye movement detection

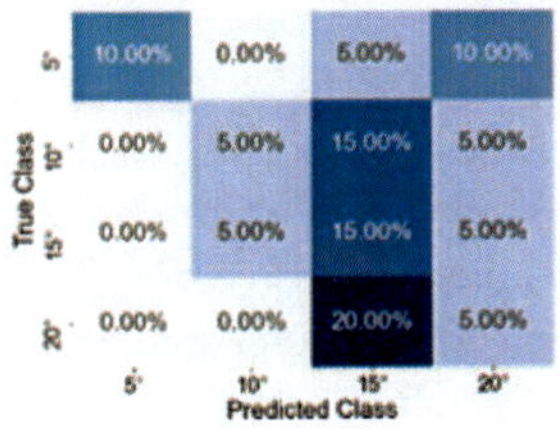

D. Vertical positive eye movement detection

Fig. 6 Normalized confusion matrices of the SVM classification for all four directions. In each direction, total of four-target class is classified

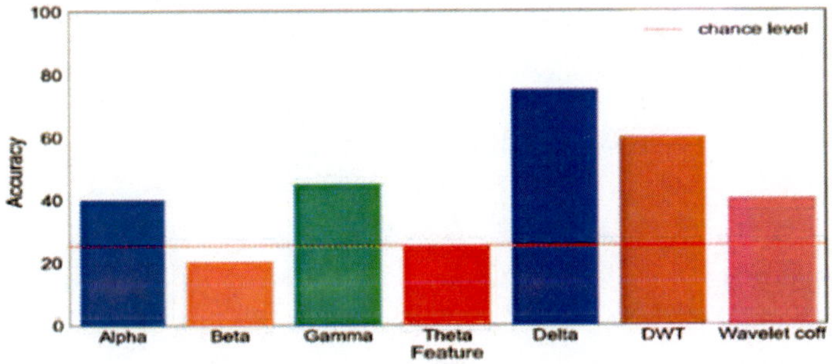

Fig. 7 EOGH positive target point classification accuracy from all the features by SMV classifier. There the delta band power from PSD gives the highest accuracy

study, we collected different angles of eye movements through EOG and measured individual distance. Here, the target visual stimuli position is the main reference point, where the classifier's distance error measures the error rate. The Euclidean distance error is calculated by given Eq. (3)

$$d_{\text{err}} = \sqrt{\sum_{i=0}^{N-1}(V_{\text{ref}(i)} - V_{\text{unknown}(i)})^2} \tag{3}$$

Table 1 Euclidean distance error of the evaluation indices

Trial	EOGH+	EOGH−	EOGV+	EOGV−
1	1.336	2.667	2.662	3.988
2	1.326	4.028	4.006	5.361
3	4.01	1.326	6.662	7.008
4	1.326	2.667	10.686	5.346
5	1.326	2.667	6.668	3.998
Avg	**1.865**	**2.671**	**6.136**	**5.141**

Proposed method shows less Euclidean distance error (cm) and achieves a Euclidean distance error of 1.865 cm on average across all trails in EOGH+ , (EOGH−, (EOGV +), and (EOGV−) direction. The mean Euclidean distance error is 1.865, 2.671, 6.136, and 5.141 cm accordingly

By calculating individual distance, we found deficient error. In most of the cases, the Euclidean error rate was not detected. Here, we found only a 1.86 cm error rate for Euclidean distance in the positive horizontal direction and 2.671 cm in the horizontal negative direction, as shown in Table 1. On average, the horizontal direction Euclidean error rate is 2.265 cm. As the vertical EOG signal does not have a smooth linear relation with the visual angel, the Euclidean distance error is higher than the horizontal direction error. Accordingly, we acquire a 6.136 cm and 5.141 cm error rate for Euclidean distance in the positive vertical direction and vertical negative direction.

The grand average Euclidean distance error rate was achieved using the proposed method in 3.95 cm, which achieved the minimum Euclidean distance error rate. A lower error rate will provide a better user experience and increase usability and extraction features at the cheapest rate. The cost of the feature will reduce, and it will be affordable for any physically impaired patients.

5 Discussion

The results of repeated eye movements in different eye angles using the SVM classifier demonstrated that the system's precision was consistent across participants. The results have shown that our system correctly measured the angle and eye movements. The experiment involved a wide range of participants, each with a unique angle of eye movement, which increased the precision of our distance measurement. Each angle of the eye movements was determined separately for the Euclidean distance after the Euclidean distance was successfully measured. We examined each eye movement and validated the patterns collected from the subjects. We independently checked each subject's records. The findings of the tests show that the SVM model is well suited for locating and recognizing EOG signals for detecting eye motions. In our research, we demonstrated how to reduce the error rate in a gaze-tracking device. For each of the four ways, we normalized the SVM classification confusion matrices.

In addition, we have shown the Euclidean distance error of the evaluation indices in Table 1. Figure 7 depicts the precision of all implemented functions in identifying horizontal positive visual angles. When it comes to extracting features from EOG, we have applied it to three different types of features. To begin, consider the power spectral density (PSD).

The delta band in the PSD had a greater power at a higher gaze angle. Then, we thought about the next ball that appeared in the center. It then went on for 5-, 10-, 15-, 25-, and 30-degree angles. The discrete wavelet transform (DWT) was then extracted, and the wavelet coefficients were then interpreted. Furthermore, EOG signals were reported spectrographically and had a high power in the low-frequency portion. In this study, the EOG signal was used to distinguish between the subjects' distinct eye motions. Every trial has been carefully monitored. We successfully measured every eye movement using the EOG signal in our research. Eye gestures based on EOG have several benefits. The eye motions can be tracked in real time by the EOG machine. EOG can track all eye movements in a short period in this way. EOG is a low-cost alternative to other systems for detecting eye movement, and it offers a reliable signal operation. Previous experiments have shown that using EOG for eye movements can be helpful for extracting a variety of features. The feature extracted from EOG can be easily implemented in every other application.

Furthermore, EOG is the most affordable signal-processing device. Each angle calculation can be compared frame by frame with SVM classification. Precision and recall were used for any individual participant's particular outcome. In this case, the SVM classifier produces results in the shortest amount of time. We achieved a maximum mean accuracy of 75% in this region. For the distance, we got a 3.95 cm average Euclidean error rate. This study's other distance yields an error-free Euclidean distance outcome.

6 Conclusion

The critical challenge in this research was to correctly detect eye movements at various angles and reduce the error rate. Minimizing the distance error rate by using Euclidean distance is difficult in this research study. We have computed the distance and determined the Euclidean error rate for each eye angle position. We discovered a 3.95 cm Euclidean error in our study.

Furthermore, EOG has a better signal transmission technique for detecting eye movements. The EOG system will be more accurate if the error rate is reduced, and features are extracted using the EOG system. As a result, reducing the error rate will increase the efficiency and accuracy of our study. The minimum error rate has been shown, and that will yield effective results over other systems.

7 Future Works and Limitations

There are many branches to explore about the eye-tracking system. In our future study, we will implement this method. We will apply multiple methods for the eye-tracking system. We will represent the comparison between different methods. We will add which method is the best work for the eye-tracking system. The limitations of our study are limited experimental subjects. In our future study, we will add different age participants in our study.

References

1. Chen Y, Newman WS (2004) A human-robot interface based on electrooculography. IEEE Int Conf Robot Autom 1:243–248
2. Zhang Ma JY, Cichocki A, Matsuno F (2015) A novel EOG/EEG hybrid human-machine interface adopting eye movements and ERPs: application to robot control. IEEE Trans Bio-Med Eng 62(3):876–889
3. Úbeda A, Iáñez E, Azorín J (2013) An integrated electrooculography and desktop input bimodal interface to support robotic arm control. IEEE Trans Human-Mach Syst 43:338–342
4. Paul G, Cao F, Torah R, Yang K, Beeby S, Tudor J (2014) A smart textile based facial EMG and EOG computer interface. IEEE Sens J 14:393–400
5. Paul G, Cao F, Huang QT, Wang HS, Gu Q, Zhang K, Shao M, Li Y (2018) An EOG-based human-machine interface for wheelchair control. IEEE Trans Biomed Eng 65:2023–2032
6. Iáñez E, Úbeda A, Azorín J (2011) Multimodal human-machine interface based on a brain-computer interface and an electrooculography interface. In: Annual international conference of the IEEE engineering in medicine and biology society. pp 4572–4575
7. Khushaba RN, Kodagoda S, Lal S, Dissanayake G (2011) Driver drowsiness classification using fuzzy wavelet-packet-based feature-extraction algorithm. IEEE Trans Biomed Eng 58:121–131
8. Torres-Valencia CA, Álvarez MA, Orozco-Gutiérrez ÁA (2014) Multiple-output support vector machine regression with feature selection for arousal/valence space emotion assessment. In: 36th annual international conference of the IEEE engineering in medicine and biology society, pp 970–973
9. English E, Hung A, Kesten E, Latulipe D, Jin Z (2013) EyePhone: a mobile EOG-based Human-Computer Interface for assistive healthcare. In: 6th international IEEE/EMBS conference on neural engineering (NER). pp 105–108
10. Khalighi S, Sousa T, Oliveira D, Pires G, Nunes U (2011) Efficient feature selection for sleep staging based on maximal overlap discrete wavelet transform and SVM. In: Annual international conference of the IEEE engineering in medicine and biology society. pp 3306–3309
11. Korkalainen H, Aakko J, Nikkonen S, Kainulainen S, Leino A, Duce B, Afara IO, Myllymaa S, Toyras J, Leppanen T (2020) Accurate deep learning-based sleep staging in a clinical population with suspected obstructive sleep apnea. IEEE J Biomed Health Inform 24:2073–2081
12. Zhang B, Zhou W, CaiH, Su Y, Wang J, Zhang Z, Lei T (2020) Ubiquitous depression detection of sleep physiological data by using combination learning and functional networks. IEEE Access 94220–94235
13. Lin C, King J, Bharadwaj P, Chen C, Gupta A, Ding W, Prasad M (2019) EOG-based eye movement classification and application on HCI baseball game. IEEE Access 7:96166–96176
14. Wu SL, Liao LD, Lu SW, Jiang WL, Chen SA, Lin CT (2013) Controlling a human-computer interface system with a novel classification method that uses electrooculography signals. IEEE Trans Bio-Med Eng 60:2133–2141

15. Lee KR, Chang W, Kim S, Im C (2017) Real-time "Eye-Writing" recognition using electrooculogram". IEEE Trans Neural Syst Rehab 25:37–48
16. Puttasakul T, Archawut K, Matsuura T, Thumwarin P, Airphaiboon S (2016) Electrooculogram identification from eye movement based on FIR system. In: 9th biomedical engineering international conference (BMEiCON). pp 1–4
17. Nugrahaningsih N, Porta M, Ricotti S (2013) Gaze behavior analysis in multiple-answer tests: an eye tracking investigation. In: 12th international conference on information technology based higher education and training. pp 1–6
18. Cai H, Ma J, Shi L, Lu B (2011) A novel method for EOG features extraction from the forehead. In: Annual international conference of the IEEE engineering in medicine and biology society. pp 3075–3078
19. Breuer A, Elflein S, Joseph T, Termöhlen J, Homoceanu S, Fingscheidt T (2019) Analysis of the effect of various input representations for LSTM-based trajectory prediction. IEEE Intell Transp Syst Conf (ITSC) 2728–2735
20. Jin L, Guo B, Jiang Y, Wang F, Xie X, Gao M (2018) Study on the impact degrees of several driving behaviors when driving while performing secondary tasks. IEEE Access 65772–65782
21. Kang M, Yoo C, Uhm K, Lee D, Ko S (2018) A robust extrinsic calibration method for non-contact gaze tracking in the 3-D space. IEEE Access 48840–48849
22. Nasrin F, Ahmed NI, Rahman MA (2020) Auditory attention state decoding for the quiet and hypothetical environment: a comparison between bLSTM and SVM. In: 2nd international conference on trends in computational and cognitive engineering (TCCE-2020). vol 1309. pp 292–301
23. Hasan MJ, Badhan AI, Ahmed NI (2018) Enriching existing ontology using semi-automated method. Future of Inf Commun Conf 886:468–478
24. Nasrin F, Yasmin A, Ahmed NI (2021) Anomaly detection method for sensor network in under water environment. In: International conference on information and communication technology for sustainable development (ICICT4SD). pp 380–384
25. Sumit SS, Watada J, Nasrin F, Ahmed NI, Rambli DRA (2021) Imputing missing values: reinforcement bayesian regression and random forest. In: Kreinovich V, Hoang Phuong N (eds) Soft computing for biomedical applications and related topics. Studies in Computational Intelligence vol 899. Springer, Cham
26. Jialu G, Ramkumar S, Emayavaramban G, Thilagaraj M, Muneeswaran V, Rajasekaran MP, Hussein AF (2018) Offline analysis for designing electrooculogram based human computer interface control for paralyzed patients. IEEE Access 6:79151–79161

Eye-Gaze-Controlled Wheelchair System with Virtual Keyboard for Disabled Person Using Raspberry Pi

Partha Chakraborty, Md. Mofizul Alam Mozumder, and Md. Saif Hasan

Abstract Eye-gaze technology enables people to operate a system by gazing at instructions on a screen. It can assist paralyzed people who are unable to communicate with others due to their inability to speak or use their muscles well enough to write on paper or type on a keyboard. So, this paper presented an eye-gaze-driven wheelchair with a virtual keyboard for paralyzed people. The main aim was to make it easier for individuals with impairments to use wheelchairs and type on a keyboard, as well as to eliminate the need for assistance for the impaired person. The system integrates an eye-controlled wheelchair with a virtual keyboard using a low-cost system. In this system, the wheelchair was controlled by eye movement and the virtual keyboard was controlled by eye blinking. The camera was positioned in front of the disabled person to take real-time images and track eye movement. The wheelchair travels left, right, and forward according to eye movement. The whole system was controlled by a Raspberry Pi.

Keywords Eye gaze · Eye blink · Eye movement · Virtual keyboard · Python · Raspberry Pi · Wheelchair

1 Introduction

The eye is regarded as a wonderful tool for understanding human communication, capable of processing information associated with the surrounding view as well as providing a correct response. It may also be a blessing for severely disabled people. Because the human body is heavily restricted from making regulated movements in any organ and perhaps even in the head due to a variety of illnesses, such as complete paralysis, lock-in syndrome, Parkinson's disease, arthritis, multiple sclerosis, and spinal cord injury. Approximately 132 million individuals with disabilities require a wheelchair [1]. Conventional wheelchairs do not adequately support severely disabled people. Some people are unable to drive a wheelchair even with

P. Chakraborty (✉) · Md. Mofizul Alam Mozumder · Md. Saif Hasan
Department of Computer Science and Engineering, Comilla University, Cumilla 3506, Bangladesh
e-mail: partha.chak@cou.ac.bd

V. Skala et al. (eds.), *Machine Intelligence and Data Science Applications*, Lecture Notes on Data Engineering and Communications Technologies 132,
https://doi.org/10.1007/978-981-19-2347-0_5

a joystick. In order to easily and safely operate the wheelchair, researchers have examined different communication technologies, such as eye tracking, electrooculography (EOG), brain–computer interfaces based on electroencephalogram (EEG), and speech recognition systems. Each technology, however, has limitations that restrict its application in daily life. Brain–computer interfaces, for example, are useful for patients in the most advanced stages of illness [2], but they are sensitive to movement and interruptions, and they are unable to implement regular tasks. Although voice recognition allows individuals to type quickly, they struggle to navigate the wheelchair and are unreliable in crowded surroundings [3]. Eye tracking is effective in controlling the wheelchair and does not require any physical contact. In addition, many disabled people have healthy eyes. A system can be set up to move the wheelchair to the target. Eye-tracking technology can be used to improve communication, mobility, and independence of disabled people. Eye tracking has been progressively implemented in various applications, like the human attention prediction system [4–6], eye-controlled wheelchair [7], automatic student attendance system [8], and eye-controlled keyboard [9–11]. The eye movement-based wheelchair control system was first introduced in 2007 with the installation of an eye tracker inside a head-mounted display (HMD) [12]. The computer processes video, measures the user's face and eye direction, and moves the wheelchair to that location. A number of challenges have hindered the comprehensive use of these technologies, such as acceptable eye detection and tracking quality [13]. In addition to a wheelchair, disabled individuals also require a virtual keyboard to convey their feelings. They cannot drive a wheelchair and cannot even write since they cannot regulate their hand motions. The goal of this study was to design an eye-gaze-driven wheelchair system with a virtual keyboard. We have presented a system that will assist disabled individuals in navigating to their destination as well as typing on a virtual keyboard utilizing their eyes. The eye-gaze-controlled wheelchair with a virtual keyboard can help a disabled person to move the wheelchair and type text without touching the keyboard. Our proposed system can minimize the effort required to move wheelchairs safely and easily in the desired direction and design a virtual keyboard for disabled and physically challenged people. Eye-gaze-controlled wheelchair and virtual keyboard technology can assist paralyzed people to live a regular life.

2 Literature Review

Eye tracking mostly determines eye-ball movement and eye-gaze direction and uses two approaches: video-based and EOG-based. An EOG signal-based wheelchair control system was proposed in [14]. Low accuracy and lack of robustness are limitations of EOG. So, researchers tried to find alternatives. Video-based eye tracking is divided into two approaches: image-based approaches as in [15–22] and infrared approaches as in [23]. A novel algorithm for object detection using HOG feature matching was introduced in [24–27]. A new wheelchair control system was presented by Nguyen and Jo [28] utilizing a head-free assumption where a three-dimensional directional sensor is fitted for monitoring head position. The proposed solution was complex and involved a large measurement attempt. Researchers were trying to merge eye-gaze motives with facial orientations as a novel hand-free demonstration of wheelchair control [29]. Jia and Hu demonstrated a wheelchair that was guided by head gestures [30]. The device cannot be used by individuals who cannot move their heads. A conventional wheelchair with a joystick controller and a conventional keyboard is not an option for handicapped individuals who are unable to move their muscles. In such circumstances, eye-gaze tracking-based wheelchairs and keyboards are useful for disabled people.

3 Methodology

The system's primary objective was to assist disabled people to navigate a wheelchair while typing on a virtual keyboard through their eyes. To meet the objective, we used face and eye region detection, eye movement, and eye-gaze tracking techniques. We utilized two options on the screen to get started with the system: one was to use the wheelchair and the other was to use the virtual keyboard. If we choose the wheelchair option with eye blinking, we can control the wheelchair movement with eye movement. We can view our virtual keyboard and type on it by blinking our eyes if we pick the virtual keyboard option. Initially, the system used the Raspberry Pi camera to capture real-time frame-by-frame pictures from a video stream. The initial step was to determine the exact region of the face, followed by the precise position of the eyes. Figure 1 depicts the proposed system's flowchart, which demonstrates the system's overall techniques: The Python Dlib library was used to locate facial landmarks using a group of regression trees using Kazemi and Sullivan's face alignment in one millisecond [31]. In this approach, face and eye recognition is accomplished using facial landmarks. The system can detect up to 68 specific facial features using face landmark recognition technology. The system uses facial training sets to determine where the specific points are in the facial structure. Then, the system draws the same points in case of interest in another image. The draw program estimates the possible distance between the main points [32]. The Dlib library displays a 68-point plot from the input image in Fig. 2.

3.1 *Face and Eye Detection*

Face detection is the process of identifying human faces in a digital image or video. It is being used in a variety of applications. The Dlib library is extensively used to locate faces using a variety of approaches. The HoG face detector from the Dlib library was utilized to recognize faces in this system. We can identify 68 unique facial landmarks using the facial landmark detection technique. A separate index is supplied for each of the 68 points. To detect eyes, we simply need to identify the

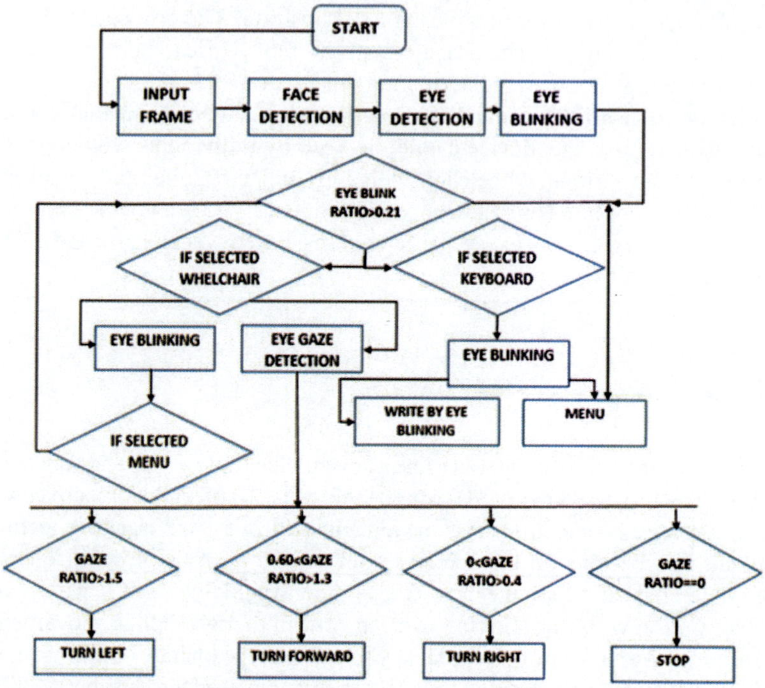

Fig. 1 Flowchart of the system

Fig. 2 Dlib facial landmark plot

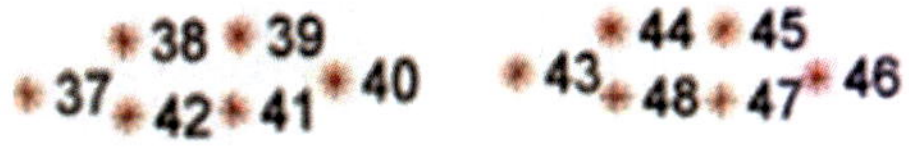

Fig. 3 Facial landmarks' eye points

points of the eyeballs. The following are the point indices for the eyes: points of the left eye: 37, 38, 39, 40, 41, and 42 and points of the right eye: 43, 44, 45, 46, 47, and 48. (Fig. 3)

3.2 Eye Blinking Detection

Blinking of the eyes can be detected by looking at important facial features. We must examine points 37–48, which represent the eyes, while considering eye blinks. Soukupova´ and Jan Cech used an equation to calculate the aspect ratio of the eyes when the eyes blink in real time using facial landmarks in [33]. The eye's aspect ratio is a measurement of the condition of the eye opening. The equation below can be used to determine the eye aspect ratio (EAR).

$$EAR = \frac{(\|p2 - 6\| + \|p3 - 5\|)}{(2\|p1 - 4\|)}$$

where p1-p6 are two-dimensional facial landmarks. The aspect ratio remains constant when the eyes are open. The eye aspect ratio quickly declines while the eyes are closed. When the aspect ratio goes below a particular threshold, the system can detect whether a person's eyes are closed or open. Applying facial landmarks to detect eye movement, the landmarks 37, 38, 39, 40, 41, and 42 relate to the left eye area. The landmarks 43, 44, 45, 46, 47, and 48 belong to the right eye area. After obtaining the left eye's coordinates, we can prepare a mask to accurately extract the region of the left eye while excluding all other inclusions. The right eye is treated in the same way. Now, the eyes are separated from the face. The picture may be sliced into a square form, so the eyes' end points (top left and bottom right) are used to get a rectangle. We also determined a threshold value to detect eyes. The eyes are gazing to the right if the sclera is more visible on the left. The eyes were then changed to gray levels, and a threshold was added. Then, we separated the left and right white pixels. The eye ratio was then calculated.

3.3 Eye-Gaze Detection and Eye Tracking

The gaze ratio indicates where the particular eye is looking. Usually, both eyes look in the same direction. Therefore, the gaze ratio of one eye can indicate the gaze direction of both eyes. To get a more accurate gaze direction, find the gaze ratio of both eyes and calculate the average value. If the user only looks to the right on one side, the user points the pupils and iris in the right direction, while the rest of the eye is completely white. The others are irises and pupils, so it is not completely white, and, in this case, it is very easy to find how it looks. If the user sees from the other side, the opposite is true. The white area between left and right is balanced when the user looks at the center. We take the eye area and make a threshold to separate the iris and pupils from the whites of the eyes. In this way, we can track eye movements.

3.4 Virtual Keyboard

We designed an on-screen virtual keyboard. The keys of the keyboard are arranged in the sequence Q, W, E, R, T, Y, U, I, O, P, A, S, D, F, G, H, J, K, L, Z, X, C, V, B, N, M, and some signs. Here, we also have SPACE, BACKSPACE, and CLEAR keys. For directly going to the next row or previous row of the keyboard, we have the NEXT and PREV keys. To return to the main menu, we have the MENU key. Eye blink is used to press the keys of the virtual keyboard. In the blink of an eye, the keys on the screen keyboard are typed over time. There are 48 characters on the keyboard.

3.5 System Model Design

The system was completely independent, and all modules operated independently. This system requires a power supply for each component, with conventional power being used for the Raspberry Pi, motors, Pi camera, and sensors. Figure 4 depicts the system's effectiveness. The system works according to the eye pupil position and moves the wheelchair left, right, and forward. When the pupil of the eye moves to the left, the wheelchair motor runs on the left side, and when the eye moves to the right, the right side of the motor will move [32]. If the eye is in the middle, the motor will also move forward. If problems are found, the system stops working. Eye blink logic is applied to starting and stopping wheelchair systems [33]. Figure 4 depicts a block diagram of a wheelchair system. The Raspberry pi camera was installed in front of the user. The distance between the eyes and the camera is fixed and must be in the range of 15 to 20 cm.

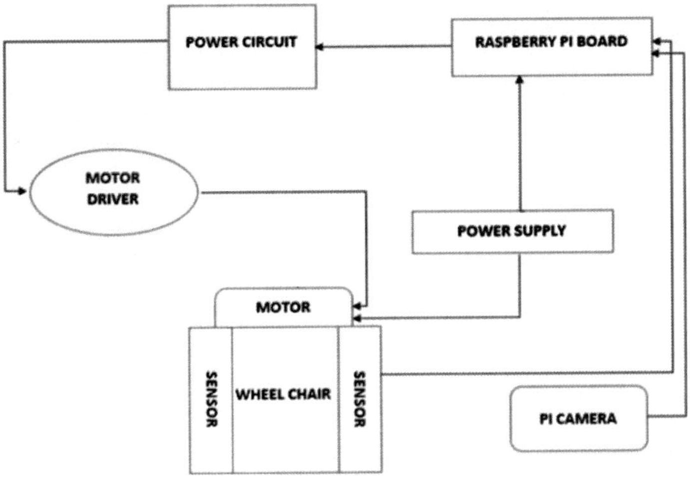

Fig. 4 Design model

3.6 System Environment

1. Raspberry Pi: The Linux operating system was installed on the Raspberry Pi. It controlled the motor driver circuit which activated the Raspberry Pi's GPIO pin.
2. 2 Pi camera: The Pi camera module is a small, portable camera that works with the Raspberry Pi. It communicates with the Raspberry Pi via the MIPI camera serial protocol.
3. Ultrasonic sensors: Obstacles in the way of wheelchair systems were detected using ultrasonic sensors. The ultrasonic sensor was connected to the Raspberry Pi.

It collected information and calculated the distance between the wheelchair and the impediments. The motor stopped operating the wheel if an impediment was sensed near the wheelchair.

4. DC motors: Two 12 V DC motors drive the wheelchair forward, backward, left, and right. The Raspberry Pi board communicates with the L298N motor driver. The L298N is a twin Hi-bridge driver that simultaneously controls the direction and speed of DC motors. DC motors with a voltage range of 5–35 V and a peak value of 2 amps may be driven by this module.
5. Putty software: Putty is a terminal emulator application for transferring network files. The Raspberry Pi was connected to the PC using the Putty program.
6. OpenCV library: OpenCV computer vision issues were solved using the Python development project library. It supports Windows, Linux, Mac OS, iOS, and Android and includes C, C++ , Java, and Python interfaces. This system uses the Python OpenCV library.

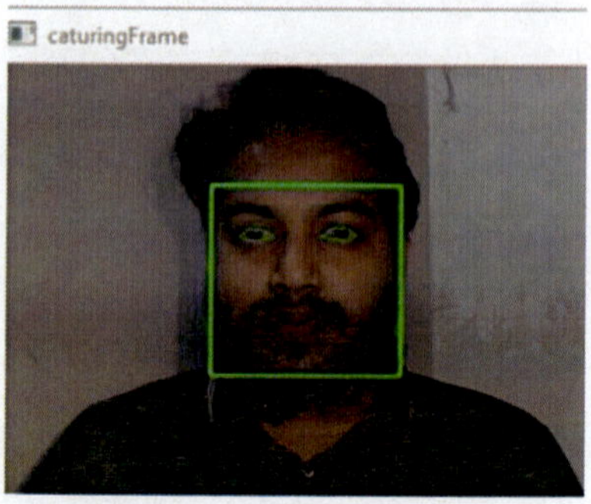

Fig. 5 Face detection and eye region detection

4 Experimental Results

Our system was executed in real time. The outcomes of the implementation are shown with the appropriate figures in this section.

4.1 Face and Eye Detection

We used a modification to a standard method for object detection to initialize Dlib's pre-trained face detector for face detection. We used Dlib's pre-trained facial landmark classifier to discover 68 particular facial landmarks for eye detection. Then, we detected corresponding landmark indexes that represent the eyes. The face region was initially extracted frame by frame from the video. Then, from the facial region, the eye regions were extracted. Then, to identify the eyeball more precisely, we used a threshold value (Fig. 5).

4.2 Eye-Gaze and Blinking Detection

We measured the eye-gaze ratio for both eyes and then calculated the average gaze ratio of the eyes to identify eye gazing. We can figure out where the user is gazing by looking at the value of the eye-gaze ratio. To distinguish between normal and desired eye blinking, we computed the eye blinking ratio (Fig. 6).

4.3 Wheelchair Navigation and Keyboard Typing

After starting the application, we were provided with the main menu. A wheelchair and a virtual keyboard were two options on the main menu. By blinking our eyes,

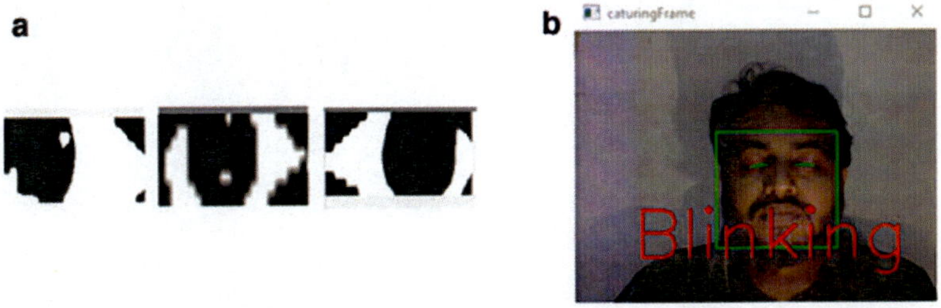

Fig. 6 **a** Eye-gaze detection **b** Eye blinking detection

Fig. 7 Wheelchair navigation **a** turn left, **b** go forward, **c** turn right

we may choose the options. We utilized eye-gaze ratio to navigate in a wheelchair. The wheelchair turned left as we gazed to the left. The wheelchair turned right as we gazed to the right. When we looked at the forward wheelchair, it moved forward in a straight line. When we provided instructions to the wheelchair, it immediately worked. When we provide commands to the wheelchair, it immediately responds. Every key was gradually illuminated. We were mostly illuminating each key for 30 frames before switching on to the next. We immediately closed our eyes for a few moments to perform blinking when our desired key was enlightened. The key would be picked and shown on the whiteboard by eye blinking, as illustrated in Fig. 8. We implemented a while loop to sequentially glow the options in white while selecting the main menu. When picking the main menu options, the white color option will be selected if the blinking ratio is greater than 0.21. We designed a loop cycle to successively glow the keys with white color for virtual keyboard typing. If the blinking ratio is more than 0.21 while typing on the virtual keyboard, a white color key from the keyboard will be written and displayed on the whiteboard. This virtual keyboard has a small amount of latency. The keyboard will be typed by eye blinking if we close our eyes for a short amount of time (Fig. 7).

4.4 Result Accuracy

This experiment was carried out in a lighter area. The wheelchair system received video processing results that were generated based on the position of the eyeball.

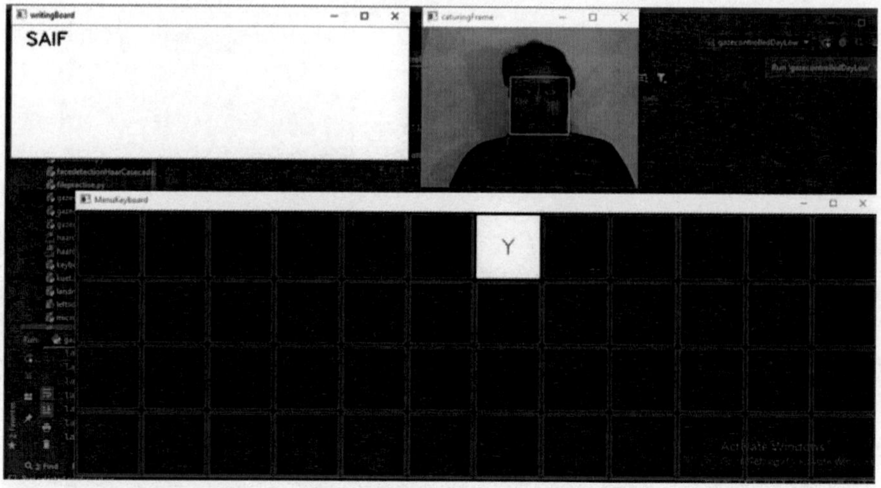

Fig. 8 Key selection and writing by eye blinking

The motor controlling circuit was attached to the Raspberry Pi, the motors were powered by the battery, and the motor-driving IC was regulated by the relay. The system generates instructional signals to execute the necessary operations, such as left, right, and front, and stops to activate GPIO pins constantly as a response to video processing. Obstacles were detected using ultrasonic sensors. It also calculated the distance between a wheelchair and an impediment. When an impediment was too near to the wheelchair, the motor triggered and the wheelchair grinded to stop. The wheelchair went ahead in all the required directions with a good response in this manner.

5 Conclusion

This study presented an eye-controlled wheelchair system with a virtual keyboard. The system allows handicapped and physically challenged people to operate their wheelchairs in the direction they desire. In order to create an interactive system for the user, we will add more sensors in the future. The system's functionality is totally dependent on the movement of paralyzed patients' eyes. This system works perfectly in normal day-light environments, but it will be challenging to work perfectly in low-light areas. So, we will work to overcome these challenges in the future. We will work on a good dataset in the future to improve the accuracy of our system (Table 1).

Simultaneously, some simple but accurate methods are employed in this work, making the system simple to implement. The system has already been proven efficient as well as valid through numerous trials, demonstrating that it has a high level of

Table 1 Result accuracy

Total wheelchair movement events	1260
Total keyboard typing events	1200
Total wheelchair movement errors	25
Total keyboard typing errors	36
Wheelchair movement accuracy	98.03%
Keyboard typing accuracy	97%

accuracy, a simple approach, and a relatively affordable cost. So, it is extremely beneficial for disabled people.

References

1. Dahmani M, Chowdhury ME, Khandakar A, Rahman T, Al-Jayyousi K, Hefny A, Ki-ranyaz S (2020) An intelligent and low-cost eye-tracking system for motorized wheelchair control. Sensors 20(14):3936
2. Hochberg LR, Serruya MD, Friehs GM, Mukand JA, Saleh M, Caplan AH, Bran-ner A, Chen D, Penn RD, Donoghue JP (2006) Neuronal ensemble control of prosthetic devices by a human with tetraplegia. Nature 442(7099):164–171
3. Acero A, Deng L, Kristjansson T, Zhang J (2000) Hmm adaptation using vector taylor series for noisy speech recognition. In: sixth international conference on spoken language processing
4. Chakraborty P, Ahmed S, Yousuf MA, Azad A, Alyami SA, Moni MA (2021) A human-robot interaction system calculating visual focus of human's attention level. IEEE Access
5. Wang M, Maeda Y, Naruki K, Takahashi Y (2014) Attention prediction system based on eye tracking and saliency map by fuzzy neural network. In: 2014 joint 7th international conference on soft computing and intelligent systems (SCIS) and 15th international symposium on advanced intelligent systems (ISIS). IEEE, pp 339–342
6. Chakraborty P, Yousuf MA, Rahman MZ, Faruqui N (2020) How can a robot calculate the level of visual focus of human's attention. In: Proceedings of international joint conference on computational intelligence. Springer, pp 329–342
7. Patel SN, Prakash V (2015) Autonomous camera based eye controlled wheelchair system using raspberry-pi. In: 2015 international conference on innovations in information, embedded and communication systems (ICIIECS). IEEE, pp 1–6
8. Chakraborty P, Muzammel CS, Khatun M, Islam SF, Rahman S (2020) Automatic student attendance system using face recognition. Int J Eng Adv Technol (IJEAT) 9:93–99
9. Al-Kassim Z, Memon QA (2017) Designing a low-cost eyeball tracking keyboard for paralyzed people. Comput Electr Eng 58:20–29
10. Chakraborty P, Roy D, Zahid Z, Rahman S (2019) Eye gaze controlled virtual keyboard. Int J Rec Technol Eng (IJRTE) 8(4):3264–3269
11. Cecotti H (2016) A multimodal gaze-controlled virtual keyboard. IEEE Trans Human-Mach Syst 46(4):601–606
12. Duchowski AT, Duchowski AT (2017) In: Eye tracking methodology: theory and practice. Springer
13. Sugano Y, Matsushita Y, Sato Y, Koike H (2015) Appearance-based gaze estimation with online calibration from mouse operations. IEEE Trans Human-Mach Syst 45(6):750–760
14. Al-Haddad A, Sudirman R, Omar C (2011) Gaze at desired destination, and wheelchair will navigate towards it. new technique to guide wheelchair motion based on eog signals. In: 2011 first international conference on informatics and computational intelligence. IEEE, pp 126–131

15. Das TR, Hasan S, Sarwar S, Das JK, Rahman MA (2021) Facial spoof detection using support vector machine. In: Proceedings of international conference on trends in computational and cognitive engineering. Springer, pp 615–625
16. Sayeed S, Sultana F, Chakraborty P, Yousuf MA (2021) Assessment of eyeball movement and head movement detection based on reading. Recent Trends in Signal and Image Processing: ISSIP 2020 1333:95
17. Chakraborty P, Nawar F, Chowdhury HA (2022) Sentiment Analysis of Bengali Facebook Data Using Classical and Deep Learning Approaches. In: Mishra M, Sharma R, Kumar Rathore A, Nayak J, Naik B (eds) Innovation in Electrical Power Engineering, Communication, and Computing Technology. Lecture Notes in Electrical Engineering, vol 814. Springer, Singapore. https://doi.org/10.1007/978-981-16-7076-3_19
18. Arai K, Mardiyanto R (2011) Autonomous control of eye based electric wheel chair with obstacle avoidance and shortest path finding based on dijkstra algorithm. Int J Adv Comput Sci Appl 2(12):19–25
19. Sayed A, Khatun M, Ahmed T, Piya A, Chakraborty P, Choudhury T (2022) Performance Analysis of OFDM System on Multipath Fading and Inter Symbol Interference (ISI) Using AWGN. In: Das AK, Nayak J, Naik B, Dutta S, Pelusi D (eds) Computational Intelligence in Pattern Recognition. Advances in Intelligent Systems and Computing, vol 1349. Springer, Singapore. https://doi.org/10.1007/978-981-16-2543-5_3
20. Rajpathak T, Kumar R, Schwartz E (2009) Eye detection using morphological and color image processing. In: Proceeding of florida conference on recent advances in robotics. pp 1–6
21. Feroz M, Sultana M, Hasan M, Sarker A, Chakraborty P, Choudhury T (2022) Object Detection and Classification from a Real-Time Video Using SSD and YOLO Models. In: Das AK, Nayak J, Naik B, Dutta S, Pelusi D (eds) Computational Intelligence in Pattern Recognition. Advances in Intelligent Systems and Computing, vol 1349. Springer, Singapore. https://doi.org/10.1007/978-981-16-2543-5_4
22. Chakraborty P, Yousuf MA, Rahman S (2021) Predicting level of visual focus of human's attention using machine learning approaches. In: Proceedings of international conference on trends in computational and cognitive engineering. Springer, pp 683–694
23. Bingham A, Hadoux X, Kumar DK (2014) Implementation of a safety system using ir and ultrasonic devices for mobility scooter obstacle collision avoidance. In: 5th ISSNIP-IEEE biosignals and biorobotics conference (2014): biosignals and robotics for better and safer living (BRC). IEEE, pp 1–5
24. Muzammel CS, Chakraborty P, Akram MN, Ahammad K, Mohibullah M (2020) Zero-shot learning to detect object instances from unknown image sources. Int J Innov Technol Explor Eng (IJITEE) 9(4):988–991
25. Sultana M, Ahmed T, Chakraborty P, Khatun M, Hasan MR, Uddin MS (2020) Object detection using template and hog feature matching. Int J Adv Comput Sci Appl 11(7). https://doi.org/10.14569/IJACSA.2020.0110730, https://doi.org/10.14569/IJACSA.2020.0110730
26. Faruque MA, Rahman S, Chakraborty P, Choudhury T, Um JS, Singh TP (2021) Ascer-taining polarity of public opinions on bangladesh cricket using machine learning techniques. Spatial Inf Res 1–8
27. Sarker A, Chakraborty P, Sha SS, Khatun M, Hasan MR, Banerjee K (2020) Improvised technique for analyzing data and detecting terrorist attack using machine learning approach based on twitter data. J Comput Commun 8(7):50–62
28. Nguyen QX, Jo S (2012) Electric wheelchair control using head pose free eye-gaze tracker. Electron Lett 48(13):750–752
29. Nakazawa N, Kim I, Mori T, Murakawa H, Kano M, Maeda A, Matsui T, Yamada K (2012) Development of an intuitive interface based on facial orientations and gazing actions for auto-wheel chair operation. In: 2012 IEEE RO-MAN: the 21st IEEE international symposium on robot and human interactive communication. IEEE, pp 173–178
30. Jia P, Hu H (2005) Head gesture based control of an intelligent wheelchair. In: Proceedings of the 11th annual conference of the chinese automation and computing society in the UK [CACSUK05]. pp 85–90

31. Kazemi V, Sullivan J (2014) One millisecond face alignment with an ensemble of regression trees. In: Proceedings of the IEEE conference on computer vision and pattern recognition. pp 1867–1874
32. Rosebrock A (2017) Facial landmarks with dlib, opencv, and python. Retrieved on-line at https://www.pyimagesearch.com/2017/04/03/faciallandmarks-dlib-opencv-python
33. Chakraborty P, Sultana S (2022) IoT-Based Smart Home Security and Automation System. In: Sharma DK, Peng SL, Sharma R, Zaitsev DA (eds) Micro-Electronics and Telecommunication Engineering. Lecture Notes in Networks and Systems, vol 373. Springer, Singapore. https://doi.org/10.1007/978-981-16-8721-1_48

SATLabel: A Framework for Sentiment and Aspect Terms Based Automatic Topic Labelling

Khandaker Tayef Shahriar, Mohammad Ali Moni, Mohammed Moshiul Hoque, Muhammad Nazrul Islam, and Iqbal H. Sarker

Abstract In this paper, we present a framework that automatically labels latent Dirichlet allocation (LDA) generated topics using sentiment and aspect terms from COVID-19 tweets to help the end-users by minimizing the cognitive overhead of identifying key topics labels. Social media platforms, especially Twitter, are considered as one of the most influential sources of information for providing public opinion related to a critical situation like the COVID-19 pandemic. LDA is a popular topic modelling algorithm that extracts hidden themes of documents without assigning a specific label. Thus, automatic labelling of LDA-generated topics from COVID-19 tweets is a great challenge instead of following the manual labelling approach to get an overview of wider public opinion. To overcome this problem, in this paper, we propose a framework named `SATLabel` that effectively identifies significant topic labels using *top unigrams features of sentiment terms and aspect terms clusters from LDA-generated topics* of COVID-19-related tweets to uncover various issues related to the COVID-19 pandemic. The experimental results show that our methodology is more effective, simpler, and traces better topic labels compare to the manual topic labelling approach.

Keywords Data-driven framework · LDA · Sentiment terms · Aspect terms · Unigrams · Soft cosine similarity · Topic · Automatic labelling

---

K. T. Shahriar (✉) · M. M. Hoque · I. H. Sarker
Department of Computer Science and Engineering, Chittagong University of Engineering & Technology, Chittagong 4349, Bangladesh
e-mail: u17mcse008p@student.cuet.ac.bd

I. H. Sarker
e-mail: iqbal@cuet.ac.bd

M. A. Moni
Faculty of Health and Behavioural Sciences, Artificial Intelligence & Digital Health Data Science, School of Health and Rehabilitation Sciences, The University of Queensland St Lucia, St Lucia, QLD 4072, Australia

M. N. Islam
Department of Computer Science and Engineering, Military Institute of Science and Technology, Dhaka 1216, Bangladesh

V. Skala et al. (eds.), *Machine Intelligence and Data Science Applications*, Lecture Notes on Data Engineering and Communications Technologies 132,
https://doi.org/10.1007/978-981-19-2347-0_6

1 Introduction

Twitter nowadays is considered as one of the most important social media platforms to explain the characteristics and predict the status of the pandemic [9]. In Wuhan, at the end of 2019, a novel coronavirus disease that causes COVID-19 was reported by the World Health Organisation (WHO). The declaration of COVID-19 as an international concern of public health emergency by WHO was reported on January 30, 2020 [1]. During the pandemic, the use of Twitter increases immensely and plays a critical role by reflecting real-time public panic and providing rich information to raise public awareness through posts and comments. However, text mining and analysis of data from social media platforms such as Twitter have become a burning issue to extract necessary information. Moreover, it is a great challenge to extract meaningful topic labels by machines instead of following diverse human interpretations of the manual labelling approach [13]. Hence, in this paper, we propose `SATLabel`, a framework that effectively identifies key topic labels of tweets automatically from the huge volume of the Twitter dataset to reduce the human effort of cumbersome topic labelling tasks.

A large number of labelled datasets is required for traditional supervised methods. Obtaining such a labelled dataset for topic labelling purposes is very difficult and expensive. In this paper, we use LDA [3], which is an unsupervised probabilistic algorithm for text documents. Thus, `SATLabel` does not need any labelled dataset for topics. A set of topics available in the documents is discovered by LDA. Sentiment terms express emotions from tweets, and aspect terms describe features of an entity [20]. We create sentiment terms cluster and aspect terms cluster for each LDA-generated topic. However, Unigram is a probabilistic language model that is extensively used in natural language processing tasks and text mining to exhibit the context of texts. `SATLabel` uses the top unigrams features from sentiment terms cluster and aspect terms cluster, respectively, and create attribute tags concatenating the two top unigrams features (first sentiment term and then aspect term). We select the attribute tag which has the highest soft cosine similarity value with respect to the tweets of the same topic to assign a meaningful label for that LDA-generated topic. Our experimental results show that the label generated by `SATLabel` has a high soft cosine similarity value with the tweets of the same topic than the manual labelling approach. The main contributions of this paper can be summarized as follows:

- We effectively utilize sentiment terms and aspect terms of tweets to produce significant topic labels.
- We propose a new framework named `SATLabel` that is useful to extract topics from COVID-19 tweets and labels them automatically instead of following the manual method.

- SATLabel effectively reduces the human effort for difficult topic labelling tasks of tweets.
- We have shown the effectiveness of SATLabel comparing with the manual labelling approach by conducting a range of experiments.

The organization of the rest of the paper is as follows. Related works are reviewed in Sect. 2. In Sect. 3, we present the methodology of the proposed framework. In Sect. 4, we assess the evaluation results of our framework by conducting experiments on the Twitter dataset. Next, we present the discussion, and finally, we conclude this paper and highlight the direction of future work.

2 Related Work

COVID-19 tweets can be helpful for identifying meaningful topic labels to highlight user conversation and understand ideas of people's needs and interests. Many researchers used the LDA algorithm to extract hidden themes of documents. Patil et al. [12] proposed a paper using the frequency-based technique to extract topics from people's reviews without mentioning the proper labelling techniques for describing the topics. Hingmire et al. [7] proposed a paper to construct LDA based topic model, but the expert association is required to assign the topic to the class labels. Hourani [8] proposed a paper to classify articles according to their topics for which labelled dataset is required. Asmussen and Møller [2] proposed a topic modelling method for researchers, but topic labelling depends on the researcher's view without having any automatic method. Wang et al. [19] proposed a paper that minimizes the problem of data sparsity without labelling key topics specifically. Zhu et al. [21] presented the change of the number of texts on topics with respect to time by following the manual topic labelling approach. Satu et al. [16] proposed a framework that extracts topics from the best cluster of sentiment classification having a manual explanation of topic labels tends to misinterpretation. Kee et al. [10] used LDA to extract higher-order arbitrary topics, but only 61.3% clear collective themes were evaluated. Maier et al. [11] presented accessibility and applicability of communication researchers using LDA-based topic labelling approach which depends manually on broader context knowledge. In our previous work, we only considered the top unigram feature of aspect terms cluster to identify the key topics with labels by implementing LDA [18]. Elgesem et al. [5] presented an analysis about the discussion of the Snowden affair using a manual topic labelling approach. Guo et al. [6] compared dictionary-based analysis and LDA analysis using a manual topic labelling approach.

The summary of the above works describes that most of the works considered a manual topic labelling approach to categorize documents and get an overview which is expensive, time-consuming and requires cumbersome human interpretations. Hence, an automatic and effective topic labelling approach would be helpful to reduce human effort and save time. Thus, in this paper, we consider the development of a framework named SATLabel to generate significant topic labels automatically to highlight users' conversations on Twitter.

3 Methodology

In this section, we present SATLabel that is a framework to label LDA-generated topics automatically as shown in Fig. 1. For analysing and mining textual data like tweets, text preprocessing is one of the most essential steps to advance in further processing steps. The working principle and overall steps to generate automatic topic labels from the Twitter dataset are shown in Algorithm 1. After preprocessing of highly unstructured and non-grammatical tweets, several processing steps are followed to produce the expected output.

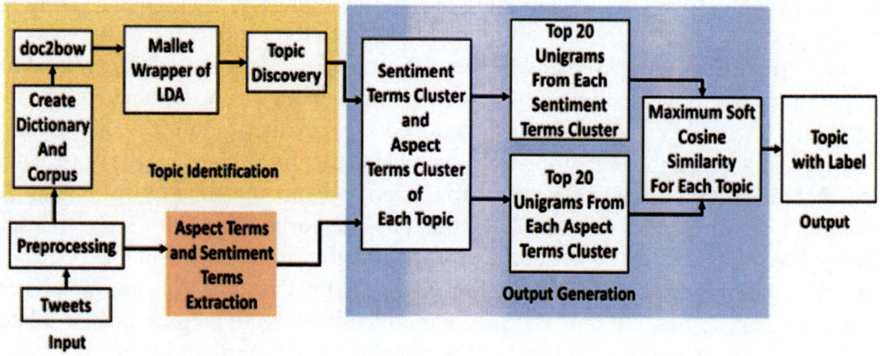

Fig. 1 SATLabel: Proposed framework for automatic topic labelling

Algorithm 1: Automatic Topic Labelling

Input: T: number of tweets in dataset
Output: Topic Label (T_{Label}).

for ***each*** $t \in \{1,2,...,T\}$ **do**
  $T_p \leftarrow Preprocess(t)$;
for ***each*** $t_p \in \{1,2,...,T_p\}$ **do**
  // Corpus Development
  $C \leftarrow Create\_Corpus(t_p)$;
  // Sentiment Terms and Aspect Terms Extraction
  $S_{T_p} \leftarrow Sentiment\_Terms(t_p)$;
  $A_{T_p} \leftarrow Aspect\_Terms(t_p)$;
// Topic Discovery
$K \sim Mallet(LDA(Doc2bow(C)))$;
for ***each*** $k \in \{1,2,...,K\}$ **do**
  for ***each*** $t_p \in \{1,2,...,T_p\}$ **do**
    $k_{dominant,t_p} \sim dominant\_topic(t_p, k)$;
    // Create Clusters from Topic
    $C_S \sim Cluster(S_{T_p} \rightarrow k_{dominant,t_p})$;
    $C_A \sim Cluster(A_{T_p} \rightarrow k_{dominant,t_p})$;
for ***each*** $k \in \{1,2,...,K\}$ **do**
  $U_S \sim max\_count(Top\_Unigrams(C_S \rightarrow k))$;
  $U_A \sim max\_count(Top\_Unigrams(C_A \rightarrow k))$;
  $T_{Label} \leftarrow max\_soft\_cosine\_similarity(U_S + U_A, k)$

3.1 Sentiment and Aspect Terms Extraction

Sentiment terms carry the tone or opinion of the text. Usually, adjectives and verbs of sentences are considered as sentiment terms that indicate expressed opinion of the text. Noun and noun phrases are considered as aspects terms of text. Objects of verbs are often regarded as aspect terms that describe the features of an entity, product, or event [20]. We follow precise parts of speech tagging which is an efficient approach to extract sentiment terms and aspect terms from texts. Examples of sentiment terms and aspect terms of sample tweets are shown in Table 1.

3.2 Topic Identification Using LDA

LDA is a popular topic modelling algorithm to discover hidden topics available in the corpus from unlabelled dataset [15]. But the challenge is how to assign significant

Table 1 Example of sentiment terms and aspect terms

Sample tweet	Sentiment terms	Aspect terms
Please read the thread	Read	Thread
To enjoy and relax for your dinner it is a great place	Enjoy, relax, great	Dinner, place
Links with info on communicating with children regarding COVID-19	Communicate, covid	Links, info, children
The retail store owners right now	Retail, right	Owners, store

labels to LDA-generated topics. The steps for topic discovery that we follow in `SATLabel` framework are discussed below:

1. *Creating Dictionary and Corpus*: A systematic way of creating a number of lexicons of a language is supported by a dictionary, and a corpus generally refers to an arbitrary sample of that language. A document corpus is built with words or phrases. In Natural Language Processing (NLP) paradigm, the corpus of a language plays a vital role in developing a knowledge-based system and mining texts. In the proposed framework, we create a dictionary and develop a corpus from the preprocessed text.
2. *Creating a BoW Corpus*: Corpus contains the word id and its frequency in every document. Documents are converted into Bag of Words (BoW) format by applying Doc2bow embedding. Each word is assumed as a normalized and tokenized string.
3. *Topic Discovery*: BoW corpus is transferred to the mallet wrapper of LDA. The presence of a set of topics in the corpus is discovered by LDA. Mallet wrapper of LDA runs faster and provides precise division of topics using Gibbs sampling technique [4]. LDA generates the most prominent words in a topic. Thus, by using the word probabilities, one can manually find dominant themes in the documents. To overcome the complex manual labelling approach, our framework `SATLabel` generates automatic topic labels using sentiment terms and aspect terms of documents without any human interpretation. Based on the topic coherence score, we choose a model that discovers 20 optimal number of topics itself. Then, we enumerate the dominant topic for each tweet to understand the distribution of topics across the tweets in the dataset.

3.3 Output Generation

The steps to generate significant topic labels automatically as output from the topics extracted by LDA are discussed below:

1. *Generation of Sentiment Terms and Aspect Terms Cluster*: We create clusters of sentiment terms and aspect terms independently from tweets corresponding to each LDA-generated topic. Thus, we get 20 sentiment terms cluster and 20 aspect terms cluster from the discovered topics by LDA.
2. *Labelling Topic Using Top Unigrams*: A unigram is a one-word sequence of n-gram. The use of unigrams can be observed in NLP, cryptography, and mathematical analysis. Soft cosine similarity considers the similarity of features in vector space model [17]. We extract the top 20 unigrams from sentiment terms cluster and aspect terms cluster, respectively, for each topic. Then, we concatenate all the possible combinations of top unigrams of sentiment terms and top unigrams of aspect terms of topic. We select a combination that has the highest soft cosine similarity value with respect to the tweets of that topic to assign with a significant topic label. We use a sentiment and aspect term tag to label each topic because that feature tag presents an attribute to describe that topic of tweets.

In the section of the methodology of this paper, we present a framework called `SATLabel` to detect key topic labels from the tweets automatically as shown in Table 2. We compare the quality of topic labels generated by `SATLabel` with the manually assigned topic labels in the experiment section. To categorize a tweet with a specific topic label from test data, we search the topic number that has a greater impact of percentage on that tweet.

4 Experiments

4.1 Dataset

We collect the Twitter dataset from the Web site at https://www.kaggle.com/datatattle/covid-19-nlp-text-classification. There are two csv files in the dataset. One is Corona_NLP_train.csv, and another is Corona_NLP_test.csv. Tweets available in the dataset are highly unstructured and non-grammatical in syntax. There are 41,157 and 3798 COVID-19-related tweets are available in Corona_NLP_train.csv and Corona_NLP_test.csv files, respectively. We apply a series of preprocessing functions to get the normalized form of noisy tweets for further processing.

4.2 Data Preprocessing

Handling ill-formatted, noisy and unstructured twitter data is one of the most important tasks for us. We preprocess the Twitter dataset to get the normalized form using the functions of transforming words into lowercase, replacing hyperlinks, mentions, and hashtags with empty string, dealing with contractions, replacing punctuation with space, striping space from words, removing words less than two characters, removing stop words, handling unicode and non-English words.

4.3 Finding the Optimal Number of Topics for LDA

We create a function to return several LDA models with multiple values of a number of topics (k) to find the optimal number of topics. The interpretable topics can be found by selecting a 'k' that identifies the end of a quick rise of topic coherence score. Sometimes, we get more granular sub-topics by choosing a higher value of topic coherence score. We pick the model giving the highest coherence value before flattening out considering better sense while the coherence score seems to keep growing as shown in Fig. 2. For the next steps, we choose the model having 20 topics itself.

4.4 Selection of Top Unigrams Features from Clusters

We create sentiment terms cluster and aspect terms cluster of tweets for each topic. We find the top counted 20 unigrams from each cluster. Figures 3 and 4 show the top 20 unigrams from sentiment and aspect terms clusters of topic no. 12, respectively. Then, we detect the topic label depending on the highest soft cosine similarity value of sentiment and aspect term tag with respect to the tweets of that topic.

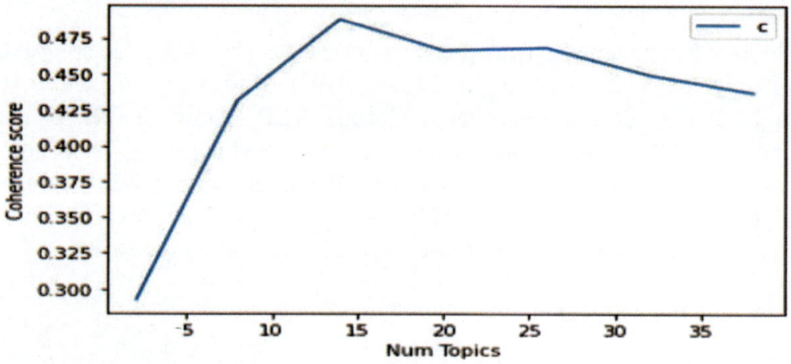

Fig. 2 Selection of the optimal number of LDA topics

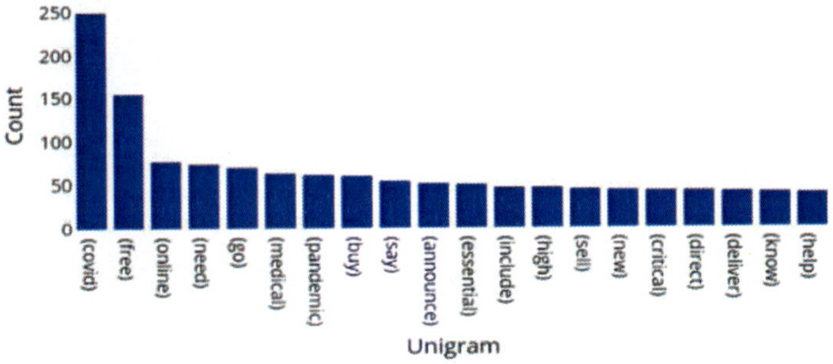

Fig. 3 Top 20 unigrams from sentiment terms cluster of topic no. 12

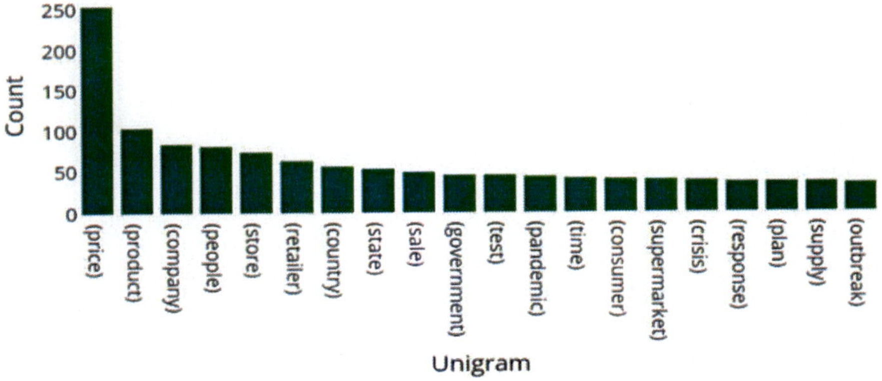

Fig. 4 Top 20 unigrams from aspect terms cluster of topic no. 12

4.5 Qualitative Evaluation of Topic Labels

An expert annotator assigns the topic labels manually using the word probabilities in LDA-generated topics to a randomly selected set of tweets. In Table 2, we present a portion of set of tweets assigned by the `SATLabel` generated topic labels. Table 2 shows that `SATLabel` generated topic labels are well-aligned and closely coherent with the descriptions of tweets. We can extract useful information related to a topic, simply by categorizing the tweets using the key label generated by `SATLabel` of that topic.

4.6 Effectiveness Analysis

In this experiment section, we calculate the soft cosine similarity (SCS) values of detected topic labels by `SATLabel` and manual approach for LDA-generated 20

Table 2 Example of topics detected on tweets

Sample tweet	Topic no.	Detected topic label (SATLabel)
Due to the COVID-19 virus and the global health pandemic, we will be closed at our retail location until further notice	17	Shut location
Dubai becomes cheaper to live in	9	Drop cost
Covid-19 is already affecting the online shopping, ok somebody slap meee plsss?	16	Online shopping
You guys still can buy food during lockdown then why need to do panic buying?	0	Hoard food
I'm going to try patenting my world-famous vegetable phall as a killer of covid-19	14	Learn scam
It's not covid 19. It's due fall in global oil prices oil cost 30 barrel …	15	Drop barrel
Here's a buying guide our community set up for the neighbourhood supermarket. Feel free to use it as a template	12	Covid product
The Consumer Financial Protection Bureau today announced that it is postponing some data collection from the financial industry	3	Learn insight
Food demand in poorer countries is more linked to income …	6	Covid food

topics. SCS is used to detect the semantic text similarities between two documents. A high SCS value provides a high similarity index, and similarity is smaller for unrelated documents. We train the word2vec embedding model to use SCS. We show the comparison of `SATLabel` and manual labelling approach for all LDA-generated topics in terms of SCS value in Fig. 5. We get SCS values generated by proposed `SATLabel` for topic nos. 4, 8, 10, 14 which are 0.77, 0.61, 0.53, 0.64, respectively, while manual approach generates 0.09, 0.06, 0.02, 0.07 SCS scores for those topics which are very low. Diverse human interpretation of topics is the possible reason for the high difference of SCS scores between proposed `SATLabel` and manual approach. For topics nos. 3, 6, 9, 12, 16, 18, we get the same SCS values for SATLabel and manual approach because of identical topic labels generated by

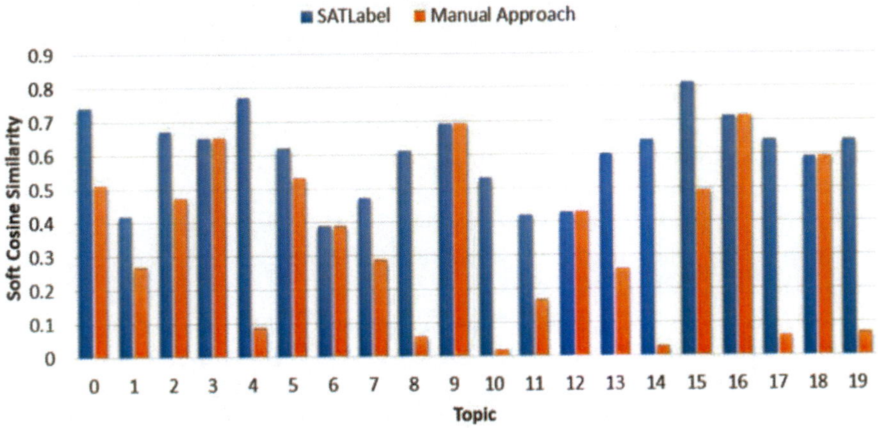

Fig. 5 Comparison of SATLabel with manual approach

both approaches. From Fig. 5, we can observe that the topic labels generated by the proposed framework SATLabel produce high SCS values for a maximum number of topics compared with the manual labelling approach. Hence, our proposed framework is more effective and traces better topic labels from unlabelled datasets to reduce the cumbersome task of the human manual labelling approach.

5 Discussion

Automatic labelling of LDA-generated topics of the tweets of social media platforms like Twitter is helpful to understand people's ideas and feelings by going through meaningful insights rather than following traditional strategies like the manual labelling approach. In this paper, we use LDA, a popular probabilistic topic modelling algorithm, to extract hidden topics from tweets. We then effectively use sentiment terms and aspect terms of tweets to create clusters. After that, we select top unigrams from the clusters to produce significant topic labels using the maximum soft cosine similarity values. Our proposed framework SATLabel helps to produce semantically similar topic labels of tweets to highlight the user's conversations and notice several COVID-19-related issues.

Overall, SATLabel is a data-driven framework for topic labelling purposes for mining texts to provide helpful information from the dataset of Twitter related to COVID-19. We firmly believe that SATLabel can be effectively used in other domains of applications like agriculture, health care, education, business, cybersecurity, etc., and also can be used to generate target class from unlabeled datasets to train deep learning models [14]. These types of contributions allow the researchers and experts in relevant departments to take necessary actions in critical situations like the COVID-19 pandemic by efficiently utilizing social media platforms.

6 Conclusion and Future Work

In this paper, we propose a new framework named SATLabel that effectively and automatically identifies key topic labels from COVID-19 tweets. Our framework saves time and reduces the human effort to minimize the overhead of difficult topic labelling tasks from the huge volumes of data to get an overview of broader public opinions on social media platforms like Twitter. We believe that SATLabel will help the reformists to discover various COVID-19-related issues by analysing automatically extracted topic labels.

In the future, we want to increase our scope of experiments by integrating the proposed framework with sentiment classification tasks using hybridization of deep learning methods. We will also implement our proposed framework to other social media platforms on different events to generate significant topic labels to handle the overload of ever-increasing data volume.

References

1. Adhikari SP, Meng S, Wu YJ, Mao YP, Ye RX, Wang QZ, Sun C, Sylvia S, Rozelle S, Raat H et al (2020) Epidemiology, causes, clinical manifestation and diagnosis, prevention and control of coronavirus disease (covid-19) during the early outbreak period: a scoping review. Infect Dis Poverty 9(1):1–12
2. Asmussen CB, Møller C (2019) Smart literature review: a practical topic modelling approach to exploratory literature review. J Big Data 6(1):1–18
3. Blei DM, Ng AY, Jordan MI (2003) Latent Dirichlet allocation. J Mach Learn Res 3:993–1022
4. Boussaadi S, Aliane H, Abdeldjalil PO (2020) The researchers profile with topic modeling. In: 2020 IEEE 2nd International conference on electronics, control, optimization and computer science (ICECOCS). IEEE, pp 1–6
5. Elgesem D, Feinerer I, Steskal L (2016) Bloggers' responses to the Snowden affair: combining automated and manual methods in the analysis of news blogging. Comput Support Coop Work (CSCW) 25(2–3):167–191
6. Guo L, Vargo CJ, Pan Z, Ding W, Ishwar P (2016) Big social data analytics in journalism and mass communication: comparing dictionary-based text analysis and unsupervised topic modeling. J Mass Commun Q 93(2):332–359
7. Hingmire S, Chougule S, Palshikar GK, Chakraborti S (2013) Document classification by topic labeling. In: Proceedings of the 36th international ACM SIGIR conference on research and development in information retrieval, pp 877–880
8. Hourani AS (2021) Arabic topic labeling using naïve bayes (nb). In: 2021 12th International conference on information and communication systems (ICICS). IEEE, pp 478–479
9. Jahanbin K, Rahmanian V et al (2020) Using twitter and web news mining to predict covid-19 outbreak. Asian Pac J Trop Med 13(8):378
10. Kee YH, Li C, Kong LC, Tang CJ, Chuang KL (2019) Scoping review of mindfulness research: a topic modelling approach. Mindfulness 10(8):1474–1488
11. Maier D, Waldherr A, Miltner P, Wiedemann G, Niekler A, Keinert A, Pfetsch B, Heyer G, Reber U, Häussler T et al (2018) Applying LDA topic modeling in communication research: toward a valid and reliable methodology. Commun Methods Measures 12(2–3):93–118
12. Patil PP, Phansalkar S, Kryssanov VV (2019) Topic modelling for aspect-level sentiment analysis. In: Proceedings of the 2nd international conference on data engineering and communication technology. Springer, Berlin, pp 221–229

13. Sarker IH (2021) Data science and analytics: an overview from data-driven smart computing, decision-making and applications perspective. SN Comput Sci 2(5):1–22
14. Sarker IH (2021) Deep learning: a comprehensive overview on techniques, taxonomy, applications and research directions. SN Comput Sci 2(6):1–20
15. Sarker IH (2021) Machine learning: algorithms, real-world applications and research directions. SN Comput Sci 2(3):1–21
16. Satu MS, Khan MI, Mahmud M, Uddin S, Summers MA, Quinn JM, Moni MA (2021) TClustVID: a novel machine learning classification model to investigate topics and sentiment in covid-19 tweets. Knowl Based Syst 226:107126
17. Sidorov G, Gelbukh A, Gómez-Adorno H, Pinto D (2014) Soft similarity and soft cosine measure: similarity of features in vector space model. Computación Sistemas 18(3):491–504
18. Tayef Shahriar K, Sarker IH, Nazrul Islam M, Moni MA (2021) A dynamic topic identification and labeling approach of covid-19 tweets. In: International conference on big data, IoT and machine learning (BIM 2021). Taylor and Francis
19. Wang B, Liakata M, Zubiaga A, Procter R (2017) A hierarchical topic modelling approach for tweet clustering. In: International conference on social informatics. Springer, Berlin, pp 378–390
20. Wang W, Pan SJ, Dahlmeier D, Xiao X (2017) Coupled multi-layer attentions for co-extraction of aspect and opinion terms. In: Proceedings of the AAAI conference on artificial intelligence, vol 31
21. Zhu B, Zheng X, Liu H, Li J, Wang P (2020) Analysis of spatiotemporal characteristics of big data on social media sentiment with covid-19 epidemic topics. Chaos Solitons Fractals 140:110123

Deriving Soft Biometric Feature from Facial Images

Mazida A. Ahmed, Ridip D. Choudhury, Vaskar Deka, Manash P. Bhuyan, and Parvez A. Boruah

Abstract This paper is based on deriving race information from face images. The work is carried out in two scenarios: first, for images taken in controlled laboratory settings and second, in natural unrestricted environment. Several small datasets are combined for the first case while BUPT dataset is selected for the second case. Linear features such as Principal Component Analysis, Linear Discriminant Analysis and Local Binary Pattern are investigated for the purpose. Experiments show these features have satisfactory discriminative ability in the first scenario with average classification accuracy of 91% while for the BUPT data the performance drops down by an average of 15%. It proves the sensitivity of these features toward variation in background, illumination, pose and expression. An overall comparison also shows that local features like LBP have more differentiating power than the holistic feature like PCA and LDA.

Keywords PCA · LDA · LBP · Nearest neighbor

1 Introduction

Soft biometric traits like gender, race (or ethnicity), age, hair color and scar complement the hard-core biometric information though are not individually sufficient for determining a person's identity. Ethnicity estimation has gained attention in the recent years along with the analysis of other soft biometric traits like gender, age, emotion, height, weight, hair color and many other features. Ethnicity is more driven toward the cultural and social terms rather than having a concrete and rigid definition to it. And this seems to be a major challenge in collecting data and thus research in this direction in hindered due to data unavailability. Nevertheless, today's data

M. A. Ahmed (✉) · V. Deka · M. P. Bhuyan · P. A. Boruah
Department of Information Technology, Gauhati University, Guwahati, India
e-mail: 14mazida.ahmed@gmail.com

R. D. Choudhury
Krishna Kanta Handiqui State Open University, Guwahati, India

V. Skala et al. (eds.), *Machine Intelligence and Data Science Applications*, Lecture Notes on Data Engineering and Communications Technologies 132,
https://doi.org/10.1007/978-981-19-2347-0_7

driven technologies have been able to solve this shortage to a considerable extent. Also, researchers have taken pain in collecting and annotating images in laboratory settings in creating digital repositories to facilitate research for better understanding the separability among the different recognized race groups.

Anthropologists have recognized three broad categories of race namely: Caucasian, Asian and African. Population belonging to these groups have high inter-race separable features in terms of face structure, complexion, hair type, height and many more apparent characteristics along with other micro-level biological characteristics such as gene type, blood and more. This paper is concerned with facial features derivable from face images. Figure 1 shows samples belonging to these three groups while Table 1 gives a brief overview about the differentiating facial features.

Automatic ethnicity identification has been trending as it finds applications in fields of surveillance as well as law enforcement. One such possible usage could be reduction in search space if the racial background of the fugitive is known. Our work is aimed at deriving the race information of a subject from his/her facial image and investigating the usage of the feature descriptors in scenarios where images are taken in (i) controlled and (ii) natural settings.

Our work involves comparing ethnicity identification results by using three methods: Principal Component Analysis (PCA), Linear Discriminant Analysis

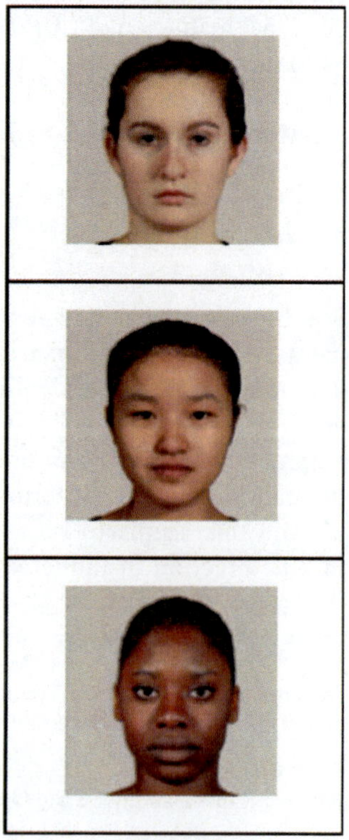

Fig. 1 Samples from three race groups. *Source* [1]

Table 1 Some prominent differentiating facial features

Features	Caucasian	Asian	African
Nose form	Narrow	Medium	Board
Face height	High	High	Low
Face width	Narrow	Wide	Narrow
Nasal profile	Straight	Concave	Concave
Skin color	Pale reddish white	Saffron to reddish brown	Black to black brown
Nasal opening	Narrow	Medium	Wide
Eye form	Medium	Narrow	Wide

(LDA) and Local Binary Pattern (LBP). PCA is a statistical tool used for analyzing the structure of data and gives the directions (vectors) of variance present in that set. The directional vectors derived are orthogonal and hence can be used as a new set of basis vectors. Projecting the data on the new set of bases will free it from correlations between any two variables. PCA not only deduces the directions but also the amount of variance along those directions. And so only those vectors are retained which have higher variances while dropping the others. By doing this, the dimension of the basis vectors along with the projected data is greatly reduced.

This paper is organized as follows: Sect. 2 discusses about the related works done beforehand; Sect. 3 describes the dataset and the methods; Sect. 4 mentions about the experiments carried out and their results obtained while Sect. 5 concludes the paper.

2 Related Work

Work on race estimation has commenced since 2004 when [2] identified the Asian and Non-Asian subjects from a database combined out of four other databases. They used posterior probability on the LDA features derived out of the 2-D images at multiple scales with an ensemble to classify them. A similar method, PCA was also employed for feature extraction used for classification of two countries population: Myanmar and Non-Myanmar by Tin and Sein [3–5].

Face texture with feature extractors like Local Binary Pattern (LBP) and Gabor are widely used for this task and are reported in many works such as Hosoi et al. [6], Qiu et al. [7], Lin et al. [8], Duan et al. [9], Saei et al. [10], Carcagnì et al. [11] and Momin and Topamo [12]. Skin color, lip color and ratio of forehead to face as are used as features for ethnicity estimation by Roomi et al. [13] on the FERET database. Another texture feature called Biologically Inspired Features (BIF) are used by Han et al. [14] and Guo and Mu [15] which shows promising results even on natural face images. Xie et al. [16] proposed a method, Kernel Class-dependent Feature Analysis (KCFA), along with facial color for ethnicity classification on two databases MBGC and a custom database created by the authors. They showed that the periorbital region consisting of the eyes and nose contain discriminating information about ethnicity.

Lu et al. [17] tracked the less explored way with 3-D scans to categorize Asian and Non-Asian classes. Weber Local Descriptors (WLD) with Kruskal–Wallis algorithm is used by Muhammed [18] on the FERET dataset.

3 Material and Methods

3.1 Dataset

Our experiments are performed for the classification of African, Caucasian and Asian. We perform the experiments with two data groups: (i) an amalgamated set of different datasets and (ii) BUPT dataset; the former holds images taken in controlled settings while the latter has unrestricted images of celebrities taken in-the-wild. The amalgamated set consists of the following datasets:

1. Yale: The Yale Face dataset [19] was developed at Yale Center for Computational Vision and Control at 1998. The samples are grayscale and vary in expression and illumination and contribute to the Caucasoid samples. The dataset is a collection of 165 images of 15 individuals with 11 images per subject with different expression and conditions like happy, sad, sleepy, surprised, with and without glasses, center, right and left light and wink.
2. ORL: Olivetti Research Ltd (ORL) dataset introduced in [20], now called 'The Database of Faces' developed at ATandT laboratory, contains a set of 92×112 sized grayscale face images of 40 subjects, with 10 samples per person, taken between April 1992 and April 1994 at the laboratory. Samples vary in pose and expression and the subjects are of Caucasoid descent.
3. MR2: It is a small **mu**lti-**r**acial, **me**ga-**r**esolution dataset [1] of frontal face images with neutral expression and no variation in illumination or pose, developed by Nina Strohminger. This is an ethnically balanced dataset consisting of images of Caucasian, Asian and African images almost equally.
4. CFD: This Chicago Face Database (CFD) [21], developed at University of Chicago, provides mega-resolution face images of subjects of four different ethnic backgrounds.
5. CUHK and JaffeDB: Chinese University of Hong Kong (CUHK) [22] and Japanese Female Facial Expression (JAFFE) [23] datasets consists of 188 and 213 face images, respectively, all of which are Asian. CUHK samples are neutral in pose, illumination and expression while JAFFE has grayscale images with different expressions of an individual. Though CUHK dataset has both sketch and images of its students, we use only the real pictures for our experiment.

Table 2 shows the contribution of the different datasets in forming the individual groups while Fig. 2 displays some samples from the constituent datasets.

Table 2 Shares of the constituent datasets

	Yale	ORL	MR2	CFD	CUHK	JAFFE	Total
Caucasian	85	386	20	76	–	–	**567**
Asian	–	–	20	109	188	213	**530**
African	–	10	31	526	–	–	**567**

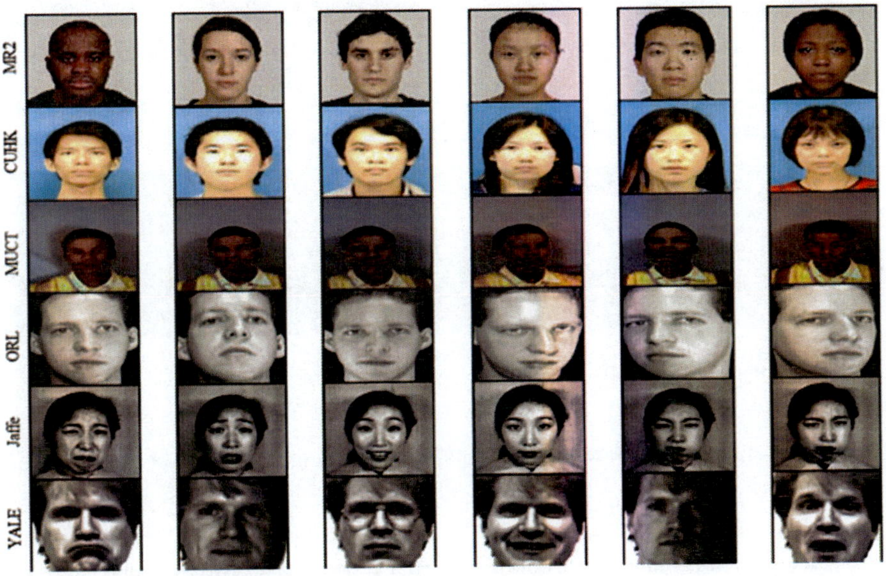

Fig. 2 Sample data from various datasets

BUPT Dataset: To test the feasibility of the linear features on natural images without the restricted laboratory settings we adopted the Beijing University of Posts and Telecommunication (BUPT-Balanced) [24] dataset is an enormous collection of 1.3M images of four racial origins namely, African, Asian, Caucasian and Indian with equal proportion. For every race group, it holds samples of 7000 individual. The samples display a wide variation in illumination, pose and expression. Figure 3 shows some of the samples from this set.

Train and Test Sets: The composite dataset is split in the ratio 7:3 such that 70% is reserved for training the model(s) while 30% is used for testing the model's performance.

Data Preprocessing: Aligning the face images is an important task before using them for model training. Preprocessing the images removes the unwanted background chaos, aligns the image so that the face area is brought to a specific location and does the necessary steps to prepare the samples for the next phase. The preprocessing

Fig. 3 Samples from the BUPT dataset

sub-tasks followed for these experiments are according to Ahmed et al. [25] which are:

1. Face detection and alignment with dlib implementation of HoG and SVM-based face detector and Kazemi and Sullivan's [26] method for 64 facial landmark detection, respectively.
2. Face area cropping.
3. Resizing the image to 40 × 40 grayscale.

3.2 Principal Component Analysis (PCA)

Let $\{x_1, x_2, \ldots, x_N\}$ be a set of images '*vectorized*' where each image vector is of size p $(= m \times n)$. From p-dimensional space, PCA derives a k-dimensional subspace $(k \ll p)$ such that the variance is maximized. Let X be the data matrix of images with N rows and p columns,

$$X = \begin{bmatrix} x_{11} & x_{12} & . & . & x_{1p} \\ x_{21} & x_{22} & . & . & x_{2p} \\ . & . & . & . & . \\ x_{N1} & x_{N2} & . & . & x_{Np} \end{bmatrix}$$

and W be the desired optimal transformation matrix. Then the projections are given by $Y = XW$. To perceive the mathematical background for deriving W, let us assume

$k = 1$ for simplicity. The variance is then given by:

$$\sigma_w^2 = \frac{1}{N}\sum_i^N (\vec{x_i}.\vec{w})^2$$

where $\frac{1}{N}\sum_i^N x_i = 0$ that is their mean is subtracted.

$$\begin{aligned} &= \frac{1}{N}(XW)^{\mathrm{T}}(XW) \\ &= \frac{1}{N}W^{\mathrm{T}}X^{\mathrm{T}}XW \\ &= W^{\mathrm{T}}\left(\frac{X^{\mathrm{T}}X}{N}\right)W \\ &= W^{\mathrm{T}}VW \end{aligned} \tag{1}$$

where V is the covariance matrix of X given by $S = \frac{1}{N}\sum_{i=1}^N (x_i - m)(x_i - m)^{\mathrm{T}}$ where $m = \frac{1}{N}\sum_{i=1}^N x_i$. As per the requirement, $\vec{w}$ should be a *unit* vector that maximizes variance and thus should satisfy the following constraint:

$$\vec{w}.\vec{w} = 1 \text{ or } W^{\mathrm{T}}W = 1.$$

Adding λ times the constraint equation to the objective function, where λ is the Lagrange's multiplier, the projection problem becomes

$$\mathcal{L}(w, \lambda) = \sigma_w^2 - \lambda\left(W^{\mathrm{T}}W - 1\right)$$

$$\frac{\delta L}{\delta \lambda} = W^{\mathrm{T}}W - 1 \tag{2}$$

$$\frac{\delta L}{\delta w} = 2VW - 2\lambda W \tag{3}$$

Setting the derivatives to zero at optimum, we get

$$\begin{aligned} &(2) \Rightarrow W^{\mathrm{T}}W = 1 \\ &(3) \Rightarrow VW = \lambda W \end{aligned}$$

(2) and (3) shows that W is an eigenvector of V. The eigenvectors are thus the principal components of V and their corresponding Eigen values indicate the amount of variance along those directions. As V is symmetric, its Eigen vectors are also orthogonal and thus their projections will be uncorrelated. Transforming the original data onto the new set of bases composed of k principal components would project

the p-dimensional data to a new compact and uncorrelated data with k-dimension where $k \ll N$.

3.3 Linear Discriminant Analysis (LDA)

LDA is another dimension reduction method which maximizes class separability. LDA finds the projection in such a way that the between-class separation is highest and the samples belonging to one class are close to each other. For a dataset of N samples with c classes and n dimensions, LDA finds the $(c - 1)$ most discriminative features. LDA is based on the Fisher Linear Discriminant Analysis which tries to maximize the criterion $J(w)$. For a projection matrix, W

$$J(w) = \frac{W^{\mathrm{T}} S_b W}{W^{\mathrm{T}} S_w W} \tag{4}$$

where S_w is the within-class scatter matrix and S_b is the between-class scatter matrix. S_w is calculated as

$$S_w = \sum_{i=1}^{c} \sum_{j=1}^{N_i} \left(x_{ij} - \overline{x_i}\right)\left(x_{ij} - \overline{x_i}\right)^{\mathrm{T}}$$

where x_{ij} is the jth sample of the ith class, $\overline{x_i}$ is the mean of the ith class and N_i is the number of samples belonging to the ith class. If X_i is the data matrix formed by the samples belonging to the ith class, then S_w and S_b are given by

$$S_w = \sum_{i=1}^{c} (X_i - \overline{x_i})(X_i - \overline{x_i})^{\mathrm{T}} \tag{5}$$

$$S_b = \sum_{i=1}^{c} N_i (\overline{x_i} - \overline{x})(\overline{x_i} - \overline{x})^{\mathrm{T}} \tag{6}$$

where $\overline{x}$ is the grand mean of the dataset. Note that the rank of S_w is at most $(N - c)$ and that of S_b is at most $(c - 1)$. Finding the maximum of $J(w)$, (4) boils down to solving the following generalized Eigen value problem:

$$J(w) = S_w^{-1} S_b W \tag{7}$$

$$W^* = \arg\max_W \left\{ \frac{W^{\mathrm{T}} S_b W}{W^{\mathrm{T}} S_w W} \right\} = [w_1, w_2, \ldots, w_t]$$

where w_i, $i = 1, 2, \ldots, t$ are the eigenvectors corresponding to the t largest Eigen values of S_w and S_b or $S_w^{-1} S_b$.

3.4 Nearest Neighbor (NN) Classifier

Nearest neighbor is a simple classification method which is easy to understand as well as implement. It is based on the idea that similar samples are close to each other, whereas dissimilar ones are far apart. It is a run-time-based method where a test sample is assigned the label of the item which is nearest to it. Thus NN does not involve any pre-training as such and classification of test items is done on-the-fly and so its run-time is relatively higher than most of the other methods.

Instead of considering only the nearest item, a variant of NN, involves k-neighbors (KNN) and its class labels and assigns the label of the majority class to the test sample. The neighbors are dependent on the distance metric used for calculating the distances between the test sample and its neighbor(s). And so, the performance of the algorithm also varies with the metric used for distance calculation. Some common distance metrics used are described briefly below.

Minkowski: Let $x = (x_1, x_2, \ldots, x_n)$, $y = (y_1, y_2, \ldots, y_n) \in R^n$, then the distance between them is calculated as:

$$d(x, y) = \left(\sum_{i=1}^{n} |x_i - y_i|^p \right)^{\frac{1}{p}} \tag{8}$$

Euclidean: Euclidean distance is the most common. It is a special case of Minkowski Distance formula when $p = 2$. This updates the distance to

$$d(x, y) = \sqrt{\sum_{i=1}^{n} |x_i - y_i|^2} \tag{9}$$

Euclidean distance formula can be used to calculate the distance between two data points in a plane.

Manhattan: It is the distance between two points in a block-like path. For $x = (x_1, x_2, \ldots, x_n)$, $y = (y_1, y_2, \ldots, y_n) \in R^n$, the Manhattan distance between x and y is calculated using an absolute sum of difference between its Cartesian co-ordinates as

$$d(x, y) = \sum_{i=1}^{n} |x_i - y_i| \tag{10}$$

Mahalanobis: Mahalanobis Distance is used for calculating the distance between two data points in a multivariate space. It is a measure of distance between a point p and a distribution $\boldsymbol{D}$. The idea behind measuring is how many standard deviations away p is from the mean of $\boldsymbol{D}$, $\vec{\mu}$. The benefit of using this method of distance measuring is that it takes covariance in account which helps in measuring the similarity between two different data objects. The distance between an observation and the mean is calculated as

$$D_M(\vec{x}) = \sqrt{(\vec{x} - \vec{\mu})^T S^{-1} (\vec{x} - \vec{\mu})} \quad (11)$$

where S is the covariance matrix.

Cosine similarity: It is a similarity index rather than a concrete distance measure between two points or vectors. It is quantified by the cosine of the angle (θ) made by two vectors ($\vec{x}$, $\vec{y}$). It gives an idea about the orientation of the vectors irrespective of its magnitude. Two vectors having same orientation have a cosine similarity of 1, -1 for opposite directions and a value of 0 for orthogonal vectors. The following expression is a measure for cosine similarity:

$$\text{Similarity}(\vec{x}, \vec{y}) = \text{Cosine}(\theta) = \frac{\vec{x}.\vec{y}}{||\vec{x}||||\vec{y}||} \quad (12)$$

Chi-square: This distance is commonly used for comparing histograms. If $x = (x_1, x_2, \ldots, x_n)$, $y = (y_1, y_2, \ldots, y_n)$ are two vectors corresponding to two histogram values, then the Chi-square distance between them is calculated as follows:

$$\chi^2(x, y) = \frac{1}{2} \sum_{i=1}^{n} \frac{(x_i - y_i)^2}{x_i + y_i} \quad (13)$$

4 Experiments and Results

4.1 Experiment 1: Race Identification with PCA

Following the steps described in Sect. 3.2, the Eigen values with its corresponding Eigen faces are deduced. Figure 4a shows the top ten Eigen faces and Fig. 4b shows the plot of the Eigen values. From the graph, it can be clearly seen that among the 1600 Eigen values only a few of them have a considerable value while the most of them are negligible as they tend to zero. Therefore, we have selected the first 50 Eigen vectors as the new bases on which the data would be projected.

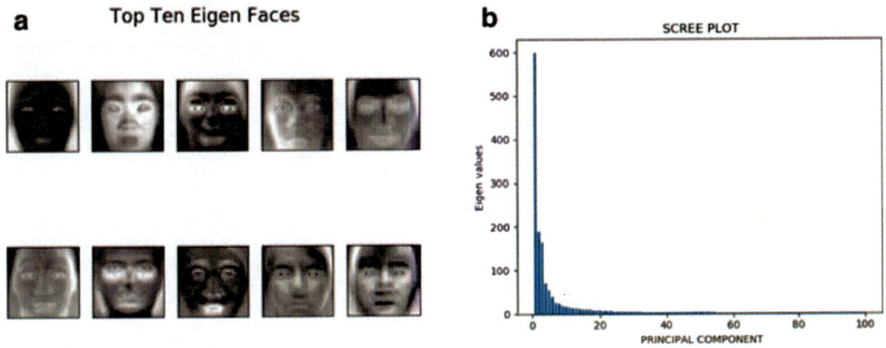

Fig. 4 **a** Top ten Eigen faces. **b** Scree plot of Eigen values

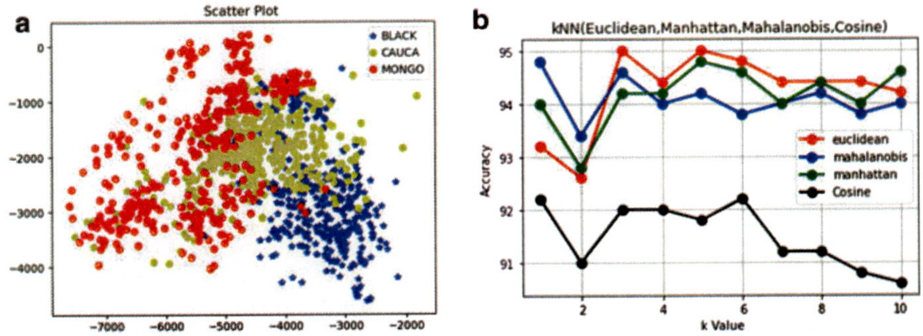

Fig. 5 **a** Scatter plot of the projected samples. **b** Accuracies for different values of k

The scatter plot of the three ethnic samples based on its new co-ordinate values is shown in Fig. 5a. Only the co-ordinates corresponding to the first two Eigen vectors are chosen for the scatter plot. The plot shows that the samples are separable. Table 3 shows the performance of the k-nearest neighbor classifier for different values of k. The table values are depicted as a graph in Fig. 5b for better readability. The model achieves the best result of 95% for Euclidean distance for $k = 3$ and 5.

Results on BUPT Data: Since BUPT is an enormous dataset with 1.3M images, applying KNN on such a set would be very expensive. So, for the train set, we select 3000 samples with 1000 samples from African, Asian and Caucasian group, ignoring the Indian race for comparability with the previous experiment. For the test set, 200 samples from each class were taken such that the images of the individuals involved in the train set does not overlap with the test set. Figure 6 shows the scatter plot of the PCA projections of the BUPT samples.

After experimenting with different distance metrics and k values, Table 4 shows the best values obtained with Mahalanobis distance metric for $k = 9$ on BUPT data. PCA features are very sensitive to the inherent variations associated with a facial image like the background, illumination, pose and expression. This is evident from

Table 3 PCA accuracy (in %) with different distance metrics in KNN classifier

k value	Distance metric			
	Euclidean	Manhattan	Mahalanobis	Cosine similarity
1	93.2	94	94.8	92.2
2	92.6	92.8	93.4	91
3	95	94.2	94.6	92
4	94.4	94.2	94	92
5	95	94.8	94.2	91.8
6	94.8	94.6	93.8	92
7	94.4	94	94	91.2
8	94.4	94.4	94.2	91.2
9	94.4	94	93.8	90.8
10	94.2	94.6	94	90.6

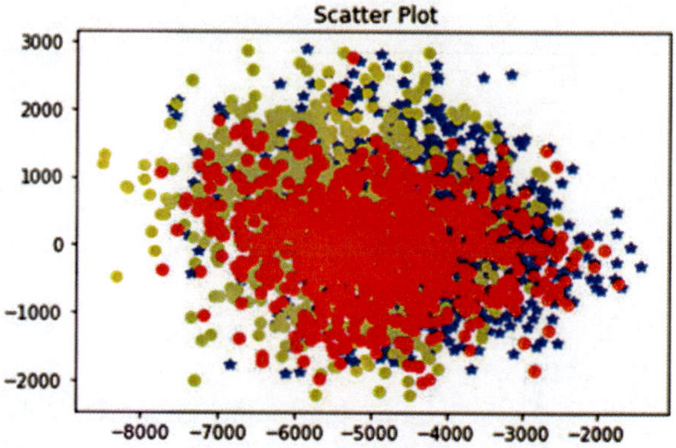

Fig. 6 Scatter plot for BUPT projected samples

Table 4 Result statistics of PCA features on BUPT dataset

	Precision	Recall	F_1-score
African	0.71	0.85	0.78
Caucasian	0.70	0.84	0.76
Asian	0.97	0.76	0.76
Average	0.79	0.76	0.76
Accuracy		0.76	

the scatter plot in Fig. 6 which shows the overlapping characteristics of the projected BUPT samples unlike Fig. 5 which clearly shows the separability among the groups for images taken in controlled environment. The decline in accuracy for natural images is, thus obvious.

4.2 Experiment 2: Race Identification with LDA

The training samples are of size 40 * 40 which equals to a dimension of 1600, whereas the cardinality of the training set for the assembled set is 1163 comprising of 70% of the total data. This poses the *Small Sample Size (SSS) Problem* wherein S_w becomes singular as not being full-rank. To solve it, we have adopted the method proposed by Rui Huang in [27]. In this method, the common null space of S_w and S_b is removed without losing the useful discriminative information and finally a subspace of S_w is derived having $(c - 1)$ most discriminative vectors. The procedure as described in [27] used for addressing SSS problem is:

Step 1: Let, $S_t = S_b + S_w$.

Step 2: Remove the common Null space of S_t by Eigen analysis. It can be achieved by forming a matrix, U whose columns are Eigen vectors corresponding to non-zero Eigen values of S_t.

Step 3: Let,

$$S'_w = U^{\mathrm{T}} S_w U \tag{14}$$

$$\text{and } S'_b = U^{\mathrm{T}} S_b U. \tag{15}$$

Step 4: Calculate null space, Q, of $S_w{}'$.
So, $S''_w = Q^{\mathrm{T}} S_w{}' Q = 0$.
From (14),

$$\begin{aligned} S''_w &= Q^{\mathrm{T}} U^{\mathrm{T}} S_w U Q = 0 \\ &= (U Q)^{\mathrm{T}} S_w (U Q) = 0. \end{aligned}$$

Thus, UQ is a subspace of the null space of S_w.
Similarly, let $S''_b = Q^{\mathrm{T}} S_b{}' Q$.
From (15), $S''_b = (U Q)^{\mathrm{T}} S_b (U Q)$.

Step 4: Remove null space of S''_b, if it exists. Let V be the matrix formed by the Eigen vectors of S''_b corresponding to the non-zero Eigen values, then the final LDA transformation matrix, W is given by

$$W = U Q V.$$

Fisher's Linear Discriminant Analysis maximizes the scatter between the classes and reduces the within-class scatter. So in the LDA space, samples belonging to a class are tightly coupled with one another. This is evident from Fig. 7. For the assembled data, all the African samples of the assembled data are projected onto the point (23.45, 18.89) with difference among them found to be less than 10^{-10}. Likewise, the Caucasian and Asian samples are also projected onto their respective points. So, the results of KNN remain unaffected by the different values of k. Also

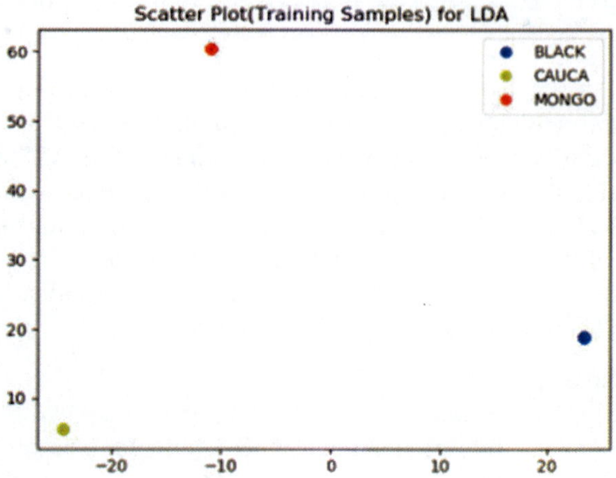

Fig. 7 Scatter plot of the projected samples in LDA

Table 5 LDA accuracy (in %) with different distance metrics in KNN classifier

k-NN distance metric	Accuracy (%)	
	Assembled data (controlled)	BUPT (avg. accuracy)
Euclidean	92.01	75.4
Manhattan	92.61	74.5
Mahalanobis	91.82	40

the KNN algorithm can be greatly made efficient in terms of time as all the training data projects to c points, where c is the number of classes. So, instead of finding the distance of the test samples from all the training samples, only c distances are evaluated and the class associated with the nearest distance be assigned to the test sample. Doing so, the time complexity reduces from $O(DN)$ to $O(Dc)$ where D and N ($D \ll N$) are the dimension and the number of training samples, respectively. For BUPT data, the projection points for the respective groups do not exactly coincide to a single point as in the above case but the differences among them lie in the range [0, 0.5]. Table 5 shows the classification results for both the assembled and BUPT data.

As in PCA, the decline in performance for LDA features is also noted for BUPT samples.

4.3 Experiment 3: Race Identification Using LBP Locally on Face Parts

To analyze texture properties for race classification, the face area is divided into its component parts (Fig. 8) namely:

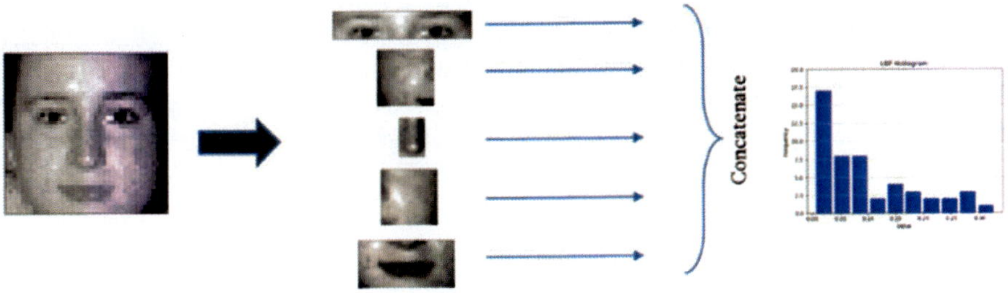

Fig. 8 Face components used for deriving LBP features

- periocular region
- nose
- right cheek
- left cheek and
- chin

To extract these parts from a face image, 64 facial landmarks are localized by the dlib's implementation of Kazemi and Sullivan's method. By choosing appropriate co-ordinates the regions are cropped. For each face part, LBP texture feature and its corresponding 10-bin histogram is computed. For each face sample, the histograms does obtained from its component parts are then concatenated to form a combined feature of size 50.

The LBP features are then supplied to the nearest neighbor classifier for classification. Since the features are actually histogram values, so chi-square distance is the most appropriate metric for similarity measure. Table 6 shows the results of classification using LBP features with face parts on the assembled data as well as on the BUPT dataset.

Table 6 Accuracy (%) for different *k* values with chi-square distance

K	Accuracy (%)	
	Assembled data	BUPT
1	97.6	100
2	92.8	82
3	92	79.2
4	88.4	78.9
5	88.4	78.4
6	86.4	76.8
7	88.4	76
8	84.4	74.6
9	85.6	72.8
10	86.4	74

As compared to the assembled set, there is a decline in performance for the BUPT set which is obvious. But LBP being a local feature proves to have more discriminating power than the holistic features such as PCA and LDA. Among the three experimented features, only LBP could achieve accuracy more than 80% (for $k = 1$ and 2) as compared to a best value of 76% for both PCA and LDA on the BUPT set.

5 Conclusion

Race or ethnicity identification, itself a conflicting concept for humans, is a challenging task for machines. Linear features like PCA, LDA and LBP are capable of differentiating the race groups in ideal situations when images have uniform background, very limited variation in pose, expression and illumination but the features become less discriminative for unrestricted environment which is definitely a weak point to look into. It is however noteworthy that local feature such as LBP possesses higher discriminative power than the simple holistic features like PCA or LDA. The experiments conducted prove this fact. Therefore, more powerful local descriptors or non-linear features should be investigated as future work for better predictive ability especially for images taken *in-the-wild*.

References

1. Strohminger N, Gray K, Chituc V, Heffner J, Schein C, Heagins T (2016) The MR2: a multi-racial, mega-resolution database of facial stimuli. Behav Res Methods 48(3):1197–1204. https://doi.org/10.3758/s13428-015-0641-9
2. Lu X, Jain A, Ethnicity identification from face images. In: Proceedings of SPIE, pp 114–123. https://doi.org/10.1117/12.542847
3. Tin H, Sein M (2011) Race identification for face images. ACEEE Int J Inf Technol 01(02), 35–37. http://hal.archives-ouvertes.fr/hal-00753272/
4. Fuentes-Hurtado F, Diego-Mas J, Naranjo V, Alcañiz M (2019) Automatic classification of human facial features based on their appearance. PLoS ONE 14(1):1–20. https://doi.org/10.1371/journal.pone.0211314
5. Logan A, Munshi T (2017) International conference on cyberworlds. https://doi.org/10.1109/CW.2017.27
6. Hosoi S, Takikawa E, Kawade M (2004) Ethnicity estimation with facial images.In: Proceedings of the sixth IEEE international conference on automatic face and gesture recognition, pp 195–200. https://doi.org/10.1109/AFGR.2004.1301530
7. Qiu X, Sun Z, Tan T (2006) Global texture analysis of iris images for ethnic classification. In: International conference on biometrics, pp 411–418. https://doi.org/10.1007/11608288_55
8. Lin H, Lu H, Zhang L (2006) A new automatic recognition system of gender, age and ethnicity. In: Proceedings of the World congress on intelligent control and automation (WCICA), vol 2(3), pp 9988–9991. https://doi.org/10.1109/WCICA.2006.1713951
9. Duan X, Wang C, Liu X, Li Z, Wu J, Zhang H (2010) Ethnic features extraction and recognition of human faces. In: Proceedings of the 2nd IEEE international conference on advanced computer control, ICACC, pp 125–130. https://doi.org/10.1109/ICACC.2010.5487194

10. Saei M, Ghahramani M, Tan Y (2010) Facial part displacement effect on template-based gender and ethnicity classification. In: 11th International conference on control, automation, robotics and vision, ICARCV, pp 1644–1649. https://doi.org/10.1109/ICARCV.2010.5707882
11. Carcagnì P, Coco MD, Cazzato D, Leo M, Distante C (2015) A study on different experimental configurations for age, race, and gender estimation problems. Eurasip J Image Video Process 2015(1):1–22. https://doi.org/10.1186/s13640-015-0089-y
12. Momin H, Tapamo J (2016) A comparative study of a face components based model of ethnic classification using Gabor filters. Appl Math Inf Sci 10(6):2255–2265. https://doi.org/10.18576/amis/100628
13. Roomi S, Virasundarii S, Selvamegala S, Jeevanandham S, Hariharasudhan D (2011) Race classification based on facial features. In: Proceedings of the 2011 3rd national conference on computer vision, pattern recognition, image processing, and graphics NCVPRIPG, pp 54–57. https://doi.org/10.1109/NCVPRIPG.2011.19
14. Han H, Jain A (2014) Age, gender and race estimation from unconstrained face images. Technical report MSU-CSE, pp 1–9. http://en.wikipedia.org/wiki/Young_adult
15. Guo G, Mu G (2010) A study of large-scale ethnicity estimation with gender and age variations. In: 2010 IEEE Computer society conference on computer vision and pattern recognition—workshops, CVPRW 2010, pp 79–86. https://doi.org/10.1109/CVPRW.2010.5543608
16. Xie Y, Luu K, Savvides M (2012) A robust approach to facial ethnicity classification on large scale face databases. In: 2012 IEEE 5th international conference on biometrics: theory, applications and systems, BTAS 2012, pp 143–149. https://doi.org/10.1109/BTAS.2012.6374569
17. Lu X, Chen H, Jain A (2006) Multimodal facial gender and ethnicity identification. In: Lecture notes in computer science (including subseries Lecture notes in artificial intelligence and Lecture notes in bioinformatics), LNCS 3832, pp 554–561. https://doi.org/10.1007/11608288_74
18. Muhammad G, Hussain M, Alenezy F, Bebis G, Mirza AM, Aboalsamh H (2012) Race classification from face images using local descriptors. Int J Artif Intell Tools 21(5):1–24. https://doi.org/10.1142/S0218213012500194
19. Yale face dataset. http://vision.ucsd.edu/content/yale-face-database. Accessed on 26-6-2021
20. Samaria F, Harter A (1994) Parameterisation of a stochastic model for human face identification. In: IEEE Workshop on applications of computer vision—proceedings, pp 138–142. https://doi.org/10.1109/acv.1994.341300
21. Ma D, Correll J, Wittenbrink B (2015) The Chicago face database: a free stimulus set of faces and norming data. Behav Res Methods 47(4):1122–1135. https://doi.org/10.3758/s13428-014-0532-5
22. Wang X, Tang X (2009) Face photo-sketch synthesis and recognition. IEEE Trans Pattern Anal Mach Intell 31(11):1955–1967. https://doi.org/10.1109/TPAMI.2008.222
23. Lyons M, Akamatsu S, Kamachi M, Gyoba J (1998) Coding facial expressions with Gabor wavelets. In: Proceedings of the 3rd IEEE international conference on automatic face and gesture recognition, FG 1998, pp 200–205. https://doi.org/10.1109/AFGR.1998.670949
24. Wang M, Deng W (2019) Mitigating bias in face recognition using skewness-aware reinforcement learning. https://doi.org/10.1109/CVPR42600.2020.00934
25. Ahmed M, Choudhury R, Kashyap K (2020) Race estimation with deep networks. J King Saud Univ Comput Inf Sci. https://doi.org/10.1016/j.jksuci.2020.11.029
26. Kazemi V, Sullivan J (2014) One millisecond face alignment with an ensemble of regression trees. In: Proceedings of IEEE computer society conference on computer vision and pattern recognition, pp 1867–1874. https://doi.org/10.1109/CVPR.2014.241
27. Huang R, Liu Q, Lu H, Ma S (2002) Solving the small sample size problem of LDA. In: Proceedings of the international conference on pattern recognition, vol 16(3), pp 29–32. https://doi.org/10.1109/icpr.2002.1047787

Retinal Disease Classification from Retinal-OCT Images Using Deep Learning Methods

Archana Naik, B. S. Pavana, and Kavitha Sooda

Abstract Retinal diseases are the damage caused to any part of the retina. OCT images are used to diagnose retina related diseases. Cross sectional view of the retina is obtained through Optical Coherence Tomography (OCT). Medical disease prediction through images is time consuming and manual process. Computer vision technology and its progress have given a solution for the prediction and classification of medical diseases through images. The major aim of this research is to present a new deep learning-based classification model for automatically classifying and predicting different retinal disorders using OCT data. In this paper, retinal disease classification is performed using convolution neural networks (CNN). CNV, DME, DRUSEN, NORMAL images are the four classifications investigated in this study. Proposed Retina CNN architecture is compared with existing architecture for the detection of retinal disease. The best accuracy and better model prediction are obtained for the Retina CNN architecture with Adam optimizer with the training accuracy is 99.29 and 97.55% validation accuracy.

Keywords CNN · InceptionV3 · Prediction · VGG19 · Retina CNN

1 Introduction

Retinal diseases (RD) cause severe visual symptoms. The retina, a thin layer of tissue on the eye's wall, can be affected by RD in any portion. The retina is made up of light-sensitive cells (rods and cones) as well as other nerve cells that receive and

A. Naik · B. S. Pavana (✉)
Department of Computer Science and Engineering, Nitte Meenakshi Institute of Technology, Yelahanka, Bangalore, India
e-mail: pavanabs9@gmail.com

A. Naik
e-mail: archana.naik@nmit.ac.in

K. Sooda
Department of Computer Science and Engineering, B.M.S. College of Engineering, Basavanagudi, Bangalore 560019, India

V. Skala et al. (eds.), *Machine Intelligence and Data Science Applications*, Lecture Notes on Data Engineering and Communications Technologies 132,
https://doi.org/10.1007/978-981-19-2347-0_8

process visual data. Through the optic nerve, the retina transmits this information to the brain, allowing one to see. For RD, there is treatment. Treatment can be used to stop or reduce the progression of the illness, as well as to improve or restore eyesight, depending on the RD status. If untreated, RD can result in visual loss or perhaps blindness. The diagnosis process involves medical professionals checking the abnormalities in the retina from observing the OCT scan images. The technology of OCT imaging has been utilized to diagnose RD. This process is a lengthy and manual one [1]. A different approach to this challenge is to create an algorithm that aids medical practitioners in detecting RD.

Overview of this paper consists of Sect. 2: Literature Survey, Sect. 3: Proposed Methodology, Sect. 4: Implementation Details, Sect. 5: Result and Analysis, Sect. 6: Conclusion and Sect. 7: Future Work.

2 Literature Survey

B-scan attentive convolution neural network (BACNN) method uses classification module, attention module through a self-attention mechanism and the features are aggregated from the module called feature extraction module of the B-scans to form an attentive vector for obtaining better classification of diseases [2]. Data augmentation is conducted by flipping on the training folds of the OCT images of B-scans for effective training. The proposed method uses CNN for detecting ocular diseases which have eight types [3]. The data is then submitted to the network for categorization after some conventional preprocessing.

The goal is to use retinal images to create a good retinal image classifier [4]. The public STARE color picture collection was used in this study. The network configuration utilized with the size of the input image with the tiny resolution of 31 × 35 pixels yielded the maximum test accuracy of 80.93%. Here OCT-NET is based on CNN for the RD classification based on OCT volumes [5].

There were four types of OCT images utilized [6]. The support vector machine (SVM) with tenfold cross-validation classification model was utilized on two transfer learning architectures and with categorical hinge loss to categorize retinal disorders. The models of long short-term memory (LSTM) and convolutional neural networks (CNN) are merged and developed [7]. In this case, the CNN is utilized to extract layers of interest and acquire edges, while the LSTM is used to trace the layer's borders. The model is trained with minimal data on normal and AMD disease images.

In order to automatically identify retinal pictures using convolutional neural networks, the surrogate-assisted technique is applied (CNNs) [8]. Denoising is used to minimize noise. The picture masks are created using morphological dilation and thresholding. The denoised pictures and masks of the image are utilized to generate a number of surrogate images that may be used to train the CNN model. The test image prediction is determined from the surrogate images that were trained on the CNN model. The pre-processed OCT scan pictures of the retina are classified using the basic CNN architecture [9]. The neural network is fed pre-processed photos that have been scaled to 128 × 128 pixels. Within the training phase, the network converged to a 98% accuracy on the train set and an 85% validation accuracy.

Neural Architecture Search (NAS) is used for the layer segmentation of the retina that uses OCT scans [10]. The U-Net model is incorporated in the NAS framework as its backbone which is used for the retina layer segmentation in the collected and pre-processed OCT image dataset. For the classification layer guided convolutional neural network is used [11]. ReLayNet is used for layer segmentation. ReLayNet generates the maps of the layer segmentation and the probability maps were extracted from the two lesion-related layers. Then, the LGCNN is used for the integration of the lesion-related layer information for classification.

The retina layers are segmented using the U-Net model [12]. The ImageNet dataset is utilized to pre-train the U-Net model with a ResNet-based encoder. For color fundus pictures, VGG19 employs a transfer learning technique for Retinal Disease Classification [13]. The merits of computer vision and machine intelligence in biomedical picture categorization problems are discussed in this overview of the literature. From the above studied literature survey, the implementation includes the CNN architecture using transfer learning. In this study pre-trained models were used such as Inception V3 and VGG16 [14].

3 Proposed Methodology

The proposed methodology is based on CNN architecture using an existing model with transfer learning and Retina CNN architecture. Pre-trained networks are trained on the ImageNet data. It is used due to the advantages like it is simple to incorporate and the models achieve good performance. Figure 1 shows the general workflow of the proposed methodology. The processes of the proposed system are: collecting the data and sending it as input, preprocessing of images by resizing them to 224 × 224. Construct the CNN architecture, train and test the model for the prediction of RD.

4 Implementation

4.1 System Requirements

Intel Core2 Duo Processor, 4 GB of RAM. The following software is required: Windows 8.1 operating system. The Python 2.7 version is used as the programming language and Google colaboratory as the IDE.

Fig. 1 General workflow of the proposed methodology

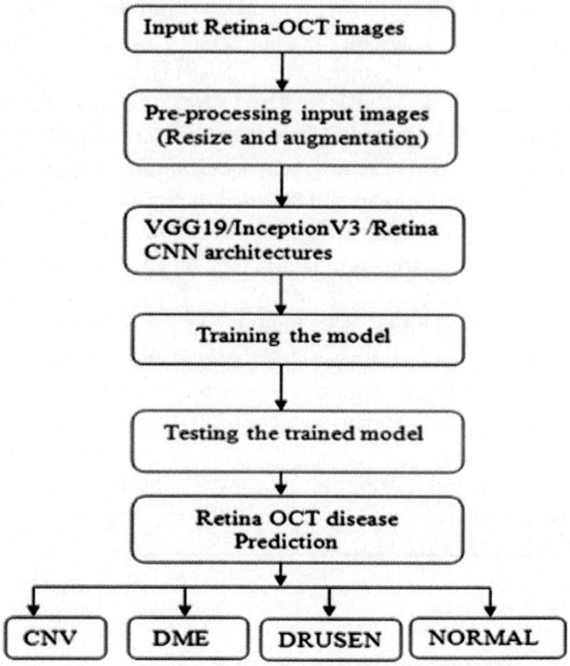

4.2 Dataset

Dataset is obtained from Kaggle website. Total images used: 4846 images. The dataset contains retinal-OCT images [15]. There are four classes to be classified they are:

1. CNV (choroidal neovascularization).
2. DME (diabetic macular edema).
3. DRUSEN.
4. NORMAL.

4.3 Implementation Details

Randomly selected 4846 images. 60% of dataset is divided for training, 20% for testing and 20% for validation. Loading of Retinal-OCT images and the images are resized to (224, 224). Since the images in the dataset are of different image size so this resizing is done on the entire dataset. Keras functions are used to load the images. Pre-trained network models on the ImageNet dataset were employed in this investigation. Data augmentation is applied to the dataset.

Transfer learning is performed by supplying the weights of the network trained on ImageNet, for the detection of RD. Deep neural networks such as VGG19 and

InceptionV3 are trained for the input pictures, and transfer learning is achieved by giving the weights of the network trained on ImageNet. Following that the trained model will be tested, and then it will be utilized to identify RD. To improve the output of neural network models, fine-tuning is accomplished. Dropout layers, batch normalization and dense layers are used. In the proposed methodology, existing model is compared with proposed model that is VGG19 and InceptionV3 with Retina CNN architecture were trained on 4846 images from Kaggel Retina OCT images.

4.4 Retina CNN Architecture

Figure 2 shows the architecture of Retina CNN. InceptionV3.This architecture has 13 convolutional layers with ReLu activation. Eleven batch normalization layers, 5 layers of max-pooling. Also includes GlobalAveragePooling2D, flatten layer and the softmax activation is used in final year for prediction of retina disease. The proposed model contains 34 layers. Training accuracy and other evaluation metrics are used to assess the proposed model's performance, also the prediction of the trained models is checked using confusion matrix.

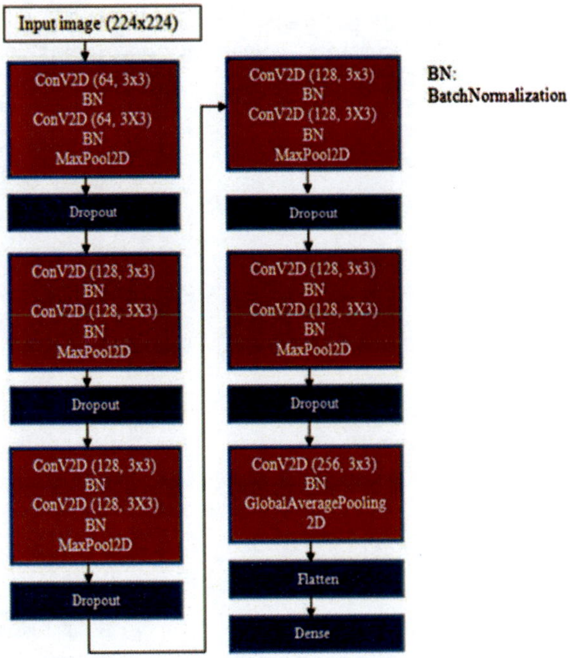

Fig. 2 Retina CNN architecture diagram

5 Result and Analysis

In this research, we find the train and test accuracy and also the confusion matrix which gives the number of correctly classified and misclassified test images. The images are mainly classified into four classes they are DRUSEN, CNV, DME and NORMAL.

Experiment has been conducted using two optimizers.

They are Adam and SGD on the above mentioned architectures. Dataset has been trained for 10 epochs for pre-trained architectures. The following are the outcomes of the implementation. The results of the Retina CNN architecture with Adam and SGD optimizer are shown in Fig. 3. Figure 4 shows the results obtained from VGG19 with Adam and SGD optimizer. Figure 5 shows the results obtained from InceptionV3 with Adam and SGD optimizer. Tables 1 and 2. Here the training and validation accuracies for CNN Architectures with Adam and SGD optimizers. Retina CNN Architecture gives good accuracy with Adam optimizer compared to other architectures.

Figure 6 shows the confusion matrix for Retina CNN architecture where 263 images are correctly classified as CNV. Two hundred twenty-five images are correctly classified as DME, 226 images are correctly classified as DRUSEN and 223 images

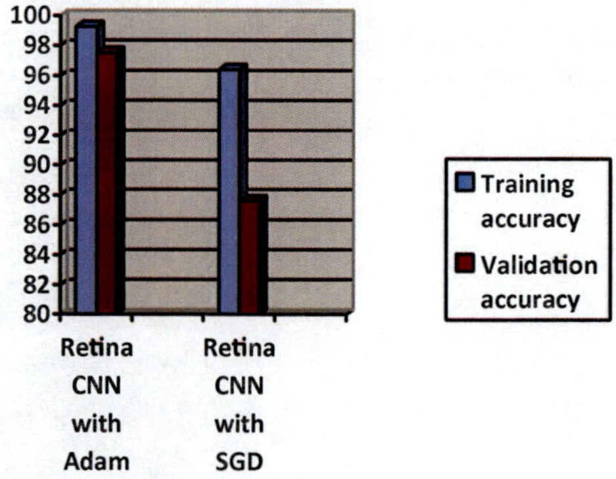

Fig. 3 Retina CNN architecture results for training accuracy versus validation accuracy

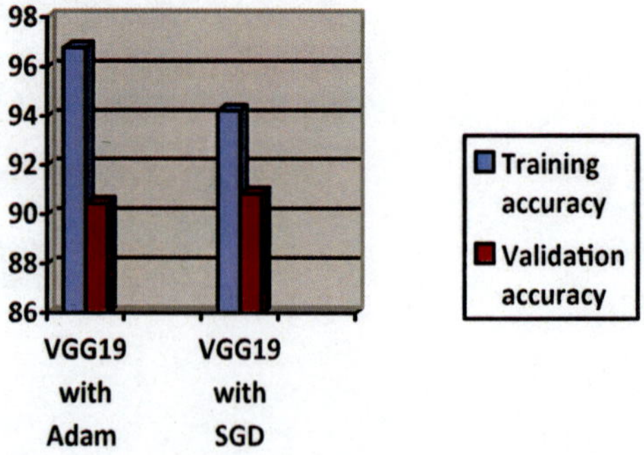

Fig. 4 VGG19 results for training accuracy versus validation accuracy

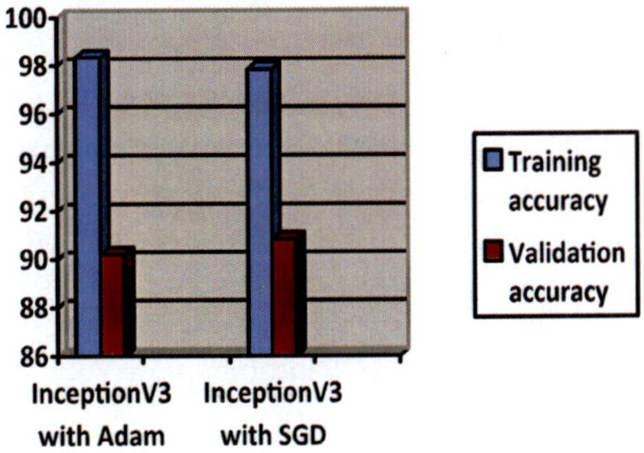

Fig. 5 InceptionV3 results for training accuracy versus validation accuracy

Table 1 Train and validation accuracies for CNN architectures with Adam optimizer

CNN architectures with Adam optimizer	Training accuracy	Validation accuracy
Reina CNN	99.29	97.55
VGG19	96.77	90.46
Inception V3	98.35	90.21

Table 2 Train and validation accuracies for CNN architectures with SGD optimizer

CNN architectures with SGD optimizer	Training accuracy	Validation accuracy
Retina CNN	96.42	87.63
VGG19	94.23	90.85
Inception V3	97.84	90.85

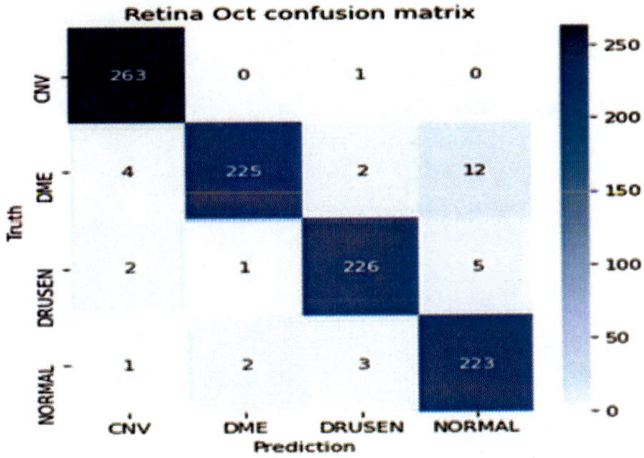

Fig. 6 Confusion matrix for Retina CNN architecture

are correctly classified as NORMAL. Figure 7 shows the confusion matrix for VGG19 architecture where 253 images are correctly classified as CNV. Hundred ninety five images are correctly classified as DME, 201 images are correctly classified as DRUSEN and 208 images are correctly classified as NORMAL. Figure 8 shows the confusion matrix for InceptionV3 architecture where 234 images are correctly classified as CNV. Two hundred eight images are correctly classified as DME, 193 images are correctly classified as DRUSEN and 204 images are correctly classified as NORMAL This shows how accurately the model can predict using Retina CNN architecture, VGG19 and InceptionV3.

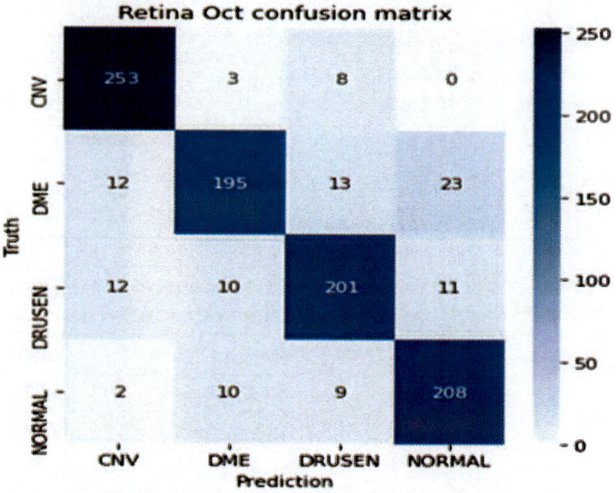

Fig. 7 Confusion matrix for VGG19 architecture

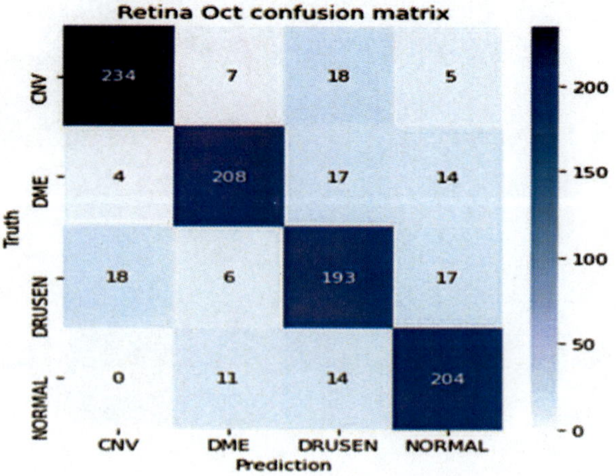

Fig. 8 Confusion matrix for InceptionV3 architecture

6 Conclusion

Retina OCT dataset included four distinct types of classes for categorization. From the result and analysis it can be concluded that Retina CNN is giving good training, validation accuracies and better prediction of the model than the existing architectures. Using the Retina CNN architecture with Adam optimizer, it was trained on 4846 pictures over 25 epochs, with the best training accuracy of 99.58% and validation accuracy of 97.55%. The automatic detection of retinal-OCT disease helps the clinicians for diagnosing the disease.

7 Future Work

A complex CNN network for retina OCT disease prediction and hyper-parameter variation to achieve the most accurate results both in terms of training and validation can be added in future construction. Increase the data size so that the training will be achieved better.

References

1. Jaffe GJ, Caprioli J (2004) Optical coherence tomography to detect and manage retinal disease and glaucoma. Am J Ophthalmol 137(1):156–169. https://doi.org/10.1016/s0002-9394(03)00792-x. PMID: 14700659
2. Das V, Prabhakararao E, Dandapat S, Bora PK (2020) B-scan attentive CNN for the classification of retinal optical coherence tomography volumes. IEEE Signal Process Lett 27
3. Islam Md T, Imran SA et al (2019) Source and camera independent ophthalmic disease recognition from fundus image using neural network. In: 2019IEEE International conference on signal processing, information, communication & systems (SPICSCON)
4. Triwijoyo BL, Heryadi Y (2017) Retina disease classification based on colour fundus images using convolutional neural networks. In: 2017 International conference on innovative and creative information technology (ICITech)
5. Perdomo O, Otálora S, González FA, Meriaudeau F, Müller H (2018) OCT-NET: a convolutional network for automatic classification of normal and diabetic macular edema using SD-OCT volumes. In: 2018 IEEE 15th international symposium on biomedical imaging (ISBI 2018), pp 1423–1426. https://doi.org/10.1109/ISBI.2018.8363839
6. Adel A, Soliman MM, Khalifa NEM, Mostafa K (2020) Automatic classification of retinal eye diseases from optical coherence tomography using transfer learning. In: 2020 16th International computer engineering conference (ICENCO), pp 37–42. https://doi.org/10.1109/ICENCO49778.2020.9357324
7. Gopinath K, Rangrej SB, Sivaswamy J (2017) A deep learning framework for segmentation of retinal layers from OCT images. In: 2017 4th IAPR Asian conference on pattern recognition (ACPR), pp 888–893. https://doi.org/10.1109/ACPR.2017.121
8. Rong Y et al (2019) Surrogate-assisted retinal OCT image classification based on convolutional neural networks. IEEE J Biomed Health Inform 23(1):253–263. https://doi.org/10.1109/JBHI.2018.2795545

9. Najeeb S, Sharmile N, Khan MS, Sahin I, Islam MT, Hassan Bhuiyan MI (2018) Classification of retinal diseases from OCT scans using convolutional neural networks. In: 2018 10th International conference on electrical and computer engineering (ICECE), pp 465–468. https://doi.org/10.1109/ICECE.2018.8636699
10. Gheshlaghi SH, Dehzangi O, Dabouei A, Amireskandari A, Rezai A, Nasrabadi NM (2020) Efficient OCT image segmentation using neural architecture search. In: 2020 IEEE International conference on image processing (ICIP), pp 428–432. https://doi.org/10.1109/ICIP40778.2020.9190753
11. Huang L, He X, Fang L, Rabbani H, Chen X (2019) Automatic classification of retinal optical coherence tomography images with layer guided convolutional neural network. IEEE Signal Process Lett 26(7):1026–1030. https://doi.org/10.1109/LSP.2019.2917779
12. Matovinovic IZ, Loncaric S, Lo J, Heisler M, Sarunic M (2019) Transfer learning with U-Net type model for automatic segmentation of three retinal layers in optical coherence tomography images. In: 2019 11th International symposium on image and signal processing and analysis (ISPA), pp 49–53. https://doi.org/10.1109/ISPA.2019.8868639
13. Das A, Giri R, Chourasia G, Bala AA (2019) Classification of retinal diseases using transfer learning approach. In: 2019 International conference on communication and electronics systems (ICCES), pp 2080–2084. https://doi.org/10.1109/ICCES45898.2019.9002415
14. Berrimi M, Moussaoui A (2020) Deep learning for identifying and classifying retinal diseases. In: 2020 2nd International Conference on computer and information sciences (ICCIS), pp 1–6. https://doi.org/10.1109/ICCIS49240.2020.9257674
15. Kermany D, Zhang K, Goldbaum M (2018) Labeled optical coherence tomography (OCT) and chest X-ray images for classification. Mendeley Data V2. https://doi.org/10.17632/rscbjbr9sj.2

Stock Recommendation BOT for Swing Trading and Long-Term Investments in Indian Stock Markets

Samarth Patgaonkar, Sneha Dharamsi, Ayush Jain, and Nimesh Marfatia

Abstract Despite being the second most populous country, the Indian stock market is underpenetrated compared to markets in developed countries. However, the recent Indian stock market trends have depicted higher interests from institutional investors as well as from the Indian retail investors. Even though information is widely available for participants to do stock market analysis, stock market traders and investors still get influenced by stock recommendations from friends, financial advisors, trading community, and various other financial blogging platforms. To avoid bias, institutional traders are showing interest towards robotic or algorithmic trading approaches which largely lead to automated decision-making. In this paper, the human bias is eliminated by assisting the automatic algorithmic trading approach by application of models like artificial neural network (ANN), random forest (RF), gradient boosting (GB), *K*-nearest neighbours (KNN), and logistic regression (LR) on stock market data. "Recommendation model" is built using the above-mentioned techniques to provide a "buy" recommendation to swing traders and investors in Indian equity market. From a short-term trading perspective (up to 3 months), the model will recommend "buy" for stocks which have a potential to provide 10% or more return on investment. From a long-term investment perspective (up to 3 years), the model will recommend a "buy" recommendation for a stock, which has a potential to provide 50% or more return on investment. As per our research, random forest (RF) model is recommended as the best model for short-term predictions, and artificial neural network (ANN) model is recommended as the best model for long-term predictions.

Keywords Stock market · Swing trading · Short-term investment · Long-term investment · Buy signal · Machine learning

---

S. Patgaonkar (✉) · S. Dharamsi · A. Jain · N. Marfatia
Great Learning, Mumbai, India
e-mail: samarth.patgaonkar2018@gmail.com

V. Skala et al. (eds.), *Machine Intelligence and Data Science Applications*, Lecture Notes on Data Engineering and Communications Technologies 132,
https://doi.org/10.1007/978-981-19-2347-0_9

1 Introduction

Investment is done for various purposes like earning a regular income, earning extra income, to out-perform inflation, creation of wealth, retirement planning, etc. All these purposes can be combined under a single term called "financial goals". Financial goals help in the selection of an asset class and direct investor's systematic investments towards achieving those goals. These goals may include achieving financial freedom, funding children's education, meet health-related expenses, buying of new assets, repayment of liabilities, building retirement corpus, etc.

As investment is aimed towards wealth creation, there is always a certain level of risk associated with it. Sometimes, an investment might not be able to generate returns as anticipated and may lose value over time.

Of all the investment avenues, the most volatile, but also the most luring, is investment into stocks through equity markets. An equity market is a market in which shares or stocks of the companies are issued and traded through exchanges. BSE (Asia's oldest stock exchange) and NSE are the two main stock exchanges in India.

Over time, traditional stock advisory and investing styles are being replaced with self-research and low-cost techno-brokerage houses. In today's world, investors have access to a variety of free historical stock data through public financial Websites. Many investors believe that only the data availability is enough for making knowledgeable decisions. In comparison, it is the capability to comprehend, analyze, and predict that enhances the chances of future profits and minimizes losses.

The fast paced environment of today leaves people with little time to go over the detailed research reports; some do not understand the technical analysis, so they tend to follow the advice of either professional advisors or get influenced by their fellow peers for investments. In addition, there are many advisory firms going online with their various stock portfolios, custom made to suit the needs of different investors. All these research methods have one thing in common: the application of human intelligence when studying historical stock data and then applying his understanding of this data to predict future movement. Through our research, as seen and demonstrated in this paper, data driven models are used to predict the movement for the future. Our solution automates the extraction and processing of data from the financial Websites including treatment of missing values to minimize the data feeding error to the model.

2 Literature Review

Budiharto [1] proposes a model based on long short-term memory (LSTM) for stock prices forecasting in Indonesian stock markets. The paper uses the Indonesian stock market data for visualizations using data science to predict and simulate the important features of stocks such as open, high, low, and closing (OHLC) with various parameters. The paper demonstrates that the usage of the most recent 1 year's data for model training seems to be more efficient than training the model with 3 years data.

Nayak et al. [2] propose two alternative models. The first model predicts the next day stock market price by performing sentiment analysis using Twitter as the medium and using the recent stock specific news availability. Other techniques are also used for monthly data to prove that stock market trends for one month are least correlated with stock market trends for another month.

Chen et al. [3] propose another LSTM model to predict the Chinese stock returns. The historical data of Chinese stock market were transformed into 30-days-long sequences with 10 learning features and 3-day earning rate labelling. The model was fitted by training on 900,000 sequences and tested using the other 311,361 sequences.

Zolotareva [4] proposes gradient boosting models to recognize stock markets long-term upward and downward trends based on historical Standard and Poor's (S&P) 500 stock market data. The research states that the proposed model can be used by both individual and institutional investors. Model produces "buy" and "sell" signals when starting or endpoints of trends are identified.

Vignesh [5] proposes SVM and LSTM models to calculate future value of stock traded on a stock exchange. The research is based on Yahoo and Microsoft's stock price data from Jan 2011 to Dec 2015.

These models helped us understand how data driven approaches were built in the past for solving problems related to stock markets. These papers gave us confidence that with good quality of data, problems related to complex areas like the stock market, can be addressed with decent accuracy and precision. There was an opportunity to gather stock data related to technical and fundamental parameters and build a data driven model using it.

3 Materials and Method

3.1 Data Collection and Processing

3.1.1 Data Pre-processing Approach for Short-Term Model

As part of short-term modelling approach, data related to 37 stocks which were part of the BSE 500 index were collected. Data collection was performed using an automated Python-based coding approach which efficiently dealt with 1.1 lakh rows of observations. Independent variables for short-term model were based on stock wise daily volume and related moving averages, stock closing price and related moving averages (MA), delivery percentages and key technical parameters relative strength index (RSI), moving average convergence divergence (MACD), money flow index (MFI), commodity channel index (CCI), average directional index (ADX), parabolic

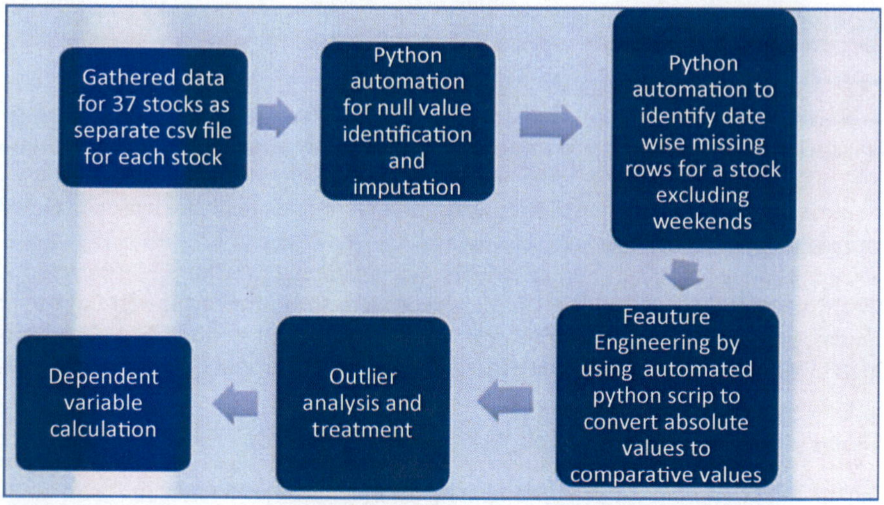

Fig. 1 Short-term model data pre-processing

stop and reverse indicator (PSAR), Bollinger Bands, William and stochastic indicators. Data pre-processing activities which included null value imputation, identification of date wise missing data for each stock, and application of feature engineering, i.e. converting the raw absolute column data into meaningful comparative data were also automated using the Python scripts. Figure 1 highlights the complete data pre-processing approach for short-term data.

Short-term indicator is the dependent variable of the short-term data set. Value true (value *1* in models) of *short-term indicator* denotes that historically stock price was able to achieve swing trade of 10% or more within 3 months period, whilst false (value *0* in models) denotes historically stock price was unable to achieve 10% stock price movement within the 3 months period.

3.1.2 Data Pre-processing Approach for Long-Term Model

The independent variables for long-term model are based on fundamental parameters which are available on an annualized basis. Data from year 2010 to 2020 were gathered. Post various elimination techniques and as per data availability, 217 stocks data were finalized for our long-term model building. Figure 2 highlights the complete data pre-processing approach for long-term data.

Long-term indicator is the calculated dependent variable on the long-term data set. Value true (value *1* in models) of *long-term indicator* denotes that historically stock price was able to achieve investment returns of 50% or more within 3 years period, whilst false (value *0* in models) denotes historically stock price was unable to achieve 50% or more returns within the 3 years period.

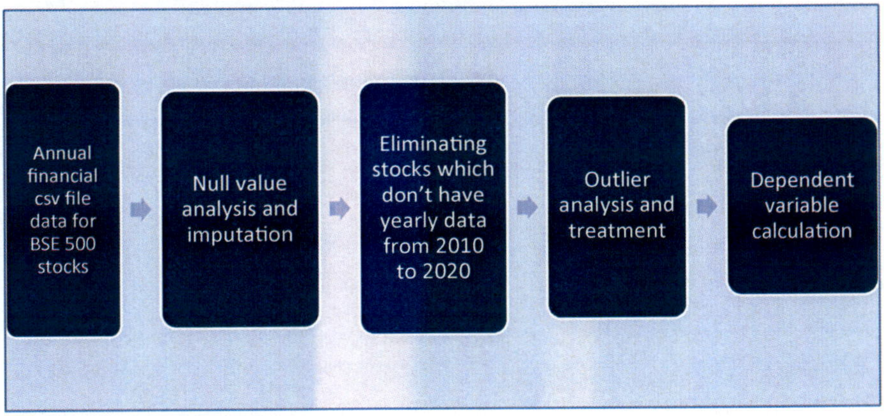

Fig. 2 Data pre-processing steps for long-term model

3.2 Data Analysis

3.2.1 Short-Term Data Set

When EDA was performed on individual 37 stocks; many prominent trends were seen which were repeating across different stocks with some minor variations and the built model ultimately confirm these trends, e.g. consider HPCL stock, Table 1 shows how closing price (CP) plays an important role with different day wise closing price-based moving averages. This type of pattern between closing price and moving averages is seen repeating across all 37 stocks in consideration.

Further, drill down using the category plots gives clearer picture about the trend. Moving average of a stock considering last 8, 13, 50, and 200 days closing price is referenced as DMA8, DMA13, DMA50, and DMA200, respectively. Figure 3 shows that for DMA8 and DMA13, if the closing price (CP) is greater than 10% in comparison with DMA8 and DMA13; the probability is quite high to get 10% upswing. This situation basically indicates buying interest has recently developed

Table 1 Closing price and moving average relationship in predicting stock price movements

Condition	Possibility of 10% swing movements
Closing price > 8 day moving average	Higher
Closing price > 13 day moving average	Higher
Closing price > 50 day moving average	Higher
Closing price < 200 day moving average	Higher

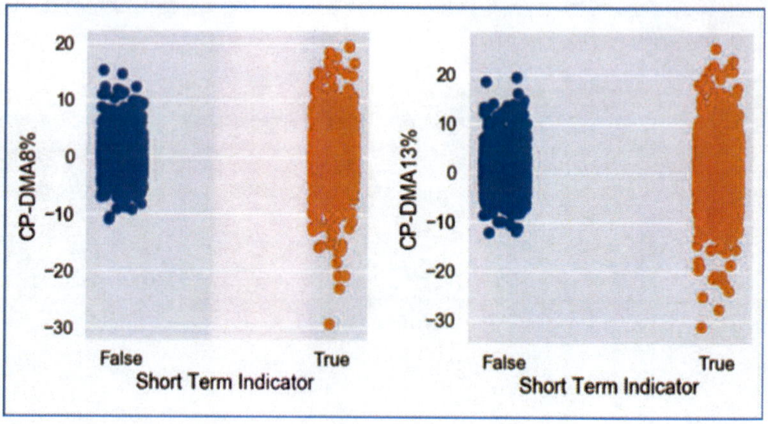

Fig. 3 CP-DMA8% and CP-DMA13% versus short-term indicator

and generally can continue for some more time before correction kicks in. Figure 4 shows that for DMA50 and DMA200; the situation is occurring at higher proportion. Also, a very clear insight obtained from Fig. 3 is that with respect to DMA8 and DMA13, if the closing price (CP) is below 10%, there is almost a pullback due in the stock price and every time the stock is giving a swing movement of 10% or above. Figure 4 also shows that when closing price (CP) is 15% and 20% below DMA50 and DMA200, respectively, swing trade of 10% movement is prominent. Such oversold conditions should be kept in mind and with any positive news flow in the limelight; once can initiate a swing trade in such situations.

A very strong trend within Bollinger Band technical parameter (upper–lower band%) was seen. First, swarm plot for HPCL stock in Fig. 5 clearly shows that on the higher side of the values for upper–lower band%, i.e. as the Bollinger Band

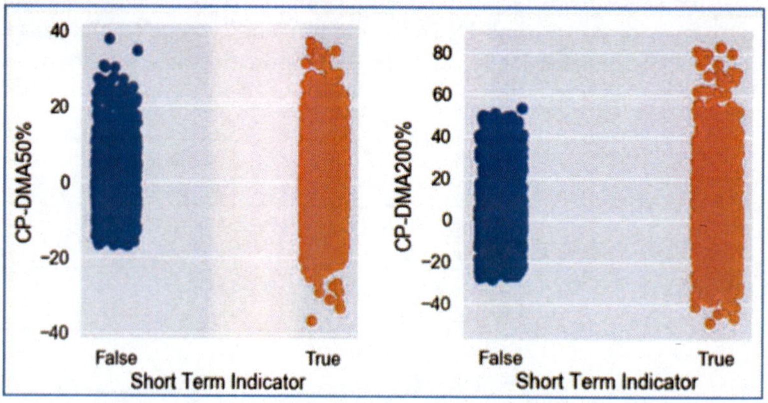

Fig. 4 CP-DMA50% and CP-DMA200% versus short-term indicator

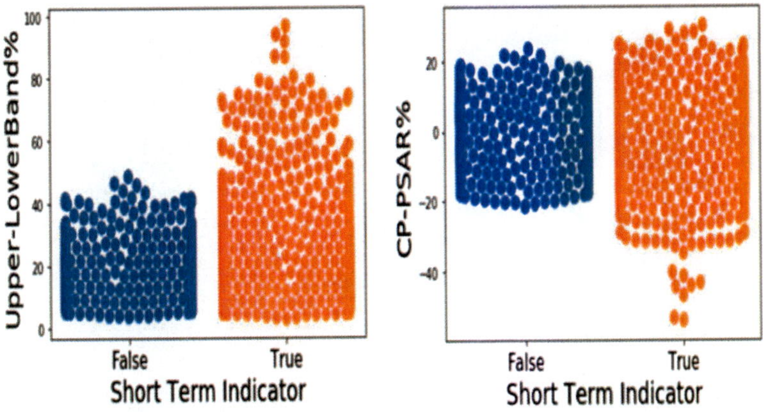

Fig. 5 Upper–lower band% (HPCL) and CP-PSAR% (BPCL) versus short-term indicator

widens, probability swing movement of 10% is almost sure. The second swarm plot for BPCL stock in Fig. 5 shows strong trend for another indicator CP-PSAR%. This is an independent variable which gives how much is the closing price, below or above PSAR technical value of stock; in % terms. The output clearly shows that when closing price (CP) is below 20% as compared to its PSAR value; BPCL stock is always bouncing back and giving a swing trade of 10% or more. These trends are seen in similar fashion across other stocks which are in consideration.

3.2.2 Long-Term Data Set

A very simple yet very strong insight derived from 217 stocks which were considered for long-term model build is that the probability of stock giving 50% or more return in a span of 3 years is more than twice the probability of stock not giving a 50% return. Our long-term model further works to explore this clear trend to pick best possible stocks.

Now, next simple but strong insight can be derived by splitting the above concept across stocks in scope. Three stocks are used in Fig. 6 to demonstrate this finding.

Figure 6 shows that stock Aarti Industries Ltd. has given a return of 50% or more in next 3 years for every financial year's data in scope. The 3M India Ltd. stock has provided 50% returns in 3 years 7 times, whilst 2 times, it has failed to do so. For Bharat Heavy Electricals Ltd., only 3 times the company has delivered 50% returns in 3-year time frame, whilst 6 times, it has failed to do so. Similar findings can help in making informed decisions across different stocks.

Next insight can be drawn using the swarm plots shown in Fig. 7.

It can be clearly seen that if the asset turnover ratio is 2 or above; return on assets (ROA) is 25 or above, and return on equity (ROE) is 50 or above for a financial year

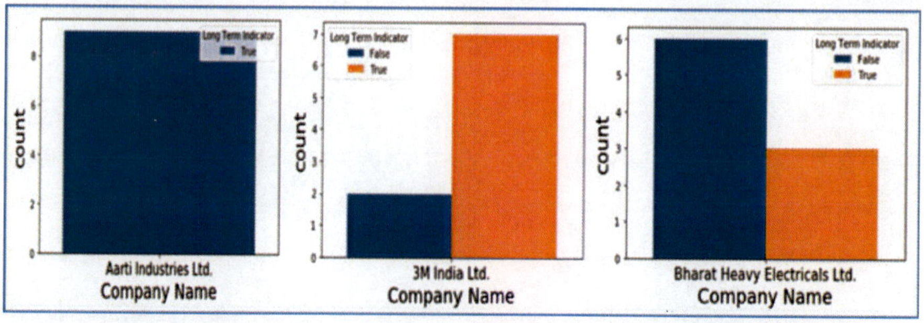

Fig. 6 Count versus company name (Hue as long-term indicator)

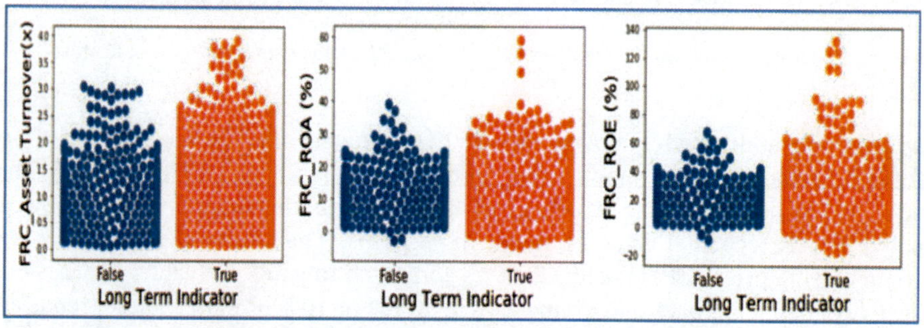

Fig. 7 Asset turnover, ROA, and ROE versus long-term indicator

for a stock in consideration; there is very strong probability of getting 50% returns for that stock in the span of next 3 years.

3.2.3 Modelling

Pre-processing of the collected data resulted in variables as seen in Table 2.

Standard data split methodology of 70:30 into training and test was used on both the short-term and long-term data sets.

Data scaling was performed on the short term as well as the long-term data sets.

It was verified if SMOTE, cross validation or any other adjustments such as modification of cutoff thresholds for classification were able to boost the model performance measures. SMOTE and cross validation techniques were not found to

Table 2 Short-term and long-term variable details

Model	No. of independent variables	Dependent variable
Short term	57	Short-term indicator
Long term	26	Long-term indicator

contribute positively but modification of probability cutoff thresholds contributed positively in long-term data set modelling.

Artificial neural network (ANN), random forest (RF), gradient boosting (GB), logistic regression (LR), and *K*-nearest neighbours (KNNs) classification techniques were used for model building. Based on model parameters comparison, the best model was selected for short-term and long-term data sets.

Model building and tuning were done using Python scripts.

4 Results

4.1 Short-Term Data Set

Random forest (RF) predicted better results for the short-term model, i.e. predicting the 10% swing trading for stocks in 3 months timeframe. Optimum hyperparameters for the tuned RF model are *max_depth* as *15*, *estimators* as *750*, *min_samples_split* as *37,* and criterion as *Gini*. Optimum probability cutoff based on thorough research is 0.6.

The evaluation statistics for the RF model on test data is shown in Fig. 8.

From a recommendation system point of view; precision parameter, i.e. ability to predict true positives (TPs) effectively was most important. The model scores a precision value of 78% in this area with overall test accuracy around 71%. Thus, it is expected that the stocks which our model will shortlist and recommend; 78% of them should end in achieving the desired results. Comparative analysis of different models is seen in Fig. 9.

Figure 10 shows comparative analysis on the short-term test data using AUC-ROC curve.

RF Test Data Classification report

	precision	recall	f1-score	support
0	0.56	0.54	0.55	14011
1	0.78	0.79	0.78	28237
avg / total	0.71	0.71	0.71	42248

Fig. 8 Random forest (RF) evaluation parameters on short-term test data

	Train_Precision	Test_Precision	Train_Accuracy	Test_Accuracy	Train Recall	Test Recall	Train_ROC-AUC	Test_ROC-AUC
ANN	0.74	0.73	0.64	0.63	0.71	0.71	0.6	0.59
KNN	0.82	0.78	0.66	0.62	0.63	0.59	0.67	0.63
LR	0.68	0.68	0.65	0.65	0.9	0.9	0.52	0.52
RF	0.87	0.78	0.8	0.71	0.83	0.79	0.79	0.67
GB	1	0.87	0.99	0.8	0.99	0.82	1	0.79

Fig. 9 Comparative analysis of short-term models

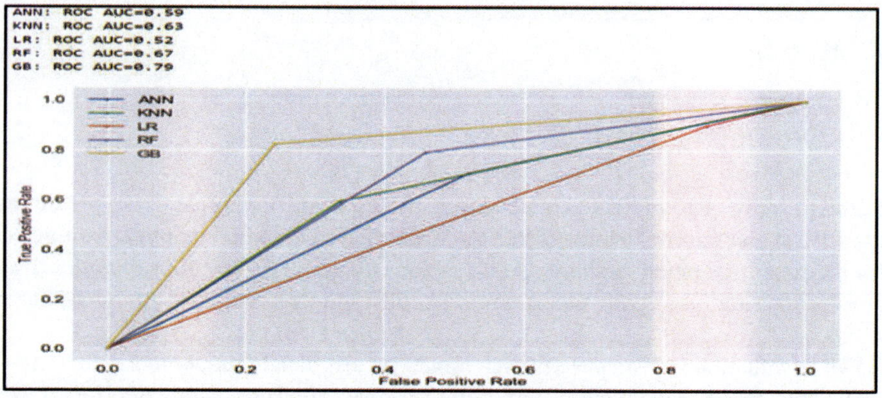

Fig. 10 AUC-ROC curve for short-term test data using RF model

4.2 Long-Term Data Set

Artificial neural network (ANN) predicted better results for the long-term model, i.e. predicting the 50% movement in a stock price in 3 years timeframe. Optimum hyperparameters for the tuned ANN model are *neurons* as *900*, *max_iterations* as *5000*, *solver* as *Adam*, *activation function* as *tanh*, and *tolerance* as *0.001*. Optimum probability cutoff based on thorough research is 0.65. The evaluation statistics for ANN model using test data are shown in Fig. 11.

Even though, the precision for predicting true negatives is weak for this model; success of the model actually depends on precision to predict true positive results.

ANN Test Data Classification report

	precision	recall	f1-score	support
0	0.36	0.39	0.38	165
1	0.75	0.72	0.74	421
avg / total	0.64	0.63	0.64	586

Fig. 11 Artificial neural networks (ANNs) evaluation parameters on long-term test data

	Train_Precision	Test_Precision	Train_Accuracy	Test_Accuracy	Train_Recall	Test_Recall	Train_ROC-AUC	Test_ROC-AUC
ANN	0.77	0.75	0.67	0.63	0.76	0.72	0.61	0.56
KNN	0.76	0.76	0.66	0.65	0.76	0.74	0.6	0.58
LR	0.76	0.75	0.66	0.63	0.77	0.73	0.59	0.55
RF	0.9	0.77	0.83	0.63	0.86	0.7	0.82	0.58
GB	0.9	0.77	0.83	0.64	0.85	0.71	0.81	0.58

Fig. 12 Comparative analysis of long-term models

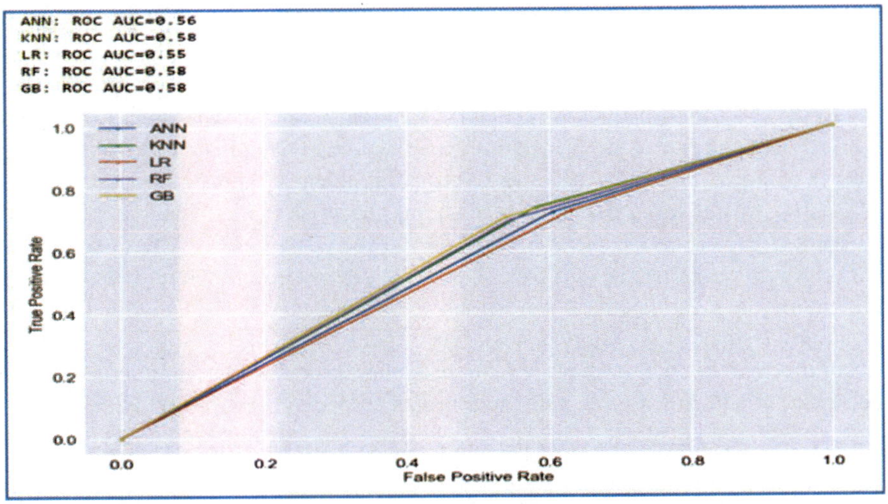

Fig. 13 Long-term data set modelling comparison using AUC-ROC curve

With precision of 0.75, it is expected that of all the stocks which model shortlists and recommend; 75% of them should end in achieving the desired results. Overall accuracy of the test model stands at 64%. Comparative analysis of different models is seen in Fig 12.

Figure 13 shows comparative analysis on the long-term test data using AUC-ROC curve.

5 Conclusion

This study is important in several ways.

Stock market accurate predictions are definitely difficult and complex as there are lot of external factors which influence the stock market movements especially the short-term movements. A short-term prediction model was built using 57 predictor variables which related to moving averages or various orders, delivery percentage, and technical parameters. Similarly, a long-term prediction model is built using 26 predictor variables based on stocks' fundamental annual parameters. Python automation was used as much as possible to reduce human errors in data extraction and to reduce repetitive human efforts. Finally, after comparing model results of multiple classification models; the best model was finalized for short-term and long-term data sets. Our research paper proves that with application of data science strategies, the short-term and long-term movements of the stocks can be predicted with reasonable accuracy and precision. From recommendation point of view, precision for true positive predictions is the most important evaluation parameter, and model outcomes show that short-term stock movements of 10% gains can be predicted successfully with precision of 78%, whilst long-term stock movements to achieve 50% gains can be predicted with precision of 75%.

Short-term model is prepared by considering 37 stocks, whilst long-term model is prepared by considering 217 stocks data. This gives scope of expansion by addition of other stocks in the future. Since our model is already automated, increase in the number of stocks does not pose any additional problems. With a suitable front-end built on top of this, even first-time investors in the stock market can take benefit of our work.

Long-term model studies 50% stock price achievement for the previous years by referencing the start date as 1st April for the reference year. Thus, whilst recommending the stock; it is assumed stock hits or goes below its 1st April price in later part of the year. Even though price remains higher; predictions will remain valid for the stock but with stock price gains reduced as per the existing difference.

References

1. Budiharto W (2021) Data science approach to stock prices forecasting in Indonesia during Covid-19 using long short-term memory (LSTM). J Big Data 8, Article number: 47. Accessed 10 Apr 21. https://doi.org/10.1186/s40537-021-00430-0
2. Nayak A, Manohara Pai MM, Pai RM (2016) Prediction models for Indian stock market, twelfth international multi-conference on information processing-2016 (IMCIP-2016). Accessed 10 Apr 21. https://www.sciencedirect.com/science/article/pii/S1877050916311619
3. Chen K, Zhou Y, Dai F (2015) A LSTM-based method for stock returns prediction: a case study of China stock market. In: 2015 IEEE International conference on big data (big data), INSPEC Accession Number: 15679536. Last accessed on 10 Apr 21. https://ieeexplore.ieee.org/abstract/document/7364089/citations#citations
4. Zolotareva E (2021) Aiding long-term investment decisions with XGBoost machine learning model. https://www.researchgate.net/publication/350991820_Aiding_Long-Term_Investment_Decisions_with_XGBoost_Machine_Learning_Model
5. Vignesh CK (2020) Applying machine learning models in stock market prediction. EPRA Int J Res Dev (IJRD) 395–398. https://eprajournals.com/jpanel/upload/138am_82.EPRA%20Journals-4361.pdf

Issues in Machine Translation—A Case Study of the Kashmiri Language

Nawaz Ali Lone, Kaiser J. Giri, and Rumaan Bashir

Abstract Machine translation has emerged as a promising research area, attracting considerable interest from researchers all over the world. Many techniques have been developed for machine translation of various languages around the world, which can be used alone or in combination. Besides that, there are several languages, including Kashmiri, in which little work has been done in terms of machine translation. The purpose of this paper is to provide an overview of the Kashmiri language, including its history, character set, and writing systems. This paper presents a brief introduction to machine translation and core issues associated with machine translation, with Kashmiri as a participating language. The study presented in this paper will be extremely useful to researchers involved in machine translation of the Kashmiri language.

Keywords Machine translation · Dataset · Kashmiri · Language processing · Character set · Syntax · Semantics

1 Introduction

Machine translation (MT) is the task of automatic or machine-assisted translation between different natural languages which does not change their meaning or context. It is a branch of applied research that combines ideas from several disciplines, including computer science, linguistics, statistics, translation theory, artificial intelligence, etc. The languages involved are referred to as source and target languages. Machine translation is not simply word-for-word translation/substitution, but rather

N. A. Lone · K. J. Giri (✉) · R. Bashir
Department of Computer Science, Islamic University of Science & Technology, Kashmir, India
e-mail: kaiser.giri@islamicuniversity.edu.in

N. A. Lone
e-mail: lonenawazali@gmail.com

R. Bashir
e-mail: rumaan.bashir@islamicuniversity.edu.in

V. Skala et al. (eds.), *Machine Intelligence and Data Science Applications*, Lecture Notes on Data Engineering and Communications Technologies 132,
https://doi.org/10.1007/978-981-19-2347-0_10

the careful consideration of various language-related aspects such as syntax (sentence structure), semantics (meanings), and so on must be taken into consideration during the development of translation algorithms.

As a result, machine translation is becoming an important and useful tool in the exchange of information between different types of people. The shared information could be of a business or non-business nature, thereby bridging the language barrier between people who speak different languages. Machine translation makes it feasible to translate large amounts of information that would otherwise require a significant amount of time and resources.

Machine translation has been around since 1951, [1]. Since then, researchers have been working to improve the efficiency of machine translation across a variety of metrics. The most recent approach is neural machine translation (NMT), which has become the dominant paradigm for machine translation in academic research as well as commercial use, with good results for many language pairs, which can be attributed to end-to-end training of the entire system [2]. The approaches are broadly classified as rule-based, corpus-based, and hybrid machine translation approaches. For every language pair, a rule-based machine translation approach is based on multiple translation rules as well as bi-lingual dictionaries. The corpus-based approach makes use of a large and structured set of sentence-aligned texts written in two different languages (participating languages). By encoding translation examples between the languages involved, translation knowledge rather than rules is created. The hybrid machine translation approach combines or integrates multiple approaches into a single system. It employs a variety of techniques to supplement or compensate for the shortcomings of other individual approaches. Recently, several techniques have been developed to deal with the resource-constrained nature of the languages used in machine translation [3–5].

2 Kashmiri Language

Kashmiri, also known as Koshur [6], is an Indo-Aryan language belonging to the Dardic subgroup of languages. Figure 1 below depicts the descent of the Kashmiri language as per Grierson.

It gets its label/name from the Kashmir region, from which it is mostly communicated. As per the 2011 census, there are approximately 7 million speakers in Jammu and Kashmir, mainly in the Kashmir Region, Chenab valley [7], Neelam valley, Leepa valley, and the Haveli district [8, 9].

The Kashmiri language is written using three orthographic systems: the Sharada script, the Devanagari script, and the Perso-Arabic script. Informally, the Roman script is now commonly used to write Kashmiri, especially online. Only after the eighth century, A.D. was the Kashmiri language written in the Sharada script. This script, however, is no longer widely used outside of religious ceremonies performed by Kashmiri Pandits [10]. Kashmiri is now primarily written in Perso-Arabic and Devanagari (modified) scripts [11]. Kashmiri is one of the Perso-Arabic script

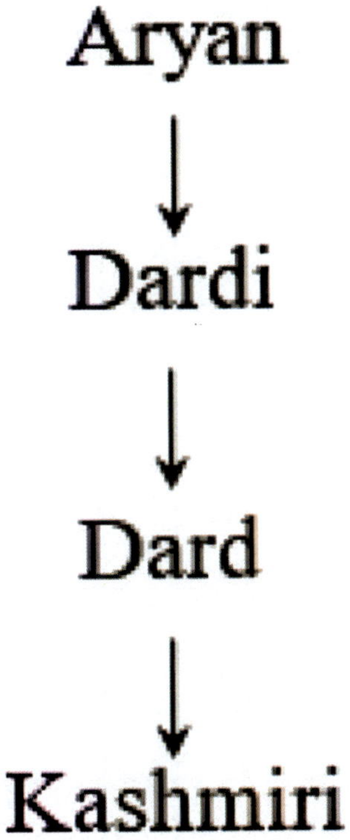

Fig. 1 Kashmiri language tree

languages, which represents all vowel sounds. The Perso-Arabic script has come to represent the Kashmiri Muslim community, while the Devanagari script has come to represent the Kashmiri Hindu community.

The Kashmiri (Perso-Arabic) script [12] is a superset of Arabic because it contains 13 more letters than Arabic, which only has 28. Kashmiri has more consonants and vowels than Arabic and other languages. Figures 2 and 3 depict the character set of the Kashmiri language.

3 Associated Challenges

NLP is one of the most successful computational technology-driven application fields, with systems outscoring humans in certain language processing activities. These developments, however, are generally dependent on the availability of large training datasets and other pre-processing tools. In terms of resources required for most tasks such as translation, sentiment analysis, or any other NLP activity, the Kashmiri language falls short.

Subsequent paragraphs, however, are indented because every NLP activity, including machine translation, is heavily reliant on available resources, such as large

ب	[b]	ج	[d͡ʒ]	ر	[r]	ض	[z]	کھ	[kʰ]	ے	[j]
پ	[p]	چ	[t͡ʃ]	ڑ	[ɽ]	ط	[t]	گ	[g]		
پھ	[pʰ]	چھ	[t͡ʃʰ]	ز	[z]	ظ	[z]	ل	[l]		
ت	[t]	ح	[h]	ژ	[t͡s]	ع	[Ø]	م	[m]		
تھ	[tʰ]	خ	[x], [kʰ]	ژھ	[t͡sʰ]	غ	[g]	ن	[n], [◌̃]		
ٹ	[ʈ]	د	[d]	س	[s]	ف	[f], [pʰ]	ں	[◌̃]		
ٹھ	[ʈʰ]	ڈ	[ɖ]	ش	[ʃ]	ق	[k]	و	[w]		
ث	[s]	ذ	[z]	ص	[s]	ک	[k]	ہ	[h]		

Fig. 2 Consonants

datasets. However, no bi-lingual dataset containing Kashmiri written text is currently available.

Because every NLP activity, including machine translation, is heavily reliant on available resources, such as large datasets. However, no bi-lingual dataset containing Kashmiri written text is currently available.

Fig. 3 Vowels

اٗ	[ə]	اۆ	[o]
اٲ	[əː]	او	[oː]
اِ	[i]	اوٚ	[ɔ]
ای	[iː]	اوا	[ɔː]
اٟ	[ɨ]	اےٚ	[e]
اٖ	[ɨː]	اے	[eː]
اَ	[a]	اُ	[u]
آ	[aː]	اوٗ	[uː]

3.1 Scarcity of Digital Resources

The majority of languages lack a parallel dataset as a main machine translation resource, one of which is the Kashmiri language, as there is no parallel dataset available with Kashmiri as a participating language.

3.2 Problems with Standardization

The Kashmiri language is very sensitive to language constructs, and it is rich in phonology, for example, through the use of diacritic marks. However, due to a lack of standardization, certain issues arise, such as incorrect spelling and the uncommon use of diacritic marks. For example, the word "گرٕ" meaning home is often written as "گر" there leaving scope for misspelling this word written differently at different places. In another example, word "نوٗر" meaning divine light is often written without diacritic mark as "نو ر". As a result, there is a good chance for misinterpretation, particularly when it comes to spoken language, as diacritic marks play an important role in preserving the phonetic richness of the Kashmiri language.

3.3 Word Order Issue

The most significant structural difference between English and the majority of Indian languages is word order [17]. In English, the subject-verb-object order is used, whereas, in Kashmir, the subject-object-verb/or free word order language is used, which presents another challenge for automatic translation.

E.g. sentence "کوٗڈ چھ ژونٹھ کھیٚون" implies that a girl is eating an apple has (subject + auxiliary + object + verb) structure, which can also be written as "ژونٹھ کھیٚو ڈ کو" having the same meaning but with a different structure (object + verb + subject).

3.4 Lack of Pre-processing Tools

It is hard to achieve automatic translation before basic pre-processing activities. But the Kashmiri language lacks many basic pre-processing tools, such as morph analyzers, tokenizers, and so on.

3.5 Multiple Dialects

The Kashmiri language is spoken in a variety of dialects. Dialect is a territorial, geographical, or sociocultural variation of recognizing differences through pronunciation, sentence structure, or vocabulary; specifically, a variation of sounds. Multi-dialect languages become even more complex to translate.

4 Conclusion

Machine translation, which has emerged as an extensive research area in recent years, now has a significant impact on the translation work of many popular major languages, e.g. English, French, Spanish, Chinese, etc., and it is of immense use in the areas of healthcare, finance, etc. The fundamental concept of retaining semantic, syntactic, and stylistic correlation between the source and target languages whilst translating renders translation a challenging task [18]. Before eliciting appropriate equivalences in the target language, it is therefore critical to consider all types of linguistic as well as resource quirks related to machine translation. This paper discusses various aspects of the Kashmiri language as well as the challenges associated with machine translation. Some of the core issues can be avoided by having good resources in terms of standard datasets, developing pre-processing tools, reordering Kashmiri sentences before translation [19] to account for structural differences, and emphasizing standardization of basic language constructs; however, some issues may persist because languages are inherently ambiguous. It is critical to start the translation-related activities in stages, where the creation of dataset and basic pre-processing tools will act as a base for machine translation.

References

1. Nye MJ (2016) Speaking in tongues: science's centuries-long hunt for a common language. Distillations 2(1):40–43
2. Dabre R, Chu C, Kunchukuttan A (2020) A survey of multilingual neural machine translation. ACM Comput Surv (CSUR) 53(5):1–38
3. Goyal V, Kumar S, Sharma DM (2020, July) Efficient neural machine translation for low-resource languages via exploiting related languages. In: Proceedings of the 58th annual meeting of the association for computational linguistics: student research workshop, pp 162–168
4. Rubino R, Marie B, Dabre R, Fujita A, Utiyama M, Sumita E (2020) Extremely low-resource neural machine translation for Asian languages. Mach Transl 34(4):347–382
5. Kuwanto G, Akyürek AF, Tourni IC, Li S, Wijaya D (2021) Low-resource machine translation for low-resource languages: leveraging comparable data, code-switching and compute resources. arXiv preprint arXiv:2103.13272
6. Kashmiri language. https://en.wikipedia.org/wiki/Kashmiri_language. Last accessed 26 Nov 2020
7. Koshur. An introduction to spoken Kashmiri. https://www.koshur.org/. Last accessed 23 Nov 2020
8. Snedden C (2015) Understanding Kashmir and Kashmiris. Oxford University Press
9. Shakil M. Languages of erstwhile state of Jammu Kashmir
10. The Sharada script: origin and development. http://www.koausa.org/Languages/Sharda.html. Last accessed 23 Nov 2020
11. Kashmiri (कॉशुर / کأشُر). https://omniglot.com/writing/kashmiri.htm. Last accessed 12 May 2021
12. Bashir R, Quadri S (2013, December) Identification of Kashmiri script in a bilingual document image. In: 2013 IEEE Second international conference on image information processing (ICIIP-2013). IEEE, pp 575–579
13. Kashmiri language: roots, evolution and affinity. http://www.koausa.org/Languages/Shashi.html. Last accessed 21 Nov 2020
14. Constitutional provisions relating to Eighth schedule. https://www.mha.gov.in/sites/default/files/EighthSchedule_19052017.pdf. Last accessed 14 June 2021
15. The Jammu and Kashmir official languages bill, 2020. http://164.100.47.4/BillsTexts/LSBillTexts/PassedLoksabha/124-C_2020_LS_Eng.pdf. Last accessed 10 Dec 2021
16. Kashmiri made compulsory subject in schools. https://www.oneindia.com/2008/11/01/kashmiri-made-compulsory-subject-in-schools1225558278.html. Last accessed 11 May 2021
17. Patel R, Pimpale P, Sasikumar M (2019) Machine translation in Indian languages: challenges and resolution. J Intell Syst 28(3):437–445. https://doi.org/10.1515/jisys-2018-0014
18. Sinha K, Mahesh R, Thakur A (2005) Translation divergence in English-Hindi MT. In: Proceedings of the 10th EAMT conference: practical applications of machine translation
19. Murthy VR, Kunchukuttan A, Bhattacharyya P (2018) Addressing word-order divergence in multilingual neural machine translation for extremely low resource languages. arXiv preprint arXiv:1811.00383

Bangladeshi Land Cover Change Detection with Satelite Image Using GIS Techniques

Kazi Atai Kibria, Al Akram Chowdhury, Abu Saleh Musa Miah, Md. Ragib Shahriar, Shahanaz Pervin, Jungpil Shin, Md. Mamunur Rashid, and Atiquer Rahman Sarkar

Abstract High-quality land use and land cover data are important for monitoring and analyzing environmental changes in the background of global warming. Bangladesh is a high-density populated country; as a consequence, high-quality land has been reducing every year. So, it is very important to monitor the locations and distributions of land cover and associated changes over time in Bangladesh. In the research work, we have assessed and examined the 17 most recent composed land cover products with the International Geosphere-Biosphere Programme (IGBP) and 15 most recent composed land cover products with the University of Maryland (UMD). We evaluated the model with the land cover of Bangladesh using 500 m multi-temporal MODIS images (MOD13Q1 v006) from the year 2001 to 2018. We have classified layer-1 and layer-2 of the MCD12Q1v006 area into 17 and 15 specific classes subsequently. Accuracy table and figure have shown that permanent wetland, natural vegetation cropland, Borland, and grasslands increase 01% meanwhile savannas land area reduced 3% in this year. Also, water bodies were increased by 01%, but cropland has fluctuated but remains the same in 2018. These land cover changes prove the significant human interventions such as the expansion of croplands with increased population pressure. Our research work is very much helpful for meteorologists for tracking change and understanding in wetlands in Bangladesh.

Keywords Remote sensing · MODIS · Land cover classification · Land cover mapping

K. A. Kibria · A. A. Chowdhury · Md. R. Shahriar · S. Pervin · Md. M. Rashid · A. R. Sarkar
Bangladesh Army University of Science and Technology, Saidpur, Bangladesh

A. S. M. Miah (✉) · J. Shin
School of Computer Science and Engineering, The University of Aizu, Aizuwakamatsu, Fukushima 965-8580, Japan
e-mail: abusalehcse.ru@gmail.com

J. Shin
e-mail: jpshin@u-aizu.ac.jp

V. Skala et al. (eds.), *Machine Intelligence and Data Science Applications*, Lecture Notes on Data Engineering and Communications Technologies 132,
https://doi.org/10.1007/978-981-19-2347-0_11

1 Introduction

Modern and early-stage people have not understood the necessity of the importance of the environment. So, the environment of the world is changing day by day. Through these changes, our climate, our weather, and overall, the geometry of the earth are changing. By the analysis from the previous years, it is observed that significant increase in the tropics by the conversion of forest, field, and grassland into farming and field land caused by various reasons like—deforestation, urbanization, and natural disasters. By observing the imagery from the modern utilization of satellite, we can clearly identify the rate and extent of the land change coverage and land usage and analyze it for any kind of changes [1]. Land spread data are significant for some applications like flood displaying, perception of rural dry spell, environmental change demonstrating, and checking of ecological variations like farming, water disasters, tsunami, and observing of carbon discharge because of cutting down forests and timberland corruption [2]. Climate characteristics are changing because of global warming. Specifically, forests are sharply decreasing. The information for gathering such data to achieve from changes over spatial and temporal extents for persuading various land changing problems and for future predictions [3, 4]. Remote sensing technology using satellites has come to our reach as the ability to observe and clarifying modifications in the situation of the surface area in the presence of time [5]. With the modern implementations of satellite sensors, now for detecting consistent and repeatable measurements from any kind of spatial scale for determining changes caused by human or nature disturbances [6, 7]. There are three types of variation occurs in the climate—seasonal change, gradual change, abrupt change. An investigation of land covers elements and its main impetus, in reality, gives the ideal establishment to the maintainability of natural resource systems [8]. Technology is also improving day by day. Nowadays, technology makes the world so closer to us. Remote sensing is the latest technology through which we can easily detect the change. There are many methods to detect the change of a specific land cover area using remote sensing [9]. For spatial, consistent, and temporal comparisons of global vegetation conditions, the moderate resolution imaging spectro-radiometer (MODIS) came with a great advantage [10].

So our motive is to use this technology to detect the land cover changes of Bangladesh for informed decision-making regarding policies about the landscape. The necessity of this analysis is to identify changes in vegetation, cropland, barren, permanent wetland, grassland, and savannas. This study provides a land cover map that will help government authorities and stakeholders set up land use plans for Bangladesh [11]. This observation presents the land cover change of Bangladesh for 18 years (odd years, 2001–2018). In this study, we analyzed the composite satellite image from MODIS to detect the land cover change of Bangladesh on the basis of International Geosphere-Biosphere Programme (IGBP) and University of Maryland (UMD) legend and class descriptions. The main purpose of this paper is to compare with the two global 1 km resolution land cover products, which both derived from the advanced very high-resolution radiometer (AVHRR) named as

International Geosphere-Biosphere Programme (IGBP) produced by U.S. Geological Survey and the second by the University of Maryland (UMd) in the whole map of Bangladesh [12]. We have collected 36 MCD12Q1 HDF files from the NASA open-source Website for 18 years from 2001 to 2018 with two layers [13]. Layer type-1 HDF satellite images contain 17 classes of land information, and layer type-2 HDF satellite images contain 15 classes of land information. Then, we transformed the 36 HDf files into 18 TIF files for both layers by using ArcGIS software with create raster dataset and mosaic tools. These TIF image files are composed of geographical information from different countries. But here, we are working only on Bangladeshi geographical data. To extract only the Bangladeshi geographical region, we have masked this TIF with a Bangladeshi map shapefile. After applying the "extract by mask" tools of ArcGIS software, we have got 18 TIF images for both layers, which contain only Bangladeshi geographical information. Each image is composed of 17 classes of land area for layer type-1 and 15 classes of land area for layer type-2. To classify the Bangladesh land into 17 categories, we have applied here IGBP, and to classify into 15 classes, we have used here UMD [14]. This paper is organized as follows: Sect. 2 literature review, Sect. 3 materials and resources, in Sect. 4 about methodology, and Sect. 5 result and discussion, finally a conclusion.

2 Literature Review

We found very limited work on the land cover of Bangladesh. Just for evaluating the signature class properly, the maximum likelihood classification method was considered to be good. Ran et al. [15] made an evaluation of four remote sensing-based land cover products over China using IGBP, UMD, GLC2000, and MODIS global land by Boston University, where it was done only for China, which made a good impact on their own country. So, such work over Bangladesh can also bring revolutionary changes on various implementations [15]. Similarly, work regarding the Singapore land cover classification was done by remote sensing methods by et al. Sidhu 2017, where using this classification, that lacks for Bangladesh is greatly determined for Singapore [16]. By the work of et al. Mengistu 2008, where they determined the land classification using percentile data to provide enough information regarding Nigeria, which is not properly available specifically for Bangladesh [17]. Again by the work et al. Cihlar 2000, here the importance of land cover mapping for a bigger area using satellite has been emphasized [18]. The research possibilities and its current status regarding its future work. Even in content of "remote sensing of land use and land cover principles and applications" by Loveland et al., Chandra P. Giri shown the implementation of the land cover mapping, which can help any government regarding land mapping decisions for its country [19]. Even in the work of et al. Loveland 2000, the development of a global land cover was characterized only for 1 km AVHRR data, where we determined the data for 500 km.

3 Materials and Resource

3.1 Working Platform

There is also ArcGIS which is considered to be commercial software [20]. ArcGIS is geospatial software that is used to view, manage, update, and analyze the geographical dataset. This software enhanced its performance by adding some feature templates with better snapping and editing, such as image analysis toolbar, which assists us in clip rasters, mosaic, perform NDVI, and image classification. Time series animation for temporal data and data management tools is one of them.

3.2 Satelite Images

There are several formats in which satellite images are stored. GeoTIFF norm is implemented by an OGC honour [21]. GeoTIFF depends on the TIFF design and is utilized as an exchange design for geo-referred to raster symbolism. GeoTIFF is thoroughly used in the NASA Earth science information framework. As of not long ago, there has been no forward-thinking detail for the GeoTIFF record design. The open geospatial consortium (OGC) distributed variant 1.1 of the OGC GeoTIFF norm in September 2019 [22]. Rendition 1.1 is in reverse viable with the first GeoTIFF 1.0 detail of 1995.

There are other logical records organizes that are grounded inside the NASA people group, e.g. HDF5 and net CDF. However, it is proceeded with interest and assertion for the GeoTIFF document design, for the most part as a conveyance design for satellite or airborne photography symbolism yet additionally for different sorts of information like digital elevation model (DEM) information and digital ortho quadrangle information [23, 24]. NASA, in 1999 launched the imaging sensor named the moderate resolution imaging spectro-radiometer (MODIS) by Santa Barbara remote sensing. The instruments capture data in 36 spectral bands ranging in wavelength from 0.4 to 14.4 μm and at varying spatial resolutions (2 bands at 250 m, five bands at 500 m, and 29 bands at 1 km). It was able to determine data in spectral bands ranging up to 36 bands, which came in wavelength from 0.4 to 14.4 μm and varying spatial resolutions. We found no study that uses this satellite to map the land cover of Bangladesh. This study will map the land cover change of Bangladesh using MODIS and ArcGIS [10, 25–27].

3.3 Different Classes Classification Procedure from Satelite Image

Satelite image contains many information, and we can classify those information with various technique. By observing Fig. 1, we can see it shows the reflectance and frequency investigations of various land cover types, where it is known that taking flood mapping comes into play by particular spectral reflectance similar to specific colour code combinations in RGB combinations. Investigating frequency properties, test regions from pictures could be chosen and prepared to recognize pixels with comparative unearthly reflectance/assimilation and appoints them to the gatherings referenced before. The mechanism is called semi-automated land cover mapping [28]. From Fig. 1, the LC scheme is clearly produced, which was used to show the nested classifications. All the parent nodes show a singular demonstration of the classification with the children nodes from the output classes, which each set of arrows generated. The surface hydrology and land use layers contain unexpected data in comparison with land cover; however, cross-over in specific definitions. The

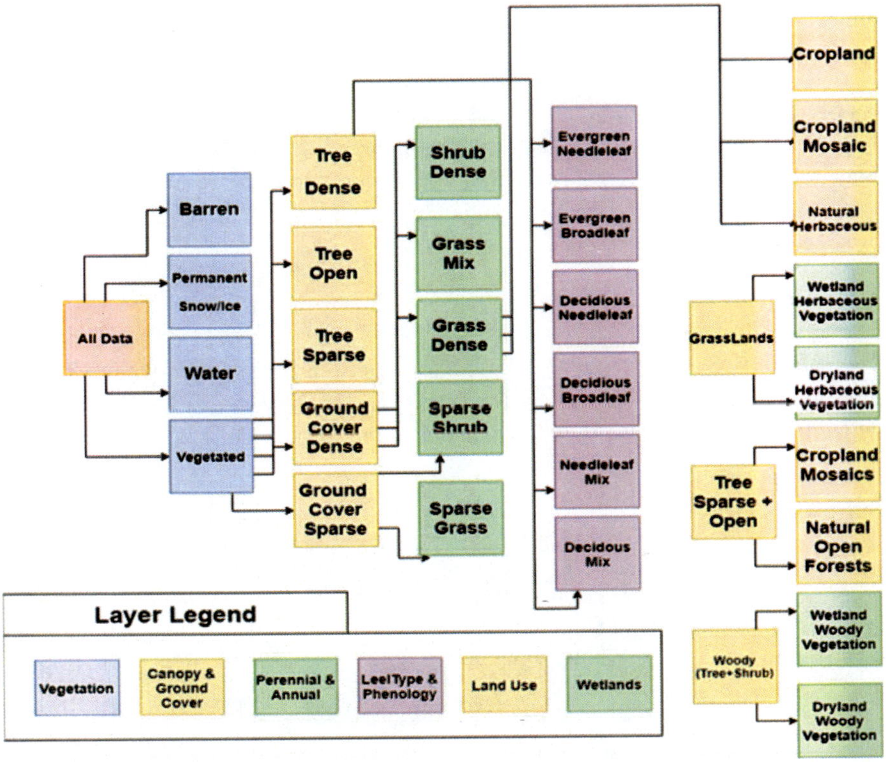

Fig. 1 Classification of data [29, 30]

various levelled nature takes into account clients to make their custom legends [29, 30].

3.4 Area and Land Cover Products

The analysis that will be conducted here is on the map of Bangladesh, where we downloaded HDF file from the open-source link: ladsweb.modaps.eosdis.nasa.gov from 2001 to 2018, each year containing two HDF files to cover the whole Bangladesh map [13]. Then, we define the projection of both HDF files each year, where the mapping is done. Bangladesh being the least-urbanized area in South Asia, the data retrieved in 2018 were 36.6% being the urban area. In our analysis, the urban land and rural land will be defined individually from the green vegetation to the savannas, water bodies, and many more. After the analysis, the land mapping of Bangladesh will be clearly defined for demonstration.

4 Proposed Methodology

Figure 2 demonstrates the rudimentary characteristics of the work that we have done to get the land cover mapping basic demonstration over Bangladesh. Here, the extraction consisted of IGBP and UMD land cover mapping.

4.1 Dataset

We have collected 36 MCD12Q1 HDF files from the NASA open-source Website (ladsweb.modaps.eosdis.nasa.gov) for 18 years from 2001 to 2018 with two layers with 2400 * 2400 * 3 pixels size.

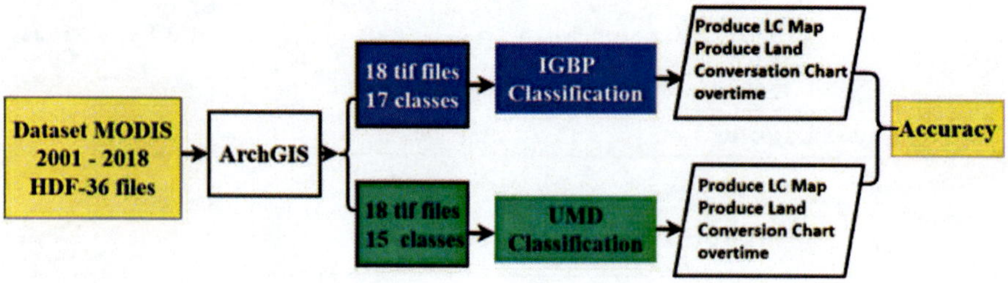

Fig. 2 Proposed methodology

4.2 Preprocessing Using ArcGIS

Then, we transformed the 36 HDF files into 18 TIF files for both layers by using ArcGIS software with create raster dataset and mosaic tools. These TIF image files are composed of geographical information from different countries. But here, we are working only on Bangladeshi geographical data.

To extract only the Bangladeshi geographical region, we have masked this TIF with a Bangladeshi map shape file. Extracting the Bangladeshi geographical region from the map are shown in Figs. 3 and 4. After applying the "extract by mask" tools of ArcGIS software, we have got 18 TIF images for both layers, containing only Bangladeshi geographical information. Each image is composed of 17 classes of land area for layer type-1 and 15 classes of land area for layer type-2. Now, at the end of preprocessing, we have gotten 18 TIF images for layer type-1 with 17 classes land and 18 TIF images for layer type 2 with 15 classes. In Fig. 4, yellow colour shows

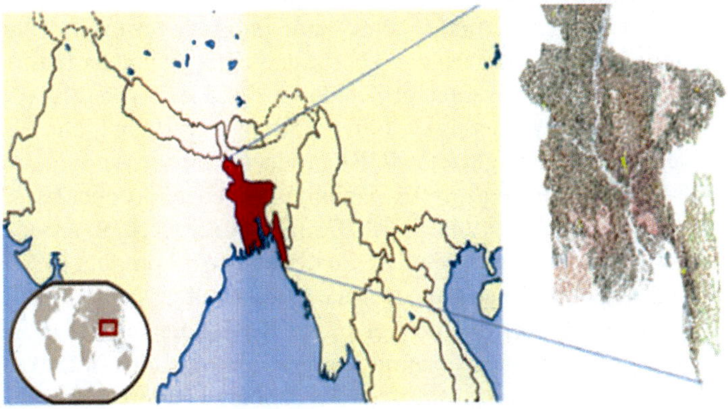

Fig. 3 Masking procedure of layer type-1 of 17 classes for IGBP [31]

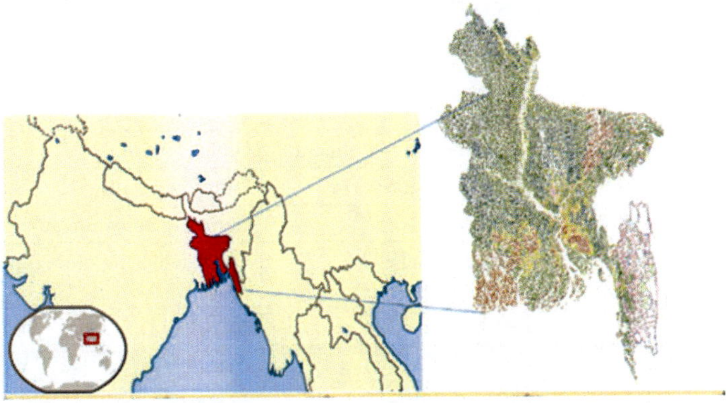

Fig. 4 Masking procedure of layer type-2 of 15 classes for UMD [31]

the deciduous broadleaf forests, red colour shows savannas, deep green colour shows the grassland, green colour shows the cropland, brown colour shows the urban and built-up lands, off-white colour shows the natural vegetation mosaics, violet colour shows the barren, and blue colour shows the water bodies. In this section, we present 18 figures representing the land covers of Bangladesh from year 2001 to 2018.

4.3 Mapping Method

Now, to classify the 18 TIF images of both layers, we have employed here 2 mapping methods such as (1) International Geosphere-Biosphere Programme (IGBP) for layer type-1 images and University of Maryland (UMD) for layer type-2. These are also called land cover classification algorithm (Table 1).

International Geosphere-Biosphere Programme (IGBP) Methodology
The IGBP land cover classification scheme came into play for the first time, which used the following criteria in the selection of the class like ground biomass (perennial vs annual), leaf longevity (evergreen or deciduous), and leaf type (broad or needle) was developed during many meetings of IGBP land cover working group held through 1995. DISCover, a system designed the IGBP product, which was used to cover up all kinds of IGBP-related science projects. A global land cover product based on 1 km AVHRR data was developed by the IGBP DIS land cover working group (LCWG), later on where it was combined in the DISCover land cover product [33, 34].

When the meeting came to an end, the final assessment was that it would be used to generate 17 classes to meet the desires of the IGBP core science projects. With the help of these three criteria with some possible combinations in six rudimentary classes like evergreen broadleaf, evergreen needleleaf, deciduous broadleaf, deciduous needleleaf, broadleaf annual, grass. With some modifications, these classes

Table 1 Similarities and differences of the IGBP DISCover and University of Maryland global land cover products [12, 32]

Product characteristics	IGBP DISCover	University of Maryland
Classification technique	Unsupervised clustering	Supervised classification tree
Processing sequence	Continent-by-continent	Global
Classification scheme	IGBP (17 classes)	Simplified IGBP (15 classes)
Refinement/update schedule	Annual	Currently being updated
Validation	September 1998	Evaluated using other digital datasets
Sensor	AVHRR	AVHRR
Time of data collection	April 1992–March 1993	April 1992–March 1993
Input data	12 monthly NDVI	41 metrics derived from NDVI and composites bands 1–5

came into play to be compatible with classification systems used at the time for environmental modelling, represent landscape mixtures and mosaics, and provide land use implications. We have employed IGBP here to classify Bangladeshi land images into 17 portions or classes. The name of each 17 types of the land of Bangladeshi geographical maps is shown in Table 2.

University of Maryland (UMD) Methodology
We have employed UMD developed by Hansen et al. [35] to classify the Bangladeshi geographical information into 15 different land classes for layer type-2 images. UMD [36] uses the 12 monthly NDVI and 29 auxiliary data derived from AVHRR 5 channels. UMD even uses a supervised technique, a decision tree, using the ground data derived from high-resolution satellite data. Most of the training data were collected through an overlay of co-registered coarse-resolution and interpreted high-resolution data. However, restricted by the computing resources, a subset of the entire dataset was extracted from the full resolution dataset—roughly every fifth pixel was sampled across each row and column. The name of the 15 classes for layer type-2 which classified using UMD is shown in Table 3.

Table 2 International Geosphere-Biosphere Programme (IGBP) legend and class details [12, 32]

Land cover class	Class	Land cover class	Class
Evergreen needleleaf forest (127)	1	Grasslands (106)	10
Evergreen broadleaf forests (201)	2	Permanent wetlands (11)	11
Deciduous needleleaf forests (16)	3	Croplands (240)	12
Deciduous broadleaf forests (52)	4	Urban and built-up lands (40)	13
Mixed forests (111)	5	Cropland/natural vegetation mosaics (70)	14
Close scrublands (24)	6	Permanent snow and ice (12)	15
Open scrublands (78)	7	Barren (110)	16
Woody savannas (69)	8	Water bodies (50)	17
Savannas (56)	9	Unclassified	18

Table 3 MCD12Q1 University of Maryland (UMD) legend and class descriptions [12, 14, 35]

Land cover class	Class	Land cover class	Class
Evergreen needleleaf forest	1	Savannas	9
Evergreen broadleaf forests	2	Grasslands	10
Deciduous needleleaf forests	3	Permanent wetlands	11
Deciduous broadleaf forests	4	Croplands	12
Mixed forests	5	Urban and built-up lands	13
Closed scrublands	6	Cropland/natural vegetation mosaics	14
Open scrublands	7	Non-vegetated lands	15
Woody savannas	8	Unclassified	255

5 Result and Discussion

We have classified our Bangladeshi geographical information into 17 land cover classes and 15 land cover classes, subsequently using IGBP and UMD.

5.1 IGPB

The guide utilizes the IGBP arrangement conspire; IGBP plots short the classes of lasting wetlands, cropland/normal vegetation mosaic, and ice and snow. For wide vegetation types, centre regions map likewise, whilst progress zones around centre regions contrast essentially. Table 4 shows the class cropland (12) from 2001 to 2008. The factuation can be seen from 2001 where it changes from top to bottom. In the specific years, it was changing with some percentage. Also, we can observe that the class cropland/natural vegetation mosaics (14) flatulating up and down as well around the years 2001–2008. Thus, we can see the changes on the table by determining the percentage ratio. Permanent wetland, natural vegetation cropland, Borland, and grasslands are increase 1% meanwhile savannas land area reduced 3% in this year. Also, water bodies were increased 1% but cropland has fluctuated but remain same at 2018. Also Table 5, we can determine that in the class cropland (12) from 2009 to 2018. The factuation can be seen from 2009 where it changes from top to bottom. In the specific years, it was changing with some percentage.

Table 4 Land cover of Bangladesh over (2001–2008) from class (1–17) in IGBP

V	2001 (%)	2002 (%)	2003 (%)	2004 (%)	2005 (%)	2006 (%)	2007 (%)	2008 (%)
1	0	0	0	0	0	0	0	0
2	1	1	1	1	1	1	1	1
4	0	0	0	0	1	1	0	0
5	0	0	0	0	0	0	0	0
8	3	3	3	3	3	3	3	3
9	11	11	11	12	11	11	11	11
10	2	2	2	2	2	2	2	2
11	10	10	10	9	9	9	10	10
12	47	46	48	47	47	48	48	48
13	1	1	1	1	1	1	1	1
14	18	18	17	17	17	16	16	16
15	0	0	0	0	0	0	0	0
16	1	1	1	1	1	1	1	1
17	6	7	6	7	7	7	7	7

Table 5 Land cover of Bangladesh over (2009–2018) from class (1–17) in IGBP

V	2009 (%)	2010 (%)	2011 (%)	2012 (%)	2013 (%)	2014 (%)	2015 (%)	2016 (%)	2017 (%)	2018 (%)
1	0	0	0	0	0	0	0	0	0	0
2	1	1	1	1	1	1	1	1	1	1
4	1	0	0	1	1	1	1	1	1	1
5	0	0	0	0	0	0	0	0	0	0
8	3	3	3	3	3	3	4	4	4	4
9	10	10	10	10	10	10	9	9	8	8
10	2	2	2	2	2	2	2	2	2	3
11	10	10	10	10	10	10	10	11	11	11
12	48	49	49	49	49	49	48	48	46	47
13	1	1	1	1	1	1	1	1	1	1
14	16	16	16	15	16	16	16	16	18	17
15	0	0	0	0	0	0	0	0	0	0
16	1	1	1	1	0	0	1	0	1	0
17	7	7	7	7	7	7	7	7	7	7

Also, we can observe that the class **cropland/natural vegetation mosaics (14)** flatuating up and down as well around the years 2009–2018. Thus, we can see the changes on the table by determining the percentage ratio. Here are some outputs created by the analysis with the map output with the bar diagram. The bar graph illustrates the amount of kinds of spreads (17 classes) which were consumed over 18 years from 2000 to 2018. Units are measured in percentage. The line graph illustrates the amount of four kinds of spreads (cropland, natural vegetation, permanent wetland, and savannas) which were consumed over 18 years from 2000 to 2018. Units are measured in percentage mentioned in Figs. 5 and 6.

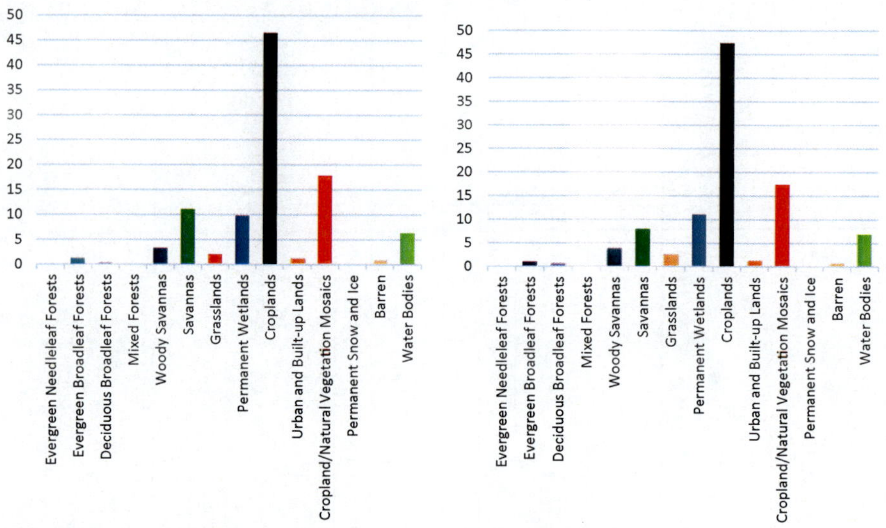

Fig. 5 Bar graph of Bangladeshi land 17 classes for 2001 and 2018

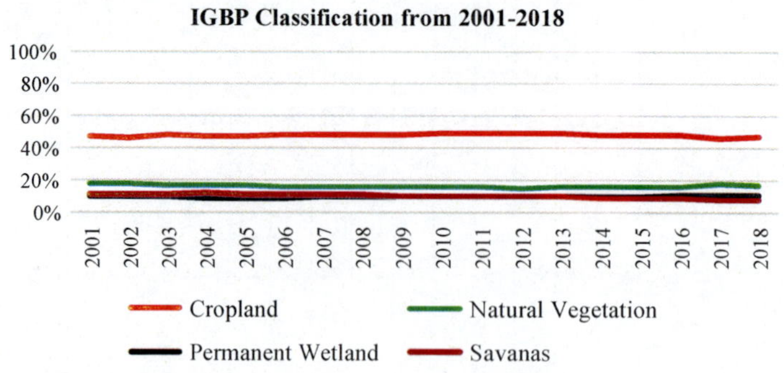

Fig. 6 Summary of pixel percentage (%) distribution for (IGBP), from year 2001 to 2018

Here, the data are completely figurative and shown above with the proper percentile from 0 to 100%. Overall, the consumption of cropland, natural vegetation, permanent wetland, and savannas over the period given, whilst for both it got reduced spreads and it also rose within 2000–2018. So we can gradually give a better perspective over the increase/decrease over the land by various changes. However, the urban and built-up land is also increasing year by year [37]. A possible explanation is that the spatial resolution of 250 m × 250 m for a pixel is very low, which fails to capture the pixels which had a proportionately low built-up area.

5.2 UMD

Table 6 shows the accuracy using UMD method. From the accuracy table, we can determine that in the class cropland (12) from 2001 to 2008 mentioned in Fig. 7.

The factuation can be seen from 2001 where it changes from top to bottom. In the specific years, it was changing with some percentage. Also, we can observe that the class cropland/natural vegetation mosaics (14) flatuating up and down as well around the years 2001–2008. Thus, we can see the changes on the table by determining the percentage ratio. Table 7 shows the class Cropland (12) from 2009 to 2018. The factuation can be seen from 2009 when it changes from top to bottom. In the specific years, it was changing with some percentage. Also, we can observe that the class cropland/natural vegetation mosaics (14) flatuating up and down as well around the years 2009–2018. Thus, we can see the changes on the table by determining the percentage ratio.

Here are some outputs created by the analysis with the map output with the bar diagram. The bar graph illustrates the amount of four kinds of spreads (15 classes) that

Table 6 Land cover of Bangladesh over (2001–2008) from class (1–15) in UMD

V	2001 (%)	2002 (%)	2003 (%)	2004 (%)	2005 (%)	2006 (%)	2007 (%)	2008 (%)
1	0	0	0	0	0	0	0	0
2	1	1	1	1	1	1	1	1
4	0	0	0	0	0	0	1	1
5	0	0	0	0	0	0	0	0
8	3	4	4	4	4	4	4	4
9	12	12	12	12	12	12	12	11
10	2	2	2	2	2	2	2	2
11	11	10	10	10	10	10	10	10
12	50	50	51	51	51	51	51	52
13	1	1	1	1	1	1	1	1
14	19	19	18	18	18	18	17	17
15	1	1	1	1	1	1	1	1

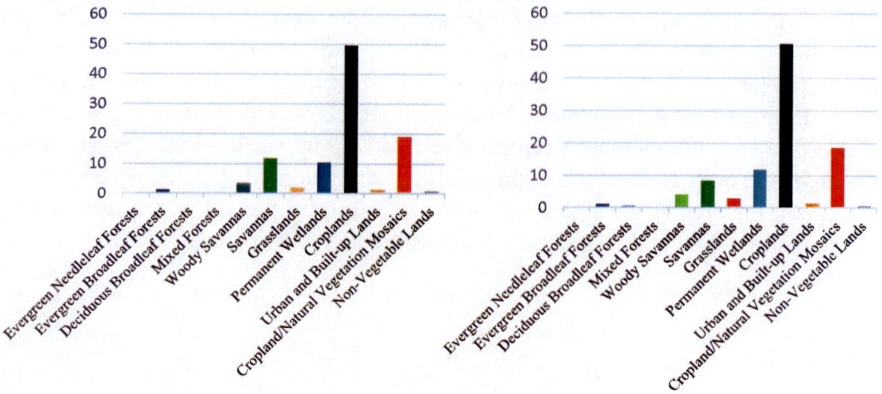

Fig. 7 Bar graph of Bangladeshi land 15 classes for 2001 and 2018

were consumed over 18 years from 2000 to 2018. Units are measured in percentage. Here, the data are completely figurative and shown above with the proper percentile from 0 to 60%. Figure 8 shows the line graph to illustrate the amount of four kinds of spreads (cropland, natural vegetation, permanent wetland, and savannas) which were consumed over 18 years from 2000 to 2018. Units are measured in percentage. Here, the data are completely figurative and shown above with the proper percentile from 0 to 100%.

5.3 *Comparative Discussion*

Figure 9 illustrates the amount of four kinds of spreads (cropland) which were consumed over 18 years from 2000 to 2018. Units are measured in percentage mentioned in Figs. 10, 11, and 12.

Here, the data are completely figurative and shown above with the proper percentile from 0 to 100%. The line graph illustrates the amount of four kinds of spreads (natural vegetation) which were consumed over 18 years from 2000 to 2018. Units are measured in percentage. Here, the data are completely figurative and shown above with the proper percentile from 0 to 100%. The line graph illustrates the amount of four kinds of spreads (permanent wetland) which were consumed over 18 years from 2000 to 2018. Units are measured in percentage. Here, the data are completely figurative and shown above with the proper percentile from 0 to 100%. The line graph illustrates the amount of four kinds of spreads (savannas) which were consumed over 18 years from 2000 to 2018. Units are measured in percentage. Here, the data are completely figurative and shown above with the proper percentile from 0 to 100%. Here, we can observe that the data analysis using IGBP and UMD, and the diagram

Table 7 Land cover of Bangladesh over (2009–2018) from class (1–15) in UMD

V	2009 (%)	2010 (%)	2011 (%)	2012 (%)	2013 (%)	2014 (%)	2015 (%)	2016 (%)	2017 (%)	2018 (%)
1	0	0	0	0	0	0	0	0	0	0
2	1	1	1	1	1	1	1	1	1	1
4	1	1	1	1	1	1	1	1	1	1
5	0	0	0	0	0	0	0	0	0	0
8	4	4	4	4	4	4	4	4	4	4
9	11	11	11	11	10	10	10	10	9	8
10	2	2	2	2	2	2	2	2	3	3
11	10	10	10	10	11	11	11	11	12	12
12	5	52	52	52	52	52	52	51	50	51
13	1	1	1	1	1	1	1	1	1	1
14	17	17	17	17	17	17	17	18	19	18
15	1	1	1	1	1	1	1	1	0	1

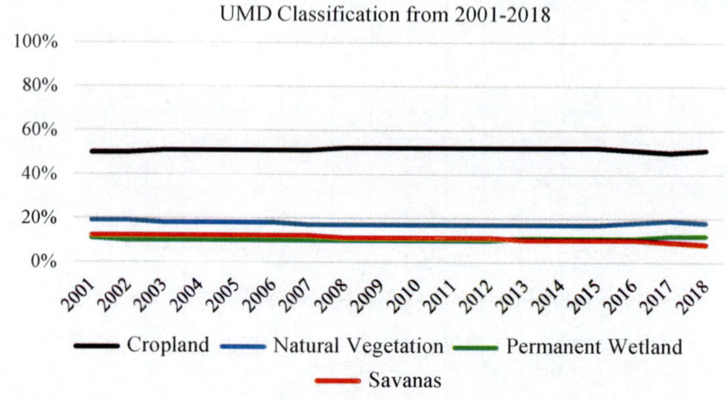

Fig. 8 Summary of pixel percentage (%) distribution for (UMD), from year 2001 to 2018

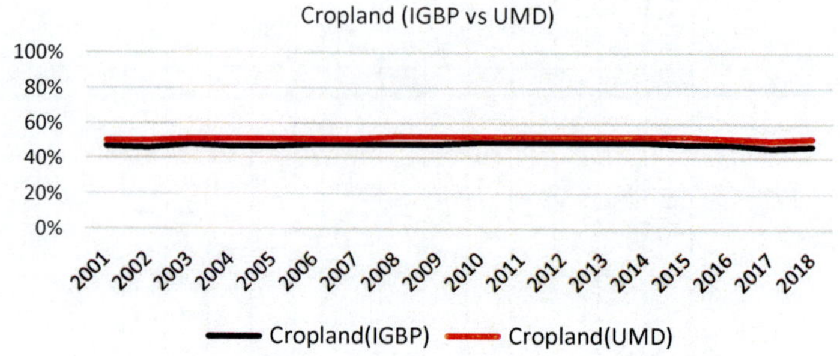

Fig. 9 Pixel percentage (%) difference for IGBP and UMD for cropland, from year 2001 to 2018

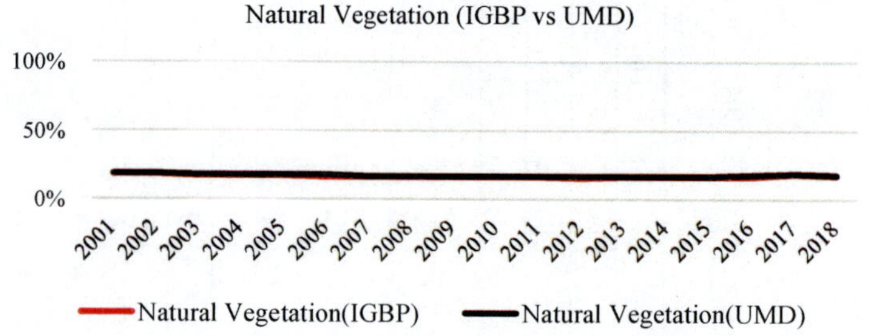

Fig. 10 Pixel (%) difference for IGBP and UMD for natural vegetation, from year 2001 to 2018

above clearly shows the data are close to each other for all the croplands, natural vegetation, permanent wetland, savannas. Thus, by which we can say that our analysis shows the data are quite a reassurance for it.

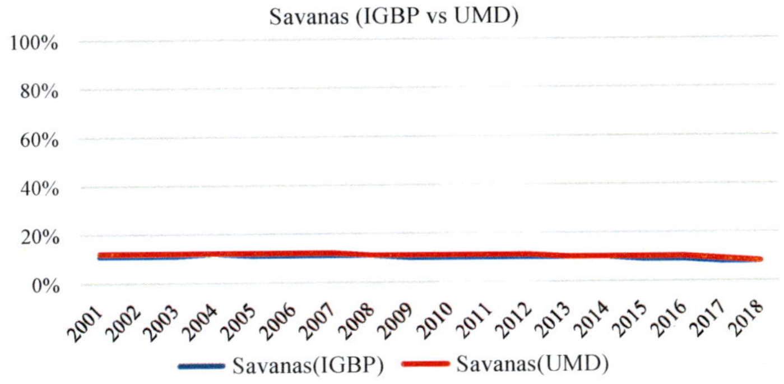

Fig. 11 Pixel (%) difference for IGBP and UMD for permanent wetland, from 2001 to 2018

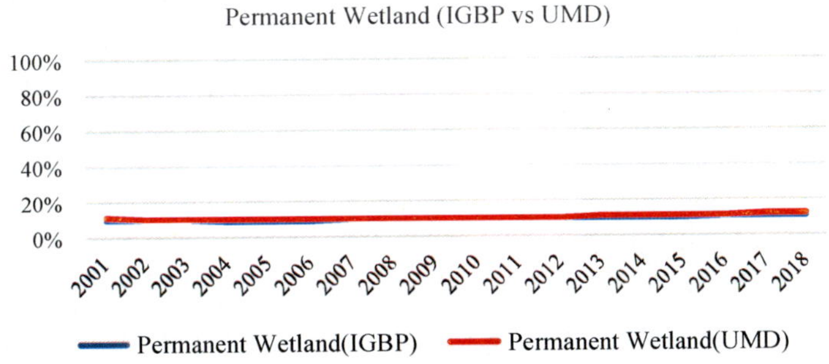

Fig. 12 Pixel (%) difference for IGBP and UMD for savannas, from year 2001 to 2018

6 Conclusion

In this study, we have been worked on finding out of IGBP and UMD from remotely sensed images for understanding its effect on cropland, natural vegetation, permanent wetland, savannas in Bangladesh from 2001 to 2018. We determine that comparisons between IGBP and UMD were given data attributes values which almost provide the same result as we expected. The drawback has been identified based on only one satellite (MODIS), and only 2 land products were utilized through a low goal of 250 m × 250 m per pixel at yearly intervals. The work that is done here is in the range of 500 m which is the latest work done so far for the land mapping of Bangladesh. The main limitation of the work is we did not collect the data for 2018–2021. In the future, we desire to work with MOD13Q1.006 Terra Vegetation Indices 16-Day Global 250 m which will be able to provide us time series climatology records and surveying vegetation conditions that will be provided every 16 days land cover changes explanation [38]. Also, a comparison between different satellites considering land cover demands further investigation. If anyone likes to work in this field, then this work will definitely give a boost to their work and open up various gates for another kind of research work related to this.

References

1. Giashuddin M, Islam M, Sheikh I, Rahman M (2011) Use of GIS and RS in agriculture of Bangladesh: present status and prospect
2. Rahman Md, Lateh H (2015) Climate change in Bangladesh: a spatio-temporal analysis and simulation of recent temperature and rainfall data using GIS and time series analysis model. Theor Appl. Climatol. https://doi.org/10.1007/s00704-015-1688-3
3. Rahman MR, Lateh H (2016) Spatio-temporal analysis of warming in Bangladesh using recent observed temperature data and GIS. Clim Dyn 46:2943–2960
4. Siraj MA, Neema MN, Shubho H, Tanvir Md (2013) Impacts of climate change on food security in Bangladesh—a GIS-based analysis. Asian Trans Eng 3:13–18
5. Dewan AM, Yamaguchi Y (2009) Using remote sensing and GIS to detect and monitor land use and land cover change in Dhaka Metropolitan of Bangladesh during 1960–2005. Environ Monit Assess 150:237. https://doi.org/10.1007/s10661-008-0226-5
6. Hossain MS, Wong S, Chowdhury M, Shamsuddoha Md (2009) Remote sensing and GIS application to mangrove forest mapping in the meghna deltaic islands of Bangladesh. Bangladesh J Mar Sci Fish 1:81–96
7. Verbesselt J, Hyndman R, Newnham G, Culvenor D (2010) Detecting trend and seasonal changes in satellite image time series. Remote Sens Environ 114(1):106–115. ISSN 0034-4257. https://doi.org/10.1016/j.rse.2009.08.014
8. Joshi BP (2014) Assessment of phosphorus loss risk from soil—a case study from Yuqiao reservoir local watershed in north China. Department of Chemistry, Faculty of Mathematics and Natural Sciences, University of Oslo, 01/2014
9. Kairu EN (1982) An introduction to remote sensing. Geo J 6(3). https://doi.org/10.1007/BF00210657
10. Justice CO, Townshend JRG, Vermote EF, Masuoka E, Wolfe RE, Saleous N, Roy DP, Morisette JT (2002) An overview of MODIS Land data processing and product status. Rem Sens Environ 83(1-2):3–15. https://doi.org/10.1016/S0034-4257(02)00084-6
11. Islam Md, Hasan Md, Farukh M (2017) Application of GIS in general soil mapping of Bangladesh. J Geograph Inf Syst 09:604–621. https://doi.org/10.4236/jgis.2017.95038
12. Hansen MC, Reed B (2000) A comparison of the IGBP DISCover and University of Maryland 1 km global land cover products. Int J Remote Sens 21(6–7):1365–1373. https://doi.org/10.1080/014311600210218
13. NASA (nd) Level-1 and atmosphere archive & distribution system distributed active archive center - LAADS DAAC. NASA. Retrieved May 9, 2022, from https://ladsweb.modaps.eosdis.nasa.gov/
14. Ran Y, Li X, Lu L (2010) (2010) Evaluation of four remote sensing based land cover products over China. Int J Rem Sens 31(2):391–401. https://doi.org/10.1080/01431160902893451
15. Ran Y, Li X, Lu L (2010) Evaluation of four remote sensing based land cover products over China. Int J Remote Sens 31(2):391–401. https://doi.org/10.1080/01431160902893451
16. Sidhu N, Pebesma E, Wang Y-C (2017) Usability study to assess the IGBP land cover classification for Singapore. Remote Sens 9:1075. https://doi.org/10.3390/rs9101075
17. Mengistu D, Salami A (2008) Application of remote sensing and GIS in land use/land cover mapping and change detection in a part of south western Nigeria. Afr J Environ Sci Technol 1:99–109
18. Cihlar J (2000) Land cover mapping of large areas from satellites: status and research priorities. Int J Remote Sens 21(6–7):1093–1114. https://doi.org/10.1080/014311600210092
19. Loveland TR, Reed BC, Brown JF, Ohlen DO, Zhu Z, Yang L, Merchant JW (2000) Development of a global land cover characteristics database and IGBP DISCover from 1 km AVHRR data. Int J Remote Sens 21(6–7):1303–1330. https://doi.org/10.1080/014311600210191
20. Khan S, Mohiuddin K (2018) Evaluating the parameters of ArcGIS and QGIS for GIS applications 7:582–594

21. GW, By, Geospatial World http://geospatialmedia.net Geospatial World by Geospatial Media!, Geospatial World, & Media!, GW by G (2018, March 16). GeoTIFF - A standard image file format for GIS applications. Geospatial World. Retrieved May 9, 2022, from https://www.geospatialworld.net/article/geotiff-a-standard-image-file-format-for-gis-applications/
22. Wikimedia Foundation (2022, February 24) Open geospatial consortium. Wikipedia. Retrieved May 9, 2022, from https://en.wikipedia.org/wiki/Open_Geospatial_Consortium
23. Balasubramanian A (2017) Digital elevation model (DEM) IN GIS. https://doi.org/10.13140/RG.2.2.23976.47369
24. Greenfeld J (2001) Evaluating the accuracy of digital orthophoto quadrangles (DOQ) in the context of parcel-based GIs. Photogram Eng Rem Sens 67
25. Islam A, Bala S, Haque A (2009) Flood inundation map of Bangladesh using MODIS surface reflectance data
26. Huete A, Didan K, Miura T, Rodriguez EP, Gao X, Ferreira LG (2002) Overview of the radiometric and biophysical performance of the MODIS vegetation indices. Remote Sens Environ 83:195–213
27. Hansen MC, Defries RS, Townshend JRG, Sohlberg R (2010) (2000) Global land cover classification at 1 km spatial resolution using a classification tree approach. Int J Rem Sens 21(6–7):1331–1364. https://doi.org/10.1080/014311600210209
28. Mora B, Tsendbazar NE, Herold M, Arino O (2014) Global land cover mapping: current status and future trends. In: Manakos I, Braun M (eds) Land use and land cover mapping in Europe. Remote sensing and digital image processing, vol 18. Springer, Dordrecht. https://doi.org/10.1007/978-94-007-7969-3_2
29. Justice CO et al (2002) An overview of MODIS land data processing and product status. Remote Sens Environ 83:3–15
30. Sulla-Menashe D, Friedl MA, User guide to collection 6 MODIS land cover (MCD12Q1 and MCD12C1) product
31. FollowthePIN.com (2021, April 19). Retrieved May 9, 2022, from https://followthepin.com/where-is-bangladesh/
32. Achugbu IC, Olufayo A, Balogun IA, Adefisan EA, Dudhia J, Naabil E (2021) Modeling the spatiotemporal response of dew point temperature, air temperature and rainfall to land use land cover change over West Africa. Model Earth Syst Environ
33. Loveland TR, Reed BC, Brown JF, Ohlen DO, Zhu Z, Yang L, Merchant JW (2000) Development of a global land cover characteristics database and IGBP DISCover from 1 km AVHRR data. Int J Remote Sens 21:1303–1330
34. Kaul HA, Sopan I (2012) Land use land cover classification and change detection using high resolution temporal satellite data. J Environ 1:146–152
35. Hansen MC, Potapov PV, Moore R, Hancher M, Turubanova SA, Tyukavina A, Thau D, Stehman SV, Goetz SJ, Loveland TR, Kommareddy A, Egorov A, Chini L, Justice CO, Townshend JRG (2013) High-resolution global maps of 21st-century forest cover change. Sci 342(6160):850–853. https://doi.org/10.1126/science.1244693
36. Chen H, Zhuang D, Li S, Shi R, Wang Y, Xinfang Yu (2005) A comparison of university of Maryland 1 km land cover dataset and a land cover dataset in China. In: Proceedings. 2005 IEEE international geoscience and remote sensing symposium, 2005. IGARSS'05., 2005, pp 2440–2443. https://doi.org/10.1109/IGARSS.2005.152547
37. Zubayer S, Wakil Md, Baksh A, Afroz F (2018) Urban growth and land use change analysis using GIS and remote sensing: a case study of Rajshahi metropolitan city (RMC)
38. MOD13Q1 V006. LP DAAC - MOD13Q1. (nd) Retrieved May 9, 2022, from https://lpdaac.usgs.gov/products/mod13q1v006/

Genetic Algorithm-Based Optimal Deep Neural Network for Detecting Network Intrusions

Sourav Adhikary, Md. Musfique Anwar, Mohammad Jabed Morshed Chowdhury, and Iqbal H. Sarker

Abstract Computer network attacks are evolving in parallel with the evolution of hardware and neural network architecture. Despite major advancements in network intrusion detection system (NIDS) technology, most implementations still depend on signature-based intrusion detection systems, which cannot identify unknown attacks. Deep learning can help NIDS to detect novel threats since it has a strong generalization ability. The deep neural network's architecture has a significant impact on the model's results. We propose a `genetic algorithm`-based model to find the `optimal number of hidden layers` and the `number of neurons` in each layer of the `deep neural network (DNN) architecture` for the `network intrusion detection` binary classification problem. Experimental results demonstrate that the proposed DNN architecture shows better performance than classical machine learning algorithms at a lower computational cost.

Keywords Genetic algorithm · Deep neural network · Hidden layer · Optimal architecture · Intrusion detection

1 Introduction

Network intrusion detection detects any unauthorized access to the computer network. Intrusion detection systems (NIDS) are classified into two types: signature-based and anomaly-based [9]. Signature-based intrusion detection uses pattern matching techniques of known attacks. With increasing cyberattacks, new unknown

S. Adhikary · I. H. Sarker (✉)
Department of Computer Science and Engineering, Chittagong University of Engineering and Technology, Chattogram 4349, Bangladesh
e-mail: iqbal@cuet.ac.bd

Md. M. Anwar
Jahangirnagar University, Dhaka, Bangladesh

M. J. M. Chowdhury
La Trobe University, Melbourne, Australia

V. Skala et al. (eds.), *Machine Intelligence and Data Science Applications*, Lecture Notes on Data Engineering and Communications Technologies 132,
https://doi.org/10.1007/978-981-19-2347-0_12

attacks are emerging. A signature-based system uses pattern matching technique that cannot detect unknown attacks. An anomaly-based NIDS uses machine learning techniques to analyze and learn the normal network behavior of a system. Then it can be used to detect unknown attacks by analyzing the deviations from normal traffic behavior.

With a good generalization ability, deep learning can enable NIDS to detect unknown attacks. One of the main topics of NIDS research in recent years has been the application of machine learning and deep learning techniques [7, 8]. Deep learning approaches show better results than conventional machine learning algorithms [5, 6]. Most deep learning architectures in the literature are selected using the trial-and-error method which is prone to error.

To overcome the neural network architecture selection problem, we used the genetic algorithm to determine the optimum number of hidden layers and neurons in each hidden layer for the intrusion detection problem. The fitness function is a linear function that combines loss, parameter, and specificity. The search for neural network architecture is a time-consuming process. To find the optimal deep neural network architectures, the genetic algorithm is used. The best neural architecture found in the search process is used to evaluate performance and compared with other classical machine learning algorithms. The proposed deep neural network showed promising results compared with previously deployed machine learning and deep learning algorithms. The contributions of this research work are as follows:

- Finding the optimal number of hidden layers and number of neurons in each layer for intrusion detection problem using genetic algorithm.
- To design an optimal deep learning model for detecting network intrusions.

The rest of the paper is organized as follows. The related work on intrusion detection and neural architecture search is discussed in Sect. 2. Section 3 gives an outline of the suggested solution. The experimental findings are presented in Sect. 4. Section 5 analyzes the limitations of the suggested solution and finishes with future research prospects.

2 Related Works

Research on cybersecurity gains popularity in recent years. Many researchers use machine learning techniques for anomaly-based network intrusion detection [7, 14, 20]. Using publicly available datasets, Vinayakumar et al. [15] conducted a systematic study of DNN and other machine learning algorithms. From a limited collection of architecture search spaces, the authors selected the DNN topology by trial-and-error method. In [12], the authors proposed a convolutional neural network for intrusion detection problems. A comparative study of various settings is used to find the hyperparameters. A systematic study of seven deep learning strategies was proposed by Ferrag et al. [5] with the deep neural network performing best. In order to choose both subsets of features and hyperparameters in one operation, the authors proposed a double algorithm based on particle swarm optimization (PSO) [3]. A deep neural network-based network intrusion detection system needs to be trained regularly with new data from network traffic which is computationally costly [1]. With the optimal architecture of the deep neural network, computational costs can be lessened. The architecture of the deep neural network greatly influences the performance of the model. As mentioned above many researchers use the trial-and-error method to find an optimal network architecture that needs extensive knowledge of deep learning and is also prone to error. In [19], different network architecture search methods were discussed. An algorithm for optimizing a multilayered artificial neural network is proposed which prunes neurons from the hidden layers as much as possible while retaining the same error rate [17]. Starting from an overly large network, pruning algorithms remove unimportant neurons before the optimum network emerges [16]. In [2], Bergstra et al. showed trials on a grid are less effective for hyper-parameter optimization than trials chosen at random. In [11], the authors used a genetic algorithm to find an optimal network topology and showed comparative analysis with other network architecture search algorithms. A hybrid genetic algorithm with a stochastic layer showed promising results in affordable computation cost by [6]. In [4], a modified evolutionary strategy was used to find optimal architecture for retinal vessel segmentation.

In the world of automation, it is important to automate the architecture search process. Deep neural network architecture search using the trial-and-error method used in most literature requires extensive expertise in the field and may not yield the best results at a reasonable computation cost. A genetic algorithm is used in this study to automate the optimum network architecture search procedure, which is both reliable and fast.

3 Methodology

The performance of a deep learning model which has three or more hidden layers depends on the network architectures and other parameters such as activation function used in each layer and optimization algorithm. Genetic algorithm is used to find the optimal neural network parameters. Our approach allows us to find the number of hidden layers and the number of possible neurons in each hidden layer. A combination of all those components creates a multidimensional search space of all possible neural network architectures. Genetic algorithms is one of the most utilized approaches in the studies of the evolution of neural network architectures [19]. It is a population-based search algorithm that can search over complex and multidimensional search space and reach global optimum or near optimum solution reliably and fast.

The block diagram of the neural network architecture search using the genetic algorithm is shown in Fig. 1. Encoded binary strings that represent the neural network architectures are initialized randomly. Each string is decoded to build a neural topology and trained using a prepossessed data set. The algorithm evaluates the model and calculates fitness using a fitness function. The goal of the algorithm is to

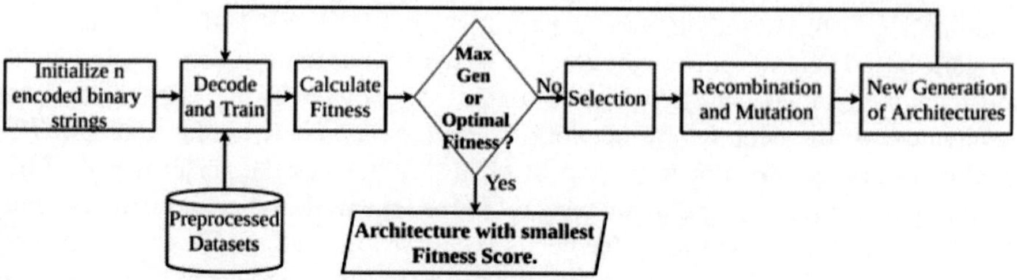

Fig. 1 Neural network architecture search using genetic algorithm

minimize the fitness value. From the fitness value of the population of architectures, a certain number of chromosome strings is selected for further process. Those strings which are not selected are discarded for the population pool. Selected binary strings are recombined and mutated to generate a new set of binary strings that represent neural architecture. This new generation of chromosome strings is used to build new models. The whole process is repeated until the maximum generation is reached or a satisfactory fitness value is achieved.

3.1 Random Set of Neural Architectures

Representing the candidate solutions is the most important issue to consider in the evolution of architecture tasks. The genetic algorithm is utilized in this study to determine the number of hidden layers and the number of neurons in each hidden layer. This information is encoded into a binary string using direct binary encoding. 50 encoded binary strings that represent the neural network architectures are initialized randomly. The neural network architectures have the 5 highest possible hidden layers and 1024 maximum possible neurons. Each chromosome is a binary string of length 50 which represents the hidden layer architecture. The maximum possible hidden layer consists of 5 hidden layers having 1023 neurons in each layer. The minimum possible architecture consists of 3 hidden layers having 32 neurons in each layer.

3.2 Build and Train

Backpropagation is used to train each neural network on the training dataset, and performance is evaluated using a test dataset. For optimization, Adam is used as an optimizer with the default parameter. Binary cross-entropy is used as a loss function. Activation function ReLU is used for inner layers, and sigmoid activation function is used for the output layer. Each architecture is trained for 90 epochs with a batch size of 512.

3.3 Fitness Function

The fitness function calculates the fitness of each candidate solution. A good intrusion detection model should detect intrusion accurately with less error rate and less computation cost. In this study, the goal of the genetic algorithm is to minimize the score calculated by the fitness function described below.

$$F = \frac{(W_l \times \text{loss}) + (W_s \times \text{specificity}) + (W_p \times \text{Norm}_{\text{parameter}})}{(W_l + W_s + W_p)} \tag{1}$$

where

$$\text{Norm}_{\text{parameter}} = \frac{(p_{\text{max}} - p_{\text{min}})}{(W_l + W_s + W_p)} \tag{2}$$

Parameter is the number of weights and bias in a deep artificial neural network. In Eq. 2, p_{max} and p_{min} represent the parameter of maximum and minimum possible neural network architecture of our proposed solution, respectively. The overall objective of using p_{min} and p_{max} is to normalize the complexity of each topology tested to the interval (0,1) so that it can be smoothly incorporated into the fitness function.

$$\text{parameter} = \sum_{l=1}^{l=\text{Layer}} (\text{Neurons}_l \times \text{Neurons}_{l-1} + \text{Bias}_l) \tag{3}$$

For loss binary cross-entropy function is used. Binary cross-entropy loss is defined as

$$\text{loss} = -(y \times \log(y^{\text{pred}}) + (1 - y)\log(1 - y^{\text{pred}})) \tag{4}$$

where y is the true binary value and y^{pred} is the predicted value.

Specificity is the metric that evaluates a model's ability to predict the true negatives of each available category. In our proposed solution, *specificity* specifies the attack class detection rate to all attacks. This fitness function tries to maximize the specificity value to get a good attack class detection rate.

$$\text{Specificity} = \text{TN}/(\text{TN} + \text{FP}) \tag{5}$$

W_l, W_s, W_p are three user-defined constants that represent the weight value of the loss, normalized parameter, and specificity to have more control over fitness function. The free parameters W_l, W_s, and W_p of the fitness function defined in the previous section are important to converge into optimal fitness. Due to computational constraints, all three free parameters are set as $W_l = 1$, $W_s = 2$, and $W_p = 0.01$. These three values are used to control the impact of loss, specificity, and parameter over fitness value. As specificity value represents the ability to detect attack class, specificity value should have the most impact on fitness value.

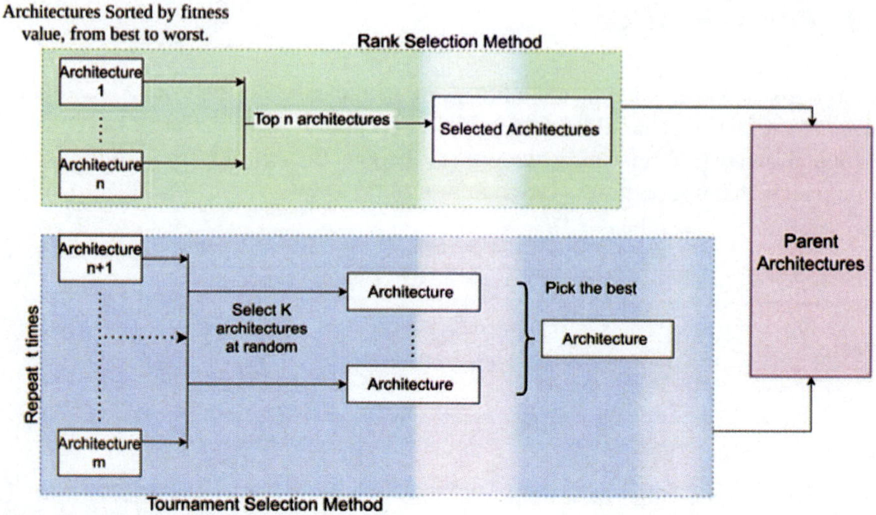

Fig. 2 Selection process of the best architecture

3.4 Selection

The selection process as shown in Fig. 2 uses a hybrid method using the rank selection and the tournament selection method. The algorithm selects several candidate solutions with the lowest fitness value. In the tournament selection method as described in Goldberg [18], the algorithm randomly selects two architectures and compares their fitness value. Architectures with the lowest fitness value are selected for recombination. This hybrid method ensures the diversity and effectiveness of the neural network architecture population pool.

3.5 Recombination

The selected architecture candidates are combined to get all possible groups of architecture candidates of size two called parents. From the pool of parents, 15 parents are selected randomly for recombination. For recombination, a two-point crossover is used as prescribed in [13]. In a two-point crossover, two randomly chosen crossover points are used to exchange a string segment that falls between the two points. The recombination process generates 10 new chromosomes. The two-point crossover recombination process is shown in Fig. 3.

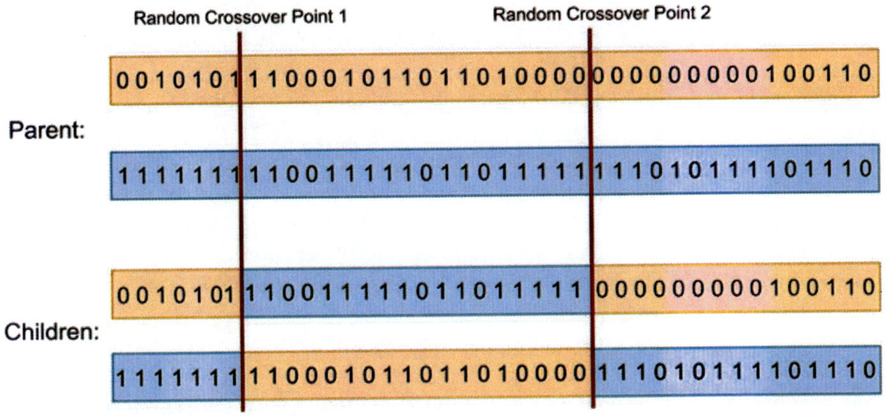

Fig. 3 Two-point crossover of parent architectures

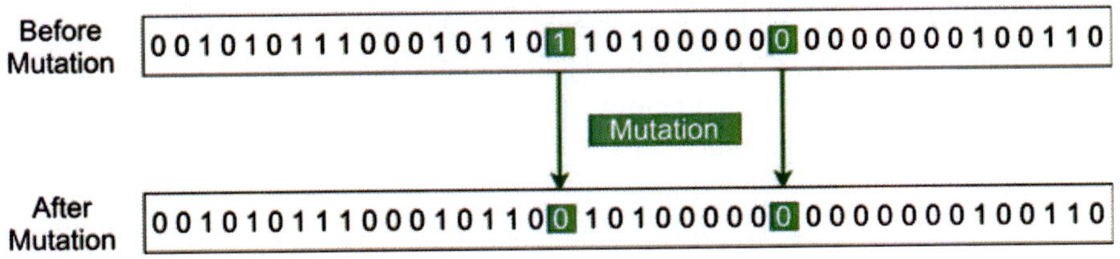

Fig. 4 Mutation process of selected architecture

3.6 Mutation

In mutation, a random binary string of the chromosome is complemented as shown in Fig. 4. A new generation is evolved by mutating the candidate strings from the recombination process. After 82 generations, the new best architecture learning rate slows down, and the network architecture search is terminated due to computational constraints. So, the total number of architecture evaluated is 1690 from architecture search space size of 1.023×10^{15}.

4 Results and Discussion

4.1 Dataset

CSE-CIC-IDS-2018 dataset collected by the Canadian Institute for Cybersecurity is used to develop a deep learning model for intrusion detection [10]. The dataset has 80 network traffic features. It comprises seven distinct attacks such as brute-force,

Table 1 Training and testing CSE-CIC-IDS-2018 dataset

Class	Train	Test
Benign	1592052	957820
Attack	807948	478910

heartbleed, botnet, dos, DDoS, web assaults, and infiltration of the network within. Data is normalized to improve the generalization ability of DNN (Table 1).

4.2 Results of Optimal DNN Search Using Genetic Algorithm

Figure 5 shows the smallest fitness score of each generation getting smaller over generations. The slope of the curve shows that the genetic algorithm has learned to find the neural network architecture of the smallest fitness value. We introduced random architecture in every generation which causes the many ups and downs of the curve. Here, 0 is the initial generation. The individuals in the population are getting better and better during the evolution process. After 82 generations, the learning rate of the evolution algorithm became very low. Due to computation hardware constraints, we stopped at the 82nd generation. The fitness score of the 82nd nd generation is the smallest, and we take the individuals as the output of the neural network architecture search. With the ReLU activation function and neurons 172, 182, 512, 38, 74, the optimal architecture has five fully connected hidden layers. For binary classification, one neuron in the output layer has the sigmoid activation function.

4.3 Results of Proposed Deep Learning Model

The degree of convergence of the training set loss is a crucial criterion for determining whether a model is effective. If it continues to plummet, it is a better training model. Figure 6 shows the training time loss and validation curve of the selected network architectures keep decreasing which indicates an effective model. The selected architecture is trained for 300 epochs to get the best results. The attack class has been defined as the negative class in the preprocessing step. A high specificity score represents the high attack class detecting ability. High specificity is crucial for intrusion detection problems. The test results of various DNN hidden layer architectures are shown in Table 2. 174, 182, 512, 38, 74, 64 are the number of neurons used for the different layers of hidden layer architecture, respectively. In terms of accuracy and F1-score architectures of hidden layers 3, 4, and 5 performed better. The highest specificity score was found in layer 5 architecture. The proposed automated

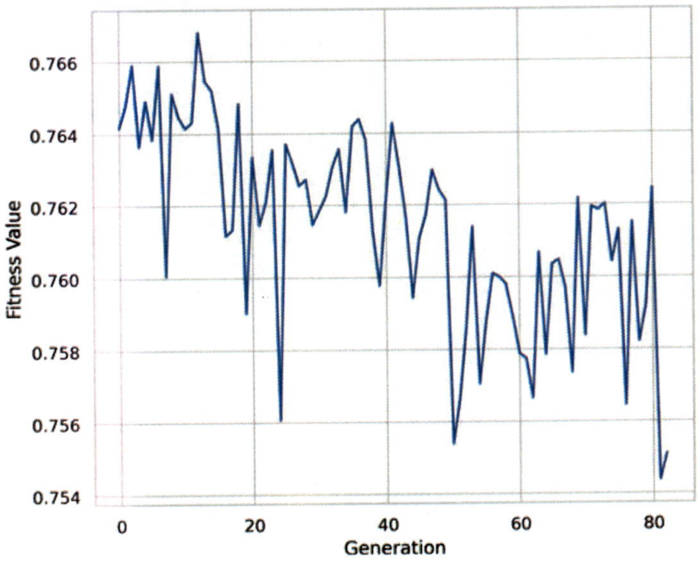

Fig. 5 Min fitness scores over generations

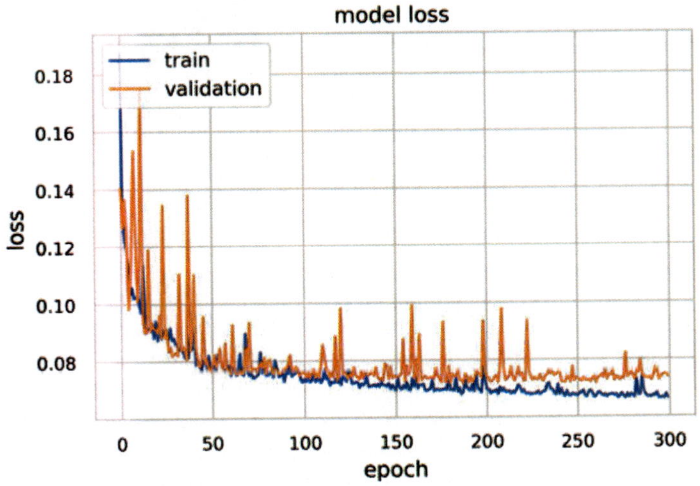

Fig. 6 Training loss of selected architecture

Table 2 Test results of DNN hidden layer architectures

No of hidden layer	Accuracy	F1 score	Specificity	Parameter
1 Layer	0.926	0.960	0.734	13,761
2 Layer	0.946	0.966	0.782	45,257
3 Layer	0.967	0.982	0.912	139,283
4 Layer	0.968	0.982	0.905	158,303
5 Layer	0.967	0.981	0.945	161,225
6 Layer	0.963	0.978	0.881	166,015

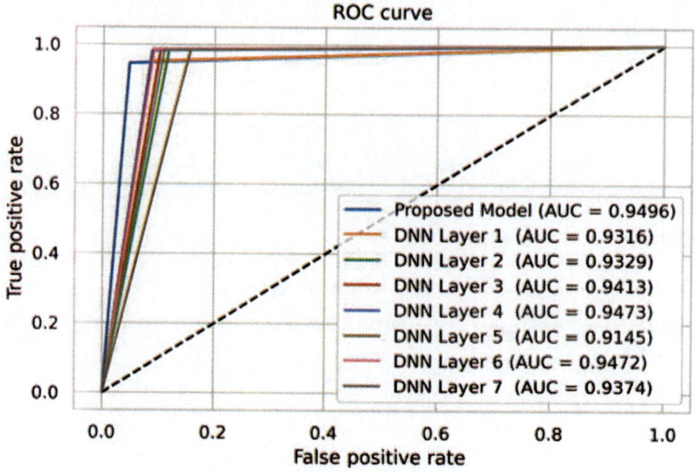

Fig. 7 ROC curves of DNN layers

method successfully finds the best architecture within a reasonable time frame. The five-hidden layer architecture has the largest area under the curve as shown in Fig. 7.

The automated search process of deep neural network architecture using the genetic algorithm found the optimal number of hidden layers and number of neurons in each layer. The performance of the proposed deep neural network model is compared with five classical machine learning algorithms as shown in Table 3 and Fig. 8.

All the models shown in Table 3 have been trained and tested on the same set of test and train datasets. Our proposed model outperforms the classical machine learning models in terms of specificity and f1 score.

Table 3 Comparison of machine learning algorithms

Algorithms	Accuracy	Precision	Recall	F1 value	Specificity
Logistic Regression	0.86	0.86	0.98	0.91	0.52
KNeighbors	0.89	0.95	0.99	0.96	0.84
Decision Tree	0.90	0.95	0.95	0.95	0.86
Gaussian Naive Bayes	0.86	0.87	0.94	0.90	0.56
Random Forest	0.91	0.95	0.99	0.97	0.85
Deep Neural Network	0.97	0.99	0.98	0.98	0.95

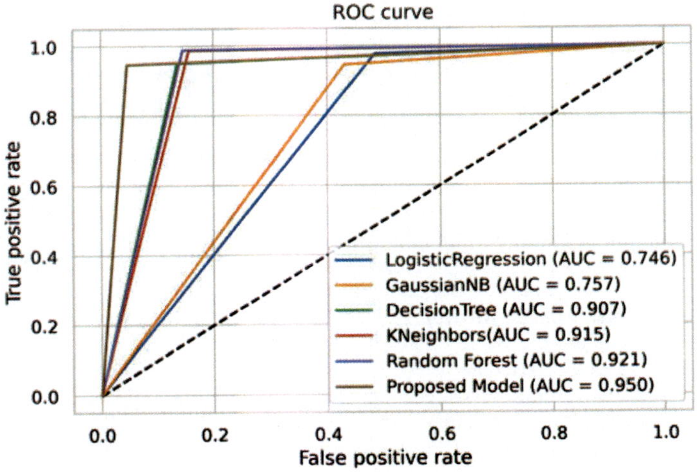

Fig. 8 ROC curves of classical machine learning algorithm and proposed model

5 Conclusion

In this study, the deep neural network model is developed by determining the optimal number of hidden layers and neurons in each layer for intrusion detection binary classification problems. The number of hidden layers and neurons in each layer is encoded in binary strings which are used with a genetic algorithm to find the optimal architectures. The proposed deep neural network architecture outperforms previously deployed machine learning in terms of attack classification ability and computational cost, to the best of our knowledge. However, using a genetic algorithm to find the optimal architecture is computationally expensive and time-intensive. This study only focused on the binary classification of network intrusion detection problems. By extending the search space on advanced hardware, the genetic algorithm's performance may improve. One of the interesting directions for future research is to use the genetic algorithm to design deep neural network architecture for other classification problems.

References

1. Ahmad Z, Shahid Khan A, Wai Shiang C, Abdullah J, Ahmad F (2021) Network intrusion detection system: a systematic study of machine learning and deep learning approaches. Trans Emerging Telecommun Technol 32(1):e4150
2. Bergstra J, Bengio Y (2012) Random search for hyper-parameter optimization. J Mach Learn Res 13(2)
3. Elmasry W, Akbulut A, Zaim AH (2020) Evolving deep learning architectures for network intrusion detection using a double pso metaheuristic. Comput Netw 168:107042
4. Fan Z, Wei J, Zhu G, Mo J, Li W (2020) Evolutionary neural architecture search for retinal vessel segmentation. arXiv e-prints pp. arXiv–2001
5. Ferrag MA, Maglaras L, Janicke H, Smith R (2019) Deep learning techniques for cyber security intrusion detection: a detailed analysis. In: 6th international symposium for ICS & SCADA cyber security research, vol 6, pp 126–136
6. Kapanova K, Dimov I, Sellier J (2018) A genetic approach to automatic neural network architecture optimization. Neural Comput Appl 29(5):1481–1492
7. Sarker IH (2021) Cyberlearning: effectiveness analysis of machine learning security modeling to detect cyber-anomalies and multi-attacks. Internet Things 14:100393
8. Sarker IH (2021) Deep learning: a comprehensive overview on techniques, taxonomy, applications and research directions. SN Comput Sci 2(6):1–20
9. Sarker IH, Furhad MH, Nowrozy R (2021) Ai-driven cybersecurity: an overview, security intelligence modeling and research directions. SN Comput Sci 2(3):1–18
10. Sharafaldin I, Lashkari AH, Ghorbani AA (2018) Toward generating a new intrusion detection dataset and intrusion traffic characterization. In: ICISSp, pp 108–116
11. Stathakis D (2009) How many hidden layers and nodes? Int J Remote Sens 30(8):2133–2147
12. Tao W, Zhang W, Hu C, Hu C (2018) A network intrusion detection model based on convolutional neural network. In: International conference on security with intelligent computing and big-data services. Springer, Berlin, pp 771–783
13. Thierens D, Goldberg D (1994) Convergence models of genetic algorithm selection schemes. In: International conference on parallel problem solving from nature. Springer, Berlin, pp 119–129
14. Tsai CF, Hsu YF, Lin CY, Lin WY (2009) Intrusion detection by machine learning: a review. expert systems with applications 36(10):11994–12000
15. Vinayakumar R, Alazab M, Soman K, Poornachandran P, Al-Nemrat A, Venkatraman S (2019) Deep learning approach for intelligent intrusion detection system. IEEE Access 7:41525–41550
16. Wagarachchi M, Karunananda A (2017) Optimization of artificial neural network architecture using neuroplasticity. Int J Artif Intelli 15(1):112–125
17. Wagarachchi NM (2019) Mathematical modelling of hidden layer architecture in artificial neural networks, Ph.D. thesis
18. Whitley D (1994) A genetic algorithm tutorial. Stat Comput 4(2):65–85
19. Wistuba M, Rawat A, Pedapati T (2019) A survey on neural architecture search. CoRR abs/1905.01392. http://arxiv.org/abs/1905.01392
20. Zaman M, Lung CH (2018) Evaluation of machine learning techniques for network intrusion detection. In: NOMS 2018-2018 IEEE/IFIP network operations and management symposium. IEEE, pp 1–5

Deep Convolutional Neural Network-Based Bangla Sign Language Detection on a Novel Dataset

Md. Jahid Hasan, S. K. Nahid Hasan, and Kazi Saeed Alam

Abstract Bangla Sign Language (BdSL) is the communication medium used by the deaf and mute people of Bangladesh. As they have speaking and hearing disabilities, they communicate with each other and with the rest of the world through sign language. It works as a communication medium and decreases the communication gap with normal people. In this research, we have worked with Bangla Sign Language detection for digit and letter signs. We have concentrated on building a complete BdSL dataset and applying various classifiers to our novel dataset. We have mostly focused on building convolutional neural networks to classify image data in this article. In addition, we have applied K-nearest neighbor (KNN), random forest (RF), support vector machine (SVM) and decision tree (DT) models to check the performance of each model on our dataset. On average, we have observed the highest accuracy of 95% for digit recognition and 91.5% for letter recognition on the CNN model. We have used 1500 images for digit and 4320 images for the letter dataset. We believe that our dataset will help researchers to carry out more in-depth studies as we have plans to make it openly accessible. Therefore, it will reduce the scarcity of the Bangla sign language dataset and expand research in this area.

Keywords Bangla sign language recognition · Sign language · Convolutional neural network · Image classification · Bangla digit dataset · Bangla letter dataset · Machine learning models

Md. J. Hasan (✉) · S. K. Nahid Hasan · K. S. Alam
Khulna University of Engineering and Technology, Khulna, Bangladesh
e-mail: jahidhasan001246@gmail.com

S. K. Nahid Hasan
e-mail: hasan1707080@stud.kuet.ac.bd

K. S. Alam
e-mail: saeed.alam@cse.kuet.ac.bd

V. Skala et al. (eds.), *Machine Intelligence and Data Science Applications*, Lecture Notes on Data Engineering and Communications Technologies 132,
https://doi.org/10.1007/978-981-19-2347-0_13

1 Introduction

About 5% of the global population are speech and hearing impaired, and they use sign language to communicate with other deaf and mute people and normal people [1]. Sign language is a visual way of communicating where someone uses hand gestures and movements, body language and facial expressions to communicate [2]. But normal people are unable to understand sign language. For making communication between normal people who do not understand sign language and deaf and mute people, a sign language interpreter or translator can be used. Every country has its own sign language. In our world, there are various sign languages like American Sign Language (ASL), British Sign Language (BSL), Indian Sign Language (ISL), etc. It varies from country to country and there is no universal sign language. In Bangladesh, Bangla sign language (BdSL) is used. Our model can be able to predict the letter or digit gesture from an image. Firstly, it gets inputs an image and preprocesses the image. Then, with the help of the training model, it predicts its actual character.

Many researchers have worked on and are still working on different sign languages projects. But in the case of Bangla sign language, only a few numbers of research works have been performed. Many researchers of Bangladesh have worked on only digit recognition systems or letter recognition systems with fewer classes. But a complete work of Bangla Sign Language is still not carried out which is the main cause of performing this research work. The dataset of BdSL detection is very limited which hinders the performance of Bangla Sign Language recognition. So, we have worked on both Bangla digits and Bangla letters, simultaneously. To get better performance we have created our own comprehensive dataset on digits and letters. We have used different Deep learning models to get a better result. Mainly we have focused on convolutional neural network (CNN), besides we used k-nearest neighbor (KNN), random forest (RF), SVM and decision tree at the end of the convolution layer of CNN to improve the result.

Deaf and mute people in any country are not a burden in society. If they can properly communicate with others, they can do all kinds of works like normal people. Using our system, the hearing- and speech-impaired people can lead a normal life and normal people can also easily understand them.

2 Related Works

At present, there are numerous forms of sign language detection research works are available like American Sign Language, Indian Sign Language, French Sign Language. However, studies on Bangla Sign Language are scarce in comparison to others [2]. Many researchers have already worked on it and obtained positive results.

Md Shafiqul Islam et al. [3] worked on Bangla Sign Language using CNN with nine random folds. They used a large dataset and worked with both letters and digits and got a decent accuracy. Shanta et al. [4] focused on sign language and used CNN for classification and SIFT for feature extraction. They acquired a total of 7600 photos of 38 Bangla letters, each of which was single-handed. They increased their performance by combining SIFT with CNN. Sanzidul Islam et al. [5] produced their own dataset 'Eshara-Lipi' in 2018, using only ten Bangla digits and a small dataset range. Based on the CNN model, they achieved 95% accuracy. Karmokar et al. published a neural network-based BdSL recognition algorithm in 2012, with a 93% accuracy. They used a dataset with the same skin tones and background, allowing for easy detection. Skin color has a significant impact on the detection of several indications. This issue is solved by employing a variety of skin tones. Many researchers have worked on real-time sign language; for example, in 2018, Oishee Bintey Hoque et al. [6] used faster R-CNN and achieved 98.2% accuracy with ten Bangla letters. They collected images with varying backgrounds and used object detection techniques.

In recent years, some other researchers have also focused on real-time detection [7]. Shahjalal et al. [8] proposed an automatic digit recognition system. They were able to reach 92% accuracy by using data augmentation. After finishing their entire process, they translated the output into Bangla speech. However, they could only work with the Bangla digits. MA Hossen et al. [9] devised a method for recognizing static hand signs of the Bengali 37 letters. They achieved an average accuracy of 84.68% here. However, in this case, they used a limited dataset and single-handed gestures. In 2019, Hossain et al. [10] proposed an approach of BdSL digit recognition using seven layers of capsule network where they used 28×28 input size of the vector and ended up with $20 \times 20 \times 256$ output shape in the first convolutional layer. They used the 'Eshara-lipi' dataset for training the model. They acquired an accuracy of 98.84%. Farhan Sadik et al. [11] investigated BdSL utilizing skin segmentation and the binary masking technique. They used an SVM classifier on a smaller dataset and achieved a standard accuracy of 99.8%. Sohrab Hossain et al. [12] worked on Bangla Sign Language utilizing CNN and achieved an accuracy of 98.75%. Various deep learning based works have also been performed in other sign language detection too which achieved overall good performance. Prangon Das et al. [13] suggested a model called recognition of American Sign Language static images that utilized CNN. They employed an ASL dataset of 26 English alphabet images for a total of 1815 images and obtained a validation accuracy of 94.34%. p. Mekala et al. [14] used CNN to recognize ASL in real time and achieved 100% accuracy. Aside from CNN, several researchers utilized alternative algorithms for sign language recognition, such as PCA and KNN, and obtained standard accuracy [15]. Similarly, some other BdSL recognition-based work was done by Roisul Islam Rumi et al. [16] and Prosenjit Roy et al. [17].

In our project, we mostly emphasized on creating a profound BdSL dataset and applied various classifiers because most of the previous works lacks the use of complete dataset. We have used both digits and letters in our dataset. The advantage is that we produced our own datasets by taking hand motions of volunteers in varied lighting and skin conditions. To increase performance, we have used different camera angles when taking pictures. We have worked with 46 classes at once (ten for

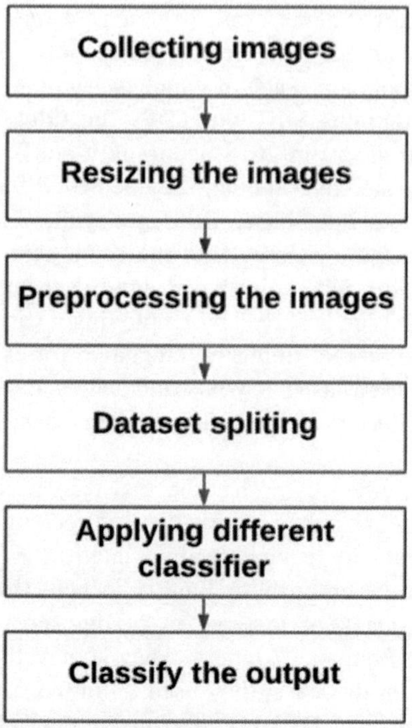

Fig. 1 Bangla Sign Language detection workflow

digits and 36 for letters). We have used 150 pictures per class in the digit part (total 10*150=1500) and 120 images per class in the letter portion (total 36 * 120 = 4320). We have primarily concentrated on convolutional neural networks, but we have also applied several Machine Learning algorithms at the end of the CNN convolution layer to improve accuracy. We have attempted to compare several machine learning techniques to find which model is superior in different conditions.

3 Methodology

In our system, we have used our own created dataset to get better performance. The entire procedure of dataset creation, dataset preprocessing, neural network application using here are described below which is shown in Fig. 1.

Table 1 Dataset summary

Class	Total class	Image per class	Train image	Test image	Total image
Digit	10	150	1200	300	1500
Letter	36	120	3456	864	4320

3.1 Dataset Description

As we have worked on our project for two segments one is for digit classification and another is for letter classification, we have separated our dataset into two main segments. In the digit section, we have a total of 10 classes, and in the letter section, we have a total of 36 classes.

For digits, we have captured about 160 images per class and filtered out some of them which have signs that are not in the right style. After filtering, we have kept about 150 images per class. In total, it would be 1500 images for digit (Table 1).

In case of letters, we have collected 130 pictures per class and after filtered out some bad quality pictures finally collected 120 images per class, and in total, we have collected $36 \times 120 = 4320$ images. Figure 2 shows 10 Bangla sign digits and 36 Bangla sign letters.

We have collected these images from different people that helped us to create different skin colors as it would be helpful to understand the performance of the models perfectly. We have also tried to change lighting effects. Rotating hand gestures in different angles, zoom in, zoom out, left-handed, right-handed combination, etc., would be considered to make a variety of images that helped us to improve our model.

At the time of training, we have split our dataset 80% for the training section and 20% for the validation section. Out of 4320 images, 3456 images have been used for training and 864 images have been used for validation. For digits, out of 1500 images, 1200 images have been used for training purposes, and 300 images have been used for validation purposes.

3.2 Dataset Preprocessing

The training images need to be preprocessed before using in CNN or any other model as raw pictures will increase complexity and thus will impact the overall performance. We have captured our images as 1600px by 1600px in RGB format. We have converted our images from RGB to a grayscale format that converts three-dimensional images into one-dimensional images which has reduced the image size. Then we have applied thresholding which helps to segment the image into different parts. It helps to remove the unnecessary background noise. Our binary threshold image contains only two colors, black and white. In Fig. 3, the preprocessing steps are shown.

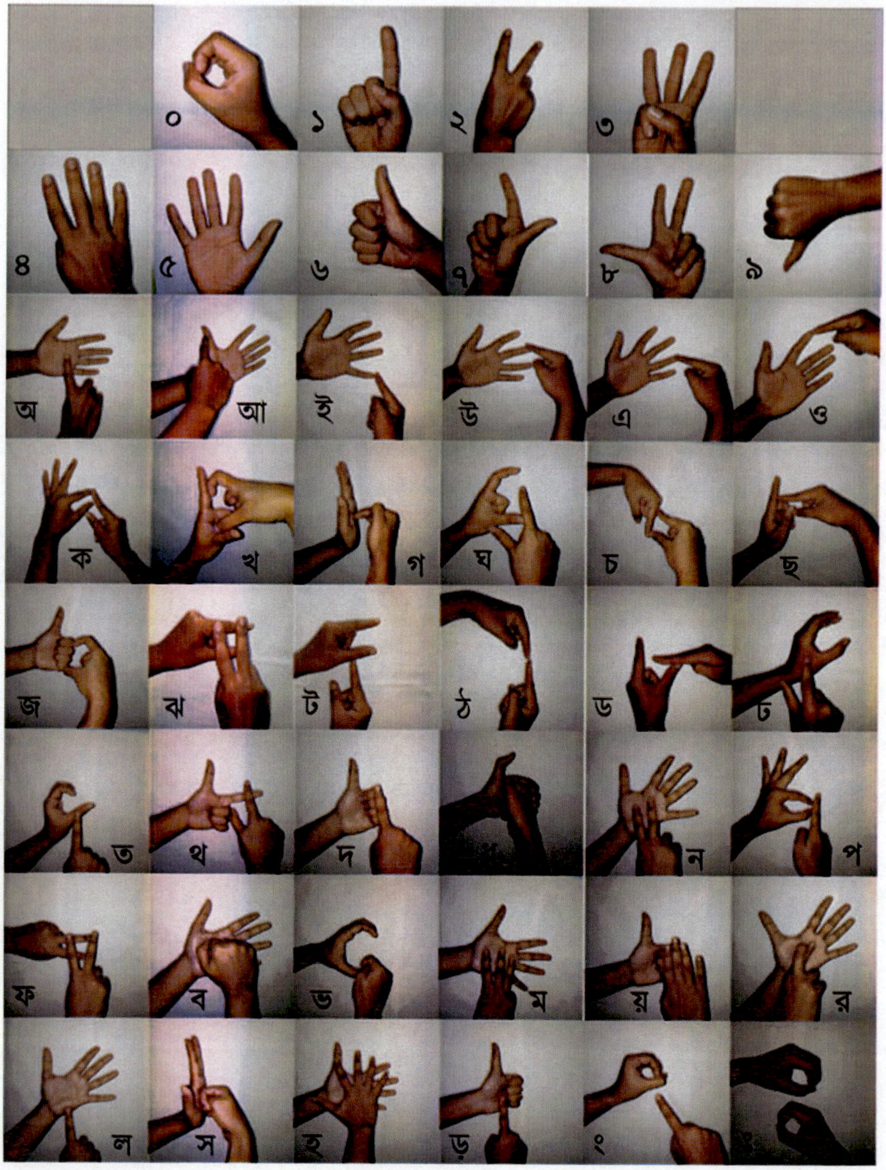

Fig. 2 Bangla digit and letter sign

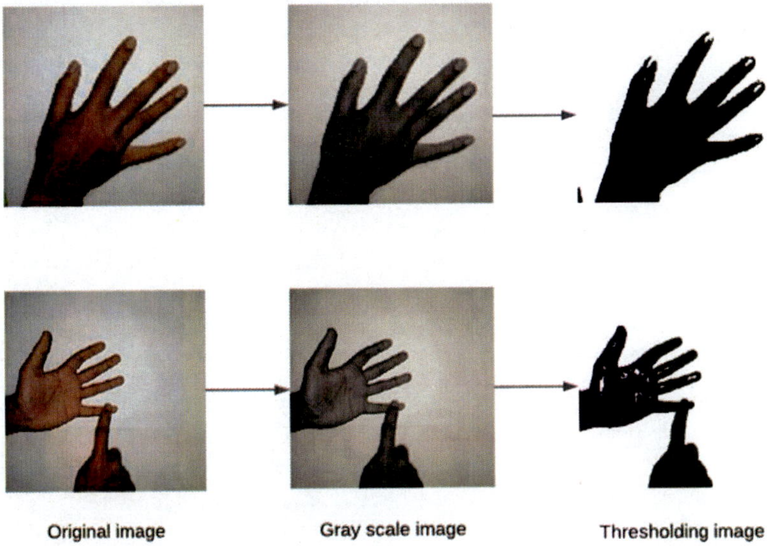

Fig. 3 Image preprocessing steps

Our original images are too large for their height and width. So, we have converted it to 128px by 128px and all the images have been labeled to their corresponding class. Finally, we have normalized our images dividing them by the highest pixel value 255.

3.3 Proposed Methodology

Convolutional Neural Network (CNN) Multi-layer convolutional neural networks with interconnected layers have been employed to recognize our digits and characters as seen in Fig. 4. In our stacked architecture, the first layer is a 'convolutional' layer with 'Rectified Linear Unit' (ReLU). The ReLU function is defined by equation is

$$ReLu(x) = Max(0, x) \tag{1}$$

In the first layer, we have used the kernel of 16 filters and each filter has a 3×3 size with a stride size of 2. We have added the activation layer to make it non-linear. It drops all the images less than the value of 0 and allows all the images greater than the value of 0. Then, with a pool size of 2 and a stride size of 2, we have created a 'max-pooling' layer. After that, a second convolutional layer of 32 filters has been added with a max-pooling layer of pool size 2. Similarly, we have added two more convolutional layers with max-pooling layers. Then we have a fully connected layer. For this, we have to flatten the output to produce a one-dimensional vector. We

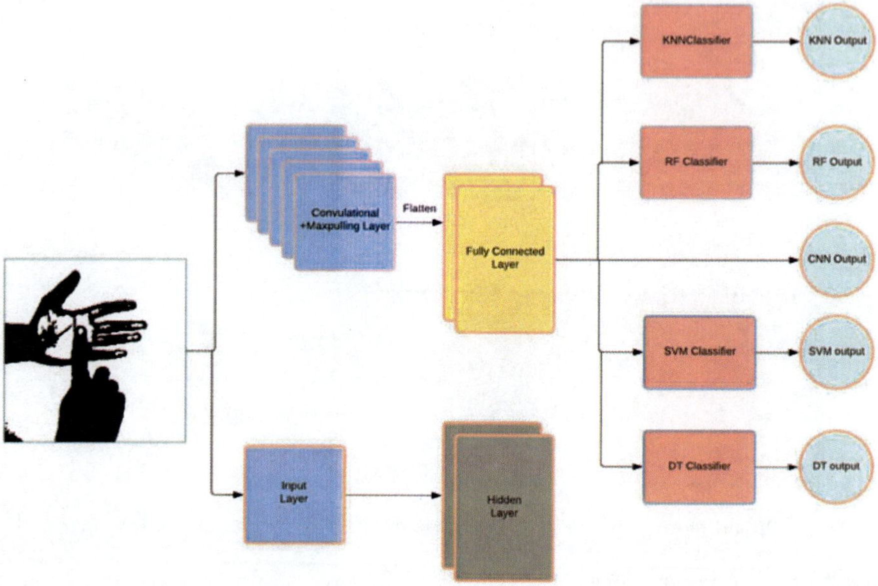

Fig. 4 Overall structure of different classifiers

have used a 25% dropout layer before flatting the output and a 50% dropout layer before the fully connected layer to reduce over-fitting. With the activation function ReLU, we have used two fully connected layers. Because of having multiple classes, a 'softmax' (2) function has been used in the last layer for final output. We have compiled the model using the 'Adam' optimizer and the 'categorical_crossentropy' loss function. We fit the model with 50 epochs and a default batch size 32.

$$S(y_i) = \frac{e_i^y}{\sum je_i^y} \tag{2}$$

Support Vector Machine (SVM) In our CNN model, we have added our fully connected layer with 256 and 128 units. We have added SVM in the out layer. We have applied a linear activation function with 'l2' regularization for it. The model was built upon using the 'Adam' optimizer and the 'squared hinge' loss function.

Decision Tree This algorithm is a graphical representation in the shape of a tree of all possible solutions to a decision. It divides a dataset into smaller and smaller subsets while also building an associated decision tree. Then, when predicting the output value of a set of features, it will predict the output based on which subset the collection of features belongs to. We have used the convolutional layer as an input of this model. We have also implemented our decision tree model with the help of Python scikit-learn toolkit.

K- Nearest Neighbor (KNN) KNN is a slow and non-parametric learning method. It is a simple method that saves all available cases and categorizes incoming data using a similarity measure. According to KNN, the data points close to each other belong to the same class. It starts by selecting a number k to be the nearest neighbor to the data point to be processed. In our proposed model convolution layers are used as input to the KNN classifiers for the prediction. We have used the Python scikit-learn toolkit for the implementation of KNN. The distance can be calculated by Eq. 3.

$$d = \sqrt{\sum_{i-1}^{k}(x_i - y_i)^2} \quad (3)$$

Random Forest (RF) Classic CNNs can face difficulties because of over-fitting issues. Random forest model have been used because Over-fitting is less likely when many trees are used. It runs efficiently on a large dataset. It generates highly accurate predictions for large datasets. We have used the convolutional layer as an input of this random forest model. Thus, our model acts in a generalized way by reducing the possibility of over-fitting. Fifty trees have been used in our random forest model.

4 Experimental Results

We have built our dataset with 5820 images (1500 for digits and 4320 for letters) that have a wide range of variations, making it a unique dataset. We have employed five different classifiers to classify our model and test its performance in our dataset. We have evaluated our model after placing 80% of total images (3456 images for letter and 1200 images for digit) into the training set and 20% of total images (864 images for letter and 300 images for digit) into the validation set. We have got accuracy around 95% in our CNN model on the digit dataset and 91.4% on the letter dataset. Other classifiers have noted this performance as well. Table 2 displays our overall accuracy for different classifiers of both letter and digit datasets. The decrease in the training loss and enhanced accuracy clearly showed that models have been trained perfectly on the dataset without over-fitting and under-fitting after analyzing few epochs. We have calculated the accuracy by dividing the total number of estimates by the number of right estimates. We have also calculated precision (P), recall (R), area under curve (AUC) and F1-measure as our classification report.

According to Table 2, we have achieved over 91.5% accuracy for CNN, the most among all classifiers for the letter dataset and 95.0% accuracy for CNN and KNN for our digit datasets, which is also the highest among all classifiers. We used a convolutional layer to extract features and a dropout layer to reduce over-fitting to improve the performance of our model. Using a variety of classifiers has a significant impact on improving accuracy. All of our classifiers have a standard accuracy, indicating the validity of our dataset. We have also applied various preprocessing techniques that

Table 2 Accuracy, F1 score and AUC of different models over digit and letter dataset

Model	Digit			Letter		
	Accuracy	F1 Score	AUC	Accuracy	F1 score	AUC
CNN	95	95	96.9	91.5	92	95.2
KNN	95	95	97.1	91	91	95.6
RF	93.8	94	96.4	91.2	91	95.5
SVM	87.8	88	89	85.2	85	92.3
DT	93.5	94	96.2	91.3	91	95.7

Table 3 Comparison between our dataset and Ishara lipi dataset

Model	Ishara lipi dataset		Our own dataset	
	Digit	Letter	Digit	Letter
CNN	92	70.5	95	91.5
KNN	91.1	70.2	95	91
RF	91.6	69.2	93.8	91.2
SVM	93.5	65.7	87.8	85.2
DT	91.1	69.7	93.5	91.3

have a positive impact on the performance of our model. Our digit dataset accuracy is always greater than our letter dataset, as shown in the table above. Aside from that, there are a few more reasons. Our digit dataset contains fewer classes than our letter dataset, and there are many differences between each digit. Our letter dataset, on the other hand, has more classes, and there are many similarities among some of them, causing interclass confusion. Furthermore, our letter pictures are double-handed, there is an issue with hand overlapping throughout the preprocessing processes. We also compare our dataset to that of other datasets. Table 3 compares the performance of our dataset and the Ishara-Lipi dataset using our suggested algorithm.

After examining the overall outcomes, we have decided that the CNN model might be used as the best model for sign language detection. The graphical representation of the accuracy of different models is shown in Fig. 5.

5 Conclusion and Future Works

Sign language detection is considered one of the most difficult problems in machine learning in the recent decade. We have proposed a sign language recognition system model based on different deep learning techniques that produce text output from images. It helps to eliminate the communication barriers between speech and hearing disabled people and normal people. We had difficulty managing a comprehensive dataset at the start of our research because wide ranges of complete BdSL datasets

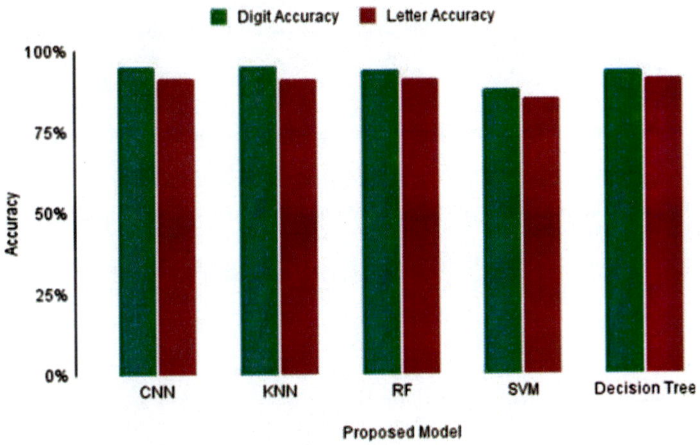

Fig. 5 Performance summary of all the classifiers

are rarely available. As a result, we have decided to construct our own dataset of entire Bangla Sign Language letters and digits. In our work, our main work was to create a wide range of novel dataset and applying different classifiers to evaluate the performance of our model on our dataset. We have used different classifiers to classify the actual class of different sign digits and letters. The development of this dataset and public sharing of it might minimize the restriction of future studies on it. We wish to use our dataset and model as a platform for standardizing Bangla Sign Language. All of our classifiers have a reasonable level of accuracy, indicating that our dataset is valid and suitable to use. But our model has some limitations. We have captured our dataset with white background. So, our model cannot work with different backgrounds. In the future, we want to capture our images with different backgrounds. Our model also cannot work in real-time. So, in the future, we want to build such a model and dataset that can also work in real time. We do not use any cross-validation techniques because our dataset is balanced, and we have an identical number of photos in each class. However, in the future, we had like to use cross-validation to examine the variation in accuracy across all classifiers. We also want to add some features so that we can convert Bangla signs into Bangla words and sentences.

References

1. According to WHO: Over 5% of the world's population—or 430 million people—require rehabilitation to address their 'disabling' hearing loss (432 million adults and 34 million children), https://www.who.int/news-room/fact-sheets/detail/deafness-and-hearing-loss. Last accessed: 14 June 2021

2. Karmokar BC, Alam KR, Kibria Siddiquee M (2012) Bangladeshi sign language recognition employing neural network ensemble. Int J Comput Appl 58: 43–46
3. Islalm MS, Moklesur Rahman M, Rahman M, Arifuzzaman M, Sassi R, Aktaruzzaman M (2019) Recognition Bangla sign language using convolutional neural network. In: 2019 international conference on innovation and intelligence for Informatics, Computing, and Technologies (3ICT), 1–6
4. Shanta SS, Anwar St, Rayhanul Kabir M (2018) Bangla sign language detection using SIFT and CNN. In: 2018 9th International Conference on Computing, Communication and Networking Technologies (ICCCNT), 1–6
5. Islam S, Mousumi SSS, Shahariar Azad Rabby AKM, Hossain SA, Abujar S (2018) A potent model to recognize Bangla sign language digits using convolutional neural network. Proc Comput Sci 143: 611–618
6. Hoque OB, Jubair M, Saiful Islam M, Akash A, Paulson AS (2018) Real time Bangladeshi sign language detection using faster R-CNN. In:2018 International Conference on Innovation in Engineering and Technology (ICIET), 1–6
7. Urmee PP, Abdullah Al Mashud M, Akter J, Jameel ASMM, Islam S (2019) Real-time Bangla sign language detection using Xception model with augmented dataset. In: 2019 IEEE International WIE Conference on Electrical and Computer Engineering (WIECON-ECE), 1–5
8. Ahmed S, Islam M, Hassan J, Ahmed M, Ferdosi BJ, Saha S, Shopon M (2019) Hand sign to Bangla speech: a deep learning in vision based system for recognizing hand sign digits and generating Bangla speech. ArXiv abs/1901.05613. n. pag
9. Hossen MA, Govindaiah A, Sultana S, Bhuiyan A (2018) Bengali sign language recognition using deep convolutional neural network. In: 2018 Joint 7th International Conference on Informatics, Electronics & Vision (ICIEV) and 2018 2nd International Conference on Imaging, Vision & Pattern Recognition (icIVPR), pp 369–373
10. Hossain, T, Shishir FS, Shah FS (2019) A novel approach to classify Bangla sign digits using capsule network. In: 2019 22nd International Conference on Computer and Information Technology (ICCIT), 1–6
11. Sadik F, Subah MR, Dastider AG, Moon SA, Ahbab SS, Fattah S (2019) Bangla sign language recognition with skin segmentation and binary masking. In: 2019 IEEE International WIE Conference on Electrical and Computer Engineering (WIECON-ECE), 1–5
12. Hossain, S, Sarma D, Mittra T, Alam MZ, Saha I, Johora FT (2020) Bengali hand sign gestures recognition using convolutional neural network. In: 2020 second International Conference on Inventive Research in Computing Applications (ICIRCA), pp 636–641
13. Das P, Ahmed T, Ali MF (2020) Static hand gesture recognition for American sign language using deep convolutional neural network. In: 2020 IEEE region 10 symposium (TENSYMP), pp 1762–1765
14. Mekala P, Gao Y, Fan J, Davari A (2011) Real-time sign language recognition based on neural network architecture. In: 2011 IEEE 43rd southeastern symposium on system theory, pp 195–199
15. Haque P, Das B, Kaspy NN (2019) Two-handed Bangla sign language recognition using principal component analysis (PCA) and KNN algorithm. In: 2019 international conference on Electrical, Computer and Communication Engineering (ECCE), 1–4
16. Rumi RI, Hossain SM, Shahriar A, Islam E (2019) Bengali hand sign language recognition using convolutional neural networks
17. Roy P, Uddin S, Arifur Rahman M, Musfiqur Rahman M, Shahin Alam M, Mahin MSR (2019) Bangla sign language conversation interpreter using image processing. In: 2019 1st International Conference on Advances in Science, Engineering and Robotics Technology (ICASERT), 1–5

Avvanna: A Website for Women's Safety

Riya Sil, Avijoy Das Bhowmick, Bharat Bhushan, and Ayan Sikdar

Abstract Indian women—the better half of our Indian Society—are becoming the most vulnerable segment as per the safety, and security of women is concerned. Everyday there are many cases related to violence against women including rape cases, trafficking, sexual assaults, domestic violence, and many more. Thus, woman safety has become a major concern in India. But not many of the women know about their rights or whom to approach at the time of scarcity and problem. According to a report of the National Crime Records Bureau, the reported incidents related to women have increased by 44% as compared to last year. In this paper, the authors have proposed a website named 'Avyanna' that provides online legal support for those women who are in need. It is a one-stop solution for women where they can consult advocates, get details of cases and Indian Penal Code (IPC) sections related to women, and also get safety instructions for women to stay alert and accordingly take precautions.

Keywords Website · Legal consultation · Law · Women's safety · Offline service

1 Introduction

Over the time, in India, violence against women has augmented to many folds [1]. Even though previously there were too much of restriction for women to stay within the walls of the house, the advancement of society and globalization has provided them with ample of opportunities to stand for themselves. Though the patriarchal mind-set of the society has changed to some extents, still there are people with the same mind-set that does not allow women to go out and work making them a tool for domestication [2]. There are various ways like domestic violence against women,

R. Sil (✉) · A. Das Bhowmick · A. Sikdar
Department of Computer Science and Engineering, Adamas University, Kolkata, India
e-mail: riyasil1802@gmail.com

B. Bhushan
Department of Computer Science and Engineering, Sharda University, Greater Noida, India

V. Skala et al. (eds.), *Machine Intelligence and Data Science Applications*, Lecture Notes on Data Engineering and Communications Technologies 132,
https://doi.org/10.1007/978-981-19-2347-0_14

eve-teasing, rape, sexual harassment, and others that male dominated society use to prove their domination over females [3]. These are some of the reasons for which violence in India is increasing at an alarming rate and women's safety has become a major concern in India [4].

In this paper, authors have designed a website named Avyanna for guiding women in their daily life by providing awareness about their legal rights and the possible way they can proceed to get their deserved right and fair justice. The website consists of different sections for all the articles and clauses of the constitution related to women's right, information about the advocates, different sections for the news related to women, a separate blog section, and safety tip-related section to deal with situation threats.

Section 1 states the origin of the work. Section 2 focuses on the basics of online women's safety system. Section 3 focuses on related research on women's safety system. Section 4 thoroughly explains our proposed model, and the performance analysis is stated. Section 5 concludes our paper with a direction to future scope of this research work.

2 Online Women's Safety System

In India, women's safety is a major issue. For resolving this issue, non-government organizations (NGOs) and Government of India both are equally trying hard and giving it the most priority. Nowadays, many women's safety applications and websites are there to help women in critical conditions [5]. As India being a digitized nation, use of Internet is almost available at all locations of the country. By using this Internet facility, many safety mobile applications are trying to provide help by reaching out to the victims specially women as much as possible. This will improve women's safety issues in India. Ministry of Home Affairs has taken a government initiative for women's safety named 'Women Safety Division' that is responsible for coordinating, formulating policy, planning to assist states/union territories, anti-human trafficking, and similar cases [6–8]. The process includes increased usage of technology in criminal justice system, thus providing supportive eco-system for crime records and forensic sciences. Another similar 70-year-old organization named CARE is a non-profit organization based on social injustice and alleviating poverty [9]. It focuses on comprehensive projects in livelihoods, education, health, and disaster preparedness and response. The overall target of the company is to improve the livelihood of women.

3 Related Work

In this section, authors have investigated some of the existing websites that provide similar service to the society. This survey would help people, specially women to get

Table 1 Comparison of various women's safety websites

S. No.	Website name	Description
1	MHA/Women Safety Division [6]	Ministry of Home Affairs has taken initiative for women's safety by designing 'Women Safety Division'. It is responsible for coordinating, formulating policy, planning to assist states/union territories, anti-human trafficking, and similar cases. The process includes increased usage of technology in criminal justice system, thus providing supportive eco-system for crime records and forensic sciences
2	SheHeroes [7]	SheHeroes is developed to empower young women to dream big, explore their interests, and passionately pursue non-traditional careers. It is done through online content and video profiles of successful women role models across all fields
3	Girl power talk [8]	Girl power enables women, men, and non-binary to feel empowered and achieve heights based on their merit opportunities. It focuses to develop leadership qualities in girls. It concentrates on education, integrates the strengths of especially abled communities, and promotes gender equality
4	CARE [9]	CARE is a 70-year-old non-profit organization. It is based on social injustice and alleviating poverty. It focuses on comprehensive projects in livelihoods, education, health, and disaster preparedness and response. The overall target of the company is to improve the livelihood of women
5	CORO [10]	CORO aims to create a society with the most marginalized community by creating equality and social justice. From these communities, it facilitates leadership to steer rights for social change

a log file of the related websites that are beneficial for the safety of women (Table 1).

4 Proposed Model

In this section, the authors have discussed about the proposed website named Avyanna for the safety of women to provide them with online legal advices. It is a one-stop solution for women's safety and empowerment which provides knowledge of laws

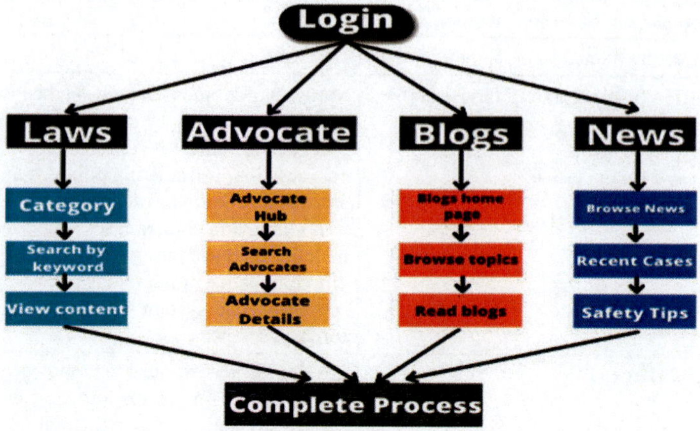

Fig. 1 Proposed model of Avyanna

regarding crimes against women, advocate details for further legal approaches, and safety tips for women to stay alert and take precautions [11].

4.1 *Operating Model of Avyanna*

Avyanna provides three key features for the women's society: (i) online legal support, (ii) advocate details, and (iii) safety tips. Also, the website has recent cases and news sections [12]. The website consists of women-related Indian Penal Code (IPC) sections that will help in understanding what legal steps should be taken against any crime. There is an advocate hub where users can also browse the advocates' profiles and can contact with them. As crime rates against women is increasing day by day, user can also visit the safety tips section [13]. As the proposed website does not have any login/sign-up procedure therefore, users can browse easily in this website for legal support. Lastly, there is a section for recent cases and news that will keep users updated about the society. Figure 1 shows the proposed model of Avyanna. Figure 2 shows the case model of the website.

4.2 *Technology Used*

The proposed website is dynamic and easy to use [14]. The below mentioned technologies have been used for creating the website that includes front-end development for which the following has been used: (i) HTML 5, (ii) CSS 3, and (iii) JavaScript, and for back-end the following has been used: (i) SQL, (ii) MySQL, and (iii) PHP, along with XAMPP Apache HTTP Server for Hosting as a Local Host and Git and

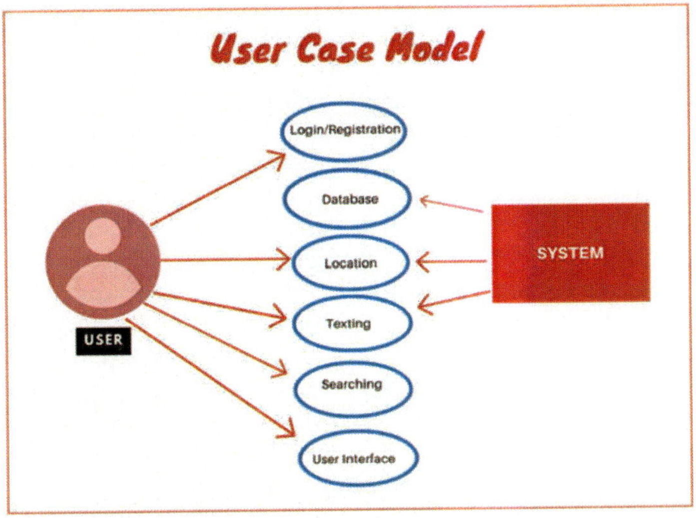

Fig. 2 User case model of Avyanna

GitHub for Version Control [15]. A search engine with a proper SEO for searching laws and crimes using keywords has also been used in this model [16–22].

4.3 Functioning of the Model

For proper functioning of the proposed model, the following are required:

(i) Project manager: Controls all the departments' design plan and strategy for the future and also the final decision maker in any kind of management-level work of each department [23].
(ii) Database administrator: Manages and monitors the database performance, regular backup, and recovery processes, maintains data integrity, and helps to support client's and admin's database [24].
(iii) Web developer: Develops the website, brings innovative and futuristic ideas on the table, and deploys updates to the websites accordingly on the basis of business requirement.
(iv) Security administrator: For ensuring security access, provide protection against unauthorized access, troubleshoots network security, and delivers staff training to prevent unexpected loss.

Business head: For all the work related to marketing and sales department, needs to take approval from business head to execute the plan.

4.4 Results

In this section, the authors have discussed the different segments of the website that have been designed. It includes the details of the legal advisors along with contact information inside the advocate hub section. The rating of the advocate has also been added based on the review of the user that will be given by genuine user that has been registered to the system. So, once the user enters the website after registering and logging in, they can see four sections and two additional sections below it on the home page. The sections include Laws, Advocates, News, and 'About us' and the rest two are recent cases and safety tips. In the Laws section, there are different categories related to 'Crime Against Women' that the user can go through, and inside it are the sections that are present in the constitution under that crime. There is also a search bar where one can search through keywords for any specific case.

There also exists an Advocate section where one will be able to find the details of the advocates and also search for nearby advocates. The details of the advocate are provided that includes advocate name, area of work, specialization, contact number, email id, address, etc. Also, the users can give ratings to the respective advocate which will help others to find a good advocate. After the Advocate section, there comes the News section which will display the 'Crimes Against Women' caused in our country. Among the two sections, one which is the recent cases is also a sub-part of the News section where the recent cases will be shown so that the users can be updated regarding the most recent crimes that have occurred against women. Finally, in the safety tips section, different situations or scenarios are provided along with their solutions for women's safety. To these features, there is another section which is the Blogs section that can be found on the navigation bar. Here, the users can read different blogs about women's safety, crimes, etc. (Fig. 3).

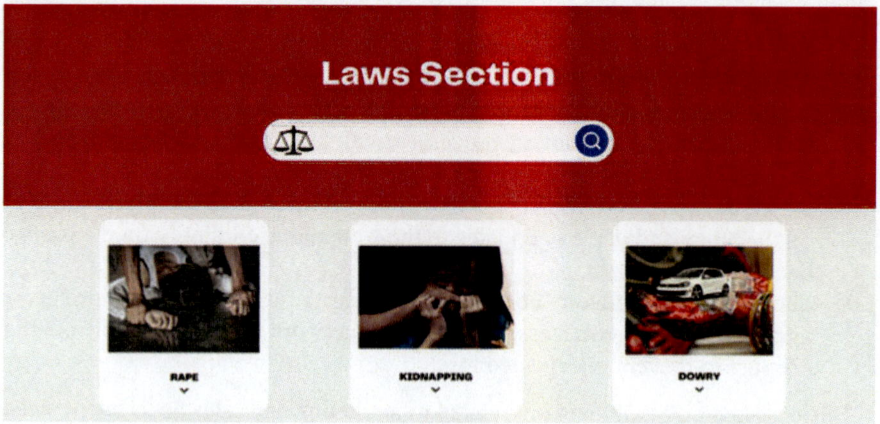

Fig. 3 Laws section

5 Conclusion

In order to improve women's safety in India, the change in mind-set of people is very essential for the safety of women. From family to educational institutions, men should be taught about respecting females. Further, their legal cases should be solved without much delay to provide justice to women. Only strict laws cannot solve the problem of women's safety in India rather the implementation of these laws in a time-bound manner can solve the issue to a large extent. Thus, to provide help to the society, the authors have designed the website that provides online legal support for those women who are in need. It is a one-stop solution where women can consult advocates and get details of cases and Indian Penal Code (IPC) sections related to women and also provides safety instructions for women to stay alert and accordingly take precautions.

Further, authors would add sections where users can provide their valuable inputs and suggestions. The website will deliver various new features along with useful contents and advocate details along with advocate rating.

References

1. Roesch E, Amin A, Gupta J, García-Moreno C (2020) Violence against women during covid-19 pandemic restrictions. BMJ m1712. https://doi.org/10.1136/bmj.m1712
2. García-Moreno C, Riecher-Rössler A (2013) Violence against women and mental health. In: Key issues in mental health, pp 167–174.https://doi.org/10.1159/000345276
3. Hlavka HR (2019) Regulating bodies: children and sexual violence. Violence Against Women 25(16):1956–1979. https://doi.org/10.1177/1077801219875817
4. Meola C (2020) Helping Aussie women online. In: Social media in legal practice, pp 130–145.https://doi.org/10.4324/9780429346088-9
5. Glass NE, Perrin NA, Hanson GC, Bloom TL, Messing JT, Clough AS, Campbell JC, Gielen AC, Case J, Eden KB (2017) The longitudinal impact of an internet safety decision aid for abused women. Am J Prev Med 52(5):606–615. https://doi.org/10.1016/j.amepre.2016.12.014
6. Women Safety Division: Ministry of Home Affairs. Ministry of Home Affairs|GoI. (n.d.). https://www.mha.gov.in/division_of_mha/women-safety-division
7. Team Archive. SheHeroes. (n.d.). https://www.sheheroes.org/team/
8. Women Empowerment in India. Girl Power Talk (2021, June 11). https://girlpowertalk.com/#aboutus
9. Samta M, Goswami MCP, Ne S, Singh VK, Rayar S, Parikh S (n.d.). Top Indian NGO: charity foundations in India for women and child education health. CARE India. https://www.careindia.org/
10. Network CRIS (n.d.) CORO India. http://coroindia.org/
11. Jewkes R, Dartnall E (2019) More research is needed on digital technologies in violence against women. Lancet Public Health 4(6). https://doi.org/10.1016/s2468-2667(19)30076-3
12. (2014) Continuing the war against domestic violence. In: Domestic violence in Indian country, pp 32–45. https://doi.org/10.1201/b17162-6
13. Wood SN, Glass N, Decker MR (2019) An integrative review of safety strategies for women experiencing intimate partner violence in low- and middle-income countries. Trauma Violence Abuse 22(1):68–82. https://doi.org/10.1177/1524838018823270
14. Langer AM (n.d.) Website design and architecture. In: Analysis and design of information systems, pp 349–369. https://doi.org/10.1007/978-1-84628-655-1_16

15. Vemula R (2017) Code version control using GitHub platform. In: Real-time web application development, pp 425–487.https://doi.org/10.1007/978-1-4842-3270-5_12
16. Cook C, Garber J (2012) HTML and CSS basics. In: Foundation HTML5 with CSS3, pp 17–36. https://doi.org/10.1007/978-1-4302-3877-5_2
17. Cal J (2009) Website design and development offshore outsourcing India—Infodreamz Technologies. SciVee. https://doi.org/10.4016/1153701
18. Deng Z, Hong Z, Zhang W, Evans R, Chen Y (2019) The effect of online effort and reputation of physicians on patients' choice: 3-wave data analysis of China's good doctor website. J Med Internet Res 21(3). https://doi.org/10.2196/10170
19. Hung C-L (2017) Online positioning through website service quality: a case of star-rated hotels in Taiwan. J Hosp Tour Manag 31:181–188. https://doi.org/10.1016/j.jhtm.2016.12.004
20. Sil R, Saha D, Roy A (2021) A study on argument-based analysis of legal model. In: Advances in intelligent systems and computing, pp 449–457.https://doi.org/10.1007/978-3-030-73603-3_42
21. Sil R, Roy A (2020) A novel approach on argument based legal prediction model using machine learning. In: International conference on smart electronics and communication (ICOSEC). https://doi.org/10.1109/icosec49089.2020.9215310
22. Saha D, Sil R, Roy A (2020) A study on implementation of text analytics over legal domain. In: Evolution in computational intelligence, pp 561–571.https://doi.org/10.1007/978-981-15-5788-0_54
23. Sil R, Alpana, Roy A (2021) A review on applications of artificial intelligence over Indian legal system. IETE J Res. https://doi.org/10.1080/03772063.2021.1987343
24. Sil R, Alpana, Roy A, Dasmahapatra M, Dhali D (2021) An intelligent approach for automated argument based legal text recognition and summarization using machine learning. J Intell Fuzzy Syst 1–10.https://doi.org/10.3233/jifs-189867

Web Page Classification Based on Novel Black Widow Meta-Heuristic Optimization with Deep Learning Technique

V. Gokula Krishnan, J. Deepa, Pinagadi Venkateswara Rao, and V. Divya

Abstract As a search engine, web page classification gives useful information which may be used to build many various applications. For effective Internet use, spam filtering, and a host of other applications, categorizing web pages is helpful. Search engines must tackle the difficulty of swiftly finding relevant results from millions of online pages. During an automatic classification process, a significant variety of elements, such as XML/HTML and text content, should be taken into account. In addition to interfering with the routine operations of the website and crawling a large amount of worthless information, present methods for listing have many disadvantages. An important goal of this study is the reduction of features needed to upsurge the speed and web page classification accuracy by dropping the number of features that need to be employed. An NBW-MHO technique was utilized in this study to choose the best features, and a deep learning-based classification ideal was proposed to learn the best structures in each web page and aid search engine listing. The project was funded by the National Science Foundation. Using only link text, side information, and header text, our suggested algorithm gives an ideal classification for websites with a large amount of web pages. When employing NBW-MHO for feature selection, classification accuracy and runtime performance are improved. Information gain and chi-square feature selection approaches were compared to the NBW-MHO algorithm.

V. Gokula Krishnan (✉)
CSIT Department, CVR College of Engineering, Mangalpally, Hyderabad, Telangana, India
e-mail: gokul_kris143@yahoo.com

J. Deepa
CSE Department, Easwari Engineering College, Chennai, Tamil Nadu, India

P. Venkateswara Rao
CSE Department, ACE Engineering College, Ghatkesar, Hyderabad, Telangana, India

V. Divya
School of Electrical and Electronics Engineering, Sathyabama Institute of Science and Technology, Chennai, Tamil Nadu, India

V. Skala et al. (eds.), *Machine Intelligence and Data Science Applications*, Lecture Notes on Data Engineering and Communications Technologies 132,
https://doi.org/10.1007/978-981-19-2347-0_15

Keywords Feature selection · Optimal classification · Link text · Spam filtering · Web page classification

1 Introduction

As a result of Web's rapid expansion, it has become increasingly difficult to identify web pages that provide useful information, as well as to filter out undesired or hazardous stuff. The last decade has seen a proliferation of web pages containing objectionable and hazardous content such as violent acts, cyber threats, and porn. Conversely, because there are so many different topics covered on the Internet, information retrieval and extraction algorithms [1] have had a difficult time generating topic-relevant results. Up until a decade ago, it seemed impossible for any computer to solve complex semantic problems due to the explosive expansion of computing performance and memory space, as well as the specialization of ML models for text and picture categorization [2]. It is used for focused crawling, assisting in the building of web directories and analyzing the subject structure of the Web.

Several retrieval and administration operations require text categorization, including information retrieval, document filtering, and the creation of hierarchical directories [3]. While text documents lack URLs, links, and HTML tags, web pages do. Web classification differs from outdated text classification because of this attribute [4]. Topic-specific online links are selected, the subject structure of the Web is analyzed, directories are created, and focused crawling is performed using categorization [5]. Web directories such as Yahoo! [6] and the Open Directory Scheme [7] are used to be built by hand, and class labels were manually applied to web documents. Hand classification requires a human effort, which makes it impractical in light of the rapid growth of the Web. Consequently, automatic web page classification systems have been in high demand [8].

There are several applications for online classification algorithms, including spam detection, web search, document organization, and cybersecurity [9]. Search engine optimization for speedy and efficient results is the focus of this study, firstly, to determine the number of classes and the attributes of each class and, secondly, to determine the probability of each document falling into each class. SVM, Naïve Bayes, and k-Nearest Neighbors (kNN) are just a few of the ML scheme that can be used for web classification [10]. There are too many features in one web page for most machine learning algorithms to be accurate. With the ML model combined with a solid feature selection approach, classification accuracy is increased.

A critical step before classification is feature selection (FS). As part of the classification process, it entails identifying a subset of pertinent features. Large document sizes make this a necessity before the classification process [11]. A few of their key advantages are that they make it easier to grasp the data and that they reduce training time. Classifier complexity and processing requirements (such as memory and desk space) will also be lowered [12]. It is the purpose of a feature selection algorithm to remove the majority of unnecessary features from a web page in order to lessen the amount of data provided into the machine learning model. It is easier for a machine learning model to learn how to correlate characteristics when the input size is small.

The remaining paper is prearranged as follows: Sect. 2 discuss the limitation of existing techniques for web page classification. The explanation of proposed practice is given in Sect. 3, where the validation of proposed method against existing techniques are described in Sect. 4. Lastly, the conclusion of the research work is given in Sect. 5.

2 Literature Review

A brief literature survey of recent investigation on link-based web classification procedures is presented in this section. A clustering method based on distance calculation stages is used to decrease the average distance between pages fit in to the same class and exploit the average distance between pages going to different classes, which is the primary purpose of these algorithms. In [13], tokenization is used to extract features from a web page. The class of a web page is then assigned based on the calculated distance among the characteristics extracted from the online page. As a result, a huge number of online links are extracted by researchers using ML algorithms such as SVM [14].

The fitness value is calculated for each iteration of the classification process in a few studies, such as [15]. Fitness function accuracy increases with each iteration. Iteration ends when fitness function accuracy does not grow in the future. This nature-inspired algorithm involves a bunch of birds searching for food at diverse locations and convergently finding it. By searching for links between documents in a group of websites, the same technique is used to categorize web documents [16, 17].

The bag of words model [18] is used in the bulk of publications in this field. Few techniques like TF-IDF are utilized to extract the more frequent words found across the web pages [19] after the document links are turned into vectors [20, 21]. When more words are substitutes of each other, it becomes difficult to signify the exact relationship; a research study [22] proposes a technique which calculates the correlation among each pair of words to solve this problem. For the Lao text, [23] provides a kNN-based classification technique that utilizes the TF-IDF metric (NLP). During their investigation, they examined seven features of the input size that they employed. Principal component analysis was applied for feature selection. Their proposed work is accurate to 71.4% using the PCA feature selection method.

A multiclass classification support vector machine is optimized by Hao et al. [24]. However, when there are more than three classes to classify, the computational cost of training increases. A hierarchical classification model was proposed in their research, in which the SVM is optimized for a multiclass classification procedure. It outperformed decision tree classifiers and kNN classifiers. For multiclass classification, the SVM classifier was subjected to a number of optimizations [25, 26].

Web page classification has never been done using the ACO until AntMiner [27]. A collection of rules that categorize web pages into categories has been developed by

Holden and Freitas using AntMiner's [27] paradigm. Neither of them has any prior assumptions about which terms in the web pages to be classified could potentially serve as potential discriminants for classification. All are taken as help in order to lessen the rarity of the data. As a result of combining WordNet generalization with title characteristics, AntMiner's classification accuracy is at 81.0%.

ACO-based feature selection technique for text classification has been introduced by Aghdam et al. [5]. The performance of the classifier and the length of the feature subset are used to evaluate the features picked by an ant. They think that classifier presentation [28] is more essential than subset length, so they provide classifier performance and subset length 80 and 20% weights, correspondingly, in the tests. kNN classifier is employed in the studies to test classification performance. On the Reuters-21578 dataset, the performance of the suggested approach is compared to that of a genetic algorithm (GA), info gain, and chi-square analysis. The ACO-based feature section was found to be superior to GA, information gain, and chi-square analysis approaches in their experimental evaluations. It was determined that 89.08% of the Reuters-21578 dataset's micro-average *F*-measure value was extracted from the bag of words.

The ACO-improved fuzzy rough feature selection developed by Jensen and Shen [29] is intended to improve the classification of web pages using fuzzy rough features. According to the TF-IDF weighting scheme, terms retrieved from web pages are weighted. In accordance with this extent and the length of the selected subgroup, pheromone values are updated. Following the selection of the best collection of features, the web pages are sorted into categories. Experimental results show that ACO-based feature selection achieves the maximum degree of feature space lessening with the least amount of information loss.

Over a 20 Newsgroup dataset, Janaki Meena et al. [30] have utilized ACO for feature selection and NB for classification. It is used as a heuristic measure to assess the quality of the features extracted using the bag-of-terms approach. Map reduce is a parallelization technique that uses map reduce. 500 ants are used in 150 repetitions of the experiment. Remember and precision are 0.94 and 0.68, respectively, for the talk.politics.middleeast dataset.

It was examined by Mangai et al. [31] Information gain and TF-IDF are used to pick features. Prior to using Ward's minimum variance metric, web page clusters with redundant attributes are identified. A subset of each cluster is kept, while the others are discarded. As a result of removing such duplicate features, categorization resources are conserved. Feature selection is done after clustering, and then, Naïve Bayes, kNN, SVM, and C4 are used to pick features [32]. For classification, five classifiers are utilized with a tenfold cross-validation. It is presented as an example of a positive website and a negative one. Comparisons are made between the suggested approach of feature selection and other commonly used methods. The suggested method outperforms most other feature selection strategies in terms of decreasing the amount of features and the preparation time for the classifier, according to experiments. kNN and SVM classifiers reach accuracy scores of 95.00% and 95.65%, respectively.

2.1 Problem Definition

A number of sub-titles are assigned to the web page categorization based on the results of the study. A few examples include the following:

The primary topic classification is as follows. Sort the web page by subject. You can think of art, business, or sports, as three good examples.

A web page's functioning mechanism is used to categorize it. Example: Home Page, Login Page, and Admin Page are some examples.

An attempt to classify emotions in order to better comprehend the author's viewpoint. Classifying web pages according to primary subjects was the goal of the program developed for this study.

However, despite their similarities, the text classification and the web page classification problems are structurally different from each other.

Primary among these are semantically and structurally associated between phrases and paragraphs in the traditional text categorization tasks. As a result, the author has complete control over the design of the text. As a result, writing has a few unique features. Applications such as decisive the author of a text can be built in this fashion, for example. In contrast to traditional textual data sources, the structure of web pages is different. Imagery, audio, and video files are all possible components of a website. There may also be no semantic or structural relationship between the textual materials. On the same web page, you may find articles on much unrelated themes.

It is possible that section titles on a website do not contain a whole sentence. The web pages may be dominated by text or fully comprised graphic elements. In contrast to the classification of classical text documents, the classification of web pages may require the analysis of imperfect textual knowledge and a greater variety of data sources. A website's HTML content is made up of tag components, as well. Visually, the users are presented with these HTML contents. The source code (raw output) of a web page is different from that of a traditional text document (browser screen). Web pages can be linked to other websites or documents using hyperlinks, which are established between web pages. Hypertexts must be analyzed while classifying a web page. In addition, the web page's links to other sources deliver some info about the web page under study. This info can be used to categorize a web page's contents. Text categorization and web page classification differ in these ways.

3 Proposed Methodology

Instead of selecting features one by one, we use NBW-MHO to select a preset number of features from the huge feature space collected from the web pages. It is based on feature groups to choose features in our feature selection approach. It is not enough to use a single feature to regulate the class of a web page; thus, NBW-run MHO's duration is increased by adding features to each spider's selected feature list. A

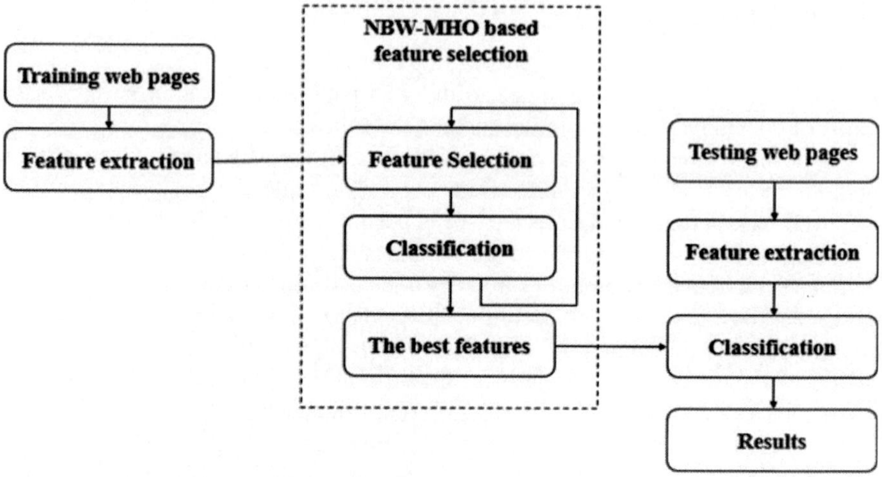

Fig. 1 Working flow of proposed method

collection of features is picked, and the relevant computations are performed simply once for each selected set. Our spiders are not blind since we feed them phrase frequency values. Before picking features, they have a notion of what terms mean. It was determined that URL features, title> tag features, tagged terms, and a bag of terms had the greatest impact on classification accuracy. The datasets we have utilized have a bigger feature space. For web page classification, we also explored which tags are more important. Figure 1 depicts the proposed method's flowchart.

3.1 Feature Extraction

For feature extraction, all tags including title, heading, and body text, as well as text comfortable and URL of web pages, are employed. We have taken all the phrases from each of the tags listed above, along with the URL of the relevant web pages included in the training set. Porter's stemming technique is employed after term extraction. Feature: Each stemmed phrase and its associated tag or URL pair. It is possible to extract different features by using the "tagged words" approach when the phrase "program" appears in the title or the "li" tag, or in the URL. This is done by grouping together terms from related HTML tags (such as "strong" and "b"), as well as "em" and I tags.

Features for each class were chosen from four diverse feature sets in this investigation. In the initial set of features, just URL addresses of web pages are used. Zweitens: For feature extraction, just the title> tags are used. All terms that appear in web pages irrespective of their HTML tag are employed as features in the third way of feature extraction. It is therefore considered to be a feature if it appears in

Table 1 Number of respect to feature extraction approaches

Class	Feature extraction method			
	Tagged terms	<title> tag	Bag of terms	URL
Course	33,519	305	16,344	479
Faculty	47,376	1502	24,641	1208
Conference	34,952	890	18,572	1115
Student	49,452	1987	22,245	1557
Project	30,856	596	15,307	686

the file regardless of its position. The "bag-of-terms" method is the name given to this approach of feature extraction. Each HTML tag described above has a feature list that includes all of the terms that appear within each of the HTML tags. To put it otherwise, a term that appears in many HTML tags is considered to be an entirely separate characteristic. Number of features vary depending on the dataset and the method used for feature extraction. Table 1 shows the number of features for each type of datasets based on the feature extraction method utilized.

Using the tagged words method on the Course class, for instance, yields 33,519 features. When only the title> tag is taken into account, the sum of features extracted for this class drops to only 305 features.

3.2 *Novel Black Widow Meta-Heuristic Optimization (NBW-MHO)*

Like any other algorithm, the proposed algorithm can develop with the initial population of spiders, so each spider is a solution. This first pair of spiders is trying to recreate a novel generation. The black widow eats the man during or after mat. He then replaces the semen kept in his sperm with egg yolk. After 11 days, the spiders lay eggs. They get pregnant numerous days a week. During this time, the brothers become cannibals. Then, they are carried through the air.

3.2.1 Initial Population

To resolve an optimization problematic, the values of the difficult must be formulated as an appropriate construction to solve the present problem. In black widow meta-heuristic optimization terminology, this structure is referred to as "chromosome" or "particle position," but here in the BWO algorithm, "widow." In the BWO algorithm, the Black Widow Spider is seen as a possible solution to any problem. Every one black widow spider represents the value of a variable problem. In this article, this structure should be seen as an array for performing test functions.

In a N_{var}–dimensional optimization problem, a widow is an array of $1 \times N_{\text{var}}$ r signifying the solution of the difficult.

To run the optimization algorithm, an entrant widow matrix of size $N_{\text{pop}} \times N_{\text{var}}$ is generated with an initial spider population. Parent pairs are then randomly selected to conduct the mating stage.

3.2.2 Procreate

As the sets of each other, they begin to mate to breed a fresh generation. At the same time, each pair mates separately from the others in their network. In the real world, about 1000 eggs are produced each time they mate, but in the end some of the stronger spider chicks survived. Well, here, in this algorithm, a matrix called Alpha must be twisted to reproduce when the widow's matrix contains random numbers. Then, the offspring are created using α with the following equation (Eq. 1), where x_1 and x_2 are parents, and y_1 and y_2 are children.

$$\begin{cases} y_1 = a \times x_1 + (1 - a) \times x_2 \\ y_2 = a \times x_2 + \left(`1 - a\right) \times x_1 \end{cases} \quad (1)$$

This process is repetitive $N_{\text{var}}/2$ times, and there is no need to duplicate randomly selected numbers. In the end, children and mothers are additional to the ranks and arranged by fitness scores. Based on the cannibal rating, some of the best people have now been added to the afresh formed population. These measures relate to all couples.

3.2.3 Cannibalism

Here, we have three types of male predators. The initial is sexual racism, in which a black widow eats her partner during or after intercourse. With these algorithms, we were able to identify women and men based on their fitness values. The second kind is related to cannibals, in which strong spiders eat their pathetic allies. In these algorithms, we set the Cannibalism Score (CR), which determines the number of survivors. In particular circumstances, a third species of ogre is often found, with small spiders eating their mother. We use fitness values to identify weak or strong spiders.

3.2.4 Mutation

At this point, we arbitrarily select Mute Pop individuals from the population. As shown in Fig. 2, each of the selected solutions arbitrarily changes two basics in the array. Intended the mute pop based on the mutation rate.

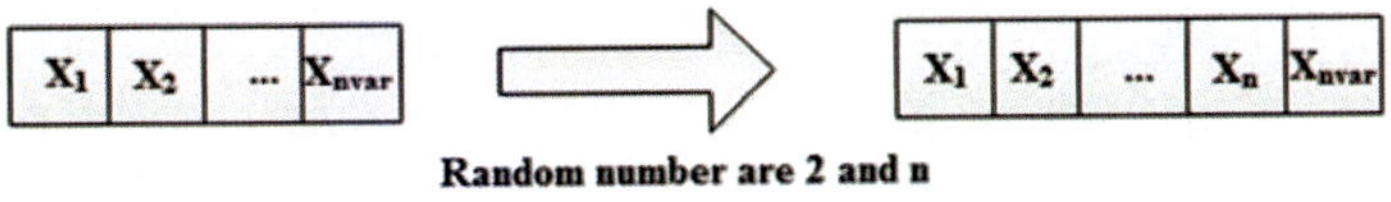

Fig. 2 Mutation

3.2.5 Convergence

Similar to other algorithms, three closing state of affairs can be deliberated: (a) a predetermined sum of repetitions, (b) the best widow compliance rating for most delegates does not change, and (c) specified accuracy. The pseudocode shown in the figure summarizes the main phases of BWO. The next section discusses some of the issues with optimizing the BWO test using. Since the best solutions for testing activities are already known, the availability of a certain level of accuracy is taken into account to determine the level of accuracy of the experimental algorithm. Also, Sect. 4 sets the maximum repetitions in the experiments as stop conditions.

3.2.6 Parameter Setting

The projected scheme has some parameters that are required to get the best results. These factors comprise purchase rate (PP), cannibalism rate (CR), and conversion rate (PM). The parameters should be adjusted accordingly so that the algorithm can find the best solutions. Better handling of a number of parameters will yield higher navigation capability on any local platform and higher ability to locate globally. Therefore, the exact number of parameters can allow to control the balance between the absorption phase and the inspection phase. The proposed algorithm has three important control factors, including PP, CR, and PM. PP is the ratio of ownership that regulates how many people are involved in the product. By regulating the production of different offspring, this parameter delivers more variety and more possibilities for a clear definition of research location. The control parameter of the CR cannibal operator is to exclude unqualified people from the population. The Prime Minister is the percentage of people who have changed. The fair value of these parameters can strike a balance between use and search. This factor can control the migration of agents in the local phase from global and lead them to better resolution.

3.3 Classification

The Forward Propagation and Back Propagation phases of the RNN model's training are clear. For example, in neural network training, Forward Propagation is responsible for computing output values. Figure 3 shows the block diagram of the RNN model.

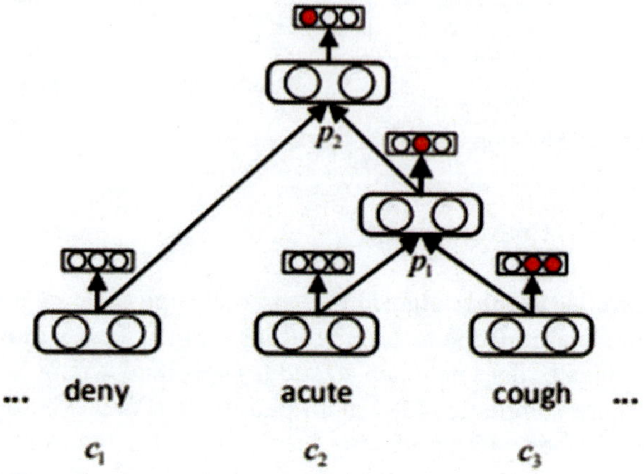

Fig. 3 RNN semantic tree models

For inductive implication tasks on complicated symbolic structures of any size, the conventional RNN is used as a traditional neural network framework. In order to compute each word's vector representation, the RNN parses a phrase into a binary semantic tree. It computes parent vectors in a bottom-up manner throughout the forward propagation training period. The following is the composition equation:

$$p_1 = f\left(W\begin{bmatrix} c_2 \\ c_3 \end{bmatrix} + b\right), p_2 = f(W\left(\begin{bmatrix} c_1 \\ p_1 \end{bmatrix} + b\right) \tag{2}$$

where f is the activation function; $W \in \mathbb{R}^{d\times 2d}$ represent as the weight matrix, and also b represent as the bias:

$$y^p = \text{softmax}(W_s.p) \tag{3}$$

where $W_s \in \mathbb{R}^{3\times d}$ is the classification matrix. The node's vector and classification result will gradually converge in this recursive procedure. The RNN can finally map the semantic representation of the complete tree into the root vector after receiving the vector of the leaf node.

4 Results and Discussion

In this section, validation of planned methodology is presented along with experimental setup, performance metrics, and description of dataset. The explanation of each section is given as follows:

4.1 Experimental Setup

During the feature extraction phase, Perl scripting was employed, and our proposed feature selection technique was applied in Java under Eclipse. Microsoft Windows 7 was used to test the procedure. As part of the research, 16 GB of RAM and an Intel Xenon E5-2643 CPU with a 3.30 GHz clock speed were used. The WebKB datasets are used to evaluate our feature selection methodology. Because after 250 iterations, there is no development in classification presentation, the algorithm is run for 250 iterations.

4.2 Dataset Description

This proposed research study makes use of the WebKB Dataset. Student, teacher, staff, department, course, project, and others are the seven categories into which these pages were manually categorized in 1997 after being collected from computer science departments at various institutions [33]. Web pages from Cornell, Texas, and Wisconsin universities as well as other university pages are included in each class's collection. It is necessary to manually classify the 8282 pages of web content into the seven categories: course (930 pages), project (504), and other (3764) pages in the class other are not considered to be "primary pages" and do not represent an instance of any of the previous six classes. 867 Cornell University web pages are included in the WebKB collection, as are 827 pages from Texas University, 1205 pages from Washington University, and 1263 pages from Wisconsin University, as well as 4120 pages from various other institutions.

Faculty, Student, and Course classes are used in this training from the WebKB dataset project. They are not considered since they have a limited number of positive examples. The WebKB project website describes how training and test datasets are produced [34]. There are relevant pages in the training set for each class, which come from three randomly chosen institutions and other classes in the dataset. During the testing phase, the fourth university's pages are consulted. In the training set, about 75% of the irrelevant pages from other classes are included. The remainder, about 25%, is included in the test set. WebKB dataset, which was employed in this work, has a number of web pages in both the train and test portions, as shown in Table 2. When it comes to the training phase, the Course class has 846 relevant and 2822 irrelevant pages, while the test phase includes 86 relevant and 942 unnecessary pages.

4.3 Performance Metrics

Evaluating the accuracy of machine learning classifiers is a key stage in data mining and information retrieval. It is usual to use error rate and F-measure to determine the

Table 2 Train/test distribution of WebKB dataset for binary class classification

Class	Test	Train
	Relevant/non-relevant	Relevant/non-relevant
Project	26/942	840/2822
Course	86/942	846/2822
Student	43/942	1485/2822
Faculty	42/942	1084/2822

accuracy of a classifier's ability to correctly classify unseen examples. As the name suggests, error rate is the percentage of test cases that were wrongly categorized in the test set. A classification model C erroneously or mistakenly classifies n occurrences in test set X, where m is the number of improperly or incorrectly classified instances. The following formula can be used to determine the correctness of C in selecting the correct classes of X instances:

$$\text{Accuracy}(C) = \frac{m}{n} \tag{4}$$

It ignores the cost of a wrong prediction in ML. F-measure is the most common solution to this problem. F-measure is considered using precision and recall. Assume, for example, that a group of texts in the test set belongs to a specific class or category. For each test text content, the ML classifier assigns a category label. These forecasts will fall into one of four groups with respect to test set S:

This is the set of text documents that were actually predicted to be in category S and, as a result, were classified as true positives (TP).

It is the set of text documents that do not fall into the category S, but were predicted to fall into a different category.

False positives (FP) are text forms that were set to be in a given category S, but are actually in a different category.

They essentially do belong to the category S while they were incorrectly projected not to. False negatives are a common occurrence (FN). When it comes to precision, it is the percentage of properly forecast texts in category S, whereas recall is the percentage of correctly forecast genuine text documents in category.

$$\text{Precision} = \frac{|\text{TP}|}{|\text{TP}| + |\text{FP}|} \tag{5}$$

$$\text{Recall} = \frac{|\text{TP}|}{|\text{TP}| + |\text{FN}|} \tag{6}$$

$$F\text{-measure} = 2 \cdot \frac{\text{Precision.Recall}}{\text{Precision} + \text{Recall}} \tag{7}$$

4.4 Performance Analysis in Terms of Proposed Feature Selection Method

In this section, the performance of the proposed NBW-MHO technique is compared with existing feature selection techniques, namely Particle Swarm Optimization (PSO), Whale Optimization Algorithm (WOA), Artificial Bee Colony (ABC), and traditional FOA schemes. Table 3 presents the performance of proposed feature selection technique.

From Table 3 and Fig. 4, it is clearly shows that the projected NBW-MHO attained better recital than existing PSO, WOA, ABC, and traditional FOA. The proposed NBW-MHO achieved nearly 95–98% on all metrics includes accuracy, precision, recall, and *F*-measure.

Table 3 Comparative analysis of proposed feature selection technique

Feature selection methodology	Parameter evaluation			
	Accuracy (%)	Precision (%)	Recall (%)	*F*-measure (%)
PSO	87.89	79.12	80.92	85.27
WOA	72.30	72.50	73.69	73.07
ABC	81.25	65.07	88.06	69.28
FOA	77.26	92.04	93.17	94.08
Proposed NBW-MHO	**95.20**	**97.64**	**98.20**	**98.67**

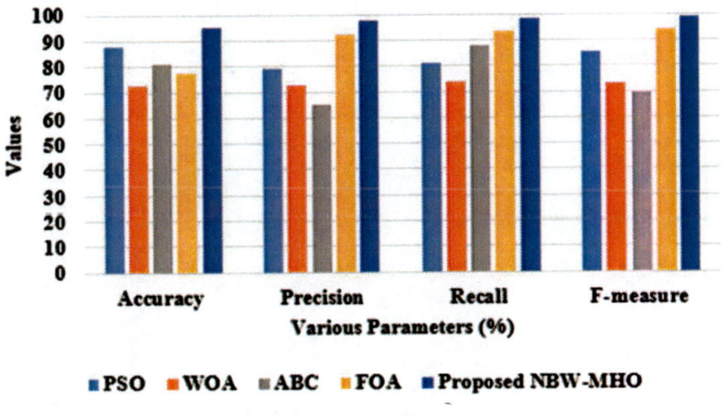

Fig. 4 Graphical representation of proposed NBW-MHO with other existing techniques

4.5 Performance Analysis in Terms of Proposed Classifiers

In this section, the proposed three classifiers such as RNN are validated with existing techniques namely CNN and recursive neural network is validated with and without feature selection technique called NBW-MHO in terms of all parameters, which is given in Table 4.

The following Fig. 5 represents the graphical results of proposed classifier without NBW-MHO in terms of all parameters.

The following Fig. 6 represents the graphical results of proposed classifier with NBW-MHO in terms of all parameters.

From the comparative study of feature selection with proposed classifier, it is clearly stated that the proposed RNN classifier achieved better results only with proposed NBW-MHO selection technique than CNN and recursive neural network.

Table 4 Comparative analysis of projected classifier with feature selection technique

Feature selection	Classifiers	Parameter evaluation			
		Accuracy (%)	Precision (%)	Recall (%)	*F*-measure (%)
Without NBW-MHO	CNN	88.14	83.56	85.75	89.23
	Recursive neural network	88.90	87.24	90.47	91.20
	Proposed RNN	**92.25**	**93.47**	**93.90**	**94.49**
With NBW-MHO	CNN	92.47	91.15	92.18	93.27
	Recursive neural network	94.80	93.70	94.74	95.49
	Proposed RNN	**97.57**	**96.49**	**97.07**	**98.15**

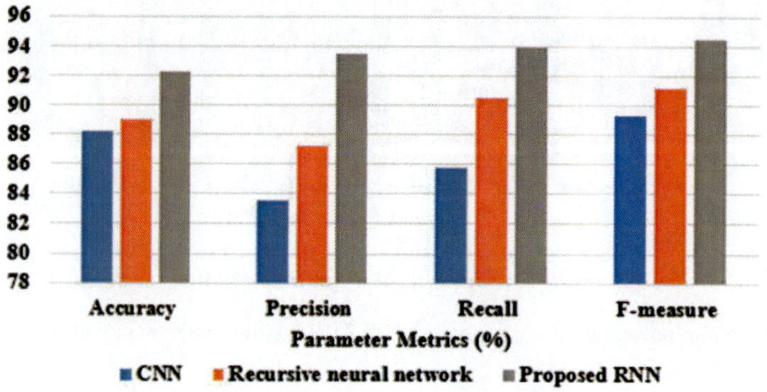

Fig. 5 Performance of proposed classifiers without NBW-MHO

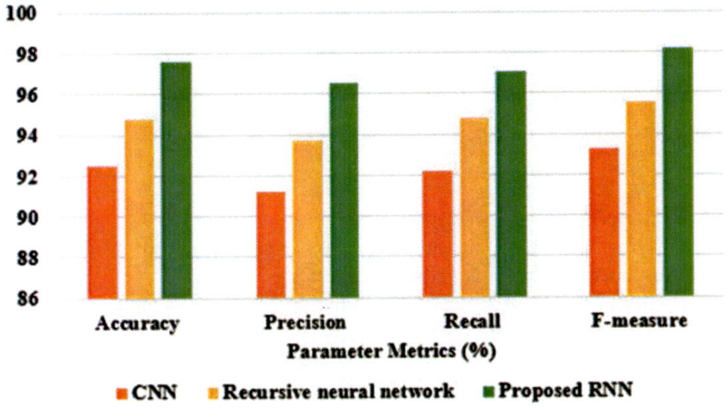

Fig. 6 Performance of proposed classifiers with NBW-MHO

5 Conclusion

A web crawler has a difficult time reading and organizing online pages because of the daily increase in the amount of web pages. As a result of this difficulty, the web classification process is becoming increasingly important. Using benchmark datasets, we built our technique in this study and found that it outperformed the existing algorithms. For forecasting the correct class for a web content, we have suggested an optimization-based feature selection approach that recommends the top *N* features to the ML model. Unlike typical text documents, web pages in a corpus are connected by hyperlinks, which provide them an added advantage in terms of classification. They serve as indicators of coupling. If there are many links between two pages, then there is a high coupling, which suggests the target class has a strong connection with the first page. This information is used in the paper's feature selection algorithm. Besides linkages, side information is also considered in the suggested method to improve classification accuracy. A conventional benchmark dataset was used to evaluate our classification algorithm. The results demonstrate that our work has a favorable beneficial effect on categorizing web content, according to the researchers. A new parameter will be added to categorization in future work: knowledge ontology and traffic links.

References

1. Hashemi M, Hall M (2019) Detecting and classifying online dark visual propaganda. Image Vis Comput 89:95–105
2. Hashemi M, Hall M (2020) Criminal tendency detection from facial images and the gender bias effect. J Big Data 7(2)
3. Qi X, Davison BD (2009) Web page classification: features and algorithms. ACM Comput Surv 41(2), article 12

4. Shang W, Huang H, Zhu H, Lin Y, Qu Y, Wang Z (2007) A novel feature selection algorithm for text categorization. Expert Syst Appl 33(1):1–5
5. Aghdam MH, Ghasem-Aghaee N, Basiri ME (2009) Text feature selection using ant colony optimization. Expert Syst Appl 36(3):6843–6853
6. Yahoo! https://maktoob.yahoo.com/?p=us
7. Open direct Project. http://www.dmoz.org/
8. Chen C, Lee H, Tan C (2006) An intelligent web-page classifier with fair feature-subset selection. Eng Appl Artif Intell 19(8):967–978
9. Altingövde İS, özel SA, Ulusoy ö, özsoyoğlu G, özsoyoğlu ZM (2001) Topic-centric querying of web information resources. Lect Notes Comput Sci 2113:699–711
10. Menczer F, Pant G, Srinivasan P (2004) Topical Web crawlers: evaluating adaptive algorithms. ACM Trans Internet Technol 4(4):378–419
11. Hamouda K (2013) New techniques for Arabic document classification
12. Ballesteros L, Larkey LS, Connell ME (2002) Improving stemming for Arabic information retrieval: light stemming and co-occurrence analysis. In: Proceedings of the 25th annual international ACM SIGIR conference on research and development in information retrieval, pp 275–282
13. Kim D, Seo D, Cho S, Kang P (2019) Multi-co-training for document classification using various document representations: TF–IDF, LDA, and Doc2Vec. Inf Sci 477:15–29
14. Bai W, Ren J, Li T (2019) Modified genetic optimization-based locally weighted learning identification modeling of ship manoeuvring with full scale trial. Futur Gener Comput Syst 93:1036–1045
15. Li L-L, Sun J, Tseng M-L, Li Z-G (2019) Extreme learning machine optimized by whale optimization algorithm using insulated gate bipolar transistor module aging degree evaluation. Exp Syst Appl 127:58–67
16. Eberhart R, Kennedy J (1995) A new optimizer using particle swarm theory. In: MHS'95. Proceedings of the sixth international symposium on micro machine and human science, pp 39–43, Nagoya, Japan
17. Xu X, Rong H, Pereira E, Trovati M (2018) Predatory search based chaos turbo particle swarm optimisation (PS-CTPSO): a new particle swarm optimisation algorithm for web service combination problems. Futur Gener Comput Syst 89:375–386
18. Francis LM, Sreenath N (2019) Robust scene text recognition: using manifold regularized twin-support vector machine. J King Saud Univ—Comput Inf Sci
19. Alaei A, Roy PP, Pal U (2016) Logo and seal based administrative document image retrieval: a survey. Comput Sci Rev 22:47–63
20. Al-Salemi B, Ayob M, Kendall G, Noah SAM (2019) Multilabel Arabic text categorization: a benchmark and baseline comparison of multi-label learning algorithms. Inf Process Manage 56(1):212–227
21. Dogan T, Uysal AK (2019) Improved inverse gravity moment term weighting for text classification. Expert Syst Appl 130:45–59
22. Yang S, Wei R, Guo J, Tan H (2020) Chinese semantic document classification based on strategies of semantic similarity computation and correlation analysis. J Web Semantics 63, article 100578
23. Chen Z, Zhou LJ, Li XD, Zhang JN, Huo WJ (2020) The lao text classification method based on knn. Procedia Comput Sci 166:523–528
24. Hao P, Chiang J, Tu Y (2007) Hierarchically svm classification based on support vector clustering method and its application to document categorization. Expert Syst Appl 33(3):627–635
25. Houssein EH, Saad MR, Hussain K, Zhu W, Shaban H, Hassaballah M (2020) Optimal sink node placement in large scale wireless sensor networks based on Harris' hawk optimization algorithm. IEEE Access 8:19381–19397
26. Houssein EH, Hosney ME, Oliva D, Mohamed WM, Hassaballah M (2020) A novel hybrid Harris hawks optimization and support vector machines for drug design and discovery. Comput Chem Eng 133, article 106656

27. Parpinelli RS, Lopes HS, Freitas A (2002) An ant colony algorithm for classification rule discovery. IEEE Trans Evol Comput 6(4):321–332
28. Holden N, Freitas AA (2004) Web page classification with an ant colony algorithm. In: Parallel problem solving fromnNature-PPSN VIII, vol 3242. Lecture notes in computer science. Springer, Berlin, Germany, pp 1092–1102
29. Jensen R, Shen Q (2006) Web page classification with aco enhanced fuzzy-rough feature selection, vol 4259. Lecture notes in artificial intelligence, pp 147–156
30. Janaki Meena M, Chandran KR, Karthik A, Vijay Samuel A (2012) An enhanced ACO algorithm to select features for text categorization and its parallelization. Exp Syst Appl 39(5):5861–5871
31. Mangai JA, Kumar VS, Balamurugan SA (2012) A novel feature selection framework for automatic web page classification. Int J Autom Comput 9(4):442–448
32. Craven M, DiPasquo D, Freitag D et al (1998) Learning to extract symbolic knowledge from the World Wide Web. In: Proceedings of the 15th national conference on artificial intelligence (AAAI'98). AAAI Press, pp 509–516
33. CMU. http://www.cs.cmu.edu/
34. WebKB. http://www.cs.cmu.edu/~webkb/

A Systematic Study on Network Attacks and Intrusion Detection System

Milan Samantaray, Suneeta Satapathy, and Arundhati Lenka

Abstract Network security is essential in all aspects of the network. The infrastructure has now been put in place throughout workplaces, educational institutions, finance companies, and everywhere. But, nearly all people in social networking media participate. While many kinds of networks security measures are being used, most of the other operations seem to be susceptible. This report provides an analysis of various kinds of network attacks and numerous IDSs, in particular. This could also present an opportunity to build new IDS to prevent various cyber threats from the network system.

Keywords Security of the network · Network assaults · Web attacks (IDS)

1 Introduction

A network with computing devices is directly linked to resource sharing. An insider or an outsider can commit a network attack. In "inner attack," a person with full authorization is attacked within the security perimeter of an entity that engages in vulnerable activities, i.e., a system resource for which an attacker has not been allowed to access some system resources [1]. To find out this kind of person is very difficult. An exogenous attack can be launched from elsewhere by an illegal or improper user interface. But outside the World Wide Web, assailants may be inexperienced or structured thieves and foreign criminals [2].

M. Samantaray (✉)
Department of ITM, Udayanath (Auto.) College of Science and Technology, Adaspur, Cuttack, Odisha, India
e-mail: milansamantaray190@gmail.com

S. Satapathy
Department of FET, Sri Sri University (SSU), Cuttack, Odisha, India

A. Lenka
Udayanath (Auto.) College of Science and Technology, Adaspur, Cuttack, Odisha, India

V. Skala et al. (eds.), *Machine Intelligence and Data Science Applications*, Lecture Notes on Data Engineering and Communications Technologies 132,
https://doi.org/10.1007/978-981-19-2347-0_16

There are two components of a computer network, namely software and hardware. The risks and vulnerabilities may exist for both of these components. Vulnerabilities to devices are easily recognized and, in place of information collected, just the machine is harmed [3, 4]. There are four types of hardware threats: physical, electrical, environmental, and maintenance. In most cases, it damages the data if the attack is made in software. Previously, the hacking program had only been written by people with high programming skills. But, by accessing Internet security tools, an individual with just a little coding skills can also become a hacker [5, 6]. Besides this, everyone intends to just use elevated software that makes it convenient to assault.

With high characteristics, safety is very likely to lack. Confidentiality, integrity, and availability are the three aims of the Software Security threat [7]. The Internet access seems to be the most widely utilized channel with several consumers. Nearly all areas of the Internet are playing an important role in connectivity. Even if an attack takes place in all types of networks, it is the web attack that is the most challenging in the broad area network. Network safety is now a region where every user seeks to defend his/her platform from invaders. The threats select a fresh texture as the Intrusion Detection System grows [8]. The advantages and disadvantages of these IDS vulnerabilities will certainly help in designing and building a system of network security. It assists the system to operate effectively.

Section 2 demonstrates inside this document the aspects of network threats and attacks. Section 3, IDS as well as its forms conducts the literature review. Section 4 and at last Section 5 are the conclusions.

2 Network Attack

2.1 Types of Network Attack

A passive or active attack can be launched. In an "active attack," the attacker will take steps to alter system resources, such as breaking or circumventing secured systems [9]. It usually leads to the disclosure of sensitive information, data modification, or, in the worst-case scenario, data loss, and Trojan horses, viruses, and worms, as well as harmful code insertion and penetration. Active attacks include stealing network data and stealing login information [10]. This kind of attack is extremely damaging to the system. Masquerade, Session Replay, Message Modification, and Denial of Service are examples of active attacks.

A "passive attack" is intended to collect or use confidential material without impacting the resources and data. The attacker uses a sniffer device throughout this form of threat and waited patiently to seize the critical material that would be used in multiple strikes, including software for traffic analyses, sniffer packets, and password filters [11]. Examples of passive attacks are the ones that follow: publish of content and evaluation of traffic.

There are the following major types of network attacks such as:

(a) **Attacks on the Internet-based Network**: The Internet-based Network Assault is a subcategory of network attacks that have their own features [12]. This makes them different from the aforementioned types of intrusion. Below are a couple of suggestions of web attacks.

An injection is a method of a web attack strategy where the individual sends web services with unknown information of the users. Injections from SQL, OS, and LDAP are forms of injection occurring whenever an intruder generates malicious scripts to a prompt or query linguist. The confrontational data of the intruder then can deceive the performer into unintended instructions or access the data without authorization [13].

(b) **Identity verification and session management broken**: This offensive might very well arise because the application functions associated with user credentials have been misimplemented to enable an attacker to hack passwords, keyboards, and other sensitive data [14]. This attack is possible in a malicious application.

Cross-Site Scripting (XSS) is a kind of intrusion that happens whenever an Internet programmer uses cross-site scripting to gather unreliable digital evidence. It uses a doughnut to transmit information to an Internet browser that is used without even an affirmation as a perpetrator browser. And this XSS allows hackers to use their ransom ware to hijack user requests and even to divert access to malicious sites on the Internet [15].

(c) **Direct References for objects which are not secure**: When the development company is not cautious, a citation to a construct validity object could unknowingly be exposed. It could be a document, a repository, or maybe even a data warehouse key [16]. This makes it possible for an assailant to directly obtain unlawful records without inspections or any other protection of availability.

(d) **Misconfiguration of security**: If your device's security measures are inadequate, an immediate or suitable visibility may be obtained by the unnamed customer. Users can participate in susceptible events by obtaining standard profiles, unutilized pages, unsecured vulnerabilities, undefended login credentials, and so forth. Large number of applications, web servers, database servers, and portals should be constructed in a secure way by any framework among many other things. This requires the design, implementation, and monitoring of the installation and configuration. In particular, the anti-malware should always be retained.

(e) **Responsive data**: This assault will likely be committed if your web application is not adequately covered by confidential information such as bank cards, revenue IDs, and authentication credentials. The intruder is primarily the right person to obtain information as well as replacement of sensitive data [17, 18]. This means the stealing of keys or concise data sets on the domain controller, attacking the person on the domain controller or even on the user's web browser. Further protection of confidential information, such as encrypted data, should indeed be provided.

(f) **Lack of Level Access Control Function**: The verification of the right to access function level is usually done before the user interface functionality is made available. However, when each function is accessed, there is no access control on the server [19, 20]. The intruder can file inquiries without appropriate authorized to view the features if the queries really are not inspected.

(g) **Forgery (CSRF) Cross-Site Request**: This attack enables a perpetrator's browser to deliver a fabricated HTTP application that includes the perpetrator's cookie and authenticating details to a web application that is susceptible. Accordingly, the application is considered legitimate [21].

(h) **Use of known vulnerabilities components**: This attack is caused by complete rights to access certain components, such as libraries, software modules, and frames. The attacker gets into his susceptible object and therefore can cause severe security breaches or procurement of the server [22]. The components in these applications undermine the safety and weaken the system.

(i) **Redirects and forward invalidated**: Web applications usually include many redirects and forward the user to various web pages and websites [23]. Without proper approval of the intended pages, the perpetrator has used some unsustainable data. It generates the physiological, malicious software or leads the user to unlawful websites.

(j) **Angriff on phishing**: This attack is now a popular assault. A false email or website like a reputable email address or a regular website has been created here. After this, they send an email to a wrong website [24] only with initials of the corrupt and immoral text. The counterfeit site looks like the regular. However, when the user starts using it, the scammer logs in by using user's login credentials, authorization, etc.

(k) **Hijack attack**: Such a threat generally happens among conversations. The aggressor participates in a release and separates the official team silently. Then, by the original disconnected party, he starts communicating with active parties [25]. The active participant does not really communicate mostly with intruder of confidential material [26].

(l) **Attack spoof**: Attacking the adversary alters the packet's source address before it gets to the target user, and so prevents the target from realizing that it is dealing with an unknown person. In this form of assault, the firewall restrictions are circumvented. When it comes to ARP, DNS, and IP address spoofing, to name a few, yes.

3 Intrusion Detection System (IDS)

The IDS intrusion detection means a computer program or device tracking fraudulent behavior or policy violation of the system or systems. The IDS intrusion detection is defined as "the detection of actions intended to impact trust, honesty, or available resources." Every identified action or breach is communicated to or efficiently gathered by the admin through the SIEM system. This SIEM brings together all outcomes from different sources to differentiate fraudulent behavior and counterfeit warning signs from alarm filters. The firewall looks completely different from the IDS [27], where both firewalls and on tribute to network safety. To stop them from occurring externally for intrusions or to avoid intrusion, a firewall limits network access. IDS may be categorized according to the place of detection (network or host) and the method of detection used.

3.1 Using Intrusion Detection Systems from the Location Network

(i) **Network Intrusion Detection Systems**: Strategic network notes or marks for network troubleshooting. It analyzes the traffic flow over the whole subnet and reflects the congestion transported to documented sub-network attacks [28]. Even when an attack or unusual conduct is discovered, the alert may be sent to the manager. For instance, snorting.

(ii) **Host detection systems for intrusion**: On network access hosts, host intrusion detection (HIDS) system operates. This HIDS monitors packets for incoming and outgoing devices as well as warnings your user or admin to suspicious activity [29]. You also need to copy the current files and folders and align the prior one. An alert will be sent to the admin in case of providing or deleting crucial file system. For instance, OSSEC, AIDE.

(iii) **Based on methods for detection**: So it is explained in two ways like,

   (a) **Signature-based**: IDS means detection mechanism, by looking for specific patterns like the network bit stream sequences or certain excellently sequences of a suspect's malicious instructions [30]. While IDS with the approval can identify common threats quickly, there can be no new intrusions, since no patterns may occur.

   (b) **Anomaly-based**: Unidentified attacks to be identified and the rapid growth of malicious software IDs based on anomalies to be addressed [31]. The fundamental idea is to build a trustworthy business strategy using machine learning and then start comparing unusual habit. While this method makes it possible to detect a new form of threat, it can also be a malicious code, but it endures from false alarms [32].

4 Literature Survey

Barghi et al. [33] have demonstrated that the Intrusion Detection System providing protection against a web attack is a huge volume of untrue, superfluous, or irrelevant alerts. That is a big disadvantage. A DARPA 1999 dataset and ShahidRajaee Port Complex dataset were proposed to use an online approach. The results showed a 94.32% reduction in the number of alerts. In certain cases, the rates were quite high and there was a very elevated alarming. And online analysis is not really suitable.

This has thus compelled the advancement and reduction of the faulty alarm rate of an innovative system with high classification accuracy.

Koo et al. [34] propose a technique for finding out whether the website is malicious or benign. First, the website content using autonomous JAVA and in addition to all the other things it will do, the software will do regular expression matches on signature verification in order to cut down the analysis time and employ a smart browser that looks up online pages. Microsoft is an excellent software tool because it is utilized for academic purposes. This system design is made up of modules: surrogate analysis, source code assessment, and a registry of behavior. Even if malevolent websites have been detected active and automated, there has always been impoverished accuracy throughout the formal verification and a static behavior of both the time and effort needed.

Friedberg et al. [35] have confirmed the use of different methods for attacks on advanced persistent threat (APT) in the initially unauthorized system. The network spreads steadily. This process aims of expanding all IDS packet-level programs to enhance their outcomes. Search models (P), events (C), hypothesis (H), and procedure rules constitute the model (R). The system presented in this paper has to deal with log lines from distributed systems and nodes having a network-wide influence. The system technique is used to examine and identify various groupings of log messages with different types of meaning. SCADA was used in conducting the test. A positive result is a genuine positive result, while a false positive is almost impossible.

Salama et al. [36] proposed the WAMID framework to combine misuses and anomalies. Web Misuse Intrusion detection and SQL Injection Attack detection algorithms are used in the field of database behavior. Initially, a profile for the valid database behavior, taken from applications of XML file association rules usually contains the SQL queries forwarded from an application, was generated mostly during the training period. The configuration of the query is then compared favorably to the valid queries during identification, and intrusion is identified as queries that differ from the standard pattern of behavior. False-positive alarms are also minimized. This work can be broadened for other attacks, such as cross-site scripts.

The botnet is a group of host (bots) of Chen and Lin [37] and is regulated via a C&C channel via a botmaster. This detection method therefore identifies the botnet attack beforehand throughout C&C. For evaluation on an organization's network, IRC traffic patterns were regarded. Resemblance and regular intervals characteristics were measured. This framework can also be used to find the malevolent Internet traffic of standard IRC customers. The studies have also shown that upwards of 90% and much less than 7% can be accomplished at a true-positive rate.

According to the paper, Kar et al. [38], who adapted a SQL injection attack from the backend database containing sensitive data, such as credit card information, developed a SQL injection technique. The team put forth a proposal to detect several SQL injection threats, including IDS-SQLiDDS (SQL Injection Detection Using Query Transformation and Document Similarity). After WHERE is used, just portion of the questions have been considered. Five honeypot web apps written in PHP and MySQL have been created for the examinations. They have been victimized by SQL, and they have detected a plethora of irregular patterns. Injector structures preserved these designs and hazelnut table. MD5 hazelnuts rescued these hazelnuts. Hierarchical clusters and modifications created a text out of them (HAC). Each cluster was keeping a separate record of every attack vector it was targeting. An SQL injection was identified, after comparing the records to the likeness of a received query.

Somwanshi and Joshi [39] report that honeypot is merely a counterfeit server which provides imitated, server-like services. Therefore, when an intruder attempts to target and is at last trapped, an intruder is diverted toward that counterfeit server. Honeypot provides the precious intruder's information. A new honeypot system was suggested in this paper. The system's components are as follows:

(i) Event Auditor–Monitor and send to IDS data exchange between nodes.
(ii) Analyzer and Alert System IDS service with two components. Compatibility analysis—the present user activity is compared to the usual behavior and the analyzer subcomponents. Knowledge Analysis—detecting known routes from certain activities remaining by a user who is also an intruder. Honeypot controls the alert system warning messages to obtain information on an assailant's behavior in order to successfully implement stronger server protection. This load balancer gets several processes done quickly while maintaining system processing performance. The honeypot system employs a load balancer. In the event of a server failure, which may or may not happen, performance will never be affected because requests are automatically directed to other servers. The flexibility of the system is increased in this way.

In the study conducted by Kaur et al. [40], a user's usage of a software automatically generated all of the user's behavior toward the weblog files. Most harmful metrics such as denial of service and brute-force assaults were avoided because of the logging system's emphasis on and attention to the system in question. It makes it easy to securely share files. The computer system is capable of distinguishing between harmful and non-malicious users. A DDoS assault indicates that the neutral threshold for a weblog file is higher than or equal to the threshold for the file. To ensure availability, brute-force attack detection must result in refused requests if information produced by the server resembles the value entered by the user. The reinforcement and accuracy were all recorded.

Kour and Sharma [41], Cross-Site Scripting Assault (XSS) is an injection code attack to exploit the web application vulnerabilities by injection. Functions of HTML tag/java showed various kinds of XSS attacks. In two steps, this system works: Firstly, to track the web application's cross-site scripting vulnerabilities. Newly created a website with PHP, hosting on your localhost (XAMPP server) and experimenting with modern browsers (Google Chrome 49, IE11, Opera15, and Firefox44.0.2). The following step is to alleviate the attack. The three steps are used to encode, purify, and use Regular Expressions Matching. The very first step is to encode the input and to erase all HTML tags from its user so that malicious code cannot be included in the dataset and regular sentences of the potentially malicious JavaScript code have also been represented in Regular Expressions Matching. Each user entry will meet all the regular phrases preset to verify that the input is accurate or not.

According to Seeber and Rodosek [42], a new way for building an IDS involves creating many IDSs, each containing their own subset of IDSs to detect intrusions from the network, which utilize core components to analyze the data. In the SDN configuration, adjust the OpenFlow properties to their appropriate values. Depending on the presence or absence of an existing or non-existent stream, OpenFlow can raise or update a flow counter when a packet arrives. To send the diverted traffic to other subnets or IPs, the use of numerous IDSs is common. The system is capable of rapidly shifting from using the cloud-based IDS solutions to the devices, which change on the fly to detect threats.

Some brute-force attacks attempt to transmit IP addresses between services without user interaction in a Saito et al. [43] time period. The team found and studied this DBF before implementing the TOPASE system. It works in two stages: first, you remove the malfunctioning parts, and then you shut down the machine. To do an in-depth analysis of login sessions from a source host to a destination host, use TOPASE. To assess TOPASE, the IDS log is evaluated and results are presented on waste time and decrease in rate.

Dual Attack Detection is devised by Ali Zardari et al. (DDBG) [44]. DDBG proposes using the two additional features of the CDS technique, including connecting dominant set (CDS) and dominant set extension (DSE), to select IDS nodes. Furthermore, IDS nodes are tested prior to their insertion into the DDBG-based IDS set to verify that they do not already exist in the blacklists. This powerful, distinct, and localized strategy, known as the CDS, is a superb method for identifying dominant ad hoc mobile network node sets in a restricted range of networks. Using the IDS nodes to get the full details on their status from their nodes is described by the IDS. The DDBG approach, which utilizes IDS nodes, enables Intrusion Detection System (IDS) nodes to assess the information gathered from the actions in order to identify erroneous nodes, and if it appears that the erroneous node is involved in the malicious activities, then it is added to the list of malicious nodes. According to these simulation results, the effectiveness of the service parameters of the technology's present routing systems grows.

5 Analysis

Table 1 shows various comparative analysis of the existing techniques and also the performance rate is depicted in Fig. 1. The Intrusion Detection System for the network platform is improvised with the implementation of the Deep Learning (DL) technique. So the DL method provides some techniques such as convolutional neural network (CNN), deep neural network (DNN), and autoencoder (AE) algorithm where these algorithms are compared with each other. If we compare the machine learning (ML) with the methods of DL, then we will get that 65% of proposed techniques are based on the DL approach, 15% of solutions use the hybrid methods of DL and ML, and 20% proposed solutions are followed by ML techniques which is shown in Fig. 2.

Table 1 Comparative analysis of the existing technique

Proposed techniques	Document reference	Data used	IDS objective	Pros	Restrictions
An ensemble approach and the best algorithms with SNORT IDS subcomponent standardization	Barghi et al. [33]	1999 data set of the DARPA and Shahid Rajaee port complex	Reduced IDS and intrusion detection large numbers of alerts	The 94.32% unimportant warnings can be reduced. With many leading IDS, it produces better results	For online analysis, it is unacceptable. Since the alarm rate is high, a system is required to lower the alarm rate
Static Java analysis—which scans Java source code for security problems using regular expressions and dynamic analysis—comprises a regex and HPC honeypot systems	Koo et al. [34]	Operating systems files for Microsoft	Malicious websites identification	It detects malicious websites actively and automated	Static analysis tests may produce results with poor precision, while dynamic analysis requires additional time and resources. A highly sensitive sensor is required. A measure must be taken to find out which technology requires the least amount of time to be detected

(continued)

Table 1 (continued)

Proposed techniques	Document reference	Data used	IDS objective	Pros	Restrictions
In this approach, system first analyzes a list of device network logs and then eliminates devices it has identified from the list	Friedberg et al. [35]	Dataset of SCADA	Continuous advanced menace (APT)	The approach can only reliably detect the evaluated anomaly through the combination of several rules in the model	As events rise in complexity, a better algorithm for hypothesis generation should be devised, and that algorithm should be given the hierarchical structure of the event class. If no relevant information about the similarities between event classes is available, a superfluous number of hypotheses build up, resulting in an overloaded model system
Misuse and anomaly detection techniques are the basic framework for WAMID	Salama et al. [36]	Logs for the database	Attack with SQL injection	It uses two different types of intrusion detection algorithms, which allows it to give the two-layered system security	The detection of other attacks may be extended

(continued)

Table 1 (continued)

Proposed techniques	Document reference	Data used	IDS objective	Pros	Restrictions
The similarity measurement and regular botnet features	Chen et al. [37]	The business network's traffic patterns	Botnet IRC-based	Detect IRC traffic abnormalities easily and find botnet activities. A truly positive rate of greater than 90% is achievable, while a false-positive rate of less than 7% is also feasible	Few points in the network still found. A new detection mechanism can be designed with encoded messages for more complicated botnet conduct
SQLiDDS	Kar et al. [38]	PHP and MySQL Web Applications	Attack of SQL Injection	98.05% overall precision. The system is error-free with respect to greater than 99% accuracy. For every false positive (FPR), there are only two false negatives (FNR). On average, the web server takes between 4.5% and 5.8% of the time to respond	A system to enhance precision and efficiency. A system for weighing down the rates of FN and FP. An algorithm for clustering that generates fewer high grade clusters
Redirection algorithm honeypot system	Somwanshi et al. [39]	Database of profile history	Security of the server	Enhances scalability. Creates fewer warnings	For improved security use can be made of the combination of Honeypot and IDS

(continued)

Table 1 (continued)

Proposed techniques	Document reference	Data used	IDS objective	Pros	Restrictions
Genetic algorithm fitness function	Kaur et al. [40]	File of web log	Attack of DDoS and brute-force	A secure platform for individual users to share their files	For encrypted data sharing, a new system can be designed
Encoding, sanitation and regular expressions technique	Kour et al. [41]	PHP web page	Scripting of cross-site (XSS)	More modern browsers compatible. Mitigate the risk of XSS attack successfully. No impact on client web browser performance	There is no need for predefined regular terms. There are also opportunities for new expressions other than predefined expressions A new method is needed, and the most practical format is one that is saved in data format rather than in HTML
OpenFlow usage IDS	Seeber et al. [42]		Attacking the network	Allows dynamic and adaptive transmission to different IDSs, including cloud-based IDS solutions	A mechanism to demonstrate the performance and reliability of cloud-based IDSs is required
TOPASE	Saito et al. [43]	Log on IDS	Brutal attacks distributed	The efficiency of the falling rate and time spent has been demonstrated	In Setting Optimal Thresholds, TOPASE cannot detect the EBF and Attenuators are unable to recognize TOPASE
An intrusion detective system in MANETs was implemented to identify black and gray hole attacks with a dual attack detection technique (DDBG)	Ali Zardari et al. [44]	MANETs' dynamic environment	Attempts on black and gray holes	DDBG is a proving and effective way to detect black and gray hole attacks	Battery life for the nodes is short because the nodes are not continually monitored. Additionally, the project presented recognizes well-known assaults, although it does not detect every single one of them

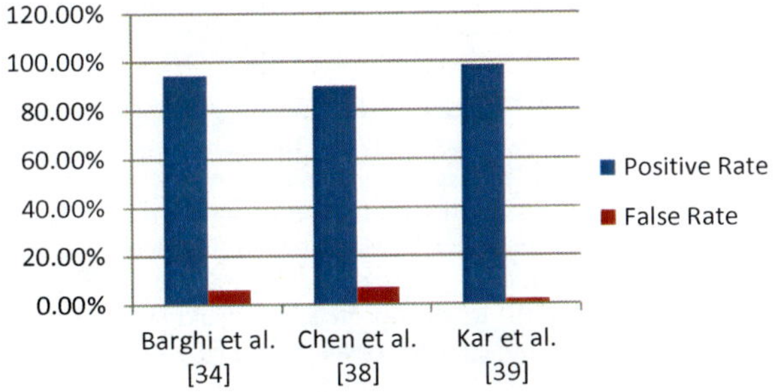

Fig. 1 Performance rate analysis

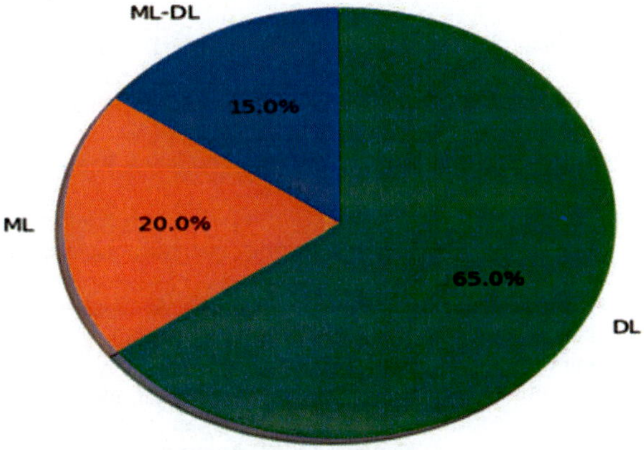

Fig. 2 Proposed methodology distribution

6 Conclusion

This paper presents the findings of a study on several Intrusion Detection Systems (IDS) methodologies offered by various authors, as well as the accompanying web attack(s) and the method by which they can be detected. The conclusion that can be derived from the aforementioned consolidated survey report is that the articles [35–38, 41, 43, 44] are solely meant to identify a specific network assault and are not intended for general use. The paper [40] discusses two types of attacks: distributed denial of service (DDoS) and brute-force. In Barghi et al. [33], an IDS is presented solely for the purpose of reducing the amount of unnecessary warnings generated by a standard IDS. Furthermore, in Refs. [34, 39, 42], there is no indication of the nature of attack. It should be noted that practically all of the techniques recommended are limited to dealing with one or two Web-based attacks. In order to develop an effective Intrusion Detection System that will defend the system from the vast majority of attacks, these features should be taken into consideration while designing a new IDS. If a single IDS is built that is capable of detecting several forms of online attacks, it will be a powerful IDS that can be utilized in a variety of network configurations and environments.

References

1. Steingartner W, Galinec D, Kozina A (2021) Threat defense: cyber deception approach and education for resilience in hybrid threats model. Symmetry 13(4):597
2. Oakley J (2018) Improving offensive cyber security assessments using varied and novel initialization perspectives. In: Proceedings of the ACMSE 2018 conference, pp 1–9
3. Adomnicai A, Fournier JJ, Masson L (2018) Hardware security threats against Bluetooth mesh networks. In: IEEE conference on communications and network security (CNS). IEEE, pp 1–9
4. Montasari R, Hill R, Parkinson S, Daneshkhah A, Hosseinian-Far A (2020) Hardware-based cyber threats: attack vectors and defence techniques. Int J Electron Secur Digit Forensics 12(4):397–411
5. Saha S, Das A, Kumar A, Biswas D, Saha S (2019) Ethical hacking: redefining security in information system. In: International ethical hacking conference. Springer, Singapore, pp 203–218
6. Samtani S, Chinn R, Chen H, Nunamaker JF Jr (2017) Exploring emerging hacker assets and key hackers for proactive cyber threat intelligence. J Manag Inf Syst 34(4):1023–1053
7. Tuma K, Calikli G, Scandariato R (2018) Threat analysis of software systems: a systematic literature review. J Syst Softw 144:275–294
8. Kim J, Kim HS (2020) Intrusion detection based on spatiotemporal characterization of cyberattacks. Electronics 9(3):460
9. Hayashi M, Owari M, Kato G, Cai N (2017) Secrecy and robustness for active attack in secure network coding. In: IEEE international symposium on information theory (ISIT). IEEE, pp 1172–1176
10. Aminuddin MAIM, Zaaba ZF, Samsudin A, Juma'at NBA, Sukardi S (2020) Analysis of the paradigm on tor attack studies. In: 8th International conference on information technology and multimedia (ICIMU). IEEE, pp 126–131
11. Jyothirmai P, Raj JS, Smys S (2017) Secured self organizing network architecture in wireless personal networks. Wireless Pers Commun 96(4):5603–5620
12. Singh K, Singh P, Kumar K (2017) Application layer HTTP-GET flood DDoS attacks: research landscape and challenges. Comput Secur 65:344–372
13. Sinha P, Kumar Rai A, Bhushan B (2019) Information security threats and attacks with conceivable counteraction. In: 2nd International conference on intelligent computing, instrumentation and control technologies (ICICICT), vol 1. IEEE, pp 1208–1213
14. Nadar VM, Chatterjee M, Jacob L (2018) A defensive approach for CSRF and broken authentication and session management attack. In: Ambient communications and computer systems. Springer, Singapore, pp 577–588
15. Sarmah U, Bhattacharyya DK, Kalita JK (2018) A survey of detection methods for XSS attacks. J Netw Comput Appl 118:113–143
16. Srinivasan SM, Sangwan RS (2017) Web app security: a comparison and categorization of testing frameworks. IEEE Softw 34(1):99–102
17. Cheng L, Liu F, Yao D (2017) Enterprise data breach: causes, challenges, prevention, and future directions. Wiley Interdiscip Rev Data Min Knowl Disc 7(5):e1211
18. Bhanipati J, Singh D, Biswal AK, Rout SK (2021) Minimization of collision through retransmission and optimal power allocation in wireless sensor networks (WSNs). In: Advances in intelligent computing and communication. Springer, Singapore, pp 653–665
19. Tourani R, Misra S, Mick T, Panwar G (2017) Security, privacy, and access control in information-centric networking: a survey. IEEE Commun Surv Tutorials 20(1):566–600
20. Biswal AK, Singh D, Pattanayak BK, Samanta D, Chaudhry SA, Irshad A (2021) Adaptive fault-tolerant system and optimal power allocation for smart vehicles in smart cities using controller area network. Secur Commun Networks 2021:13, Article ID 2147958. https://doi.org/10.1155/2021/214795
21. Rankothge WH, Randeniya SM (2020) Identification and mitigation tool for cross-site request forgery (CSRF). In: IEEE 8th R10 humanitarian technology conference (R10-HTC). IEEE, pp 1–5

22. Cheminod M, Durante L, Seno L, Valenzano A (2017) Detection of attacks based on known vulnerabilities in industrial networked systems. J Inf Secur Appl 34:153–165
23. Touseef P, Alam KA, Jamil A, Tauseef H, Ajmal S, Asif R, ... Mustafa S (2019) Analysis of automated web application security vulnerabilities testing. In: Proceedings of the 3rd international conference on future networks and distributed systems, pp 1–8
24. Franz A, Benlian A (2020) Spear phishing 2.0: how automated attacks present organizations with new challenges. HMD Praxis Wirtschaftsinformatik 57:597–612
25. Apostolaki M, Zohar A, Vanbever L (2017) Hijacking bitcoin: routing attacks on cryptocurrencies. In: IEEE symposium on security and privacy (SP). IEEE, pp 375–392
26. Biswal AK, Singh D, Pattanayak BK (2021) IoT-based voice-controlled energy-efficient intelligent traffic and street light monitoring system. In: Green technology for smart city and society. Springer, Singapore, pp 43–54
27. Pradhan M, Nayak CK, Pradhan SK (2020) Intrusion detection system (IDS) and their types. In: Securing the Internet of Things: concepts, methodologies, tools, and applications. IGI Global, pp 481–497
28. Ken FY, Harang RE, Wood KN (2017) Machine learning for intrusion detection in mobile tactical networks. In: Cyber sensing, vol 10185. International Society for Optics and Photonics, p 1018504
29. Jose S, Malathi D, Reddy B, Jayaseeli D (2018) A survey on anomaly based host intrusion detection system. J Phys Conf Ser 1000(1):012049
30. Chawla A, Lee B, Fallon S, Jacob P (2018) Host based intrusion detection system with combined CNN/RNN model. In: Joint European conference on machine learning and knowledge discovery in databases. Springer, Cham, pp 149–158
31. Zavrak S, İskefiyeli M (2020) Anomaly-based intrusion detection from network flow features using variationalautoencoder. IEEE Access 8:108346–108358
32. Biswal AK, Singh D, Pattanayak BK, Samanta D, Yang MH (2021) IoT-based smart alert system for drowsy driver detection. Wireless Commun Mob Comput
33. Barghi MN, Hosseinkhani J, Keikhaee S (2015) An effective web mining-based approach to improve the detection of alerts in intrusion detection systems. Int J Adv Comput Sci Inf Technol (IJACSIT), (ELVEDIT) 4(1):38–45
34. Koo TM, Chang HC, Hsu YT, Lin HY (2013) Malicious website detection based on honeypot systems. In: 2nd International conference on advances in computer science and engineering (CSE 2013). Atlantis Press, pp 76–82
35. Friedberg I, Skopik F, Settanni G, Fiedler R (2015) Combating advanced persistent threats: from network event correlation to incident detection. Comput Secur 48:35–57
36. Salama SE, Marie MI, El-Fangary LM, Helmy YK (2012) Web anomaly misuse intrusion detection framework for SQL injection detection. Editorial Preface 3(3)
37. Chen CM, Lin HC (2015) Detecting botnet by anomalous traffic. J Inf Secur Appl 21:42–51
38. Kar D, Panigrahi S, Sundararajan S (2015) SQLiDDS: SQL injection detection using query transformation and document similarity. In: International conference on distributed computing and internet technology. Springer, Cham, pp 377–390
39. Somwanshi AA, Joshi SA (2016) Implementation of honeypots for server security. Int Res J Eng Technol (IRJET) 3(03):285–288
40. Kaur J, Singh R, Kaur P (2015) Prevention of DDoS and brute force attacks on web log files using combination of genetic algorithm and feed forward back propagation neural network. Int J Comput Appl 120(23)
41. Kour H, Sharma LS (2016) Tracing out cross site scripting vulnerabilities in modern scripts. Int J Adv Networking Appl 7(5):2862
42. Seeber S, Rodosek GD (2015) Towards an adaptive and effective IDS using OpenFlow. In: IFIP international conference on autonomous infrastructure, management and security. Springer, Cham, pp 134–139
43. Saito S, Maruhashi K, Takenaka M, Torii S (2016) Topase: detection and prevention of brute force attacks with disciplined IPs from IDs logs. J Inf Process 24(2):217–226

44. Ali Zardari Z, He J, Zhu N, Mohammadani KH, Pathan MS, Hussain MI, Memon MQ (2019) A dual attack detection technique to identify black and gray hole attacks using an intrusion detection system and a connected dominating set in MANETs. Future Internet 11(3):61

Achieving Data Privacy Using Extended NMF

Neetika Bhandari and Payal Pahwa

Abstract Data mining plays a vital role today for decision making and analysis in education, health care, business and more. It is very important to protect the data before the mining process such that it is protected from security threats and produces correct and desirable results. Privacy-preserving data mining (PPDM) allows securing data, thus maintaining data privacy. In this paper, we have used perturbation-based methods for data transformation, making it secure before applying the data mining process. The authors have proposed extended non-negative matrix factorization (NMF), which includes the NMF method followed by double-reflecting data perturbation (DRDP) method to distort data. This gives higher protection levels compared to NMF alone based upon various privacy measures. We have used R language for the implementation of the research work. We have evaluated and compared various privacy parameters to show that the proposed method of extended NMF (NMF followed by DRDP), provides higher level of protection to nonnegative numeric data compared to NMF alone.

Keywords Data distortion · Double-reflecting data perturbation (DRDP) · Non-negative matrix factorization (NMF) · Data perturbation · Privacy-preserving data mining (PPDM)

1 Introduction

Data is a very important part of today's digital world. Every day we deal with data in one form or the other and use it extensively in education, health care, banking, business, defense and many more. This data is growing tremendously at a very fast rate. This vast data plays a vital role as it can be used to extract useful information and make important decisions in business, healthcare and various other domains. Chen et al. [1] define data mining as the process used for the fast analysis of data to extract the useful information and patterns. But, it is very important to keep this data

N. Bhandari (✉) · P. Pahwa
Guru Gobind Singh Indraprastha University, Delhi, India
e-mail: neetika.igdtu@gmail.com

V. Skala et al. (eds.), *Machine Intelligence and Data Science Applications*, Lecture Notes on Data Engineering and Communications Technologies 132,
https://doi.org/10.1007/978-981-19-2347-0_17

safe and secure before and after the data mining process. If the privacy of sensitive data is not ensured, it can lead to various security threats harming the privacy of individuals or organizations. Thus, it is important to secure the critical data and information for security purposes. This is done by privacy-preserving data mining (PPDM) techniques, which protect the sensitive data and information from being disclosed to untrusted users maintaining the utility of data [2].

Perturbation-based PPDM techniques are simple data distortion methods that transform the original data making it secure. These include methods like SVD (singular value decomposition), NMF (non-negative matrix factorization), DRDP (double-reflecting data perturbation), rotation, shearing, addition and many more [3–5]. Certain applications like banking and physics do not allow negative values, and for the analysis of these datasets we need secure non-negative numeric data only. In this paper, we have proposed to use the extended NMF method to secure data. Extended NMF method combines NMF and DRDP methods sequentially to perturb non-negative numeric data values of various datasets to improve the privacy level. NMF is a matrix factorization method which retains the non-negativity of dataset providing higher degree of privacy than simple distortion methods. It is a fast method with low computational costs. Similarly, DRDP is also a simple method with low computational costs, and it preserves the statistical features of the dataset. The proposed method of "Extended NMF" provides higher security level to the dataset at low computational costs. We have evaluated the performance of the proposed method based on various privacy measures like value difference and position difference. The performance of the proposed extended NMF method is compared with that of NMF method based on the privacy measures.

Section 2 summarizes the literature survey. In Sect. 3, we explain perturbation-based PPDM methods such as NMF and DRDP. In Sect. 4, various privacy measures used for evaluation purposes have been explained. Section 5 covers the methodology and results. In Sect. 6, the conclusion and future scope of the research work have been defined.

2 Literature Survey

Data distortion is used to perturb data and ensure security. Research shows that various methods are used to distort data for perturbation-based PPDM. Xiao-dan et al. [3] have introduced the use of additive data perturbation, multiplicative data perturbation and rotation data perturbation to distort and hide data effectively. They have used computational cost as evaluation parameter to evaluate performance, efficiency, quality and accuracy for evaluating effectiveness of the algorithm. Maheswari and Revathi [4] have used QR decomposition method to provide data privacy maintaining the utility of data. Further, they have applied hierarchical clustering to measure the effectiveness, i.e., proportion of data points clustered in the same cluster after applying data perturbation. Manikandan et al. [5] have used the shearing method to add noise to original data. They have combined shearing with translation and scaling to get various modifications by using K-means clustering to compare the results of original and distorted data.

Various other methods used for data distortion are singular value decomposition (SVD), non-negative matrix factorization (NMF) and discrete wavelet transform (DWT). Kabir et al. [6] have proposed the use of NMF with sparseness constraints for data distortion. The evaluation measures used by them include privacy parameters (VD, RP, RK, CP, CK) and utility parameters (accuracy of KNN). Bhandare [7] has proposed the Tanh normalization for PPDM and has used privacy parameters VD, RP, RK, CP, CK and utility parameter (J48) for evaluation. Peng et al. have proposed various combinations of SVD, NMF and DWT for data distortion and have used the privacy measures to compare the results [8]. Zhang et al. in [9] have compared NMF and SVD for data distortion with noise addition. Another method for perturbation has been proposed by Li and Zhang [10] in which DRDP and rotation-based transformation methods are combined for privacy. They have used the degree of privacy and misclassification error for comparison of results.

Lee and Seung [11] have explained NMF, and Li et al. have used it for perturbation-based PPDM in [12] by combining with random perturbation. The analysis of results has been done based on the mean absolute error. Wang et al. [13] have used NMF in an iterative manner for privacy protection and also have used privacy measured for evaluation. Nagalakshmi and Rani in [14] have combined NMF with PCA to provide a higher level of data distortion. They have applied K-means clustering on the final data to evaluate the degree of privacy and clustering quality. Li and Xi [15] proposed an improved NMF to preserve privacy using privacy measures to compare the results. Xu et al. in [16] have proposed the use of sparsified SVD for data distortion. They have proposed various parameters to measure privacy that include value difference measures (VD) and position difference measures (RP, RK, CP and CK). Afrin et al. [17] have combined NMF and SVD to perturb a dataset, and the utility of this perturbed dataset is measured with respect to query accuracy. They have measured privacy for the perturbed dataset using metrics VD, RP, CP, RK and CK. In [18], Koushika and Premlatha have used SVD and then repeated 3D rotation data perturbation for preserving the data privacy. Further, they have used different classifiers on both perturbed data and original data to measure the percentage of accuracy.

3 Perturbation-Based PPDM

Privacy-preserving data mining (PPDM) techniques are used to protect sensitive data from unauthorized access without decreasing the data utility. According to Xu et al. [2], PPDM works at two levels: protection of sensitive data before being sent for mining and protection of mining results before being sent for decision-making purposes. There are various PPDM techniques like perturbation-based PPDM, anonymization-based PPDM, randomization-based PPDM, condensation-based PPDM and cryptography-based PPDM [19–21], which are dependent upon the method used for hiding the data. Bhandari and Pahwa [22] have evaluated the advantages and disadvantages of these methods.

Perturbation-based PPDM methods are simple and efficient data transformation methods that are used before sending the data for analysis purposes. These methods are suitable for centralized as well as distributed databases. These methods allow treating each attribute individually securing the data while preserving its statistical properties. Various perturbation methods exist based upon the methodology used for data distortion. Perturbation methods can be additive like noise addition and can be multiplicative like rotation and shearing. They can be geometric, allowing sequence of geometric transformations [3]. Apart from these, there are some perturbation methods that involve matrix decomposition like SVD, NMF and QR decomposition [4, 6, 7]. There are other perturbation techniques as well, namely DRDP, DWT and many more [8, 10].

3.1 Nonnegative Matrix Factorization (NMF)

Datasets are usually represented as high-dimensional matrices. There are different matrix decomposition techniques as discussed by Zhang et al. [9], which can help secure data efficiently. NMF or non-negative matrix factorization is a matrix decomposition method, which works on non-negative values for attributes. This method is a fast method with low computational cost. It maintains the non-negativity of the datasets. It is explained as follows:

Let V be a $m \times n$ non-negative data matrix. NMF factorizes V into two matrices W and H such that W and H are non-negative and have size $m \times k$ and $k \times n$, respectively, such that

$$V \approx WH \tag{1}$$

k is chosen such that $k < \min(m, n)$ and $(m + n)k < mn$ [12]. W is generally considered to be the basis matrix, which is used to combine each vector of coefficient matrix H to optimize V. The result of NMF depends on the value of k, and thus, it is more secure than other matrix decomposition methods as it does not give unique results [11–15]. The values of matrices W and H are selected such that the error between V and WH is minimized. The error function used by Wang et al. [13] is defined as:

$$E(W, H) = 1/2 \sum_i \sum_j \left(V_{ij} - (WH)_{ij}\right)^2 \tag{2}$$

3.2 Double-Reflecting Data Perturbation (DRDP)

DRDP or double-reflecting data perturbation is a simple data distortion method used to perturb all or some confidential attributes of a dataset [10, 23]. It is an easy-to-implement method that transforms data based on the operation shown below:

$$\mathrm{op}_j = \rho V_j + (\rho V_j - v_j) = 2\rho V_j - v_j \tag{3}$$

where V_j is the confidential attribute and v_j is an instance of V_j. Here ρV_j is defined as:

$$\rho V_j = \left|\max(V_j) + \min(V_j)\right|/2 \tag{4}$$

4 Privacy Measures

Various privacy measures/parameters are proposed to analyze the performance of different data distortion methods. These measures are evaluated using the value of datasets before and after the distortion is done. In this paper, we have used the privacy measures proposed by Xu et al. [16] as defined below:

4.1 Value Difference

Value difference (VD) represents the relative difference in the value of datasets V before distortion and V' after distortion. It is measured as the ratio of Frobenius norm of difference of V and $|V'|$ to Frobenius norm of V.

$$\mathrm{VD} = \left\| V - |V'| \right\|_F / \| V \|_F \tag{5}$$

A good data distortion method has a higher value for the value difference parameter VD.

4.2 Position Difference

Just like the value of data changes after distortion, similarly the relative order of data also changes. The position difference metrics are used to measure this change. These are explained below:

RP represents the average change of order for all the attributes. It is calculated as:

$$\text{RP} = \frac{1}{mn}\sum_{i=1}^{m}\sum_{j=1}^{n}\left|\text{Ord}_j^i - \overline{\text{Ord}}_j^i\right| \tag{6}$$

where dataset V has m attributes and n instances. Ord_j^i is the ascending order of jth element in attribute i in V and $\overline{\text{Ord}}_j^i$ is the ascending order of jth element in attribute i in V'.

RK represents the percentage of elements that keep their order same in their respective columns post the distortion process. It is calculated as:

$$\text{RK} = \frac{1}{mn}\sum_{i=1}^{m}\sum_{j=1}^{n}\text{Rk}_j^i \tag{7}$$

where Rk_j^i is 1 if the element keeps its position in the column after distortion also, otherwise it is 0.

After the distortion process, the order of average value of each attribute changes just like the change of order of values in each column.

CP is used to measure the change of order of average values of the attributes after distortion. It is calculated as:

$$\text{CP} = \frac{1}{m}\sum_{i=1}^{m}\left|\left(\text{OrdAv}_i - \overline{\text{OrdAv}_i}\right)\right| \tag{8}$$

where OrdAv_i and $\overline{\text{OrdAv}_i}$ are ascending order for attribute i based on their average values before and after distortion, respectively.

CK represents the percentage of attributes for which the average value is in the same order after distortion as before distortion. It is evaluated as:

$$\text{CK} = \frac{1}{m}\sum_{i=1}^{m}\text{Ck}^i \tag{9}$$

where Ck^i is 1 if attributes have the same order of average value after distortion, else it is 0.

A good data distortion method ensures more privacy if RP and CP are increased while RK and CK are decreased.

5 Methodology and Results

The experimental work is performed in R language. R is an open-source tool, which is free of cost, easily available for download and can be used on different operating systems and hardware. It deals with statistical computations of data for analysis.

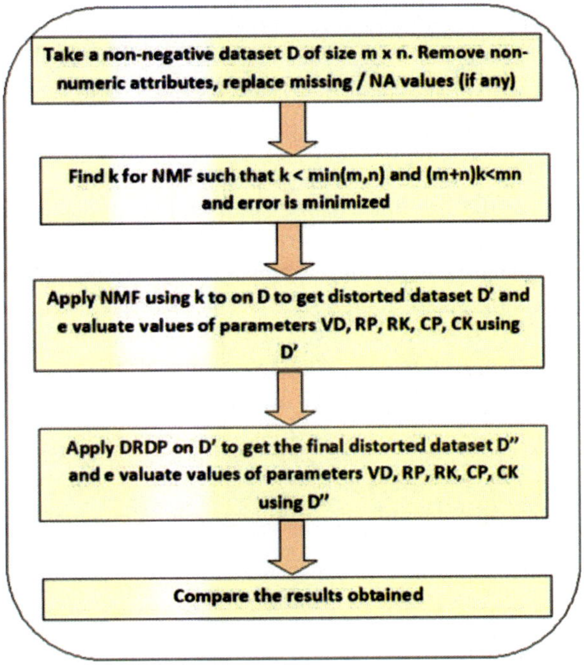

Fig. 1 Flowchart of the work done

R has various repositories of packages like the CRAN "Comprehensive R Archive Network," which has around 13,695 available packages currently. In our work, we have used the NMF package (version 0.20.6) of R [24]. This package provides a framework to perform nonnegative matrix factorization (NMF).

We have performed the experimental work on 5 datasets from the UC Irvine Machine Learning Repository [25]. The methodology followed is explained below, and the flowchart is shown in Fig. 1.

1. Non-negative dataset D was taken, and non-numeric attributes were removed.
2. Missing or NA values were replaced with mean.
3. NMF was applied to get distorted dataset D' using Eq. (1) where the value of k was chosen such that $k < \min(m, n)$ and $(m + n)\,k < mn$ so as to minimize the error in Eq. (2).
4. Privacy parameters, i.e., value difference (VD) in Eq. (5) and position difference parameters (RP, RK, CP, CK) in Eqs. (6), (7), (8) and (9) were evaluated for distorted dataset D'.
5. Final distorted dataset D'' was generated by applying DRDP on dataset D' obtained in step 3 using Eqs. (3) and (4).
6. Privacy parameters, i.e., value difference (VD) and position difference parameters (RP, RK, CP, CK), were evaluated for the final distorted dataset D'' obtained in step 5.
7. The results obtained in steps 4 and 6 were compared to evaluate the performance of proposed method of extended NMF with that of NMF alone.

Results of values of various parameters for five datasets are shown below.

5.1 IRIS Dataset

This is the Iris Plant dataset [26] with 150 instances and 5 attributes. Four out of the 5 attributes are numerical which represent sepal length, sepal width, petal length and petal width, while 1 attribute is the class attribute. For experimental purposes, four numeric attributes have been considered, while the 5th attribute representing the class has been removed. The screenshots of the IRIS dataset after removing the non-numeric attribute, after performing NMF and after NMF followed by DRDP, are shown in Tables 1, 2 and 3, respectively.

Table 1 IRIS dataset after removing non-numeric attribute

V_1	V_2	V_3	V_4
5.1	3.5	1.4	0.2
4.9	3	1.4	0.2
4.7	3.2	1.3	0.2
4.6	3.1	1.5	0.2
5	3.6	1.4	0.2
5.4	3.9	1.7	0.4
4.6	3.4	1.4	0.3
5	3.4	1.5	0.2
4.4	2.9	1.4	0.2
4.9	3.1	1.5	0.2
5.4	3.7	1.5	0.2

Table 2 IRIS dataset after applying NMF

V_1	V_2	V_3	V_4
5.113612	3.484247	1.389459	0.2022147
4.803223	3.115013	1.469385	0.2226879
4.699577	3.200659	1.300202	0.2002391
4.628351	3.071801	1.455843	0.2734002
5.029893	3.569006	1.359493	0.2606411
5.419527	3.879062	1.676759	0.4341626
4.639207	3.354516	1.367840	0.3437485
5.026599	3.371554	1.468017	0.2401053
4.406707	2.895185	1.381310	0.2433567
4.862765	3.152169	1.491968	0.2276014
5.401819	3.700578	1.486751	0.2352824

Table 3 IRIS dataset after applying the extended NMF

V_1	V_2	V_3	V_4
7.239236	2.761682	6.580751	2.47943609
7.557718	3.115841	6.513207	2.45241053
7.654101	3.043131	6.673460	2.48103893
7.684480	3.225783	6.524554	2.45256327
7.238407	2.788313	6.627479	2.52175205
6.821024	2.519600	6.297478	2.37383426
7.599932	3.043070	6.618867	2.46968712
7.341253	2.853090	6.504117	2.42553087
7.949543	3.344238	6.593816	2.43284975
7.499620	3.076324	6.489994	2.44554194
6.957445	2.534493	6.489833	2.44435086

Table 4 Privacy parameters for IRIS dataset

S.No.	Parameters	NMF	Extended NMF
1.	VD	0.02038	0.5409
2.	RP	4.81333	74.68
3.	RK	0.1217	0.00167
4.	CP	0	0
5.	CK	1	1

To perform the experiment, the value of k for NMF has been set to 3 for this dataset. The results of various parameters are shown in Table 4.

It can been seen from Table 4 that the values of VD and RP have increased for the proposed sequential implementation of NMF followed by DRDP and that the value of RK has reduced. The values of CP and CK remain unchanged. Thus, for the IRIS dataset it is clear that the proposed perturbation method is better in privacy than NMF method alone.

5.2 *Absenteeism_At_Work Dataset*

This dataset [27] represents the absenteeism at work data at a courier company in Brazil. It has 740 records and 21 attributes, and all of them are numerical. These attributes represent Individual identification, Reason for absence, Month of absence, Day of the week, Seasons, Transportation expense, Distance from Residence to Work (kilometers), Service time, Age, Work load Average/day, Hit target, Disciplinary failure, Education, Son (number of children), Social drinker, Social smoker, Pet (number of pets), Weight, Height, Body mass index and Absenteeism time in hours.

Table 5 Privacy parameters for Absenteeism_At_Work dataset

S.No.	Parameters	NMF	Extended NMF
1.	VD	0.01171	0.50203
2.	RP	101.5006	336.2687
3.	RK	0.013835	0.00097
4.	CP	0	1.0476
5.	CK	1	0.52381

Table 6 Privacy parameters for Final_Grades dataset

S.No.	Parameters	NMF	Extended NMF
1.	VD	0.02126	0.98399
2.	RP	9.42115	27.7599
3.	RK	0.17025	0.00538
4.	CP	0.2222	1.7778
5.	CK	0.7778	0.1111

The value of k for NMF is taken as 20 based on the required conditions. Table 5 shows the values of various parameters for this dataset.

Table 5 shows that values of VD, RP and CP have increased, while the values of RK and CK have decreased for the proposed sequential NMF-DRDP method making it better in terms of privacy than NMF.

5.3 *Final_Grades Dataset*

This dataset represents the final grades obtained by students for a course in the final examination [25]. This dataset has 62 instances and 18 attributes. First attribute represents student ID, and remaining attributes represent the grades scored by the students. The value of $k = 13$, based on the required conditions. The results of all parameters are shown in Table 6.

It can be seen from Table 6 that values of parameters VD, RP and CP have increased, while those of CK and RK have decreased for the proposed method.

5.4 *Wholesale Customers Dataset*

This dataset [28] from the UCI repository represents the annual expense on various products for a wholesale distributor. It contains 440 instances and 8 attributes. First 2 attributes in this dataset represent the customer channel and region. The next six attributes represent the annual spending on fresh products, milk products, grocery products, frozen products, detergents and paper products and delicatessen products.

Table 7 Privacy parameters for wholesale customers' dataset

S.No.	Parameters	NMF	Extended NMF
1.	VD	0.01441	6.38811
2.	RP	34.9403	206.3432
3.	RK	0.3599	0.00028
4.	CP	0	0.25
5.	CK	1	0.75

For the experimental work, value of k is taken as 7 for NMF. The results for this dataset are shown in Table 7.

It is clear from Table 7 that VD, RP and CP have increased and RK and CK have decreased for this dataset also for the proposed sequential method of NMF and DRDP compared to NMF only.

5.5 *Breast_Cancer_Wisconsin Dataset*

This dataset was obtained from the University of Wisconsin Hospitals, Madison from Dr. William H. Wolberg [29]. It is the breast cancer dataset representing two possible classes, namely benign and malignant. It contains 699 instances and 11 attributes. Sixteen instances had missing values that were replaced with the mean of the attribute. The value of k was chosen to be 10 for NMF based on the requirements.

In Table 8, we can see that values of VD, RP and CP have increased and values of RK and CK have decreased. From the tables shown above, graphs for various parameters are given in Figs. 2, 3, 4, 5 and 6.

Figures 2, 3 and 5 show that the values of parameters VD, RP and CP, respectively, have increased for all the datasets for extended NMF in comparison with NMF alone. It can be seen from Figs. 4 and 6 that values of RK and CK have decreased for extended NMF compared to NMF alone for all the datasets.

Thus, it is clear that the values of privacy measures VD, RP and CP have increased in extended NMF (sequential implementation of NMF followed by DRDP) than in NMF alone. Also, the values of RK and CK have reduced in the proposed combined method than NMF alone.

Table 8 Privacy parameters for Breast_Cancer_Wisconsin dataset

S.No.	Parameters	NMF	Extended NMF
1.	VD	7.8223e−06	9.251683
2.	RP	214.428	245.170
3.	RK	0.0848	0.0013
4.	CP	0.18182	0.36364
5.	CK	0.81818	0.72727

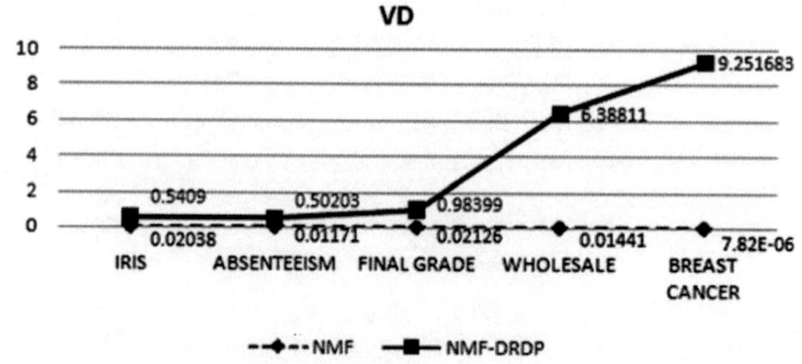

Fig. 2 Graph of values of parameter VD for all the datasets for NMF and extended NMF

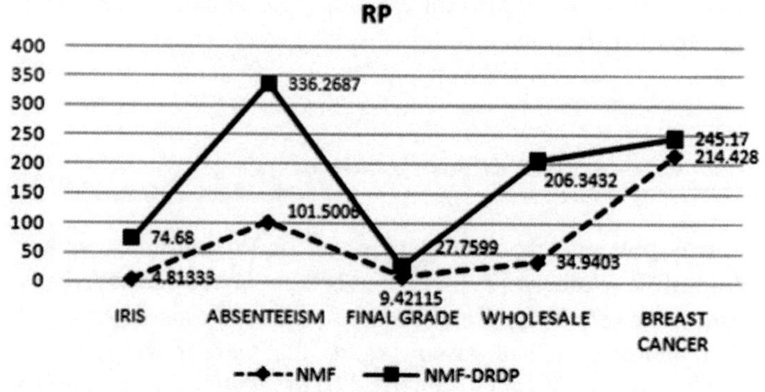

Fig. 3 Graph of values of parameter RP for all the datasets for NMF and extended NMF

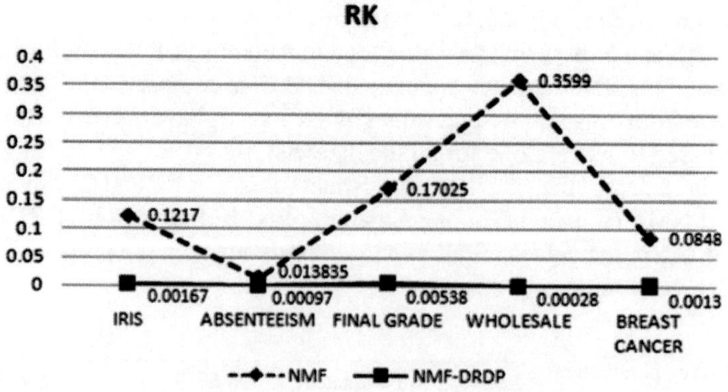

Fig. 4 Graph of values of parameter RK for all the datasets for NMF and extended NMF

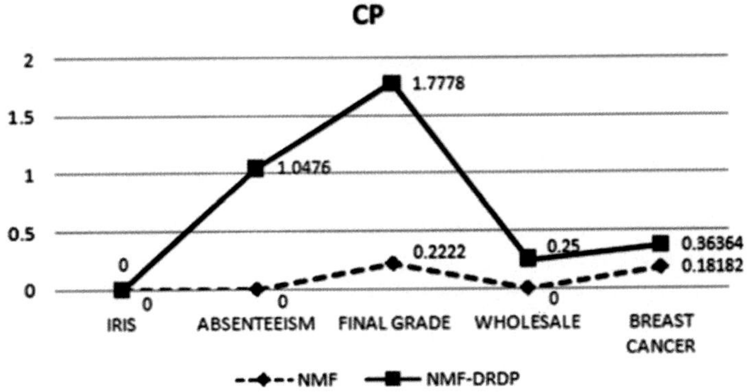

Fig. 5 Graph of values of parameter CP for all the datasets for NMF and extended NMF

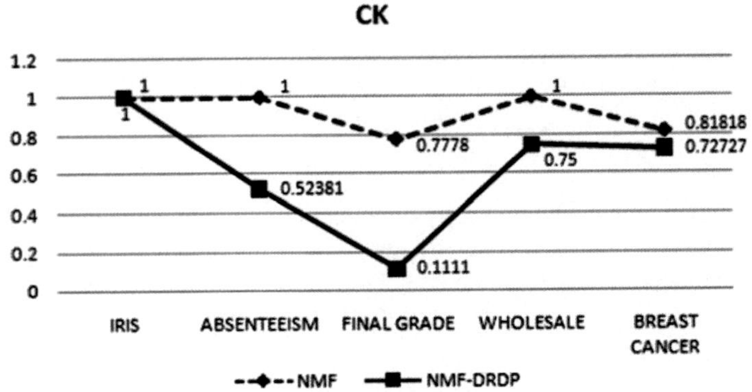

Fig. 6 Graph of values of parameter CK for all the datasets for NMF and extended NMF

6 Conclusion

Data privacy is a very sensitive and vital issue in today's data-oriented world. It is very important to keep the data safe before using it for any information retrieval purpose. Perturbation-based PPDM allows to ensure data privacy without changing the utility and statistical properties of the data.

In this paper, two perturbation methods, namely NMF and DRDP, are sequentially executed one after the other to produce a distorted dataset. The privacy parameters for the resultant distorted dataset from extended NMF (NMF followed by DRDP) are compared with the privacy parameters evaluated from distorted dataset obtained from NMF alone. The results show that the proposed method of sequential implementation of NMF followed by DRDP has increased values for privacy measures such as VD, RP and CP and decreased values for privacy measures such as RK and CK for different non-negative numeric datasets. Thus, it can be concluded that our proposed perturbation method of extended NMF (NMF followed by DRDP) provides better privacy levels as compared to NMF alone to non-negative numeric datasets.

Data is of no use until it is mined and information is extracted from it to help organizations make important and crucial decisions. Clustering is one of the data mining methodologies that allow to extract important information. So in the future, it is of interest to evaluate the performance of various clustering algorithms and compare which clustering algorithm performs better with secure perturbed data from extended NMF compared to distorted data from NMF.

References

1. Chen MS, Han J, Yu PS (1996) Data mining: an overview from a database perspective. IEEE Trans Knowl Data Eng 8:866–883
2. Xu L, Jiang C, Wang J, Yuan J, Ren Y (2014) Information security in big data: privacy and data mining. IEEE Access 2:1149–1176
3. Xiao-dan W, Dian-min Y, Feng-li L, Yun-feng W, Chao-Hsien C (2006) Privacy preserving data mining algorithms by data distortion. In: International conference on management science and engineering, Lille, pp 223–228
4. Maheswari N, Revathi M (2014) Data security using decomposition. Int J Appl Sci Eng 12(4):303–312
5. Manikandan G, Sairam N, Sudhan R, Vaishnavi B (2012) Shearing based data transformation approach for privacy preserving clustering. In: 2012 Third international conference on computing, communication and networking technologies, ICCCNT'12, Coimbatore, pp 1–5
6. Kabir SMA, Youssef AM, Elhakeem AK (2007) On data distortion for privacy preserving data mining. In: 2007 Canadian conference on electrical and computer engineering, Vancouver, BC, pp 308–311
7. Bhandare SK (2013) Data distortion based privacy preserving method for data mining system. Int J Emerg Trends Technol Comput Sci 2
8. Peng B, Geng X, Zhang J (2010) Combined data distortion strategies for privacy-preserving data mining. In: 3rd International conference on advanced computer theory and engineering, ICACTE, Chengdu, V1-572–V1-576
9. Zhang J, Wang J, Xu S (2007) Matrix decomposition based data distortion techniques for privacy preserving in data mining. Technical report, Department of Computer Science, University of Kentucky, Lexington. Retrieved from https://www.academia.edu/7981302/Matrix_Decomposition-Based_Data_Distortion_Techniques_for_Privacy_Preservation_in_Data_Mining
10. Li L, Zhang Q (2009) A privacy preserving clustering technique using hybrid data transformation method. In: 2009 IEEE International conference on grey systems and intelligent services, GSIS 2009, Nanjing, pp 1502–1506
11. Lee DD, Seung HS (2000) Algorithms for non-negative matrix factorization. In: Proceedings of 13th international conference on neural information processing systems, NIPS'00, pp 535–541
12. Li T, Gao C, Du J (2009) A NMF-based privacy-preserving recommendation algorithm. In: 2009 First international conference on information science and engineering, Nanjing, pp 754–757
13. Wang J, Zhong W, Zhang J (2006) NNMF-based factorization techniques for high-accuracy privacy protection on non-negative-valued datasets. In: Sixth IEEE international conference on data mining—workshops, ICDMW'06, Hong Kong, pp 513–517
14. Nagalakshmi M, Rani KS (2013) Privacy preserving clustering by hybrid data transformation approach. Int J Emerg Technol Adv Eng 3
15. Li G, Xi M (2015) An improved algorithm for privacy-preserving data mining based on NMF. J Inf Comput Sci 3423–3430

16. Xu S, Zhang J, Han D, Wang J (2005) Data distortion for privacy preservation in terrorist analysis System. In: Proceedings of IEEE international conference on intelligence and security informatics, ISI 2005, vol 3495, Atlanta, GA, USA
17. Afrin A, Paul MK, Sattar AHMS (2019) Privacy preserving data mining using non-negative matrix factorization and singular value decomposition. In: Proceedings of 4th international conference on electrical information and communication technology, EICT, pp 1–6
18. Koushika N, Premlatha K (2021) An improved privacy-preserving data mining technique using singular value decomposition with three-dimensional rotation data perturbation. J Supercomput 1–9
19. Malik MB, Ghazi MA, Ali R (2012) Privacy preserving data mining techniques: current scenario and future prospects. In: 2012 Third international conference on computer and communication technology, Allahabad, pp 26–32
20. Li X, Yan Z, Zhang P (2014) A review on privacy-preserving data mining. In: 2014 IEEE International conference on computer and information technology, Xi'an, pp 769–774
21. Vaghashia H, Ganatra A (2015) A survey: privacy preserving techniques in data mining. Int J Comput Appl 119
22. Bhandari N, Pahwa P (2019) Comparative analysis of privacy-preserving data mining techniques. In: Bhattacharyya S, Hassanien A, Gupta D, Khanna A, Pan I (eds) International conference on innovative computing and communications. Lecture notes in networks and systems, vol 56. Springer, Singapore. (Proceedings of ICICC, Delhi, India, vol 2, 2018)
23. Balajee M, Narasimham C (2012) Double-reflecting data perturbation method for information security. Orient J Comput Sci Technol 5:283–288
24. Gaujoux R (2018) An introduction to NMF package version 0.20.6. Retrieved from https://cran.r-project.org/web/packages/NMF/vignettes/NMF-vignette.pdf
25. UCI Machine Learning Repository. Available online: https://archive.ics.uci.edu/ml/index.php
26. Fisher RA (1988) Iris. UCI Machine Learning Repository
27. Martiniano A, Ferreira R (2018) Absenteeism at work. UCI Machine Learning Repository
28. Cardoso M (2014) Wholesale customers. UCI Machine Learning Repository
29. Wolberg WH (1992) Breast cancer Wisconsin (original). UCI Machine Learning Repository

Challenges of Robotics: A Quest for an Integrated Action

Md. Toriqul Islam, Ridoan Karim, and Sonali Vyas

Abstract Because of the inherent potentials, robotics has grown increasingly popular in many workplaces throughout time. Unless malfunctioning, a robot can perform its assigned jobs non-stop, perfectly, and quickly. It can perform in extreme conditions, such as deactivating explosives; exploring mines; finding sunk shipwrecks; and rescuing survivors. These large-scale uses of robotics inescapably cause tremendous ethical, social, and legal challenges in the contemporary world, which need to be redressed. The main focus of this article is to analyze those challenges encompassing AI and robotics and shed light on the prospective solutions thereof. This paper argues that the challenges in the field cannot be solved by a single effort; rather, an integrated action is needed from all stakeholders. Hence, a joint action plan, accelerated by national–international collaboration and cooperation and led by the United Nations, might be the chosen alternative.

Keywords Emergence of robotics · Challenges · Prospects · Regulations · Suggestions

Md. T. Islam
Faculty of Law, University of Malaya, 50603 Kuala Lumpur, Federal Territory of Kuala Lumpur, Malaysia
e-mail: toriqul@siswa.um.edu.my

R. Karim
Department of Business Law and Taxation, School of Business, Monash University Malaysia, Jalan Lagoon Selatan, 47500 Bandar Sunway, Selangor Darul Ehsan, Malaysia
e-mail: ridoan.karim@monash.edu

S. Vyas (✉)
School of Computer Science, University of Petroleum and Energy Studies, Dehradun, Uttarakhand 248007, India
e-mail: vyas.sonali86@gmail.com

V. Skala et al. (eds.), *Machine Intelligence and Data Science Applications*, Lecture Notes on Data Engineering and Communications Technologies 132,
https://doi.org/10.1007/978-981-19-2347-0_18

1 Introduction

Last few decades, we have witnessed a remarkable development in robotics and increased applications thereof in domestic, commercial, or military spheres. The widespread human–robot interactions give rise to several ethical, social, and legal challenges, which need to be resolved. As a result, robotics and automation attracted the attention of regulators to the legal, social, and political complexities. It will be a disaster if a robot fails or malfunctions since it links humans through various services and operations. In general, robotics is considered a complex and interdisciplinary science requiring the attention of a rising amount of social, ethical, and legal issues due to its diversity in design, production, applications, and operations.

A single viewpoint will never manage the intricacies that have evolved in the field; instead, a multidisciplinary strategy needs to be established [1]. Additionally, there should have uniformity in the regulation of robotics to avoid the plea of jurisdictional limitations. The authors further urge that the United Nations should play the central role in combining all endeavors encompassing the robotics' regulation into a specific direction. This paper aims to offer a brief overview of the social, ethical, and legal challenges of robotics, together with the regulatory aspects thereof, while prescribing some specific suggestions and recommendations.

2 Emergence of Robotics

The term 'robot' has emerged from the Czech word 'robota' that was coined by Czech poet Karel Capek in 1924, the meaning of which is forced labor [2]. Later, in the 1940s, Isaac Asimov used the term 'robots' as one of the leading characters in his visionary novels. Subsequently, industrial robots came into reality in the 1960s by Joseph F. Engel Berger, who introduced the PUMA robot with a modern shape and efficiency [3].

The robotics history can be divided into four phases. The year 1954 can be shown as the first phase; the year 1978 represents the second phase; the year 1980 represents the third phase, while the year 1995 to this day can be shown as the fourth phase. For instance, George Devol designed the first programmed robot in 1954 terming as the Universal Automation, and subsequently, he abbreviated it to Unimation, and later, this turned into the name of the first robot company of the world in 1962. In 1978, the Unimation company developed the Puma (Programmable Universal Machine for Assembly) robot in association with General Motors.

The third phase started in 1980—when the robotics industry had expanded in leaps and bounds. Many universities and institutions started introducing programs and courses on robotics among a wide range of disciplines, from electrical engineering, and mechanical engineering to computer science. The last phase has started in 1995, and that is continuing to date—when the use of robotics has dramatically expanded and spread out everywhere as an educator, entertainer, and executioner [4].

3 Prospects and Challenges

There remains a long hope encompassing this cutting-edge technology, as robots relieve humans from doing toiling jobs while providing comfort, speed, perfection, etc. Although there are worries that automation would take over everything and strip us of our employment, there are also a lot of opportunities that are inclined with robotics. As a modern feature of today's technology, robotics expands fast enough to touch upon all the elements of our life. It is presumed that the next-generation robots will collaborate with every human effort and task to enrich efficiency. They will cohabit with humans and provide physical and psychological assistance, eventually contributing to creating a safe and peaceful civilization. Thus, a group of mathematicians and the member of the French Parliament once opined that AI has been ascended in this era to generate many hopes by executing much more crucial jobs than before. It endeavors continually for paving us the new way forward, and eventually, the black box algorithms will never be capable of stop-ping our digital society [5].

Apart from the numerous prospects, there are huge challenges, problems, and dilemmas encircling robotics. However, in this article, we will concentrate on the ethical, social, and legal challenges and analyze the regulatory measures thereof. Our focus is not to evaluate the regulatory aspects comprehensively, rather to indicate that they are sufficiently inadequate because of the lack of universal applicability, lack of cooperation, and coordination.

3.1 *Ethical Challenges*

Too much autonomy and wide expansive interaction with human beings enable robots to pose serious ethical threats nowadays. Occasionally, questions in AI and robotics are left unsolved due to ambiguity. A self-driving car, for example, may be in danger of colliding with a pedestrian or slamming into anything else. This circumstance may create additional dilemmas, such as whether it is reasonable to have multiple methods depending on the age or physical condition of the pedestrian. Many feel that robots should not handle such extreme situations [6].

Another ethical and legal issue is that a privacy breach is always likely to happen in robotic applications [7]. It is important to remember that most AI-based systems process personal data for a variety of objectives. For example, self-driving vehicles and drones process massive amounts of personal data in the name of offering a wide range of services, which may be misused for nefarious ends. There are other severe issues concerning robotics. For example, childcare robots may be very speculative, causing negative impacts on children's mental, physical, and emotional well-being [8].

It is essential to see how the law and society see and implement AI as further concerns are present with robotic prostheses. What happens if technology, such as consumer robots, become dishonest and misleading to humans, endangering, and invading the privacy of individuals? A robust and practical regulatory system based on ethical, social, and legal norms and values is needed in this environment. Ethical issues may play a critical role in the absence of potential legal safeguards or social aversion toward the sector.

3.2 *Social Challenges*

Robots can also cause many societal issues, such as high unemployment. According to Oxford academics, robots may be able to nullify up to 47% of US employment [9]. On the other hand, Frey and Osborne anticipated a bit more, estimating that up to half of the American job market might be lost to robots in the next 20 years [10]. Müller and Bostrom found that half of the human vocations will be replaced by robotics in 2040–2050, with the number rising to 9 out of 10 by 2075 based on a poll of 170 respondents out of 549 highly qualified academics and professionals [11].

As a result, labor-intensive economies such as India, South Africa, Brazil, China, Russia, and Bangladesh may suffer severe societal consequences due to the inevitable unemployment concerns generated by robots. Even law professionals may be replaced by AI technologies, as ROSS Intelligence and IBM's Watson provide legal advice that is more satisfactory and accurate than that of human lawyers [12]. Meanwhile, AI is used to decide thousands of court cases in China.

Although there are predictions of significant job losses and job creations as prepared by a group of world-famous technologists and economists, it is tough to predict how many jobs will be created or lost due to the innovation and application of robotics. To predict the probable implications of robotics on future employment, the following table has been shared that is prepared following a report of the MIT Technology Review.

Realizing the risks, world-famous corporate executives, scholars, and scientists have been warning the rapid growth of robotics and AI, as well as the services they provide. Stephen Hawking has enumerated in an interview with BBC that 'human beings can never compete with AI, and they are always likely to be superseded by AI because of their slow biological growth' [13]. Bill Gates observed, 'I am in a society that is highly tensed as to superintelligence' [14]. Bill Joy of Sun Microsystems wrote, the mighty twenty-first century technologies, e.g., nanotech, genetic engineering, and robotics, are always posing threats to the human being causing an apprehension of extinction [15].

3.3 *Legal Challenges*

The increasing complexities encompassing robotics create conditions that sometimes undervalue the traditional legal doctrines, e.g., tort, negligence, or product liability.

Sometimes, the courts feel embarrassed about imposing civil, or criminal liability on robots, especially on autonomous robots. In particular, due to the lack of mens rea, it is uncomforting for the judges to inflict criminal liability on robots. Furthermore, it is absurd for the judges to treat a robot as a wrongdoer.

Similarly, it is also impossible to impose accountability upon robots [16]. Recently, the EU approached whether they could define robots precisely and attribute personhood upon robots, but the Europeans have been divided into parts on the question [17].

By an open letter, 156 AI experts from the 14 EU nations have rejected the EU Parliament's recommendation of the imposition of legal personhood to robots. The expert group opined that it is a blunder to shift the human's responsibility to the machine in the case of an occurrence or accident. This would exempt the human actors from their mistakes, which they can do in the designing, manufacturing, and operating stages [17]. Even though the imposition of legal personhood to robots seems to be a fantasy today, this will surely be true shortly for controlling the large-scale human–robot interactions [18].

4 Regulations

In the absence of a universally recognized framework, the following instruments play vital roles in robots' regulation.

4.1 *Asimov's Law*

Isaac Asimov's three ethical principles, often known as Asimov's rule of robotics, come first in all regulatory debates on robots. The summary of those principles is—a robot cannot harm a human being by any means; unless conflicting with the first principle, a robot must obey the orders of human being, and a robot must protect its existence unless that is conflicting with the first, or second principles [19].

4.2 *Code*

Another significant consideration in the robotics regulatory framework is the codes. Designers, manufacturers, and operators should respect Asimov's principles until and unless robots become self-supporting. As a result, the code is a regulatory block, a set of architectural or behavioral principles implanted in a system that forbids departure. Many contend that self-driving cars, such as the Google car, should adhere to pre-programmed traffic restrictions into the car's software [20].

4.3 Soft Law

To deal with the transnational nature of robots, international bodies, non-state players, and independent organizations create soft law instruments. This is necessary for the sector's worldwide standard, quality, and flexibility to be maintained. Many agencies, international groups, and other non-state players, such as the International Standardization Organization (ISO), are involved in the development of several soft law rules.

4.4 Social Norms and Values

Unlike the legal method, social standards are governed by society as a whole rather than by the state power. Although it looks that social punishments are less severe, this is not accurate. In certain civilizations, minor infractions such as smoking in front of youngsters, pregnant women, or at the dinner table can elicit intense communal resentment, ending in expulsion from the society with a stigma attached [21].

4.5 Formal Legislation

The EU passed the 'Civil Law Rules on Robotics' on 16 February 2017, which contains 68 principles under a broad heading 'General principles concerning the development of robotics and artificial intelligence for civil use.' This is the first legal document of this kind containing comprehensive provisions for all aspects of AI and robotics ranging from ethical to societal, and legal implications [22].

In Feb 2017, the Virginia State of the USA passed a law allowing robots for the delivery of products to the customer's door. Later the Idaho and Florida States also wished to pass similar legislation. Meanwhile, Nevada State has enacted law and framed rules allowing the root permits of autonomous trucks.

The California Consumer Protection Act, 2018, and the Vermont Data Broker Legislation, 2018, contain provisions relating to AI. Additionally, over 20 other US States have enacted laws containing provisions regarding the safety, security, liability, and accountability issues of autonomous vehicles.

Apart from the USA, the UK, France, Germany, China, India, Japan, Singapore, South Korea, Australia, and New Zealand have taken numerous guidance, published reports, set strategies, adopted model frameworks, established institutions, built networks, and proposed to enact bills concerning the AI and robotics.

The United Nations and other international and regional forums are also very much concerned about the impacts of AI and robotics. In particular, the UN has taken many efforts to address the issue, which have been evident through its continuous endeavors by many of its specialized organizations, such as UNICRI, UNESCO, UNCTAD,

ITU, and so on. In the regional spheres, the prominent endeavors are—Robolaw Project, ROBOLAW.Asia, etc.

On April 8, 2019, a high-level expert group of the EU has introduced the Ethical Guidelines for Trustworthy AI. As per the said Guidelines, a trustworthy AI should have the following three components that must be met throughout its entire life cycle: (1) an AI shall have to be lawful through compliance with all existing laws, rules, and regulations; (2) it must be ethical by ensuring its adherence to the ethical principles, standards, and values, and (3) an AI should be robust in terms of both the social and technical perspective, as it is likely to cause unintentional harm by AI despite having built-in good intentions.

4.6 Recent Legal Developments

The way we work, live, play, or entertain today is heavily dependent on AI or robotic systems. Automation or machine make life easier, faster, and smarter but poses tremendous challenges too. This technology-dependent world system requires extensive legal rules, regulations, and guidelines for establishing risk, or danger-free modern life. Nonetheless, there are legal vacuums for regulating technologies, especially for AI and robotics industries. There exist some scattered legal postulates, rules, frameworks, guidelines, etc., which are mostly enforced unevenly. The functions, operations, or implications of both AI and robotics are global in nature that entails global response for governance. Some recent activities of the United Nations, European Union, and OECD are playing significant roles toward establishing integrated global governance for AI and robotics.

On behalf of the United Nations, UNESCO is planning to release its final draft on Ethics of Artificial Intelligence at the end of 2021, the first draft of which was released in 2019. On September 30, 2021, the UN General Assembly arranged a symposium in searching for Recommendations on UNESCO's Ethics of AI. Among others, the Recommendation proposes an anticipatory, novel, and transformative framework conducive to conclusive policy actions based on universal values and principles.

The current advancement in some frontier technologies, such as AI, biotechnology, and robotics has had huge potentials for sustainable development. It is evident that these technologies currently present a market share of $350 billion, and it could rise to $3.2 trillion by 2025. Nonetheless, there exist potential risks of inequality and disparity among nations across the globe. To reduce such risks, the UN is working for years through its specialized institutions. Through the UNCTAD, the UN recently calls for strengthening international cooperation to advance innovation capabilities among developing nations; promote technology transfer; raise women's participation in IT industries; carry out technological assessments; and encourage extensive debates on the implications of these cutting-edge technologies on sustainable development.

In recent years, the European Commission has also been working intensively to facilitate and enhance cooperation encompassing AI industries; increase competence;

and secure trust in the region based on the EU values. Current major EU initiatives to develop inclusive AI strategy and AI regulation include, among others, the European Parliament resolution of 16 February 2017 with recommendations to the Commission on Civil Law Rules on Robotics (2015/2103(INL)); European Strategy on AI 2018; Guidelines for Trustworthy AI 2019; European Commission's White Paper on AI 2020, and the Proposal for AI Regulation 2021. Among all EU initiatives, the latest proposal for AI Regulation shall have immense global implications for future AI development, research, and governance.

Despite having huge potentials across industries, especially by offering new products and services; increasing productivity; enhancing performance, reducing costs, AI entails some inherent dangers too. Admitting this reality, the OECD has undertaken numerous efforts for AI governance and regulation to establish trustworthy AI. To date, it has released many publications and policy notes; attempted to classify AI systems, and especially, adopted AI Principles in 2019. It is expected that the OECD AI Principles will promote innovative and trustworthy AI; respect human rights and democratic values; set standards for practical and flexible AI, and eventually, pass the test of time. The OECD AI Principles include: (a) inclusive growth, sustainable development and well-being (Principle 1.1); (b) human-centered values and fairness (Principle 1.2); (c) transparency and explainability (Principle 1.3); (d) robustness, security and safety (Principle 1.4), and (e) accountability (Principle 1.5).

5 Suggestions

In the foregoing discussion, it appears that there is neither any uniformity nor any universality, or generalization in the understanding concerning the regulations of AI and robotics. In the absence of the self-sufficing legal rules, Asimov's law, embedded code, soft law, social norms, ethics, etc. may play vital roles, but they seem to be feeble forces against the giant monsters. In this context, we recommend the following suggestions for consideration:

5.1 Liability

Ordinarily, the liability lies upon the users, but it is still unsettled who is liable for any incident caused by algorithm-run devices, e.g., autonomous cars. Usually, devices operated by quality algorithms adhere to the best standard, and thus, manufacturers of the worst algorithms must compensate for the costs of an accident caused by their products. The Directive 85/374/EEC, for example, inflicted strict liability on the producers due to the defects of his products holding the view that they cannot disregard the due care test [23].

5.2 Personhood

Much has been talked about from the sociological and philosophical point of view of whether personality shall accord to robots, especially, in the cases where they possess artificial intelligence. Like 156 AI experts, we think that the imposition of personhood to sophisticated robots misses the mark, because it may relieve the manufacturers and producers inappropriately against their liability. The subsequent embodiment of accidental risk on the sufferers is intolerable as it destroys the duty of care of the manufacturers.

5.3 Adapting to Change

Some people may likely lose their jobs because of AI and robotics. We can, however, assume that people would be displaced by robots, not because of their lack of qualifications, but because of their reluctance to adapt to the new systems. We can share the lessons of Kodak cameras, which ignored accepting the new technology and continued operating the analogue system, being overconfident upon its superiority and reputation in the market, and subsequently, became bankrupt, whereas businesses like Nikon welcomed the emerging technology and consequently grabbed the position of Kodak. Thus, it is essential to adapt to the change; otherwise, we will die and must not forget the survival of the fittest.

5.4 Preparation for AI

AI or robotics cannot stand alone unless it is welcomed by all stakeholders in the prevailing socio-technical connections. Surely, both AI and robotics bring enormous opportunities for mankind other than their risks, but their potentials depend on how we utilize them [24]. To be benefitted from AI or robotics, researchers and manufacturers would have proper training on their liability to the AI systems, and that resulted in numerous implications to the society. Machines and algorithms should be designed in responsive to societal values. Disseminating basic AI education for all can be a fruitful effort without any doubt.

5.5 Asimov's Law

Experts in the field regard Asimov's law of robotics as one of the normative ideals, and minimum standards in the robots' production, even though these are not the law in a lawyer's sense. Observing the widespread usage, we can surely say that robots are not the mere probability, rather they are the reality. Eventually, it is likely to have some people who might wish to make money by producing faulty robots without taking care of the interests of the common users. In the absence of compelling legal rules, they can be bound to obey Asimov's simple but far-sighted principles in designing and manufacturing robots for ensuring safer, reliable, and comfortable automated machines.

5.6 *Preparing Laws for Robotics*

Considering huge potentials, and the wider implications, we cannot give up AI research and development but regulate it. For doing so, we should consider the following postulates: (i) all laws should be applied against the automation fully as long as they are applicable against their human counterparts; (ii) there must have an explicit statement from the AI that it is not a human; (iii) the automation must not keep or reveal the personal data without the express consent of the data subjects; (iv) the regulatory framework toward robots must be context-specific; (v) the existing legal rules should be sensibly applied; (vi) the robot regulation must be designed compatible with the profound normative structure of our society [25].

5.7 *Roles of the UN*

In this golden age of technological advancement, the global community is affected by the wide-scale use and applications of technologies, but there is hardly any accord in regulating the technologies, especially robotics. Even though all are affected, only a few nations have attempted to make regulations for AI and robotics. It is also mentionable that in terms of the regulatory framework, China, the USA, the EU, South Korea, and Japan hold the 5th, 4th, 3rd, 2nd, and 1st positions, respectively [26]. Among them, whereas the EU promotes AI regulation, China and the USA, on the other hand, favor the technology. In this given context, the United Nations should play the central role in combining all efforts to set a global regulation for AI and robotics. In doing so, the UN should consider the following:

a. to understand the ins and outs of AI and robotics for receiving better services by reducing the probable harms;
b. to analyze the problems and prospects of AI in every step of design, manufacturing, and operation for avoiding the irreparable losses that robots may cause later;
c. to search for the ways, means, and mechanisms to get rid of the undesired consequences;
d. to synthesize all possible means to provide a better option for application;

e. to integrate all means to have a simple, practicable, and generalized notion from the inherent intricacies of the selected tools;
f. finally, framing the global comprehensive regulation following the steps stated above.

6 Conclusion

As a dynamic field of modern science, robotics is blessed with both enormous prospects and numerous ethical, social, and legal challenges. In addressing those challenges, the regulators will have to devise workable solutions. It is admitted that there is no one-size-fits-all solution. Therefore, the scientist, manufacturers, policy-makers, and regulators should work together for striking the balance between the problems and prospects encompassing robotics. Again, the challenges in the field are not against a particular nation, or a region, but to the whole world at large; hence, international collaboration and cooperation led by the United Nations are strongly advisable.

References

1. Veruggio G, Solis J, Van Der Loos M (2011) Roboethics: ethics applied to robotics. https://doi.org/10.1109/MRA.2010.940149
2. Barthelmess U, Furbach U (2013) Do we need Asimov's laws? In: Lecture notes in informatics (LNI). Proceedings—Series of the Gesellschaft fur Informatik (GI)
3. Schweitzer G (2003) Robotics—chances and challenges of a key science. In: 17th International congress of mechanical engineering
4. Lin P (2012) Introduction to robot ethics. In: Robot ethics: ethical and social implications of robotics
5. Villani C (2017) For a meaningful artificial intelligence. AI for Humanity
6. Holder C, Khurana V, Harrison F, Jacobs L (2016) Robotics and law: key legal and regulatory implications of the robotics age (Part I of II). Comput Law Secur Rev 32. https://doi.org/10.1016/j.clsr.2016.03.001
7. Torresen J (2018) A review of future and ethical perspectives of robotics and AI. Front Robot AI 4. https://doi.org/10.3389/frobt.2017.00075
8. Lichocki P, Kahn PH, Billard A (2011) The ethical landscape of robotics. IEEE Robot Autom Mag 18. https://doi.org/10.1109/mra.2011.940275
9. Pham QC, Madhavan R, Righetti L, Smart W, Chatila R (2018) The impact of robotics and automation on working conditions and employment. IEEE Robot Autom Mag 25. https://doi.org/10.1109/MRA.2018.2822058
10. Leenes R, Palmerini E, Koops BJ, Bertolini A, Salvini P, Lucivero F (2017) Regulatory challenges of robotics: some guidelines for addressing legal and ethical issues. Law Innov Technol 9. https://doi.org/10.1080/17579961.2017.1304921
11. Müller VC, Bostrom N (2016) Future progress in artificial intelligence: a survey of expert opinion. Presented at the https://doi.org/10.1007/978-3-319-26485-1_33
12. Semmler S, Rose Z (2017) Artificial intelligence: application today and implications tomorrow. Duke Law Technol Rev 16
13. Cellan-Jones R, Stephen Hawking warns artificial intelligence could end mankind

14. Rawlinson K (2015) Bill Gates insists AI is a threat. BBC News
15. Veruggio G (2005) The birth of roboethics. In: IEEE International conference on robotics and automation. Workshop on roboethics
16. Pagallo U (2011) Killers, fridges, and slaves: a legal journey in robotics. AI Soc 26. https://doi.org/10.1007/s00146-010-0316-0
17. Floridi L, Taddeo M (2018) Romans would have denied robots legal personhood. Nature 557. https://doi.org/10.1038/d41586-018-05154-5
18. Atabekov A, Yastrebov O (2018) Legal status of artificial intelligence across countries: Legislation on the move. Eur Res Stud J 21. https://doi.org/10.35808/ersj/1245
19. Deng B (2015) The robot's dilemma. Nature 523
20. Leenes R, Lucivero F (2014) Laws on robots, laws by robots, laws in robots: regulating robot behaviour by design. Law Innov Technol 6. https://doi.org/10.5235/17579961.6.2.193
21. Lessig L (1999) The law of the horse: what cyberlaw might teach. Harv Law Rev 113. https://doi.org/10.2307/1342331
22. Nevejans N (2017) European civil law rules in robotics
23. Schütze R (2018) Directive 85/374 on the approximation of the laws, regulations and administrative provisions of the member states concerning liability for defective products. In: EU treaties and legislation. https://doi.org/10.1017/9781108624374.031
24. Dignum V (2017) Responsible artificial intelligence: designing AI for human values. ICT Discov
25. Reed C (2018) How should we regulate artificial intelligence? Philos Trans R Soc A Math Phys Eng Sci 376. https://doi.org/10.1098/rsta.2017.0360
26. Eidenmueller H (2017) The rise of robots and the law of humans. SSRN Electron J. https://doi.org/10.2139/ssrn.2941001

Developing an Integrated Hybrid App to Reduce Overproduction and Waiting Time Using Kanban Board

Akansha Yadav and Girija Jha

Abstract This paper is aimed at reducing two of the eight types of wastes under lean manufacturing. The targeted wastes are overproduction and waiting time wastages. Appendices were used to opinionate various departments in addition to the WIP data, regarding the major factors causing the waiting time delays, their impacts on performance, and the constraints to implementation of Kanban techniques under different departments. The paper also provides for a separate hybrid application developed to assist the merchandising department in its activities by overcoming the problem of unaccountability of fabrics, leading to overlapping, and ambiguous order placement leading to capital losses. The paper also provided for the basis of replacing every department's activity scheduling process from considering forecasted demands to consider actual demand and thus increasing efficiency. This is accompanied by providing a mechanism to track the raw material down the production line for better analysis. The paper provides for a central tracking platform with all the departments linked together making it possible to analyze the demand and the resources at hand, thus providing an opportunity to plan the department's future activities such that the wastages are minimum, efficiency is maximum, waiting time is reduced, and so is the lead time leading to customer satisfaction. Visual analysis of various factors in the production line and with their help makes the detection of bottlenecks quite efficient. It also is an effective management platform in terms of resources as well as manpower and highlights the problems through lead time analysis. The paper adopts the Kanban pull production technique to address these wastes. This paper also encompasses the just-in-time (JIT) concept of Toyota Production System (TPS).

Keywords Kanban pull production · WIP · Little's law · Overproduction · Waiting time delay wastages · Inventory management · Effective planning · Hybrid application

A. Yadav (✉) · G. Jha
National Institute of Fashion Technology, Hauz Khas, New Delhi 110016, Delhi, India
e-mail: akansha.yadav1@nift.ac.in

V. Skala et al. (eds.), *Machine Intelligence and Data Science Applications*, Lecture Notes on Data Engineering and Communications Technologies 132,
https://doi.org/10.1007/978-981-19-2347-0_19

1 Introduction

In a production line, wastages not only lead to decrease in production efficiency leading to losses but also increases the time between customers' orders and the shipment. Two of these wastages identified at PEE Empro Exports Ltd. were of overproduction, i.e., unnecessary production of more quantitates than required and waiting time delay wastages, i.e., wastages caused because of stoppage of work due to some reason.

Overproduction:

Blind production by workers, even when those who receive their output are either not ready or do not need the items [1].

Waiting Time Delay:

- The person next in the production line is overwhelmed.
- Occurrence of fault in a machine in the production line.
- Waiting for approved materials.
- Waiting due to slow supply of essential materials from the preceding departments required to keep the production line running smoothly [1].

Another problem that has also been contributing to increased wastages, poor efficiency, and hampering profit margins is the problem regarding fabric wastages occurring due to unaccountability leading to ambiguous and overlapping order placement by the merchandising department.

2 Need for the New Methodology

The need for the new methodology was felt after analyzing the data pertaining to the above-mentioned observations, i.e., overproduction and waiting time delay wastages.

From Table 1, it is clear that the overproduction wastages are a humungous problem as they are leading to over 50% of the products being wasted, and there is a need to address them as soon as possible.

Table 1 Overproduction wastages

Departments	% overproduction wastages	Avg. overproduction wastages
Fabric inspection department	–	**50.04%**
Cutting department	28.34%	
Sewing department	52.04%	
Finishing department	69.4%	

Table 2 Waiting time wastages

Departments	Target WIP (days)	Avg. WIP (days)
Fabric inspection department	–	**1.5**
Cutting department	1.5	
Sewing department	1.5	
Finishing department	1.5	

From Table 2, it was observed that the industry is maintaining a WIP inventory of 1.5 days to tackle the waiting time delay wastages due to any of the afore-mentioned reasons and ensure a continuous process flow at the industry with minimal disruptions. But maintaining a WIP inventory of around 1.5 days is also adding to the capital cost and is a contributing factor toward the overproduction wastages as well. Hence, it is essential to reduce this period from 1.5 days without compromising on the continuity of the workflow and tacking the delay factors effectively, thus minimizing disruptions.

From both the tables, it is visible that the industry is unable to maintain any sort of data concerning the fabric inspection department which is then leading to unaccountability of raw materials. In an audit of the fabric inspection department carried out at the company, it was found that a considerable amount of fabric rolls were kept unused in the inventory and were totally unaccounted for leading to placement of order for new fabric rolls with the same specifications and ignoring the current stock completely.

3 Objectives

- Reduction of wastages caused due to overproduction
- Reduction of wastages caused due to waiting time
- Developing an integrated and timely updated inventory that is innovated such that it accommodates all real-time operations and assists the merchandising department regarding order placement.

4 Literature Review

4.1 Lean Manufacturing

Lean manufacturing is a philosophy that focuses on reducing the time between customer orders and distribution by eliminating waste.

It is based on Toyota's Production System (TPS). Lean manufacturing aims to eliminate waste in all aspects of the manufacturing process, including customer relations, product design, supplier networks, and plant management. With fewer resources, lean means providing more value to customers. A lean organization recognizes the value of its customers and focuses its major activities on continually increasing it. Its purpose is to become highly responsive to client demand while manufacturing superior quality products in the most efficient and cost-effective manner possible by incorporating fewer human efforts, less inventory, less time to develop products, and less space [2].

To do so, lean thinking shifts management's attention away from optimizing individual technologies, assets, and vertical divisions and toward maximizing product flow across complete value streams that flow horizontally across technologies, assets, and departments to customers [3].

There are mainly 8 principles/concepts of lean manufacturing:

- Value specification
- Value stream
- Waste
- Equipment reliability
- Continuous flow
- Pull production
- Continuous improvement
- People involvement [2]

For the scope of this paper, two of these principles have been considered, such as the waste principle and the pull production principle.

Under the waste principle of lean manufacturing, there are eight types of wastes:

- Overproduction
- Inventory
- Transportation
- Motion (operation)
- Extra-processing
- Defects/quality
- Waiting
- People's skills/unused talents [4]

Another principle of lean manufacturing that has been implemented to address these wastages is the pull production principle. A pull production system regulates the flow of resources by simply replacing what has been consumed. Customer demand drives a pull mechanism. A Kanban system is a pull system that regulates upstream production and delivery flow using color-coded cards attached to parts or part containers. We can set minimum and maximum on-hand quantities for raw materials, supplies, and each assembly or product manufactured using a Kanban system [5].

The just-in-time (JIT) principle was another TPS component that was taken into account while designing the Website. It means giving each process exactly what it needs, when it needs it, and in what quantity it needs it. JIT focuses on reducing manufacturing lead times, which is generally accomplished by lowering work-in-progress levels (WIP). The outcome is a continuous flow of small amounts of items throughout the manufacturing process. These stock reductions will be accompanied by significant improvements in quality and manufacturing, resulting in previously unseen cost savings. It primarily consists of three stake holdings, all of which it seeks to reduce [2]:

- Input Material: Stocking extra raw material than what is actually required. It can be decreased by involving suppliers into the manufacturing process [2].
- Work-In-Progress (WIP): Inadequate production planning, insufficient machine capacity, poor operator skills, product variants, changing product priorities, and machine breakdowns are the main reasons of excessive WIP. To reduce WIP, production should be spread evenly over time to ensure a smooth flow between processes, as well as the Kanban system's pull signals, which instruct the operators when to build the next part [2].

Finished Goods: This refers to the excess inventory in the warehouse. There would be no finished goods in the dead stock if the Kanban pull method was used. Working capital would be reduced as a result of this [2].

4.2 Little's Law

The Little's law was given and proved by John Little in 1961. The law states:

The average number of customers in a stable system (over some time interval) is equal to their average arrival rate, multiplied by their average time in the system [2].

Despite the fact that it appears intuitively sensible, the finding is extremely astounding. In a manufacturing unit, little's law can be mathematically expressed as cycle time is equal to amount of work in process divided by the output during this time, i.e.,

$$\text{Cycle Time} = \frac{\text{Work in process (WIP) (in units)}}{\text{Output (in units) per unit time}} \tag{1}$$

Cycle time here can be referred as the time taken to complete the production cycle or the average time it takes to produce one unit [6].

As a result, if the total units throughout the work area are constant, and the output per time unit is constant, the cycle time may be easily calculated. In addition, if the WIP remains constant while the output decreases, the cycle time will increase; conversely, if the output remains constant while the WIP decreases, the cycle time will drop.

Therefore, if the industrial unit can maintain a closed control on the entire process, from the input point till the completion point, then the delivery date can be easily predicted and thus customer satisfaction can be increased [2].

But this would be an ideal case as in a realistic industrial unit, and manufacturing processes are very hard to predict. If the input is the same and the output goes down, then the WIP would build up indicating bottlenecks that would need immediate fixing. Little's law is useful in this as it gives us an approximation about how much it could raise or lower the output. As a result, whenever a product is introduced to the store floor, it is critical to take all necessary steps to keep the workflow going. It is preferable to slow or cease introducing new products into the production line if the products are stuck anywhere [2].

Thus, the most important conclusion from Little's law is that in a production system, the best way to increase the output is not to increase the input which may seem intuitive, but it is to find the bottlenecks by looking at the building up of WIP at any work area and fix them to increase the output.

4.3 Kanban

Kanban has a variety of definitions based on the stages of development and functions. Some are more developed as a component of lean, while others are closer to what Toyota uses. The following two definitions are enough to get an overall understanding of Kanban [3].

1. Demand scheduling is what Kanban is characterized as. Operators manufacture products based on real usage rather than predicted usage in Kanban-controlled systems. The Kanban schedule takes the place of the standard weekly or daily production schedule. Visual signals and established decision criteria are used to replace the schedule, allowing the production operators to schedule the line [2]. What Kanban replaces is
    - The daily scheduling operations required to run the manufacturing process.
    - The requirement for production planners and supervisors to keep a constant eye on schedule status in order to select the next item to run and when to switch over.
2. Kanban is a lean agile system that may be used to improve any software development life cycle, such as scrum, XP, waterfall, PSP/TSP, and other approaches. Its purpose is to deliver value in an effective manner [2].
    - Kanban emphasizes the lean notion of flow to provide value consistently and reliably.
    - To make activities and offer clear directions, the task and workflow are made apparent.
    - Kanban keeps WIP to a minimum to encourage quality, concentration, and completion.

The Website implements a hybrid of the two definitions, thus incorporating the benefits of both and minimizing drawbacks.

Kanban Board: The Kanban that has been employed comprises of three components:

1. *Visual Signals*: These signals can be in the form of cards, tickets, stickers, etc. Once on the Kanban board, these visual signals would help the workers and the supervisors quickly understand the workflow [7].
2. *Columns*: Each column on a Kanban board represents a specific activity that together composes a workflow. The columns used for the purpose of this paper are 'to-do', 'in-progress', and 'completed'.
3. *Work-In-Progress (WIP) Limit*: At any given time, this is the maximum number of cards that can be in one column. When the column is 'maxed-out', the team must swarm on those cards and move them forward before new cards may be added to the workflow. These WIP restrictions are crucial for identifying workflow bottlenecks, reducing overproduction, and increasing flow [7] (Fig. 1).

Kanban Cards: The Kanban card is a color paper that is used across the production process, from inventory to carton packing. Kanban cards accompany production items and identify the style number, line number, Kanban number, color, and quantity that have been issued. Kanban cards are used for both transactions and communication [2].

The Kanban cards used are digitized and created after filling a form where details can be modified as per requirements, thus removing the constraint of fixed details in a physical card. The card moves through the production process just like a physical card but provides the advantage of altering the details neatly as it moves down the line and also tracking a card centrally despite of which department issues it.

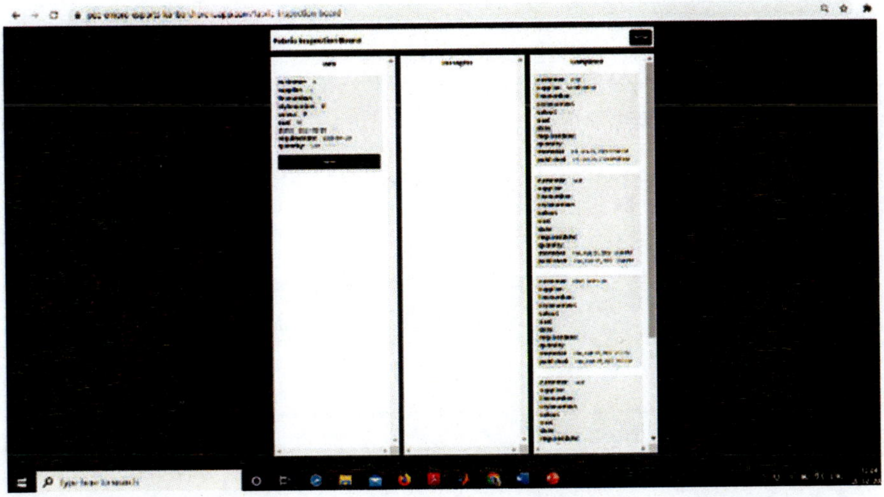

Fig. 1 Kanban board

E-Kanban System and its Advantages: The E-Kanban system provides the organization and its suppliers with the tools they need to meet the expectations that drive value and performance across the supply chain. Even Toyota, the Kanban system's originator has adapted the E-Kanban system to send external pulling signals to faraway suppliers.

The advantages of the E-Kanban system are

- Removes the issue of misplaced cards.
- The demand requirement is met on time.
- The amount of time and effort required to handle cards is reduced.
- Kanban cards may be optimized quickly and effectively.
- Ensures that material shortages are kept to a minimum.
- Transparency in the supply chain is improved.
- Aids in the evaluation of supplier efficiency [8].

4.4 Kanban Pull Production

The real consumption at the downstream process generates a signal (at finishing). As a result, the finishing department has a 'pull' on the entire manufacturing system. The pull signal continues to rise in response to actual consumption until it reaches the upstream process (i.e., fabric inspection). The 'pull' exerted by the downstream department is visually represented to the providing upstream operation in a Kanban system [9].

While a pull system schedules work based on data from within the system, it also imposes the previous limit on the work in progress in the system and allows work to upstream processes using Kanban cards [4] (Fig. 2).

5 Case Studies

The following two case studies were done to analyze and understand the working of the Kanban system and its constraints provided for the basis to build on them further.

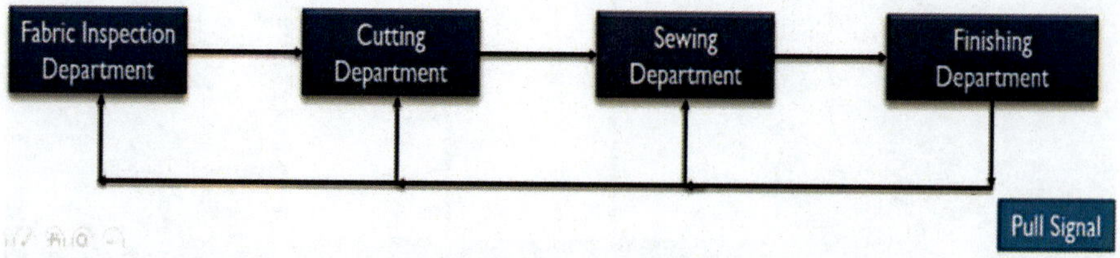

Fig. 2 Pull system [9]

5.1 Case Study 1—Toyota Production System

To structure its manufacturing operations, Toyota came up with the idea of just-in-time (JIT) production, which includes logistics, supplier management, and customer delivery. Its fundamental aim was to reduce costs by eliminating waste and optimizing machine and human capabilities [10].

The Kanban system process for TPS can be summarized as

1. Two types of Kanban cards are utilized in this Kanban system: a 'conveyance Kanban card', which is used to order production of the percentage withdrawn by the subsequent process, and a 'production Kanban card', which is used to order production of the proportion withdrawn by the subsequent process. The containers have these cards attached to them.
2. The conveyance Kanban card is withdrawn from the container when the contents of the container begin to be utilized. To pick up this part, a worker takes this conveyance Kanban card and goes to the stock point of the previous procedure. This conveyance Kanban card is then attached to the container containing this part by the worker.
3. The production Kanban card linked to the container is then removed and becomes the process' dispatching information. They manufacture the part as soon as feasible in order to replenish it.
4. As a result, the final assembly line's production activities are linked in a chain to the preceding processes or subcontractors, allowing for just-in-time manufacturing of the entire process [11].

Toyota employs the following six rules for the effective implementation of the Kanban system:

1. Never give substandard merchandize to others.
2. Take only what you require.
3. Produce the exact number of units required.
4. Increase the output.
5. Fine-tune the process
6. The process should be stabilized and rationalized [10].

5.2 Case Study 2 —An Apparel Industry in India

They use Kanban card, Kanban dashboard with Andon system, Kanban loading, and input book as part of their physical Kanban system. To facilitate the production flow, they all function together as a system:

1. Kanban Cards: To calculate the number of Kanban cards for the production scheduling process, they employ the same formula as mentioned above. They use colored papers as Kanban cards that move along the garment cartons.

- Blue Kanban: Displayed when bit parts are available for cutting.
- White Kanban: Displayed on the finishing table indicating adherence to packing list.
- Pink Kanban: Displayed on the finishing table indicating non-adherence to the packing list.
- Yellow Kanban: Displayed on the warehouse indicating garment wash after sewing [12].

The following signaling routine is followed:

- On the Kanban dashboard, a card indicates the availability of bits in the supermarket.
- The dashboard receives batch/line signals indicating the need for bits in the cutting table.
- The parts, as well as the blue Kanban card, are moved from the supermarket to the cutting table.
- The pieces are sliced according to the micro-cut plan, and a white Kanban card is then displayed on the cutting table's Kanban post, ready to be loaded back onto the manufacturing line.
- When the white Kanban card is returned to the Kanban post at the end of the production line, it is used as a visual indicator for the line's operation.
- The Kanban card is displayed until the garment is finished, and the clothes are packed into cartons.
- After receiving the cartons at the CTPAT area, the sticker linked to the Kanban card is adhered to the Kanban input book [12].

2. Kanban Dashboard: The Kanban dashboard is situated on the factory floor outside the cutting supermarket. It makes an educated guess about the state of the available cutting pieces.

   The workflow is as follows:

   - Cutting manager examines the available pieces dashboard and keeps an eye on the cutting table.
   - When bits are available on the cutting table, turn the switch off and on again, and the dashboard light blinks to alert the person in charge of issuing bits to the cutting area [12].

3. Kanban Loading and Input Book: Kanban loading and input book are a helper tool for Kanban cards that keep track of when a Kanban is loaded into a batch and when it is completed [12].

Some of these problems identified from internal studies at the industry involve

- Lack of effective management information systems
- Lack of visual aids that are simple and unambiguous to follow, including the visual board
- Standard Kanban stickers, thus avoiding confusion between various types of cards

- Excess time to fill up the Kanban books leading to mismanagement on a daily basis and back-logs, and also multiple books amplify the problem even further
- Easy manipulation of records and signals by incompetent employees
- Reworks
- Labor absenteeism
- Quality issues
- Miscommunication and mismanagement
- Non-adherence to cutting and packing plans leading to constraints on in-serial clearance

Therefore, as mentioned earlier, the paper analyzes all these factors on a digital scale via the virtue of a hybrid application achieving all the goals set up for the physical Kanban system along with some additional benefits. The above-mentioned constraints have also been kept in mind while drafting and innovating mechanisms, features, techniques, and implementation of the methodology to bridge all these gaps left wide exposed by the physical signaling system of Kanban, therefore developing a highly efficient system of Kanban aimed at achieving the objectives of this paper.

This paper aims at both the process scheduling and the production line along with targeting the administrative aspects of the system and provides much greater ease and efficiency in these.

6 Pre-implementation Findings

The following conclusions were drawn based on the findings from various departments, i.e., fabric inspection, cutting, sewing, finishing, and IE department.

1. Waiting time delay wastages are a major contributing factor in the total wastages and also a major constraint in customer satisfaction.
2. The prime reasons for waiting time delay wastages are labor absenteeism, reworks, lack/delays in supplies from other department, and machine failures and overwhelming of labor.

Therefore, to address these factors, the Website was equipped with an inventory that considers all real-time operations, and the rules for implementation were made such that there are no delays in supplies and reworks are tackled effectively. Thus, these features along with analysis of WIP blockade with the button changing feature would allow the waiting time wastages to be reduced by a substantial factor. These features would also help in reducing the waiting time WIP inventory without causing any unnecessary interruptions and thus reduce capital cost and increase customer satisfaction.

7 Methodology

7.1 Existing Methodology

The current methodology at PEE Empro Exports Pvt. Ltd. follows the following procedure:

- The cutting department sends a 'layer control and inspection form' to the fabric inspection department with the amount of fabric required mentioned on it. 5% allowance is made on the original demand of cut panels. This form is then signed by the supervisor, and the fabric inspection department releases the fabric according to the mentioned specifications.
- Based on the released fabric, the fabric inspection department maintains a register which stores the details of the fabric released along with dates and some other specifications.
- After the cutting of the panels is completed by the cutting department, the cutting department sends the cut panels to the sewing department along with a 'job card' which contains the specifications regarding the panels.
- The sewing department maintains a log of all these job cards in a register.
- The sewing department then sends the lot of garments along with a production card to the finishing department.

7.2 Working Procedure of Kanban Pull Production

- The cutting operator pulls the withdrawal Kanban card from the container/trolley/bin and sets it in the Kanban post, using items maintained near the assembly line.
- To retrieve products, the operator takes the withdrawal Kanban card to the fabric inspection department. He takes the products that correspond to 1 Kanban card and places them in the Kanban post with the production Kanban card that came with them.
- After that, the production Kanban card is placed on the Kanban board.
- The cards are issued to the units to begin creating more things as soon as the pre-specified quantity of cards collect on the production board.
- The same method will be followed for the remaining departments [9].

The number of Kanban cards to be included in the process of to-do list can be calculated by

$$\text{Number of Kanban Cards} = \frac{\{\text{Average Daily Demand} \times (\text{Order Frequency} + \text{Lead Time} + \text{Safety Time})\}}{\text{Container Quantity}} \quad (2)$$

[9]

Here, average daily demand is the average demand for a type of garment on a daily basis,

Order frequency is the how many times on a daily basis is order received for that particular type of garment.

Lead time is the total time taken for an item to go through, starting from the fabric inspection department up till finishing department (to be used in this formula, the time in hours must be divided by the number of hours a shift lasts for the company to get the lead time in term of days with respect to shift time for the company).

Safety time is the time factored in for breakdowns, quality losses, etc. It is generally taken as half of the duration of one shift but varies from company to company.

Container quantity is the number of items a container/trolley/bin can hold which is used for transporting the items from one department to another and holds the Kanban cards [9].

The formula has been derived from the TPS Kanban system in which

$$\text{Number of Kanban Cards} = \frac{\text{Average Daily Demand} \times (\text{Lead Time} + \text{Safety Time}) \times (1 + \alpha)}{\text{Container Quantity}} \quad (3)$$

Here,

α is a policy variable which is determined according to the factory's capability to manage external factors and is usually in the range [0, 0.1] [11].

7.3 *Working Mechanism of Website and Mobile Application*

To structure its manufacturing operations, Toyota came up with the idea of just-in-time (JIT) production, which includes logistics, supplier management, and customer delivery. Its fundamental aim was to reduce costs by eliminating waste and optimizing machine and human capabilities [10].

The Kanban pull production technique has been digitized by the virtue of a Website and mobile application which would not help fasten up the process but would also be a better utilization of resources. The workflow of the Website can be identified as follows (Fig. 3).

The Website can be accessed at: https://pempro-kanban.herokuapp.com/ (Fig. 4).

The mechanism of the Website is as follows:

1. The Website is made password protected so as to provide maximum control with the administrator/supervisor and ensure effective implementation without any unnecessary disruptions.
2. There are individual sections for all the four departments on the home screen.

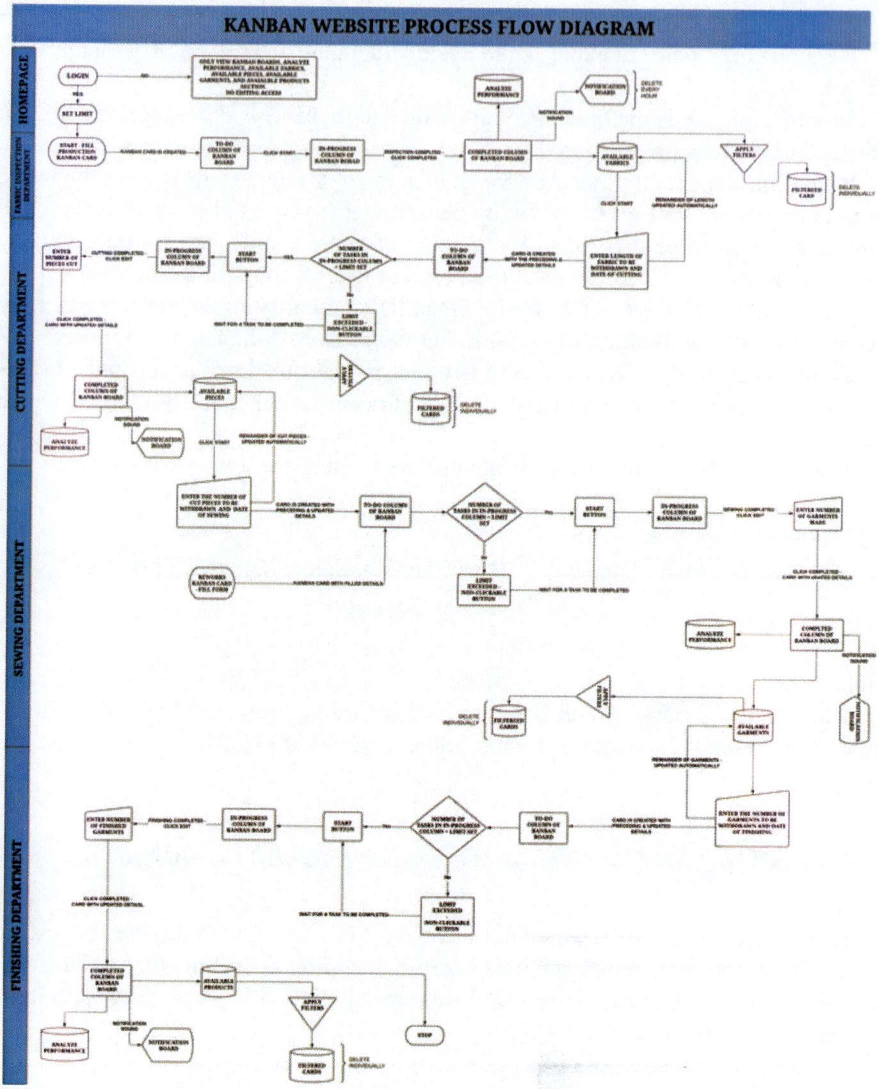

Fig. 3 Process flow diagram

3. Each department has an independent Kanban board, each containing three columns, i.e., to-do, in-progress, and completed.
4. Apart from this, there is an available products section under each department which provides for the inventory for each department and acts as a 'supermarket' for the following department. This section is equipped with filters to filter out a Kanban card based on the supervisor's requirements, and also each Kanban card in here has a 'delete' button to delete the card individually, thus

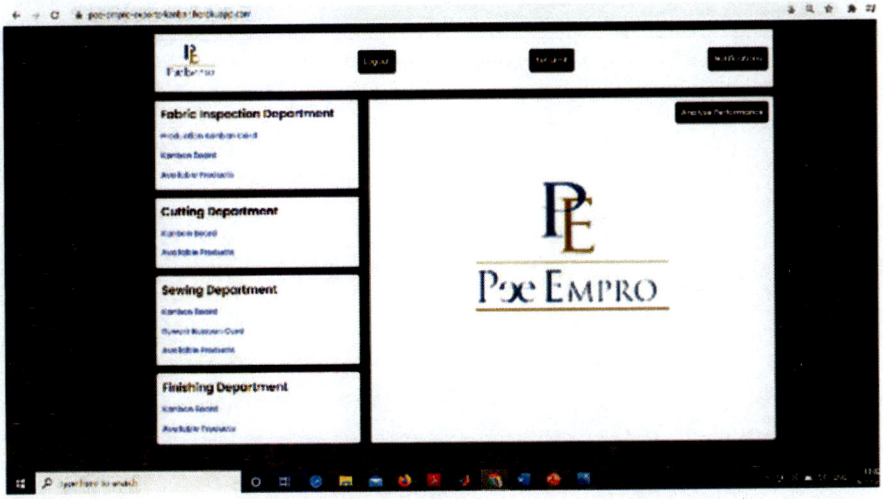

Fig. 4 Website homepage

providing the supervisor of the concerned department with an opportunity to maintain the inventory and timely update it (Fig. 5).

5. As soon as the supervisor will log into the Website with the given credentials, he/she will be able to access all the features that are there to offer. The supervisor would begin by setting up the limit to the number of tasks that can be put into the production line at a moment, by clicking on a 'set limit' button on the homepage. This limit can be changed by the supervisor at any moment of

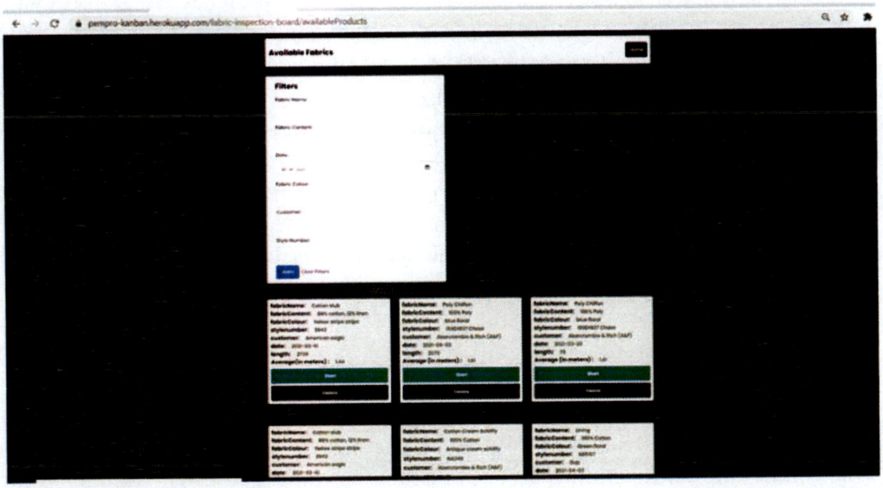

Fig. 5 Available fabrics section

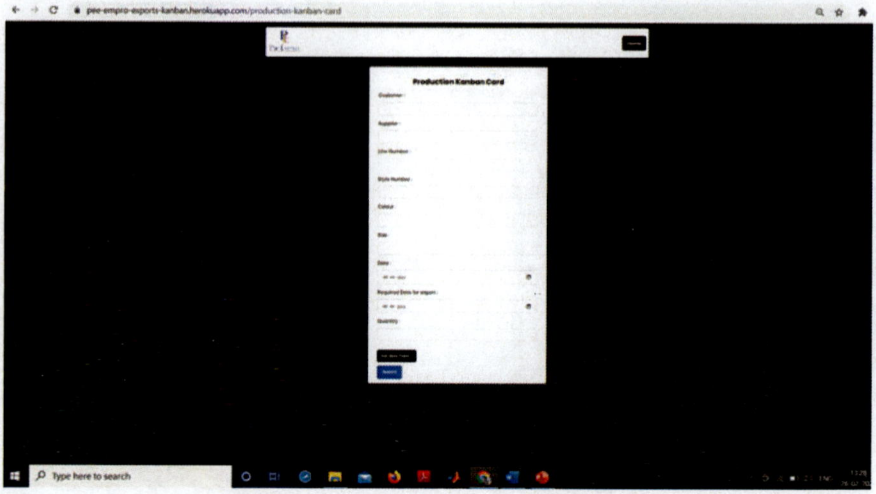

Fig. 6 Production Kanban card

time as and when required and is the key factor in controlling overproduction wastages.

6. The limit set would not affect the fabric inspection department's Kanban board as it was realized during the implementation that the fabric inspection department independently maintains its inventory, and the cutting department makes withdrawal from it as per its requirement.
7. Following this, the supervisor will be required to create a task/card for the initial department, i.e., fabric inspection department under the production Kanban card button. In this section, the supervisor would be provided with a form with some pre-set fields to be filled in and would also have the freedom to create custom fields as and when the need arises. All fields have been made optional considering the variations in the orders (Fig. 6).
8. As the task/card is created, it will move into the to-do column of the Kanban board of the fabric inspection department and the supervisor can start the task, thereby clicking on the 'start' button that will move the task/card to the in-progress column, and then, he/she can assign his/her team to complete inspection for the given number and type of fabric samples. As soon as the task will be completed, the supervisor can click the 'completed' button on the task/card and it would move into the completed column of the Kanban board.
9. As soon as the task is completed, the supervisor will get a notification regarding the completion of a task by his/her department with a sound. The notification will also be visible under the notification board that could be accessed by clicking the notifications button on the homepage. The notification board will contain the notification that would show the department that completed the task and at what time. These notifications would be made temporary so as to not slow down the Website, i.e., they would disappear after an hour. If the

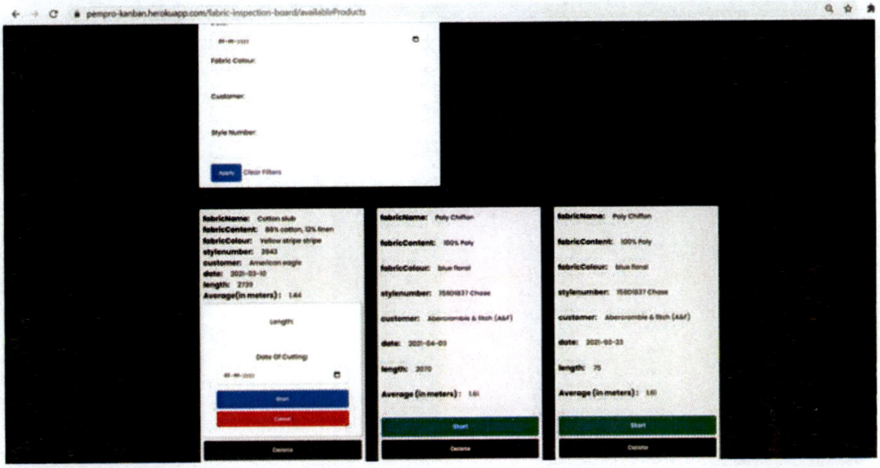

Fig. 7 Withdrawal

board looks cluttered, the supervisor will have the power delete all of them at once as and when required, by clicking on the 'delete all' button.

10. The task/card will then automatically move to the available fabrics section of the fabric inspection department. This section as mentioned before would have the feature to filter the cards out using filters and also to delete the cards in case of any discrepancy and also the available pieces, available garments, and available products section down the line would have these features, with slight modifications in filter specifications based on department specific details.
11. The cutting department's supervisor can then make a withdrawal of the required fabric from the inspected fabrics by clicking the 'start' button on bottom of every Kanban card in the available fabrics section of the fabric inspection department (Fig. 7).
12. Upon clicking 'start', the user would be prompted to enter the length of the fabric to withdraw for cutting along with the date of cutting.
13. As soon as these details are filled a Kanban card with the previous details along with the updated length and date of cutting would move to the to-do column of the cutting department's Kanban board.
14. At the same time, the length of that particular type of fabric would be updated in available fabrics section of the fabric inspection department with the remainder of the length.
15. If the need arises to alter specification for a card at any stage, the supervisor can do so by clicking the 'edit' button present on the card in the to-do column of all the departments, except the initial department, i.e., the fabric inspection department.
16. If the need arises to alter specification for a task/card at any stage, the supervisor will be able to do so by clicking the 'edit' button present on the task/card in the to-do column of all the departments, except the initial department.

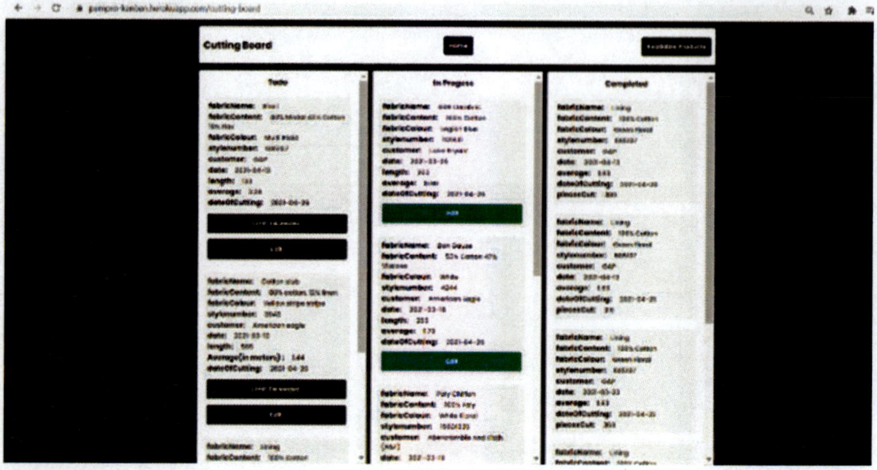

Fig. 8 Limit exceeded

17. When the set number of tasks/cards are present in the production line/in-progress column, the supervisor will be indicated this with the change of 'start' button into the 'limit exceeded' non-clickable button (Fig. 8).
18. As the cutting of the issued fabric is completed, the Kanban card/task can be completed by clicking the 'edit' button on the Kanban card in the in-progress column of the cutting department's Kanban board. This would then prompt the user to enter the number of pieces cut, and then click 'completed'.
19. As this value is updated the Kanban card with the updated detail moves into the completed column of the cutting department's Kanban board and also the available pieces section of the cutting department.
20. The same procedure will be followed under every subsequent department.
21. Considering the need for reworks under the sewing department, a section of rework Kanban card has been provided under this department so that whenever due to any defects, there is a need for reworking an item, a Kanban card could be created for it there itself without engaging the previous departments and its effect be considered in the productivity and wastage analysis. The rework task should be done on priority so as to avoid any kind of back-log.
22. There is an analyze performance section on the homepage, where the supervisor would be able analyze the performance of each department task-wise. The task/card there contains the lead time for every task for each department, i.e., the time taken by a task to be completed by a department. This is an effective tool in checking for bottlenecks in any department (Fig. 9).
23. To maintain the analyze performance section organized the tasks will automatically delete from there every 30 days, and the supervisor can instruct to store the important data in a workbook on a monthly basis if the need arises.

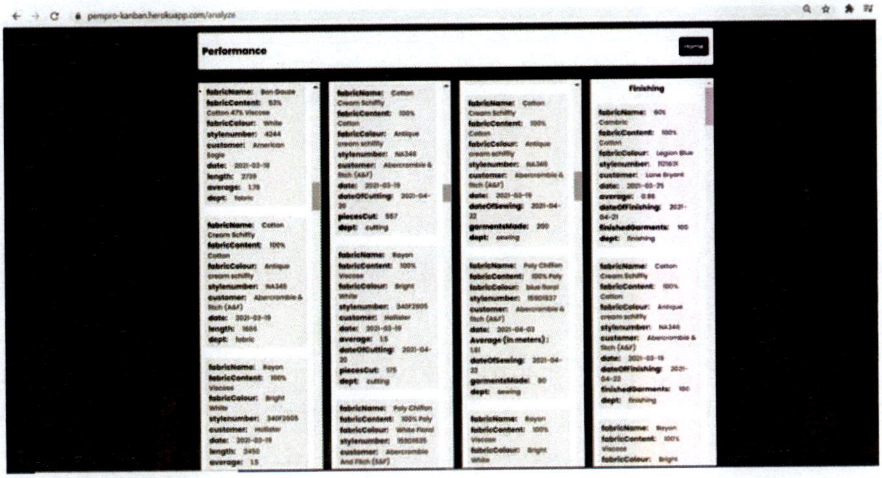

Fig. 9 Analyze performance section

24. If not logged in, the user would be able to only view the Kanban boards of every department and the analyze performance section but would not be able to make a single change in the entire workflow.
25. The Website has been developed into a mobile application as well with all the above-mentioned features, so that the operators and workers can access it even on a simple smartphone and track their progress and self-analyze. This mobile application can be installed from within the Website itself (Fig. 10).

The inventory management Website also follows a similar structural approach, and the workflow can be identified as follows (Fig. 11).

And the Website can be accessed at: https://peempro-inventory-manager.herokuapp.com/.

7.4 *Rules for Implementation in the Industry*

To in the Kanban system, both the Kanban cards and materials move in a continuous flow.

The major goal of the implementation rules is to describe how to run the Kanban and to give process operators control over the line's scheduling. The only way the production operators may take control of the line's schedule is if the regulations provide clear direction and advice [2].

The following rules have been followed to implement this digitized Kanban pull production technique

- Stick to the calculated number of Kanban cards.

Fig. 10 Mobile application

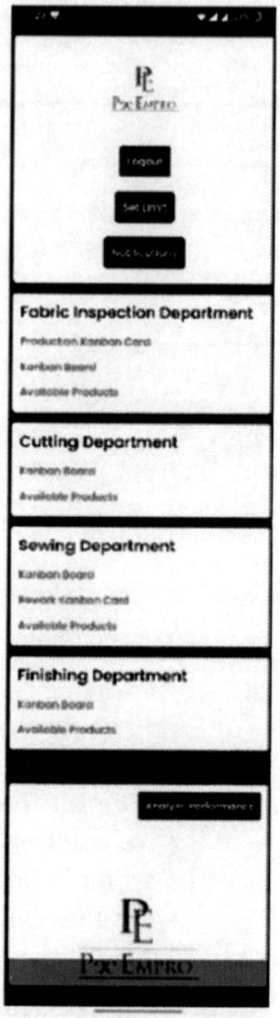

- Any department should not fill its inventory with more materials than required to for the WIP limit of tasks. For example, if the WIP limit is 3 and there are already 3 tasks in line for the cutting department, then the cutting department and the fabric inspection department supervisor should make sure that only materials pertaining to next 3 tasks are present in the cutting department's inventory. Same goes for other departments as well. This could be easily tracked with the help of features explained above like notifications, changing of 'start' button to non-clickable 'limit exceeded' button, and available products section. This would ensure that only that much amount of material is going ahead that is in demand and would minimize wastages. And would also ensure that time is not wasted in blind production rather the department moving ahead of schedule can focus on new order fulfillment.
- Delay in production of materials/clearance of tasks or cards should be tracked by supervisors using lead time and piling up of tasks in the 'to-do' section of any department, and necessary steps should be taken in such a scenario to resolve the bottleneck and maximize efficiency.

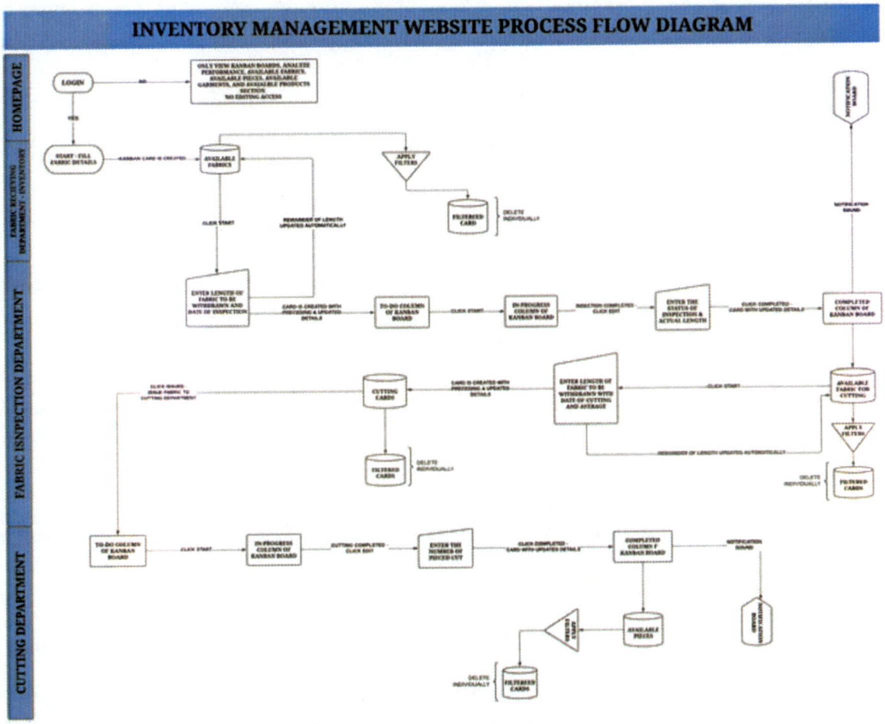

Fig. 11 Process flow diagrams—inventory management

- Delay in production of materials/clearance of tasks or cards can be tracked by supervisors using piling up of tasks in the 'to-do' section of any department, blocking of WIP shown by changing of button into non-clickable button for a significant amount of time, and necessary steps should be taken in such a scenario to resolve the bottlenecks and maximize efficiency.
- Also, action regarding the production line planning and introduction of new styles should be taken by checking upon the inventory from time to time to have an idea of the demand from the downstream departments and thus plan the future plan of action accordingly.
- Any rework task should be prioritized so that the flow of Kanban cards and the entire Kanban pull production system is not disrupted, and the rework lot is cleared with ease.
- Auditing shall be done with utmost care so that defective goods can be recognized at the departmental floor only and can be reworked and cleared with the lot rather than being identified at the packaging stage as the Kanban system would not be able to take into account such scenarios and the workflow would be disrupted by it as the entire lot would then have to be held before the defect is removed and the lot is cleared.

At the end of each shift production of each department should be analyzed under the analyze performance section to resolve issues if any according to the measures in place. For this purpose, Kanban auditors can also be assigned.

7.5 *Benefits of Proposed Approach*

- The communication between two departments would become easier and faster as they would be notified of each other's progress and demands by the virtue of Kanban cards and the notification system.
- The WIP can also be maintained effectively and in a hassle-free manner with the set limit option and its blockage can easily be detected by changing of the button into non-clickable buttons which when stays as it is for a substantial amount of time is a indication of bottlenecks.
- Easy restriction on WIP and a system of pull production generating user friendly and easily understandable pull signals along with efficient tracking would reduce the overproduction wastages.
- Each department would have access to each other's completed goods effectively and thus can plan their production accordingly without any delays.
- This workflow would also work as a check on overproduction as the departments could constantly monitor their inventory on a central platform which would deplete when the order is placed and the departments downstream make a withdrawal, thus prioritizing the production for that order, and they could also notice piling up of material for a product whose order has not been placed, thus halting its production and move to next style/product to maintain their inventory, thus reducing the waiting time delays faced by the downstream departments due to lack of supply of material. The set limit feature along with this feature would ensure only that much is produced which is demanded, and thus overproduction is minimized. Thus, based on the readily available information, the departments can plan their future activities and style plans so as to maximize efficiency and reduce wastages and also reduce delivery times leading to customer satisfaction.
- An inventory would be maintained with details of all the raw materials that are arriving in the industry and what quantity of it is being utilized. This would also be timely updated using the features provided along with the Kanban cards.
- No raw material would be unaccounted for and the production efficiency can be easily analyzed with an integrated inventory.
- Supervisors would be able to analyze the performance of all the departments and track the raw material going into every department effectively and centrally along with their outputs and therefore can also keep track of WIP and thus look for bottlenecks, thus increasing output and maximizing efficiency without increasing the input.
- The reworks can also be handled effectively and prioritized to avoid any disruptions and breakdown of the workflow.

8 Results

- From merchandising department's point of view, they now have access to a timely updated inventory that is capable of accommodating real-rime operations. Hence, at any moment of time, they have accurate data regarding the availability of a particular type of raw material which can be quickly assessed using the filters provided. Hence, they can go ahead and place orders for new raw materials based on this information which in turn is reflecting the real-time demand.
- Also, some objectives have been achieved as
  - The maxing out of the to-do and in-progress columns of the Kanban boards can be effectively observed by changing of the start button into a non-clickable limit exceeded button, and also this can be unambiguously visually interpreted, thus helping to diagnose WIP buildup, and expose and solve the bottlenecks. The bottlenecks are also signaled by unusually excessive time taken by a department to clear a set of tasks which can be interpreted from the analyze performance section by calculated lead time and appropriate action can be taken to resolve excessive lead times. This has also initiated a practice of analyzing individual department's performance and take action accordingly.
  - The raw materials can be effectively tracked from the initial department to the final department as the Kanban cards accommodate every transaction and action into themselves modifying them accordingly thus a single card can be tracked all the way down, rather than creating a card for every step which then becomes hard to track.
  - During the period of implementation, several disruptions pertaining to waiting time delays came up especially labor absenteeism accounting for the second wave of the pandemic, and some other including machine breakdowns, etc., but it was observed that due to the rule of keeping material corresponding to WIP-1 tasks into the to-do column at any moment helped in overcoming these disruptions as the workflow did not come to a halt and operations continued. Thus, this with every passing day minimized the need for excessive WIP inventory (initially 1.5 days).
  - All the supervisors concurred to the fact that they were updated regarding other department's activities with the engaging feature of the notification accompanied by notification sound and the notification board.

Hence, from the above inferences, it is evident that one of the primary objectives, and all of the secondary objectives that were set out were achieved during implementation.

Regarding the major objectives of reduction in overproduction and waiting time wastages, their effect could not be absorbed in the WIP report of the company, which was the basis for detection of these wastages because to analyze the effect of the Kanban pull production on these wastages as discussed under the review of literature requires the entire workflow to go through this system and make every step accountable on the Website which was slowed down due lockdown and lack of workforce at the industry due to the pandemic restrictions.

Due to these factors, the implementation was stuttered and hence could not indicate the desired results into the WIP report which comes out every month, and the above-mentioned factors did not allow the workflow to run smoothly for a month continuously.

And when implemented on full scale, there is enough evidential basis that very well suggests that the expected and desired results in terms of reduction of overproduction and waiting time wastages (pertaining to WIP inventory) would reduce drastically and therefore would increase the company's profit margins.

9 Scope for Future Study

The following are some limitations in this methodology which make this research open for future studies leading to better development of Kanban implementation.

- Cumbersome to analyze raw data due to lack of graphical analysis.
- No option for raising or redressal of any red flags, which might prove critical to waste reduction.
- Lack of integration with mobile OS for a sleep mode notification system.

Apart from these, the methodology is also prone to some limitations of the Kanban system as well:

- Incompetence on the part of in-line and end-line checkers, who clear garments by inspecting quality, but at the end of the line, after the last quality check, the garment is refused, delaying Kanban clearing and disturbing the entire Kanban system.
- Lack of communication between the management and the departmental floor operators regarding the flow of information might also cause a limitation to the method's effective implementation.

10 Conclusion

Not only PEE Empro Exports Pvt. Ltd., but industries all across the globe are targeting to increase productivity at lower costs and to produce with the product of best quality. The effective usage of Kanban tool is critical to improving the practices that can help us to struggle with existing garment industry, as is suggested by the Little's law as well stating that in order to increase the output, it is not advised to increase the input rather check for WIP buildup and clear the bottlenecks, and the Kanban tool with its techniques provides us a way to do so. As this is the most optimum technique to reduce overproduction and waiting time wastages which further reduced the lead time and capital costs, thus increasing the productivity, efficiency, and customer satisfaction.

Also, effective inventory management is the key to success of any industry as it not only reduces the wastages leading to decrease in the working capital of the industry but also helps the industry with the data to make base for its future activities and planning, and the Kanban technique has provided a very strong base to achieve this through its various innovative features.

The paper offers a digitalized methodology of Kanban pull production to limit the WIP, provide a central tracking platform, and a platform to identify bottlenecks easily along with maintaining timely updated, error free, and efficient database and inventory stock, providing a base for every department involved to plan its activities in the most optimum and efficient manner made possible by innovating it to accommodate the real-time operations, and also bridging the gap between departments, with the specific aim of reducing waiting time delays and overproduction wastages. Also, reduction of working capital, WIP inventory, workflow variations have been made possible with the help of this technique by achieving just-in-time (JIT), thus reaching the goals of lean manufacturing by implementing its various principles by the virtue of the Kanban technique, and also proving to be a critical project management tool. This paper has also supported the merchandising department in making a well-informed decision regarding placement of orders in accordance with the existing stock and demand, and raw material tracking down the production line has also been made possible.

Though, there are drawbacks to this tool as well, its complete success in the extensive run depends on close understanding between the management and departmental floor executives.

References

1. Ahmed S, Chowdhury SI (2018) Increase the efficiency and productivity of sewing section through low performing operators improvement by using eight wastes of lean methodology. Global J Res Eng J General Eng 18:2–45
2. Yang Z (2010) Kanban re-engineers production process in Akers Sweden AB. Department of Innovation, Design and Product Development Malardalen University, pp 14–25
3. Steven S, Bowen H (1999) Decoding the DNA of the Toyota production system. Harvard Bus Rev, 96–106
4. Liker JK (2004) The Toyota way—14 management principles from world's greatest manufacturer. McGraw-Hill
5. Lu DJ, Kyokai NN (1989) Kanban just-in time at Toyota. Japan Management Association
6. Little's Law for Production Development. http://www.shmula.com/263/littles-law-for-product-development. Last accessed 12 Mar 2021
7. What is Kanban Board? https://www.atlassian.com/agile/kanban/boards. Last accessed 27 Jan 2021
8. Mariam H, El Laila A, Abdellah A (2017) E-Kanban the new generation of traditional Kanban system, and the impact of its implementation in the enterprise. In: International conference on industrial engineering and operation management, pp1261–1270

9. Reflexive Production System: Use of Kanban. https://apparelresources.com/business-news/manufacturing/reflexive-production-system-use-of-kanban/. Last accessed 27 Jan 2021
10. Toyota Production System. https://kanbanzone.com/resources/lean/toyota-production-system/. Last accessed 28 Jan 2021
11. Sugimori Y, Kusunoki K, Cho F, Uchikawa S (1977) Toyota production system and Kanban system materialization of just-in-time and respect-for-human systems. Int J Prod Res 15(9):553–564
12. Technical Standard. https://www.scribd.com/doc/23494/kanban. Last accessed 25 Jan 2021

Implementation of Gamified Navigation and Location Mapping Using Augmented Reality

R. Janarthanan, A. Annapoorani, S. Abhilash, and P. Dinesh

Abstract From hand-drawn maps, and compass to technology-based navigation systems, people have always relied on some kind of tool to help them reach their destination. At present, there are a lot of people equipped with a smartphone which has become a part of the daily life. These smartphones have applications such as Google maps which makes use of the GPS technology to facilitate navigation in the outdoor environment. It provides a great deal of accuracy to reach our destination but the same cannot be said for indoor navigation. Indoor navigation systems are still under research and development. The present indoor navigation systems use the existing technologies such as Bluetooth, Wi-Fi, RFID, and Computer Vision for navigating through the indoor environment. In this paper, we point out some of the issues of the existing technologies for indoor navigation and propose a method for indoor navigation using the Augmented Reality technology and ARCore.

Keywords Navigation · ARCore · Augmented reality · Mobile application · Indoor environment

1 Introduction

Tools such as maps and compasses have helped travelers since the ancient times. They had an important role in the daily lives of the people of the past. Without the help of these tools, the time to reach their destination would have been extended by a large amount. In the modern times, people rely on the power of technology for navigational purposes. With the growing popularity of the smartphone technology, smartphones have become one of the prominent devices of the modern era. Almost all the people are equipped with a smartphone these days. They have inbuilt applications such as Google maps which uses the GPS technology for facilitating outdoor navigation. The accuracy of the outdoor navigation provided by this kind of application is fairly high but when the same applications are used in the indoor environment, the accuracy is

R. Janarthanan (✉) · A. Annapoorani · S. Abhilash · P. Dinesh
Centre for Artificial Intelligence, Chennai Institute of Technology, Tamil Nadu, Chennai, India
e-mail: janarthananr@citchennai.net

V. Skala et al. (eds.), *Machine Intelligence and Data Science Applications*, Lecture Notes on Data Engineering and Communications Technologies 132,
https://doi.org/10.1007/978-981-19-2347-0_20

extremely low. Buildings with complex internal environments such as universities, malls, and industries are hard to navigate for the people who are visiting for the first time. Therefore, reaching their destination becomes a time-consuming process. The greatest challenge in navigating such environments is to guide the people to their destination in the most optimal manner and with the highest possible accuracy.

Various techniques such as Bluetooth, Wi-Fi, RFID, Computer Vision, etc. were implemented to tackle the issue of low accuracy [1, 2]. These techniques were able to increase the accuracy of indoor navigation but they had their drawbacks [3–9]. Augmented Reality (AR) is a reality-altering technology that is under rapid development [10–24]. AR enables the virtual objects to be perceived along with the real-world objects. There are several Software Development Kits (SDK) that make the development of the application easier. Some of the specific SDKs for the development of AR applications include ARCore, ARKit, and Vuforia. ARCore enables AR development for android devices and ARKit enables AR development for Apple's iOS.

2 Background

2.1 Previous Technologies

Bluetooth technology focuses on using beacons for improving the accuracy of indoor navigation [3–9]. Bluetooth Low Energy (BLE) technology provides the necessary setup by mounting several beacons that send signals which are used to determine the location in an indoor environment [1, 2]. These beacons are radio transmitters that can emit signals in the range of 10–30 m. They provide an accuracy of up to one meter. Based on the detected beacon, it is also possible to determine the current floor where the user is at.

Wi-Fi technology is similar to the Bluetooth technology but instead of using beacons, it utilizes Wi-Fi hotspot points to calculate the position of the device. The Received Signal Strength Indication (RSSI) helps in determining the position of the user. The accuracy of this method is 5 to 15 m which is far lesser compared to the beacon method. The distance is calculated using the latitude, longitude, and the relative position of the device with each Wi-fi access point.

The Radio Frequency Identification (RFID) method utilizes the information saved into a tag that can track in an indoor environment. It is most commonly employed in the logistics sector where they have to keep track of their materials. It uses tags that can store a specific amount of information which can be scanned using another device and this information can be used to calculate the tag's position. Passive, Semi-passive, and Active are three types of tags available in this technology. The passive tags are not provided with any energy, they obtain the power from the readers. However, the other tags are provided with an energy storage component.

2.2 Technology Limitations

The above-mentioned technologies have greatly helped in increasing the accuracy of the indoor navigation but they have their limitations too. Bluetooth beacons have greater accuracy compared to the Wi-Fi access points but the downside to this accuracy is that it is directly proportional to the number of beacons used. Mounting beacons over a small area such as an office is easy and cost-effective but when it comes to a larger structure such as universities, it becomes impractical. The problem with the Wi-Fi access points is not only its accuracy but also the fact where failing to establish a connection to the device or a delay in the connection may result in errors and false mapping of the user's position. The RFID technology may be suitable for implementing in buildings such as a warehouse but it is not suitable for implementing in universities or malls. For higher performance and accuracy, we can use the active tags but they are very costly. Moreover, a large number of tags are required to cover such a huge area. Passive tags can be used instead of active tags but the range and efficiency of these tags pale in comparison to that of the active tags.

2.3 Augmented Reality and ARCore

In 1968, a system known as the "*Sword of Damocles*" was introduced which would later become the predecessor for the reality technologies. Augmented Reality (AR) specializes in overlaying the real-world environment with virtually generated objects. The goal of AR is to provide the clearest and most accurate representations where the user finds it hard to differentiate the virtual augmentations that are being applied. AR is used in game development, military training, simulations, designing, and in the entertainment industry. The increasing popularity of the AR technology has led to the rise of several SDKs which support the development of the AR applications. Vuforia, ARCore, ARKit are some of the SDKs for developing AR applications. Vuforia has a recognition feature for identifying various visual objects and environments. Local storage or cloud storage can be used to enable the recognition feature. ARKit was developed by Apple which can be enabled from iOS 11 and later versions. It supports the creating of AR applications for iPad and iPhone.

ARCore was developed by Google for building AR applications. It works only for a specific set of devices but the number of supported devices is increasing steadily. There are three key technologies in ARCore that allows the integration of the virtual and physical world using the device's camera.

1. Six degrees of freedom are provided which enables the device to realize the current position and track accordingly with respect to the world coordinates.
2. The understanding of the environment enables a device to track the position of the planes such as the floor plane and the top flat surface of furnitures. It can also detect the size of such surfaces.

3. The light estimation feature enables the device to detect and analyze the environmental lighting and provides various methods to act upon it.

3 Literature Survey

Corotan et al. [10] implemented an Augmented Reality system for the purpose of indoor navigation. They created an app that acted as a controller and could also perform sensing operations. Their robot had been powered by Arduino Uno. The robot had a smartphone mounted on it for navigating automatically and to carry out localization operations. The autonomous navigation had three major aspects which include object detection, localization, and routing. Q-learning algorithm was used for optimally planning the route and the blueprint which was used to navigate was limited to a single floor. They used a parser that could split a floor map into smaller states. They solved the localization issue by using the motion tracking system of ARCore along with a scaled blueprint.

Francis et al. [11] implemented a long-range indoor navigation using PRM-RL. They implemented the system by training a planner which is independent of the environment, creating a small road map for the planner followed by data query, generating trajectories, and data execution. They have evaluated the performance with the basic blueprints of the floor layers and using the SLAM maps. They have analyzed the various threshold values that were obtained from the robot and set variable velocities to test the performance. The numerical data of the physical experiments have been tabulated. The simulation environment was created from the real-world environment perspectives and the metric maps were derived from them.

Michel et al. [12] implemented a system for precise identification of attitude for indoor navigation using AR and smartphones. The variation of attitude was estimated by the combined use of the accelerometer, magnetometer, and the gyroscope of the device. An absolute quaternion is obtained along with a compensated drift. They tested the recorded measurements using iPhones and android smartphones for accuracy. They varied the sampling rate of the sensors and tested out the associated algorithms for precision according to sampling rates. Their tool and experimental protocol allowed them to confirm the parameter values that yield the best results.

Al Rabbaa et al. [13] implemented a multisensory application with a simplified cognitive effort requirement. The cyclical process involved three stages: ideation, user interface, and user experience. Using Place note SDK, the world was scanned using the camera of an iPhone. The varying depths of the surrounding space were calculated and 3D point clouds of the horizontal and vertical planes were represented. The multisensory experiences were tested with a few human participants and their feedback was used for further improvement of the system.

4 Proposed System

The goal of the system is to overcome the limitations of the previous technologies (Sect. 2) by using AR technology for indoor navigation. AR technology development has been made simpler because of the latest SDKs. It provides a simple and inexpensive solution for navigational purposes.

4.1 System Requirements

The application is focused on android mobile devices that can support ARCore. The list of supported devices has been provided by Google on their developers' site under the supported-devices section. Software such as Android Studio and Unity is required for building the application along with the Android Gradle build support. The level of the android package used should be at least 24 to support the application development. Android Studio 3.1 or later and Unity 2017.4.34f1 or later versions are the necessary for importing and integrating ARCore into the project.

5 System Overview

The system comprises two modes of operation: user mode and admin mode. The admin mode allows access to the cloud storage for storing the navigational markers that are set to real-world environments. In this mode, the device scans its environment and acquires a basic understanding based on the detected Points of Interest (PoI). Once the required PoI is detected, the ground plane detection algorithm starts estimating the size and distance of the floor. The detected floor surfaces can now be superimposed with the virtual objects (navigational markers). These markers are then hosted to the cloud storage for later retrieval using the cloud anchors. The orientation of the device, the world coordinates along with the detected features are captured when a marker is placed and these data are stored such that there is no deviation of the marker placed in the real-world environment when the application is used at a later time.

The user mode is provided to the people who seek navigation within the complex indoor environment. When the user opens the application, the recent places that have been visited pop up to make the search easier. If the required destination is not displayed in the recent bar, it is possible to search for the destination manually. When the destination is set, the interface changes to the AR navigation interface where the device scans the environment and compares the detected feature points with the ones that are stored in the cloud. Once a match is found, the navigational markers are displayed which guide the user to their destination in the most optimal manner. The state of optimality is reached by analyzing the shortest route using the shortest route detection algorithm which maps the points that are stored in the cloud and find the shortest route with the A* algorithm [25]. When the user reaches their destination, the navigation interface jumps back to the home interface and the user is able to search for a new destination.

ARCore provides the ground plane detection, light estimation, and environmental understanding features which play an important role in the implementation of this project. The ground plane detection algorithms are able to clearly define any flat surfaces such as the floor or tabletops that have well-defined feature points. The light estimation allows to react to the instances of light that are available in the surroundings. It also enables the brightening up of the scene when a specific flag is triggered. The environmental understanding enables the device to identify the PoI and then analyze the data to determine the type of the object that is detected. Making use of these features, an interactive indoor navigation system is provided to efficiently reach the destinations inside a complex indoor environment. Figure 1 represents the mobile interface.

Fig. 1 Mobile application interface

6 System Design

The home interface of the mobile application is shown in Fig. 1. From the home interface, the user mode and the admin mode are accessible. Selecting any of the destinations that are displayed in the recent bar would immediately display the user mode navigation interface. To improve the user experience, the switching of the interfaces was implemented with a smooth transition and different backgrounds were applied to the home interface. The unique destinations were provided with their own icons to make the process of finding them easier. The plane detection algorithm works when the camera of the device is focused on the ground. Since the user is unaware of this information, the user navigation interface provides a small animation indicating that the user should focus their device camera on the floor. When the user successfully completes this task, the detected planes are represented by a dot mesh on the floor. Based on the orientation of the device, it is able to identify the walls and other obstacles too but there are no visual representations for them similar to the dot mesh of the ground.

There are few instances where the system is not able to correlate the data stored in the cloud and the data that is currently being detected by the device camera. This may happen when there are insufficient feature points or when there is a problem with establishing a connection to the cloud storage. The problem of insufficient feature points can be easily solved when the device is moved around a little and camera is focused on adjacent areas as well. The problem of establishing a connection depends on the available network speed and coverage. This is a device and location-specific issue that can be solved if the user has taken the appropriate measure of securing a stable connection.

The dataflow model of a sophisticated indoor navigation system is shown in Fig. 2 which represents the sequence of execution and the various components that are involved. The project involves a simplified system where ARCore is the heart

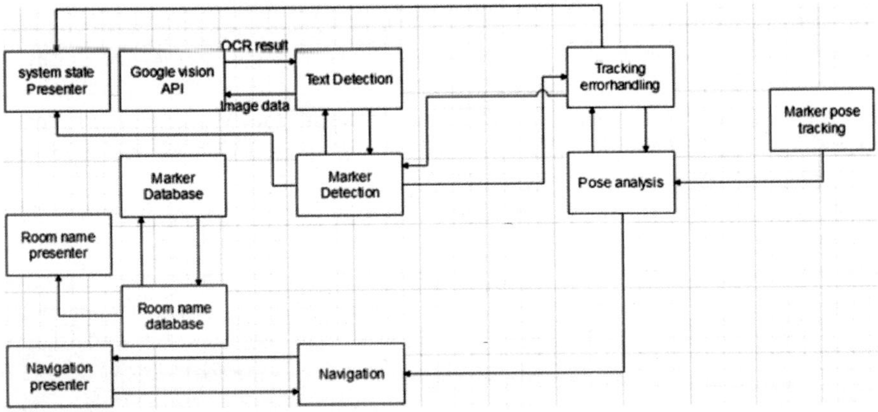

Fig. 2 Dataflow representation

of the system and the cloud storage is the database used. The detection and analysis algorithms are integrated into the ARCore such that there are no separate calls to the function. The data stored in the cloud is not related to the destinations rather it is the position data of the markers placed in the real-world environment. Figure 2 represents the dataflow representation.

7 Experimental Setup

We conducted an experiment to test the accuracy of the system. The markers were limited to a single floor and they were set to point to the various laboratories available on that floor. A white-tiled surface covered the entire region of the floor along with glass walls and doors that bounded the laboratories. Every lab was registered as a destination in the mobile application. The two participants were asked to choose different destinations and they started navigating simultaneously. Even though every location of the floor was almost identical, the system was able to detect the markers that were placed in the admin mode and were properly rendered in the user mode. They were successfully able to reach their destinations without any problems. The results have confirmed the localization, positioning, and rendering capabilities of the system in a small-scale environment.

8 Discussion and Results

In comparison with the previous technologies mentioned in Sect. 2, Augmented Reality provides a simpler method for indoor navigation. ARCore has greatly simplified the process of developing the AR application and has provided various methods that are well suited for acquiring data from the environment. In addition to being simple to develop and implement, it is also an inexpensive solution where there is no requirement of hardware to produce signals for positioning and tracking within the indoor environment. The interface provided by AR is very interactive and the user experience is greatly improved compared to the earlier technologies. The accuracy that we obtained by carrying out the above experiment was also found to be higher compared to the previous technologies.

The accuracy comparison of the various technologies used for indoor navigation is shown in Fig. 3. From the above representation, it is clearly evident that the accuracy of GPS is the lowest for indoor navigation. This is due to the fact that the reception of the satellite signal is very low in the indoor environment. Wi-Fi has a higher accuracy compared to GPS as there is hotspots set up in various locations within the indoor environment. The received strength of the signal helps in the location estimation of the navigating device. The localization method of Bluetooth beacons is the same as Wi-Fi but it offers a higher level of accuracy. The highest accuracy is offered by the AR technology in combination with ARCore. The AR technology is not based

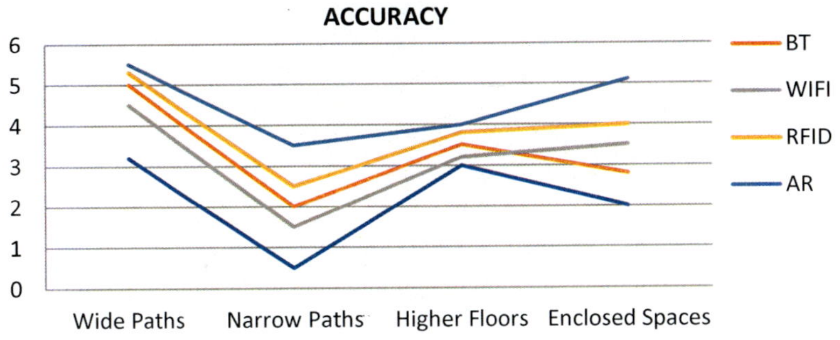

Fig. 3 Accuracy of navigation technologies

Table 1 Comparison of navigation technologies

Name	Type	Accuracy	Setup (1 = easy, 5 = hard)	Usability
Bluetooth beacons	2D	3–8 m	3	3
Compass based	2D	5–10 m	4	1
Apple indoor maps	2*D*	4–8 m	3	1
Ceiling antennas	2*D*.	10–50 cm	5	2
GPS	2*D*	5–15 m	1	3
Visual recognition/SLAM	3*D*/6*DoF*	10–30 cm	1	1
Markers/*QR* codes (AR)	3*D*/6*DoF*	5–15 cm	4	1

on signals rather it focuses on observing the indoor environment and uses the data that is stored in the cloud for positioning and localization. Table 1 represents the comparison of navigation technologies.

Earlier, the navigation technologies were used to work on a two-dimensional frame. The navigation was only based on the *X* and *Y* dimensions. The maps were represented as a planar surface and the position of the device was represented by a dot on that map. The third dimension for navigation was introduced by technologies such as AR and Computer Vision. Table 1 shows the various navigation technologies along with their dimensional characteristics, accuracy, and usability. AR provides the highest level of usability due to its super-friendly user interface and easy-to-understand navigation. The difficulty associated with the visual recognition and SLAM technology is because of the complexities in mapping the device's location and positioning it accordingly.

The AR technology is under rapid development and updates are being constantly released to improve the efficiency. The Reality-based technologies have become very popular in the recent years due to their rich user experience. These technologies are able to produce an environment where the border between virtuality and reality is broken down and it became harder to differentiate from one another. It enables a scenario where the virtual objects can be brought into the real world or the real-world objects can be brought into the virtual world. It makes the user to experience

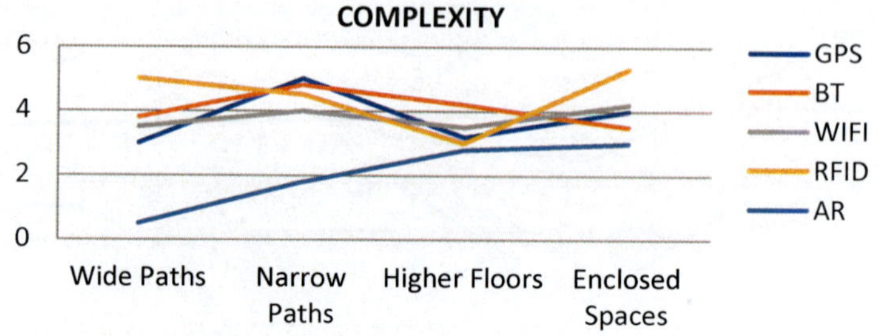

Fig. 4 Complexity of navigation technologies

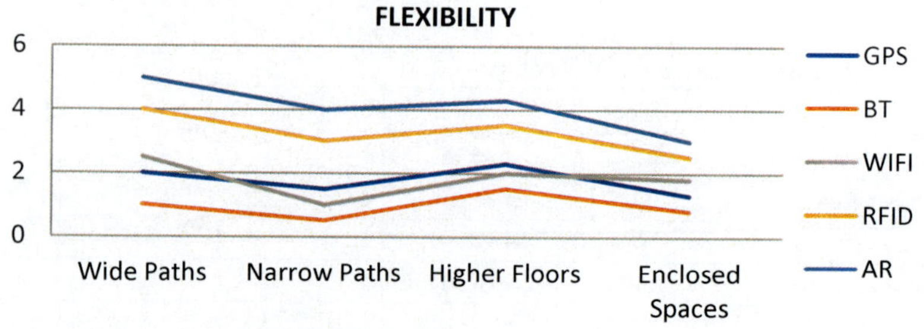

Fig. 5 Flexibility of navigation technologies

a world that is fundamentally different from what they are normally used to. These technologies have the potential to unlock unlimited possibilities which could be incorporated into every existing field of work.

They are currently being used in a wide range of fields such as video game development, military training, vehicle simulations, component design, etc. and are considered to be some of the top technologies of the future. The complexity and efficiency comparison of the various technologies used for indoor navigation is shown in Figs. 4 and 5.

9 Conclusion and Future Evolution

The solution which has been provided here is an inexpensive and convenient indoor navigation system that does not require any additional hardware. This system was particularly checked for effectiveness within a university campus and the results proved to be satisfactory. The AR-based navigation system provided an immersive experience to the users from the start of their route till they reached their destination. This immersive experience can be further improved by integrating a voice support, enhancing the details provided during the navigation, and by adding fun features such as character guides to the navigation system. This indoor navigation system uses a common cloud storage for holding all the markers that are placed in the real-world environment. However, it is inefficient while implementing this project on a large scale across various malls and universities.

This problem can be solved by allocating separate storage spaces for every different internal environment such that a memory overload does not occur when the device tries to access the marker information. This would further enhance the system by providing a quicker retrieval from the storage and rendering them smoothly on the device screen. Integrating the above-mentioned voice support system, character guidance system, and Natural Language Processing (NLP) system would provide various possibilities for the user to navigate the indoor environment. The NLP would be a great addition to the system as it allows the users to easily acquire their destination using voice input. English is one of the most spoken languages in the world. Therefore, most of the NLP systems are focused around the English language. But these systems would not be useful for the common people who are ignorant of languages other than their local language. By evolving the NLP system to include several local languages, the user experience of the system could be tremendously increased and it would become greatly useful to several communities.

10 Discussions

The major focus of future development of the indoor navigation system would be to increase the efficiency of the system's marker tracking and positioning along with efficient storage and retrieval. The system should not only cover the premises of a single university but it should be able to detect the location of any indoor environment that is integrated into the system. The position of the building is determined from the latitude and longitude coordinates obtained by using the GPS technology. This information is relayed to the server which calls the appropriate marker information from the cloud storage corresponding to the detected indoor environment. Thus, the system would be able to guide the users to their destination in various indoor environments simultaneously and efficiently.

References

1. Fischer G, Dietrich B, Winkler F (2004) Bluetooth indoor localization system. In: Proceedings of the 1st workshop on positioning, navigation and communication
2. Seichter H, Mulloni A et al (2011) Handheld augmented reality indoor navigation with activity-based instructions. In: Proceedings of the 13th international conference on human computer interaction with mobile devices and services, pp 211–220
3. Jiang L, Zhao P, Dong W, Li J, Ai M, Wu X, Hu Q (2018) An eight-direction scanning detection algorithm for the mapping robot path finding in unknown indoor environment. Sensors 18(12):4254

4. Nikoohemat S, Peter M, Oude Elberink S, Vosselman G (2018) Semantic interpretation of mobile laser scanner point clouds in indoor scenes using trajectories. Remote Sens 10(11):1754
5. Alattas A, Zlatanova S, Van Oosterom P, Chatzinikolaou E, Lemmen C, Li KJ (2017) Supporting indoor navigation using access rights to spaces based on combined use of IndoorGML and LADM Models. ISPRS Int J Geo Inf 6(12):384
6. Poux F, Neuville R, Nys GA, Billen R (2018) 3D point cloud semantic modelling: integrated framework for indoor spaces and furniture. Rem Sens 10(9)
7. Ivanov M, Sergyienko O, Tyrsa V, Lindner L, Flores-Fuentes W, Rodríguez-Quiñonez JC, Hernandez W, Mercorelli P (2020) Influence of data clouds fusion from 3D real-time vision system on robotic group dead reckoning in unknown terrain. IEEE/CAA J Autom Sinica 7(2):368–385
8. Real Ehrlich C, Blankenbach J (2019) Indoor localization for pedestrians with real-time capability using multi-sensor smartphones. Geo-spatial Inf Sci 22(2):73–88
9. Li X, Li X, Khyam MO, Luo C, Tan Y (2019) Visual navigation method for indoor mobile robot based on extended BoW model. CAAI Trans Intell Tech 2(4):142–147
10. Corotan A, Irgen-Gioro JJZ (2019) An indoor navigation robot using augmented reality. In: 5th international conference on control, automation and robotics, pp 111–116
11. Francis A, Faust H-T, Chiang L, Hsu J, Kew J, Fiser M, Lee T-WE (2019) Long range Indoor navigation with PRM RL
12. Michel T, Genevès P, Fourati H, Layaïda N (2018) Attitude estimation for indoor navigation and augmented reality with smartphones. Pervas Mob Comput 46:96–121
13. Al Rabbaa J et al (2019) MRsive: an augmented reality tool for enhancing way finding and engagement with art in museums. Springer, Cham, pp 535–542
14. Zhao Y; Azenkot S et al (2017) Understanding low vision people's visual perception on commercial augmented reality glasses. In: Proceedings of the 2017 CHI conference on human factors in computing systems—CHI '17, pp 4170–4181
15. Gerhard et al (2009) Handheld augmented reality for underground infrastructure visualization. Personal Ubiquitous Comput 13(4):281–291
16. Van Krevelen DWF, Poelman R (2010) A survey of augmented reality technologies, applications and limitations. Int J Virt Real 9(2):1–20
17. Pomberger G, Narzt W, Ferscha A et al (2005) Augmented reality navigation systems. Univ Access Inf Soc 4(3):177–187
18. Azuma RT (1997) A survey of augmented reality. Presence: Teleoperators Virtual Environ 6(4):355–385
19. Mortari F, Zlatanova S, Liu L, Clementini E (2014) Improved geometric network model (IGNM): a novel approach for deriving connectivity graphs for indoor navigation. ISPRS Ann Photogrammetry, Rem Sens Spat Inf Sci 2(4)
20. Akinci B, Tang P, Huber D, Lytle A (2010) Automatic reconstruction of as-built building information models from laser-scanned point clouds: A review of related techniques. Autom Const 19(7):829–843
21. Tran H, Khoshelham K, Kealy A, Díaz-Vilariño L (2019) Shape grammar approach to 3d modelling of indoor environments using point clouds. J Comput Civ Eng 33(1):04018055
22. Yan X, Liu W, Cui X et al (2015) Research and application of indoor guide based on mobile augmented reality system. Xi'an University of Post & Telecommunications, Xi'an, China. Pp 308–311
23. Chang CP, Chiang CS, Tu TM et al (2005) Robust spatial watermarking technique for colour images via direct saturation adjustment. Vis Image Sig Proc IEEE Proc 152:561–574
24. Boguslawski P, Zverovich V, Mahdjoubi L (2016) Automated construction of variable density navigable networks in a 3D indoor environment for emergency response 72:115–128
25. Rodenberg O (2016) The Effect of A* Path finding characteristics on the path length and performance in an octree representation of an indoor point cloud. Master's Thesis, Technical University of Delft, Delft, The Netherlands

A Survey on Autonomous Vehicles

Md. Hashim and Pooja Dehraj

Abstract Autonomous vehicles are the group of vehicles that do not or partially need a human driver to drive them. So, they are also known as driverless or robotic vehicles. Despite the name or group or size of vehicles, the purpose behind the technology is same. Besides the rapid development in technology of autonomous vehicles, the experiments have been started in early 1920 and due to lack of highly powerful computing devices, they were just controlled by radio wave technologies popularly known as remotes. Then, later on, the research and development took its speed in 1950 with the development of computers. But, nowadays with the development of highly powerful and cost-efficient computing hardware GPU the enhancement in technology becomes so rapid that it is updating day by day and becoming capable of concurring every aspect of human drivers' skills. AI and Machine Learning algorithms are being developed to train the model for autonomous driving over huge amount of recorded driving data of human drivers. As the time goes on these models started mimicking drivers' behavior and improves it day by day. So, as these Autonomous driving getting normalize and passing various parameters for driving the automobile industries saw a large potential for business and profit in autonomous vehicles leading the giant companies (Waymo Google, Tesla, Volvo, Renault, Uber, Toyota, Audi, Mercedes-Benz, Nissan, General Motors, Bosch and Continental's motors) started developing Level-3 Autonomous vehicles and released in 2020. By the rigorous development and solving challenges, autonomous vehicles are becoming our irreplaceable need for future.

Keywords Autonomous vehicles · Artificial intelligence · Machine learning

---

Md. Hashim (✉) · P. Dehraj
Noida Institute of Engineering and Technology, Greater Noida, India
e-mail: md.hashim127@gmail.com

P. Dehraj
e-mail: drpooja.cse@niet.co.in

V. Skala et al. (eds.), *Machine Intelligence and Data Science Applications*, Lecture Notes on Data Engineering and Communications Technologies 132,
https://doi.org/10.1007/978-981-19-2347-0_21

1 Introduction

With the rapid development of Artificial intelligence and sensory technologies, the researchers found an opportunity for harnessing the benefits of these technologies for making the driving decisions. After an intensive research on customers' behavior and needs the researchers came across the conclusion that the autonomous might be the best solution to satisfy the customers' need and requirements along with solving the major traffic and commuters problems like traffic jams and accidents which mainly occurs because of drivers wrong/delayed decisions and commons and repeating mistakes which can be resolved up-to an extend by implementation of artificial intelligence in vehicles in place of human drivers. Various researches claim that the traffic jams and accident rates will drastically fall to as low as zero because of right decisions made by AI. The AI models inside the autonomous vehicles will be trained over thousands of driving data of trained and professional drivers. The key hardware sensing components used in autonomous vehicles are LIDAR, cameras, RADAR, and ultrasonic sensors. AV's are more secure in many aspects in which human drivers are vulnerable and it becomes hard to have full trust in human drivers by their co-passengers, especially in case of cab services provided by cab provider organizations. Other benefits are the maintenance costs for AV's are far cheaper as the chances for misinterpretation in the control reduces to very low which results in drastically fall in wear and tear in vehicles. It will be boon for the cab provider companies as other than their driver's salaries, an impactful portion of profit goes into this maintenance which could be avoided. So, by considering these many governments have proposed for identification and development of autonomous vehicle zones/city infrastructure. So, we could say in upcoming future AV's are going to be more necessity than luxury.

Along with this, the design and development of autonomous vehicles include implementation of autonomic computing techniques in which system is made self-managed. It means the vehicle autonomically manages the internal processing with less human intervention [1–3] Autonomic computing technique is a combination of four major self-adaptive features that is self-configuration, self-healing, self-optimization, and self-protection. Among these, autonomous vehicle development process focuses mainly on self-optimization and self-configuration. Self-optimization feature handles the resource optimization work of the autonomous vehicle and self-configuration feature handles the system's configuration-related processes [4–6].

2 Design and Development of Autonomous Vehicles

2.1 Initiation Phase

In this research work, we are going to develop an intelligence software to drive the vehicles without any human interactions, i.e., autonomously. Software will be tested in a real-world simulated environment, provides by an open-source simulator named CARLA, on various parameters like time efficiency, accuracy of predictions, and reflected actions accordingly. Feasibility of this research work depends on various external parameters like Hardware's Technology, Geographical diversities, Human external interactions, etc.

2.2 Definition Phase

The efficiency and accuracy mainly depend on three-parameter deciding resources such as Hardware, Algorithm used, and real-world Training Datasets gathered by different organizations like CVPR (Conference on Computer Vision and Pattern Recognition), BDD (Berkeley DeepDrive) [7].

2.3 Design Phase

This research work will pass through four different phases till its completion.

Designing and Selection. Basic hardware system, Identification of main components of software and creating a safety measure assessment strategy, Defining the PID for longitudinal control and Path following controllers for lateral control, and testing of their control design.

State Estimation Sensing and Localization. Understanding of the key methods for parameter and state estimations. Development and use of models for typical vehicles localization sensors. Deployment of Kalman filters to the vehicle state estimation problems. Registering of point clouds from LIDAR sensor's data to 3D Maps.

Visual Perceptions for Self-driving Car. Projection of 3D points on the camera image planes. Calibration of the pinhole camera model. Integration of feature detection algorithms for localization and mapping. Development and training of neural network for objects detection and semantic segmentation.

Motion Planning. Devise trajectory rollout motion planning. Calculate the time of collision. Define high-level vehicles behavior.

2.4 Implementation Phase

Collection of Dataset. O s in Fig. 1, the dataset for training of Convolutional Neural Network (CNN), time-stamped video from the cameras (three mounted at data-acquisition car) is captured simultaneously along with the steer angle applied by the human driver, are obtained by tapping into the vehicle's Controller Area Network (CAN) bus. Training data contains single images sampled from the video, paired with the corresponding steering command (1/r); where "r" is the turning radius in meters.

Training of CNN. As in Fig. 2, after creation of dataset, images are being fed into a CNN, which computes a proposed steering command. The proposed command is compared to the desired command for that image, and the weights of the CNN are adjusted to bring the CNN output closer to the desired output. The weight adjustment is accomplished using backpropagation.

Taking Command from CNN. As in Fig. 3, once trained, the network is able to generate steering commands from the video images of a single-center camera.

Computing Hardware Requirements for Implementation.
Image Processing, Object Detection, Mapping. ASICs—Application Specific Integrated Chip. FPGAs—Field Programmable Gate Array. GPUs—Graphic Processing Unit. Synchronization Hardware. To synchronize different modules and provide a common clock.

Assumptions Aggressive Deceleration = 5 m/s*2, Comfortable deceleration = 2 m/s*2, Stopping distance — d = v2/2a.

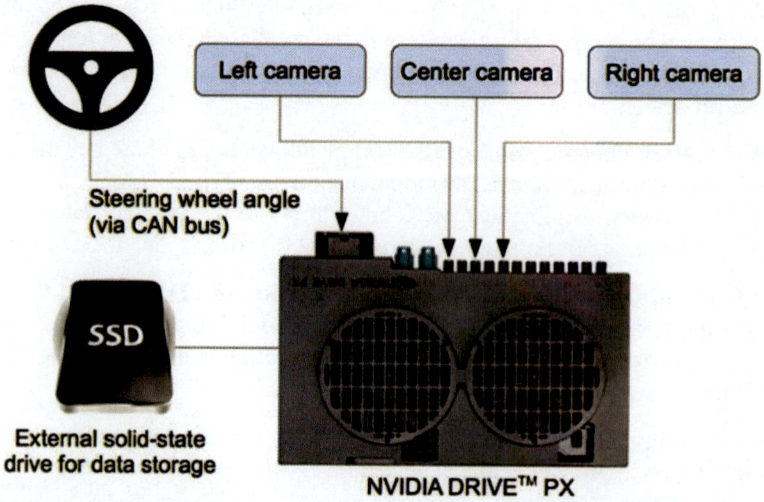

Fig. 1 Data collection system [8]

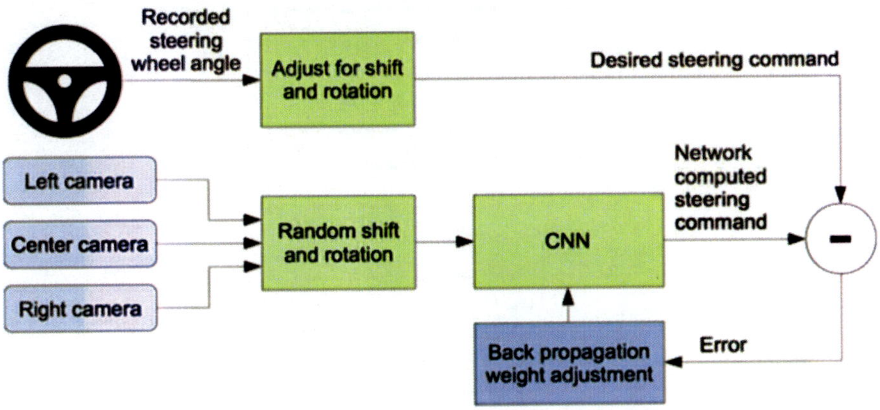

Fig. 2 Block diagram of neural network training [8]

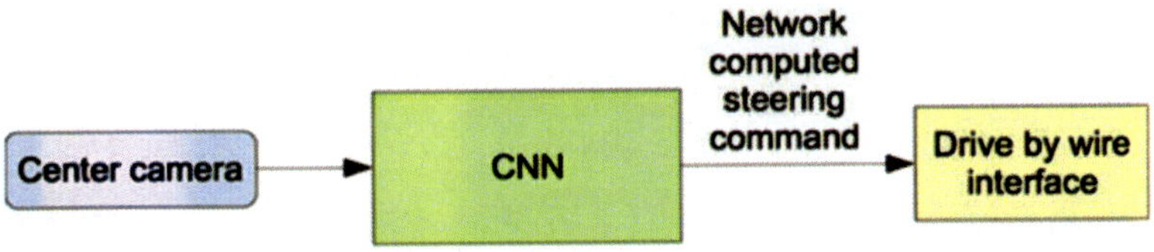

Fig. 3 Command flow diagram [8]

3 Design Architecture of Autonomous Vehicles

Architecture designing is one of the most important and crucial parts of autonomous vehicles development. It is broadly classified into hardware architecture and software architecture. As in Fig. 4, firstly cover the overall mission control planning then based on that behavioral control after that local planner and finally overall vehicle controller.

3.1 Description of Software Architecture

As in, Fig. 5 [9], Software Architecture, on taking the output data from sensors, the software map the positions of the vehicles with respect to the environment after that it perceives the environment and then the CNN starts motion planning and sends command signals to the controller and the control controls the vehicle.

Environment Perception. After collecting data from sensors, as in Fig. 6 [9], the vehicle perceives the environments with respect to vehicle like vehicle position, moving or static objects and road positioning.

Environment Mapping. The vehicle maps its positioning with respect to global maps like occupancy grid map, localization map, and detailed road map and maps

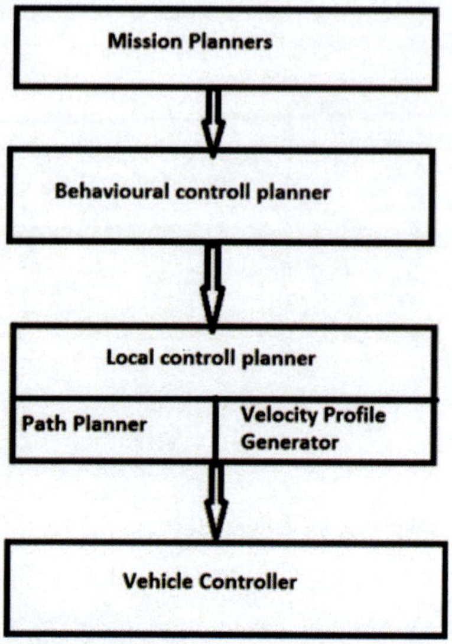

Fig. 4 Design architecture flow diagram [9]

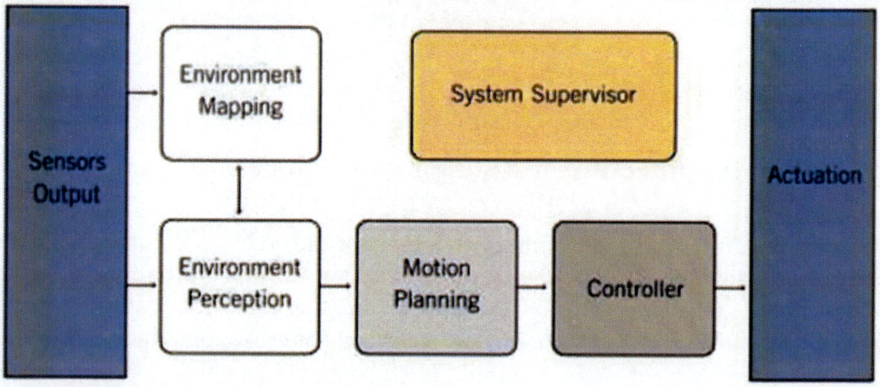

Fig. 5 Block diagram of software architecture

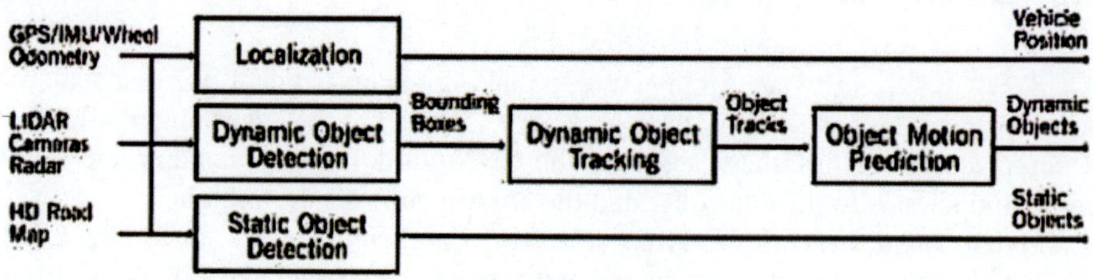

Fig. 6 Environment perception architecture diagram

different scenarios like road boundaries, sign poles, lanes, and traffic lights. As in Fig. 7.

Controller. It acts on the basis of control signal provided by CNN by environment perception and mapping to different controllers like velocity (throttle and brake) and steering controller. As shown in Fig. 8 [9].

Motion Planning. As in Fig. 9 [9], the vehicle planes its motion and behavior based on the local and global environments and its destination. It is categorized into mission planer, behavior planner, and local planner.

System Supervisor. As in Fig. 10, it is broadly divided into software and hardware supervisor. Software supervisor perceives and supervises the software modules whereas hardware supervisor supervises the hardware modules like sensors outputs.

Sensors. Sensors detect the environment and send data to CNN in different formats based on their types like heatmap, video data, ultrasonic data, etc. [10]. There are numerous types of sensors are used and mounted on the vehicle, as in Fig. 11.

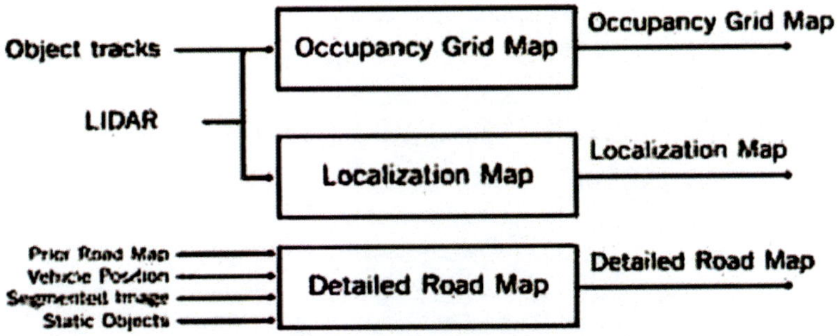

Fig. 7 Environment mapping architecture diagram [9]

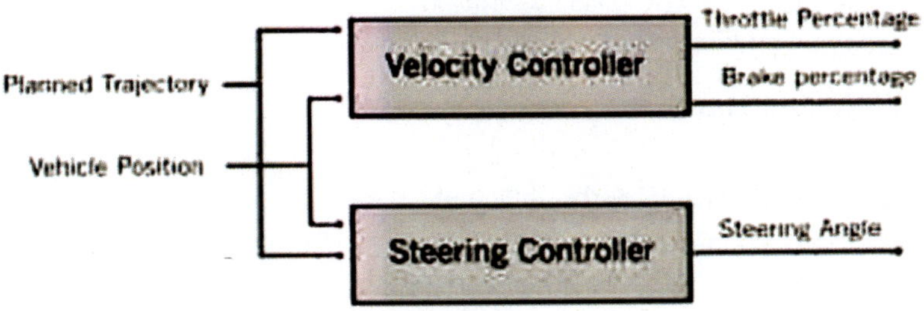

Fig. 8 Block diagram of controller architecture

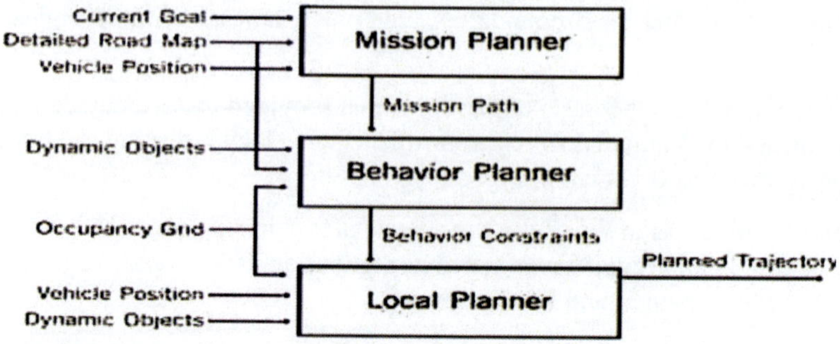

Fig. 9 Motion planner architecture diagram

Fig. 10 System supervisor architecture diagram [9]

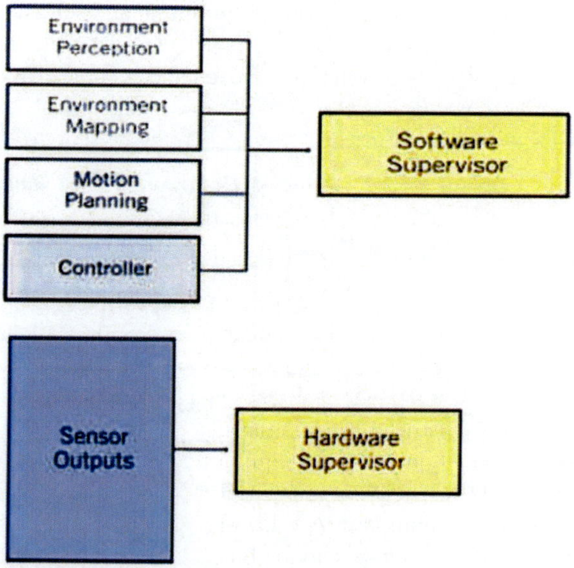

3.2 *Highway Analysis Process*

It is one of the most essential and crucial parts of autonomous vehicle design process, by the different highway analysis and congestion control to ensure the high saver mode of commute by the autonomous vehicles [12].

Emergency Stop. It is used in case of unpredicted hazards or situations.

Longitudinal Coverage. Used to calculate the stopping distance or the retardation needed to stop the vehicles within the available space ahead of vehicle, as shown in Fig. 12 [9].

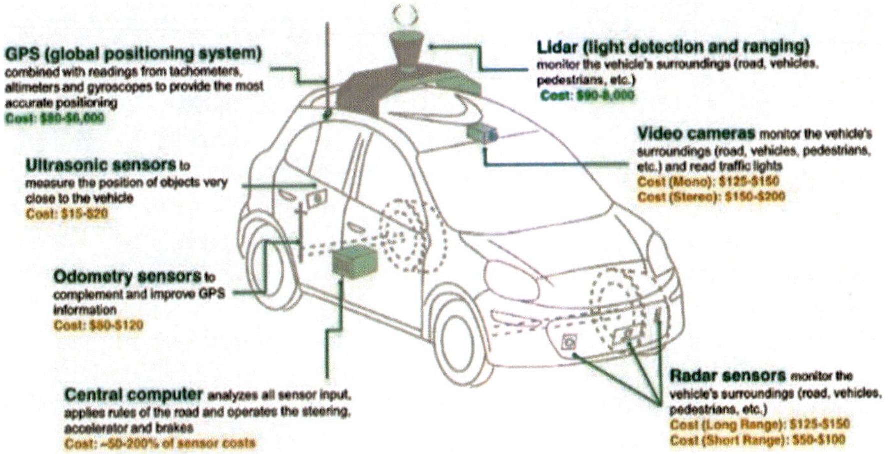

Fig. 11 Blueprint of Sensors positioning [11]

Fig. 12 Longitudinal coverage diagram

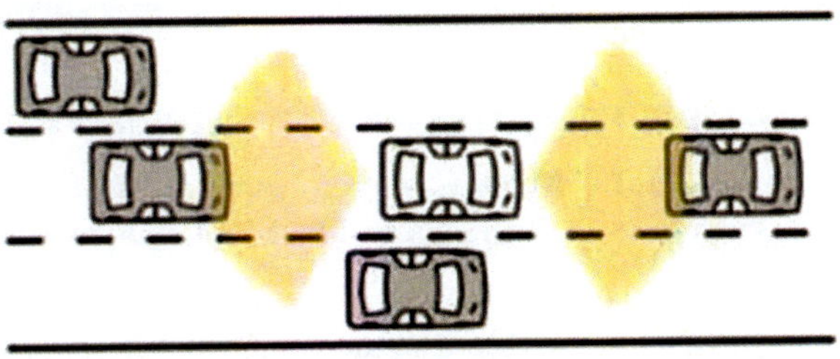

Lateral Coverage. Used to sense and measure the lateral margin or lateral space available ahead of vehicle to avoid lateral (side by side) collision, as in Fig. 13.

Lane Changing. Need to sense and calculate the longitudinal and lateral safe distance for lane change of a vehicle. And also needs to analyze and predict the behavior of ahead and behind vehicles to avoid uncalculated collisions, as in Fig. 14 [9].

Fig. 13 Lateral coverage diagram [9]

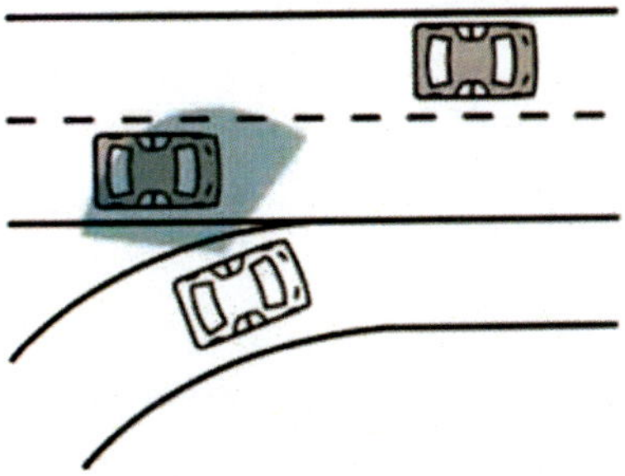

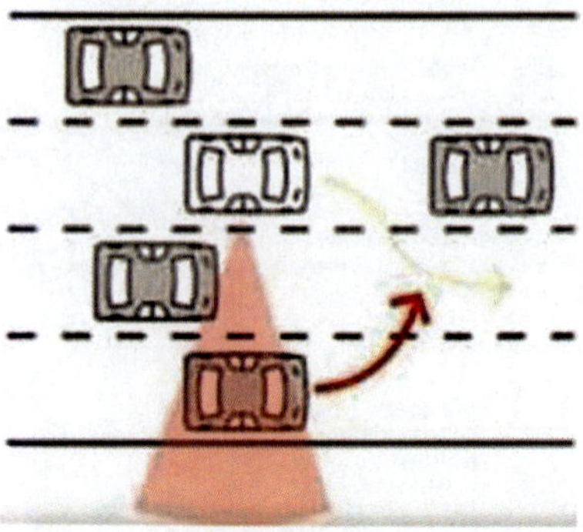

Fig. 14 Lane changing diagram

3.3 Urban Analysis

It is a step ahead of highway analysis including the different additional scenarios like unfamiliar turning and crossings and round bounds where the decision-making capabilities of autonomous vehicles needs to extend to a new level by many additional parameters like ABS [13].

Emergency Stop. It is used in case of unpredicted hazards or situations.

Longitudinal Coverage. Used to calculate the stopping distance or the retardation needed to stop the vehicles within the available space ahead of vehicle, as shown in Fig. 15 [9].

Lateral Coverage. Used to sense and measure the lateral margin or lateral space available ahead of vehicle to avoid lateral (side by side) collision, as in Fig. 16.

Lane Changing. Need to sense and calculate the longitudinal and lateral safe distance for lane change of a vehicle. And also needs to analyze and predict the behavior of ahead and behind vehicles to avoid uncalculated collisions, as in Fig. 17 [9].

Overtaking. While trying overtake a moving or parked vehicle, it needs to detect oncoming traffic ahead and should be beyond predicted time and point of return to its own lane, as shown in Fig. 18 [9].

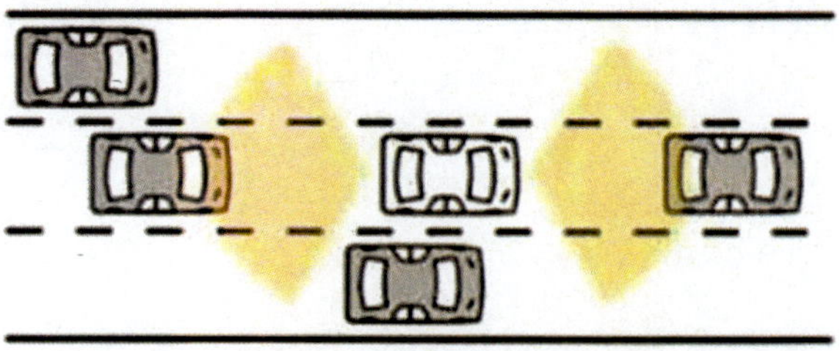

Fig. 15 Longitudinal coverage diagram

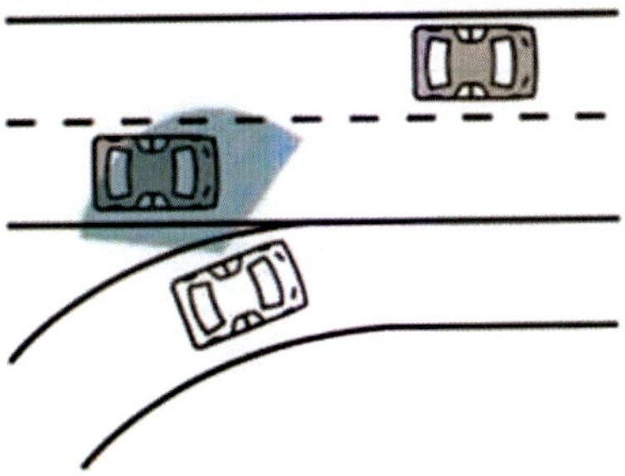

Fig. 16 Lateral coverage diagram [9]

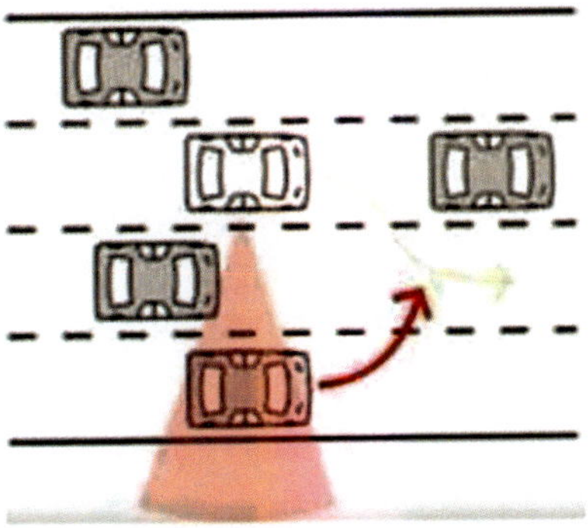

Fig. 17 Lane changing diagram

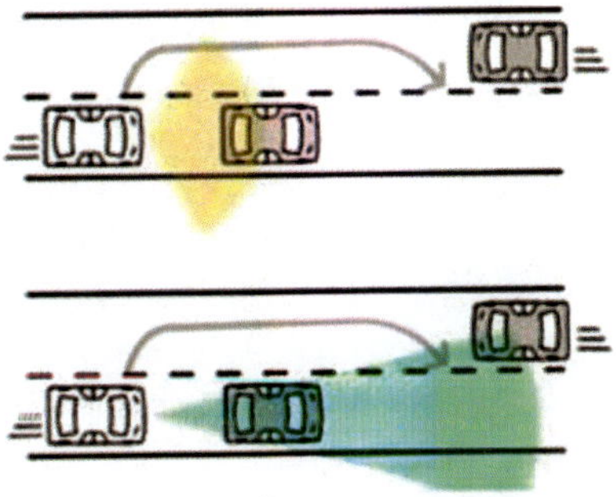

Fig. 18 Overtaking controller diagram

Turning, Crossing at Intersections. Vehicles have to observe and predict beyond intersection points for approaching vehicles, pedestrian crossings, and clear exit lanes. It requires near omnidirectional sensing for arbitrary intersection angles, as shown in Fig. 19 [9].

Passing Roundabouts. Due to the shape of the roundabout, the vehicles need a wider field of view and predictions to avoid collisions, as shown in Fig. 20.

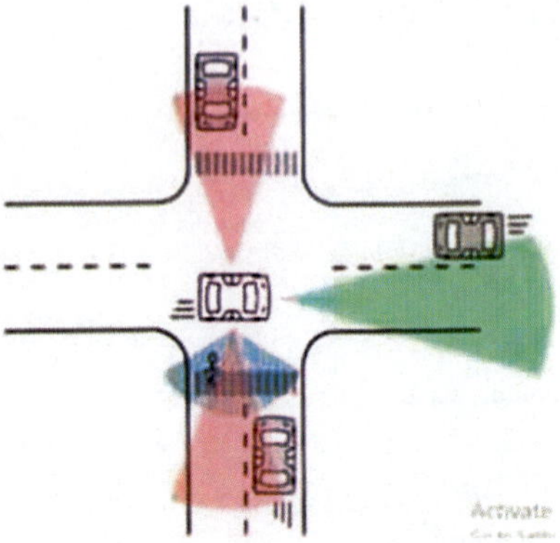

Fig. 19 Decision-making for crossing at intersections

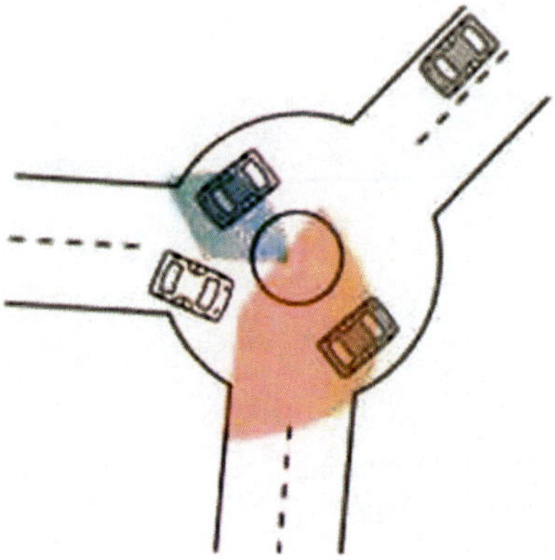

Fig. 20 Decision for passing roundabouts [9]

3.4 Sensors for Perceptions

- Detailed 3D scene geometry from LIDAR point cloud.
- Comparison metrics: Number of beams, Points per second, Rotation rate, Field of view
- Configurations: WFOV (short-range) and NFOV (long-range)
- Sensors needed for perception: Camera, Radar, Lidar, Ultrasonic, GNSS/IMU, Wheel Odometry

Detection Process. Different sensors such as Cameras, radar, and lidar which provides rich data about vehicles' surrounding to central control unit of the vehicle. So, much like the human brain processes visual data taken in by the eyes, an vehicle should be able to make sense of this constant flow of information.

Feature Detection. TFeatures are points of interest in an image that is needed for computation and analysis. Points of interest should have the following characteristics [14]:

Fig. 21 Detection of traffic signs and signals [8]

- Saliency-distinctive, identifiable, and different from its immediate neighborhood
- Repeatability—can be found in multiple images using same operations
- Locality—occupies a relatively small subset of image space
- Quantity—enough points represented in the image
- Efficiency—reasonable computation times.

Traffic Sign and Signal Detection. Autonomous vehicles are needed to be highly trained toward traffic sign and signal detection for safer and smooth driving, as in Fig. 21. Traffic signs and signals appear smaller in size compared to cars, two-wheelers, and pedestrians. Traffic signs are highly variable with many classes to be trained. Traffic signals have different states that are required to be detected. In addition, traffic signals change state as the car drives.

4 Testing and Validations

4.1 Testing

After the development phase, the vehicle goes through various phased testing which tests the performance and decision-making intelligence to the next level considering different real scenario test conditions and requirements.

- Performance testing at different levels
- Requirement validation of components, levels
- Fault injection testing of safety-critical functionality
- Intrusive testing, such as electromagnetic interference, etc.
- Durability testing and simulation testing.

4.2 Safety Thresholds

The vehicles are equipped with two key safety thresholds-

Fail Safes—there is redundant functionality (second controllers, backup systems, etc.) such that even if primary systems fail, the vehicle can stop normally.

SOTIF—All critical functionalities are evaluated for unpredictable scenarios.

4.3 Safety Processes

It consists of different design and analysis processes steps and parameters for the vehicles to withstand different tough scenarios which occur while driving.

- Deductive Analysis
- Fault tree analysis
- Inductive Analysis
- Design & Process FMEA
- Exploratory Analysis
- HAZOP—Hazard and Operability Study
- Safety through CRM (Comprehensive Risk Management) and deep integration
- Identity and address risks, validate solutions
- Prioritize elimination of risks, not just mitigation
- All hardware, software systems meet
- Self-set standards for performance, reliability, crash protection, security, safety, and serviceability,
- Address all 12 elements of NHTSA Safety Framework.

4.4 Levels of Testing to Ensure Safety

Different levels of testing are considered to ensure highest possible safety and nullify the chance of errors.

- Simulation testing
- Test rigorously with simulation, thousands of variations, fuzzing of neighboring vehicles
- Closed-course testing
- Focus on four most common crashes-
- Rear-end, intersection, road departure, lane change
- Real-world driving
- Start with smaller fleet, expand steadily
- Already testing thousands of vehicles, with more on the way.

4.5 RANSAC Algorithm

It is a learning technique that estimates parameters of a model by random sampling of observed data for all the given datasets whose data element contains both inliers and outliers, by using the voting scheme to find the optimal fitting result [15].

Initialization.

Given a model, find the smallest number of samples, from which the model can be computed.

Main Loop.

- From your data, randomly select M samples
- Compute model parameters using the selected M samples
- Check how many samples from the rest of your data actually fit the model. We call this number the number of inliers C
- If C > inliers ratio threshold or maximum iterations reached, terminate and return the best inliers set. Else, go back to step 2.

Final Step.

Re-compute model parameters from entire best inliers set.

5 Conclusion

So, by the advancement in AI technologies and Machine Learning algorithms, the autonomous vehicles are becoming smarter day by day with high pace of improvements and advancements. They are going to change the future and way of transportation in upcoming future and going to be an irreplaceable part of our daily commute. By the current result of tests and prediction, it is giving us a sense of assurance of safer, economical, and hassle-free mode of commute. For every automobile industry, it will be necessity to invest in research and development of autonomous vehicles to be sustainable and profitable in their business. Also for its potential customers means commuters must have to adopt the changes and experience from the conventional way of human driving to AI-driven autonomous vehicles. Besides from that the govt. need to transform transportation infrastructure and rules made for automobile for the smoothness increment of use and advantages of autonomous vehicles.

References

1. Dehraj P, Sharma A (2020) An approach to design and develop generic integrated architecture for autonomic software system. Int J Syst Assur Eng Manage 11(3):690–703

2. Dehraj P, Sharma A (2021) A review on architecture and models for autonomic software systems. J Supercomput 77(1):388–417
3. Dehraj P, Sharma A (2019) Autonomic provisioning in software development life cycle process. In: Proceedings of international conference on sustainable computing in science, technology and management (SUSCOM), Amity University Rajasthan, Jaipur-India
4. Dehraj P, Sharma A (2020) A new software development paradigm for intelligent information systems. Int J Intell Inf Database Syst 13(2–4):356–375
5. Dehraj P, Sharma A (2019) Guidelines for measuring quality of autonomic systems. In: Proceedings of international conference on sustainable computing in science, technology and management (SUSCOM), Amity University Rajasthan, Jaipur-India
6. Dehraj P, Sharma A, Grover PS (2019) Maintenance assessment guidelines for autonomic system using ANP approach. J Stat Manage Syst 22(2):289–300
7. Faisal A et al (2016) Understanding autonomous vehicles. J Trans Land Use 12(1):45–72
8. NVIDIA https://developer.nvidia.com/blog/deep-learning-self-driving-cars/. Accesed: June 2021
9. Zhihu.com: https://zhuanlan.zhihu.com/p/379250208. Accesed: June 2021
10. Rojas-Rueda D et al (2020) Autonomous vehicles and public health. Ann Rev Publ Health 41:329–345
11. wired.com https://media.wired.com/photos/59372bb59a93607bd17ca79a/master/w_1920,c_limit/sensor_info.jpg
12. Schwarting W, Alonso-Mora J, Rus D (2018) Planning and decision-making for autonomous vehicles. Ann Rev Cont Robot Auton Syst 1:187–210
13. Wiseman Y (2021) Revisiting the anti-lock braking system. Technical Report
14. Janai J et al (2020) Computer vision for autonomous vehicles: Problems, datasets and state of the art. Found Trends® in Comput Graph Vis 12(1–3):1–308
15. Kuutti S et al (2019) Deep learning for autonomous vehicle control: algorithms, state-of-the-art, and future prospects. Synth Lect Adv Automot Technol 3(4):1–80

Impact of Innovative Technology on Quality Education: The Resourceful Intelligence for Smart Society Development

Suplab Kanti Podder, Benny Thomas, and Debabrata Samanta

Abstract Quality education is the systematic learning and execution road map that build confidence among the learners and develop employability skills. Innovative techniques are the central facilitators of providing quality education for the younger generation. The economists, scientists, management experts, and research initiators are putting their efforts to develop a certain sustainable system in quality education through innovative technology. This is about digital equity, customized education, activity-based classroom, where the young mind is to be in synchronizing with the technology to explore new possibilities of learning and accomplishment. The research initiative reveals the system of implementing the innovative technology for quality education that has a direct impact on smart society development. The principal outcomes of the research initiative include the innovative ideas that transform the traditional education system into a dynamic education framework. The framework includes the integration of tools and techniques for standard mode of operations that reflects the productive and realistic education system. The researchers gracefully interconnected the concepts, methods, and applications of a quality education system that will open up new vistas for future research initiatives in the area of digital education, industry-institution collaboration, developing smart society, and economy of a nation at large. There is significant level of impact of innovative technology on quality education that leads to independent employability skills, creative, and innovative projects for facilitating future generation. All the influencing factors of resourceful intelligence together have great impact on smart society development

S. Kanti Podder
Department of Management and Commerce, Dayananda Sagar College of Arts, Science and Commerce, Bangalore, India
e-mail: skpnext@gmail.com

B. Thomas
Department of CSE, CHRIST Deemed to be University, Bangalore, India
e-mail: frbenny@christuniversity.in

D. Samanta (✉)
Department of Computer Science, CHRIST Deemed to be University, Bangalore, India
e-mail: debabrata.samanta369@gmail.com

V. Skala et al. (eds.), *Machine Intelligence and Data Science Applications*, Lecture Notes on Data Engineering and Communications Technologies 132,
https://doi.org/10.1007/978-981-19-2347-0_22

that leads to provide modern facilities for the residents of smart society and create favorable environment for the future generations.

Keywords Quality education · Innovative technology · Resourceful intelligence · Smart society

1 Introduction

Invention and innovation are the two important factors of sustainable development that facilitate overcoming future threats and challenges [1]. The innovative techniques upgrade the teaching and learning methods that describe the new way of developing the modern education system. Through the research work, we tried to find out the impact of quality education and innovative entrepreneurship on the development of a smart society [2]. The impact of innovative technology on quality education is the productive invention that ensures the systematic process of implementing modern tools and techniques of education [3]. Figure 1 shows conceptual framework related to the impact of innovation technology on quality education various models are given below with suitable statistical data and graphical representations: The framework of the innovative model of education instructs us to provide systematic training to familiarize ourselves with the new tools and techniques of modern education [4]. The innovative model of education requires the installation and usage of various tools like digital mode of teaching and learning, accelerate remote learning, artificial intelligence, smart classroom, and flipped classrooms. In this age of digitalization, every student must have an equal opportunity to digital education. Education now should adopt an asynchronous mode where students can access and use information through online platforms [5]. Artificial intelligence is the mechanism that ensures the most efficient and accurate performance for a human being. The arrangement can be

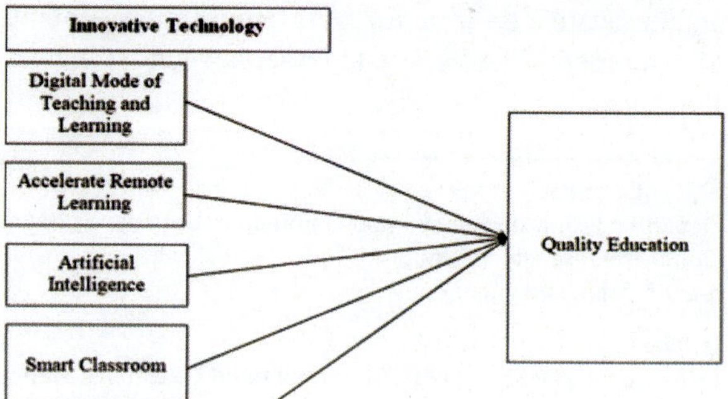

Fig. 1 Conceptual framework related to the impact of innovation technology on quality education

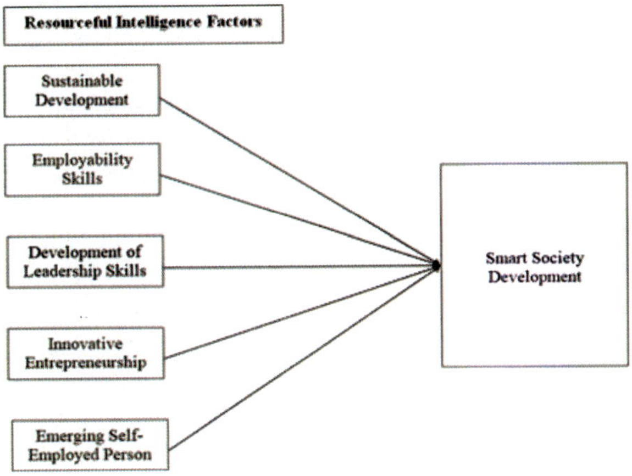

Fig. 2 Conceptual framework of resourceful intelligence factors that influence to smart society development

utilized for an innovative education system by integrating information and communication technology [6, 7]. The ICT-based teaching and learning system are implemented that consider the better understanding of conceptual and practical aspects. A smart classroom is the integration of a digital mode of operations and leads to develop the electronic content for systematic learning [8, 9]. Flipped classroom activities are the innovative teaching and learning method that encourage the students for reading and understanding the concepts at home and practice in the classroom. Figure 2 expresses conceptual framework of resourceful intelligence factors that influence to smart society development. The resourceful intelligence factors include sustainable development, employability skills, development of leadership skills, innovative entrepreneurship, and emerging self-employed person are the interconnected aspects of smart society development [10]. The residents of smart society expect the best experience for modern education, healthcare services, smart supply chain management, and sustainable development initiatives for future generations [11]. Quality education and smart society are considered as the recent development concepts and applications in modern society [12, 13]. The education system, education policy, corporate requirements, and individual expectations are changing time to time. The smart decision for quality education keeps changing as per the national and international environment and situation. The present research paper identified the research problem by considering the previous articles and expectation from future generations and their survival. The novel solutions for the identified problems have addressed by describing the innovative and realistic action plans that facilitate the systematic guidelines of utilizing technology on quality education and provide modern facilities for the residents of smart society to create favorable environment for the future generations.

2 Reviews of Literature

The innovative technology has a great impact on quality education that can play a major role in developing a smart society. The technology includes diverse fundamentals of learning environment to offer optimized learning structure and methodology to each learner in response to the smart earnings and improving the standard of living [14, 15]. This innovative model sets in a learning ecosystem that will support innovative and optimized learning. This innovative learning introduces new and improved educational digital content [16, 17].

The innovative technology helps in building innovation and an information-driven educational system where the objective and the content of courses will be problem-solving and critical-thinking centered education [18]. The innovative model has tried to promote reconstruction and reorganization of curriculum for computational thinking-oriented education [19]. The smart initiative facilitates the development of autonomous competency of educational institutions. The model also tries to establish an adaptive learning structure using intelligent information and higher technology to maximize learning efficacy.

It encourages digital learning and teaching. The technology has advised virtual and blended learning practices. Digital learning encompasses a balanced blend of cloud based and personalized learning and wide use of, AI platform and e-resources. The initiative fosters the use of digital and innovative teaching techniques and tools [20]. The instructional practices and the assessment methods will also be digital tools and technology-oriented strategies for the educational field. Digital learning will enhance the learning experience and even transform the teaching practices and assessment methodology.

The curriculum would produce complete literacy including technology literacy, social literacy, and human values literacy [21]. The abilities and skills acquired by the learners will be conceptual and trade knowledge, knowledge of advanced technology, and proficiency in strategies for future global culture. The curriculum will build up students that will possess creative learning and risks taking. An innovative curriculum would work toward making the learners learn and apply the learning. It is an application-based curriculum [22]. The modern education expects the latest technology upgradation and application of modern education tools like Edmodo, Socrative, Projeqt, Thinglink, TED-Ed, cK-12, and ClassDojo that facilities smart education and creative implications. The ICT-based education system can improve the teaching and learning ability and develop more employability skills.

3 Contribution

Based on the statement of problems and curiosity of the researcher the following, objectives were identified:

1. To identify the impact of innovative technology on quality education.

2. To find out the resourceful intelligence factors that influence smart society development.

4 Research Objectives

The hypothesis formulation can be summarized as follows:

$H01$: There is no significant level of impact of innovative technology on quality education.

Or,

Mathematically, $H01$: [the significant level of impact of innovative technology on quality education] = 0.

$H02$: There is no significant level of influence of resourceful intelligence on smart society development.

Or,

Mathematically, $H02$: [the level of resourceful intelligence factors' influence to smart society development.] = 0.

5 Results and Discussions

The innovative techniques of quality education consist of a digital mode of teaching and learning, accelerate remote learning, artificial intelligence, smart classroom, and flipped classrooms. Table 1 shows the impact of innovative technology on quality education. Summary of multiple regression analysis comprises the *t*-test, ANOVA test, and *p*-value (significance value) to analyze the responses of primary sources. The calculated *p*-value is 0.000 which is less than 0.05 at the level of significance that indicates the rejection of the null hypothesis. Figure 3 shows the summary of multiple regression analysis with respect to standard coefficient, *t*-test value, and *p*-value.

The above graphical representation indicates that there is a significant level of impact of innovative technology on quality education. All the influencing factors of innovative technology together have a great impact on quality education that leads to independent employability skills, creative, and innovative projects for facilitating future generations.

The resourceful intelligence factors include sustainable development, employability skills, development of leadership skills, innovative entrepreneurship, and emerging self-employed person are the interconnected aspects of smart society development. Summary of multiple regression analysis comprises the *t*-test, ANOVA test, and *p*-value (significance value) to analyze the responses of primary sources. The calculated *p*-value is 0.000 which is less than 0.05 at the level of significance that indicates the rejection of the null hypothesis. Table 2 describes the resourceful intelligence factors that influence smart society development.

Table 1 Impact of innovative technology on quality education

Highlights of analysis					
DepVar: Quality education, N: 400, Multiple R: 0.561, Squared multiple R: 0.315, Adjusted squared multiple R: 0.306, Standard error of estimate: 0.836					
The result of MRA					
Effect	Coeff	Std error	Std Coeff	t	Sig
(Constant)	0.648	0.938		0.690	0.490
Digital mode of teaching and learning	−0.385	0.033	−0.768	−11.752	0.000
Accelerate remote learning	0.335	0.037	0.5	8.972	0.000
Artificial intelligence	0.763	0.074	0.775	10.342	0.000
Smart classroom	0.365	0.061	0.485	5.944	0.000
Flipped classrooms	0.195	0.052	0.208	3.745	0.000
Significant at 0.05 level					
ANOVA					
Source	Sum-of-Squares	df	Mean-Square	F-ratio	Sig
Regression	126.777	5	25.355	36.251	0.000
Residual	275.583	394	0.699		
Significant at 0.05 level					
Durbin–Watson D Statistic = 2.169; First Order Autocorrelation = 0.131					

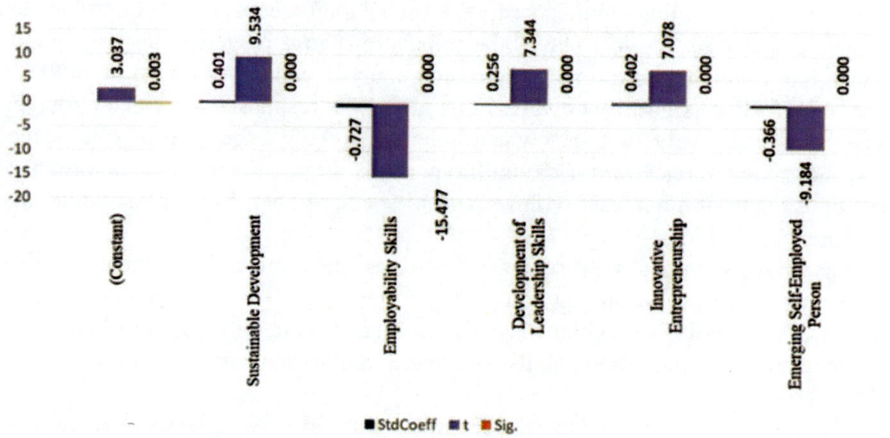

Fig. 3 Summary of multiple regression analysis with respect to standard coefficient, *t*-test value, and *p*-value

Table 2 Resourceful intelligence factors that influence smart society development

Highlights of analysis					
DepVar: Smart Society, N: 400, Multiple R: 0.788, Squared multiple R: 0.620, Adjusted squared multiple R: 0.615, Standard error of estimate: 0.310					
The result of MRA					
Effect	Coeff	Std Error	Std Coeff	t	Sig
(Constant)	0.586	0.193		3.037	0.003
Sustainable development	0.452	0.047	0.401	9.534	0.000
employability skills	−0.461	0.03	−0.727	−15.477	0.000
Development of leadership skills	0.351	0.048	0.256	7.344	0.000
Innovative entrepreneurship	0.281	0.04	0.402	7.078	0.000
Emerging self-employed person	−0.415	0.045	−0.366	−9.184	0.000
Significant at 0.05 level					
ANOVA					
Source	Sum-of-Squares	df	Mean-square	F-ratio	Sig
Regression	61.803	5	12.361	128.713	0.00
Residual	37.837	394	0.096		
significant at 0.05 level					
Durbin–Watson D Statistic = 1.551; First-order autocorrelation = 0.121					

The graphical representation indicates that there is a significant level of influence of resourceful intelligence on smart society development. All the influencing factors of resourceful intelligence together have a great impact on smart society development that leads to provide modern facilities for the residents of smart society and create a favorable environment for future generations. Figure 4 shows the summary of multiple regression analysis with respect to standard coefficient, *t*-test value, and *p*-value.

6 Conclusion

The standard mode of operations is the orderly combination of digital mode of teaching and learning, accelerate remote learning, project-based learning, digital mode of examination and assessment, and field-specific experience. Digital mode of teaching and learning activities is performed through usages of electronic devices for efficient learning. All the tools and techniques are used for supporting modern teaching and learning activities. Accelerate remote learning is the initiative to open up

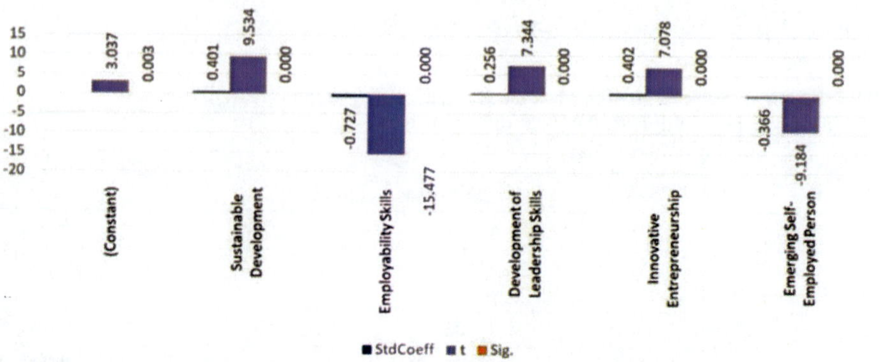

Fig. 4 Summary of multiple regression analysis with respect to standard coefficient, *t*-test value, and *p*-value

new learning experiences. The system encourages social learning and taking ownership of creativity and innovation. Educational institutions should teach and encourage imparting theoretical knowledge using digital means. The students will learn the theoretical concepts using digital platforms and media. Project-based learning is the student engagement procedure in which social issues are addressed in the form of problem-solving initiatives. At first, the students create a team of few members and start work for identifying the problems. Students will be taught through a dynamic classroom approach. They would be taking up real-time projects and explore the real-time issues and challenges of the corporate world and acquire a deep understanding of the concepts through the projects.

References

1. Askling B, Stensaker B (2002) Academic leadership: prescriptions, practices and paradoxes. Tert Educ Manag 8(2):113–125
2. Sridevi KB (2020) Filling the quality gaps for a futuristic management education. J Econom Adm Sci ahead-of-print(ahead-of-print)
3. The University of Learning: Beyond Quality and Competence
4. Hedberg C-J, von Malmborg F (2003) The global reporting Initiative and corporate sustainability reporting in Swedish companies. Corp Soc Responsib Environ Manag 10(3):153–164
5. Podder SK, Samanta D (2019) Impact of climate changes on rural livelihoods and re-orienting the situation through human resources towards a sustainable society. (ID 3355347), March 2019.
6. Delpech Q, Rundell E (2014) Outsourcing labor rights struggles: the internationalization of the US trade union repertoire of action in central America. Critique Inter 64(3):33–46
7. Podder SK, Samanta D (2019) Factors that influence sustainable education with respect to innovation and statistical science. Int J Recent Technol Eng 7:373–376
8. Crowther D (2000) Corporate reporting, stakeholders and the internet: mapping the new corporate landscape. Urban Stud 37(10):1837–1848
9. Podder SK, Samanta D, Gurunath R (2020) Impact of business analytics for smart education system and management functions, 479–488

10. Croppenstedt A, Demeke M, Meschi MM (2003) Technology adoption in the presence of constraints: the case of fertilizer demand in Ethiopia. Rev Dev Econ 7(1):58–70
11. Park S-T, Jung J-R, Liu C (2020) A study on policy measure for knowledge-based management in ICT companies: focused on appropriability mechanisms. Inf Technol Manage 21:03
12. Biswas M, Podder SK, Shalini R, Samanta D (2019) Factors that influence sustainable education with respect to innovation and statistical science. Int J Recent Technol Eng 7:373–376
13. Pandey S (2006) Para-teacher scheme and quality education for all in India: Policy perspectives and challenges for school effectiveness. J Educ Teach 32:319–334
14. Dyllick T, Hockerts K (2002) Beyond the business case for corporate sustainability. Bus Strateg Environ 11(2):130–141
15. Podder SK, Samanta D (2022) Green computing practice in ICT-based methods: innovation in web-based learning and teaching technologies. Int J Web-Based Learn Teach Technol (IJWLTT) 17(4):1–18
16. Evidences E (2012) Revisiting the impact of integrated internet marketing on firms' online performance. Procedia Technol 5:418–426
17. Shahzaib Khan J, Zakaria R, Aminudin E, Adiana Abidin NI, Mahyuddin MA, Ahmad R (2019) Embedded life cycle costing elements in green building rating tool. Civ Eng J 5(4):750–758
18. Darabpour MR, Darabpour M, Sardroud JM, Smallwood J, Tabarsa G (2018) Practical approaches toward sustainable development in Iranian Green construction. Civ Eng J 4(10):2450– 465
19. Tilak JBG, Varghese NV (1991) Financing higher education in India. High Educ 21(1):83–101
20. Podder SK (2019) Information and communication technology in sustainable reporting towards paradigm shift in business. Int J Emerg Technol Innovat Res 240–250
21. Hart SL, Milstein MB (2003) Creating sustainable value. Acad Manag Perspect 17(2):56–67
22. Hart SL (1997) Beyond greening: strategies for a sustainable world. Harvard Bus Rev

A Review on Virtual Reality and Applications for Next Generation Systems and Society 5.0

Dishant Khosla, Sarwan Singh, Manvinder Sharma, Ayush Sharma, Gaurav Bharti, and Geetendra Rajput

Abstract Virtual reality is no myth to anyone as of now. As we have seen in movies like Star Trek where a holographic panel is projected for the shuttle's operation, which was a fascinating sight for anyone. But now, this is possible for anyone to use and create such technologies and which are creating new market for its growth and evolution. In the following work, we will be discussing about the introduction of virtual reality in the gaming sector which can also be referred in the entertainment criteria. The convergence of virtual reality with the physical space through society 5.0 is discussed which enables the analysis of big data through artificial intelligence which is fed back to the humans in the real space. This adds value to the economic development and social resolutions. The growth of this technology in the market and its future scope is also mentioned. How it will create a change for the international Esports players and introduction of new tournaments. The most awaited Apple VR is also discussed which will come in play around 2021 code named as t288.

Keywords Augmented reality · Virtual reality · Society 5.0 · Compound annual growth rate · Virtual reality modeling language · Military simulations · Cupertino

1 Introduction

Virtual reality is a user interface which immerses a person in a digital 3D environment with computer-generated imagery which aims at simulating the real world through senses. It can also be stated as a three-dimensional environment in which the person

D. Khosla (✉) · M. Sharma
CGC Group of Colleges, Landran, Mohali, Punjab, India
e-mail: dishant.coeece@cgc.edu.in

S. Singh
NIELIT, Chandigarh, India
e-mail: sarwan@nielit.gov.in

A. Sharma · G. Bharti · G. Rajput
CGC College of Engineering, Landran, Mohali, Punjab, India

V. Skala et al. (eds.), *Machine Intelligence and Data Science Applications*, Lecture Notes on Data Engineering and Communications Technologies 132,
https://doi.org/10.1007/978-981-19-2347-0_23

Fig. 1 User immersed in a virtual environment

can explore and interact with them virtually. The virtual reality systems use VR headsets or multi-projection environment to generate realistic sounds, images, and other sensations. The effect of VR is created by a head-mounted display (HMD) in the headset which has a small screen in front of the user [1]. This effect can also be created in specially designed rooms which has multiple large screens. The virtual reality environment is not just only about the graphics, but out of all the senses of human, the hearing and visual senses are the central ones. The humans react more swiftly to the audio cues than the visual cues. In order to create a perfect VR experience, the accurate environmental sounds are a must [2]. To accomplish this task, the headphones should be a perfect noise-cancelation device which should not hinder with the audio of virtual and real world. In 1982, an Australian science fiction, Damien Broderick used the term "virtual reality" in a science fiction context The Judas Mandala [3, 4]. In Fig. 1, it presents the view of a user immersed in the virtual environment by putting on the head-mounted display (HMD).

1.1 VR in Society 5.0

Society 5.0 is defined as a human centric society in which the social problem resolutions are balanced by the advancement in economy. There is a significant integration of physical space and virtual reality (cyberspace) in society 5.0. In the fifth Basic Plan of Science and Technology, society 5.0 was being proposed. It is followed by society 1.0 (hunting society), society 2.0 (agricultural society), society 3.0 (industrial society), and society 4.0 (information society). The information and knowledge shared cross-sectionally in society 4.0 which is the information society were not enough. It used to be a burden to find the information because of the labor limitation which leads to the restricted scope of action [5, 6]. The adequate response was restricted because of different restrictions such as aging population and decrease in the birth rate. There is a scope of forward-looking society achieved by innovations in society 5.0. This society has members who have mutual respect and understanding for each other and where enjoyable and happy life can be led by every person in the society through virtual reality. The virtual reality (cyberspace) and the real space (physical space) are significantly converged in society 5.0. The databases and cloud services were accessed in the virtual reality in society 4.0 (information society) through Internet, and further, information was analyzed [7, 8]. Figure 2 shows the evaluation of society 5.0 from the information society.

In society 5.0, the sensors collect the information in the physical space which is then stored in the virtual reality or the space. Artificial intelligence (AI) analyzes the big data in the cyberspace. Here, the results are analyzed and fed back in the real space to the humans in different forms. In society 4.0 (information society), the information was collected through the network via Internet and was analyzed by the humans. But in society 5.0, there is an interconnection between the things, people, and system in the cyberspace. AI collects the optimal results which are fed in the physical space to the humans. The society and industry add new value which was not possible previously [9, 10]. It balances the economic advancement with the resolution of social problems by incorporating new technologies such as IoT, robotics, AI, and big data in all industries and social activities which provides goods and services. The economic development is balanced by society 5.0, and the social issues are solved [11, 12].

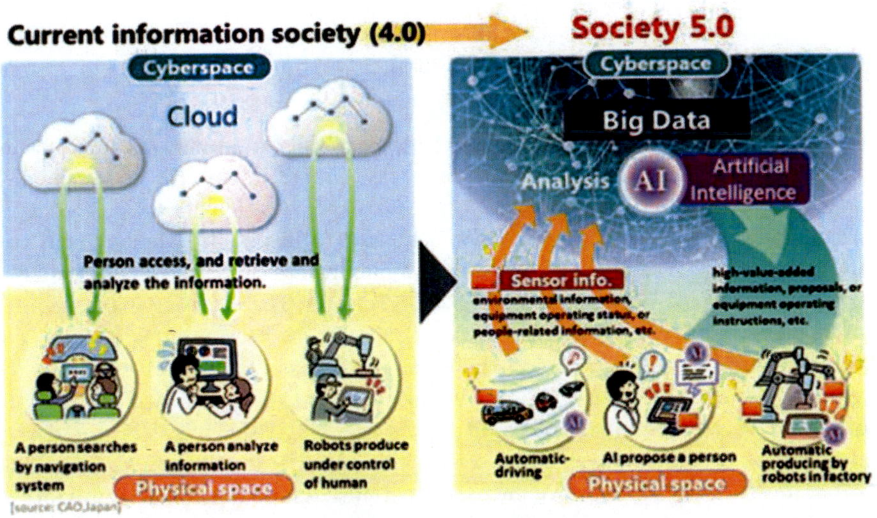

Fig. 2 Evaluation of society 5.0 from the information society

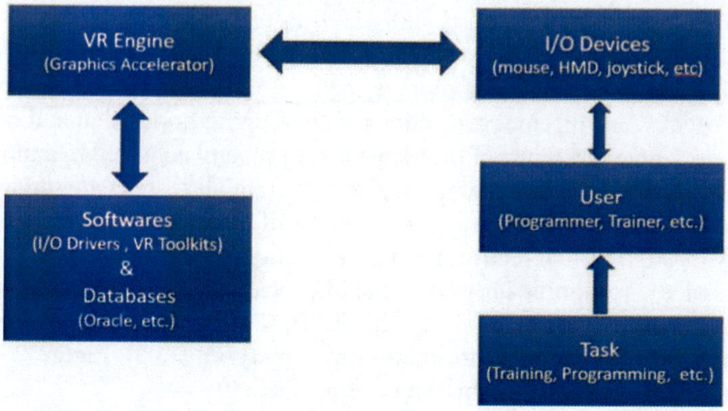

Fig. 3 Process of VR in gaming

1.2 Working of VR in Gaming Devices

In Fig. 3, we can tell the process of how a virtual reality compatible game runs. Out of all the present sources, the most important one is the user, and it can be a human or a bot/dummy which operates the system. The task to be performed by the user-end is conveyed by it to the I/O devices like the head-mounted display (HMD) [13, 14]. Then, the process of converting the 2D images into 3D images is carried out with the help of VR engine which uses the help of software like virtual reality modeling language (VRML) and some other toolkits. To store all the data of the task performed, the use of database takes place.

1.3 Advantages of Virtual Reality

With the introduction of VR in gaming, the user can go beyond the screen, i.e., the interaction of user with virtual world will get more interactive. Better gaming experience with visual display is incentive. Gamers can orientate their perspective base on the head motion. Due to its popularity among the people, it helps to push the technological advancements of VR. The communication of the players or called as "in-game voice chat" will get more refined and clearer. The detailing in the graphics is very good and creates a wonderful experience for the user in terms of video and audio. It is used to make the demonstration more impressive and catchier. It provides user an experience which user normally will not be able to experience in a real life. Virtual reality creates a realistic world for the user [15, 16]. Thus, VR would open the vast area of the gaming field. User can see and feel the shaped surface under his/her fingertips which improve the user experience within a game. VR provides an efficient way for gaining of the knowledge and increased learning of the students. Oculus offers low latency, fast head-tracking performance. VR provides adjustable focus for near-or far-sighted gamers. VR is used for the training and the simulation not only in the gaming field but also for the public services such as vehicle components and bus or flight services. It will create job opportunities not only for the developers but also the underdog industry of Esports will grow. It will train the humans to make quick and precise decisions as soon as possible because the majority of the games available are fast-paced and based on survival skills [17, 18].

1.4 Effects of Using VR Extensively

With the usage of VR going to be the next eminent phase of upcoming technologies, the number of users of such technology will increase and so will their involvement with it. As there is a saying that, "excess of anything is never good." Similarly, there are going to be some after-effects of using VR technology for long hours. The problems can be related to physical, mental, or social life of that person. For example, if the game is too violent (as in the sense of extreme not in a gross or negative manner), the user might also get violent which will reflect on his real life somewhere. This will create a mental imbalance on him/her leading to a change in their characteristic behavior and because of that change in their behavioral pattern, this might also affect their social life as well. There are also some chances where it may also harm/effect their relational matters with other people [19, 20]. As the VR games requires total concentration and involvement of the user, it also needs the user to perform certain actions to overcome the challenges faced by them in the game. Holding certain positions and performing quick-reflexive movements can at times cause muscle injury which can permanently hamper the physical structure of the body. In an example of a combat game where the user as a solider is slowly crouching toward the target and suddenly he is required to get up and sprint toward a safe spot to avoid the grenade thrown at him can lead to make the user use quick reflex movement and in doing so might tear a muscle ligament of his thigh which will cripple him for life [21, 22]. The only thing which will conclude to the statements given above are to not completely get depended on such addictive technologies. It should not matter which genre of game the user will play, one thing which they should remember is that it is for entertainment purpose, and the actions performed in them are nowhere associated to real-life entities. Keeping oneself physically fit is one of the best solutions in any matter. If the person is fit physically and mentally, nothing can be bad for him [23, 24].

2 Introduction of VR in Gaming Industry

The introduction of virtual reality for the humans is all about realistic and immersive simulation of a three-dimensional environment. It is a growing sector which is playing a major role in every other field from healthcare to architecture, military simulations to agricultural purpose and also to the gaming industry. According to a study done by a Website, marketresearchfuture.com, they predict that the growth of virtual reality market will expand to $14.6 billion (USD) by the end of 2023. The entertainment sector is quite enormous where the user is willing to spend lots and lots of money for their leisure and entertainment such as video games, amusement parks, and social networks. Introducing the VR in the gaming sector has given a new twist to the video game business [25, 26]. The users/gamers are getting benefited with the VR in certain ways like the gamers are able to take players inside the game in real time, and it is more appealing for the professional gamers who compete at international level and enhances the engagement of the user with an immersive experience and also enriches the gaming environment with cutting-edge capabilities. The growth of VR gaming market is based on the development of more mobile and affordable VR headsets by the hardware production leaders like Samsung, Oculus, HTC, Sony, Google, etc. Innovative technologies of VR headset provide the freedom of movement and recreation of 5 basic feelings of the humans. The prices of the VR game products of Nintendo, Microsoft, and Sony are predicted to get lower in the coming years [27]. It consists of a head-mounted display (HMD) which is compatible with smartphones which includes gyroscope and also motion sensors which can track body, hand, and head positions. Also small screens (HD) for stereotypic displays are provided with small, fast, and lightweight processors. Some of the challenges faced by hardware of VR are data security, freedom of movement, mobility, quality content, and 5G speed Internet (or higher). The language used to describe a three-dimensional (3-D) image is known as virtual reality modeling language (VRML). It enables the possibility of the user interaction with the 3D environment. With the help of VRML, a sequence of visual images can be built into Web settings with which a user can interact by rotating, viewing, moving, and otherwise interacting with an apparently 3D scene. Some of the best VR games available in the market are Hover Junkers, Star Trek: Bridge Crew, Elite Dangerous, No Limits 2, Lone Echo, L.A. Noire: Tthe VR case files and Pavlov VR. The current situation signifies that the interaction between the brain and the virtual reality has triggered the advancements of VR in gaming and entertainment and other fields as well [28]. Also, a certain number of researches points to the fact that the market size of the virtual reality globally in gaming and entertainment will grow from $4.15 billion (USD) in 2018 and reach $70.57 billion (USD) by 2026 exhibiting a CAGR of 40.1% as mentioned in Fig. 4.

Although, these statistics may vary differently on comparing with a number of researches and surveys done by different organizations, so we can get on the point stating that the future of VR in gaming and entertainment sector is bright and will flourish at rapid pace [29, 30].

3 Apple in the VR Industry

As for the Android smartphone users, the introduction of Samsung Gear and Oculus and many other VR headsets by HTC, Google, and many more has increased the

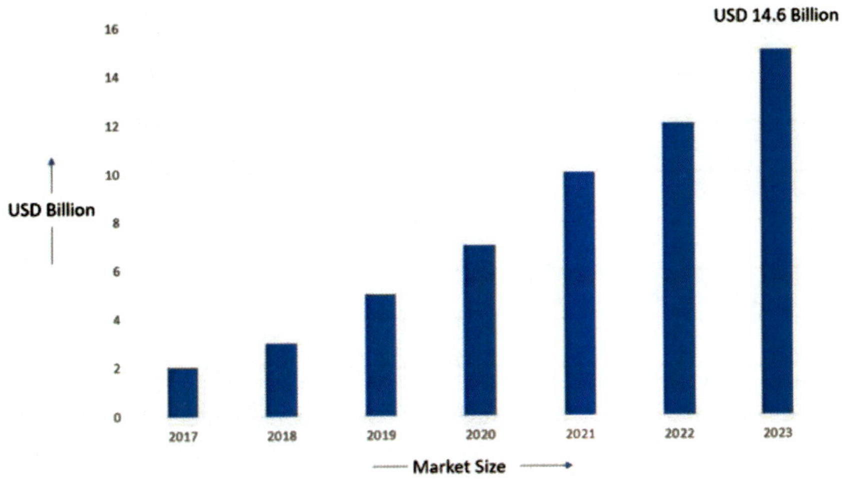

Fig. 4 Global market size of VR

VR market. But what about the iPhone users? They do not need to worry about it because very soon Apple is going to introduce their AR/VR headsets compatible with the iPhones. The rumors are that they are planning to launch their first AR/VR supported device by the second half of 2020. It is reported that the launch of a new iPad Pro will feature a module featuring two camera sensors and a small hole for the 3D system which will let people create three-dimensional reconstruction of rooms, objects, and people [31]. It will have a new 3D sensor system to make it possible. By 2021 or 2022, Apple aims to release a combination of AR and VR headset which is focused more on gaming, streaming videos, and virtual meetings. It is also rumored that the research on the VR/AR technology in the Cupertino has been going for almost 10 years. A team of skilled engineers was up to provide a technology which will be more advanced than the face-ID of the Apple products. The team consists of highly skilled engineers from the company itself and also consists of ex-NASA engineers, former game developers, and graphics experts [32]. They are also working on a patent which will make statement in the virtual reality technology at mass. It would let the AR and VR viewers to watch the streamed video from any angle of their choice by composting multiple streams. It is basically beneficial for the viewers and content creators who stream live videos on the YouTube of gaming tournaments (e.g., PUBG Mobile Club Open, or PMCO). This project of theirs is code named as t 288. The future aim is to add this hardware component as a mandatory accessory for the iPhone products which will lead to produce slimmer and lightweight iPhones and iPads. Plans are that it will have an 8 K display for each eye which is higher resolutions as compared to the televisions. The release of AR/VR glasses is expected to be around 2021. However, it can still change or scrap the plan and reschedule their launch. The much-awaited Apple VR will create a hype among the users and the other competitive brands in the VR technology to produce more fine

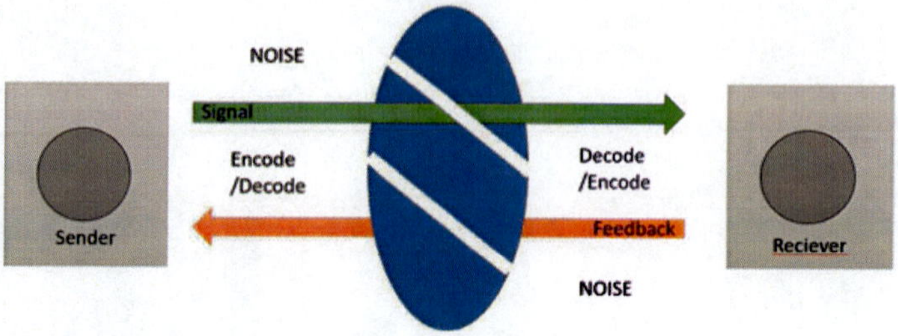

Fig. 5 Software technology for VR

technology and hardware system for the virtual experience [33, 34]. Figure 5 shows the description of software technology for VR. In short, behind the scenes view of the virtual environment of how the actions produced by the user are converted into the signals which are received at the other end and how it shows its effect on the environment.

4 Conclusion

The future of the virtual reality industry for the gaming sector is going to expand at an exponential rate. The introduction of new hardware devices for the gamers will not only change the way of gaming but it will also lead to a new market for these gadgets. It has also been observed that the upcoming startups and entrepreneurs in the market for VR have also increased, and new software and hardware systems are being introduced on day-to-day basis. There has been a tremendous significance of VR in society 5.0 as it integrates virtual space with the physical space which enables the analysis of big data through artificial intelligence. The burden on the human labor is significantly reduced and thus optimizing the entire organizational and the social system. Society 5.0 is not merely monitored by robots and AI, and each person in the society is given the due importance. It balances the economic advancement with the resolution of social problems by incorporating new technologies such as IoT, robotics, AI, and big data in all industries and social activities which provides good and services. The society and industry add new value which was not possible previously. Initially, the VR head-mounted displays were only made for entertainment purposes, but now with the advancements in the technology, the applications of VR HMDs have stepped up and expanded to marketing, retail, military, healthcare, education, and fitness. It does not matter which brand makes the advanced version of the present VR whether it is Samsung, HTC, Google, or Apple, what it shows is that the interaction of humans to the virtual world will get more comfortable, and the evolution of the human race will climb one more step ahead and will define the entire civilization as whole.

References

1. Burdea GC, Coiffet P (2003) Virtual reality technology. Wiley
2. Ryan M-L (2001) Narrative as virtual reality. Immersion Interactivity Lit
3. Sveistrup H (2004) Motor rehabilitation using virtual reality. J Neuro Eng Rehabil 1(1):10
4. Biocca F, Levy MR (2013) Communication in the age of virtual reality. Routledge
5. Sharma M, Singh S, Khosla D, Goyal S, Gupta A (2018) Waveguide diplexer: design and analysis for 5G communication. In: 2018 fifth international conference on parallel, distributed and grid computing (PDGC), IEEE, pp 586–590
6. Fukuyama M (2018) Society 5.0: aiming for a new human-centered society. Jap Spotlight 1:47–50
7. Sharma M, Khosla D, Pandey D, Goyal S, Gupta AK, Pandey BK (2021) Design of a GaN-based flip chip light emitting diode (FC-LED) with au bumps and thermal analysis with different sizes and adhesive materials for performance considerations. Silicon, pp 1–12
8. Fukuda K (2020) Science, technology and innovation ecosystem transformation toward society 5.0. Int J Prod Econom 220:107460
9. Sharma M, Pandey D, Palta P et al (2021) Design and power dissipation consideration of PFAL CMOS V/S conventional CMOS based 2:1 multiplexer and full adder. SILICON. https://doi.org/10.1007/s12633-021-01221-1
10. Salgues B (2018) Society 5.0: industry of the future, technologies, methods and tools. Wiley
11. Kaur SP, Sharma M (2015) Radially optimized zone-divided energy-aware wireless sensor networks (WSN) protocol using BA (bat algorithm). IETE J Res 61(2):170–179
12. Ferreira CM, Serpa S (2018) Society 5.0 and social development. Manage Org Stud 5:26–31
13. Sharma M, Sharma B, Gupta AK, Khosla D, Goyal S, Pandey D (2021) A study and novel AI/ML-based framework to detect COVID-19 virus using smartphone embedded sensors. In: Sustainability measures for COVID-19 pandemic. Springer, Singapore, pp 59–74
14. Zyda M (2005) From visual simulation to virtual reality to games. Computer 38(9):25–32
15. Sharma M, Gupta AK (2021) An algorithm for target detection, identification, tracking and estimation of motion for passive homing missile autopilot guidance. In: Mobile radio communications and 5G networks. Springer, Singapore, pp 57–71
16. Porras DC, Siemonsma P, Inzelberg R, Zeilig G, Plotnik M (2018) Advantages of virtual reality in the rehabilitation of b alance and gait: systematic review. Neurology 90:22:1017–1025
17. Singla BS, Sharma M, Gupta AK, Mohindru V, Chawla SK (2020) An algorithm to recognize and classify circular objects from image on basis of their radius. In: The international conference on recent innovations in computing. Springer, Singapore, pp 407–417
18. Morel M, Benoît Bideau JL, Kulpa R (2015) Advantages and limitations of virtual reality for balance assessment and rehabilitation. Neuro Phys Clinique/Clinical Neurophysi 45(4–5):315–326
19. Khosla D, Malhi KS (2019) Rectangular dielectric resonator antenna with modified feed for wireless applications. Int J Cont Aut 12(5):487–497
20. Regan C (1997) Some effects of using virtual reality technology. In Virtual Reality, Training's Future. Springer, Boston, MA, pp 77–83
21. Khosla D, Malhi KS (2018) Investigations on designs of dielectric resonator antennas for WiMax & WLAN applications 2018 fifth international conference on parallel, distributed and grid computing (PDGC). Solan Himachal Pradesh, India, pp 646–651. https://doi.org/10.1109/PDGC.2018.8745754
22. Riva G, Mantovani F, Capideville CS, Preziosa A, Morganti F, Villani D, Gaggioli A, Botella C, Alcañiz M (2007) Affective interactions using virtual reality: the link between presence and emotions. CyberPsychol Behav 10(1):45–56
23. Minhas S, Khosla D (2017) Compact size and slotted patch antenna for WiMAX and WLAN. Indian J Sci Technol 10(16):1–5. https://doi.org/10.17485/ijst/2017/v10i16/102762
24. Lee EA-L, Wong KW (2008) A review of using virtual reality for learning. In: Transactions on edutainment I. Springer, Berlin, Heidelberg, pp 231–241

25. Sharma N, Khosla D (2018) A Compact two element U shaped MIMO Planar Inverted-F Antenna (PIFA) for 4G LTE mobile devices 2018 fifth international conference on parallel, distributed and grid computing (PDGC). Solan Himachal Pradesh, India. pp 838–841
26. Cowan B, Kapralos B (2011) GPU-based acoustical diffraction modeling for complex virtual reality and gaming environments. In: Audio engineering society conference: 41st international conference: audio for games. Audio Engineering Society
27. Halton J (2008) Virtual rehabilitation with video games: a new frontier for occupational therapy. Occup Ther Now 9(6):12–14
28. Goude D, Björk S, Rydmark M (2007) Game design in virtual reality systems for stroke rehabilitation.“ In MMVR, pp 146–148
29. Miller KJ, Adair BS, Pearce AJ, Said CM, Ozanne E, Morris MM (2014) Effectiveness and feasibility of virtual reality and gaming system use at home by older adults for enabling physical activity to improve health-related domains: a systematic review. Age Ageing 43(2):188–195
30. Lange BS, Requejo P, Flynn SM, Rizzo AA, Valero-Cuevas FJ, Baker L, Winstein C (2010) The potential of virtual reality and gaming to assist successful aging with disability. Phys Med Rehabil Clin 21(2):339–356
31. Yoffie, DB (1992) Apple Computer–1992
32. Zhu YY (2013) Quick time virtual reality technology applies to practical teaching recording system. TELKOMNIKA Indonesian J Electr Eng 11(11):6315–6320
33. Anthes C, García-Hernández RJ, Wiedemann M, Kranzlmüller D (2016) State of the art of virtual reality technology. In: 2016 IEEE aerospace conference. IEEE, pp 1–19
34. Chen SE (1995) Quicktime VR: An image-based approach to virtual environment navigation.“ In Proceedings of the 22nd annual conference on computer graphics and interactive techniques, pp 29–38

Facial Recognition with Computer Vision

Vishesh Jindal, Shailendra Narayan Singh, and Soumya Suvra Khan

Abstract The cutting-edge age of innovation has developed at high speed to make our lives quiet. The progression in security administrations of the advanced world has prompted the working of frameworks and gadgets all the more effectively and precisely by giving the greatest to most extreme wellbeing and security. The significant method of getting our frameworks and gadgets is passwords or passphrases however this has a few bugs which can undoubtedly be broken or hacked, this prompted the appropriation of further developed procedures, i.e., biometric finger impression scanner, this gives sufficient measure of safety to frameworks like opening the telephone and section in schools or workplaces. In any case, examining the current situation of this gigantic pandemic Covid-19, some control is needed for the protection of individuals which advances touchless registration into workplaces, schools, and universities. This examination paper denotes the use of Computer Vision innovation by carrying out python programming language and its libraries.

Keywords Face recognition · Haar-cascade classifiers · Local binary patterns histogram

1 Introduction

This is a time when technology has boosted progressively, so it is important and necessary to keep all of our security [1, 2] systems in check. So, one of the chief reasons for security in this digital era is Identification and verification where everything is secured with the help of biometrics. Fingerprint Scanner provided decent

V. Jindal · S. Narayan Singh
Department of Computer Science, Amity School of Engineering and Technology, Amity University, Noida, Uttar Pradesh 201313, India
e-mail: snsingh36@amity.edu

S. Suvra Khan (✉)
Department of Computer Science and Engineering, Meghnad Saha Institute of Technology, Kolkata, West Bengal 700150, India
e-mail: soumyasuvra@msit.edu.in

V. Skala et al. (eds.), *Machine Intelligence and Data Science Applications*, Lecture Notes on Data Engineering and Communications Technologies 132,
https://doi.org/10.1007/978-981-19-2347-0_24

security; however, it lacks in providing full protection from hackers. Keeping in mind the full-fledged security, we introduced more advanced technology, i.e., Facial Recognition System [3, 4], combining this technology with the fingerprint scanner provides ultimate security to the system and devices [5, 6].

A facial Recognition System is that innovation which can distinguish and check the individual or an item from a computerized picture. This innovation can distinguish and perceive the appearances and attempt to measure their facial highlights and surfaces like size or state of the eyes, nose, jaw, cheekbones, skin tone, and so on and afterward coordinating with them with the put away faces in the dataset for acknowledgment reason. This system can also be regarded as a Biometric Artificial Intelligence-based system [7, 8].

2 Computer Vision and Machine Learning

In layman's terms, Computer Vision is fundamentally the copy of human vision, i.e., it impersonates human vision and attempts to do some usefulness with the framework by applying various calculations. Computer Vision Technology is a significant field of Artificial Intelligence. This innovation is utilized for show machines the best way to gather data from the pixels and change over into PC meaningful configuration and ten preparing the model with the assistance of various calculations [9, 10]. With the latest trends in this digital world, digital security plays an important role in safeguarding our data, and Computer Vision Technology is one field of AI that fulfills the needs of security [11, 12].

Now talking about Machine Learning which is all about analyzing the given situation and working accordingly by the machines. It is also the major field of Artificial Intelligence and is widely used across various industries. The main ideology behind machine learning is using optimization methods for observing, processing, and identifying patterns given within the dataset [7, 13]. Figure 1 illustrates the basic principles of VJ and ML algorithms.

3 Related Work

Systems that are accessible within the merchandise are really helpful during this digital era and a huge pandemic of COVID-19 to make everyone's life easy which promotes touchless check-ins and check-outs. Facial Recognition System is one among those systems which is very much used for having digital security [14].

There are some add-, i.e., when the character of the individual is not affirmed there will be a signal sound of typical recurrence and for a term of 10–15 s. by indicating unknown face [5, 7].

The authors in the paper [15] have proposed a computer vision-based methodology to find transform functions to play chess by facial readings. Another approach can

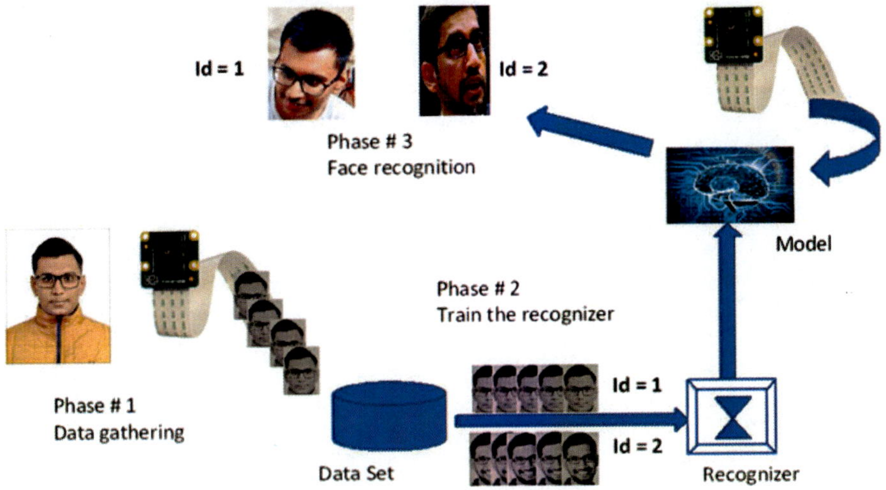

Fig. 1 CV and ML algorithms

be seen in [16] to illustrate the attendance control by face recognition. The security system is recommended in [17] where the authors have used computer vision for face detection. Network information security is also developed using face detection proposed in [18]. The authors in [19] have introduced the Principal Component Analysis (PCA) by face recognition. PCA reduces the time cost by improving the data processing speed is shown in [20, 21]. The Eigenface classification algorithm is proposed in [22] for PCA-based facial extraction. Linear Discriminate Analysis (LDA) is proposed in [23]. The authors in [24] have used LDA in face classification. Another approach can be shown in [24] where Support Vector Machine (SVM) is proposed especially for small dataset high-definition facial recognition problems. In [26] face recognition is shown using extracted face features and SVM. Another idea has been proposed in [27] to integrate different classifiers to increase overall performance. Adaboost [28] has improved the performances of face classifiers in [27] by partial boosting algorithms. The authors in [39] proposed Haar classifier which is a cascading approach of Adaboost. In [30] Howland et al. proposed a combined method. It combines the linear discriminant analysis with generalized singular value decomposition (GSVD) to solve the small samples size problem.

A novel approach can be seen in [31] where a fine-grained discriminator and a wavelet-based discriminator have been constructed to enhance visual quality. The recognition accuracy of a face recognition algorithm is easily affected by occlusion, light, and, distance. The work in [32] presents a convolutional neural network (CNN) that was trained to advance the accuracy of face detection with the ability to capture facial landscapes. The proposed procedure incapacitates the situation when the face accompanies by occlusion. We can see the advancement of the face recognition algorithm in [33, 34] where the authors have proposed masked and unmasked face recognition. In [35] the authors have proposed intelligent face recognition on edge computing using neuromorphic technology. In [36] face recognition has been used in communication APP for deaf and mute people. Other applications of face recognition can be seen in [37–39].

4 Methodology

In Fig. 2, the system methodology is going to explain different approaches and ways which is required in achieving the final result.

4.1 Python Programming Language

Python is a simple, elegant, and high-level programming language which has a clear syntax that automatically leads to having a wide amount of documentation. It supports an object-oriented approach where programmers can write clear and logical code and can easily work on small as well as big projects. This is best suited for Computer Vision, Machine Learning [5, 13].

4.2 Libraries Used

NumPy: It is an open-source python library that is majorly used in the field of Science and Technology. NumPy stands for Numerical Python, and it is faster than python lists. It is used for creating multi-dimensional arrays, i.e., array, we can also specify a data type for the array [7, 10]. NumPy arrays are the foundation of modern python data science.

OpenCV: Open CV library provides infrastructure for live Computer Vision, Machine Learning, and Image Processing. This library is used for processing images and videos and trying to figure out or identify the faces, objects, and handwriting of human beings and contains over 2500 optimized algorithms. This library also supports frameworks of deep learning like PyTorch, TensorFlow, Keras, etc. [14, 40].

Pillow: Pillow library is a subdivision of the Python Imaging Library (PIL). This library is used in our project which is used for image detection and creating thumbnails and also for converting an image from one format to another. It supports jpeg, png, gif, ppm, etc. [41].

Matplotlib: This library is a graph plotting library that is used for visualization. This library is used in my project because it displays the [42] images in the format of Blue Green Red (BGR) color channels.

Winsound: This module is used for creating sound by the machinery which supports the Windows platform. The function of this module which is used in my project is Winsound. beep (f, t) where f denotes the frequency of sound in Hertz (Hz)

Fig. 2 Flow diagram of the proposed methodology

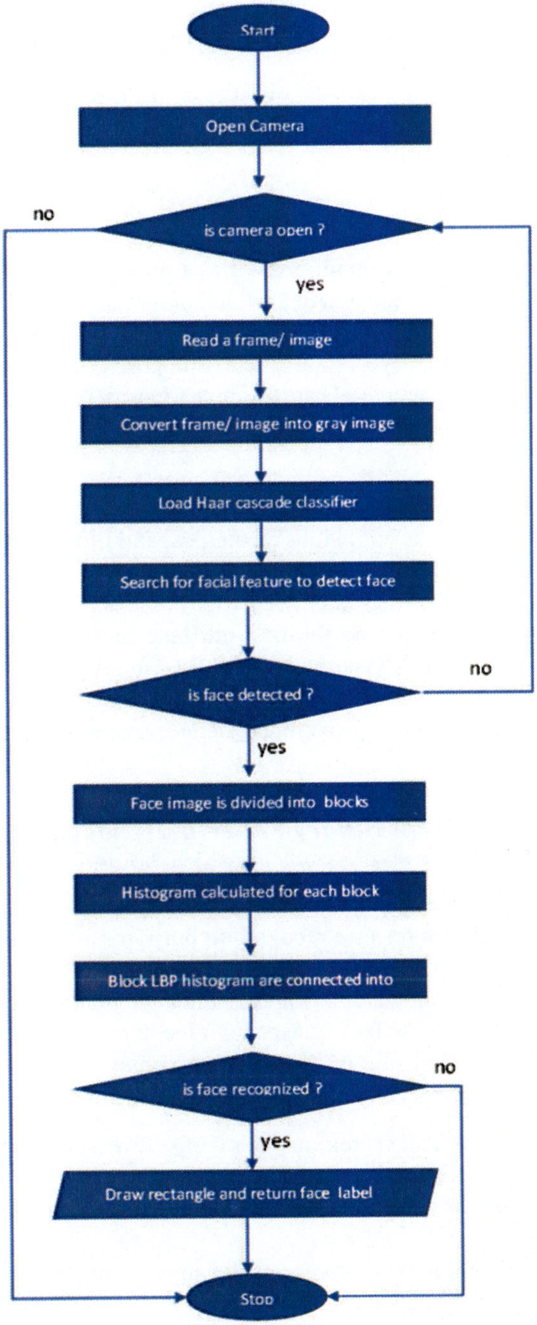

and t denotes the duration of sound in milliseconds. This library is used to produce a beep sound whenever the face is not recognized, i.e. when it shows an unknown face.

4.3 Haar-Cascade Classifier

Computer Vision libraries offer a variety of things for the execution of projects in the domain of detection of any kind of feature, one such important and valuable classifier is "Haar-Cascade Classifier" which attributes to all kinds of detection like an object, face, smile, eyes, mouth detection [10, 17]. Majorly the best detection comes out in grayscale images. Haar-Cascade Classifiers contain three terminologies to get trained and require a variety of positive and negative images of the face. These terminologies are as follows:

- The first step is of collecting various calculations, i.e., Calculating features of Haar-like edge, four rectangles, and line features [8, 15].
- Next is boosting up the speed of Haar features which can be done by creating Integral Images [11].
- Then choosing the best attribute and training the classifiers accordingly. This training is known as Adaboost Training. In this, we are detecting the objects by creating strong classifiers from weak classifiers.
- haarcascade_frontalface_default.xml which detects the frontal part of the face.

4.4 Local Binary Patterns Histogram (LBPH)

Local Binary Pattern Histograms Algorithm is one of the easiest and most popular algorithms for face recognition purposes and was introduced in the year 1996. This algorithm extracts the result of an image as a binary number by labeling the pixels of the image, i.e., by thresholding the neighborhood of each pixel. This algorithm represents the face images with the data vector [9, 43].

The first step involves understanding the parameters that are being used by the Local Binary Pattern Histograms algorithm which are:

- Grid X: It represents the no of cells that are there in the horizontal direction.
- Grid Y: It represents the no of cells that are there in the vertical direction.
- Radius: It is mainly set to 1 and shows a radius that is around the center of the pixel [8].
- Neighbors: It represents the no of points that are required to build a circular local binary pattern.

The next step is where we need to train the algorithm by creating a dataset folder on the pc which should contain the face pictures of the person whom we want to recognize along with setting the specific id number Last step is performing face

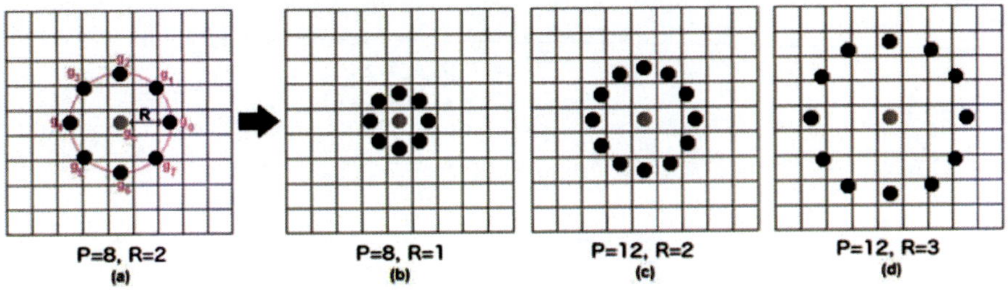

Fig. 3 Performing LBPH algorithm

recognition where a histogram is created for each image in dataset folder. Euclidean distance (D) is the square root of the summation of the difference between the squares of two histograms, if the nearest match is found, the face is recognized with the specified name [8]. Figure 3 mentioned below explains the execution of the Local Binary Patters Histogram Algorithm.

5 Experimental Analysis

This project has six phases ranging from learning some basic concepts and implementing them in phases to getting hold of concepts and technologies involved in the final execution. Different phases are as follows:

5.1 *Phase-I: Understanding NumPy Arrays*

```
In [1]: import numpy as np
```

This phase constitutes holding a good grip on NumPy arrays by understanding the logic behind them. Firstly, importing the necessary library, i.e., NumPy which is required for having understanding and working of NumPy arrays.

After importing the NumPy library, Fig. 4 depicts creating a list and then converting that list into NumPy arrays by using np.array(). It now prints the list into NumPy arrays and then finds the shape of the array which is 5 and the type which is NumPy.ndarray.

In Fig. 5 there is an array matrix of ones with two rows and four columns and it prints ones in the format of floating-point numbers depicted in Fig. 4. Array matrix of zeros with same two rows and five columns with arrange () method which is used

Fig. 4 Creating NumPy arrays

```
In [2]: # converts list into array
        li=[1,2,3,4,5]
        arr=np.array(li)
        arr

Out[2]: array([1, 2, 3, 4, 5])

In [3]: # shape of array
        print(np.shape(arr))
        type(arr)

        (5,)

Out[3]: numpy.ndarray

In [4]: print(arr.min()) # min value
        print(arr.max()) # max value
        print(arr.argmax()) # max value index
        print(arr.argmin()) # min value index

        1
        5
        4
        0
```

to give all the numbers ranging from start value till stop value minus one in sequence [44].

5.2 *Phase-II: Images and NumPy*

In this phase, there is the use of NumPy, matplotlib, and Python Imaging Library by working with the images. Every single digit of the image can be represented as an array.

In the above code declaration of a variable pic, i.e., Fig. 6 which stores the image vj2.jpg that which is opened from faces_vj folder on the desktop and printing the pic which shows my image [10, 41].

Now the image shown in Fig. 7 depicts the use matplotlib library which is used to plot the image with show() method having all the color channels, i.e., RGB in the image.

Image. plots the image with the red intensity and color mapping of gray along with the shape which has height and width of 1440 and only one color channel red.

```
print(np.ones((2,5))) # printing matrix of ones
print("\n")
print(np.zeros((2,5))) # printing matrix of zeros
```

```
[[1. 1. 1. 1. 1.]
 [1. 1. 1. 1. 1.]]


[[0. 0. 0. 0. 0.]
 [0. 0. 0. 0. 0.]]
```

```
a=np.arange(0,100,1) #printing elements from 0 to 100 with step value 1
print(a)
print("\n")
b=a.reshape(10,10) # reshaping the array into matrix of 10 rows and 10 coloumns
print(b)
print(b[1,3]) # getting element at row 1 and column 3
```

```
[ 0  1  2  3  4  5  6  7  8  9 10 11 12 13 14 15 16 17 18 19 20 21 22 23
 24 25 26 27 28 29 30 31 32 33 34 35 36 37 38 39 40 41 42 43 44 45 46 47
 48 49 50 51 52 53 54 55 56 57 58 59 60 61 62 63 64 65 66 67 68 69 70 71
 72 73 74 75 76 77 78 79 80 81 82 83 84 85 86 87 88 89 90 91 92 93 94 95
 96 97 98 99]


[[ 0  1  2  3  4  5  6  7  8  9]
 [10 11 12 13 14 15 16 17 18 19]
 [20 21 22 23 24 25 26 27 28 29]
 [30 31 32 33 34 35 36 37 38 39]
 [40 41 42 43 44 45 46 47 48 49]
 [50 51 52 53 54 55 56 57 58 59]
 [60 61 62 63 64 65 66 67 68 69]
 [70 71 72 73 74 75 76 77 78 79]
 [80 81 82 83 84 85 86 87 88 89]
 [90 91 92 93 94 95 96 97 98 99]]
13
```

Fig. 5 Performing operations

Fig. 6 Actual image

```
pic=Image.open('C:/Users/Dell/Desktop/faces_vj/vj2.jpg')
pic
```

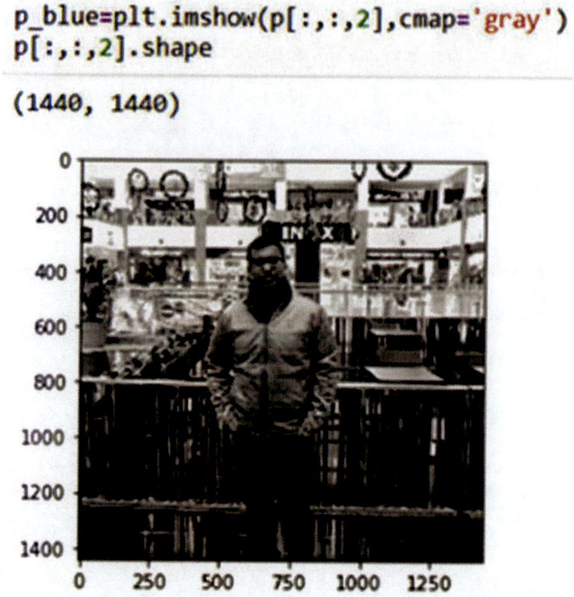

Fig. 7 Modified version of actual image

5.3 *Phase-III: Working with OpenCV Library*

In this part of the phase, there is a very good comparison between matplotlib library and OpenCV library that matplotlib library displays the image in the format of Red Green Blue (RGB) whereas OpenCV displays the image in the format just reverse of matplotlib, i.e., Blue Green Red (BGR) [8, 12].

Reading from the faces_vj folder and storing the image in img variable and plot it along with finding the type of image, i.e., numpy.ndarray, Figs. 8, 9 and 10 show the image (img) in the format of BGR.

The above code explains about drawing a rectangle in Fig. 11, circle, line on the blank image, and also displaying text on a blank image. Displaying it, after that creating various shapes using cv2.rectangle(), cv2.circle(), cv2.line() [11, 12].

5.4 *Phase-IV: Connecting to Webcam with OpenCV and Python*

In this phase, there is an implementation on how to connect the webcam for capturing and saving the video file shown in Fig. 11 [14, 44]. Then, the width is 640 × 480 pixels of course.

Next, declared the while loop which will run for the condition to be true, after that it displays the resulting frame using cv2.imshow() method. Now stated with the

```
img=cv2.imread('C:/Users/Dell/Desktop/faces_vj/vj1.jpg')
type(img)
```

```
numpy.ndarray
```

```
plt.imshow(img)
```

```
<matplotlib.image.AxesImage at 0x19b5ca033c8>
```

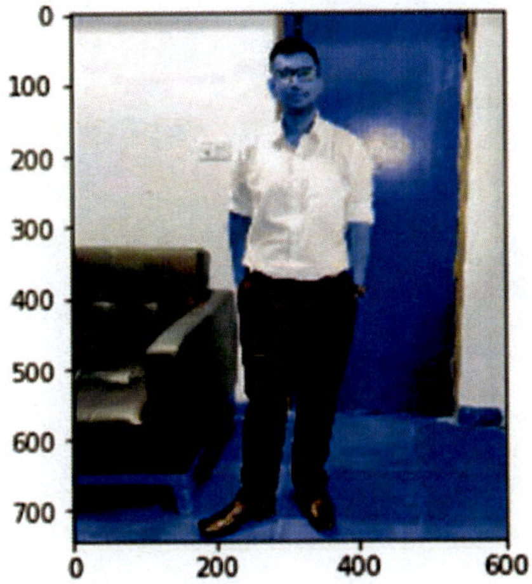

Fig. 8 Image in BGR

if the condition that cv2.waitKey(1) which shows the output for 1 ms and because it is running in the loop [8, 40].

5.5 *Phase-V: Face Detection*

In this phase performing in real-time, face detection in Fig. 13 and face detection will be there by taking images as an input from the desktop shown in Fig. 12 with the help of OpenCV and python.

Haarcascade_frontalface_deafult.xml file which is used to detect the frontal part of the face in the image or live webcam [40, 44]. Then there is a loop while the condition is True and ensured whether the camera is opened or not using ret and read the image and stored it in a frame and also declared face_count variable to count the no of faces detected [9].

```
new_img=cv2.cvtColor(img,cv2.COLOR_BGR2RGB)
```

```
plt.imshow(new_img)
```

```
<matplotlib.image.AxesImage at 0x19b20063a20>
```

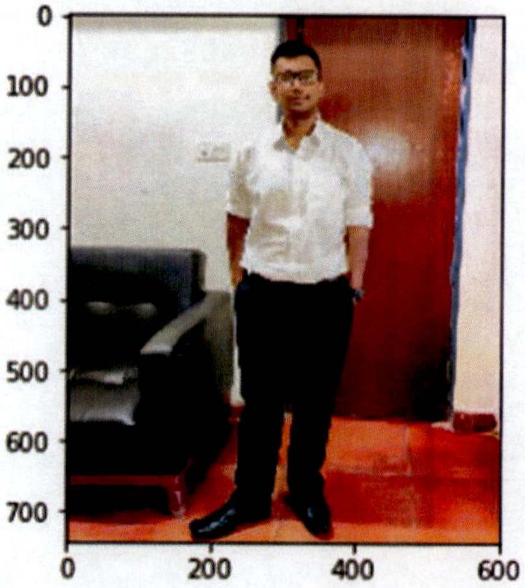

Fig. 9 Converting in RGB

5.6 *Phase-VI: Facial Recognition with Computer Vision Technology*

Now comes the final part of working where there is a full-fledged Facial Recognition System that will identify and recognize the faces and return the names and identities whether confirmed or not. This execution involves three steps which are described as follows:

5.7 *Step-1: Detection and Gathering*

Video Capture () object is used which is utilized to open up the camera to catch the feed, and afterward set the width and tallness of the edge as 640 × 480 pixels and proclaimed the haar course classifier which is utilized to identify the front-facing part of the face in the live webcam [14]. From that point forward, taking the ID of the individual to be perceived as information and pronounced the check variable which holds no face pictures to be caught as shown below in Figs. 14 and 15 show the dataset folder.

```
plt.imshow(cv2.flip(new_img,0))
```

<matplotlib.image.AxesImage at 0x193db841f28>

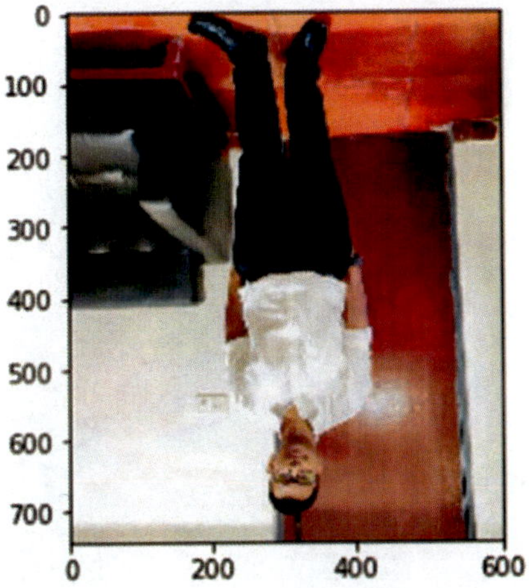

Fig. 10 Flipping the image

Fig. 11 Drawing shapes

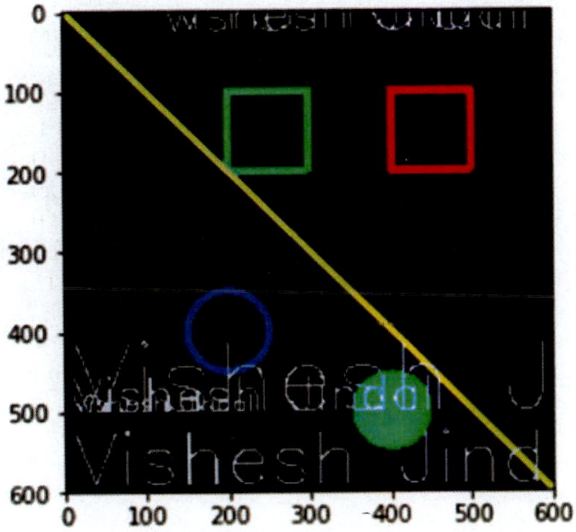

5.8 Step-2: Training

In this step, saved images are going to be trained for recognition purposes using the LBPH algorithm.

Fig. 12 Face detection from image

```
Total no.of Faces::  5
Facial Coordinates::
 [[ 311  641  123  123]
 [ 596  657  121  121]
 [ 879  661  129  129]
 [1218  617  140  140]
 [ 111  675  132  132]]

<matplotlib.image.AxesImage at 0x2631a91b4a8>
```

Fig. 13 Real-time face detection

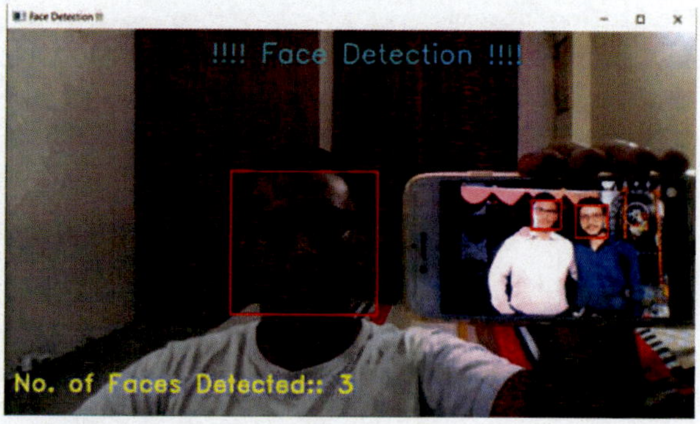

Fig. 14 Putting the ID of the face

```
Enter the ID of person for recognition----->:5
```

After storing five faces in the dataset folder, now training those faces for recognition by extracting the path which consists of images and labels, and after that using the Local Binary Pattern Histograms (LBPH) algorithm to recognize the frontal face of the image stored with the live image [13, 44] shown in Fig. 16.

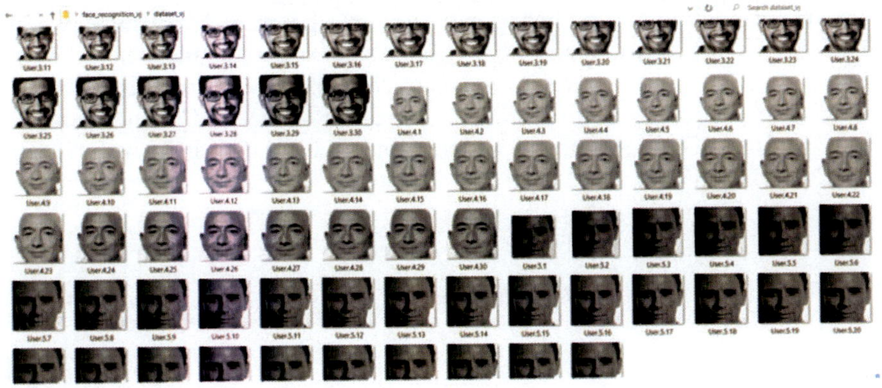

Fig. 15 Dataset folder

Fig. 16 Faces trained

```
5 faces trained !!!!!
```

5.9 Step-3: Real-Time Recognition

This is the end part of the code where real-time face recognition takes place by printing their names along with the identities whether confirmed or not based on training the model shown in Fig. 17.

This explains loading the recognizer from train_dataset_vj yml file and creating the cascade classifier for detection and declaring the id which starts from index 0 and also creating the list of face_names and placing the names according to the input id which was earlier fed into training [12, 42]. Now capturing using the VideoCapture () object and justifying the width and height of the frame to be 1280 × 1280 pixels. Declaring the frequency of beep which is in Hertz along with the duration of beep in milliseconds.

Starting with the while loop which gets the image and converts it into grayscale, now it detects the face in the object and after that recognizes the id using

Fig. 17 Real-Time Face Recognition with three faces detected

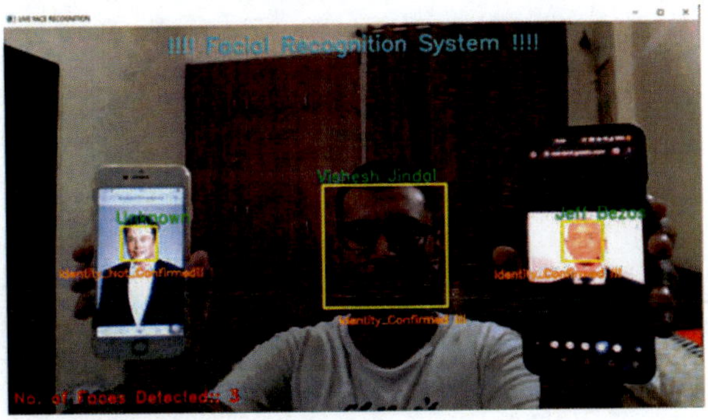

face_recognize. predict () method from the data which is available in trainer.yml file and returns to the owner along with the percentage concerning the matched owner. If the percentage is less than 100, it will return the name of a person with probable id along with printing "Identity Confirmed", otherwise will return Unknown face with the beep sound of the above-mentioned frequency and duration along with "Identity Not Confirmed" [7].

6 Conclusion and Future Scope

Utilization of both the advances that are Computer Vision and Machine Learning together which comprises the significant field of Artificial Intelligence gave a productive outcome in the field of computerized security and helps in getting this advanced time more got with the headways and fixing the bugs that happen alongside the taking care of the issue of the biometric unique mark scanner in the current circumstance of enormous pandemic COVID-19 which advances the use of touchless registration and registration.

Till now there is an establishment of the framework in the field of Artificial Intelligence by effectively constructing this undertaking by gaining from fundamentals till cutting edge in Computer Vision Technology and applying Machine Learning calculations. There will be an attempt to make enhancements and headways in this venture which increment the extent of his innovation on the lookout.

The future of facial recognition systems is bright as various companies are pitching this software and are accommodating it. This holds from retail to policing. This technology will be going to generate huge revenues in the coming time.

References

1. Narang S, Nalwa T, Choudhury T, Kashyap N (2018)An efficient method for security measurement in internet of things. In: 2018 international conference on communication, computing and internet of things (IC3IoT), pp 319–323. https://doi.org/10.1109/IC3IoT.2018.8668159
2. Srivastava R, Tomar R, Sharma A, Dhiman G, Chilamkurti N et al (2021) Real-time multimodal biometric authentication of human using face feature analysis. Comput Mater Cont 69(1):1–19
3. Sarishma D, Sangwan S, Tomar R, Srivastava R (2022) A review on cognitive computational neuroscience: overview, models, and applications. In: Tomar R, Hina MD, Zitouni R, Ramdane-Cherif A (eds) Innovative trends in computational intelligence. EAI/Springer Innovations in Communication and Computing. Springer, Cham. https://doi.org/10.1007/978-3-030-78284-9_10
4. Dhamija J, Choudhury T, Kumar P, Rathore YS (2017)An advancement towards efficient face recognition using live video feed: for the future. In: 2017 3rd international conference on computational intelligence and networks (CINE), pp 53–56. https://doi.org/10.1109/CINE.2017.21
5. Prakash RM, Thenmoezhi N, Gayathri M (2019) Face recognition with convolutional neural network and transfer learning. In: 2019 International conference on smart systems and inventive technology (ICSSIT), IEEE, pp 861–864

6. Calder J, Young AW (2005) Understanding the recognition of facial identity and facial expression. Nature Rev Neurosci 6(8):641–651
7. Yip, Sinha P (2002) Role of color in face recognition. Perception 31:995–1003
8. Canedo D, Neves AJR (2019) Facial expression recognition using computer vision: a systematic review. Appl Sci 9:4678
9. Robertson DJ, Noyes E, Dowsett AJ, Jenkins R, Burton AM (2016) Face recognition by metropolitan police super-recognisers. PLoS One. 11:e0150036. pmid:26918457
10. Ghazi MM, Ekenel HK (2016) A comprehensive analysis of deep learning-based representation for face recognition. In: 2016 IEEE conference on computer vision and pattern recognition workshops (CVPRW). IEEE, Las Vegas, NV, pp 102–109
11. Young W, Hellawell D, Hay DC (1987) Configurational information in face perception. Perception 16:747–759
12. Bobak AK, Dowsett AJ, Bate S (2016) Solving the border control problem: Evidence of enhanced face matching in individuals with extraordinary face recognition skills. PLoS One. 11:e0148148. pmid:26829321
13. Ding C, Tao D (2017) Trunk-branch ensemble convolutional neural networks for video-based face recognition. IEEE Trans Pattern Anal Mach Intell 40:1002–1014. pmid:28475048.
14. He K, Zhang X, Ren S, Sun J, Deep residual learning for image recognition. In: 2016 IEEE conference on computer vision and pattern recognition (CVPR). IEEE, Las Vegas, NV, pp 770–778
15. Silver D, Schrittwieser J, Simonyan K, Antonoglou I, Huang A, Guez A, Hubert T, Baker L, Lai M, Bolton A et al (2017) Mastering the game of go without human knowledge. Nature 550(7676):354
16. Manjula VS, Santhosh Baboo Lt Dr S et al (2012) Face detection identification and tracking by prdict algorithm using image database for crime investigation. Int J Comput Appl 38(10):40–46
17. Lander K, Bruce V, Bindemann M (2018) Use-inspired basic research on individual differences in face identification: implications for criminal investigation and security. Cognitive Res: Principles Implications 3(1):1–13
18. Hu Y, An H, Guo Y, Zhang C, Li Y (2010) The development status and prospects on the face recognition. In: 2010 4th international conference on bioinformatics and biomedical engineering (iCBBE)
19. Gottumukkal R, Asari VK (2004) An improved face recognition technique based on modular pca approach. Pattern Recogn Lett 25(4):429–436
20. Hoyle DC Rattray M (2003) Pca learning for sparse high-dimensional data. Epl
21. Vijay K, Selvakumar K (2015) Brain fmri clustering using interaction k-means algorithm with pca. In: 2015 International conference on communications and signal processing (ICCSP)
22. Li J, Zhao B, Hui Z, Jiao J (2010) Face recognition system using svm classifier and feature extraction by pca and lda combination. In: International conference on computational intelligence and software engineering, 2009. CiSE 2009
23. Chintalapati S, Raghunadh MV (2013) Automated attendance management system based on face recognition algorithms. In: 2013 IEEE international conference on computational intelligence and computing research. IEEE, pp 1–5
24. Juwei L, Plataniotis KN, Venetsanopoulos AN (2003) Face recognition using lda-based algorithms. IEEE Trans Neural Netw 14(1):195–200
25. Cortes C, Vladimir V (1995) Support-vector networks. Machine Learning
26. Sun A, Lim E-P, Liu Y (2009) On strategies for imbalanced text classification using svm: a comparative study. Decis Support Syst 48(1):191–201
27. Freund Y, Iyer R, Schapire RE, Singer Y, Dietterich TG (2004) An efficient boosting algorithm for combining preferences. J Mach Learn Res 4(6):170–178
28. Ratsch G (2001) Soft margins for adaboost. Mach Learn 42(3):287–320
29. Xiang-feng L, Wei-kang Z, Xin-yuan D, Kun L, Dun-wen Z (2019) Vehicle detection algorithm based on improved adaboost and haar. Measurement and Control Technology
30. Howland P, Wang J, Park H (2006) Solving the small sample size problem in face recognition using generalized discriminant analysis. Pattern Recogn 39(2):277–287

31. He R, Cao J, Song L, Sun Z , Tan T (2020) Adversarial cross-spectral face completion for NIR-VIS face recognition. IEEE Trans. Pattern Anal Mach Intell 42(5):1025–1037. https://doi.org/10.1109/TPAMI.2019.2961900.
32. Tsai C, Ou Y -Y, Wu W-C, Wang J-F (2020)Occlusion resistant face detection and recognition system. In: 2020 8th international conference on orange technology (ICOT), pp 1–4. https://doi.org/10.1109/ICOT51877.2020.9468767
33. Ejaz MS, Islam MR, Sifatullah M, Sarker A (2019) Implementation of principal component analysis on masked and non-masked face recognition. 2019 1st International conference on advances in science, engineering and robotics technology (ICASERT), pp 1–5. https://doi.org/10.1109/ICASERT.2019.8934543
34. Malakar S, Chiracharit W, Chamnongthai K, Charoenpong T (2021)Masked face recognition using principal component analysis and deep learning. In: 2021 18th International conference on electrical engineering/electronics, computer, telecommunications and information technology (ECTI-CON), pp 785–788. https://doi.org/10.1109/ECTI-CON51831.2021.9454857
35. Kim J-W, Nwakanma CI, Kim D-S, Lee J-M (2021) Intelligent Face recognition on the edge computing using neuromorphic technology. Int Conf Inf Network (ICOIN) 2021:514–516. https://doi.org/10.1109/ICOIN50884.2021.9333967
36. Tao Y, Huo S , Zhou W (2020)Research on communication APP for deaf and mute people based on face emotion recognition technology. In: 2020 IEEE 2nd international conference on civil aviation safety and information technology (ICCASIT, 2020, pp 547–552. https://doi.org/10.1109/ICCASIT50869.2020.9368771
37. Harikrishnan J, Sudarsan A, Sadashiv A, Ajai RAS (2019) Vision-face recognition attendance monitoring system for surveillance using deep learning technology and computer vision. Int Conf Vis Towards Emerg Trends Commun Netw (ViTECoN) 2019:1–5. https://doi.org/10.1109/ViTECoN.2019.8899418
38. Chen Z –B, Liu Y (2020) Application of Face Recognition in Smart Hotels. In: 2020 IEEE eurasia conference on iot, communication and engineering (ECICE), pp 180–182 https://doi.org/10.1109/ECICE50847.2020.9302014
39. Min WY, Romanova E, Lisovec Y, San AM (2019) Application of statistical data processing for solving the problem of face recognition by using principal components analysis method. In: 2019 IEEE conference of russian young researchers in electrical and electronic engineering (EIConRus), , pp 2208–2212. https://doi.org/10.1109/EIConRus.2019.8657240
40. Deng J, Guo J, Xue N, Zafeiriou S (2019) ArcFace: additive angular margin loss for deep face recognition. In: 2019 IEEE/CVF conference on computer vision and pattern recognition (CVPR). IEEE, Long Beach, CA 2019. pp 4685–4694
41. Jenkins R, Burton A (2008) 100% accuracy in automatic face recognition. Science 319(5862):435
42. Ahonen T, Hadid A, Pietikainen M (2006) Face description with local binary patterns: application to face recognition. IEEE Trans Pattern Anal Mach Intell 28(12):2037–2041
43. Benson PJ, Perrett DI (1991) Perception and recognition of photographic quality facial caricatures: implications for the recognition of natural images. Eur. J. Cognitive Psychol. 3(1):105–135
44. Pike G, Kemp R, Brace N (2000) The psychology of human face recognition," IEE Electronics and Communications: Visual Biometrics, 00/018 (2000).

Community Strength Analysis in Social Network Using Cloud Computing

A. Rohini, T. Sudalai Muthu, Tanupriya Choudhury, and S. Visalaxi

Abstract In large cloud services, the social network domain has provided the services of data analysis efficient and faster. The study has designed a Facebook application for the development of a programming interface environment. In the context of social cloud to access the cloud service is more expensive; this design consideration is presented to get the data from the callback service. The Edge Learning Web miner coordinates the links from the user request from different services, the user has given the request to generate and link the users and return the user request and response. In the Weight-Based Dynamic Link Prediction. The node tendency has increased and influenced the community. Experimental results show that strength analysis of the prediction accuracy is 97.66% in comparison with state-of-the-art solutions.

Keywords Link prediction · Social network · Edge learning

1 Introduction

For the everyday part of people, social network sites have been used for real-world relationships. It provides a platform to support communication and sharing between the nodes in the network. The development of social network analysis has been focused on the social relationship of linking individuals. The study has analyzed the

A. Rohini (✉)
Miracle Educational Society Group of Institutions, Vizianagaram, Andhra Pradesh 535216, India
e-mail: rohinaruna@gmail.com

T. Sudalai Muthu
Hindustan Institute of Technology and Science, Chennai, Tamilnadu, India

T. Choudhury (✉)
School of Computer Science, University of Petroleum and Energy Studies (UPES), Dehradun, Uttarakhand 248007, India
e-mail: tanupriya@ddn.upes.ac.in

S. Visalaxi
Research Scholar, Hindustan Institute of Technology and Science, Chennai, Tamilnadu, India

V. Skala et al. (eds.), *Machine Intelligence and Data Science Applications*, Lecture Notes on Data Engineering and Communications Technologies 132,
https://doi.org/10.1007/978-981-19-2347-0_25

link between the nodes called the structure. It examines how various kinds of objects interact together to form different kinds of links to make a homogeneous community. The study of the structural approach is based on the interaction among the nodes or an actor which is called social network analysis. The relationships are linked with individual objects or nodes. It may be demographics, organization, or actors. The approach of the social network is the ground truth of the pattern of the social ties in which nodes have been embedded with the significant values for those actors or nodes. It exposes the various kinds of patterns and determines the conditions under which patterns are to expose the consequences of the relationship between the nodes. The social network structure is dynamic virtualization between the relationships between the nodes. The abstraction of storage computation in a cloud environment normally provides low-level computations. It is a complimentary act of building blocks in the high-level service stacks. Storage is often extended the capabilities of the physical layer. Web 2.0 is the interface in cloud computing [1, 2] for delivering the services. It encompasses facilitating the service of information sharing and propagation and composition of applications. This sequential access of information has transformed into the runtime environment development of Web 2.0. It captures and developed the delivery services through the target nodes and service applications. This interface brings the flexibility of interaction in the social networking sites and provides the user access flexibility of functions in the desktop applications. It integrates the collection of standards which are AJAX, JavaScript, and XML. That software allowed us to build the applications. Furthermore, the diffusion process has increased the dynamism for the technological key element. It integrates the constant rate of usage trend in the community. Loose coupling is another fundamental property. New applications can be "synthesized" simply by composing existing services and integrating them, thus providing added value. This way it becomes easier to follow the interests of users. Finally, Web 2.0 applications aim to leverage the "long tail" of Internet users by making themselves available to everyone in terms of either media accessibility or affordability. With the emergence of social network [3] analysis, researchers are used to conducting structural research on social phenomena through the combination of the following four approaches:

(a) By the systematic analysis of collected nodes of data that has built the pattern.
(b) Basic structural instincts are clarifying and extended.
(c) Develop the graphical visualization of node pattern links, and.
(d) Evaluated the computational complexity or Graph metrics of social patterns.

Cloud [4] storage service providers such as Amazon, Google App Engine, and Azure. It provides scalable resources like storage, computations, and applications through posted predominantly. Social networking platforms, in particular, gain access to huge user communities, can exploit existing user management functionality, and rely on pre-established trust formed through user relationships.

2 Related Works

Hosting social networks in a cloud platform is a scalable application in the structure of the social network. There are many instances in the integration of cloud computing and social network. The scalable cloud base application of Facebook has been hosted by the amazon web services [5]. Related methods have been depending on the cloud infrastructure in the social networking sites of authentication and user management, Automated Provisioning environment approach has integrated with web 2.0, cloud computing, and social networking applications in the Facebook user community. Some similar efforts have been taken into the grid community environment in the concepts of detection mechanisms [6]. The social network field has developed community decision-making, problem-solving, community, group, influence propagation, influence diffusion, and coalition [7]. Social Network Sites (SNS) have altered the level and style of societal bonding. It has affected the trend of interactions and sharing information, from the microscopic local level to the macroscopic global level [8]. Analyzing the current trends, it is obvious that the years to come will witness major changes in terms of connectivity, as the social web continues to evolve and adapt to our habits and technology [9]. Dynamic scenario drives the research community toward the development of innovative techniques to provide an enhanced amiable and hassle-free social interactive environment [10]. With the widespread attention in digital education (2019) the authentic social network has been categorized into friends, close friend's followers, and acquaintances [11]. Temporal method used in the regularity of interpersonal communication in the social network. The closeness of links in the complex structure was grouped into clusters. Each cluster had homogeneous. Prediction algorithms for local data scheduling could recognize that the homogeneous links were used to find the closeness of links in the identified groups. The data network analysis algorithm was efficiently predicting the nodes to incorporate the changes in the social network [12]. Optimized the influence propagation was forwarded to all the nearby nodes to explore the structure of the social network. Each node examined the presence of the requested node [13, 14]. The node intensity was not present; the request was forwarded to other nearby nodes until its time reached. The simulation result showed that better efficiency in hit success rate and response time than the other strategies [15].

3 Design

The architecture has been designed for a Facebook application for the development of an application programming interface environment (Fig. 1).

In a social network site cloud has been mapped into a particular user to identify the Facebook users. ex., the user could find the ego friends in the cluster. These interactions have been analyzed the variety of patterns in the structure to identify the global and local influential entities of network dynamics. The social network field has developed community decision-making, problem-solving, community, group, influence propagation, influence diffusion, and coalition. The Social Network Sites (SNS)

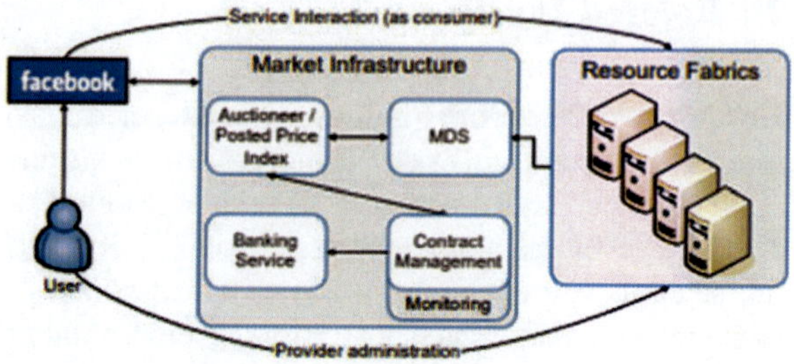

Fig. 1 Social Cloud Facebook application architectural design

has altered the level and style of societal bonding. It has affected the trend of interactions and sharing information from the microscopic local level to the macroscopic global level.

4 Materials and Methods

4.1 Facebook Application

By the REST interface methods of Facebook application to get the data from the user page. It retrieves the information of groups, insights of the user, photos, comments, tags, and friend-of-friend. Facebook Markup Language has the proprietary extension of creating and integrate the application. Facebook JavaScript is the parser to create the virtual application when the page is loaded. The application has exposed the interface methods to get a collection of data which are insights of profiles, photos, friends, and groups. Facebook Markup Language (FBML) has enabled the creation of the application to integrate with the Facebook runtime environment. Facebook JavaScript and sandboxing have been used to create the virtual application. FBJS has been used to load the page in the virtual environment. The application is hosted independently and a URL has been created to access the users and map the user to remotely host. The rendering process of the application has shown in Fig. 2.

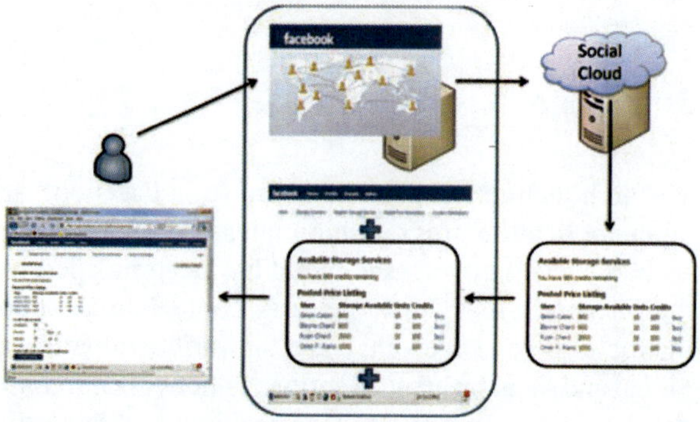

Fig. 2 Cloud web application generates page content of Facebook to create an end-user page

The Facebook user page has been requested through the canvas URL (http://apps.facebook.com/socialcloud/), the server has forwarded the request to define the URL callback service. The page has created based on the request and return to the social network site application of Facebook. The user page is parsed to the specific content and added to the FBML instruction page and return to the user. In the context of social cloud to access the cloud service is more expensive, this design consideration presents the callback services to get the data. FBJS has been used asynchronously in specific services without routing in the application server.

4.2 Web Services

The Graph aimed to evaluate optimization strategies that would eventually be used in the Link Optimization, so the structure of the target links had to mimic as closely as possible the structure of the learning pattern. However, not all components were important or affected by the optimization strategies. So it modeled with only the relevant ones as associated with node management (Gemmel et al., 2016). Each user's intensity had a Message on its initiative (MOI), an Influence User (IU), and a Previous message from a User (PU). The MOIs were submitted to Users via a Web Service of Edge Learning Web miner (WSELWMN) (Fig. 3).

The Edge Learning Web miner coordinates the links from the user request from different services, the user has given the request to generate and link the users and return the user request and response. The user's request is in the structured form of data; it has obtained the queries in the database which stores the user's request. It will be called Web services among the WS-Classifications, Associator, and WS-Clustering. WS-Associator and Web Service-Social Network Analysis (WS-SNA) depend on what content the user has requested. Edge Learning-Web Miner (EL-WM) has used the WS-SNA to generate the information and send it to the WS-Classification to make the prediction task. The selected services will return the user's need and

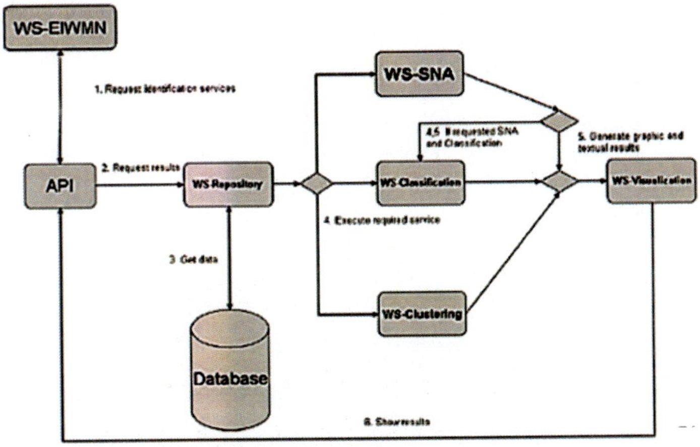

Fig. 3 Generalized model of social network architecture

send it to the WS-Visualization, which will return the result graphically. Finally, the influence node is to be identified and make the community in the social network.

The MOI interacted with an IU to obtain the required information for interacting. Each IU contained an individual optimizer that used a link optimization strategy to make predict the decisions about data files. These files were stored in WS-ELWM and might be predicted on different sites. The IU was used to move the edges or nodes between the communities. The Link Location Service consisted of a Catalog to hold logical to physical node mappings.

The components of analyzing the edges and vertices could be categorized into Timing, Network, Node Selection, Link Access Patterns, Scheduling Policies, and Link Optimization Strategies.

(i) Timing: The time has measured by the social network analysis graph packages of igraph.
(ii) Network: The model of the underlying network along with how it simulated the background of Link analysis.
(iii) Node Selection and Access Rate: Modeling different types of online social network users.
(iv) Learning Link Access Patterns: The order in which links are accessed within some nodes.
(v) Scheduling Policies: The implementation of the policies to decide where to find the influential target node.
(vi) Link Optimization Strategies: The various strategies implemented in igraph.

4.3 Node Proximity

The Node Proximity Clustering algorithm has defined the proximity functions between the nodes. In the dimensional space, the proposed algorithm has measured the closeness between the data points. In the ensemble of K-means clustering algorithm is used to find similar nodes. It has grouped by the unsupervised method of topical perspective. If the neighboring nodes were in the same group. The bipartite graph returns from the algorithm. Each Node or Vertex has visited the properties of neighboring nodes once in the cluster. The adjacency matrix has been used to storing the nodes in the structure. Thus it takes time to traverse all the vertices and their neighbors in the graph. The result in the complexity has O (V2). In this, each node in the graph needs to be traversed once. Hyperlink Induced Topic Distillation iterates until the node has converged and predicted the links in the cluster.

5 Result and Discussion

Community Strength analysis of Weight [16, 17]-Based Dynamic Link Prediction and Threshold Dynamic Link comparison has shown in Fig. 5. It has analyzed with five attributes of betweenness, Degree, Closeness, Coreness, and eccentricity. In the Weight-Based Dynamic Link Prediction, the node tendency has increased and influenced the community. In the comparative analysis of five attributes, the degree attains 15% in the WBDLP. The relative importance of centrality metrics of vertices for betweenness has yielded the value like 21%, where the negative influential links have gone down in the community. The closeness of egocentric nodes yielded as 22% of diffusion in the community, due to bridges and gaps in the structure. The iteration process takes to influence of a node in the network to highly raised value in the Coreness yielded as 30% of edges has influenced in the network (Fig. 4).

By the analysis of Coreness and degree have attained more closeness of the vertices in the structural network. The links are also strongly connected in the betweenness and closeness of the nodes. The eccentricity has yielded the threshold values of 15%, respectively. The WBDLP and TDL algorithms were simulated with 100, 500, 900, 1300, 1700, 2100, and 2500 links for Measuring the Spread of Information propagation validation. The WBDLP algorithm yielded as 525,032.5 s, 719,769 s, 1,008,768 s, 1,316,165 s, 1,632,420 s, 2,001,818 s and 2,360,484 s in 100, 500, 900, 1300, 1700, 2100 and 2500 number of links, respectively.

The TDL algorithm yielded Measuring Spread of Information propagation as 575,000 s, 790,000 s, 1,110,000 s, 1,450,000 s, 1,800,000 s, 2,210,000 s and 2,610,000 s on 100, 500, 900, 1300, 1700, 2100 and 2500 number of links.

The Binomial prediction and Node influence metrics had predicted the value of the link using the Binomial distribution and graph distribution, respectively. (1) The number of times that link was requested, (2) the frequency of link was being requested, (3) the last time that the node-link was requested, and (4) the size of the structure was considered as the parameters. The parameters affect the "importance value" of the node and links. The numbers of times, frequency of information propagation are important factors to predict future requests in the social network structure. The relationship between two nodes was an important factor in impacting the performance of structural size. The size of the structure also played a role in the relationship. Hence, the size was also taken as a factor for Link prediction.

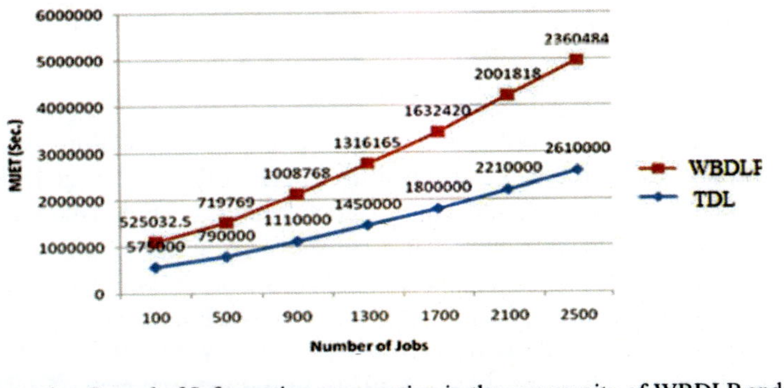

Fig. 4 Measuring Spread of Information propagation in the community of WBDLP and TDL

Table 1 Link prediction accuracy of 34 attributes with 600 nodes

Confusion matrix		Predict	
		Negative	Positive
Actual	Negative	$a = 128$	$b = 9$
	Positive	$c = 5$	$d = 458$

5.1 Performance

The performance of such systems is normally evaluated using the data in the matrix. Table 1 shows the confusion matrix for a two-class classifier.

Confusion Matrix for prediction of links using 34 attributes with 600 nodes.

From Table 1 the confusion matrix entries have the following meaning in the context of our study:

- *a*—Number of accurate predictions that an instance is negative,
- *b*—Number of incorrect predictions that an instance is positive,
- *c*—Number of incorrect predictions that an instance negative and.
- *d*—Number of accurate predictions that an instance is non-negative links.

Several standard terminologies defined for the class 2 matrix are:

The accuracy (A) is the proportion of the total number of predictions that are correct and is calculated using following equation:

$$A = (a + d)/(a + b + c + d).$$

The correlation between network metrics and the prediction accuracy was 97.66%. The effectiveness was high when the link interprets the ego network with an average shortest path distance of 5.317 to accessing the node. It was strongly connected with high betweenness centrality in the community.

6 Conclusion

The presented work can be used in information propagation on *Online Social Media Networks* combining knowledge from different networks. The social network is representing the friendship relationships among Facebook users and the geographical network representing the physical location of the set of users. In particular, it is to verify the most efficient way to choose a small subset of nodes of the network from which information is spread such that to maximize a certain function. The future direction of the research work will focus on designing an algorithms using machine learning techniques to perform access quickly in the link patterns of unstructured network. Limitation of the study to find the interpretation of links in the ego nodes.

References

1. Sharma TC, Kumar P (2018) Health monitoring & management using IoT devices in a Cloud Based Framework, In: 2018 International Conference on Advances in Computing and Communication Engineering (ICACCE), pp 219–224. https://doi.org/10.1109/ICACCE.2018.8441752
2. Dewangan BK, Jain A, Choudhury T (2020) GAP: hybrid task scheduling algorithm for cloud. Revue d'Intelligence Artificielle 34(4):479–485. https://doi.org/10.18280/ria.340413
3. Sarishma, Tomar R, Kumar S, Awasthi MK (2021) To beacon or not?: Speed based probabilistic adaptive beaconing approach for vehicular Ad-Hoc networks. In: Paiva S, Lopes SI, Zitouni R, Gupta N, Lopes SF, Yonezawa T (eds) Science and technologies for Smart Cities. SmartCity360° 2020. Lecture Notes of the Institute for Computer Sciences, Social Informatics and Telecommunications Engineering, vol 372. Springer, Cham. https://doi.org/10.1007/978-3-030-76063-2_12
4. Wilkison S et al (2016) Storj: a peer-to-peer cloud storage network. https://storj.io/storj.pdf
5. Alemany J (2019) Metrics for privacy assessment when sharing information in online social networks. IEEE Access
6. Babu SS (2020) Earlier detection of rumors in online social networks using certainty-factor-based convolutional neural networks. In: Social network analysis and mining
7. Muthu CT (2020) Review on graph feature learning and feature. arxiv
8. Carley MK (2012) Trends in science networks: understanding structures and statistics of scientific networks. In: Social network analysis and mining, pp 169–187
9. Carley MK (2014) An incremental algorithm for updating betweenness centrality and k-betweenness centrality and its performance on realistic dynamic social network data. In: Social network analysis and mining, p 235
10. Chakraborty A (2018) Application of graph theory in social media. Int J Comput Sci Eng
11. Chakraborty K (2020) A survey of sentiment analysis from social media data. IEEE Transaction
12. Chekkai N (2019) Weighted graph-based methods for identifying the most influential actors in trust social networks. Int J Netw Virtual Organisations 101–128
13. Chen P-Y (2019) Identifying influential links for event propagation on Twitter: a network of networks approach. IEEE Trans Signal Inf Process Over Netw 5(1)
14. Kumar S, Tomar R (2018) The role of intelligence in space exploration. In: 2018 International Conference on Communication, Computing and Internet of Things (IC3IoT), pp 499–503. https://doi.org/10.1109/IC3IoT.2018.8668161
15. EitanMullera R (2019) The effect of social networks structures on innovation performance: a review and directions for research. Int J Res Mark 3–19
16. George G, Sankaranarayanan S (2019) Light weight cryptographic solutions for fog based blockchain. In: Proceedings of International Conference Smart Structure Systems (ICSSS), pp 1–5
17. Iqbal R, Butt TA, Afzaal M, Salah K (2019) Trust management in social Internet of vehicles: factors, challenges, blockchain, and fog solutions. Int J Distrib Sensor Netw 15(1). Art. no. 1550147719825820

Object Detection on Dental X-ray Images Using Region-Based Convolutional Neural Networks

Rakib Hossen, Minhazul Arefin, and Mohammed Nasir Uddin

Abstract In dentistry, Dental X-ray systems help dentists by showing the basic structure of tooth bones to detect various kinds of dental problems. However, depending only on dentists can sometimes impede treatment since identifying things in X-ray pictures requires human effort, experience, and time, which can lead to delays in the process. In image classification, segmentation, object identification, and machine translation, recent improvements in deep learning have been effective. Deep learning may be used in X-ray systems to detect objects. Radiology and pathology have benefited greatly from the use of deep convolutional neural networks, which are a fast-growing new area of a medical study. Deep learning techniques for the identification of objects in dental X-ray systems are the focus of this study. As part of the study, Deep Neural Network algorithms were evaluated for their ability to identify dental cavities and a root canal on periapical radiographs. We used tensor flow packages to detect dental caries and root canals in X-rays. This method used faster R-CNN technology. For this reason, the proposed method is accurate at 83.45% which is 10% greater than previous research.

Keywords Deep learning · Faster R-CNN · Convolutional neural network · Dental X-ray image

1 Introduction

In recent decades, dentistry has improved. However, as dentistry has grown, so has the number of people with dental issues. Dentists are often required to see a large number of patients in a single day. A significant number of dental X-ray [1] films are taken every day as an important diagnostic tool for assisting dentists. A majority of

R. Hossen · M. Arefin (✉) · M. Nasir Uddin
Jagannath University, Dhaka, Bangladesh
e-mail: minhazularefin21@gmail.com

M. Nasir Uddin
e-mail: nasir@cse.jnu.ac.bd

V. Skala et al. (eds.), *Machine Intelligence and Data Science Applications*, Lecture Notes on Data Engineering and Communications Technologies 132,
https://doi.org/10.1007/978-981-19-2347-0_26

dentists perform film reading work, which takes up important clinical time and can lead to mis- or under-diagnosis because of personal aspects like fatigue, emotions, and poor skill levels. Intuitive dental disease detection technologies [2] may minimize the labor load of dentists and the incidence of misdiagnoses, therefore improving the quality of dental care. For smart health care in this context, automated object recognition in dental X-ray systems is a critical duty to be completed on time. It is a form of structural deterioration that causes teeth to develop cavities or holes. Cavities, on the other hand, are the consequence of tooth decay, which is caused by bacterial infections. Root canal is being due to inflammation or infection or external fracture in the roots of a tooth. They are common infectious oral diseases [3]. Many individuals, from teenagers to adults suffer from caries and root canal which leads to severe pain for a lifetime and even tooth loss. Further, the Sever problem may lead to oral cancer. In 2016, the WHO estimated that about 50% of the world population [4] is affected by dental caries and in some Asian-Pacific countries, the incidence of oral cancer is within the top three of all cancers. Caries and a root canal can be difficult to detect and diagnosis due to internal damage confused lesions that cannot be identified with naked eyes. So far, for caries and root canal detection, several image segmentation and classification techniques such as level set method, morphological image processing, and machine learning (ML) based classification methods have been improved. But they could not achieve any significant results.

Dental caries and root canals have been detected using a variety of techniques in recent decades. As a result of tooth anatomy and restoration shapes, there has not been a significant improvement. Particularly when deep fissures, strong interproximal contacts, dark areas, and subsequent lesions are present, identification becomes difficult. With the help of standard machine learning and image processing techniques, these methods concentrated on extracting hand-crafted features, and building effective classifiers for the identification and recognition of objects and people. Such techniques are restricted by the fact that they typically require medical specialists to create useful features since vital visual characteristics for clinical choices [5] may be combined with irrelevant pixels in these approaches. Aside from that, each component of the detection pipeline is improved independently, resulting in a suboptimal detection pipeline overall. As a side note, these characteristics are typically restricted and useless, resulting in incorrect ROI and edge detection. That means that both expert systems and medical science will gain from this research. Using this framework does not require a lot of experience on the part of the end-users.

2 Background Studies

Most teens and adults in the globe suffer from dental caries, a persistent infectious oral illness. For the identification of dental caries and root canals, a number of studies have been conducted. These attempts to identify tooth cavities and root canals on the basis of dental X-ray images [6] are discussed here. It was proposed by Rad et al. [7] that a method of automated segmentation and feature extraction be used for the detection of caries in dental radiographs. A wide range of image processing techniques was employed by the author, including picture enhancement and segmentation [8] as well as feature extraction, detection, and classification. To improve the contrast of the structure of interest, picture enhancement methods are utilized. A k-mean clustering technique is then used to segment the picture in order to identify any errors. Individual tooth areas have been separated using an integrated projection method. If there are caries in the area of interest (ROI), characteristics such as entropy, intensity, mean, and energy are retrieved from the ROI and compared to nearby areas. For caries detection, dental experts analyze the characteristics of the caries region. As a result of this method's effectiveness, segmentation is often challenging owing to changes in intensity area, and tooth extraction issues might lead to incorrect results.

To help in the diagnosis of proximal dental caries in periapical pictures, Choi and colleagues [9] developed an automated detection method. Proximal dental caries was investigated in regions near the crown borders, according to the authors. The crown areas were initially divided using a level set approach in order to identify them. Second, the authors used the identified crown regions to filter out areas outside of crowns. Then, crown borders were used to reduce the risk of dental pulp cavities. To decrease computing expenses, the scientists scaled all of the experimental pictures to the same size. Five folds were randomly assigned to the pictures. Four folds were used to train the CNN, and one fold was used to test the CNN. Test folds are altered one at a time to ensure that all periapical pictures have been checked. HAPT (horizontal alignment of photographed teeth), probability map creation, crown extraction, and refining are the four components of the system.

An improved caries detection system was suggested by Rad et al. [10] using a novel segmentation approach and detection methodology. It consists of three primary phases: preprocessing, segmentation, and analysis, which are described in detail below. Initial contour creation and intelligent level set segmentation finish the segmentation process. In this technique, features are extracted manually, which might lead to incorrect IC synthesis. Despite the author's attention to the segmentation process, a sufficient amount of research has not been done to identify caries. The BPNN also suffers from local minima, therefore this technique does not give good convergence. As the learning rate must be set carefully, it is possible that the detection of caries will be inaccurate. In this technique, features are extracted manually, which might lead to incorrect IC synthesis. Despite the author's attention to the segmentation process, a sufficient amount of research has not been done to identify caries.

The BPNN also suffers from local minima, therefore this technique does not give good convergence. As the learning rate must be set carefully, it is possible that the detection of caries will be inaccurate. According to Lee and colleagues [11] deep CNN algorithms may be used to identify and diagnose dental caries on periapical radiographs. Rotation, width and height shifting, zooming, shearing, and horizontal flipping were used to augment the training dataset ten times at random. The datasets were preprocessed before being trained with transfer learning using a GoogleNet Inception v3 CNN network. The v3 architecture's inception has learned roughly 1.28 million photos with 1000 object categories. To extract distinct scale properties, a convolutional filter with 22 deep layers can be utilized. It is possible to achieve a variety of scale features by combining convolutional filters of different sizes inside the same layer. The nine inception modules used included an auxiliary classifier, two fully-connected layers, and softmax functions. With a learning rate of 0.01 and 1000 epochs of training, the dataset was randomly partitioned into 32 batches for each epoch. To improve the detection of dental caries, fine-tuning was done to optimize weights and increase output power by modifying hyperparameters. This method uses a sophisticated and expensive algorithm to calculate the sliding window for caries detection. Detecting premolar caries may be challenging as a result of this procedure.

On the basis of dental radiograph image categorization, Yang et al. [12] presented an automated root canal treatment quality evaluation technique. For the root canal filling therapy, the scientists employed an automated apical foreman-area recognition technique based on dental scans. Researchers in this study employed a labeled dataset of periapical radiography pictures obtained before and after therapy to develop their findings. Root canal filling was identified using image subtraction techniques at first. The ROIs—the apical area in image preprocessing—was extracted here. It was first necessary to discover feature points between the two images using the SIFT and SURF algorithms. The authors then used the minimal grayscale difference approach to get the best ternary matching spots. These best ternary points were then used to construct an affine matrix, which was then applied to images in order to do an image subtraction. As a result of this, the authors were able to determine the apical foreman and its surrounding region using the filling area. The authors then input this portion of the pictures to the CNN training software in order to train their classification models on this component of the dataset. The authors employed six layers of deep neural networks to tackle the overfitting problem. Two convolution layers were added to the Inception structure in order to minimize the amount of parameters while maintaining a high level of performance. Because ROI detection is so imprecise, this research has a major flaw. There are a few CNN layers used in this technique, which limits the performance. A similar method does not function well with molar pictures. The discrepancies between certain pictures are so great that even manual calibration, which relies solely on the affine matrix, does not work effectively. As a result, new ways for dealing with these picture pairings are needed.

3 Proposed Methodology

This study technique follows the comparable Faster R-CNN, a state-of-the-art deep learning system for object detection. This research method follows a similar deep learning framework. The first element is a Regional Proposal Network (RPN) that generates a list of areas that are likely to contain objects or ROIs; the second part is a Fast R-CNN that categorizes a section of the picture as objects (and background) and reinforces their borders. The convolution layers used for feature extraction in both components have identical values, allowing object detection tasks to be completed at a reasonable speed. This research paper proposed a new model for object detection

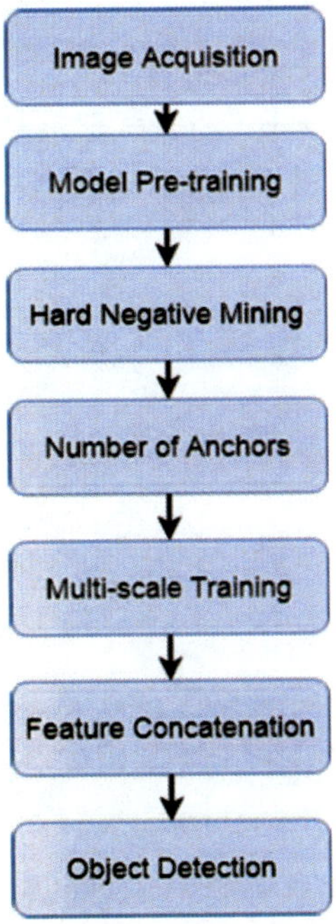

Fig. 1 Proposed methodology of object detection on dental X-ray images using R-CNN

that uses a deep learning technique. It is based on the Faster Region-based Convolution Neural Networks (Faster R-CNN), made up of the Regional Proposal Network (RPN) & (Fast R-CNN). We will describe the process flow with a high-level view in the first place. Figure 1 shows the top-level view of the proposed model.

3.1 Image Acquisition

The Digital Dental Periapical X-Ray Dataset was utilized to evaluate the suggested model. The primary objective of this research is to recognize and mark each pixel for the item category designated (e.g., implant and root canal, etc.). Over the years, the Computer Vision Community has produced several benchmark data sets in order to evaluate academics and enthusiasts' models. A small segment of the dataset is shown in Fig. 2.

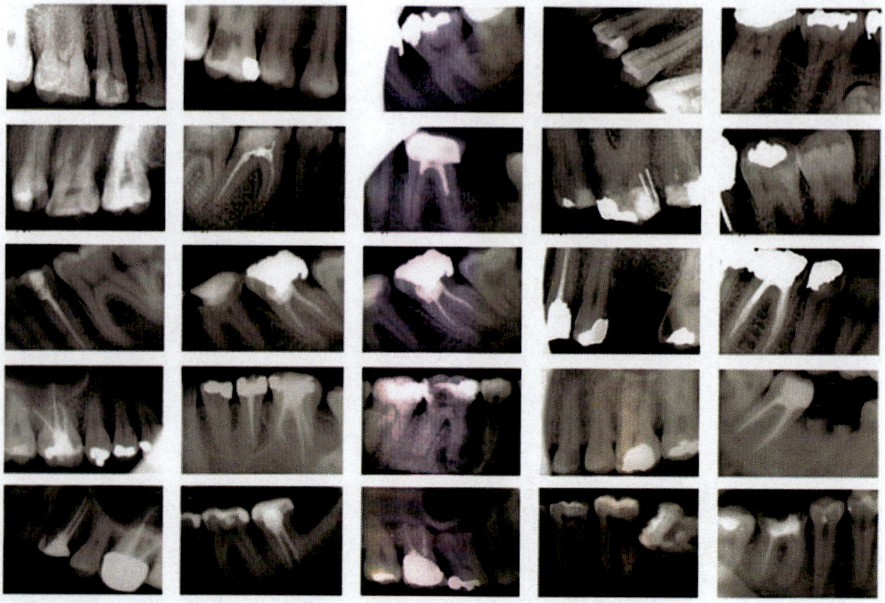

Fig. 2 Segment of digital dental periapical X-ray dataset

3.2 Model Pre-training

We have decided to fine-tune a pre-trained *faster r-cnn inception v*2 model using a COCO dataset to adapt the Faster R-CNN to object detection. The Digital Dental Periapical X-Ray Database is a widely known dataset featuring unconstrained objects. However, it may not be a good choice to merely refine this data set, as it is a very tiny dataset with just two items in 120 pictures. We have pre-trained our model in a COCO dataset, which is a lot bigger object dataset with many more challenging instances before we finalize the digital dental periapical X-ray database. In order to manage those tough instances that might interrupt the process of convergence, attention should be paid to the discovery of certain training data where the experimental section contains information. Hard negative mining is also required in this dataset before training in order to limit the amount of false-positive results created. This technique shows the intricate architecture in the Fig. 3.

3.3 Hard Negative Mining

Hard negative mining has been demonstrated as an efficient technique to increase the performance of deep education, particularly for the identification of objects. The concept behind this procedure is that the places in which the network has failed to

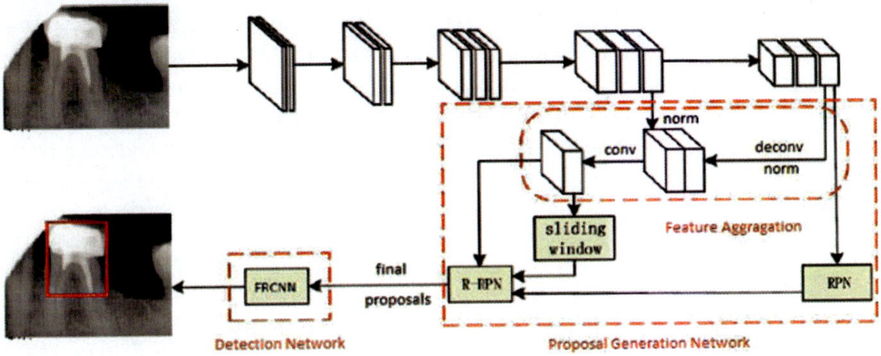

Fig. 3 Architecture of faster R-CNN

forecast correctly are harsh negative. Hard negatives are therefore sent back into the network to enhance our trained model. The ensuing training procedure can then enhance our model to reduce false-positive effects and increase classification performance. When the classifier incorrectly classifies a negative example, the model will automatically identify it as a negative example and place it in the training set for retraining. This phase is akin to having the improper set of questions. If the same question is answered incorrectly many times, the student is incorrect. During the first stage of our training process, hard negatives are extracted from the pre-trained model. Then we regard a region to be hardly negative if it is less than 0.5 at its intersection between the union (IoU). We intentionally add these harsh negatives to RoI's throughout the hard negative training process in order to finalize the model. It also completes the ratio of background and foreground near about 1:3. It is the same ratio we used in the first step.

3.4 *Number of Anchors*

In this research, the Faster R-CNN architecture has set up numerous critical hyper-factors, in which the most important of these parameters seem to be the number of anchors in the RPN portion. Usually, Faster R-CNN utilizes nine anchors that are able to remember tiny things. Smaller items such as implants seem to be quite frequent in object detection tasks. In the faster R-CNN, we use two conditions to assign a positive label to an anchor.

$$L(p_i, t_i) = \frac{1}{N_{\text{cls}}} \sum_i L_{\text{cls}}(p_i, p_i^*) + \lambda \frac{1}{N_{\text{reg}}} \sum_i p_i^* L_{\text{reg}}(t_i, t_i^*) \tag{1}$$

A mini-batch has an anchor i, and p_i is a projected probability that anchor i will be a real item. We assume the anchor in the figure is positive. For that reason, the ground-truth label p_i* is 1, else it is 0. There are four parameterized coordinates in t_i, and t_i is the ground-truth bounding box. It is associated with a positive anchor in t_i. L_{cls} is a logarithm base loss across two classes. For the regression loss, we use $L_{reg}(t_i, t_i^*) = R(t_i, t_i^*)$ where R is denoted as the loss function (smooth L_1). A regression loss $L_{reg}(t_i, t_i^*) = R(t_i, t_i^*)$ is only enabled for positive anchors ($p_i^* = 1$) and deactivated for negative anchors ($p_i^* = 0$). As a result, the outputs of *cls* and *reg* are p_i and t_i, respectively.

3.5 Multi-scale Training

Technique for all the training pictures, the Faster R-CNN architectures use a fixed scale. The detector can learn features over a broad variety of sizes, thereby enhancing its performance toward a level invariance, by scaling pictures to a random scale. We allocate one of three scales for each image randomly in this study prior to it being included in the network. Our experimental section provides details. Our empirical results demonstrate that our model is more resilient to diverse dimensions and improves the performance of detection on benchmarks. We compute the bounding-box regression by parameterized the four coordinates of the detected object. It is shown in Eq. 2:

$$\begin{aligned} t_p &= (p - p_a)/r_a, \; t_q = (q - q_a)/s_a, \\ t_r &= \log(r/r_a), \qquad t_s = \log(s/s_a), \\ t_p^* &= (p^* - p_a)/r_a, \; t_q^* = (q^* - q_a)/s_a, \\ t_r^* &= \log(r^*/r_a), \qquad t_s^* = \log(s^*/s_a), \end{aligned} \tag{2}$$

where, p, q represents the width and r, s represent the height of the box. Variable p represents the predicted box. Variable p_a denotes the anchor box. Variable p^* is for ground-truth box.

3.6 Feature Concatenation

The RoI pooling on the final functional map layer is used for the classic Fast R-CNN networks to create regional characteristics that are subsequently examined by the classification component of your network. With this innovative architecture, the classification network may use the characteristics computed by RPN to save many needless calculations. This method is however not always ideal and occasionally certain crucial characteristics may be missed since features of the deeper convolution layer output have broader receptive fields and are more granular. The reverse function computed the partial derivative of the loss function. It worked in the RoI pooling layer.

For calculating it works with respect of each input variable p_i. Using argmax quation we find that in Eq. 3:

$$\frac{\delta L}{\delta p_i} = \sum_r \sum j[i = i^*(r, j)] \frac{\delta L}{\delta q_{rj}} \tag{3}$$

Here, r is denoted as each mini-batch RoI and y_{rj} represents each pooling output unit. The partial derivative $\frac{\delta L}{}$ is accumulated if i is the argmax selected for q_{rj} by max pooling. Because there is no concept of a ground truth in the bounding box for background ROIs, L_{loc} is ignored. We utilize the loss, for bounding-box regression Eq. 4:

$$L_{\text{loc}}\left(t^u, v\right) = \sum_{i \in p,q,r,s} \text{smooth}\left(t_i^u - v_i\right) \tag{4}$$

in which,

$$\text{smooth}_{L1}(p) = \begin{cases} 0.5p2 & \text{if} |p| < 1 \\ |p| - 0.5 & \text{otherwise,} \end{cases} \tag{5}$$

Here, L_1 is used, cause it is sensitive to the image outliers and L_2 is used in R-CNN. They both are loss functions. In this research, we want to enhance the RoI pooling by merging many convolution-layered characteristic maps, which include both low- and high-level functions, to capture more finely-grained RoI information. In order to get the final bonding features for detection tasks, we propose to concatate the output of several convolution feature maps. In other words, combined with the final feature map in RPN, we use certain intermediate findings, which we combine to produce the final pooling functions. Specifically, features are RoI-pooled and L2-normalized from several lower-level convolution layers accordingly. These characteristics are then concatenated and rescaled as if the original scale of the features had not been adopted. A 1 1 convolution is done to match the original network's number of channels. This technique shows the intricate architecture in Fig. 4.

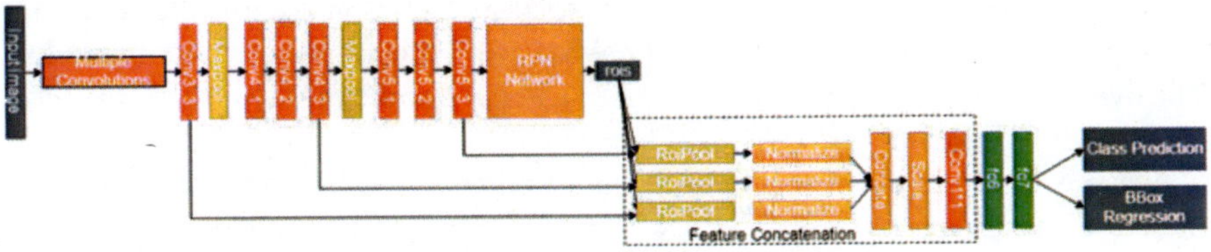

Fig. 4 Architecture of feature concatenation

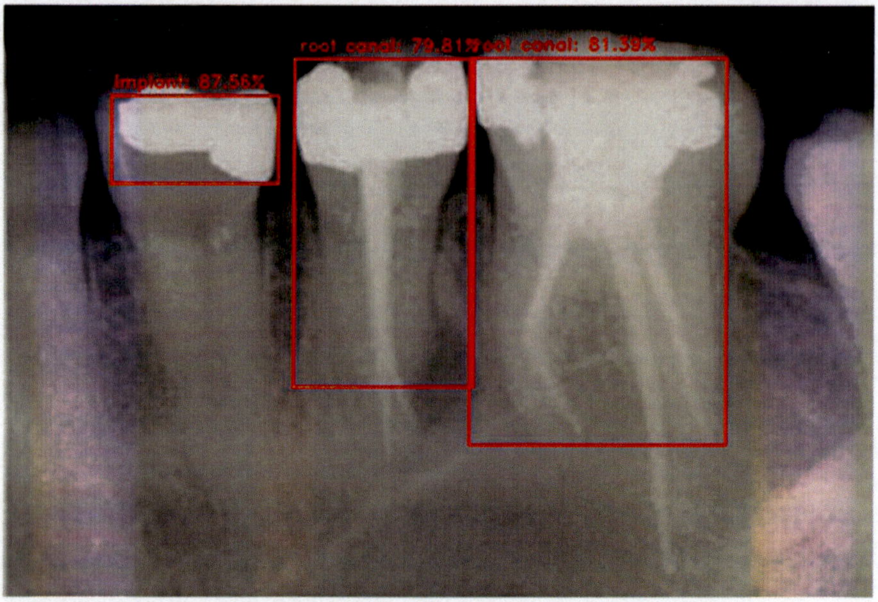

Fig. 5 Detected object on dental X-ray image

3.7 *Object Detection*

Object detection is the last step of this process where usually we assign some levels to an object based on its features. In the last step of this methodology, using softmax classifier this method can finally detect objects in dental X-rays images successfully which is shown in Fig. 5.

4 Result

In order to detect objects, there are a number of metrics that are variations of intersection over union (IoU). The overlap between two borders is measured by the IoU. The overlap between our anticipated border and the ground reality is then calculated (the real object boundary).

Let $n_{i,j}$ be the number of pixels of class i predicted to belong to class j, where there are $n_c l$ different classes, and let $t_i = \Sigma_j \text{ n}_{ij}$ be the total number of pixels of class i.

Mean IoU is defined as:

$$\text{Mean}_{\text{IoU}} = \left(\frac{1}{n_{\text{cl}}}\right) \sum_i \frac{n_{ii}}{t_i + \text{sum}_j n_{ji} - n_{ii}} \tag{6}$$

To determine if an item is present in the image, the mean average precision (mAP) is utilized as a measure of accuracy. It is important to know the accuracy and recall of a classifier in order to comprehend the AP. To train RPNs, we assign each anchor a binary class label (whether it is an object or not). Two types of anchors are given a good designation. Intersection over Union (IoU) anchors having the highest overlap with a ground-truth box, or any anchor with an IoU overlap greater than 0.7 with a ground-truth box, are highlighted. Assigning positive labels to numerous anchors is possible with a single ground-truth box. All ground-truth boxes with an IoU ratio less than 0.3 will be labeled as negative. Following the multi-task loss in Fast R-CNN, we minimize an objective function with these definitions in place (Fig. 6).

The result shows that the proposed methodology can find out 83.45% accurate dental object detection with an inaccuracy of 16.55%. The results comparing the proposed method with three others are shown in Table 1. It shows that the mAP of R-CNN is better than the earlier results. If we increase the size of the training data, more mAP can be obtained. The existence of a large volume of data is always useful for the R-CNN model.

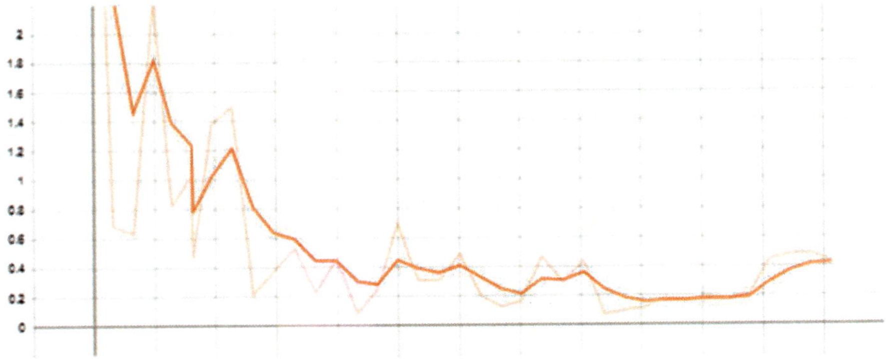

Fig. 6 Total Loss function per epoc (in thousand)

Table 1 Comparison with existing models

Models	mAP (Digital dental periapical X-ray database) (%)
RFCN Resnet 101	68.3
GoogleNet Inception V3	55.01
SSD Inception V1	73.56
Our study	83.45

5 Conclusion

As well as reducing the expense of oral health care, accurate identification of dental decay and root canals improves the chance of natural tooth preservation in the long run. In this research, deep learning techniques are used to recognize and classify dental X-ray items. In order to recognize generic objects, we have utilized the Faster R-CNN framework which features feature concatenation, multi-scale training, hard negative mining, and correct tuning of anchor sizes for RPN, among other things. Because this framework combines a variety of approaches, it is able to overcome many of the limitations of single methods. Both dental cavities and root canals have been detected. As a result of the classification accuracy, our technique has been proven to be reliable and efficient. When training and testing were done on the dataset, classification accuracy increased. 83.45% of caries and root canals were detected using our technique, which is better than any other method. A better local minimum is achieved with this strategy compared to a conventional random initialization of network weight. Our approach may be utilized to detect dental caries and root canals from dental X-rays, therefore in conclusion, we can conclude that it works. In future we will try to use a better deep learning method with multiple datasets.

References

1. Lakshmi MM, Chitra P (2020) Classification of dental cavities from X-ray images using deep CNN algorithm. In: 2020 4th International Conference on Trends in Electronics and Informatics (ICOEI)
2. Lee JH, Kim DH, Jeong SN, Choi SH (2018) Diagnosis and prediction of periodontally compromised teeth using a deep learning-based convolutional neural network algorithm. J Periodontal Implant Sci 48(2):114–123
3. Al Kheraif AA, Wahba AA, Fouad H (2019) Detection of dental diseases from radiographic 2d dental image using hybrid graph-cut technique and convolutional neural network. Measurement 146:333–342
4. Oral health. https://www.who.int/news-room/fact-sheets/detail/oral-health. Last accessed 19 Aug 2021
5. Lee J-H, Kim D-H, Jeong S-N, Choi S-H (2018) Detection and diagnosis of dental carries using a deep learning-based convolutional neural network algorithm. J Dent 77:106–111
6. Silva G, Oliveira L, Python M (2018) Automatic segmenting teeth in X-ray images: trends a novel data set benchmarking and future perspectives. Expert Syst Appl 107:15–31
7. Rad AE, Amin IBM, Rahim MSM, Kolivand H (2015) Computer-aided dental carries detection system from x-ray images. In: Computational intelligence in information systems. Springer, Heidelberg, pp 233–243
8. Ali M, Khan M, Tung NT (2018) Segmentation of dental Xray images in medical imaging using neutrosophic orthogonal matrices. Expert Syst Appl 91:434–441
9. Choi J, Eun H, Kim C (2018) Boosting proximal dental caries detection via combination of variational methods and convolutional neural network. J Signal Process Syst 90(1):87–97
10. Rad AE, Rahim MSM, Kolivand H, Norouzi A (2018) Automatic computer-aided carries detection from dental x-ray images using intelligent level set. Multimedia Tools Appl 77(21):28843–28862

11. Lee J-H et al (2018) Detection and diagnosis of dental caries using a deep learning-based convolutional neural network algorithm. J Dent 77:106–111
12. Yang Y, Xie L, Liu B, Xia Z, Cao, Guo C (2018) Automated dental image analysisby deep learning on small dataset. In: IEEE 42nd Annual Computer Software andApplications Conference (COMPSAC), vol 1. IEEE, pp 492–497

Real-Time and Context-Based Approach to Provide Users with Recommendations for Scaling of Cloud Resources Using Machine Learning and Recommendation Systems

Atishay Jain

Abstract The recent paradigm of IT services has shifted from desktop-based infrastructure to cloud-based infrastructure. One of the most essential services provided by the cloud is infrastructure-as-a-service (IaaS). IaaS allows customers to use infrastructure such as servers, storage, and processing units of a provider on a pay-as-you-go basis. Customers can scale resources up and down as per the requirement. There are many infrastructure providers in the market available today, but if a customer can be recommended about the requirement of resources in the upcoming future based on the context-driven recommendation system using the historic data of requests in various periods, then this can help customers scale up their resources in the time of need. By the use of this recommendation system, customers can scale up their resources when their usage requirements are about to increase without bearing the loss in business; moreover, customers can take easy calls with the support of this recommendation system as to when they should increase their requirements.

Keywords Scaling · Cloud resources · Recommendation system · Contextual · Real time · Machine learning · Preventing business loss

1 Introduction

The recent advancements in the field of cloud computing have led the computing technology from standalone devices to cloud-based systems. The delivery of various services over the Internet is known as cloud computing [1]. These resources include data storage, servers, databases, networking, and software, among other tools and applications. Cloud computing turns IT infrastructure into a utility by allowing you to connect to it via the Internet and use computer resources without having to build or maintain them on-premises. Using cloud computing over traditional hardware-based computing systems overcomes many shortcomings like limited geographic access, outdated off-the-shelf software, dedicated in-house IT support, and data

A. Jain (✉)
Dell EMC, Bangalore, India
e-mail: atishayjain3098@gmail.com

V. Skala et al. (eds.), *Machine Intelligence and Data Science Applications*, Lecture Notes on Data Engineering and Communications Technologies 132,
https://doi.org/10.1007/978-981-19-2347-0_27

storage devices [2]. Increased use of cloud computing has helped mitigate such issues and has provided various other benefits like lower IT costs, improve agility and time-to-value, scale more easily and cost-effectively, automatic upgrades, and protection from data loss [3]. Such benefits have helped cloud computing gain so much popularity and widespread adaption from traditional systems.

Through cloud computing, various types of services can be provided to the user over the Internet. Software-as-a-service (SaaS), infrastructure-as-a-service (IaaS), and platform-as-a-service (PaaS) are the three most common models of cloud services. SaaS (also known as cloud-based software or cloud apps) is cloud-based application software that you access and use through a Web browser, a dedicated desktop client, or an API that connects with your desktop or mobile operating system [4]. SaaS users typically pay a monthly or annual membership fee, while some may provide "pay-as-you-go" pricing depending on actual usage. PaaS offers software developers an on-demand platform—a complete software stack, hardware, infrastructure, and even development tools—for operating, creating, and managing applications without the complexity, expense, or inflexibility that comes with maintaining that platform on-premises. The cloud provider hosts everything, including networks, operating system, storage, servers, software, middleware, and databases, with PaaS [5]. IaaS gives pay-as-you-go access to basic computing resources such as real and virtual servers, networking, and storage through the Internet. End users may expand and decrease resources as required using IaaS, which eliminates the need for large, upfront capital expenditures, unneeded on-premises or "owned" infrastructure, and overbuying capacity to accommodate periodic spikes in demand. Users can minimize the cost and complexity of purchasing and managing real servers and datacenter equipment by using IaaS. Each resource is supplied as a separate service component, and you only pay for it for as long as you require it. The infrastructure is managed by a cloud computing service provider, while a user buys, installs, configures, and maintains their software, such as operating systems, middleware, and apps [6].

IaaS provisions users to scale the resources required by their application up and down based upon the requirement and business demands. This scaling of resources can be much more efficient if a user is notified beforehand when their requirements will increase or decrease for the needed resources such that they can scale up or down. This can help prevent any business loss to the user, as the user will be aware in advance when their requirement will be higher or lower than the current demand. This can help a user take an easy call when to scale up or scale down the resources. The approach discussed here talks about a system that will act as a monitoring and recommending system, which will monitor the requests coming from the user, and based upon the requests and various other factors, the recommendation system will provide recommendations to the user on when to scale resources up or down. Through this system, users can take an easy and informed decision in advance such that the user doesn't suffer any business loss due to resources [7].

A recommendation engine is a data filtering tool that uses machine learning algorithms to suggest the most relevant things to a certain user or client. It works on the premise of identifying patterns in customer behavior data, which may be gathered either implicitly or explicitly. Recommendation systems use algorithms and data analysis techniques to recommend the most relevant product/items to a particular user [8]. A recommendation system generally has four phases that is collection of data, storing of data, analyzing the data, and filtering of data. A recommendation system can analyze the real-time data and based on the various factors and then provide real-time contextual recommendations for a system [9].

2 Literature Review

The major advancements in cloud computing technology have paved the way for the adaption of cloud-based services over traditional systems. Cloud provides IaaS that is infrastructure-as-a-service which provided various resources like servers, networks, and databases as a service, which are used in the development and running of the application. These resources are used on a pay-as-you-go model that is the number of resources currently being used is only paid for. The requirement of these resources can vary based on the demand so to manage that scaling of resources from the cloud is required. [10] proposes an approach for resource auto-scaling in the cloud environment according to Markov decision process (MDP). The author uses reinforcement learning and also considers the Markov model and concluded that the approach used is efficient but faces difficulty in having minimum cost. Marinescu [11] brings this concept to its logical conclusion by proposing a framework for enabling self-managing resources in the cloud. It introduces the idea of coalitions, consisting of collaborating resources that work together for service delivery. It implies that limiting user-cloud service provider contact to a well-defined services interface might be beneficial. It demonstrates how clouds may be seen as engines for delivering a suitable collection of resources in response to service requests. Kriushanth et al. [12] the author provides an overview of cloud computing, and it emphasizes auto-scaling; they discuss the recent trends and services available in auto-scaling in different levels of scaling. Sadooghi and Raicu [13] The author discusses the design and implementation of a scheduling and execution system (CloudKon). This system has three main features that is first, it is built to get the best performance and utilization out of cloud resources by utilizing a range of cloud services (Amazon SQS and DynamoDB); second, it is completely distributed and capable of running large-scale applications; third, it can handle a variety of workloads at the same time, including MTC and HPC applications. Rana [14] the author proposes a scalability enhancement in cloud-based applications using software-oriented models. This approach shows that software architectural improvement approaches are preferable choices for scalability enhancement as compared to adding more resources physically. Recommendations systems are also used in scaling resources in the cloud. In [15], author proposes a recommender system, a new declarative approach for selecting cloud-based infrastructure services which automate the mapping of users-specified application requirements to cloud service configurations. Guo et al. [16] the author proposes a service recommendation method based on requirements. In this approach, user communities are built by clustering to narrow the range; then, the reported QoS values and the evaluation QoS values are combined to forecast user QoS requirements, and then, the degree of user-to-service matching is then calculated. Then finally, based on their similarity, the difference in their matching degrees, and the ratings of services by the target user's neighbors, service ratings are predicted for the target user to obtain the suggestion list.

3 Methodology

The approaches discussed above do not provide real-time context-based recommendations for scaling resources up and down in a cloud environment. The approach discussed in this paper tries to rectify this challenge by providing users with real-time contextual recommendations for scaling resources up or down in cloud infrastructure such that there is no business loss for users (Fig. 1).

The above figure describes the proposed approach where the interaction between the client devices and the cloud provider is illustrated. The proposed approach is divided into four stages as described below:

1. **The connection between user and cloud**: In the first step, the request and response are shared between the client devices and the load balancer. Load balancer manages the request received by the cloud and sends the response back to the client.
2. **API Gateway Service**: The load balancer sends the received request from the client to the API gateway service. API gateway service manages which

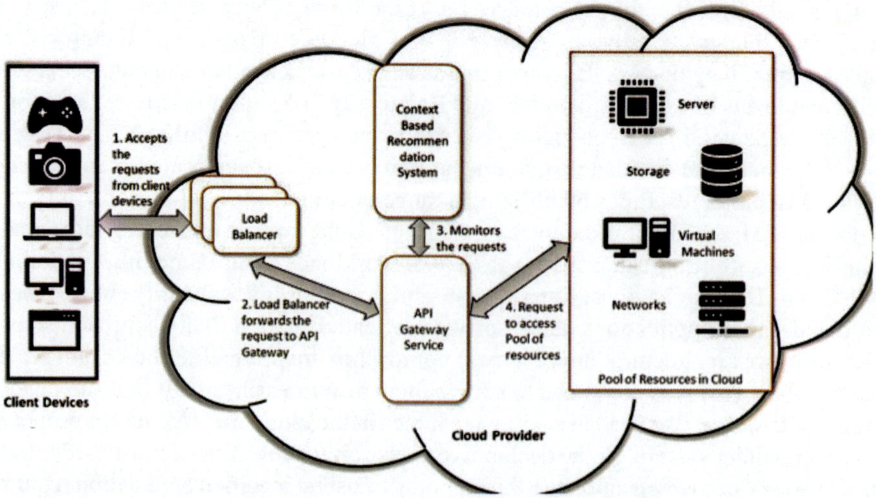

Fig. 1 Overview of methodology for context-based recommendation system

request is to be sent to which API. This service sends the request to two places simultaneously, first, the pool of resources, second, the recommendation system.

3. **Recommendation System**: The recommendation system receives the request from the API gateway service and monitors all the other requests. The recommendation system is used for analyzing the type of request, number of requests, and various other factors. Then, this data are fed to a recommender system model inside a processing engine, which analyzes the data and provides the information that if there is a need for scaling of resources for various users. The recommendation system is an independent service from the flow; it just receives the real-time request and stores them in its lightweight database but doesn't send any response immediately. Once the recommendation system analyzes the requests, it directly sends the mail to the user using the mailing engine highlighting the need for the resources and whether scaling of resources is required or not.
4. **The pool of resources**: API gateway service sends the same requests to the pool of resources that are accessible through APIs; the request is processed as usual here, and the request is sent back to the client.

4 Methodology

To understand the implementation of the above-discussed approach, let's first discuss the component diagram of the recommendation system and understand the functionality of each component in the proposed approach. The five basic components in the model used are as follows:

1. API Handler: This object of the recommendation system accepts the requests coming from the API gateway service of the cloud provider. The main use of this object is to accept the incoming requests and re-route them to the monitoring engine. There is no response sent back from the API handler of the recommendation system, as the recommendations are directly sent via mail to the customer.
2. Monitoring Engine: It is a utility that accepts requests from the API handler. Its main function is to monitor the requests of the client. This utility breakdowns the request into meaningful data. This utility refines any unnecessary data present in the request coming from the customer. This utility also pulls static factors like upcoming sale data or launch date of a new product from various sources for each customer. These static factors help the recommendation system present at the processing engine in analyzing and generating meaningful recommendations. The refined data from requests and the static data pulled are together sent to the database.
3. Lightweight Database: This component is the storage unit of the recommendation system; all the refined request data are stored in the database over a while for each user. This data are stored for further use to analyze trends and patterns from the data and in providing the customer with the recommendation for scaling of

resources. This database component stores various static factors for each user. This data are analyzed by the processing engine.

4. Processing Engine: It is the central processing unit (or brains) of the entire system. The processing engine is loaded with a pretrained recommendation machine learning model. This engine pulls the data for each user in a Round-Robin fashion from the database to analyze the data. The machine learning model deployed in the engine consumes this data and provides recommendations considering various contextual and real-time data. If the factor of recommendation is above a threshold, then that recommendation is eligible to be sent to the user. Once the recommendation is prepared, it is sent to the mailing server.
5. Mailing Server: It is a utility that runs an SMTP mailing server. The mailing server is responsible for mailing the users with the recommendations.

Each of the components as described above works together to provide real-time context-based recommendations to the user. As illustrated in Fig. 2, the entire process for providing recommendations is divided into six sub-processes; each of the processes is described below:

(1) In the first process, the request received by the cloud from the client devices is sent to the recommendation system by the API gateway service of the cloud. The request is received in the API handler of the recommendation system.

(2) The API handler re-routes the request to the monitoring engine. There are two major tasks of the monitoring engine that is first, to monitor the request received from the cloud and refine then by removing any unnecessary data from the request, and second is to pull various static factors for each user like upcoming sale data, the launch of a new product, and offers. The monitoring

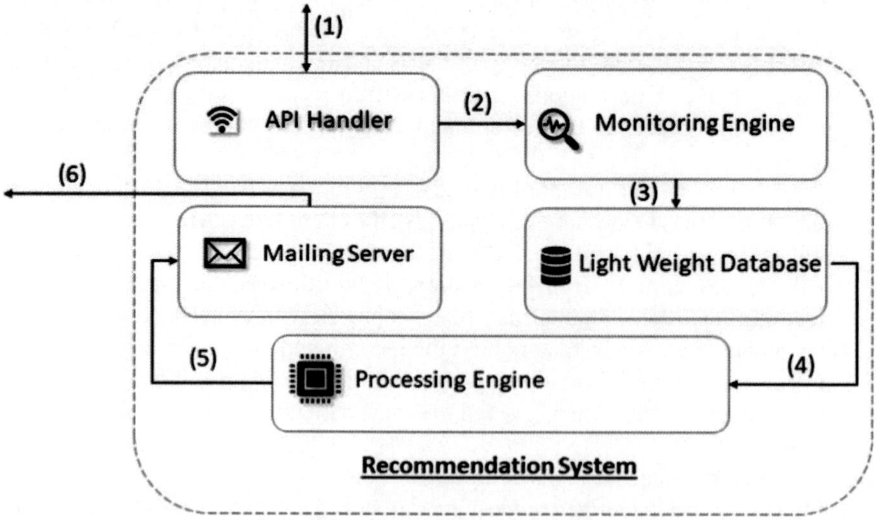

Fig. 2 Overview of implementation for context-based recommendation system

engine combines this data into a meaningful readable format and sends it to the database.

(3) The lightweight database stored the data received from the monitoring engine; this data are stored with an ID of the user being the key. The data from the database are pulled from the processing engine of the recommendation system.

(4) The recommendation system is the centralized and powerful processing unit of the entire system which is deployed on a server; the processing unit is loaded with a machine learning recommendation model, which in Round-Robin format pulls the data for each user from the database and feeds it to the recommendation system; based on the machine learning model, the real-time context-based recommendation is generated from the model. These generated recommendations have a score that signifies the depth and value of the generated recommendation. This score is compared to a threshold score, and if the score is higher than the threshold, then the generated recommendation is sent to the mailing server; otherwise, the recommendation is discarded.

(5) After processing in the processing engine, the eligible recommendations are sent to the mailing server. The mailing service is a running SMTP server that generated an email and sends it to the user. This email helps the user in making an informed decision whether to scale the cloud resources or not well in advance.

(6) Once the email is generated, it is sent out to the respective user through the mailing server APIs.

5 Conclusion

The approach described in this paper solves a major setback of existing models that is current models do not provide real-time and contextual recommendations to users on whether or not to scale cloud resources up or down even before the requirement has come. The discussed work proposes an approach to provide these recommendations well in advance by using the recommendation systems in the cloud environment. The model used in this paper monitors the real-time requests coming from the user and stores them in a database. This model also considers various other contextual factors like upcoming sale data, offers data, and new product launch data. These factors help the model to generate context-based recommendations. The user is informed about the generated recommendation via emails. This approach helps the user in taking an easy call to prevent any business loss well in advance so that when the demand increases/decreases, the user is ready with scaled resources to handle those requests.

References

1. "SP 800-145, The NIST Definition of Cloud Computing | CSRC." https://csrc.nist.gov/publications/detail/sp/800-145/final. Accessed 06 Aug 2021

2. Jansen WA. Cloud Hooks: security and privacy issues in Cloud Computing
3. Müller SD, Holm SR, Søndergaard J (2015) Benefits of cloud computing: literature review in a maturity model perspective. Commun Assoc Inf Syst 37(1):851–878. https://doi.org/10.17705/1CAIS.03742
4. Srivastava P, Srivastava P, Khan R (2018) A review paper on Cloud Computing. Int J Adv Res Comput Sci Softw Eng 8(6):17–20. https://doi.org/10.23956/ijarcsse.v8i6.711
5. Kim S-T (2011) Leading Korea's e-government advances. J E-Governance 34(4):165–165. https://doi.org/10.3233/GOV-2011-0271
6. Sala-Zárate M, Colombo-Mendoza L (2012) Cloud computing: a review of PAAS, IAAS, SAAS services and providers. Lámpsakos 7:47. https://doi.org/10.21501/21454086.844
7. Shahzadi S, Iqbal M, Qayyum ZU, Dagiuklas T (2017) Infrastructure as a service (IaaS): a comparative performance analysis of open-source cloud platforms. IEEE Int Work Comput Aided Model Des Commun Links Networks, CAMAD. https://doi.org/10.1109/CAMAD.2017.8031522
8. Bhatt B, Patel PJ, Gaudani H (2014) A review paper on machine learning based recommendation system. Int J Eng Dev Res 2:2321–9939. Accessed 06 Aug 2021. (Online). Available: www.ijedr.org
9. Fanca A, Puscasiu A, Gota DI, Valean H (2020) Recommendation systems with machine learning. In: Proceedings of 2020 21st international Carpathian Control conference ICCC 2020, Oct 2020. https://doi.org/10.1109/ICCC49264.2020.9257290
10. (12) (PDF) An Efficient Approach for Resource Auto-Scaling in Cloud Environments. https://www.researchgate.net/publication/305180276_An_Efficient_Approach_for_Resource_Auto-Scaling_in_Cloud_Environments. Accessed 06 Aug 2021
11. Marinescu DC, Paya A, Morrison JP, Olariu S (2017) An approach for scaling cloud resource management. Cluster Comput 20(1):909–924. https://doi.org/10.1007/S10586-016-0700-8
12. Kriushanth M, Arockiam L, Mirobi GL (2013) Auto scaling in Cloud Computing: an overview. Int J Adv Res Comput Commun Eng 2. Accessed 06 Aug 2021. [Online]. Available: www.ijarcce.com
13. Sadooghi I, Raicu I. Scalable resource management in cloud computing committee members
14. Rana ME, Farooq U, Rahman WNWAB (2019) Scalability enhancement for cloud-based applications using software oriented methods. Int J Eng Adv Technol 8(6):4208–4213. https://doi.org/10.35940/IJEAT.F8869.088619
15. Zhang M, Ranjan R, Nepal S, Menzel M, Haller A (2012) A declarative recommender system for cloud infrastructure services selection. Lect Notes Comput Sci (including Subser Lect Notes Artif Intell Lect Notes Bioinformatics) 7714:102–113. https://doi.org/10.1007/978-3-642-35194-5_8
16. Guo L, Luan K, Zheng X, Qian J (2021) A service recommendation method based on requirements for the cloud environment. J Control Sci Eng. https://doi.org/10.1155/2021/6669798

Traffic Density Estimation Using Transfer Learning with Pre-trained InceptionResNetV2 Network

Md. Nafis Tahmid Akhand, Sunanda Das, and Mahmudul Hasan

Abstract Traffic jam is a major problem in urban areas. It is a global-scale problem, and almost every country and every city face this to some extent. Traffic jam is a man-made problem. That means lack of proper planning is one of the root causes of this problem. And, that is why the problem of traffic jams can be avoided with proper planning and optimizing the flow of traffic. An intelligent traffic management system tries to manage the traffic flow of the area it covers most efficiently. Such a system needs to measure the traffic density of the roads and use that data to optimize the traffic flow throughout the area. This study attempts to solve the first problem. To accomplish the task, transfer learning and different pre-trained backbone networks such as ResNet-50, ResNet-101, MobileNet, Xception, and InceptionResNetV2 have been used in our study. The proposed method determines the traffic density of the roads using images from traffic cameras. The necessary data for this study were obtained from an open-source API provided by Land Transport Authority, Singapore. Various deep learning models were investigated for determining traffic density. Different backbone networks were tried, and among them, InceptionResNetv2 achieved the best accuracy of 84.69% and top 2 accuracy of 97.50%.

Keywords Image processing · Traffic data analysis · Transfer learning · Deep learning

1 Introduction

Transfer learning is a very popular method in deep learning. This approach is generally used in natural language processing and computer vision-related works where a pre-trained model trained on a different but related dataset is used as the starting point [1]. The weights from the previous dataset are saved; then, the researchers use

Md. N. T. Akhand · S. Das (✉) · M. Hasan
Department of Computer Science and Engineering, Khulna University of Engineering & Technology, Khulna 9203, Bangladesh
e-mail: sunanda@cse.kuet.ac.bd

V. Skala et al. (eds.), *Machine Intelligence and Data Science Applications*, Lecture Notes on Data Engineering and Communications Technologies 132,
https://doi.org/10.1007/978-981-19-2347-0_28

those weights and this pre-trained model on their own dataset as a backbone network [2]. This approach is used to overcome the hassle of developing neural networks from scratch and to acquire better results. Transfer learning can help to build on previous experiences which can aid in acquiring a robust feature extraction. That is why transfer learning has been used in our study for estimating the traffic density of the roads.

A smart traffic control system estimates traffic density on the roads using the input which is taken from various sensors. By using these data, the system is able to provide an optimal traffic signal control system [3]. Such an intelligent system considers both local traffic density and the condition of the vehicles of a whole city or entire region it covers, and thus, it provides the most robust and efficient traffic management system to manage a constant flow of cars and is able to reduce traffic congestion [4]. Hilmani et al. [5] showed that an intelligent system not only controls the traffic but also provides suggestions to the drivers which can help them reach their destination more quickly by avoiding heavy traffic congestion. In other words, it can help them to find the best route for their journey.

Most developing country in the world suffers from traffic congestion on both financial and social levels. According to a recent survey, five million hours of labor are lost every year and Tk 370 billion is squandered per annum due to traffic jams [6]. Furthermore, people face a wide range of issues, from being late for crucial meetings to more severe issues such as being trapped in rush hour gridlock with a patient or an expectant mother in an emergency ambulance [7]. Even it is not unheard of for pregnant women to give birth within an ambulance when stopped in traffic [8]. As a result, building an intelligent system that is able to maintain an effective flow of traffic while avoiding traffic congestion is badly needed for a developing country like Bangladesh as well as any other country.

Building up a smart traffic management system is divided into two sections. The first section is to determine the road's traffic density. And, the second section is to regulate traffic flow effectively by using the evaluated data. The first section of this problem, determining the density of traffic on the roadways, is addressed in this study. To obtain the system, different backbone networks were used. These networks were trained on the collected dataset. Finally, the proposed model with InceptionResNetV2 obtained the highest accuracy. Images taken from traffic cameras are used as input to determine the number of incoming cars. It is possible to determine the incoming road's traffic density quite accurately by using this approach.

2 Related Works

As traffic jam is one of the major problems in all big cities, many methods for estimating traffic density have been presented to resolve this issue.

Chowdhury et al. [9], Patekar et al. [10], and Abbas et al. [11] considered a particular direction of the road for evaluating traffic density. Their proposed system worked by comparing the real time frame of live video with the reference image and by searching vehicles only in the region of interest (i.e., road area) to reduce the traffic congestion. Chowdhury et al. [9] collected video samples, and then background subtraction was applied; in both Chowdhury et al. [9] and Patekar et al. [10], the former also applied foreground detection using the Mixture of Gaussian (MoG) algorithm. Abbas et al. [11] divided their model into four parts. It included processing video signal and images acquisition, cropping the images, and object detection, and then, finally traffic density was counted. The numbers of cars in each lane were counted based on their line of centroid at the exit region. MATLAB was used for processing video signals and image cropping. Binarization of different images along with erosion, dilation, and morphological closure operations were used to enhance the presence of objects.

Uddin et al. [12] and Walid et al. [13] applied image processing technique for counting vehicle's number for the measurement of traffic density. The method of Uddin et al. [12] is mainly divided into three steps, and they are empty road analysis, traffic road analysis, and making the decision. The cropping image tool was used to remove unwanted errors. The canny method was applied for detecting the edges of vehicles. Walid et al. [13] used support vector machine (SVM) for analyzing their result. They only estimated low and high density. They defined their processes as three phases. At first, suitable images from roadside cameras were selected for analysis. They used erosion and dilation for image processing. For counting vehicles in the image object detection technique was applied. Finally, the SVM classification algorithm is used for result representation. A threshold value was set here to compare the high and low density. Traffic management system, roadway system, MATLAB, and SVM classifier algorithm were needed for their proposed model. The model gained 60% accuracy.

Nubert et al. [14] proposed a traffic density estimation model using convolutional neural network. The dataset they used for making the model is from monitoring traffic data provided by the Land Transport Authority (LTA), Singapore. They divided their algorithm into two parts. At first, the model receives live camera images for determining the traffic density of every lane, and then, it sets up the optimal traffic light state based on traffic density. They designed the problem as a multi-class classification problem. The camera images were classified into five different traffic density classes. They got 71.35%, 73.0 %, and 74.3% accuracy for basic CNN algorithm, basic CNN with class imbalance method, and basic CNN with CI measures and masking method, respectively. InceptionV3 was also applied, and they obtained 66.38% accuracy. They also evaluated these classifiers and methods on F1-scores and top 2 accuracies matrices. Aimsun and SUMO were used for simulation.

Ikiriwatte et al. [15] proposed a traffic signal management system using CNN. They used traffic images that were collected from Kirulapone junction, Sri Lanka. They used YOLOV3 for vehicle density estimation. For estimating the crowd density at a junction, they used TensorFlow object detection API. They also identified the vehicles violations and the crowdsourced application. Finally, they designed an optimal traffic lights controlling system at a four-way junction based on vehicle density estimation and crowd density estimation using Q-Learning. SUMO was used for testing the proposed model's accuracy. They trained the models using ResNet-152,

ResNet-101, DarkNet-53, and DarkNet-19. Then, some performance metrics were used to evaluate those models. In their study, the model with DarkNet-53 attained better results compared to the other models.

In this study, transfer learning is applied with the help of InceptionResNetV2 to correctly estimate traffic density.

3 Methodology

Images from traffic cameras mounted at the intersections are loaded into the system for determining the density of the oncoming lanes. The framework of the whole method is illustrated in Fig. 1.

Here, all the pre-trained backbone networks were trained using the ImageNet dataset. Some adaptations had to be made for using these models to recognize the traffic density classes. First, all the upper layers of these pre-trained models were frozen. Then, new flattening and dense layers were added at the end. Lastly, these newly adapted models were trained with our train dataset and the recognition models were created. Then, our test dataset was used on these recognition models to test their accuracy in predicting traffic density based on the number of vehicles on the road.

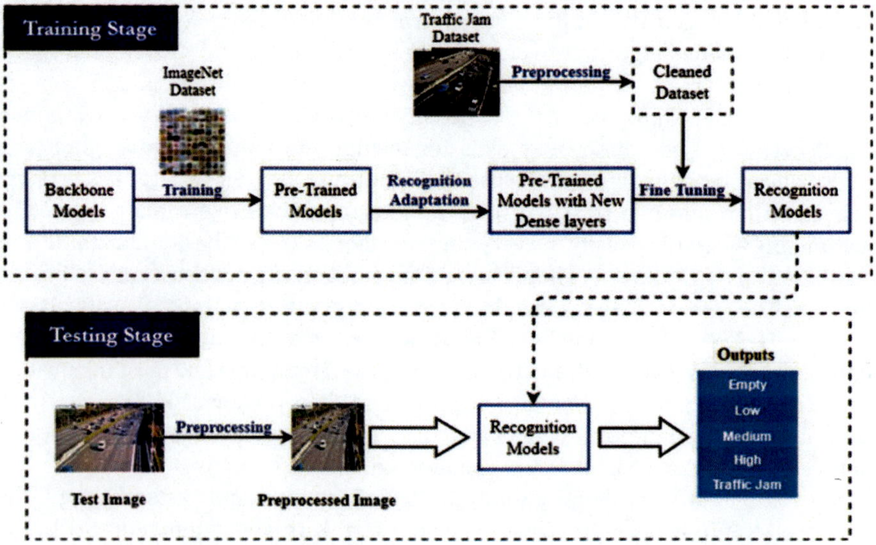

Fig. 1 Framework of the proposed method

Table 1 Traffic density class definitions

Classes	Explanation	Definition
Empty	Almost empty street	0–5 Cars
Low	Street has a few vehicles	6–15 Cars
Medium	There are some vehicles on the street	16–30 Cars
High	Almost filled street	31–50 Cars
Traffic jam	The vehicles on the road almost aren't moving	>50 cars

3.1 Dataset

Live images from traffic cameras mounted at the road intersections are taken as input in our system. So, the model had to be trained with such images. From an open-source API provided by LTA, Singapore, these necessary images for training the model were collected [16]. They have images from 87 cameras situated at different intersections of the city. Among them, 20 cameras were selected for our dataset based on image clarity and variation in traffic density. Our dataset contains a total of 4048 images. These were randomly divided as 80% for the training set and 10% for the validation and testing sets each. Images from both day and night times as well as different traffic densities were collected. The dataset was labeled into five different classes which are defined in Table 1.

3.2 Image Augmentation and Preprocessing

Overfitting is a common problem for deep learning models. So, there are various ways to solve this problem. A very common way of solving this problem is image augmentation. Augmenting the training images ensures that there is less class imbalance in the training dataset as artificial data instances are created on the fly with the help of augmentation. It also provides a diverse set for the model to train. In this way, the model is better suited to be applied in real-life situations where its input would certainly be varied. For our dataset, various data warping techniques like horizontal flip, changing the brightness of the image, zooming, shearing, etc., were used. Figure 2 illustrates the performed image augmentations on our dataset in detail. It can be seen that random images were zoomed by 50%. Shearing was performed on some random images about 20% of their original state. Horizontal flipping was performed randomly, and also, the brightness of some random images was increased up to two times.

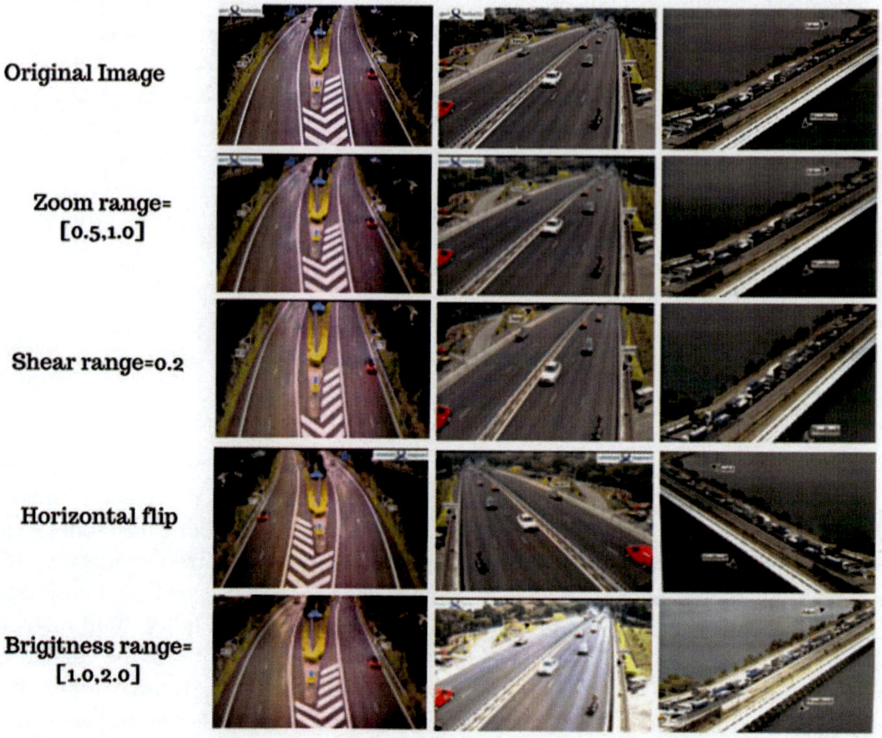

Fig. 2 Illustration of different augmentation techniques used in our method

The images were also preprocessed to some extent to fit the model. All the images were rescaled by diving them with 255. The color mode was set to RGB, and reshaping of the images was done as 150 × 150. Also, the images were randomly shuffled to avoid the overfitting of the model.

3.3 Model Overview

The system takes RGB images as input in the form of a 150 × 150 array. Then, those images are fed through some pre-trained models to see their performance. Finally, they give an output which represents the previously defined five classes of the system. Figure 3 shows the block diagram of the method. And, related information about backbone networks is shown in Table 2.

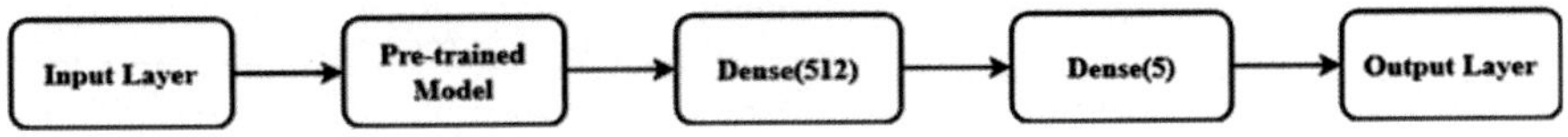

Fig. 3 Block diagram of the method

Table 2 Related information about backbone networks

Backbone networks	Input shape	Total parameters
ResNet-50	150 × 150 × 3	26,213,253
ResNet-101	150 × 150 × 3	43,709,829
MobileNet	150 × 150 × 3	4,805,829
Xception	150 × 150 × 3	21,913,133
InceptionResNetV2	150 × 150 × 3	55,126,245

3.4 Evaluation Metrics

Some evaluation metrics were used to assess the performance of the models. These are accuracy, top 2 accuracy, precision, recall, and F1-score. The precision–recall curve and receiver operating characteristic (ROC) curve were generated. All of these evaluation metrics and curves use various ways to show the ratios of true-positive (TP), true-negative (TN), false-positive (FP), and false-negative (FN) values predicted by the models [17]. The confusion matrix also illustrates these values with the help of a 2D matrix of $l \times l$.

$$\text{Precision} = \frac{\sum_{i=1}^{l} \text{TP}_i}{\sum_{i=1}^{l} \text{TP}_i + \text{FP}_i} \tag{1}$$

$$\text{Recall} = \frac{\sum_{i=1}^{l} \text{TP}_i}{\sum_{i=1}^{l} \text{TP}_i + \text{FN}_i} \tag{2}$$

$$\text{F1-score} = \frac{2 * \text{Precision} * \text{Recall}}{\text{Precision} + \text{Recall}} \tag{3}$$

$$\text{Accuracy} = \sum_{i=1}^{l} \frac{\text{TP}_i + \text{TN}_i}{\text{TP}_i + \text{FN}_i + \text{FP}_i + \text{TN}_i} \tag{4}$$

Here, i denotes the matrix's row number and l represents the total number of classes.

Top 2 Accuracy: Instead of predicting direct true or false values, softmax predictions predict in the form of probability distribution. The class with the highest probability in the distribution is chosen as the predicted class. In top N accuracy, if the probability of a class falls under the top N values of the distribution, then that class is deemed to be correct instead of just the top one class. Finally, the ratio of correct predictions (classes that fall under the top n values of the softmax distribution) over incorrect predictions (classes that do not fall under the top n values of the softmax distribution) is calculated and that is called top N accuracy. In case of top 2 accuracy, the value of $N = 2$.

4 Result Analysis

Five different backbone networks were used. These models were trained with our dataset to achieve the coveted method. Among them, method InceptionResNetV2 achieved the highest accuracy of 84.69%. The loss versus epoch and accuracy vs. epoch graphs of the five different models are presented in Fig. 4a, b respectively. After careful inspection of Fig. 4, it is evident that although the method with MobileNet achieved the lowest loss and highest accuracy during the training process, after few epochs, the validation loss starts to increase, as well as the validation accuracy remains approximately constant which indicate the overfitting aspects of the model. The other four models, on the other hand, are free of overfitting since loss decreases and accuracy increases with each epoch during both the training and validation phases.

The confusion matrices of all five models are presented in Fig. 5. In Fig. 5e, the confusion matrix of the method (InceptionResNetv2) can be seen. Among the 64 images of the traffic jam class, the model accurately predicts 59 of them as true positive. Among the other five images, one is predicted in medium traffic density and the other four are classified in the high traffic density class which is the closest class to the traffic jam class. Similarly, the true-positive, false-positive, true-negative, and false-negative values of the other four classes can be interpreted. Figure 5a–e illustrates the confusion matrices of the other four models.

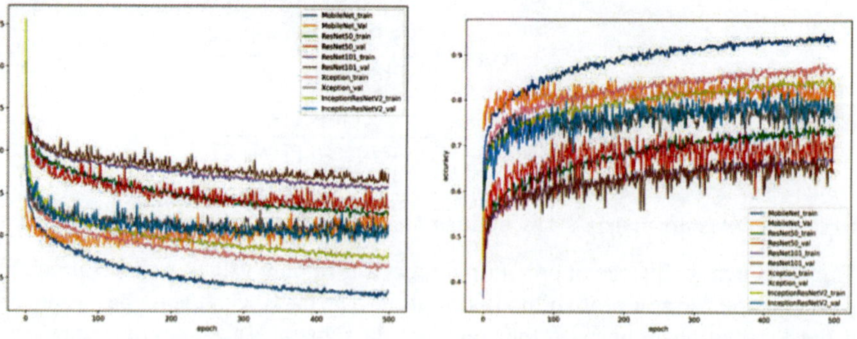

(a) loss vs epoch graph (b) accuracy vs epoch graph

Fig. 4 Loss vs epoch graph and accuracy vs epoch graph during training and validation phase for five different models

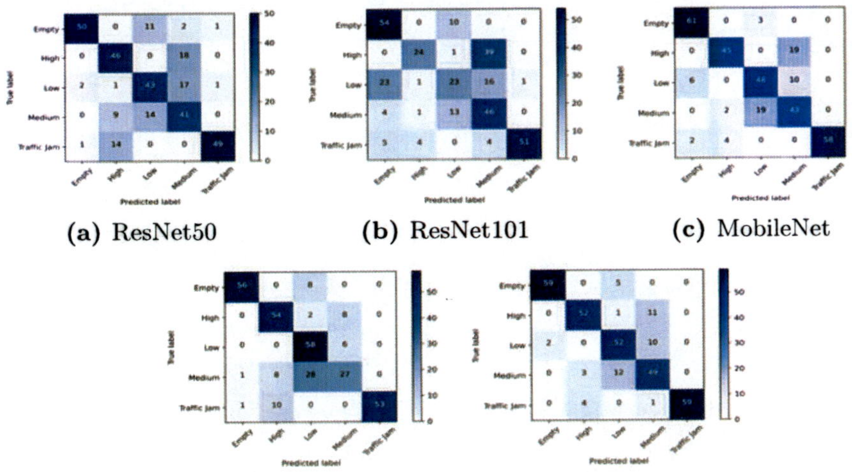

(a) ResNet50 (b) ResNet101 (c) MobileNet

(d) Xception (e) InceptionResNetV2

Fig. 5 Confusion matrix of five different models

The five models were evaluated on a few metrics: precision, recall, accuracy, F1-score, and top 2 accuracy. The individual class performance for the precision, recall, and F1-scores of the five models is shown in Table 3. Method InceptionResNetV2 achieved higher and better results on all the evaluation metrics over the other four models as well as the other previous works done on this topic. The results are presented in Table 4.

Figure 6a–e illustrates the receiver operating characteristic (ROC) curves of the five different models. Figure 6e shows the ROC curve produced by the method InceptionResNetV2. Along the x-axis of the curve, the false-positive values of the individual classes are plotted, whereas the true positives along with micro and macro-average are plotted along the y-axis. The area under curve (AUC) shows promising results as almost all of them are above 0.9 and very close to 1. It means most of the images provided for testing were correctly predicted by the method.

Figure 7a–e illustrates precision–recall curves of the five different models. Figure 7e illustrates the precision–recall curve of method InceptionResNetV2 where along the x-axis the recall values of individual classes are plotted. And, along the y-axis, precision values along with the micro-average values are plotted. The classes have a high micro-average value of 91.3%.

Table 3 Precision, recall, and F1-score for the five different classes of the five different backbone networks

Methods	Classes	Precision (%)	Recall (%)	F1-score (%)
ResNet50	Empty	94.00	78.00	85.00
	Low	63.00	67.00	65.00
	Medium	53.00	64.00	58.00
	High	66.00	72.00	69.00
	Traffic Jam	96.00	77.00	85.00
ResNet-101	Empty	63.00	84.00	72.00
	Low	49.00	36.00	41.00
	Medium	44.00	72.00	54.00
	High	80.00	38.00	51.00
	Traffic Jam	98.00	80.00	88.00
MobileNet	Empty	88.00	95.00	92.00
	Low	69.00	75.00	72.00
	Medium	60.00	67.00	63.00
	High	88.00	70.00	78.00
	Traffic Jam	100.00	91.00	95.00
Xception	Empty	97.00	88.00	92.00
	Low	60.00	91.00	72.00
	Medium	66.00	42.00	51.00
	High	75.00	84.00	79.00
	Traffic Jam	100.00	83.00	91.00
InceptionResNetV2	Empty	97.00	92.00	94.00
	Low	74.00	81.00	78.00
	Medium	69.00	77.00	73.00
	High	88.00	81.00	85.00
	Traffic Jam	100.00	92.00	96.00

5 Conclusion

Traffic jam is a major problem for Bangladesh. As more and more urbanization took place throughout the country in the last few decades, traffic jam also increased rapidly with that. It is not only a problem in Bangladesh; rather almost every country suffers from it to some extent. That is why many pieces of research have been done and are currently ongoing throughout the world to solve this problem. This study attempts to provide an effective solution to the traffic density estimation problem where the proposed method with InceptionResNetV2 pre-trained on 'ImageNet' outperformed previous similar methods and was able to generate a better result with respect to accuracy, precision, recall, F1-score, and top 2 accuracy. The highest accuracy of 84.69% was achieved by the method InceptionResNetV2. It had a top 2 accuracy of 97.50%. It means, the traffic density of a road can be predicted accurately more than

Table 4 Performance comparison between different models

Methods	Precision (%)	Recall (%)	Accuracy (%)	F1-score (%)	Top 2 accuracy (%)
Basic CNN model[14]	–	–	73.0	80.52	93.23
Basic CNN with CI measures & masking[14]	–	–	74.3	81.3	94.00
ResNet-50	74.40	71.60	71.56	72.40	95.00
ResNet-101	66.80	62.00	61.88	61.20	89.38
MobileNet	81.00	79.60	79.69	80.00	97.19
Xception	79.60	77.60	77.50	77.00	96.25
InceptionResNetV2	**85.60**	**84.60**	**84.69**	**85.20**	**97.50**

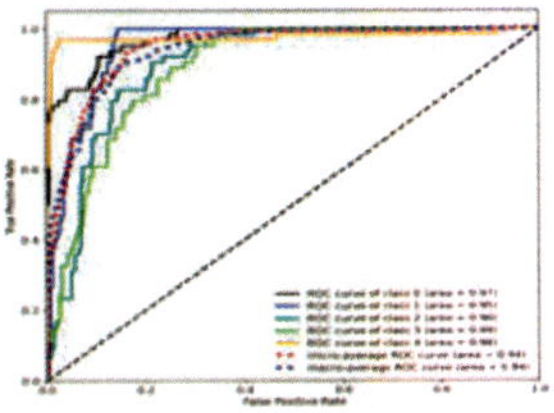

(a) ResNet50

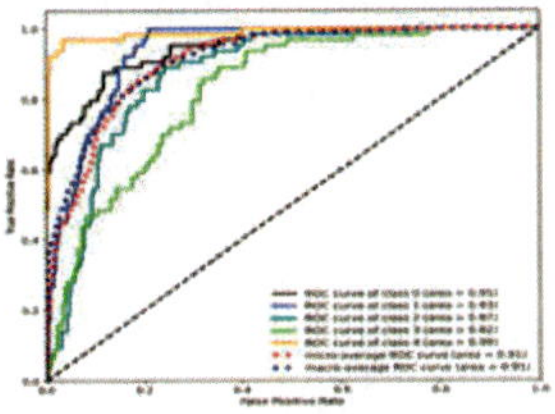

(b) ResNet101

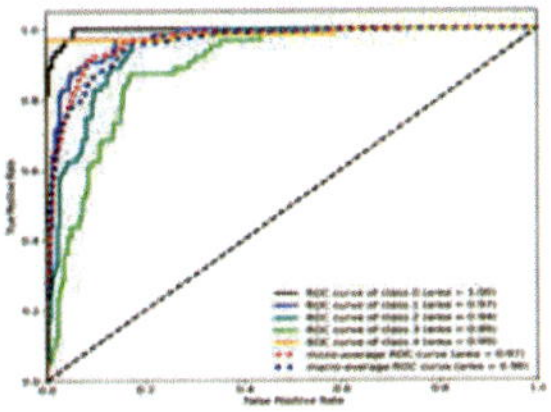

(c) MobileNet

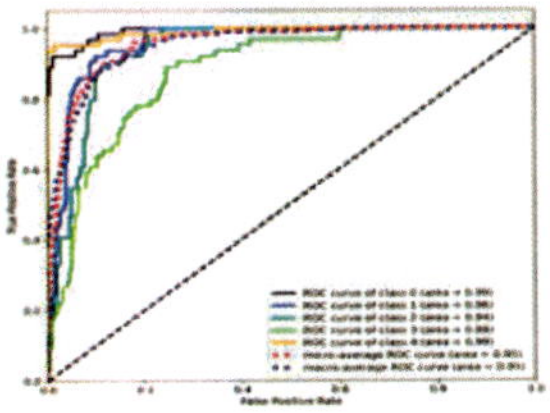

(d) Xception

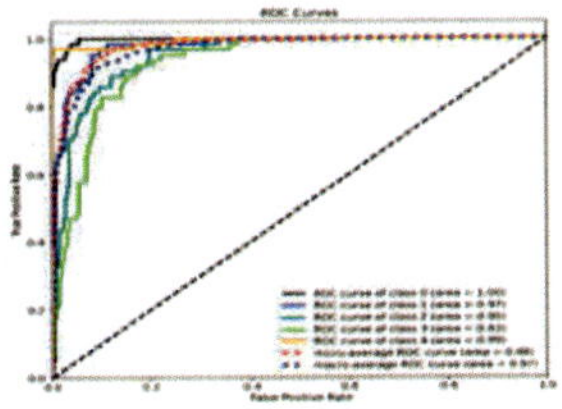

(e) InceptionResNetV2

Fig. 6 ROC curve of five different models

84% of the time by the model, and in its top 2 predictions, it can predict the traffic density more than 97% of the time. If a model predicts traffic jam as high traffic density or vice versa, then these predictions are close enough that they can work in a real-life scenario. And, a top 2 accuracy of 97.50% signifies just that. So, it can be said that the proposed model is quite dependable and can predict traffic density fairly accurately.

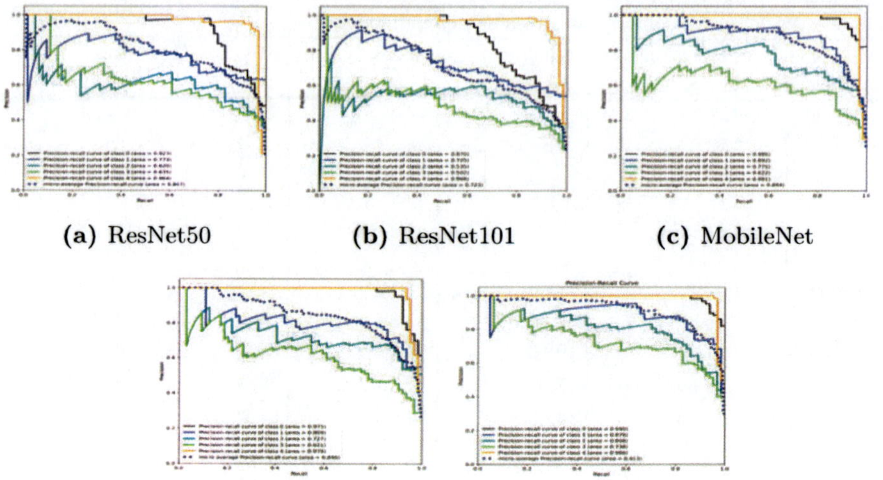

(a) ResNet50 (b) ResNet101 (c) MobileNet

(d) Xception (e) InceptionResNetV2

Fig. 7 Precision–recall curve of five different models

References

1. Weiss K, Khoshgoftaar TM, Wang D (2016) A survey of transfer learning. J Big Data 3(1):1–40
2. Pan SJ, Yang Q (2009) A survey on transfer learning. IEEE Trans Knowl Data Eng 22(10):1345–1359
3. Khanna A, Goyal R, Verma M, Joshi D (2018) Intelligent traffic management system for smart cities. In: International conference on futuristic trends in network and communication technologies. Springer, Berlin, pp 152–164
4. Mandhare PA, Kharat V, Patil C (2018) Intelligent road traffic control system for traffic congestion a perspective. Int J Comput Sci Eng 6(07):2018
5. Hilmani A, Maizate A, Hassouni L (2020) Automated real-time intelligent traffic control system for smart cities using wireless sensor networks. In: Wireless communications and mobile computing
6. Kibria A. The heavy cost of traffic congestion. https://thefinancialexpress.com.bd/views/the-heavy-cost-of-
7. Chowdhury PP. Caught between life and traffic. https://www.thedailystar.net/star-weekend/city/news/caught-between-life-and-traffic-1798837
8. Stuck in tailback, woman gives birth on road in tangail. https://www.thefinancialexpress.com.bd/public/national/country/stuck-in-tailback-woman-gives-birth-on-road-in-tangail-1559655312
9. Chowdhury MF, Biplob MRA, Uddin J (2018) Real time traffic density measurement using computer vision and dynamic traffic control. In: 2018 Joint 7th International Conference on Informatics, Electronics & Vision (ICIEV) and 2018 2nd International Conference on Imaging, Vision & Pattern Recognition (icIVPR). IEEE, pp 353–356
10. Patekar AR, Dewan JH, Umredkar SA, Mohrir SS (2018) Traffic density analysis using image processing. Int J Comput Appl
11. Abbas N, Tayyab M, Qadri MT (2013) Real time traffic density count using image processing. Int J Comput Appl 83(9):16–19

12. Uddin MS, Das AK, Taleb MA (2015) Real-time area based traffic density estimation by image processing for traffic signal control system: Bangladesh perspective. In: 2015 International Conference on Electrical Engineering and Information Communication Technology (ICEEICT). IEEE, pp 1–5
13. Hosne-Al-Walid NA, Tubba U, Akter L, Asfad Z (2015) Traffic density measurement using image processing: an svm approach. Traffic 3(6)
14. Nubert J, Truong NG, Lim A, Tanujaya HI, Lim L, Vu MA (2018) Traffic density estimation using a convolutional neural network. arXiv preprint arXiv:1809.01564
15. Ikiriwatte A, Perera D, Samarakoon S, Dissanayake D, Rupasignhe P (2019) Traffic density estimation and traffic control using convolutional neural network. In: 2019 International Conference on Advancements in Computing (ICAC). IEEE, pp 323–328
16. Authority LT. Traffic images. https://data.gov.sg/dataset/traffic-images
17. Sokolova M, Lapalme G (2009) A systematic analysis of performance measures for classification tasks. Inf Process Manage 45(4):427–437

A Mobile Application for Sales Representatives: A Case Study of a Liquor Brand

Sharon Xavier, Saroj Kumar Panigrahy, and Asish Kumar Dalai

Abstract In this paper, an efficient system is presented which can be used by sales representatives as a mobile application for accurate and efficient collection of data. The mobile application created can place sales orders, track inventory stock, and conduct surveys. It makes use of GPS location which tracks the user's location and maintains logs for each one of them. Using the GPS location of the user, the store details are automatically filled into the forms from the database to avoid any errors. The barcode scanner allows the user to enter or view product information by scanning the item's barcode. Various reports are also generated from the data collected by the application which can be used for devising marketing strategies. The application can also automatically place sales orders when a product quantity is below a certain requirement. In addition to this, the application also stores the contact information of the company's customers. The online feedback form is used to collect information about the end-user's preferences about the product like—preferred taste, price, quality, quantity, etc. This information helps the company to manufacture products which are much more desired by the consumers, thereby increasing the sales and profits of the company. Sales forecasting is one of the crucial aspects when it comes to sales and marketing as it helps to understand the consumer's buying patterns. Hence, machine learning algorithm is used to predict the future sales of various brands from the data collected.

Keywords Mobile application · Sales · Prediction · Forecasting · Machine learning · SARIMAX

1 Introduction

The advent of mobile computing has opened up applications of this technology in several areas of the enterprise. The ability to check for information on the fly or

S. Xavier · S. K. Panigrahy (✉) · A. K. Dalai
VIT-AP University, Amaravati, Andhra Pradesh 522237, India
e-mail: saroj.panigrahy@vitap.ac.in

V. Skala et al. (eds.), *Machine Intelligence and Data Science Applications*, Lecture Notes on Data Engineering and Communications Technologies 132,
https://doi.org/10.1007/978-981-19-2347-0_29

execute transactions and display analytics for decision-making has made mobile-based applications, a powerful tool in all sectors of the industry. Sales personnel needs to check stock to promise customers whether they can deliver on a particular date, check for invoices which need to be collected when they are visiting an area. Purchasing department can use it to trigger a workflow to approve purchase orders (POs), and the very approval itself can be done on the app. Quality personnel can check materials and update their usage decision (passed or failed) using their mobile app wherever they are. Plant maintenance technicians can trigger a notification if anything breaks down or if there is an emergency situation. Employees can use a mobile app to file a leave application, and the manager can use it to approve it. These are all valid scenarios where mobile apps deliver value by giving the users the flexibility to carry on business as usual and providing a user interface, leveraging the power of the mobile application.

This paper addresses one area of the enterprise detailed above, i.e., sales and distribution, and attempts to provide a solution to solve some of the problems faced by the sales personnel in the field. While this paper has been inspired by a real case business scenario of the tender requirements of a liquor manufacturing company, the application can be very easily applied with minor edits to real-time scenarios such as retail, manufacturing, or service industries.

Angostura products were not just used for medicinal purposes but also for cooking and a base to many cocktails [1]. With increase in demand, Angostura implemented an IOS-based SAP Mobile Sales Solution, in the year 2013. This seemed useful for a while but with the IOS update and lack of vendor commitment, the solution could no longer be maintained. The company has since adopted a manual process to ensure that the business continues to service its customers. Salespersons either submit a sales order form (via e-mail or manually) or call the customer service representatives who takes their orders. While this approach works, it has resulted in a reduction in the efficiency of the sales order process. Additionally, the lack of a mobile solution results in error-prone entries, loss of real-time processing of orders, and inaccurate information.

Due to the above drawbacks, development of a mobile application for the sales representatives was implemented. The following are the objectives of the same:

- To design an application which allows only users with valid credentials to access it.
- To design a system which is capable of storing the data collected.
- To provide a list of stores closest to the user so that he/she can schedule their visits accordingly.
- To record the date and time of the sales representative when they visit a store.
- To display the distance and time necessary to reach the store from the user's current location.
- To maintain a list of wholesalers.
- Allow the user to add new contacts.
- Allow the sales representative to place sales/return orders.
- Ability to track in-store inventory.

- Allow for in-store customer survey.
- Generate reports based on the collection and analysis of data.
- Enable the store to automatically place a sales order when the product quantity is below the minimum requirement level.
- Ability to scan a product by its barcode and display/save the product information.
- Help to forecast the sales based on the data collected.

The rest of the paper is organized as follows. Section 2 describes the background and related works. Section 3 explains the proposed system and working methodology. Section 4 describes the software and hardware details of the mobile application system. Section 5 discusses the results obtained after the project was implemented. Finally, Sect. 6 concludes the paper with a scope for future development.

2 Background and Related Works

Kurniawan has implemented the representational state transfer (REST) Web service for mobile-based sales order and sales tracking [2]. It consists of a mobile and Web application. The Web application has been developed using ASP.NET model, view, and controller (MVC), and the mobile application has been developed using Android. The mobile app consists of the following: a check-in feature, barcode reader, and global positioning system (GPS) locator. The check-in feature is used to determine whether a salesperson has visited a particular store or not. The barcode reader is used to place sales orders by reading barcodes, and the GPS locator is used to track the location of the user. The Web application is mainly developed for the company to keep track of the salesperson's monthly visit to store and to monitor sales orders and transactions.

Khawas and Shah have studied the application of Firebase in Android application development [3]. Their study shed light on the features and simplicity of using Firebase which is mobile platform of Google that helps in developing high-quality apps quickly [4]. The study also compared Firebase with other known databases on various parameters. Firebase is a cloud-based database which makes use of JavaScript Object Notation (JSON) text. The data stored in Firebase are unstructured when compared to the structured data in relational database management systems (RDBMS). Daniel Pan has showed how an Android app is connected to Firebase [5]. He also showed the basics of designing the structure in Firebase.

Nari et al. have developed an application of seasonal autoregressive integrated moving averages with exogenous regressors (SARIMAX) model to forecast daily sales in food retail industry [6]. The authors have taken the basic SARIMAX model [7] which is a machine learning (ML) algorithm used for data with seasonal trends and tried to improve it by considering other factors such as seasonality, holidays effects, and price reduction effects.

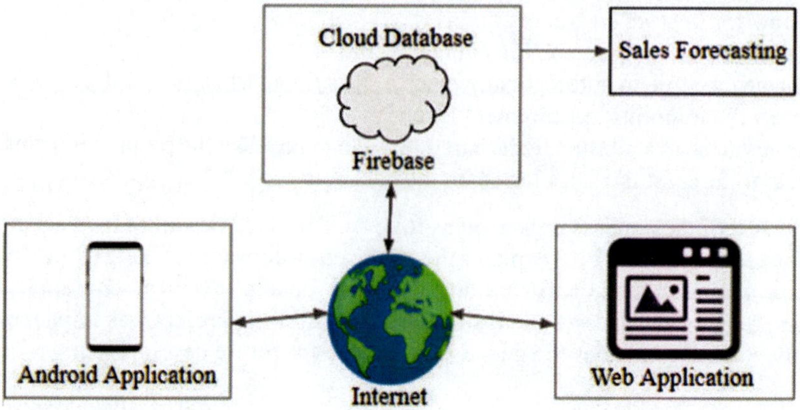

Fig. 1 System block diagram

3 Mobile Application for a Sales Representative

This section describes the proposed system, working methodology, and standards details.

3.1 Proposed System

This system consists of one main section which includes the software. The software includes a mobile application, a Web page, and sales forecasting using machine learning. The hardware part is in the form of a mobile phone and is used for deploying and testing the app. The system architecture of the project is depicted in Fig. 1.

3.2 System Architecture

The mobile application is made using Android Studio [8]. Multiple forms created are used for collecting various informations like order details, inventory, competitor prices, etc. The data entered via these forms are stored in firebase which can be retrieved for analysis by the marketing team. The data collected are also used for generating various reports such as—overdue sales report, delivery report, and inventory report which gives an overview of the sales and consumer behavior.

The Webpage is an online feedback form which is created using Microsoft Visual Studio [9]. The form allows the end-users to give feedback regarding their preferred products like taste, smell, quantity, alcohol content, etc. With this information, company officials can decide what kinds of products to make so that their sales and profits can be maximized.

The sales forecasting for different brands of liquors is done using the SARIMAX method which is a time series forecasting method used with data containing trends and seasonality. The dataset is first tested to see if any seasonal trend is present. The values which give the lowest Akaike information criteria (AIC) values are selected and are considered as inputs into the SARIMA model for predicting values. A graph is plotted with the predicted values to determine whether there is an increase or decrease in the sales. Apart from sales, this data can also be used by manufacturers to determine how much quantity of a product should be produced in the coming months.

4 Mobile Application Implementation Details

This section describes the software and hardware details for the implementation of the proposed system.

4.1 Software Details

The main software in this application includes an Android app, Web page, and Firebase. In addition to this, a machine learning algorithm is used to predict the sales.

Android Application: The Android application is built using Android Studio which is the official integrated development environment (IDE) for development of Android applications. It allows a developer to create applications which are compatible with various mobile devices. The application consists of the following modules—login page, menu screen, contacts, GPS location, logs, sales order form, inventory form, survey form, reports, and barcode scanner. GPS shows the user's current location and also the stores in the vicinity of the user. The scanner scans a product's barcode and displays information regarding the product such as name, manufacturer, and price. A similar scanner present at the store's end deducts the item's quantity from the inventory when purchased and automatically places a sales order for that particular item when it is below the minimum requirement level.

Web Application: The Web application is developed using Visual Studio and hosted in localhost using Apache XAMPP server [10].

Sales Forecasting: Future sales prediction of various brands of liquors is done by SARIMAX model using Google Colab [11].

Firebase: In this application, both mobile and Web application are connected to Firebase. Firebase is used for storing information from the app and the Web page. It is also used for displaying data on the application.

4.2 Integration of Hardware and Software

This project is completely software based, and the only hardware used is a mobile phone with Android operating system which is used for deploying and testing the application. To deploy our application in the mobile device, the mobile device is connected to the development machine and is integrated to run the app and store data.

5 Results and Discussion

This section describes the results of sales prediction and forecasting of different liquor products after implementing the mobile application.

5.1 Results of Sales Forecasting

The data required for carrying out a time series forecasting are supposed to be large as the trends are considered on a yearly basis. A dataset called "Iowa liquor sales" [12] published by the state of Iowa is taken for this purpose which contains information about the name, price, kind, quantity, and location of sales of individual containers or packages of containers of alcoholic beverages. The required data are extracted from this dataset and are used for predicting the sales [13]. Figure 2 depicts the sales of three different liquor brands.

Trends in rum, bitters, and coffee liqueurs are plotted in Figs. 3, 4, and 5, respectively. Then, it is checked that whether the data are having any seasonal trend. Once

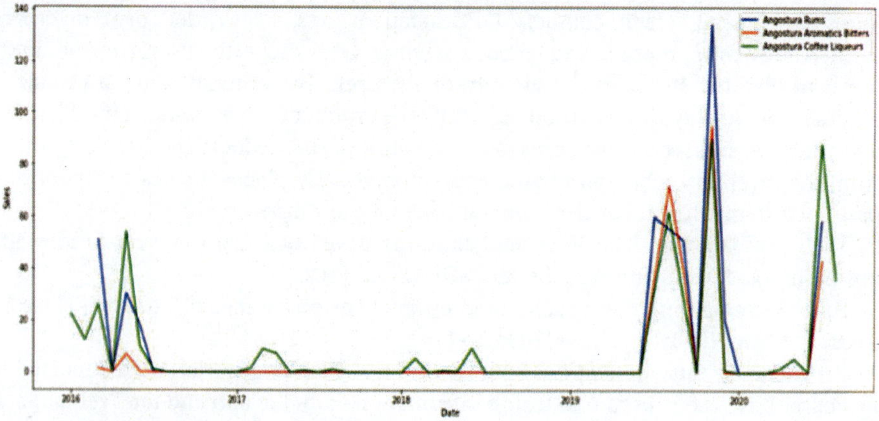

Fig. 2 Sales of liquor brands

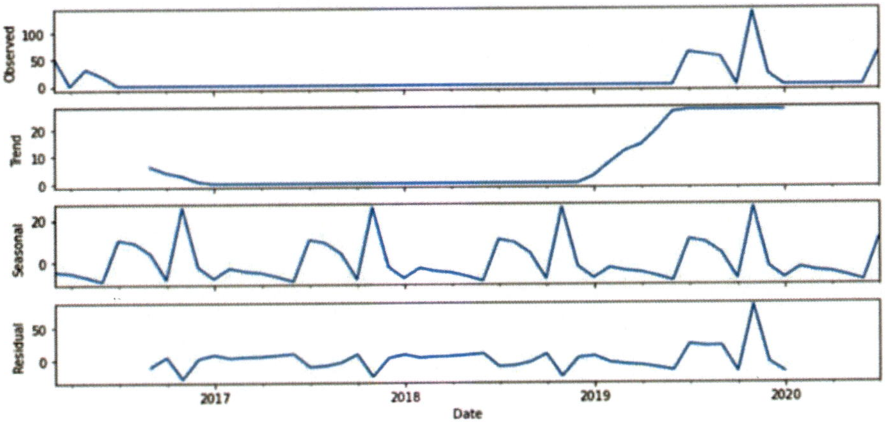

Fig. 3 Trend in rum

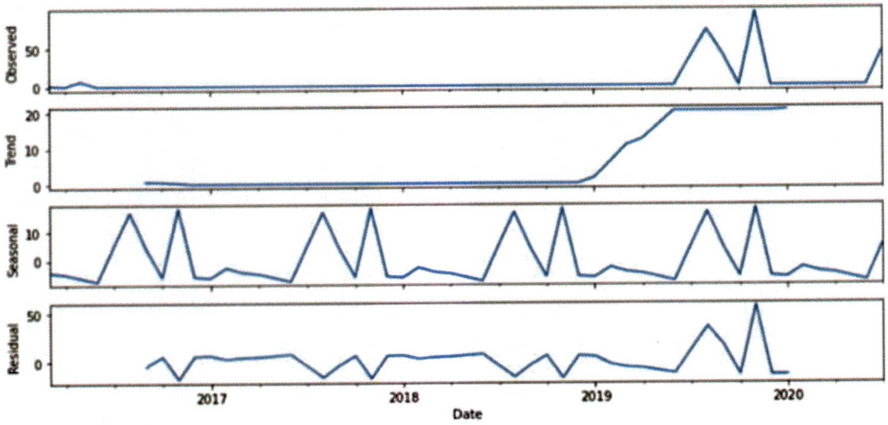

Fig. 4 Trend in bitters

a seasonal trend has been identified, the SARIMAX algorithm is applied. As it is required to predict the sales of three different brands of liquors, the algorithm is run three times separately for each of the brands. The predicted values are then checked to see whether they are in line with the given values. Figures 6, 7, and 8 represent the sales prediction of rums, bitters, and coffee liqueurs, respectively. The 20-months sales forecasting for rums, bitters, and coffee liqueurs is shown in Figs. 9, 10, and 11, respectively.

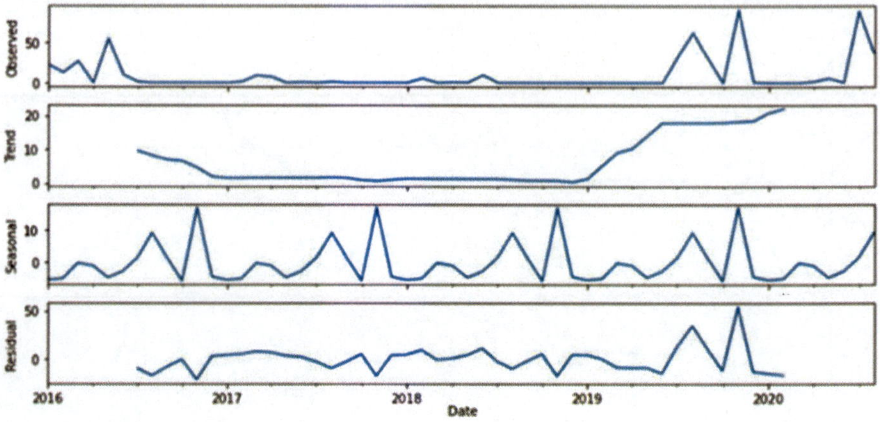

Fig. 5 Trend in coffee liqueurs

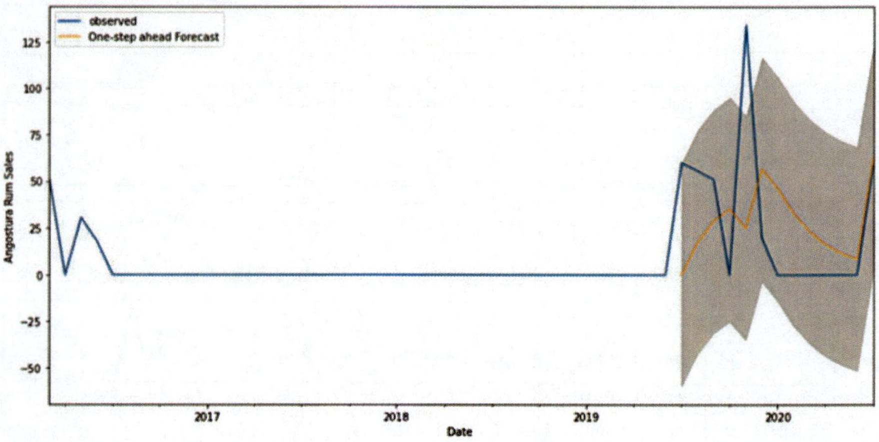

Fig. 6 Sales prediction of rum

6 Conclusion and Future Work

The job of a sales personnel is vital to a company as he acts as a point of communication between the company officials and customers. The data collected by the salesperson are very important to a company as it can determine the sales and profits of the company. As such, making use of a mobile-based application which makes the job of the sales rep easier will be highly beneficial, as the sales rep can survey more stores in less time. The above proposed system not only takes orders and records inventory but also makes use of a GPS system. In addition to this, it makes use of a barcode reader which automatically enters the product details on scanning the

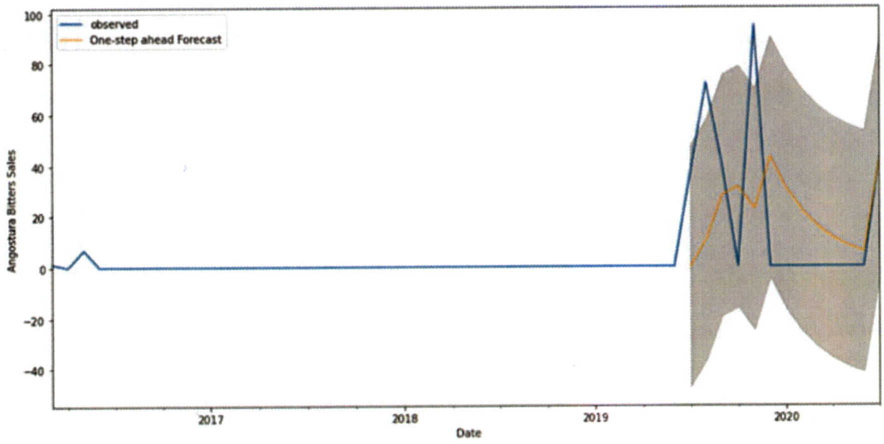

Fig. 7 Sales prediction of bitters

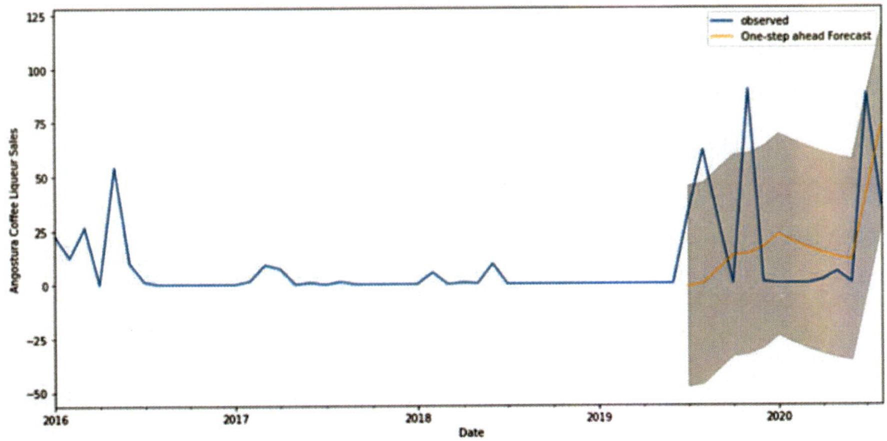

Fig. 8 Sales prediction of coffee liqueur

product's barcode. The app also generates various reports which gives the company an overview of the sales.

The data gathered from the feedback form can be used to determine the type of product likely to be sold the most in the market. Using the sales forecasting model, the future sales can be predicted which is not only useful to marketing managers but also to manufacturers. The current application works only from a sales perspective and does not include other modules such as delivery, production, and finance. It can be further developed to include these modules for a clearer understanding of the process. Another development could be that the marketing strategies developed after analyzing the data could be automated. The application can also be integrated with

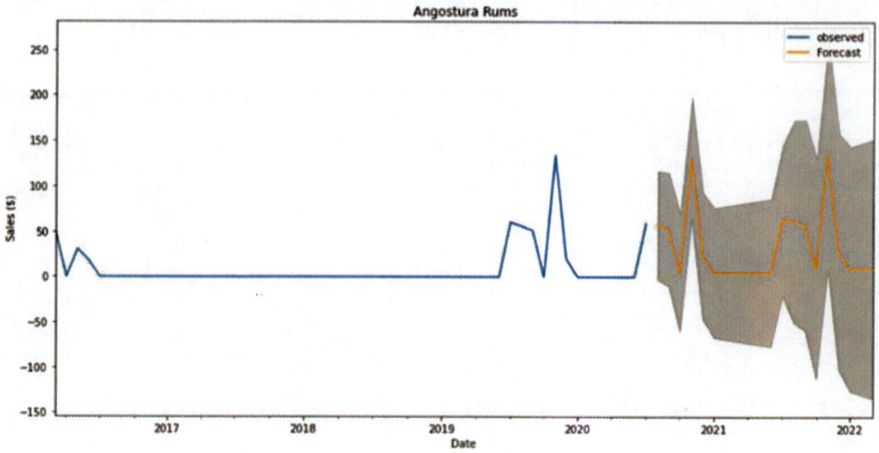

Fig. 9 Sales forecasting (20 months) of rum

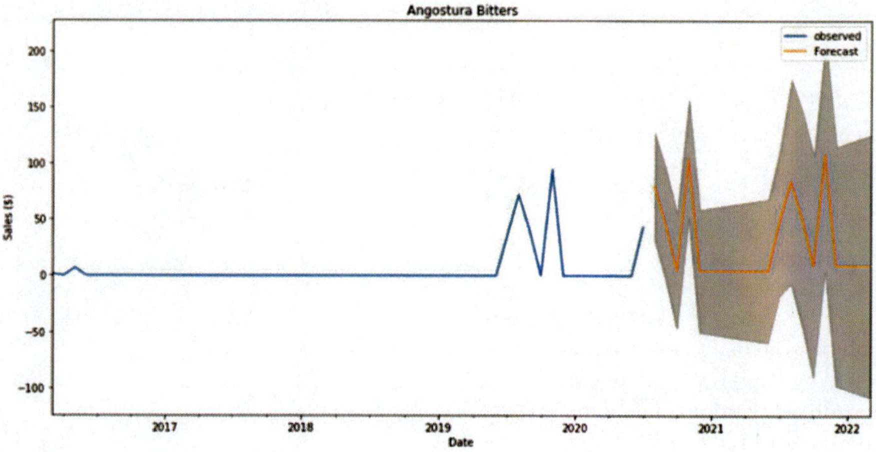

Fig. 10 Sales forecasting (20 months) of bitters

SAP at a future date. The data from Firebase can be downloaded and fed into the SAP database where it can be made use of accordingly.

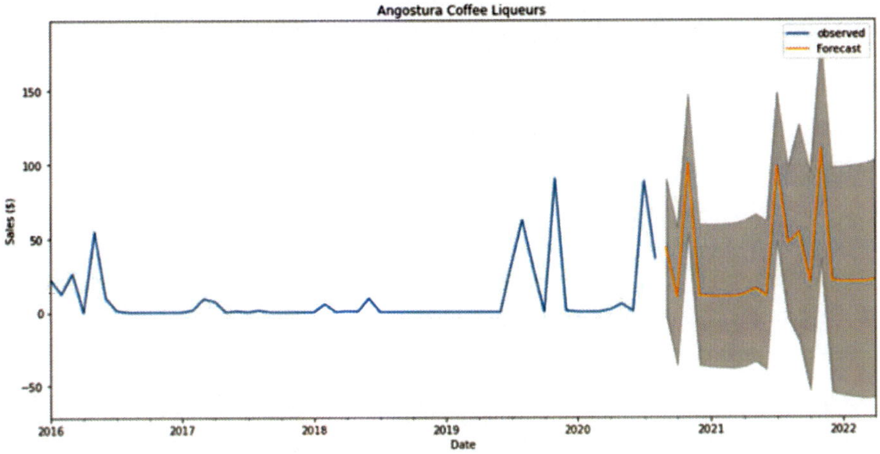

Fig. 11 Sales forecasting (20 months) of coffee liqueurs

References

1. The House of Angostura. http://angosturabitters.com. Last accessed 2021/05/01
2. Kurniawan E (2014) Implementasi REST web service Untuk sales order dan sales tracking Berbasis mobile. EKSIS J 7(1):1–12
3. Khawas C, Shah P (2018) Application of firebase in android app development—a study. Int J Comput Appl 179(46):49–53
4. Google Firebase. https://firebase.google.com. Last accessed 2021/05/01
5. Daniel Pan: Firebase Tutorial. https://www.andrew.cmu.edu/user/gkesden/ucsd/classes/fa16/cse110-a/applications/ln/firebase.pdf (2016)
6. Arunraj NS, Ahrens D, Fernandes M (2016) Application of SARIMAX model to forecast daily sales in food retail industry. Int J Oper Res Inf Syst 7(2):1–21
7. SARIMAX Introduction. https://www.statsmodels.org/dev/examples/notebooks/generated/statespace_sarimax_stata.html. Last accessed 2021/05/01
8. Griffiths D, Griffiths D (2017) Head first android development: a brain-friendly guide, 2nd edn. O'Reilly Media
9. du Preez OJ (2019) Visual studio 2019 in depth, 1st edn. BPB Publications, India
10. XAMPP Installers and Downloads for Apache Friends. https://www.apachefriends.org. Last accessed 2021/04/30
11. Google Colab. https://colab.research.google.com. Last accessed 2021/05/01
12. Iowa Liquor Sales. https://www.kaggle.com/residentmario/iowa-liquor-sales
13. Store Item Demand Forecasting Challenge. https://www.kaggle.com/c/demand-forecasting-kernels-only/notebooks

Wireless Sensor Networks (WSNs): Toward an Energy-Efficient Routing Protocol Design

Nitish Pathak, Neelam Sharma, and Harshita Chadha

Abstract Emerging as a class of network, wireless sensor network (WSN) remains highly resource-constrained. The primary concern revolving around WSN systems is the energy consumption. The current study employs a cross-layer design approach toward the development of an energy-efficient routing protocol to address this concern. The proposed is titled the PRRP (Position Responsive Routing Protocol) system. The design of the proposed methodology is such that it serves to minimize the energy that is consumed in the respective distance nodes relative to the given networks. The study also involves a critical evaluation of how the proposed PRRP protocol behaves and performs in relation to the parameters of the network's energy consumption, throughput, and network lifetime, focusing on the individual basis for the respective data packets. From the results, the research has established that, upon being analyzed and also benchmarked relative to well-known models such as CELRP and LEACH protocols, PRRP comes with significant improvements in the WSN systems concerning energy efficiency. As such, PRRP is confirmed to steer marked improvements in WSNs' overall performance.

Keywords Wireless Sensor Network (WSN) · Energy efficiency · Routing protocol · PRRP (Position Responsive Routing Protocol) · CELRP protocol · LEACH protocol · Network lifetime · Throughput

1 Introduction

In contemporary society, the wireless sensor technology continues to play a very crucial role in most commercialized industrial automation procedures, as well as certain real-life applications [1–3]. Specifically, the technology plays a critical role

N. Pathak
Bhagwan Parshuram Institute of Technology, New Delhi, India

N. Sharma · H. Chadha (✉)
Maharaja Agrasen Institute of Technology, New Delhi, India
e-mail: harshitaachadha@gmail.com

V. Skala et al. (eds.), *Machine Intelligence and Data Science Applications*, Lecture Notes on Data Engineering and Communications Technologies 132,
https://doi.org/10.1007/978-981-19-2347-0_30

when it comes to harsh environment applications in which network infrastructural deployment is nearly impossible or proves difficult [4]. Examples of such environments include high thermal environments, hazardous chemical plants, and battlefields [5–7]. Thus, in most cases, crucial systems such as security and surveillance applications have employed sensor-based applications [8]. Also, previous studies indicate that most of the preferred sensors are those that are economical and small-sized [9].

Imperative to note is that all sensor networks exhibit certain sensing mechanisms through which data is collected from targeted physical contexts, either by the event triggering method or a time-driven technique [10]. Through these techniques, sensors engage in the conveyance of the sensed data to destinations or sinks, as well as multiple sinks or destinations through routing frameworks or algorithms. Some of the previously proposed routing algorithms include the Directed Diffusion Routing Protocol (DDRP) and the Minimum Cost Forwarding Algorithm (MCFA), as well as cluster-based routing protocols [11]. Given their small-sized nature, sensor nodes tend to be developed with finite battery power capacity, small memory storage, and limited computational capacity [12, 13].

For typical WSN nodes, therefore, they exhibit four crucial components in terms of power units, processing units, analog-to-digital converters, and sensing elements [14]. Of importance to note is that WSNs remain resource-constrained networks, whereby the main parameter that shapes operation ability involves energy efficiency, which determines the life span of the battery of the sensor nodes [15]. Thus, the unique parameters tend to be considered (including data packet routing activities) while seeking to address challenges and issues concerning variables such as application areas/environment, network deployment, node mobility energy consumptions or efficiency, node administration, node distribution, runtime topologies management, and network coverage [16, 17].

Indeed, in WSNs, nodes tend to be memory, computation, and energy-constrained [18]. Hence, there tends to be a need for scholarly examinations of resource-aware and low-computation WSN algorithms, especially in relation to highly resource-constrained and small embedded sensor nodes. From previous investigations, one of the most important factors in WSN applications entails energy consumption, leading to the design of some hardware and algorithms based on the achievement of energy awareness or energy efficiency as a central point of focus [19, 20]. This study focuses on WSN energy efficiency enhancement relative to communication routing protocols. Particularly, a new routing protocol in the form of PRRP (Position Responsive Routing Protocol) is proposed. The study also strives to offer a comparative analysis of the performance outcomes of the proposed PRRP protocol with CELRP and LEACH protocols, upon which the efficacy of the proposed model relative to system performance enhancement and energy efficiency realization or improvement might be confirmed or otherwise.

2 Methodology

The study strives to establish the PRRP protocol, a new and energy-efficient routing protocol. The motivation lies in the need to have WSN energy issues addressed, as well as have the WSN's energy efficiency enhanced. For the proposed PRRP protocol, it comes with a novel way through which WSN cluster heads are selected. For the existing CELRP and LEACH, cluster head selection occurs randomly relative to all nodes in relation to their associated residual energy. For the case of the proposed PRRP protocol, different variables are considered. They include the average distance between the candidate cluster head node and the neighboring nodes, the energy level, and the distance from the sink.

Also, in the proposed PRRP protocol, there is the division of the WSN into cells and grid and then into various tiers. For the sink, it is assumed to be placed on the topology's center. Also, there is the random fashion distribution of the nodes, assuming further that they exhibit awareness of their location or position via local means such as the Global Positioning System (GPS). The gateways, thus, form nodes closer to sinks. Additionally, there is the selection of gateways alongside other tiers based on the factors of the number of nodes in the neighbor, position from the sink, and the node energy level. The eventuality is that the operation of PRRP occurs in various phases, beginning with the selection of the gateways up to the point of data transfer. There is also the formation of tree rooted at the sinks before allowing for data collection via TDMA scheduling. The assumption of the proposed routing protocol, at this point, is that it is only when a neighbor is available in the same tier that a node can join it. Thus, there is the ensuring of the minimum distance, allowing further for energy saving in the process of data transfer, as the respective nodes only select parents from closest neighbors, avoiding data transfer in long distances. The proposed PRRP routing protocols also focus on short node transmission ranges, with the respective nodes listening to the transmission of nodes available in the same tier or close to them. Hence, the study strives to unearth a technique that might assure minimum distance among interacting nodes in the course of data transfer, translating further into a beneficial effect of energy saving in WSN applications.

3 Results and Discussion

In this simulation study, some of the parameters that were evaluated included the data packet size, control packet size, initial energy in the respective nodes, the amplifier energy, and the electronics energy. Others included maximum communication between nodes, delay for retransmission, time-out constant, routine data source probability, unusual event sources, the radio range, sink location, sensor distribution, number of sensors, and the network area. The central assumption was that there was a reliable physical channel and that there was no message loss. Other assumptions entailed open area distribution of nodes and the antenna being omnidirectional, as well as circular direction coverage of the radio.

Initially, therefore, there was a comparative analysis pitying LEACH and the proposed PRRP routing protocol's performance in relation to various variables. Imperatively, PRRP was found to exhibit some degree of similarity with LEACH in such a way that a given node could communicate with the sink and that there was a time-based schedule arrangement when it came to the transmission mechanism or data collection process. However, the research revealed several differences in functionality, operation, and performance between PRRP and LEACH. For example, in LEACH, the cluster heads and clusters would be formed in the respective clusters for the transmission of data, but in PRRP, there would be the construction of trees rooted in the sinks. Also, in LEACH, data collection from associated clusters would be achieved by cluster heads before being transferred to the sink, but in PRRP, gateways were found to play the role of data transfer to the sinks. In PRRP, also, the number of children was found to be less when compared to the case of LEACH implementation.

3.1 Proposed PRRP Versus LEACH

Regarding PRRP versus LEACH's performance evaluation, with variables or parameters such as the average energy used per packet for the respective periods of data transmission, the network throughput, the number of live nodes, and energy efficiency, it was in LEACH that cluster heads and non-head nodes that emerged as two sensor node types, but in PRRP, three sensor node types emerged and they included non-leaf nodes, leaf nodes, and gateway nodes. With several data transmission periods elapsed, a comparison of the two protocols saw PRRP depict significant improvements as shown in the figures below. Collectively, with all rounds considered, simulation outcomes revealed that PRRP nodes exhibit the capability to stay longer and also have the maximum possible energy utilized for a longer time period than the case of LEACH. A specific illustration confirming these observations was the case of a 10-round test run. From the figures, it can be seen that it is after 275 s that LEACH protocol's last node dies, but it is after 350 s that PRRP protocol's last node dies (Fig. 1).

Regarding the total energy consumed, in PRRP, as the tree formation phase sets in, it was established that there could be signal transmission by the nodes with the least energy level because of the selection of closer nodes by the parents, leading to distance reduction; hence, node energy saving. For LEACH, distance or location of the nodes is not considered, as it is assumed that the respective nodes are capable of hearing all others in the given network. Thus, in PRRP, the transmission distance proved to be shorter. A specific example confirming these results was that in which, at $t = 150$ s, with 10 periods of data transmission, a value of about 220 J was found to be the total consumed energy for the case of LEACH, but the value stood at 120 J when it came to the case of PRRP. Thus, PRRP was associated with significant

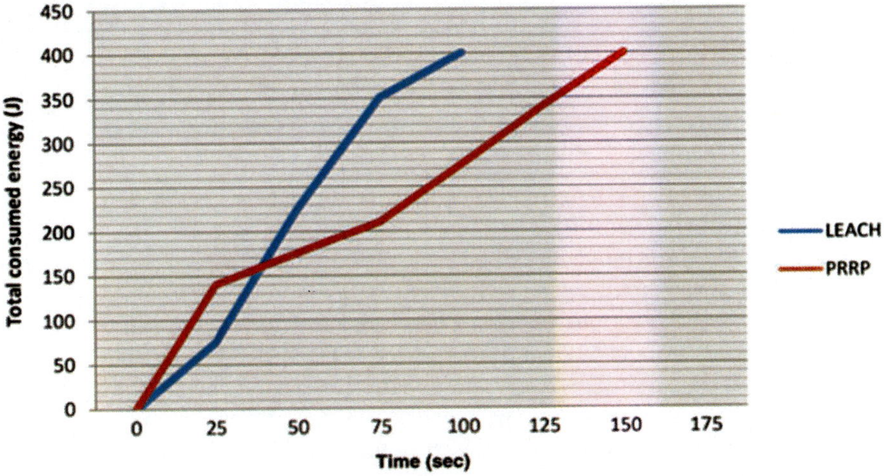

Fig. 1 Total consumed energy (J—*Y*-axis) versus time (sec—*X*-axis) (PRRP versus LEACH)

improvements in the overall energy consumed up to the network lifetime's end, with higher throughput also observed.

Lastly, the comparison of PRRP with LEACH focused on per-packet average consumed energy. Here, PRRP exhibited superior performance because of its associated consumption of less energy as shown in Fig. 2. Important to note is that as the initial transmission periods set in, PRRP exhibited higher average consumed energy.

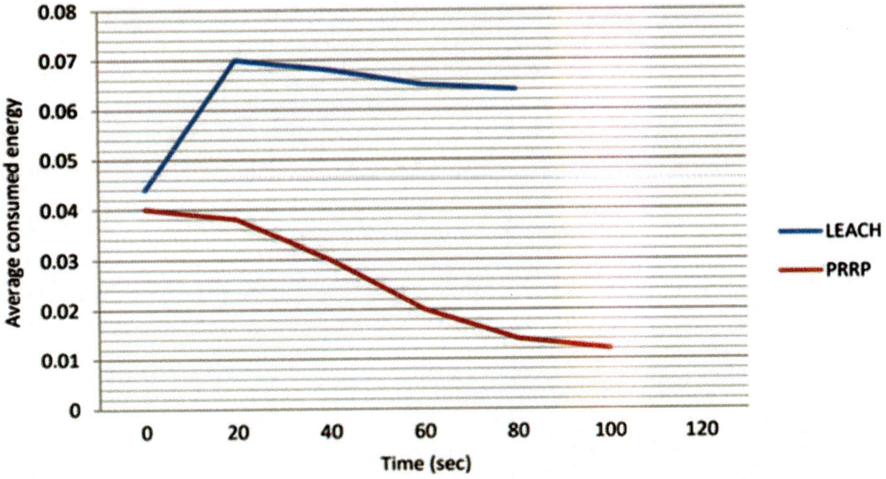

Fig. 2 Average consumed energy for each packet (*Y*-axis) versus time (sec—*X*-axis) (PRRP versus LEACH)

Table 1 PRRP versus LEACH on throughput trend

Data transmission period	1	5	10
LEACH	3607.00	13,045.00	25,640.00
Proposed PRRP	15,834.00	53,950.00	64,100.00

However, with more and more rounds of transferring data, a comparison of PRRP with LEACH revealed that PRRP exhibited higher efficiency.

Regarding the throughput, for all rounds, PRRP yielded marked improvements in the network throughput, saving significant energy amount as the phase of transmitting data set in for various periods. There was also further increase in network throughput with increased number of rounds regarding PRRP implementation, with significant performance improvements noted concerning parameters of the utilization of maximum possible energy of sensor network nodes, the energy efficiency variable, and the factor of network lifetime. The table below highlights these outcomes (Table 1).

3.2 *Proposed PRRP Versus LEACH*

When it comes to the total energy consumed, the proposed PRRP routing protocol was found to exhibit strength in terms of data transmission, effective TDMA scheduling, and tree building phases. In PRRP, for example, as tree building proceeded, nodes were found to transmit signals with the least energy level because parents would select closer nodes. Hence, distance between the child and parent node was reduced, saving significant energy during the transmission of energy. Also, PRRP had its sink located at the network's center, reducing the overall distance for the respective sensor nodes, inclusive of gateways, non-leaf nodes, and the leaf nodes. For CELRP, nodes acted as cluster head leaders, cluster heads, and normal nodes, with the sink occurring outside the distance. With CELRP associated with longer transmission distances, therefore, the associated system's energy loss was more than that linked to PRRP.

Concerning the factor of per-packet average consumed energy, good performance was initially depicted by CELRP when compared to the case of PRRP. However, after some short time span, which involved increased data periods, PRRP exhibited significant improvement on this parameter. For PRRP, the higher average energy consumed initially was attributed to three initial phases required for setup, with there being no data transmission at these stages.

Lastly, there was the comparison of PRRP with CELRP's performance based on the factor of throughput. Here, simulation outcomes depicted significant network throughput improvement for all rounds, but PRRP was found to exhibit the capacity to save more energy amounts in the course of data transmission for multiple periods. Overall, the investigation established and confirmed PRRP's significant performance

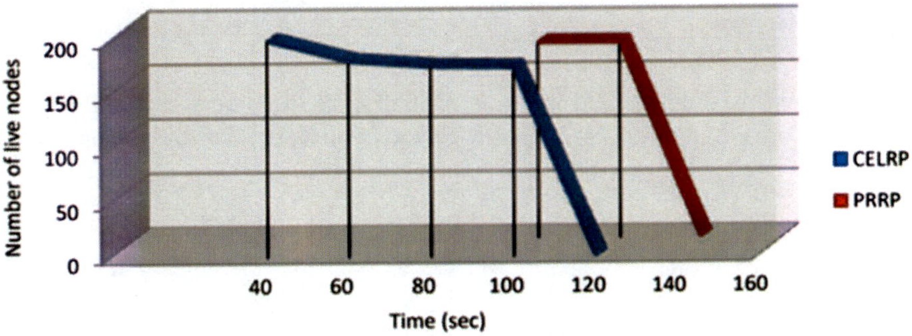

Fig. 3 PRRP versus CELRP—on the number of live nodes against time

improvement relative to factors of network throughput, the utilization of maximum possible energy, energy efficiency, and the network lifetime (Figs. 3 and 4; Table 2).

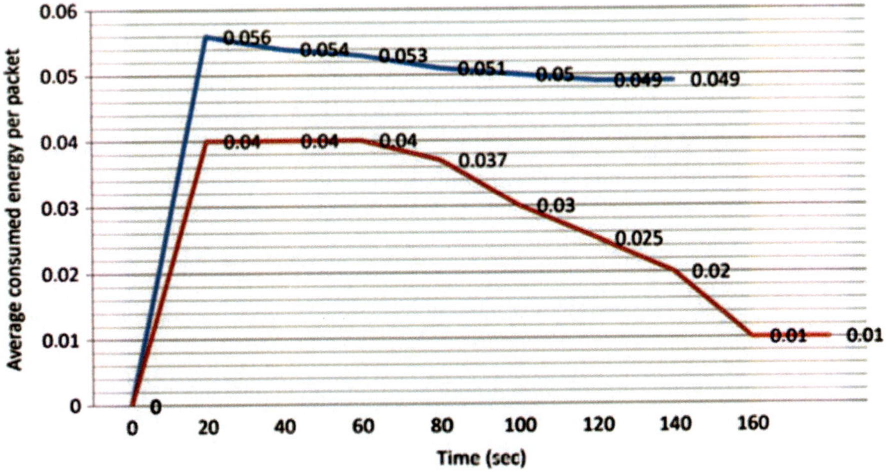

Fig. 4 PRRP (blue) versus CELRP (red)—on the average consumed energy per packet

Table 2 PRRP versus CELRP—on the factor of network throughput

Data Transmission Period	1	5	10
CELRP	5950.00	19,985.00	35,710.00
Proposed PRRP	16,733.00	54,555.00	65,305.00

4 Conclusions

This study focused on the performance evaluation of the proposed PRRP routing protocol relative to the variables of average energy consumption and the network's average lifetime, both on per-packet basis and individually, as well as the behavior of the proposed algorithm relative to the parameter of network throughput. The results were compared with the performance of previously proposed models of CELRP and LEACH. From the findings, PRRP was associated with superior results than LEACH, especially in terms of the capacity to steer improvements in the network lifetime through reductions in energy consumption, with a similar trend observed regarding the network throughput variable. Particularly, there was improvement of over 50% in the network lifetime and 2.5 times increase in the throughput. When compared to CELRP, the proposed PRRP model improved the network lifetime by over 35%, also increasing the network throughput by 1.82 times, yielding further energy consumption reductions by over 35%. Thus, the superiority of the proposed model over CELRP and LEACH was confirmed.

References

1. Zhang W, Wei X, Han G, Tan X (2018) An energy-efficient ring cross-layer optimization algorithm for wireless sensor networks. IEEE Access 6:16588–16598
2. Sobrinho JL (2017) Correctness of routing vector protocols as a property of network cycles. IEEE Trans Netw 25:150–163
3. Mouapi A, Hakem N (2018) A new approach to design autonomous wireless sensor node based on RF energy harvesting system. Sensors 18:133
4. Zhang Y, Liu M, Liu Q (2018) An energy-balanced clustering protocol based on an improved CFSFDP algorithm for wireless sensor networks. Sensors 18:881
5. Bahbahani MS, Alsusa E (2018) A cooperative clustering protocol with duty cycling for energy harvesting enabled wireless sensor networks. IEEE Trans Wirel Commun 17:101–111
6. Zheng HF, Guo WZ, Xiong N (2017) A kernel-based compressive sensing approach for mobile data gathering in wireless sensor network systems. IEEE Trans Syst Man Cybern Syst 48:2315–2327
7. Shen J, Wang A, Wang C, Hung PCK, Lai C (2017) An efficient centroid-based routing protocol for energy management in WSN-assisted IoT. IEEE Access 5:18469–18479
8. Sohn I, Lee J, Lee SH (2016) Low-energy adaptive clustering hierarchy using affinity propagation for wireless sensor networks. IEEE Commun Lett 20:558–561
9. Zhao Z, Xu K, Hui G, Hu L (2018) An energy-efficient clustering routing protocol for wireless sensor networks based on AGNES with balanced energy consumption optimization. Sensors 18:3938
10. Roy NR, Chandra P (2018) A note on optimum cluster estimation in LEACH protocol. IEEE Access 6:65690–65696
11. Hosen A, Cho G (2018) An energy centric cluster-based routing protocol for wireless sensor networks. Sensors 18:1520
12. Sharma D, Bhondekar AP (2018) Traffic and energy aware routing for heterogeneous wireless sensor networks. IEEE Commun Lett 22:1608–1611
13. Kaur T, Kumar D (2018) Particle swarm optimization-based unequal and fault tolerant clustering protocol for wireless sensor networks. IEEE Sens J 18:4614–4622

14. Behera TM, Samal UC, Mohapatra SK (2018) Energy-efficient modified LEACH protocol for IoT application. IET Wirel Sens Syst 8:223–228
15. Alnawafa E, Marghescu I (1863) New energy efficient multi-hop routing techniques for wireless sensor networks: static and dynamic techniques. Sensors 2018:18
16. Jadoon R, Zhou W, Jadoon W, AhmedKhan I (2018) RARZ: Ring-zone based routing protocol for wireless sensor networks. Appl Sci 8:1023
17. Tanwar S, Tyagi S, Kumar N, Obaidat MS (2019) LA-MHR: learning automata based multilevel heterogeneous routing for opportunistic shared spectrum access to enhance lifetime of WSN. IEEE Syst J 13:313–323
18. Priyadarshi R, Singh L, Singh A, Thakur A (2018) SEEN: stable energy efficient network for wireless sensor network. In: Proceedings of the 2018 5th international conference on Signal Processing and Integrated Networks (SPIN), Noida, India, 22–23 Feb 2018, pp 338–342
19. Kang J, Sohn I, Lee S (2019) Enhanced message-passing based LEACH protocol for wireless sensor networks. Sensors 19:75
20. Cheng HJ, Su ZH, Xiong NX, Xiao Y (2016) Energy-efficient nodes scheduling algorithms for wireless sensor networks using Markov random field model. Inf Sci 329:461–477

Empirical Analysis of Machine Learning and Deep Learning Techniques for COVID-19 Detection Using Chest X-rays

Vittesha Gupta and Arunima Jaiswal

Abstract Due to the Coronavirus (COVID-19) cases growing rapidly, the effective screening of infected patients is becoming a necessity. One such way is through chest radiography. With the high stakes of false negatives being potential cause of innumerable more cases, expert opinions on x-rays are high in demand. In this scenario, Deep Learning and Machine Learning techniques offer fast and effective ways of detecting abnormalities in chest x-rays and can help in identifying patients affected by COVID-19. In this paper, we did comparative analysis of various Machine Learning and Deep Learning techniques on chest x-rays based on accuracy, precision, recall, f1 score, and Matthews correlation coefficient. It was observed that improved results were obtained using Deep Learning.

Keywords COVID-19 · Coronavirus · Machine learning · Deep learning

1 Introduction

The severe acute respiratory syndrome coronavirus 2 (SARS-CoV-2) causes a highly infectious disease known commonly as coronavirus. The initial clinical sign which allowed case detection of the SARS-CoV-2 related disease coronavirus was pneumonia [1, 2]. There was no independent feature of coronavirus pneumonia on analysis of chest radiographs that was specific or diagnostic by itself, but a combination of multiple foci peripheral lung changes of ground-glass opacity and/or consolidation was present [3]. These factors along with the study of previous radiographs can determine if the patient has coronavirus [4]. The effective screening of chest radiographs of the infected patients is a necessary step to control and prevent further spread while assessing the severity of coronavirus infection in the patient [5]. With the

V. Gupta (✉) · A. Jaiswal
Department of Computer Science and Engineering, Indira Gandhi Delhi Technical University for Women, Delhi, India
e-mail: vittesha110btcse19@igdtuw.ac.in

A. Jaiswal
e-mail: arunimajaiswal@igdtuw.ac.in

V. Skala et al. (eds.), *Machine Intelligence and Data Science Applications*, Lecture Notes on Data Engineering and Communications Technologies 132,
https://doi.org/10.1007/978-981-19-2347-0_31

high stakes of false negatives being the potential cause of innumerable more cases, expert opinions on radiographs are high in demand. Manually checking and classifying large quantities of radiographs is a tedious and cumbersome task. For a faster diagnosis, various Deep Learning (DL) and Machine Learning (ML) techniques can be applied to train computers on existing data of chest radiographs of coronavirus positive patients and can be used by doctors as an aid to identify coronavirus cases. We did a comparative analysis of ML and DL techniques by training various techniques on images of chest radiographs and performing binary image classification. The dataset was classified into two classes- Normal and coronavirus cases. In this work, we have implemented various ML techniques- K-Nearest Neighbors (KNN), Naïve Bayes (NB), Random Forest (RF), Support Vector Machine (SVM), Decision Tree (DT), Logistic Regression (LR), Extreme Gradient Boosting (XGB) and Light Gradient Boosting (LGBM), along with various DL techniques- Recurrent Neural Network (RNN) and Convolutional Neural Network (CNN). We trained the above techniques and then tested them on the Coronavirus Radiography Dataset developed by researchers belonging to Qatar University in Doha, Qatar, and the University of Dhaka in Bangladesh, in collaboration with medical doctors [6–8]. The database consists of chest radiographs images for coronavirus positive as well as healthy patients (see Fig. 1).

3000 images of coronavirus positive and normal chest radiographs in total have been used for analysis where 2250 images were used to train and 750 were used to test these techniques. Accuracy, precision, recall, f1 score, and matthews correlation coefficient (MCC) scores were used as a performance measures to compare these techniques.

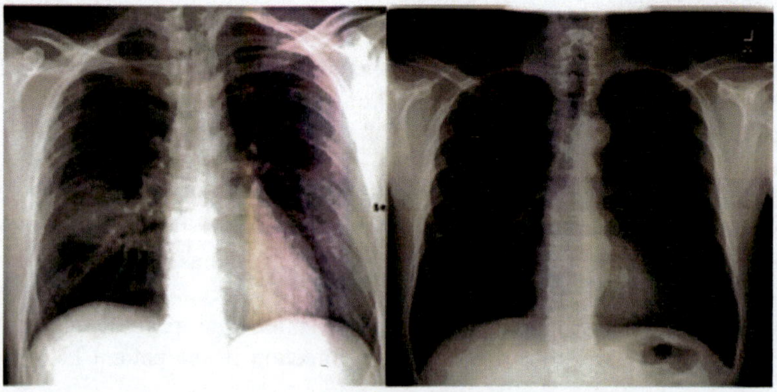

Fig. 1 Chest radiograph of coronavirus positive (left) and normal (right) patient

2 Related Work

Since the early days of the Coronavirus pandemic, researchers in various fields have actively participated in trying to combat the growing cases of coronavirus by developing techniques to detect the infected people at the earliest and prevent the spread of the virus [9, 10]. Many researchers have worked on the use of chest radiographs [11, 12] for the detection of coronavirus. There has been working using CT scans, along with blood tests for COVID-19 diagnosis [13]. New ML techniques have also been developed for obtaining better results [14, 15]. Some researchers have also taken into account several variables such as time from symptoms occurrence and subsequent submission, the count of white blood cells of the patient, and their levels of oxygen saturation for risk prediction of coronavirus positive patients [16]. Researches have also trained DL techniques such as simple CNN with a pretrained AlexNet model on CT scans and x-rays images [17]. During the early stages of the pandemic, researchers worked with limited publicly available datasets of CT scans and x-ray images of coronavirus positive patients, for training DL techniques which they overcame by assembling their own datasets [17] and using smaller trainable parameters [18]. The relevance of DL technique applications for detection has also been researched where the researchers have discussed the applications of computer vision and natural language processing [19]. The possible limitation of DL applications have been discussed and in conclusion, it was seen that the DL techniques are useful and important for fighting coronavirus. New methods for coronavirus detection have also been proposed, such as a method with a total of four phases- data augmentation, pre-processing, and stage 1 and stage 2 of DL model designing [20], along with other methods [21, 22]. Analysis of CT images through transfer learning using pretrained DL techniques such as resnet, googlenet, alexnet, and xception [21, 23]. Use of pretrained DL techniques like ResNet50, ResNet152, InceptionV3, Inception-ResNetV2 and ResNet101 [24, 25]. Detection of the severity of coronavirus infection has also been considered an important feature for the prevention of any more casualties due to this virus [26].

3 Application of Techniques

In this dataset, the chest radiograph images were distinguished and well labeled. We used supervised learning-based traditional ML techniques for comparison and then implemented ensemble techniques involving better performing ML techniques to get improved results [25–28] as shown in Table 1. The measures used to compare performance are precision, recall, f1 score, and MCC. Precision gives a measure of out of all the predicted radiographs as coronavirus positive, how many are actually positive. This is an important measure to take into consideration to avoid false positives as they can be detrimental to our aim. Recall gives a measure of out of all actual coronavirus positive cases, how many cases are correctly predicted as positive by

Table 1 Machine learning techniques

Technique	Details
KNN	The inputs consist of k closest training data points in the datasets. The data points are mapped onto points in space and the training examples closest to that data point are known as its neighbors. The output of this technique is determined by assigning the class most common among its k-nearest neighbors
SVM	It maps training data to points in k-dimensional space (where k is the number of categories) so as to maximize the gap between different categories. Its objective is to find a hyperplane that draws a definite decision boundary in n-dimensional space between data points belonging to different classes. The hyperplane depends on the dimensional space, i.e., line for two-dimensional and plane for three-dimensional. We predict the category of data points based on where they fall with respect to the hyperplane
NB	It is a probabilistic classifier. Probabilistic classifiers use conditional probability distribution, which is the probability of assigning a class to an example given a specific set of features. It creates a probability distribution of all classes in the dataset and then decides which class is to be assigned for every example
DT	It constructs a hierarchical structure consisting of nodes and directed edges based on questions asked to each training example in the dataset. The series of questions and answers are asked till a class label can be assigned
RF	It is an ensemble method which works by merging a collection of independent decision tree classifiers to get more accurate prediction. It uses majority vote for making the final predictions
LR	It computes the class probability for every data point which lies between 0 and 1. This probability is used to classify the data
XGB	It is an ensemble method of several DT trained on training data subsets
LGBM	It is another ensemble technique that has automatic feature selection which helps in getting improved results using this technique

that technique. It is also a crucial performance measure as false negatives would be counterproductive to our aim of detecting coronavirus positive cases and deploying correct measures to curb any potential spreading of this infection. F1 score is precision and recalls' harmonic mean that takes both the measures into account while delivering the result which is imperative to our work. Apart from these commonly used performance measures, we have also taken into consideration MCC [29, 30] which gives us a measure of the correlation between the true and predicted values. We also implemented various DL techniques [31, 32] as shown in Table 2.

We have implemented the aforementioned techniques using Python.

4 Dataset Details

The dataset on the coronavirus Radiography Database was developed by researchers belonging to the University of Dhaka in Bangladesh and Qatar University in Doha,

Table 2 Deep learning techniques

Technique	Details
RNN	It is a type of artificial neural network where a directed graph along a temporal sequence is formed using connections between the nodes
CNN	It is a type of multilayer perceptron that has been regularized. Every neuron present in one layer is linked with all others of the successive layer which makes it fully connected. This makes them prone to overfitting data. It learns to optimize the kernels or filters through automated learning and hence is independent of dependency on human intervention or requiring any prior knowledge to aid in extraction of features. These features make CNNs more desirable and better than other image classification algorithms

Qatar, in collaboration with medical doctors [7, 8]. The dataset consists of 10,200 Normal images and 3616 coronavirus positive chest radiographs. As the dataset was imbalanced, we have implemented the techniques on a reduced dataset by using 3000 images for classification to prevent any skewed results. We have performed binary classification on this dataset. The categories of classification are coronavirus positive and Normal. 2250 coronavirus positive and Normal chest radiograph images were used to train and 750 were used to test the techniques (see Fig. 2).

All the chest radiographs images are present in png format with size of 256 by 256 pixels. We have preprocessed the data by resizing the images to 32 × 32 and then flattening them into a list of raw pixel intensities. A 3D color histogram was extracted from every image and the flattened histogram was used as the feature vector for that image.

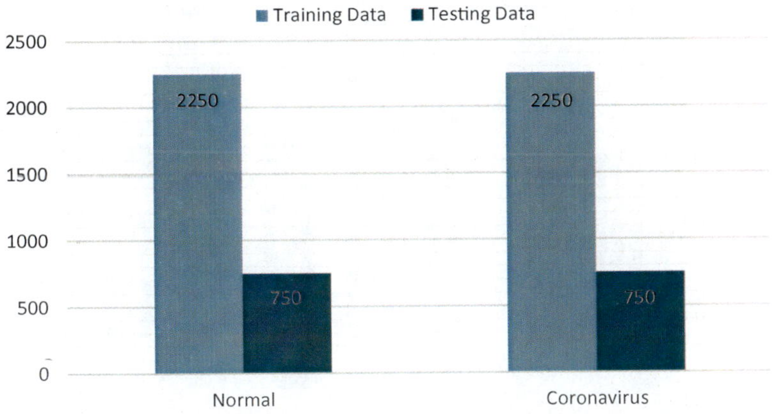

Fig. 2 Distribution of dataset into the two categories-COVID-19 positive and Normal

5 Result and Discussion

In this section, we have computed and drawn comparison between the performance of ML techniques using accuracy (see Fig. 3), precision, recall, f1 score (see Fig. 4), and MCC (see Fig. 5). The performance of DL techniques is compared using accuracy scores (see Fig. 6).

According to our performance analysis on this dataset, it was observed that among the traditional ML techniques, KNN gave better results followed by DT, followed by SVM, then LR, and lastly NB. Taking this into consideration, we implemented RF, XGB, and LGBM which are ensemble techniques made up of many trained DT to

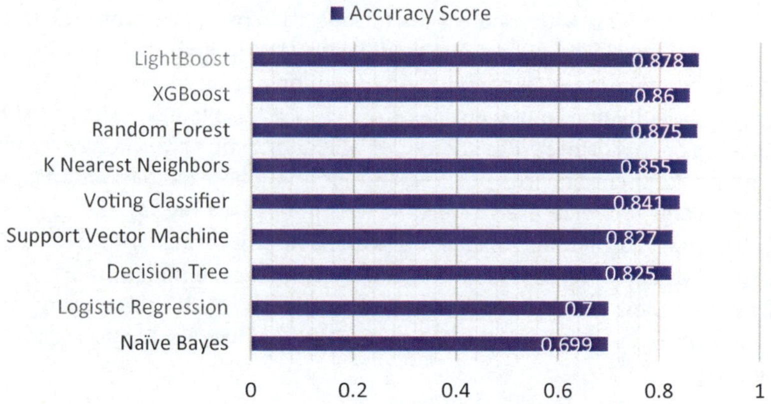

Fig. 3 Comparative analysis of different ML techniques based on accuracy scores

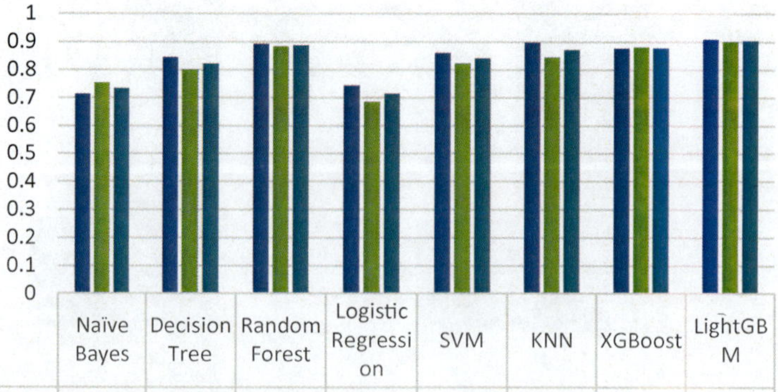

	Naïve Bayes	Decision Tree	Random Forest	Logistic Regression	SVM	KNN	XGBoost	LightGBM
Precision	0.714	0.845	0.893	0.745	0.86	0.898	0.876	0.909
Recall	0.755	0.801	0.884	0.687	0.824	0.845	0.881	0.902
F1 score	0.734	0.822	0.888	0.715	0.842	0.871	0.878	0.905

Fig. 4 Comparative analysis of different ML techniques based on precision, recall, and F1 scores

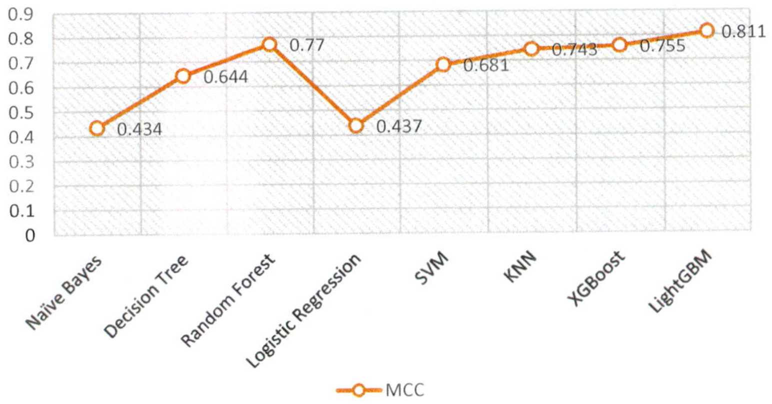

Fig. 5 Comparative analysis of various ML techniques based on their MCC scores

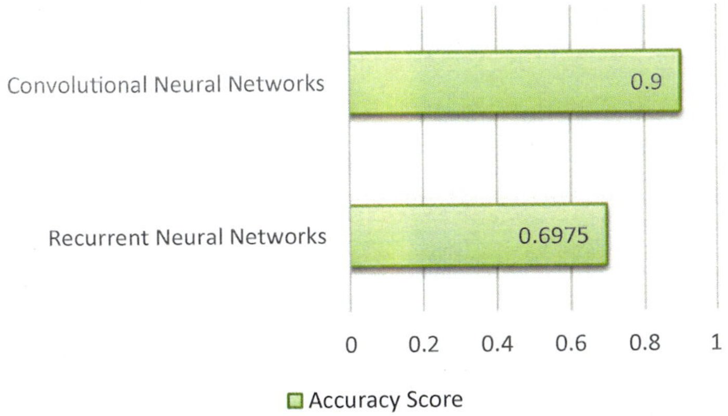

Fig. 6 Comparative analysis of different DL techniques based on accuracy scores

get improved results. The results obtained using LGBM were the highest among the tested techniques. LGBM uses automatic feature extraction for the input images and works by the development of many DT on training data subsets which increases its accuracy in prediction. LGBM shows improved results as the development of a large number of DT eradicates any errors that may occur on development of these individual DT. The DL technique with the highest accuracy was CNN with an accuracy score of 0.9 and with the lowest accuracy was RNN with an accuracy score of 0.675. CNN outperforms all the implemented techniques as it is one of the most promising techniques for pattern-recognizing problems. It extracts the relevant features from the input data automatically without the need for human supervision or intervention. The use of the pooling layer in CNN also reduces the computational power needed by the computer during classification, further increasing the desirability of this technique.

6 Conclusion and Future Scope

Coronavirus is the biggest challenge of the twenty-first century. Effective measures need to be used to control and prevent any further spread of the virus. The analysis of chest radiographs is one such measure that can help identify coronavirus-affected patients and can also be used to determine the degree of severity of infection. This paper drew a comparison between different ML techniques, namely, Naïve Bayes, Logistic Regression, Decision Tree, SVM, Voting Classifier, KNN, and Random Forest and DL techniques, namely, RNN and CNN by training them on the same dataset of chest radiographs and then using accuracy as a measure to compare these techniques. Among all the techniques implemented, CNN had the highest accuracy on this dataset with an accuracy score of 0.9. As a future direction, a further improvement in the prediction accuracy of chest radiograph image classification can be tested by implementing and evaluating other soft computing techniques like hybrid neural networks and transfer learning.

References

1. Chan JF, Yuan S, Kok KH et al (2020) A familial cluster of pneumonia associated with the 2019 novel coronavirus indicating person-to-person transmission: a study of a family cluster. Lancet S0140–6736(20):30154–30159. https://doi.org/10.1016/S0140-6736(20)30154-9
2. Jiang F, Deng L, Zhang L, Cai Y, Cheung CW, Xia Z (2020) Review of the clinical characteristics of coronavirus disease 2019 (COVID-19). J Gen Intern Med 1–5
3. Cleverley J, Piper J, Jones MM (2020) The role of chest radiography in confirming covid-19 pneumonia BMJ 2020
4. Rousan LA, Elobeid E, Karrar M et al (2020) Chest radiograph findings and temporal lung changes in patients with COVID-19 pneumonia. BMC Pulm Med 20:245
5. Cohen JP, Dao L, Roth K et al (2020) Predicting COVID-19 pneumonia severity on chest radiograph with deep learning. Cureus. 12(7):e9448
6. Chowdhury MEH, Rahman T, Khandakar A, Mazhar R, Kadir MA, Mahbub ZB, Islam KR, Khan MS, Iqbal A, Al-Emadi N, Reaz MBI, Islam MT (2020) Can AI help in screening Viral and COVID-19 pneumonia? IEEE Access 8:132665–132676
7. Rahman T, Khandakar A, Qiblawey Y, Tahir A, Kiranyaz S, Kashem SBA, Islam MT, Maadeed SA, Zughaier SM, Khan MS, Chowdhury ME (2020) Exploring the effect of image enhancement techniques on COVID-19 detection using chest radiograph images
8. Sharma TC, Kumar P (2018) Health monitoring & management using iot devices in a cloud based framework. In: 2018 international conference on advances in computing and communication engineering (ICACCE), pp 219–224. https://doi.org/10.1109/ICACCE.2018.8441752
9. Hamet P, Tremblay J (2017) Artificial intelligence in medicine. Metabolism 69:S36–S40
10. Cozzi D, Albanesi M, Cavigli E et al (2020) Chest radiograph in new coronavirus disease 2019 (COVID-19) infection: findings and correlation with clinical outcome. Radiol Med 125(8):730–737. https://doi.org/10.1007/s11547-020-01232-9
11. Ke Q, Zhang J, Wei W, Połap D, Woźniak M, Kośmider L et al (2019) A neuro-heuristic approach for recognition of lung diseases from radiograph images. Expert Syst Appl 126:218–232

12. Ai T, Yang Z, Hou H, Zhan C, Chen C, Lv W et al (2020) Correlation of chest CT and RT-PCR testing in coronavirus disease 2019 (COVID-19) in China: a report of 1014 cases. Radiology 200642
13. Cabitza F, Campagner A, Ferrari D, Di Resta C, Ceriotti D, Sabetta E, Colombini A, De Vecchi E, Banfi G, Locatelli M, Carobene A (2021) Development, evaluation, and validation of machine learning models for COVID-19 detection based on routine blood tests. Clinic Chem Lab Med (CCLM) 59(2):421–431
14. Elaziz MA, Hosny KM, Salah A, Darwish MM, Lu S, Sahlol AT (2020) New machine learning method for image-based diagnosis of COVID-19. PLoS ONE 15(6):e0235187
15. Tomar R, Tiwari R, Sarishma (2019) Information delivery system for early forest fire detection using internet of things. In: Singh M, Gupta P, Tyagi V, Flusser J, Ören T, Kashyap R (eds) Advances in computing and data sciences. ICACDS 2019. Communications in computer and information science, vol 1045. Springer, Singapore. https://doi.org/10.1007/978-981-13-9939-8_42
16. Assaf D, Gutman Y, Neuman Y et al (2020) Utilization of machine-learning models to accurately predict the risk for critical COVID-19. Intern Emerg Med 15:1435–1443
17. Maghdid HS, Asaad AT, Ghafoor KZ, Sadiq AS, Khan MK (2020) Diagnosing COVID-19 pneumonia from radiograph and CT images using deep learning and transfer learning algorithms. arXiv preprint arXiv:2004.00038
18. Oh Y, Park S, Ye JC (2020) Deep learning COVID-19 features on CXR using limited training data sets. IEEE Trans Med Imaging 39(8):2688–2700
19. Shorten C, Khoshgoftaar TM, Furht B (2021) Deep learning applications for COVID-19. J Big Data 8:18
20. Jain G, Mittal D, Thakur D, Mittal MK (2020) A deep learning approach to detect Covid-19 coronavirus with radiograph images. Biocyber Biomed Eng 40(4):1391–1405, ISSN 0208-5216
21. Rahimzadeh M, Attar A, Sakhaei SM (2021) A fully automated deep learning-based network for detecting COVID-19 from a new and large lung CT scan dataset. Biomed Sig Proc Control 68:102588. ISSN 1746-8094, https://doi.org/10.1016/j.bspc.2021.102588
22. Panwar H, Gupta PK, Siddiqui MK, Morales-Menendez R, Singh V (2020) Application of deep learning for fast detection of COVID-19 in radiographs using nCOVnet. Chaos, Solitons Fractals 138:109944. ISSN 0960-0779
23. Zhou T, Lu H, Yang Z, Qiu S, Huo B, Dong Y (2021) The ensemble deep learning model for novel COVID-19 on CT images. Appl Soft Comput 98:106885. ISSN 1568-4946. https://doi.org/10.1016/j.asoc.2020.106885
24. Narin A, Kaya C, Pamuk Z (2021) Automatic detection of coronavirus disease (COVID-19) using Radiograph images and deep convolutional neural networks. Pattern Anal Applic 24:1207–1220
25. Dewangan BK, Jain A, Choudhury T (2020) GAP: Hybrid task scheduling algorithm for cloud. Rev d'Intell Artif 34(4):479–485. https://doi.org/10.18280/ria.340413
26. Monaco CG, Zaottini F, Schiaffino S et al (2020) Chest radiograph severity score in COVID-19 patients on emergency department admission: a two-centre study. Eur Radiol Exp 4:68
27. Jaiswal A, Monika (2019) Pun detection using soft computing techniques. In: 2019 international conference on machine learning, big data, cloud and parallel computing (COMITCon). pp 5–9. https://doi.org/10.1109/COMITCon.2019.8862264
28. Kumar A, Jaiswal A, Empirical study of twitter and tumblr for sentiment analysis using soft computing techniques. In: Proceedings of the world congress on engineering and computer science, vol 1, iaeng.org
29. Chicco D, Jurman G (2020) The advantages of the Matthews correlation coefficient (MCC) over F1 score and accuracy in binary classification evaluation. BMC Genomics 21:6
30. Kumar A, Jaiswal A (2020) Particle swarm optimized ensemble learning for enhanced predictive sentiment accuracy of tweets. In: Singh P, Panigrahi B, Suryadevara N, Sharma S, Singh A (eds) Proceedings of ICETIT 2019. Lecture notes in electrical engineering vol 605. Springer, Cham

31. Jaiswal A, Malhotra R (2018) Software reliability prediction using machine learning techniques. Int J Syst Assur Eng Manag 9(1):230–244
32. Tomar R, Patni JC, Dumka A, Anand A (2015) Blind watermarking technique for grey scale image using block level discrete cosine transform (DCT). In: Satapathy S, Govardhan A, Raju K, Mandal J (eds) Emerging ICT for bridging the future—proceedings of the 49th annual convention of the computer society of India CSI Volume 2. Advances in intelligent systems and computing, vol 338. Springer, Cham. https://doi.org/10.1007/978-3-319-13731-5_10

Comparative Analysis of Machine Learning and Deep Learning Algorithms for Skin Cancer Detection

Nikita Thakur and Arunima Jaiswal

Abstract Skin cancer is a typical type of disease, and early recognition builds the endurance rate. Skin cancer is a perilous and far and wide sickness. The endurance rate is under 14% whenever analyzed in later stages. Notwithstanding, if skin cancer is recognized at the beginning phases, the endurance rate is almost 97%. This requests the early location of skin cancer. Motivated by the same we, in this research we implemented different machine learning and deep learning techniques for skin cancer detection. We performed the comparative analysis of various machine and deep learning models, implemented on a fixed dataset. The analysis was based on accuracy and it was observed that the deep learning techniques produced enhanced results. The various techniques we used in this paper are, CNN, RESNET, DECISION TREE, KNN, SVM, NAÏVE BAYERS, INCEPTION V3, VGG -16. Accuracy was used as the performance measure.

Keywords Machine learning · Deep learning · Disease detection · Image classification · Skin cancer first section

1 Introduction

Cancer is an illness wherein cells in the body outgrow abnormally. Right when cancer affects the skin, it is called skin disease. They start in the basal and squamous layers of the skin, separately. Melanoma, the third most normal kind of skin cancer, starts in the melanocytes. In the most key terms, disease implies cells that grow out of control and assault various tissues. Cells may become malignant due to the gathering of blemishes, or changes, in their DNA. Certain gained genetic defects (for example, BRCA1 and BRCA2 changes) and defilements can construct the risk of

N. Thakur (✉) · A. Jaiswal
Department of Computer Science, Indira Gandhi Delhi Technical University For Women, Delhi, India
e-mail: nikita112btcse19@igdtuw.ac.in

A. Jaiswal
e-mail: arunimajaiswal@igdtuw.ac.in

V. Skala et al. (eds.), *Machine Intelligence and Data Science Applications*, Lecture Notes on Data Engineering and Communications Technologies 132,
https://doi.org/10.1007/978-981-19-2347-0_32

malignant growth. An abnormal mass of cells is called a tumor. Tumors are further classified as benign and malignant. Benign tumors develop locally and do not spread hence, not viewed as cancer. They can in any case be hazardous, particularly if they press against imperative organs. Malignant tumors can spread and damage different tissues. There are various sorts of threats dependent on where a cancer tumor begins [1–5]. Melanoma which is a kind of skin malignant growth is by and large reparable when recognized and treated early. At whatever point melanoma has spread further into the skin or various bits of the body, it ends up being harder to treat and can be hazardous. The evaluated five-year perseverance rate for U.S. patients whose melanoma is perceived early is around near 100%. While the five-year perseverance speed of stage 3 and 4 melanomas are 63.6 and 22.5% separately. An anticipated 7180 people (4600 men and 2580 women) will fail miserably of melanoma in the U.S. in 2021.Of the 207,390 expected cases of cancer, 106,110 cases will be noninvasive and bound to the top layer of skin (in situ). An expected 101,280 cases will be obtrusive, entering into the skin's subsequent layer. As indicated by the American Cancer Society, the five-year endurance rate for Merkel cell stages 0, 1, and 2 is 78%. While it is 51% for stage 3 and 17% for stage 4 [6]. Image processing is having a very important role in the medical domain [7]. Image classification is an essential field of exploration. Expanding pace of media information, distant detecting, and web photograph display need a class of various images for the appropriate recovery of client. Different analysts apply unique approaches to image classification. Content of images like tone, surface, and shape also, size assumes a significant part in semantic image classification. Machine learning and deep technique improve the performance of image classification [8]. Different tests might be acted to affirm a cancer determination. Positron Outflow Tomography and Registered Tomography (PETCT) Outputs and other comparable tests can feature "problem areas" of cancer cells. There are a few quantities of systems which can be followed, the first can be an overall body test, A trial to inspect general signs of prosperity, including inspecting for indications of contamination, for instance, bulges or whatever else that gives off an impression of being odd. The different methodology may be, a skin checkup, a checkup of the skin for knocks or spots that look unusual in any form. Skin biopsy, all or part of the strange-looking development is removed from the skin and checked for cancer cells. There are four main types of this technique. The main types of biopsies are: shave biopsy, punch biopsy, incisional biopsy, and excisional biopsy. There are eight types of treatments that are used for the treatment of skin cancer. Each of these is explained as follows. Surgery might include a simple excision where the affected region is cut from the skin. Mohs micrographic surgery where the affected cells are cut in a thin layer, shave excision, where the affected area is shaved off, curettage and electrodesiccation, where the affected cells are removed from the skin with a tool called a curette, and cryosurgery, where a tool is used to freeze and destroy the affected tissues, laser surgery, which uses a narrow beam of light to remove affected cells, and dermabrasion, which is the removal of the top layer of skin by the use of a rotating wheel. Radiation therapy, which is the second form of treatment uses high-energy x-rays to treat the cancer cells or stop them from spreading them. Chemotherapy utilizes medicines to limit the extension of affected cells, this is done by either killing cells or limiting their division. Some type of light is used to treat the cancer cells in the technique called Photodynamic therapy. In the next treatment method, i.e., Immunotherapy, the affected person's immune system is used to fight the disease. Targeted therapy is yet another form of skin cancer treatment, where drugs or some other substances are used to find and fight the affected cells. A chemical peel is the method where a chemical mixture is applied to the skin. The other drug therapies include the usage of retinoids to fight the cancer on the skin. The risk to life due to skin cancer depends on the thickness of the tumor formed. The growth of the tumor depends on the time for which the disease goes without identification. Early detection is the only way to an increase positive recovery from the disease and lowers the death rates.

In this paper, we studied the various techniques that can be used to detect and classify skin cancer using a dataset of images of skin cancer-affected parts of the body. The various techniques we used in this paper are, CNN, RESNET, DECISION TREE, KNN, SVM, NAÏVE BAYERS, INCEPTION V3, VGG -16. Accuracy was used as the performance measure. Hence through this paper, the main aim is to tackle the problem of the long and time taking process of skin cancer detection and its further classification. The paper has been coordinated as follows. A brief idea about the past works done for image classification and image detection is given in Sect. 2. Section 3 contains the system architecture and describes its details. Whereas the following sections provide the results and conclusion, respectively.

2 Related Works

Lately, the occurrence of skin cancer cases has kept on heightening quickly and the condition currently influences a huge number of individuals around the world, basically because of the delayed treatments. In the course of recent many years or somewhere in the vicinity, numerous specialists and researchers have been drawn to foster superior programmed strategies for skin cancer discovery by the use of image classification. In [9], a programmed technique for the division of pictures of skin malignancy and other pigmented sores is introduced. This technique initially diminishes a color picture into an intensity picture and around fragments the picture by intensity thresholding. Then, at that point, it refines the division utilizing picture edges.

In [10] the highlights extricated from skin lesions are taken care of to characterize skin injuries into one of two particular classifications: cancerous or benign, by just contrasting the component boundaries and the predefined limits. [11] discusses how Machine learning has emerged as a very useful component of intelligent computer vision programs. With the advent of image datasets and benchmarks, machine learning and image processing have lately gained a lot of importance. A creative coordination of machine learning and image processing is probably going to have an extraordinary advantage to the field, which will add to a superior comprehension of mind-boggling pictures. While [12] presents that deep learning is viewed as additionally giving invigorating and precise answers for clinical imaging, and is viewed as a critical technique for future applications in the medical services. Mhaske and Phalke [3]: Different Data Pre-handling techniques, Disease Diagnosis, Maximal Frequent Itemset algorithm for preparing, K-means grouping for division, and huge regular patterns for characterization have been examined in this paper. This work proposed a powerful structure of a CAD framework for melanoma skin cancer mostly by utilization of an SVM model. Assessments on an enormous dataset of dermoscopic pictures have exhibited that the proposed system shows better execution as far as execution files of affectability, explicitness, and precision by accomplishing 98.21, 96.43, and 97.32%, separately, without losing continuous consistence. [13] talks about the execution of strategies, for example, dermoscopy or epiluminescence light microscopy (ELM) in aiding the deadliest form of skin cancer, that is, melanoma.

Jain et al. [14] published by IEEE reported the figures and the outcomes of the most important implementations existing in the field. A correct and efficient diagnosis of skin cancer is equally as important as an early diagnosis, which can be achieved by first correctly classifying the affected cells as a particular type of skin cancer. The two types of classes for the same are, Benign and Malignant. The correct classification initiates a better approach to treating the disease resulting in positive results. The work proposed in Jain et al. [14] also presented the methods for image segmentation of skin lesions, and further classify them as benign or malignant. A three-step procedure is followed as (i) Image segmentation using suitable methods (ii) Selecting the most accurate segmentation results, and (iii) classification of the lesion as malignant or benign. Maglogiannis and Doukas [15] also implemented the classification of the skin tumors by implementing a deep learning algorithm. In this study, the aim was to classify skin lesions using the CNN model [16].

3 Dataset Description

The initial step for working on the topic of research is to collect a suitable dataset. The dataset used here is a part of the ISIC (International Skin Image Collaboration) Archive. It comprises of a set of 1601 and 1304 pictures of the benign and malignant skin cells, respectively. From the total images, approximately 8% of the data is taken for testing and 91% of the data is considered for training purposes. The resolution of the pictures has been set to low (224 × 224 × 3) RGB [17]. After taking the dataset, as the next step, various libraries are imported for the model. The images from the dataset are loaded and then turned into NumPy arrays using their RGB values. Due to the already low resolution, the images are not needed to be resized. In the end, the pictures are added together to a big training set and shuffled. The values of the pictures are needed to be normalized; this is done by splitting up the Red Green Blue values by a value of 255. After these steps, the model building phase is started. Image classification has become the focal point of exploration in recent years because of its variety and intricacy of data. Through this study, different models are implemented using the given dataset for the results as classification of the skin lesions.

4 Application of Techniques

The various models implemented in this study are as follow (Table 1).

5 Experiment and Implementation Section

5.1 Libraries Used

The libraries and modules used for implementing the models are, cv2 from OpenCV, Keras, NumPy, Pandas, Random, PIL (Python Imaging Library), tqdm, Tensorflow, Matplotlib.pyplot, IPython.display, sklearn, glob, seaborn, skimage.

Opencv is an open-source library used for computer vision applications. Keras is a python library used for deep learning models. For working with the arrays in Python the numpy library is imported. Pandas library assists with data structure and data analysis tools, it runs with numpy. The Random module is used to generate random numbers. The image processing is done with the help of the PIL (PythoN Image Library). Tensorflow is an artificial intelligence library, which is used for computing of numerical data. For the visualization of data, the matplotlib library is used in python, seaborn library runs on its top. Many useful tools for machine learning are provided by sklearn library [22, 23].

5.2 Implementation

After importing the libraries and modules, the train and test data set was imported. The images are then converted into array. The feature extraction for each model and the model building followed the process. As a result, the model accuracies are attained in the end.

6 Result

In this section, results are discussed which are obtained upon application of aforesaid machine learning and deep learning techniques using accuracy as the performance measure. Accuracy is defined as the closeness of the measurements to a specific value.

Table 1 Description of techniques used

Heading level	Example
ResNet 50	The skip associations in ResNet 50 tackle the issue of disappearing inclination in profound neural organizations by permitting substitute easy routes way for the slope to move through. The alternate way that these associations help is by permitting the model to become familiar with the character capacities which guarantees that the higher layer will perform basically as great as the lower layer, and not more terrible
Logistic regression	Logistic regression is a model acquired by machine learning. It is taken from the field of measurements
Random forest	In this model numerous decision trees are made and then combined to obtain more enhanced results
K nearest neighbor	*K* Nearest Neighbor is one of the simplest Machine Learning algorithms. In this algorithm, the unidentified data points are classified by finding the most common class which is available among the *k*-closest examples
Convolutional neural network	Convolutional neural networks (CNNs) demonstrated a significant performance in various visual recognition problems [18]. Their capacity can be obliged by changing their significance and broadness, and they moreover make strong and generally right assumptions about pictures (explicitly, stationarity of estimations and domain of pixel conditions)
SVM characterization	SVM portrayal uses different planes in space to parcel data centers using planes. An SVM model is a depiction of the models as focused in space, arranged with the objective that the occurrences of the different groupings or classes are separated by a disconnecting plane that helps the edge between different classes. This is a direct result of the truth if the secluding plane has the greatest distance to the nearest getting ready data points of any class. The test centers or request centers are then arranged into that equivalent space and expected to have a spot with a class considering which side of the opening they fall on [19]
Decision tree algorithm	The decision tree algorithm basically breaks down a dataset into minor subsets and a connected decision tree is created, which has decision nodes and leaf nodes
Naive Bayers method	The naive Bayers method simply implements the Bayes theorem with the "naive" presumption that every two features are independent. Only a little portion of training data is needed for the semination for necessary parameters
Inception-V3	Inception-v3 is a convolutional neural network design that makes a few improvements including utilizing label smoothing, factorized 7×7 convolutions, and the utilization of an assistant classifier to engender name data lower down the network [20]
Vgg16	VGG16 is a convolutional neural network architecture and it is considered to be an excellent vision model [21]

It is a way of assessing any given model. In this study, the accuracy of the different Machine learning and deep learning models and algorithms are implemented. The approximate accuracies of the various machine learning (Fig. 1) and deep learning techniques (Fig. 2) for skin cancer detection and classification into benign and malignant tumors are presented below in the form of a graph (Fig. 3). The accuracies of the deep learning techniques are comparatively higher than the machine learning techniques as found from the observations. The Vgg16 model shows the least accuracy while the Inceptionv3 model shows the highest accuracy among all the given models. The least accuracy in the case of the vvg16 model might be due to a high training loss value. That is, for the given dataset, there was a greater loss during training of the dataset which resulted in a lesser accuracy. The high accuracy of the Inception V3 can be explained due to the changes implemented in the inception model. Specific changes in the previous model have led to significant improvements in the Inception v3 model.

Fig. 1 Comparative analysis of various machine learning techniques

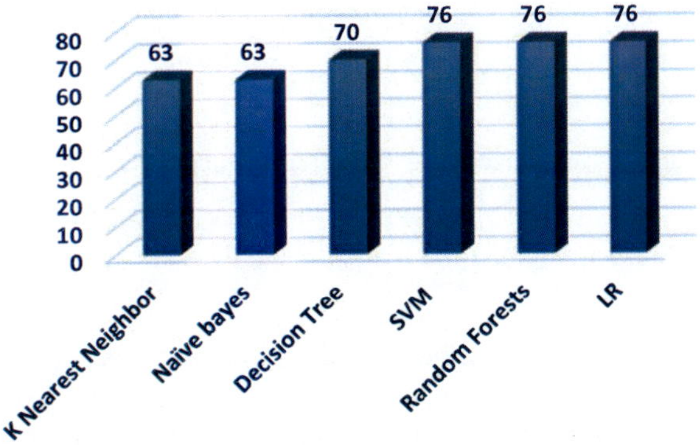

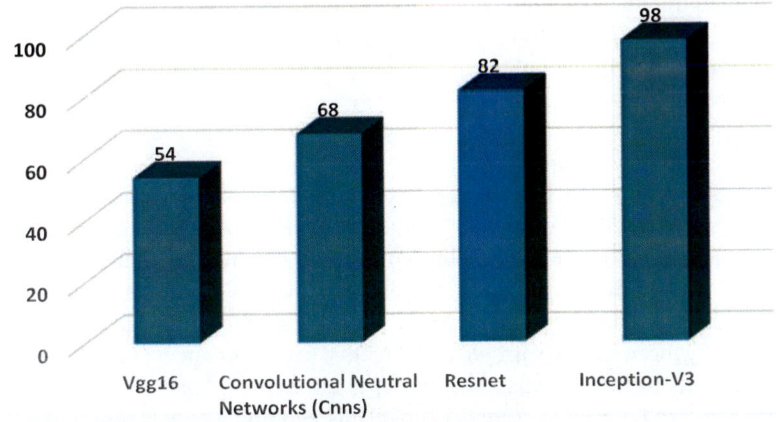

Fig. 2 Comparative analysis of various deep learning techniques

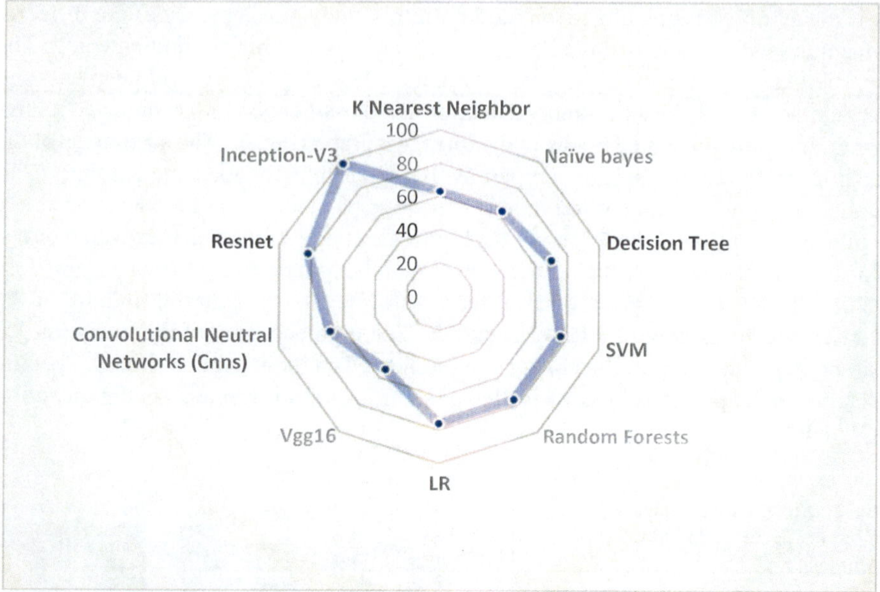

Fig. 3 Machine learning versus deep learning techniques

7 Conclusion

In this study of skin cancer detection and classification into the two categories, that is, benign and malignant tumors we implemented various machine learning and deep learning techniques and algorithms for image classification and detection on the given dataset. The accuracy of the inception v3 model, which is a deep learning model, is the highest as obtained by the study.

8 Future Scope

Due to a lack of time, a further deeper study on the various machine and deep learning models and their outcomes could not be done. All these experiments and implementations can be conducted in the future to expand the results and learning outcomes. In the future, the performance and efficiency of the models can be increased by taking a bigger dataset with more vivid and diverse skin sample images. This will make the study more globally usable and effective to detect cancer cases for providing early treatments to the patients and save lives ultimately. Apart from accuracy, the $F1$ score can also be used to compare the results and find a conclusion in the future. The $F1$ score is a better measure for comparing the various techniques as it is better to evaluate the imbalanced class distribution. Other various techniques and models can also be implemented in the future for studies and detection.

References

1. Ahmed K, Jesmin T,Early D (2013) Prevention and detection of skin cancer risk using data miningdz. Int J Comput Appl 62(4)
2. Alquran H et al (2017)The melanoma skin cancer detection and classification using support vector machine. In: 2017 IEEE Jordan conference on applied electrical engineering and computing technologies (AEECT), pp 1–5. https://doi.org/10.1109/AEECT.2017.8257738
3. Mhaske HR, Phalke DA (2013)Melanoma skin cancer detection and classification based on supervised and unsupervised learning. In: 2013 international conference on circuits, controls and communications (CCUBE), pp 1–5. https://doi.org/10.1109/CCUBE.2013.6718539
4. Hossin MA, Rupom FF, Mahi HR, Sarker A, Ahsan F, Warech S (2020) melanoma skin cancer detection using deep learning and advanced regularizer. Int Conf Adv Comput Sci Inf Syst (ICACSIS) 2020:89–94. https://doi.org/10.1109/ICACSIS51025.2020.9263118
5. Choudhury T, Aggarwal A, Tomar R (2020) A deep learning approach to helmet detection for road safety. J Sci Ind Res (JSIR) 79(06):509–512
6. arXiv:1512.00567 [cs.CV] Rethinking the inception architecture for computer vision
7. Vidya M, Karki MV (2020) Skin cancer detection using machine learning techniques. In: 2020 IEEE international conference on electronics, computing and communication technologies (CONECCT), pp. 1–5. https://doi.org/10.1109/CONECCT50063.2020.9198489
8. Kumar S, Khan Z, Jain A (2012) Int J Adv Comput Res Bhopal 2(3):55–60
9. Xu L, Jackowski M, Goshtasby A, Roseman D, Bines S, Yu C, Dhawan A, Huntley A (1999) Segmentation of skin cancer images. Image Vis Comput 17(1)
10. Bakheet S (2017) An SVM framework for malignant melanoma detection based on optimized HOG features. Computation 5:4. https://doi.org/10.3390/computation5010004t
11. Machine Learning in Image Processing Olivier Lezoray, ´ 1 Christophe Charrier,1 Hubert Cardot, 2 and Sebastien Lef ´ evre ‘ 3
12. Razzak MI, Naz S, Zaib A (2018) Deep learning for medical image processing: overview, challenges and the future. In: Dey N, Ashour A, Borra S (eds) Classification in BioApps. Lecture notes in computational vision and biomechanics, vol 26. Springer, Cham. https://doi.org/10.1007/978-3-319-65981-7_12
13. Jain S, Jagtap V, Pise N (2015) Computer aided melanoma skin cancer detection using image processing. Procedia Comput Sci 48:735–740. ISSN 1877-0509
14. Maglogiannis I, Doukas CN (2009) Overview of advanced computer vision systems for skin lesions characterization. IEEE Trans Inf Technol Biomed 13(5):721–733. https://doi.org/10.1109/TITB.2009.2017529 Epub 2009 Mar 16 PMID: 19304487
15. Krizhevsky A, Sutskever I, Geoffrey E. Hinton ImageNet classification with deep convolutional neural networks
16. Dinote A, Sharma DP, Gure AT, Singh BK, Choudhury T (2020) Medication processes automation using unified green computing and communication model. J Green Eng 10(9):5763–5778
17. https://www.isic-archive.com/#!/topWithHeader/wideContentTop/main)
18. Kim J, Kim B-S, Savarese S, Comparing image classification methods: *K*-nearest-neighbor and support-vector machines
19. Vijayalakshmi MM (2019) Melanoma skin cancer detection using image processing and machine learning. Pub Int J Trend Sci Res Dev (ijtsrd) 3(4):780–784. ISSN: 2456-6470, URL: https://www.ijtsrd.com/papers/ijtsrd23 936.pdf
20. Wang C et al (2019) Pulmonary image classification based on inception-v3 transfer learning model. In: IEEE Access, vol. 7, pp. 146533–146541. https://doi.org/10.1109/ACCESS.2019.2946000; Han SS, Kim MS, Lim W, Park GH, Park I, Chang SE (2018) Classification of the

clinical images for benign and malignant cutaneous tumors using a deep learning algorithm. J Invest Dermatol 138(7):1529–1538. doi: https://doi.org/10.1016/j.jid.2018.01.028. Epub 2018 Feb 8. PMID: 29428356.Convolutional neural networks for hyperspectral image classification, Shiqi Yua Sen Jiaa Chunyan Xub,

21. Qassim H, Verma A, Feinzimer D (2018) Compressed residual-VGG16 CNN model for big data places image recognition. In: 2018 IEEE 8th annual computing and communication workshop and conference (CCWC), pp. 169–175. https://doi.org/10.1109/CCWC.2018.8301729
22. Kumar S, Tomar R (2018) The role of artificial intelligence in space exploration. In: 2018 International conference on communication, computing and internet of things (IC3IoT), pp 499–503. https://doi.org/10.1109/IC3IoT.2018.8668161
23. Sarishma, Tomar R, Kumar S, Awasthi MK (2021) To beacon or not?: speed based probabilistic adaptive beaconing approach for vehicular Ad-Hoc networks. In: Paiva S, Lopes SI, Zitouni R, Gupta N, Lopes SF, Yonezawa T (eds) Science and technologies for smart cities. SmartCity360° 2020. Lecture notes of the institute for computer sciences, social informatics and telecommunications engineering, vol 372. Springer, Cham. https://doi.org/10.1007/978-3-030-76063-2_12

Investigating Efficacy of Transfer Learning for Fruit Classification

Vikas Khullar, Raj Gaurang Tiwari, Ambuj Kumar Agarwal, and Alok Misra

Abstract Automated artefact identification and classification are a highly coveted field of study in a wide variety of commercial fields. While humans can easily discern objects with a high degree of multi-granular similarity, computers face a much more difficult challenge. In several deep learning technologies, transfer learning have shown efficacy in multi-level subject classification. Traditionally, current deep learning models train and test on the transformed features created by the rearmost layer. The objective of this research paper is to fabricate an automated and efficient method of fruit classification using deep learning techniques. Since the algorithm is automatic, it does not require human involvement, and the mechanism is more accurate than human-involved processes. For the classification of fruits, a pre-trained deeply trained model is fine-tuned. To distinguish fruits, we used transfer learning-trained architecture such as VGG16, InceptionV3, ResNet50, DenseNet, and InceptionResNetV2 models. The Fruits-360 dataset is used to conduct the evaluation. Extensive testing reveals that the InceptionResNetV2 outperforms in comparison to other deep learning methods.

Keywords Machine learning · Fruit classification · Supervised learning · Object identification · Deep learning · Transfer learning

V. Khullar (✉) · R. Gaurang Tiwari · A. Kumar Agarwal
Chitkara University Institute of Engineering and Technology, Chitkara University, Punjab, India
e-mail: vikas.khullar@gmail.com

R. Gaurang Tiwari
e-mail: rajgaurang@gmail.com

A. Kumar Agarwal
e-mail: ambuj4u@gmail.com

A. Misra
Institute of Engineering and Technology, Lucknow, Uttar Pradesh, India
e-mail: amisra.csed.cf@ietlucknow.ac.in

V. Skala et al. (eds.), *Machine Intelligence and Data Science Applications*, Lecture Notes on Data Engineering and Communications Technologies 132,
https://doi.org/10.1007/978-981-19-2347-0_33

1 Introduction

Today, precise classification of various fruit types is a hot topic. It is a matter of importance not only in academic study, but also in commercial applications. Numerous practical applications can be developed using this classification scheme. One of the primary uses is to assist cashiers in supermarkets. Cashiers must be able to distinguish not just the species of fruit purchased by the consumer, but also the range in order to accurately price it. Such classification-based applications will automatically classify the type a consumer purchases and compare it to the right price [1].

Owing to the digitisation of world, with more and more flowing in of text data, image data, voice data, and sensor data which are most commonly unstructured, new techniques are needed to analyse them. The scenario what we had few decades back is not the same now. With the rise in technologies like cloud computing, big data, machine learning and graphics processing unit (GPU) architectures, machines are proving near human intelligence. Machine learning finds major use in applications like face recognition, speech recognition, object detection, spam detection, pedestrian detection, information retrieval, drug discovery, and many more. Classic machine learning techniques have two main components: feature extraction and classification/detection. Deep learning is a kind of machine learning technique in which the feature extraction is automated. With deep learning, the features are learned on its own with simpler concepts building the complex ones.

In recent years, the fruit industry has seen remarkable advancements due to advent of computer vision and machine learning. Convolutional neural networks (CNNs) are a form of deep learning that has been widely used in the field of computer vision in recent years for tasks such as image recognition, object detection, image captioning, handwritten digit recognition, face recognition, image segmentation, action recognition, pedestrian detection, and so on. CNNs are neural networks consisting of multiple layers which identifies the features layer by layer and constructs the feature representations. Early layers detect simple patterns which collectively generate complex patterns or abstractions in latter layers [2].

The training images are fed to the network which in every layer convolutes with some set of weights and propagates through all the layers (in the forward direction). At the end of the network, the difference is computed using loss function (errors). Based on these errors the weights are adjusted in every layer while propagating in the backward direction. The weights are adjusted by some optimisation function. A complete cycle of forward and backward propagation corresponds to a single iteration during which the weights are updated and this continues until convergence. The CNN has been implemented in a variety of areas over the last few years, including object recognition, image analysis, and video classification. Numerous studies have been performed in the last few years to determine the identification, classification, and ranking of fruits on-tree [3].

Transfer learning is a critical component of machine learning because it address the fundamental issue of inadequate training data. The aim of this approach is to take the information gained from one task (problem) and apply it to related work by solving

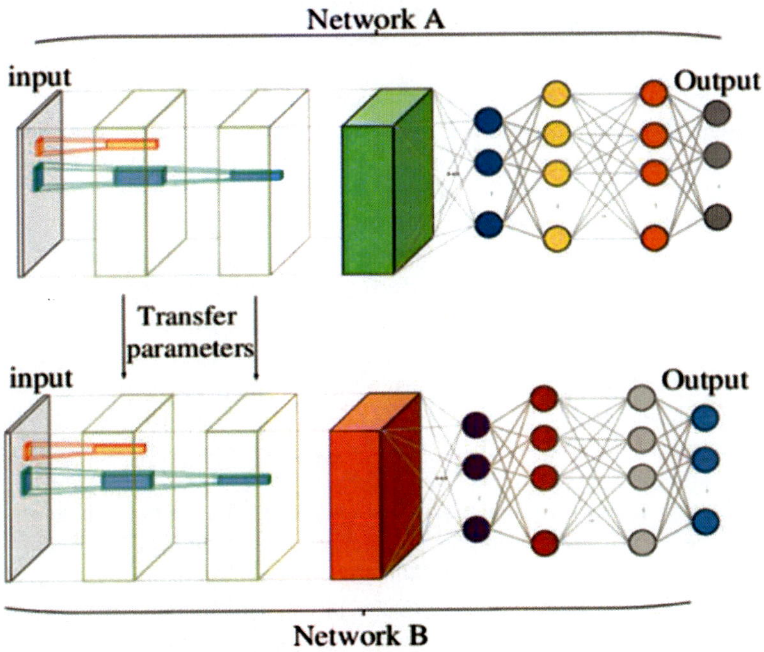

Fig. 1 Process of transfer learning [3]

isolation learning problems. This accumulated experience provides inspiration to address the issue of multiple fields that face difficulties improving as a result of insufficient or incomplete training data. Figure 1 illustrates the transfer learning mechanism [2, 3].

Over the last several years, several variants of CNN architectures have been created, including GoogleNet [3], ResNet50 [4], InceptionV3 [5], and Inception-ResNetV2 [6]. By the depth and breadth of the network, the inception architecture with residual links works more effectively. The recently released InceptionResNetV2 was selected as the basis for our proposed approach due to its superior success on ImageNet in the 2015 ImageNet Large Scale Visual Recognition Competition (ILSVRC) challenge [6].

The remainder of the paper is organised in the subsequent way. Section 2 contains information about previous studies in the same field. Section 3 discusses the suggested process. Section 4 contains the investigational findings. Lastly, in Sect. 5, the conclusions are summarised.

2 Related Studies

Numerous systems for automating the examination of fruits for faults, maturity process detection, and classification can be found in the literature. Jose Luis Rojas-Aranda [7] proposed a framework for image recognition based on lightweight CNN with the aim of accelerating the checkout procedure in shops. A new image dataset was launched that included three types of fruits, both with and without plastic bags. To improve classification performance, the CNN architecture was enhanced with additional input functionality. The RGB histogram, a single RGB colour, and the RGB centroid obtained through K-means clustering are all valid inputs. The findings indicated an average classification precision of 95% for fruits without a plastic bag and 93% for fruits with a plastic bag. Xiang et al. [8] developed a system for classifying fruit images using a lightweight neural network called MobileNetV2 and a transfer learning technique. They started with a MobileNetV2 network that had been pre-trained on the ImageNet dataset and then replaced the top layer with a classical convolutional layer and a Softmax classifier. They simultaneously applied dropout to the newly introduced conv2d to minimise overfitting. The features were extracted using the pre-trained MobileNetV2 and classified using the Softmax classifier. They learned this new model in two steps, each with a different learning rate, using the Adam optimiser. Their approach obtained a precision of 85.12% in classifying 3670 images of five fruits in image dataset. Duong [9] suggested a realistic approach to the problem of fruit recognition by using EfficientNet and MixNet, classifiers, which are two deep neural network families, to create an expert system capable of efficiently and rapidly identifying fruits. Such a framework may be implemented on low-resource devices to generate precise and well-timed endorsements. The method's success was tested using a real-world dataset comprised of 48,905 training images and 16,421 research images. The investigational findings indicate that when EfficientNet and MixNet are applied to the deliberated dataset, the inclusive accuracy significantly increases when compared to a well-established baseline. Behera [10] suggested a support vector machine (SVM) classification algorithm for 40 types of Indian fruits based on deep features derived from completely connected layer of a CNN model. Additionally, a transfer learning-based technique was recommended for the identification of Indian fruits. The tests were conducted in six of the most popular deep learning architectures available, including GoogleNet, AlexNet, ResNet-18, ResNet50, VGGNet-16 and VGGNet-19. The test findings indicated that the SVM classifier with deep learning capability outperforms its transfer learning counterparts. The deep learning capability of SVM and VGG16 results in a maximum of 100% accuracy, precision, and F1 ranking.

3 Materials and Methods

3.1 Dataset

All pictures from the Fruit-360 dataset which is available on Kaggle were selected for training and testing. The dataset comprises 90,483 images of fruits in various categories (100 × 100 pixels) [11].

3.2 Methodology

Recently, deep learning frameworks built on the transfer learning architecture have been used to solve computer vision issues. To distinguish fruits, we used deep CNN architecture-based VGG16, InceptionV3, ResNet50, DenseNet, and InceptionResNetV2 models in conjunction with transfer learning techniques. Transfer learning also aids in the management of insufficient data and model execution time. Figure 2 shows a graphical representation of the architectures used, along with pre-trained models for implementation of transfer learning.

4 Results

We tested classification model accuracy using three different deep neural network architectures and the same datasets in this study. We used deep learning models (i.e. VGG16, InceptionV3, ResNet50, DenseNet, and InceptionResNetV2) from pre-trained CNN networks to do fine-tuning based on transfer learning as shown in Fig. 3. Here, VGG and InceptionV3 responded below expectation with very low accuracy and other parameters. However, ResNet50, DenseNet, and InceptionResNetV2, resulted top parametric outcomes in terms of accuracies greater than 99% with balanced precision, recall, and area under curve and also with least validation losses. As mention in Tables 1 and 2, InceptionResNetV2 achieves the highest precision (0.99), recall (0.99), training and validation accuracy (0.99 and 0.98), and area under curve (0.99) values, thus incurring the lowest training and validation losses (0.003 and 0.066).

5 Conclusion

Recognising objects is a critical component of computer vision and artificial intelligence in general. Strong vision models are critical enablers for artificial intelligence applications that can interpret visual inputs. Very deep convolutional networks have

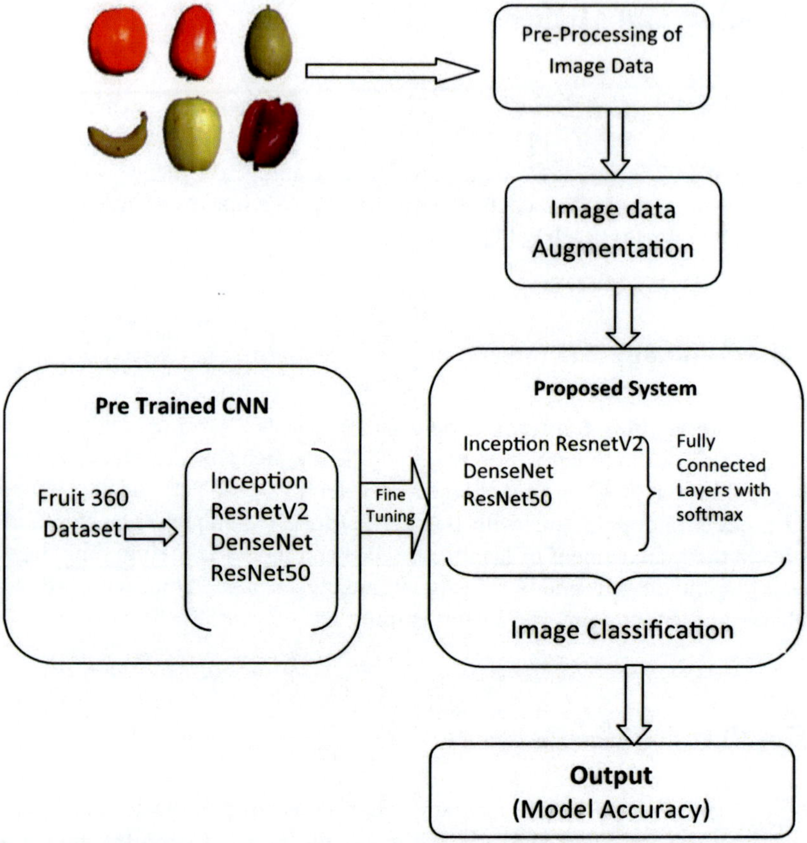

Fig. 2 Proposed framework

been instrumental in recent years' most significant improvements in image recognition efficiency. In this research work, we use VGG16, InceptionV3, ResNet50, DenseNet, and InceptionResNetV2 classification using deep features derived from transfer learning models' completely linked layers. It demonstrates that classification model outputs vary in arithmetical value with statistical implication. The test findings indicate that the InceptionResNetV2 classifier with higher accuracy, precision, recalls, and least losses. Hence, paper proposes the utilisation of transfer learning models for classification of different fruits using computer vision.

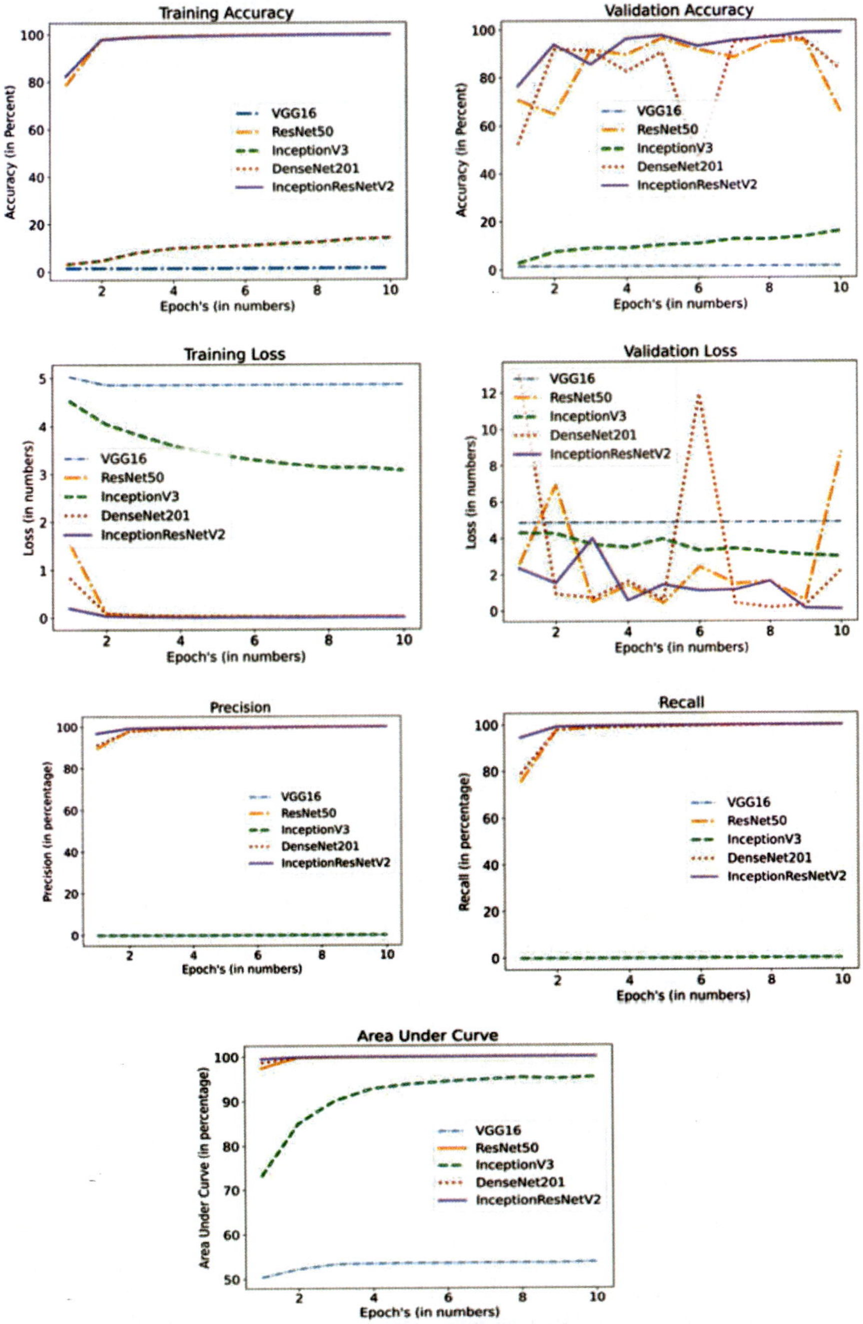

Fig. 3 Comparative results of transfer learning on fruit-360 dataset

Table 1 Transfer learning implementation results (training- validation accuracy loss) on fruit-360 dataset

Algorithm	Training accuracy	Validation accuracy	Training loss	Validation loss
VGG16	0.0145	0.0145	4.857	4.8564
InceptionV3	0.1403	0.16	3.069	2.9663
ResNet50	0.9983	0.6558	0.009	8.7435
DenseNet201	0.9977	0.8304	0.011	2.2123
InceptionResNetV2	0.9994	0.9878	0.003	0.0667

Table 2 Transfer learning implementation results (precision, recall, and auc) on fruit-360 dataset

Algorithm	Precision	Recall	Area under curve
VGG16	0	0	0.5376
InceptionV3	0	0	0.9541
ResNet50	0.9983	0.998	0.9997
DenseNet201	0.9977	0.998	0.9996
InceptionResNetV2	0.9994	0.999	0.9999

References

1. Kang H, Chen C (2020) Fast implementation of real-time fruit detection in apple orchards using deep learning. Comput Electron Agric 168:105108
2. Dandekar M, Punn NS, Sonbhadra SK, Agarwal S (2020) Fruit classification using deep feature maps in the presence of deceptive similar classes. arXiv preprint arXiv:2007.05942
3. Szegedy C, Liu W, Jia Y, Sermanet P, Reed S, Anguelov D, Erhan D, Vanhoucke V, Rabinovich A (2015)Going deeper with convolutions. In 2015 IEEE conference on computer vision and pattern recognition (CVPR), pp 1–9
4. He K, Zhang X, Ren S, Sun J (2016) Deep residual learning for image recognition. In: IEEE conference on computer vision and pattern recognition (CVPR), pp 770–778
5. Szegedy C, Vanhoucke V, Ioffe S, Shlens J, Wojna Z (2016) Rethinking the inception architecture for computer vision. In: 2016 IEEE conference on computer vision and pattern recognition (CVPR), pp 2818–2826
6. Szegedy C, Ioffe S, Vanhoucke V, Alemi AA (2017) Inception-v4, inception-resnet and the impact of residual connections on learning. In: AAAI
7. Rojas-Aranda JL, Nunez-Varela JI, Cuevas-Tello JC, Rangel-Ramirez G (2020) Fruit classification for retail stores using deep learning. In: Mexican conference on pattern recognition. Springer, Cham, pp 3–13
8. Xiang Q, Wang X, Li R, Zhang G, Lai J, Hu Q (2019) Fruit image classification based on Mobilenetv2 with transfer learning technique. In: Proceedings of the 3rd international conference on computer science and application engineering, pp 1–7
9. Duong LT, Nguyen PT, Di Sipio C, Di Ruscio D (2020) Automated fruit recognition using efficient net and MixNet. Comput Electron Agric 171:105326
10. Behera SK, Rath AK, Sethy PK (2020) Fruit recognition using support vector machine based on deep features. Karbala Int J Mod Sci 6(2):16
11. Mureşan H, Oltean M (2017) Fruit recognition from images using deep learning. arXiv preprint arXiv:1712.00580

BERT-Based Secure and Smart Management System for Processing Software Development Requirements from Security Perspective

Raghavendra Rao Althar and Debabrata Samanta

Abstract Software requirements management is the first and essential stage for software development practices, from all perspectives, including the security of software systems. Work here focuses on enabling software requirements managers with all the information to help build streamlined software requirements. The focus is on ensuring security which is addressed in the requirements management phase rather than leaving it late in the software development phases. The approach is proposed to combine useful knowledge sources like customer conversation, industry best practices, and knowledge hidden within the software development processes. The financial domain and agile models of development are considered as the focus area for the study. Bidirectional encoder representation from transformers (BERT) is used in the proposed architecture to utilize its language understanding capabilities. Knowledge graph capabilities are explored to bind together the knowledge around industry sources for security practices and vulnerabilities. These information sources are being used to ensure that the requirements management team is updated with critical information. The architecture proposed is validated in light of the financial domain that is scoped for this proposal. Transfer learning is also explored to manage and reduce the need for expensive learning expected by these machine learning and deep learning models. This work will pave the way to integrate software requirements management practices with the data science practices leveraging the information available in the software development ecosystem for better requirements management.

Keywords Software development · Requirements management · Bidirectional encoder transformers · Knowledge graphs · Transfer learning

R. R. Althar
Data Science Department, CHRIST (Deemed to be University), Bangalore, Karnataka, India
e-mail: raghavendra.a@res.christuniversity.in

Specialist-QMS, First American India Private Ltd., Bangalore, Karnataka, India

D. Samanta (✉)
Department of Computer Science, CHRIST (Deemed to be) University, Bangalore, Karnataka, India
e-mail: debabrata.samanta369@gmail.com

V. Skala et al. (eds.), *Machine Intelligence and Data Science Applications*, Lecture Notes on Data Engineering and Communications Technologies 132,
https://doi.org/10.1007/978-981-19-2347-0_34

1 Introduction

The need for a focused security management system at the requirements phase is due to the impact of security vulnerabilities in the software application. More robust analytical methods are needed as there are limitations in the current systems. Since there is a dependency on the vulnerability being exposed to the industry, not all the vulnerabilities are reported as Common Vulnerabilities Exposures (CVE) [1, 2]. In some cases, the time taken to expose the vulnerabilities may be too high and costly. CVEs also have limitations of the hackers knowing the vulnerabilities and getting ready to exploit the same. The software has many dependencies on the publicly available components; these may have vulnerabilities that are not known. Managing these vulnerabilities across these open-source components is a challenge [3].

Though security vulnerabilities are taken seriously, when it comes to managing them in the software development life cycle, it takes a back step due to the amount of effort required. One concern is the information explosion; with much of the knowledge available, catching up with that ever-growing information is a challenge. There is a need to build a system that can consume all this information periodically and grow smarter. Information from this system can feed in with crucial information and the software development community [4, 5]. Requirements manager is right at the leading stage of software development; they need to be focused while this system is built. Mostly, these players are oriented toward business and less toward technical requirements. The maintenance of these data management pipelines decides on the durability of the system. With all these factors under consideration, an exploration of building a smart requirements management system (SRMS) thought process is constructed. The paper is organized into sections for review of the literature, approach for smart requirements management system, and role of BERT [6–8]. Further section covers knowledge graph application in smart requirements management system, knowledge graphs role in the modeling industry, best practice of security for software development, understanding the landscape of requirements in title insurance domain and agile model of development. Paper concludes with leveraging transfer learning to cross-learn across applications.

2 Review of Literature

Work in [9] focuses on leveraging the abundant pattern hidden in the software source code. They intend to compare the naturalness of the code with natural language. There is a comparison of machine learning, software engineering, and programming language to derive the best aspects. Probabilistic modeling is explored in light of all critical elements of natural language and programming language. Design principles of the models are explored in this study, and a review of the literature is conducted. The authors also examine the applications of these concepts for software applications.

This work also inspires to leverage the statistical modeling of software source code for the application already in production.

Learning gathered from this can be fed back to the requirements management phase for better management of requirements. Work in [10] focuses on getting the bird's eye view of requirements engineering. The current state of practices in machine learning and software engineering practices is explored with key machine learning categories such as regression, classification, and clustering. Some standard machine learning models such as decision tree, *k*-nearest neighbor, and Naïve Bayes are explored for requirements engineering. Vector space modeling is explored for data preprocessing. Publicly available requirements data is used as a data set in this study. Some of the observations made in this study are the need for automated requirements elicitation to optimize the time involved. Classification and clustering are used to categorizing the data gathered from the public domain. Non-functional requirements (NFR) are identified later in software development life cycle (SDLC), so there is a need to identify them early in SDLC. Also, NFRs are in a scattered form that needs to be streamlined for better management. Early identification of NFRs also ensures that the risk associated with SDLC is considerably reduced. Prioritization of the requirements is another area that needs attention. Understanding security-related requirements, among other NFRs, is one of the prominent aspects. There is an emphasis on leveraging the visualization approaches to get a better insight into the requirements data. Organization of the requirements-related document is another area that needs attention. Natural language processing (NLP) is primarily used to preprocess the requirements data to be conceived by machine learning models to identify patterns from them. Requirements management in software development is intuition based mainly, and there is potential to make it objective and scientific by leveraging machine learning approaches. Parsing the requirements data and summarizing them takes a lot of time. Machine learning will have a crucial role in mining the data from the non-formal format and free-form text. The drawback of this study is that the experiments are not validated in industrial settings. Some of the critical insights that can be derived from reviewing this paper are as follows. Insurance domain-specific requirements data can be targeted for the study, and classification approaches can experiment. Identification of security-related needs from this domain can be explored. The agile model of development in SDLC can be assessed to discover how it influences requirements management. There would be a focus on enhancing the interpretability of the models.

3 Approach for Smart Requirements Management System

Requirement management optimization in software development life cycle (SDLC) involves a critical aspect of ensuring complete security-related requirements which are identified and lined up for addressing during SDLC. Some of the concerns that block this ideal situation are the lack of security focus in the requirements discussion between the customer and the requirements management team. The conversation

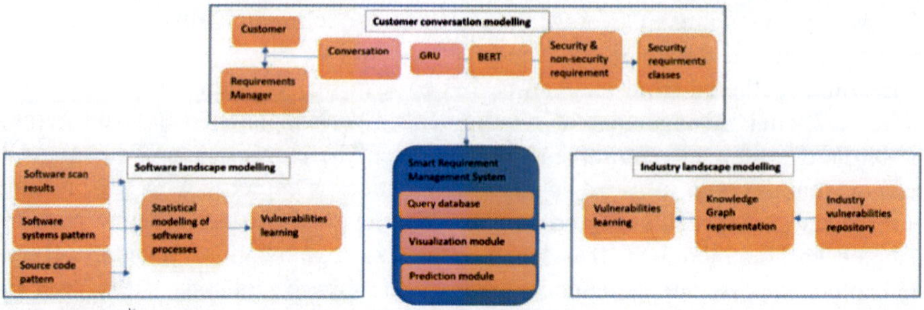

Fig. 1 Architecture of smart requirements management system

primarily focuses on the business, and related security priorities lose prominence and cause trouble only once recognized in the later part of SDLC. This situation can impact the customer and result in loss of confidence and business revenue. It will be helpful to divide the problem into smaller chunks from a machine learning perspective and act on it. Understanding customer requirements and classifying them in security and the rest of the requirements can be the first part. This classification needs a classifier to be built to execute this first step. Categorization will have two perspectives; first, it would be a labeling problem or classification problem. The labeling problem is an important aspect, as the historical requirements data would not have labels identified. In the same process of transformation of customer requirements conversation, requirements are drafted. Further refinement involves identifying the specific technical requirements in discovery mode and is more dependent on expert knowledge. A key focus in this situation is to provide the requirements management team with an intelligent system to make an informed decision as part of requirements elicitation.

As shown in Fig. 1, this system's architecture will have three components: customer conversation modeling framework, software landscape modeling framework, and industry landscape modeling framework. All three modules focus on learning the security-related vulnerabilities and feeding them back to the central system of smart requirements management system (SRMS). The customer conversation module specializes in understanding the pattern of customer conversation over time. Modeling the customer requirements to understand the construct of the same is the objective. From this construct, exploring and extracting security needs is the intent. Though security would be targeted, this can be extended to various other aspects of the software application landscape requirements. Software landscape modeling will encapsulate learning from the application, which is already running in production. Software undergoes periodic scanning for vulnerabilities that is one prominent source for this module. Source code itself is the key construct of the application landscape; this holds considerable information, leveraging this system. Statistical modeling of software source code comes into the picture here. Various other aspects of the software landscape can be explored through this module. Threat modeling is one of the main approaches used to analyze the vulnerabilities in the software development processes. This process is generally a subjective process that depends on expert knowledge. This approach can be enhanced to make it objective with the data available and cut down the subjectivity. This further exploration can be done on the statistical modeling of software processes to improve this module to leverage the software development processes' knowledge. Industry landscape modeling is another component of the system. Industry knowledge repositories like Open Web Application Security Project (OWASP) publish the top vulnerabilities information periodically; these information sources are leveraged in the industry landscape modeling module. This module is the source that will learn from the knowledge of the industry and feed into the central system. This module can be further enhanced with other security knowledge sources like International Organization for Standardization (ISO) standards that document global best practices from various domains. ISO standards also cover an information security domain. These sources of the central knowledge system of the smart requirements management system will provide the requirements management team with the required knowledge base for optimal requirements management with security being focused. This major system will provide the capability of the query database from which the requirements manager will be able to look up any critical information that is needed. Visualization features will help to visualize the requirements trends and construct. The prediction module will assist in monitoring various aspects of the requirements and project the possibilities. It would predict prospective vulnerabilities in the system and other vital parameters to enable the requirements management team. Though the knowledge graph is depicted to be used only in the industry landscape modeler, the entire system provides more opportunities to utilize the knowledge graph's capabilities. The whole system can be enabled with a knowledge graph as its key theme. Machine learning finds its significance in software development processes and [11] attempts to explore the successful prediction of android applications using neural networks. Work [7] focuses on modeling the architecture of Internet applications in run time. Software system architectures are focused in this study to understand their composition in run time.

4 Exploring Convolution Neural Network

In understanding the best approaches for modeling requirements-related information, some of the constructs of convolution neural network (CNN) are to be explored. This understanding helps to examine further the role of bidirectional encoder representation from transformers (BERT) in this discussion. The equation below shows the convolution equation, '*f*' and '*g*' are the functions, and at '*t*' point of time, ${}^{0}\tau^{0}$ converts one functional form to another.

$$(f * g) = \text{def} \int_{-\infty}^{\infty} f(\tau) g(t - \tau) d\tau \qquad (1)$$

Rectified linear unit (ReLU) is one of the activation functions that is predominantly used in CNN, same is depicted in the equation.

$$\phi(x) = \max(x, 0) \tag{2}$$

$$\sum_{i=1}^{m} = \omega_i x_i \tag{3}$$

Convolution activity makes the images convoluted and passed across feature maps; ReLu is used to make the output nonlinear if the convolution has introduced linearity. ReLu works on the function $w_i x_i$. Softmax function helps to consolidate the outputs of the last layer into two outputs in case of binary classification. 'z' will be the output that needs to be consolidated. Logit function, as shown below, is used to normalize outcomes in the range of 0–1.

$$f_i(z) = \frac{e^{zj}}{\sum_k e^{zk}} \tag{4}$$

Below equations are different forms for the cross-entropy. 'P' and 'q' values are the prediction value and prediction probability from the final prediction that gets to be fed into cross-entropy. For example, among binary predictions of 1 and 0, if the probability is 0.9 and 0.1, then 'p' gets fed with 1 and 'q' with 0.9. It is a loss function to validate the outcomes of the network in a much granular level better than classification error and mean squared error as metrics.

$$L_i = -\log\left(\frac{e^{f_{y_i}}}{\Sigma_j e^{f_j}}\right) \tag{5}$$

$$H(p, q) = -\sum_x p(x) \log q(x) \tag{6}$$

Below are the constructs of a CNN network in Table 1.

Table 1 Constructs of a CNN network

CNN network operations	Description
Convolution operation	Breaks up the image with feature detector
ReLu layer	No optimizing the linearity
Max pooling	Pooling the key features
Flattening	To align it to be fed in to neural network
Full connection	Configured network layer for modeling

Long short-term memory (LSTM) facilitates the information to be processed across time. LSTMs can be represented with the below equations. 'ϵ_t' is the output at the time 't', and 'W rec' is the weights at every point of time. Based on the value of 'W rec' over time, there is exploding or vanishing gradient that has to be managed in LSTM. To manage this variation, 'W rec' was proposed to be made as 1.

$$\frac{\partial \epsilon}{\partial \theta} = \sum_{1 \le t \le T} \frac{\partial \epsilon_t}{\partial \theta} \tag{7}$$

$$\frac{\partial x_t}{\partial x_k} = \prod_{t \ge i \ge k} \frac{\partial x_i}{\partial x_{i-1}} = \prod_{t \ge i \ge k} w_{rec}^{T} diag^{(\sigma'(x_i - 1))} \tag{8}$$

With this background, we will be able to explore BERT more effectively.

5 Role of BERT

In Fig. 2, BERT-based classifier pipeline is used; the BERT consumes labeled data for its pre-training purpose, where the BERT does generalized learning from the corpus. This pre-trained BERT model then does a specific task with supervised learning consuming the labeled data from the task. Later classifier model can sit on top of it and use BERT language modeling capability to conduct classification. Since the insurance domain is involved here, fin-BERT will be appropriate. In the proposed architecture for the SRMS system, BERT can play a prominent role. All the knowledge sources can be collected and gathered as events in the first stage of the machine learning or deep learning module. This information source can help identify the security-related events; in the next phase of the pipeline, another module can validate the identified events to highlight any possible vulnerabilities for the system. There can be a stage to validate these predictions by a human. Based on

Fig. 2 BERT-based classifier pipeline

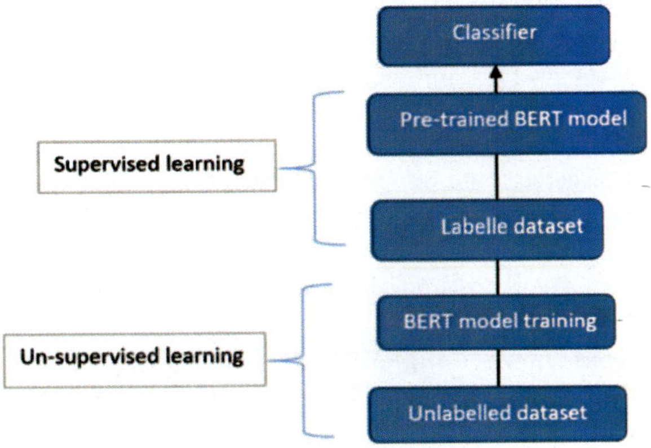

these inputs, the event collector can learn for itself and feed the machine learning pipeline with better information next time. The first stage of this module may need approaches that can provide natural language processing capabilities; in the second module, BERT can identify security vulnerabilities. Some of the probable modeling approaches that can influence this architecture are fundamental machine learning approaches like linear models, tree models, ensemble models, and stacking models. In the next stage, an experiment can include deep learning approaches like convolutional neural network (CNN), deep neural network (DNN), and long short-term memory (LSTM). Gated recurrent unit (GRU) with attention mechanism can be a good fit in processing natural language information followed with BERT providing the language understanding capabilities. (Virtanen et al.) In work [12], the author experimented with a pre-trained model for large corpora of unannotated text to demonstrate efficient transfer learning. To explore the capabilities of the approach, authors have focused on a common resource language like Finnish. In work [13], there is an exploration of BERT for conversational question answering, where BERT is used as an encoder for combining various sources of information.

6 Knowledge Graph in Action for Smart Requirements Management System

A text-based knowledge graph is an excellent information mining approach. This approach helps build a relational knowledge graph in a subject domain and learn from it for various specific knowledge associated with the domain. Work in [14] attempts to derive understanding from Wikipedia pages by building knowledge graphs. On similar lines, knowledge hidden in the software requirements management process across various data sources of the software development life cycle (SDLC) can be leveraged to enable the requirements management team to make smart decision-making. Sentence segmentation, entities extraction, and relations extraction are the critical steps associated with a simple knowledge graph building approach. Knowledge graphs (KG) are interconnected entities, which are based on the relation between these entities. An edge connecting the node holds the critical relationships between the nodes [15].

Knowledge graphs are helpful in the space of learning from the highly unstructured data from the public domain. In SRMS, learning from industry benchmark data is targeted to be done with knowledge graphs. These public data are complex and unstructured due to various data components like hyperlinks, video clips, and other elements amid text data. Making this information machine consumables is a challenge. Work [14] provides an overview of knowledge graph utilization for handling such data.

As shown in Fig. 3, two entities are represented as the node will be connected with an edge defining the relation between them. There can be multiple relations between the nodes that can also be specified. This graph complexity can be built on

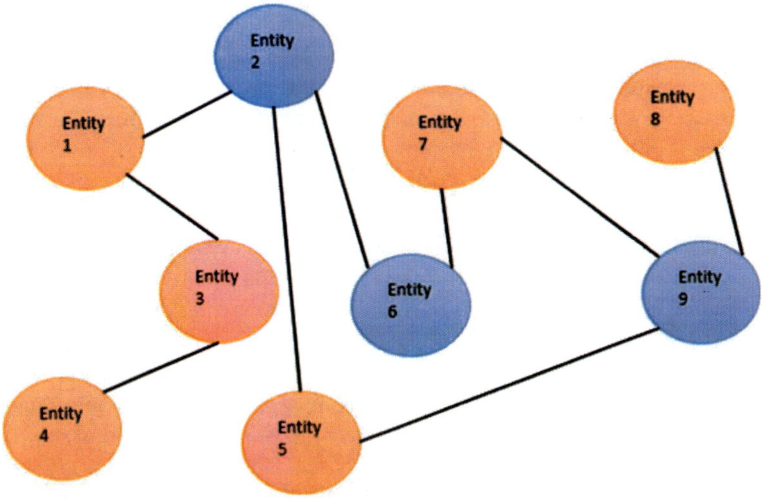

Fig. 3 Simple knowledge graph representation

adding more entities and their relation to the graph. Extracting these entities and their relationship is supposed to be done by the machine, but it needs the natural language understanding capabilities. NLP techniques like dependency parsing, entity recognition, and part of speech tagging can be applied [16]. In sentence, segmentation article is broken down into sentences. Subject and object are identified from these sentences. To determine the subject and objects entities, extraction will assist; Parts of speech (POS) tagging can be the approach. For cases where entities are spread across multiple words, dependency parsing methods can be applied. Entities extraction is the first part of the knowledge graph building; the relation between the entities needs to be established for the second part. The root of the sentence, which is the verb, provides the base for relation extraction. spaCy library of Python offers the capability of this processing. Extracted entities like subject–object pairs are the base for the creation of knowledge graphs. The resultant knowledge graph would be a complex representation of the object entity relation. Knowledge graphs need to be provided with features to query for any critical information required. In SRMS, while the knowledge graph is built on the industry knowledge repository of security vulnerabilities, there should be provision to derive only a graph related to specific themes of interest like user login vulnerabilities [17]. Requirements understanding is one of the essential parts of effective requirements management. The power of data visualization can be leveraged for this purpose. Techniques like work cloud can generate some text analytics to highlight the key aspects of the requirements. These help ensure that security-related requirements implicit within the customer requirements are called out. This visualization system will assist the requirements manager with real-time information on the requirements landscape visually. The architecture of the SRMS visualization module is intended to serve this purpose. Knowledge graphs themselves can provide the visualization capability in this module. Knowledge graph also will have a prominent role in contributing to the query database of SRMS. This interface is intended to enable the requirements manager to query the system and get different dimensions of the information to have a meaningful requirements management phase. Use of word cloud or any other similar word analysis methods to find out critical themes in the product's requirements data is objective [6]. Knowledge graph built on the requirements data set can help to use the key theme that will be derived to pull out the smaller subset of the knowledge graphs for each of these themes [18]. These smaller knowledge graphs will provide a refined requirement; if the targeted theme is certain security-related ones, the graph should identify those theme-related security requirements. Based on these subsets of knowledge graphs, understanding the pattern of security requirements in the requirements ecosystem is possible. These pattern understanding will assist in making future requirements elicitation smarter. In work [19], there is a proposal of natural language-based questioning capability built for complex knowledge graphs. Work [8] explores the exciting approach of graph neural network-based recommender system. This system can provide helpful insight for SRMS, particularly for the knowledge database intended to be built.

7 Knowledge Graphs Role in Modeling Industry Best Practice of Security for Software Development

Open Web Application Security Project (OWASP) is a nonprofit organization that works to secure the software. There are community-led open-source software projects that include thousands of developers and technologists. OWASP's top ten are the means of standard awareness among developers about web application security. Developers globally recognize this source as the industry-leading standard reference for security. This information source should be considered the starting point by the developers to ensure the security mindset in the development. Though developers are exposed to this information during the information explosion, they tend to miss vital information. Also, the requirements management team is less exposed to this information. Smart requirements management system will help focus on both concerns and create a level playing ground for all stakeholders to bring in the security mindset in software development.

OWASP's top ten vulnerabilities also provide further information around threat agents, security weakness, and impact. It includes information on when one can considers their application is vulnerable to a security threat and recommends ways to handle the threat. This information can be handy to devise a strategy for secured application development. Consuming this information and creating a smarter system for the software developers can help build a security mindset. Though this information is available handy, various permutations and combinations in which the threat may enter the system and affect it are available. A smart system needs a smart system that can study these intricacies and devise an information system for the users to act.

As depicted in Fig. 4, threat agents can potentially choose multiple pathways to attack a system, leveraging various system security weaknesses. Security controls need to understand these weaknesses and plug in the same from navigating the system and exposing them. Technical impacts would be seen on various information assets, finally leading to business impact. As part of the modeling of this system, knowledge graphs can be explored. Knowledge graphs specialize in aggregating the ecosystem's knowledge by building the relationship between various entities in the system. Knowledge graphs can help develop understanding from OWASP and other sources of wisdom from the industry and integrate them. Understanding multiple components of the security vulnerability sources would be a crucial part of building these knowledge graphs. OWASP provides information for understanding the type of attacks, possible weaknesses in the system, probable controls that can tackle these weaknesses, assets that can get exploited, and resulting impact which are the components involved. The knowledge graph can help to build an association with each of these and build knowledge repositories. This knowledge repository can continuously learn from the sources of information and get smarter. The requirements management team can leverage this module to gather information from the industry. Smart requirement management system has an internal software landscape modeling module and customer conversation modeling module. All these sources can provide an intelligent platform to understand the requirement better and ensure security which is given utmost importance in the requirements management phase. The knowledge graph's ability to recognize the information landscape's semantics is a critical attribute that helps process the information unambiguously and efficiently. Entities built into the graph will represent part of the information and help provide context to the information while they are interpreted. Knowledge graphs bring data management characteristics like database ability, knowledge base ability, and graph ability. Databases ability provides the querying possibility for the users. The formal structured information can be used within the graphs to interpret the information and derive new information. The graphical nature of the knowledge graphs helps to analyze the network data structure. The construct of the knowledge graph provides performance capability handling millions of data points. Work [20] attempts to build traceability between security vulnerabilities and software components with the knowledge graph.

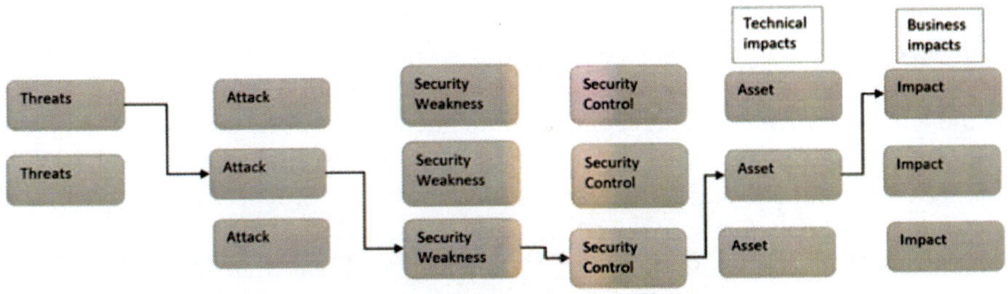

Fig. 4 Application security risks exploitation as depicted in OWASP top ten report

8 Transfer Learning to Cross-Learn Across Applications

Transfer learning is helpful to cut down on the training data needed for machine learning. Since there are variations associated with the different software applications, it will be beneficial to explore the transfer learning capabilities to reduce training data and time. These characteristics also resemble closure to how humans learn, where it is possible to use the logic of learning used in one area to be leveraged elsewhere. In traditional machine learning, every data set will provide the required knowledge to be used in the experiments. But in transfer learning, there is an accumulation of experience every time it is extended incrementally. Transfer learning can be defined as a mix of two components for a targeted domain, feature space in which the learning has to be done, and the distribution of the data in that space. There is space identified for the labels that are the output values of the input; this is label space. A predictive function binds together the feature vector and the labels; for each of the feature vectors in the target domain, there is a label associated with the same. There is a combination of the source domain, source tasks, target domain, and target tasks. Conditional probability in the target domain is learned based on the source domain's information and source tasks. In this case, the source domain and target domain are different, and the source task and target tasks are various. These come in handy in the target space's case having lesser examples to model than source examples. Transfer learning also means that different domains are involved; different feature spaces differ in their marginal distribution in other domains. In case if there are various tasks, then they have different label spaces of different conditional distributions. In the case of transfer learning, if the data has the same tasks but labels available only in the source domain, then the learning is called transductive transfer learning. For different domains, transductive learning will involve domain adaptation, and for other languages, it involves cross-lingual learning. If the tasks involved are various, with labeled data available in the target domain, then inductive transfer learning is involved. If tasks are learned simultaneously, it is called a multi-task learner, but if the tasks are learned sequentially, it is called sequential transfer learning.

Word2Vec, GloVe, ELMo, ULMFiT, GPT, and BERT have been used in the NLP pipeline, leading to good outcomes. Some of the applications of these approaches are classification, sequence labeling, and question answering. Training and adaptation are vital phases of transfer learning. The context-based representation can be done with Word2Vec and fastText. Recurrent neural network (RNN) and long short-term memory (LSTM) approaches can build language models. ULMFiT and BERT qualify as the language model that can generalize. While these models are reused, they can follow two approaches, either the entire architecture and weights are reused, or they undergo fine-tuning to suit the new domain. Work in [21] proposes a novel approach for training a convolutional neural network. Here, retraining the neural network architecture for visual pattern recognition is managed with transfer learning. This training will provide a valuable reference for improving the performance of machine learning approaches by reducing training time.

9 Research Limitation and Threats

Dynamic variation in the threat and vulnerabilities in the industry will be a challenge to handle. Some of the threats may not frequently impact, but when they do occur, they have a significant effect; making sure that area is addressed is crucial. Machine learning models need periodic calibration to keep them relevant at any point in time. Changing landscape of technology will also add to the challenge of keeping up with the security needs. Integrating multiple models will be challenging to tackle as the proposed architecture intends to provide a central system for decision-making. Data associated with the process will be sensitive, and access to data would be a challenge; utmost care may have to be taken while handling and processing the sensitive data. The explainability of the models will be a challenge to handle, as a need to convince the practitioners of the recommendations provided by the architecture. Extension of the work to other domains would involve more profound work.

10 Conclusion

The proposed system here is approached with various dimensions of the problems associated with the practitioners. Since there are natural language data involved, natural language data modeling capabilities are explored. This part of the system can be further explored to build more efficient approaches; customization can be an essential exploration. Since there is a role for various software development experts, enabling them with the correct information to make their decision-making process objective is another critical area. In these cases, entity relation building, connecting the knowledge sources is essential. Practical approaches to managing this knowledge gathering can be further explored. Combining the knowledge from various sources and aligning with the software development community's needs is the focus of this exploration. Further steps will involve experimenting with the title insurance domain's data and validating how this system can be built for this industry. Also, involving the experts to validate the effectiveness of this system is a way forward.

References

1. Tyo J, Goseva-Popstojanova K (2018) Identification of security related bug reports via text mining using supervised and unsupervised classification. In 2018 IEEE international conference on software quality, reliability and security (QRS). IEEE, pp 344–355
2. Morrison P, Oyetoyan TD (2021) An improved text classification modelling approach to identify security messages in heterogeneous projects. Softw Quality J 1(45)
3. Malik G, Cevik M, Parikh D, Başar A, Kici D (2021) A bert-based transfer learning approach to text classification on software requirements specifications. In: The 34th Canadian conference on artificial intelligence

4. Samanta D, Guha A (2021) Hybrid approach to document anomaly detection: an application to facilitate RPA in title insurance. Int J Autom Comput 18:55–72
5. Rashed AN, Boopathi CS, Amiri IS, Yupapin P, Samanta D, Sivaram M (2020) Distributed feedback laser (DFB) for signal power amplitude level improvement in long spectral band. J Optical Commun 18:55–72
6. Singh PK, Rani P, Samanta D, Khanna A, Bhushan B, Khamparia A (2020) An internet of health things-driven deep learning framework for detection and classification of skin cancer using transfer learning. Trans Emerg Telecommun Technol e3963:2317–2328
7. Colucci S, Vogli E, Grieco LA, Sciancalepore M, Mongiello M (2016) Run-time architectural modeling for future internet applications. Complex Intell Syst 2(2):111–124
8. Lu J, Jin Y, Zhang Q (2020) Artificial intelligence in recommender systems. Complex Intell Syst 1–19
9. Barr ET, Devanbu P, Sutton C, Allamanis M (2018) A survey of machine learning for big code and naturalness. ACM Comput Surv (CSUR) 1–37
10. Elahidoost P, Lucio L, Iqbal T (2018) A bird's eye view on requirements engineering and machine learning. In: 2018 25th Asia-Pacific software engineering conference (APSEC). IEEE, pp 11–20
11. Seifzadeh H, Beydoun G, Nadimi-Shahraki MH, Dehkordi MR (2020) Success prediction of android applications in a novel repository using neural networks. Complex Intell Syst (6):573–590
12. Virtanen A, Kanerva J, Ilo R, Luoma J, Luotolahti J, Salakoski T, Ginter F, Pyysalo S (2019) Multilingual is not enough: BERT for finnish. arXiv preprint arXiv:1912.07076
13. Rong W, Zhang J, Zhou S, Xiong Z, Wang Y (2020) Multi-turn dialogue-oriented pretrained question generation model. Complex Intell Syst
14. Joshi P (2019) Knowledge graph—a powerful data science technique to mine information from text (with python code)
15. Kumar SS, Karuppiah M, Samanta D, Maheswari M, Geetha S, Park Y (2021) PEVRM: probabilistic evolution based version recommendation model for mobile applications. IEEE Access. https://doi.org/10.1109/ACCESS.2021.3053583
16. Padhy N, Samanta D, Gomathy V et al (2020) Malicious node detection using heterogeneous cluster based secure routing protocol (HCBS) in wireless adhoc sensor networks. J Ambient Intell Human Comput 11:4995–5001
17. Nagaraju R, Samanta D, Sivakumar P et al (2020) A novel free space communication system using nonlinear. In: GaAsP microsystem resonators for enabling power-control toward smart cities. Wireless Netw 26:2317–2328
18. Samanta D, Althar RR (2021) The realist approach for evaluation of computational intelligence in software engineering. Innovations Syst Softw Eng
19. Zou Y, Cao Y, Xie B, Wang M (2019) Searching software knowledge graph with question. In: International conference on software and systems reuse. Springer, Cham, pp 115–131
20. Ren X, Wu Y, Chen J, Ye-W, Sun J, Xi X, Gao Q, Zhang S, Du D (2018) Refining traceability links between vulnerability and software component in a vulnerability knowledge graph. In: International conference on web engineering. Springer, Cham, pp 33–49
21. Vyas V, Anuse A (2016) A novel training algorithm for convolutional neural network. Complex Intell Syst 2(3):221–234

Hyper Parameter Optimization Technique for Network Intrusion Detection System Using Machine Learning Algorithms

M. Swarnamalya, C. K. Raghavendra, and M. Seshamalini

Abstract Cyberspace is a concept describing a widespread, interconnected digital technology with numerous users. New standards add more concerns with tremendous information gathered from various network sources, which can be utilized for focused cyber-attacks. Digital attacks in networks are getting more complex and subsequently introducing expanding difficulties in precisely distinguishing network intrusions. Inability to forestall the network intrusions could debase the validity of safety administrations, for example, information privacy, trustworthiness, and accessibility. In this work, we center around developing a Network Intrusion Detection System which is implemented utilizing Machine Learning Techniques. IDS dependent on ML techniques are successful and exact in distinguishing varied networks intrusions. Also, many of the ML-based IDS experience the ill effects of an increment in false-positive rate, leading to lower precision and accuracy. Consequently, we present an exploration of the UNSW-NB15 intrusion detection dataset that will be utilized for preparing the machine learning models. In our tests, we carry out various ML approaches namely, Support Vector Machine (SVM), Logistic Regression (LR), Random Forest Classifier (RF), Decision Tree (DT), Gradient boosted decision trees (GBDT). The results demonstrated that the hyperparameter optimization technique using Grid Search and Random Search methods and evaluation using *k*-fold cross-validation gives better optimization of model parameters and henceforth brings about better execution.

Keywords Cross-validation · Cyberattack · Decision tree (DT) · Gradient boosted decision trees (GBDT) · Hyperparameter Tuning · Logistic regression (LR) · Network intrusion detection system (IDS) · Random forest classifier (RF) · Support vector machine (SVM)

M. Swarnamalya (✉) · C. K. Raghavendra
Department of Computer Science and Engineering, B N M Institute of Technology, Bangalore, Karnataka 560070, India
e-mail: swarnamalya65@gmail.com

M. Seshamalini
Manhattan Associates, Bangalore, Karnataka, India

V. Skala et al. (eds.), *Machine Intelligence and Data Science Applications*, Lecture Notes on Data Engineering and Communications Technologies 132,
https://doi.org/10.1007/978-981-19-2347-0_35

1 Introduction

The pace and agility at which the innovations like the Artificial Intelligence, Internet-of-Things (IoT), and Robotics are progressing have led network intruders and hackers to develop with a higher speed regarding their capacities. Current networked business environments require an undeniable degree of safety to guarantee protected and confided correspondence of data between different associations. An Intrusion Detection System (IDS) goes about as a versatile defends innovation for framework security after the traditional advancements fizzle. Cyber-attacks will certainly turn out to be more refined, so it is significant that protection technologies be equipped to overcome the forthcoming dangers and menaces. Thus, Intrusion Detection Systems (IDSs) are an indispensable implementation in a network organization. Also, an IDS is equipped for reacting to any vindictive network exchanges and reporting them likewise. An IDS is thus characterized as an equipment or a product framework that screens an association's network for the approaching and for the anticipated dangers or threats. There is a wide exhibit of IDS, extending from antivirus programming to layered monitoring frameworks that follow the traffic of a whole organization. IDSs are classified into three groups, specifically, hub or host-based IDS (HIDS), network or circulated IDS (DIDS or NIDS), and Hybrid IDS (HYIDS). HIDS works on an enterprise network framework, analyzing the security and system configuration. Conversely, a DIDS monitors the multiple hosts that are connected to the network. HYIDS follows the implementation of both the above techniques.

In this paper, various ML strategies for implementing IDS are considered. Because of the huge dimensionality of dataset, it is essential to perform data processing to clean and process the raw dataset. For the work introduced in this paper, hyperparameter optimization method is employed which results in a model that achieves the best performance on a given dataset. The hyperparameter optimization also termed hyperparameter tuning utilizes the two common methods namely Grid Search and Random Search to tune the model hyperparameters. The result of a hyperparameter optimization is a single subset of well-performing hyperparameters that can be used in configuring the model. Evaluation of predictive models for a given hyperparameter is done by cross-validation. Thus, K-fold cross-validation results will show how good the model has been trained by using the training data given a set of hyperparameters in the ML model. In this research, the following ML approaches for IDS are implemented: Logistic Regression (LR), Support Vector Machine (SVM), Decision Tree (DT), Random Forest (RF), and Gradient boosted decision trees (GBDT). The hyperparameter optimization technique that uses Grid Search and Random Search methods is used in each of the above ML models. It results in the best hyperparameter of the training data model after validating using the 3-fold cross-validation. The model is then evaluated using the testing data that could be used to make predictions for unseen data. The metrics of evaluating a model include accuracy, precision, recall, $F1$ score, and AUC value of ROC (Fig. 1).

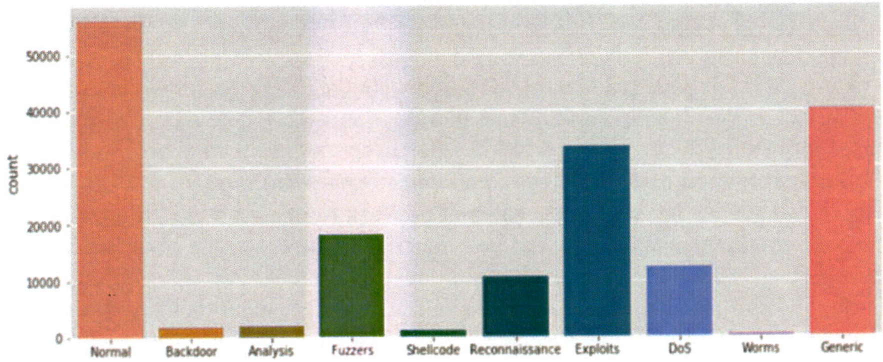

Fig. 1 Count of the list of IDS attacks contained in train data

2 Related Work

Jing [1] proposed SVM based technique for implementing the IDS system for binary and multiclass classifications. Log function scaling is deployed in the pre-processing stage followed by applying min-max normalization to obtain graded values. In the next stage, RBF SVM kernel is used that gives outstanding performance and yields an accuracy of 85.99% for binary classification.

Kasongo [2] proposed XGBoost-based feature selection technique to be adopted in various ML models. The ML models described in this paper are SVM, KNN, LR, ANN, and DT. Both binary and multiclass configurations are considered and have been noted that the filter-based feature reduction technique using XGBoost has increased the accuracy of the ML models from 88 to 91% using the above approach. The researchers used the UNSW-NB17 dataset to validate the performance of the model.

Ayo [3] has proposed a deep learning neural network for implementing a network intrusion detection system a typical hybrid feature selection (RHF-ANN) method. The detection is carried out in three stages: feature selection, classification, and prediction. The feature selection is implanted using the methods like Genetic search. Feature evaluation is first performed that returns the correlation values of the attributes. The genetic search method evaluates the weights and returns attributes with greater correlation values. The selected features serve as base for the ANN model to train and classify. The results show that the proposed method outperforms other methods with an accuracy rate of 98.8% and reduction in false alarm rate to 1.2%.

Ibor [4] proposed an approach of deep learning architecture using the activation function in the feed-forward neural network. The first component of the architecture is the capture of network traffic. The captured network (benign or malign) is manipulated using the UNSW NB15 and CICIDS2017 datasets. The dataset is then normalized using the *z*-score normalization to achieve unbiased results. The Principal Component Analysis (PCA) is used for feature engineering, which results in

p uncorrelated feature set. ReLU is the activation function used in DNN and has obtained an accuracy of 99.9%.

Khraisat [5] provides a contemporary review of the IoT IDS deployment methods and strategies involved in the implementation. The prescribed method is Anomaly-based intrusion detection system (AIDS). There are majorly two phases involved in the implementation of AIDS. Model training is first performed using the captured network. In the testing stage, new cases of network intrusions are generalized using the dataset. The results fetched, can be improvised with the usage of other machine learning-based methods. Shu [6] proposed the use of Gen-AAL algorithm that uses GAN model to detect the network intrusions. The GAN model typically consists of a generator and discriminator network. The generator comprises of encoder and decoder module which is based on Variational Autoecoder (VAE) model. GAN reduces the error resulting in a loss function to less than 0.1 and achieves the classification accuracy of 99.7%.

Sornsuwit [7] proposed a hybrid machine learning technique that detects multiple network intrusions. The methodology adopted is feature selection based on correlation combined with adaptive boosting that eliminates the unwanted features from the dataset and results in higher accuracy. The experiment is conducted on datasets such as UNSW-NB 15, KDD Cup'99 and is found to yield higher efficiency using the above-mentioned method. Feature Selection based on correlation is primarily used to weigh the relation between the features and significantly less correlated features are eliminated. The second phase involves training the various classifiers, namely, SVM, C4.5, KNN, MLP, and LDA. Each model can detect a particular genre of network attack. The third phase involves in building a Adaboost classifier.

Keserwani [8] has proposed an approach for feature selection followed by neural network learning technique employed. The precise methods used are gray wolf optimization (GWO) combined with crow search algorithm (CSA) which enables extracting more accurate features from the cloud network.

Hammad [9] has proposed four different methods namely naive bays (NB), J48, Random Forest (RF), and Zero R. Clustering or the classification of the dataset into distinct classes is performed using K-MEANS and Expectation-Maximization algorithms. First stage involves collection of data through the benchmarking dataset like UNSW-NB15. The second stage involves the feature selection using the correlation-based feature selection (CFS) that performs feature reduction and classification. The third stage is the classification process that uses the classifiers like J48, RF along with the EM clustering algorithms to obtain the predictions with higher accuracy. Finally, the accuracy and the performance of the model are assessed using the metrics like recall, precision, and f1 score. The results computed denote that the RF and J48 algorithms performed well and obtained an accuracy of 97.59 and 93.78%, respectively.

Al-Daweri [10] has approached to implement the IDS using three methods namely, rough-set theory (RST), back-propagation neural network (BPNN), and discrete variant of the cuttlefish algorithm (D-CFA). In the first stage, the correlation between the features is calculated using the RST method. In the second stage, BPNN is employed for classification task by considering the feature set as the input. Third, the feature selection is performed for multiple times to calculate the frequency of each selected feature. The validation is performed using the k-fold cross-validation method with the value of k set to 10. The result shows that the proposed approach yielded accuracy greater than 84% using the KDD99 dataset. This result is yielded when 18 features of dataset are considered with false alarm rate of 0.5%.

3 Overview of ML Approaches

The following below describes the ML approaches used in this work.

3.1 Linear Regression

Linear Regression uses a regression technique in which there exists a linear relationship between the independent and the dependent variables. The main objective of the linear regression is to consider all the given data points and construct a best fit line that fits the model in the best possible way. The linear equation [11] representing the best fit line is given as

$$Y = b_0 + b_1 x + e \tag{1}$$

Y denotes the dependent variable that is to be predicted.b_0 denotes the y-intercept, i.e., the point at which it intercepts the y-axis. b_1 denotes the slope of the line and x represents the independent variable that determines the value of Y. denotes the error in the resultant prediction.

3.2 Support Vector Machine (SVM)

Support Vector Machine is used in classification and regression tasks but predominantly in classification problems. In SVM, each point is plotted as a feature value in an n-dimensional space. The hyperplane is determined that perform classification by segregating it into different classes. The SVM algorithm is implemented using the kernel which enables transforming the input vector to the required form. Below mentioned is the list of available kernels that can be implemented [12].

- Linear Kernel: It is used to perform dot product between two feature observations. The representation of the linear kernel is as follows:

$$k(x, x_i) = \text{sum}(x * x_i) \tag{2}$$

where x and x_i represent the two vectors and hence the equation is the sum of dot product of individual vectors.

- Polynomial Kernel: It is the commonly used kernelized model with the SVM and determines the similarity of vectors in the feature space. The representation of a polynomial kernel is given as:

$$k(x, x_i) = 1 + \text{sum}(x * x_i) \wedge d \quad (3)$$

  Here d represents the degree of the polynomial being used.

- Radial Basis Function Kernel: It is the predominantly used kernel model that determines the hyperplane between the distinct classes. Its representation is given by:

$$k(x, x_i) = \exp\left(-\text{gamma} * \text{sum}\left(x - x_i^2\right)\right) \quad (4)$$

  Here, the range of gamma is specified between 0 and 1.

3.3 *Decision Tree (DT)*

Decision trees is based on supervised learning algorithm where the given data is classified based on an input parameter. The tree mainly comprises of decision nodes and leaves. The decision nodes represent the classes of data after splitting. The leaves denote the outcome of the decision tree.

3.4 *Random Forest (RF)*

It is a popular supervised learning technique used for both classification and regression problems. It adopts the approach of ensemble learning to solve the complex problems by using multiple classifiers. This results in higher accuracy and better performance of the model. It predicts the final output by taking the majority votes of the prediction given by various classifier trees.

3.5 *Gradient Boosted Decision Trees (GBDT)*

GBDT uses the method of boosting in combining the decision trees. Boosting helps in minimizing the error of previous decision trees. It follows the approach of sequentially fitting the decision trees to residuals from the previous trees. This results in minimization of the mean squared error and thus yields an optimized model.

4 Unsw-Nb15 Dataset

In this research, UNSW NB15 dataset [13] is being used for the experimental process. The dataset comprises of nine classes of attacks, namely, Analysis, Fuzzers, Backdoors, DoS, Exploits, Generic, Reconnaissance, Shellcode and Worms. In total, 49 features are generated using the algorithms that are built with the Argus and Bro-IDS tools [14]. The descriptions of these features are detailed in UNSW-NB15_features.csv file. A part of this dataset is classified as training data and is labeled under UNSW_NB15_training-set.csv. This constitutes for the data used in model training. The remaining data is labeled under UNSW_NB15_testing-set.csv and is used to validate the trained model. The total number of records that accounts for training set is 175,341 records and that of testing set is 82,332 from the different types, attack and normal.

5 Proposed Approach

The architectural design of the proposed system is described in the figure. The first block describes about the data pre-processing. The data pre-processing involves the following steps: cleaning, feature engineering, standardization, and normalization. Cleaning of dataset refers to processing of raw dataset that includes removal of null values, exclusion of incompatible data values, and fixing the datatype mismatch issues. Sequentially, feature engineering process is performed to remove the highly correlated features from the dataset. log1p () transformation is applied on the numerical columns and thereby removing the original columns. This helps to determine and drop the highly correlated features from the dataset as they are more likely to yield the same plot as already plotted with another feature. At this stage, standardization of the dataset is performed. As the range of few features in this dataset is very large, we will keep everything within certain range by using standardizing. It is performed by mean centering and variance scaling using the standard-scaler object that is trained on train data. As a result, all the features will have mean 0 and be scaled to unit variance. Finally, normalization is performed as a part of data pre-processing. This dataset consists of categorical columns with text data. But ML models cannot process text data directly as numbers. So there is a need to convert the categorical columns to numerical columns and is achieved using One Hot Encoder technique. One hot encoding creates binary columns, denoting each possible value from the original data where 1 is assigned if the value is present for the row and in other cases 0 is assigned.

Following the data pre-processing, various ML-trained models are set up using the hyperparameter optimization with cross-validation technique. Hyperparameter tuning is used to determine the well-performing set of model parameters. At least two hyperparameters need to be tuned for good performance on unseen data: a regularization constant C and a kernel hyperparameter γ. The best subset of parameters

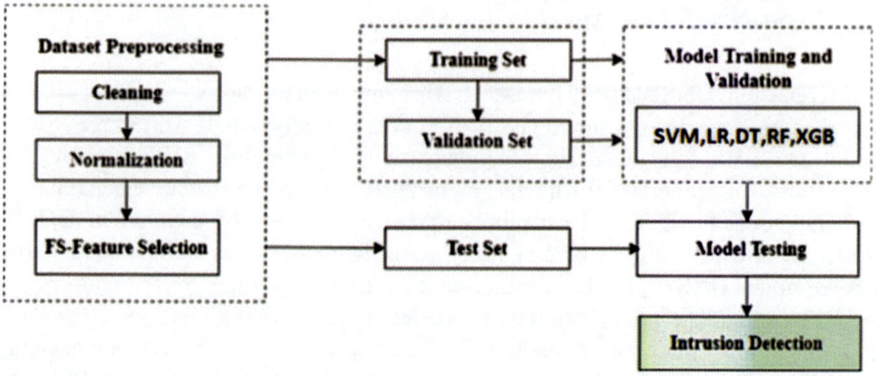

Fig. 2 Stages of system model

thus obtained is used to train the model. The *k*-fold cross-validation is used along with hyperparameter tuning. Cross-validation of a model parameter determines how good a model has been trained using those parameters. Score for each of the hyper-parameters is yielded by the cross-validation. The parameters with the best score are considered the best model parameter (Fig. 2).

5.1 Hyperparameter Tuning with k*-fold Cross-Validation*

The parameters defining a model architecture are referred to as hyperparameters and the technique of determining the most ideal model parameter is referred to as hyperparameter tuning. Thus, the hyperparameter tuning is performed using the below methods:

1. Grid Search: Grid search is the most basic hyperparameter tuning method that is used to build a simple model for each possible combination of the hyperparameter values. On evaluating each model, the architecture which produces the best results is selected.
2. Random Search: Random search differs from the grid search in which statistical distribution for each hyperparameter is provided rather than a discrete set of hyperparameter values. Using the distribution, the values are randomly sampled.

6 Performance Metrics

There prevails numerous metrics in evaluating ML-based IDS models. However, the following metrics are most used in this experimental process.

1. AUC: AUC is the area under the ROC curve. It measures the area underneath the entire ROC curve. ROC represents the probability curve and AUC represents the

measure of separability of classes. It denotes how well the model can classify into distinct classes. The ROC curve is plotted with TPR (True Positive Rate) against FPR (False-Positive Rate) where FPR is on x-axis and TPR is on y-axis.

The terminologies used in AUC and ROC curves are:

$$\mathrm{TPR} = \frac{\mathrm{TP}}{\mathrm{TP} + \mathrm{FN}} \tag{5}$$

$$\mathrm{FPR} = \frac{\mathrm{FP}}{\mathrm{TN} + \mathrm{FP}} \tag{6}$$

2. $F1$ Score: $F1$ score gives a better measure than the accuracy metric. It is calculated using the precision and recall and is computed as the harmonic mean of them.

$$\mathrm{Precision} = \frac{\mathrm{TP}}{\mathrm{TP} + \mathrm{FP}} \tag{7}$$

$$\mathrm{Recall} = \frac{\mathrm{TP}}{\mathrm{TP} + \mathrm{FN}} \tag{8}$$

$$F1\,\mathrm{Score} = 2\frac{\mathrm{Precision.Recall}}{\mathrm{Precision} + \mathrm{Recall}} \tag{9}$$

7 Experiments and Results

The experimental process illustrates the performance of various ML models that includes: SVM, LR, DT, RF, and GBDT. The below results describe the output obtained on various machine learning models.

The steps involved in training the model are:

- Step 1: Train a ML model and validate it via threefold cross-validation (CV). The CV results will show how good the model has been trained by using the training data given a set of hyperparameters in the ML model. The metrics for evaluating a model include $F1$ score, AUC value of ROC.
- Step 2: Evaluate the model by using the testing data. It will show how good the model could be used to make predictions for unseen data.

7.1 Logistic Regression (LR)

The hyperparameter considered for tuning here is "alpha" and "l2 penalty". The threefold cross-validation is used in the hyperparameter tuning for each of the eight candidates, thus totaling to 24 fits. The optimal value of alpha is found to be 1e-05 and the maximal cross-validation score obtained by the fit is 0.9991911276142402. The plot of performance versus alpha is displayed below that depicts the varying ranges of alpha and its dependency with the performance.

The hyperparameter considered for tuning in the second case is "alpha" and "l1 penalty". The fivefold cross-validation is used in the hyperparameter tuning for each of the eight candidates, thus totaling to 40 fits (Figs. 3 and 4).

The below plot depicts the performance metrics for the training and testing data (Fig. 5).

It is observed that the Logistic Regression model uses wide range of values for "alpha" from 10^−6 to 10^3 and "penalty" as l1 and l2. The score of the model is

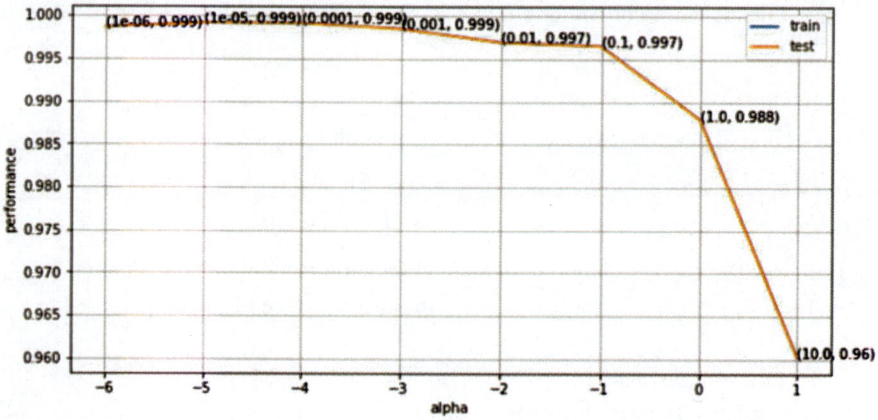

Fig. 3 Plot of performance versus alpha for l2 penalty

```
Best Score: 0.9991936630246743
Best Parameters: {'alpha': 1e-06}
```

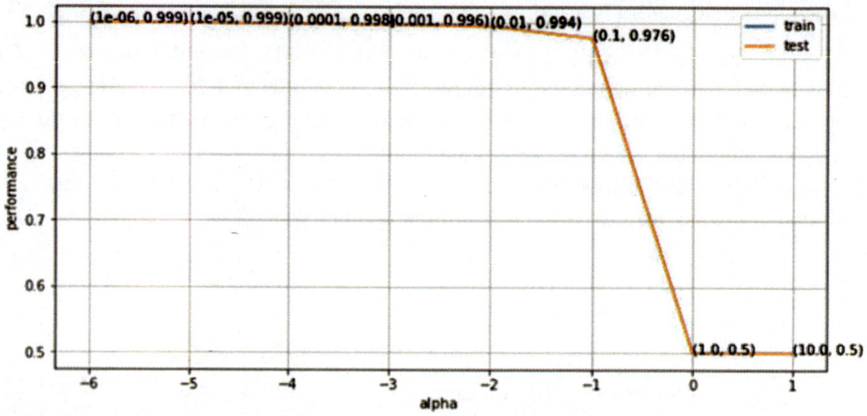

Fig. 4 Plot of performance versus alpha for l1 penalty

Dataset	Model	AUC	F1-score	False Alarm Rate
Train	LR	0.9878791093434393	0.9585322685085258	0.012120890656560756
Test	LR	0.9875316737082718	0.9581082448153768	0.012468326291728213

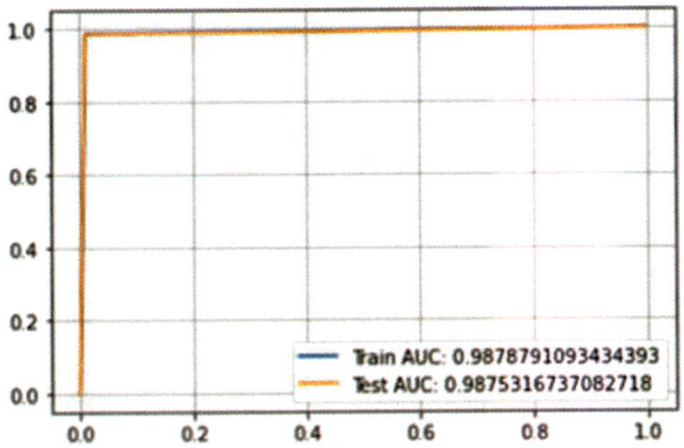

Fig. 5 Test and train data validation for LR model

high for values of alpha till 0.1. The best parameters of the model are obtained when penalty is l1 and alpha is 10^−6. Thus, these CV parameters yield a test accuracy of 0.954498 which demonstrates more accurate classification in the Linear Regression Model. The model clearly signifies that the AUC score for the train and test is very close denoting the absence of overfitting in the model. Also, the false-positive rate is very less, inferring that the model is performing well.

7.2 *Linear SVC*

The hyperparameter considered for tuning here is "alpha" and "penalty". The threefold cross-validation is used in the hyperparameter tuning for each of the 16 candidates, thus totaling to 48 fits. Thus, the best score obtained for the model is 0.9990533249115621 when alpha is 0.0001 and penalty is set to l2. The apt tuning parameters led to the higher AUC test score of 0.99 which depicts the higher performance of the model using the threefold cross-validation. The heatmap for the test and train data is obtained as below (Figs. 6 and 7).

The tuning parameters are "max_depth", "min_sampples_split" and "min_samples_leaf". The performance of the model is mostly dependent on "max_depth" and less dependent on other two parameters. Best score for the model is obtained when max_depth is 10, min_samples_split is 6, and min_samples_leaf is set to 9. The test accuracy obtained using the above CV parameters is 0.8974 which shows better distinction of the attacks into their respective classes. There is no overfitting of the model involved here owing to its better AUC score.

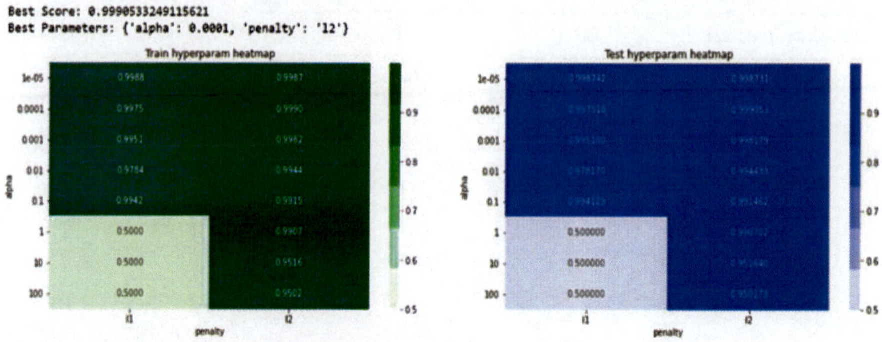

Fig. 6 Heatmap for test and train data in SVM model

Dataset	Model	AUC	F1-score	False Alarm Rate
Train	SVM	0.9907030504693602	0.955294589640227	0.009296949530639703
Test	SVM	0.9906682443423142	0.9552035837133029	0.009331755657685866

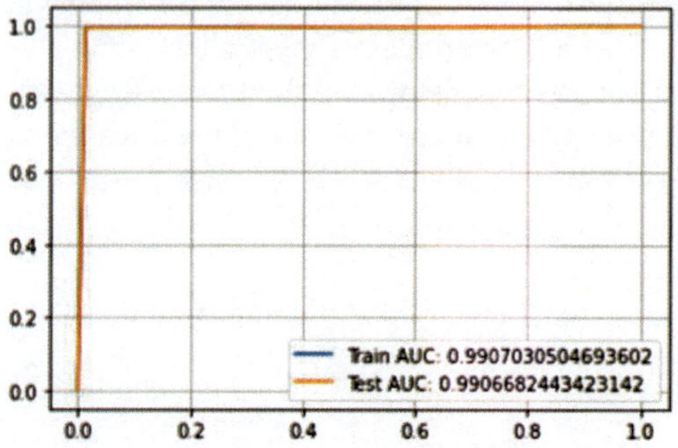

Fig. 7 Test and train data validation for SVM model

7.3 *Random Forest (RF)*

The tuning parameters involved here are "*n*_estimators", "max_depth", "min_samples_split", and "criterion". The following plot shows the model score when "criterion" is the parameter used (Fig. 8).

The plot of AUC score for the test and train data is depicted below. Best score for the model is obtained when the parameters like criterion, take the value "gini" and max_depth is set to 22, min_samples_split set to 6, and n_estimators set to 300. Further the CV parameters yield a better accuracy of 0.870731 for the Random Forest Classifier model. There is slightly significant amount of overfitting on train data in comparison with above models as the test and train AUC score is not similar (Fig. 9).

```
# Tuning Criterion
param = {'criterion':['gini', 'entropy'], 'min_samples_split':6, 'max_depth':22}
dt_clf, _, _ = cross_validation(CLF_NAME, param, 'criterion')
 50%|█████     | 1/2 [03:39<03:39, 219.52s/it]
For criterion: gini, train auc: 0.999941892813254 and test auc 0.9997759872555106
100%|██████████| 2/2 [06:49<00:00, 204.79s/it]
For criterion: entropy, train auc: 0.999923816545292 and test auc 0.9997642074606826
```

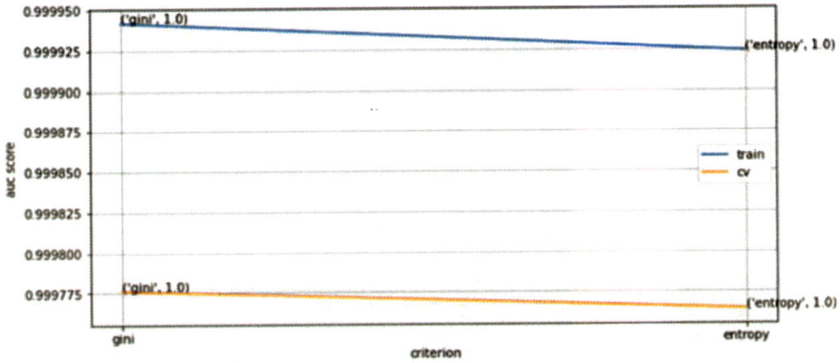

Fig. 8 Performance plot of AUC score versus criterion parameter for RF model

Dataset	Model	AUC	F1-score	False Alarm Rate
Train	RF	0.9927925575846711	0.9897532424861752	0.007207442415328972
Test	RF	0.9854768258366028	0.9767504956694146	0.01452317416339732

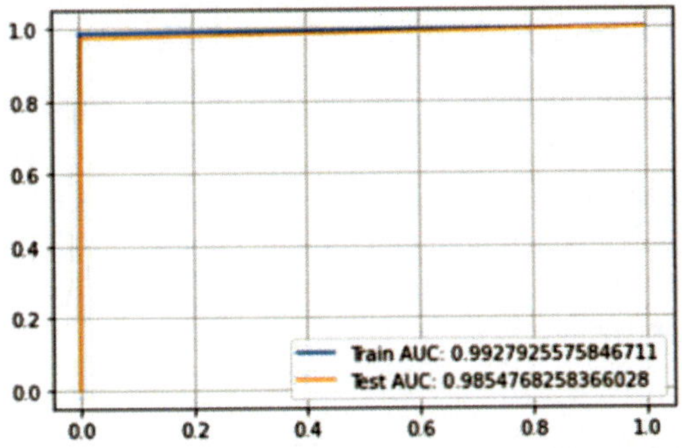

Fig. 9 Test and train data validation in RF model

7.4 GBDT

Learning rate is one of the model parameters being used with varying values. The plot of the accuracy score with the model parameters is shown Fig. 10.

The following plot shows the AUC score of test and train data (Fig. 11).

```
param = {'n_estimators':[200, 300, 400]}
xgb_clf,_,_ = cross_validation(CLF_NAME, param, 'n_estimators', x_train=x_train.iloc[train_sample], y_train=y_train.iloc[train_sample])
 33%|███      | 1/3 [10:34<21:08, 634.02s/it]
For n_estimators: 200, train auc: 0.9996954660106072 and test auc 0.9996688047624311
 67%|██████   | 2/3 [26:20<12:07, 727.79s/it]
For n_estimators: 300, train auc: 0.9997390417498165 and test auc 0.9996977639118547
100%|█████████| 3/3 [47:09<00:00, 943.19s/it]
For n_estimators: 400, train auc: 0.9997804231677948 and test auc 0.9997233525687121
```

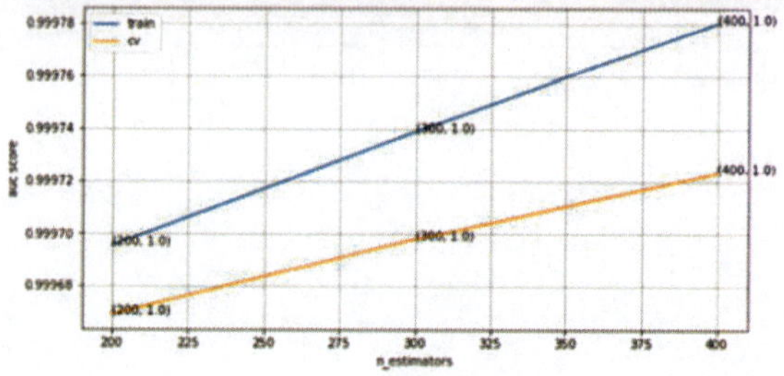

Fig. 10 Performance plot of AUC Score versus n_estimators parameter for GBDT model

Dataset	Model	AUC	F1-score	False Alarm Rate
Train	XGB	0.9953893531099667	0.9920798181106495	0.004610646890033333
Test	XGB	0.9864086030089578	0.9763625484533937	0.013591396991042138

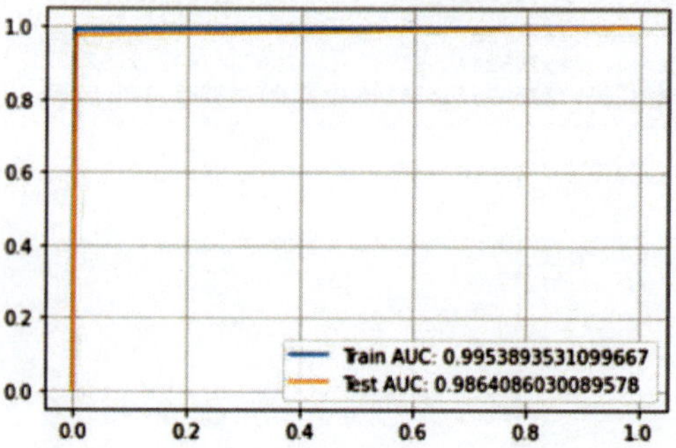

Fig. 11 Test and train data validation for GBDT model

It is observed that the performance is mostly dependent on "learning_rate" and slightly on "max_depth" and "n_estimators". The best score for the model is obtained when "n_estimators" is set to 400, "max_depth" is set to 12, "learning_rate" is set to 0.1, "colsample_bylevel" is set to 0.5and the "subsample" is set to 0.1. The CV parameters result in better accuracy of 0.862605 in the above decision tree model.

8 Model Evaluation

The ML models involved in the paperwork deploy different cross-validation parameters to obtain the best score and higher levels of accuracy. Thus, the result metrics for each of the model is depicted as shown below.

8.1 AUC Score Evaluation

The ROC curve is a metric that evaluates the performance of the classification model. The AUC score or AUC ROC (Area Under the Receiver Operating Characteristic Curve) score denotes how accurate the model can distinguish into the respective classes. Higher the AUC, better the model is at predicting the attacks. The roc_curve metric uses the predicted and the observed values of the train and test data to compute the false-positive rate (FPR) and true positive rate (TPR). The AUC metric is further computed using the FPR and the TPR values. According to the above table, SVM classifier outperforms others because it has got the highest AUC score.

8.2 F1 Score Evaluation

The $F1$ score is the evaluation metric used to show the performance of the trained model. It is used as a measure of harmonic mean of recall and precision. It is used as a better metric to evaluate the accuracy of the model. From Table 1, it can be observed that the $F1$ score is high for the Random Forest classifier model. While the other models also follow closely with the $F1$ score values of Random Forest model which makes them valid and acceptable.

Table 1 Performance comparison with various models

	Model	AUC	$F1$
0	LR	0.987532	0.958108
1	SVM	0.990668	0.955204
2	DT	0.987745	0.962851
3	RF	0.985477	0.97675
4	XGB	0.986409	0.976363

9 Conclusion

This paper involves detailed research on the application of hyperparameter tuning with cross-validation on various ML models to implement IDS with good performance. The UNSW-NB15 dataset is used to estimate and compare the accuracy metrics of the ML models namely, LR, SVM, DT, RF, XGB. In the preliminary stage, an exploratory literature survey was done, and the different techniques applied to the UNSW-NB15 dataset were reviewed. We scrutinized the detailed summary of the performance scores obtained in the literature review and compared them with the proposed methodology. Notably using the feature engineering techniques and hyperparameter tuning with cross-validation summarizes better performance in the ML models as shown below.

References

1. Jing D, Chen H-B (2019) SVM based network intrusion detection for the UNSW-NB15 dataset. In: 2019 IEEE 13th international conference on ASIC (ASICON)
2. Kasongo SM, Sun Y (2020) Performance analysis of intrusion detection systems using a feature selection method on the UNSW-NB15 dataset. J Big Data 7
3. Ayo FE, Folorunso SO, Abayomi-Alli AA, Adekunle AO, Awotunde JB (2020) Network intrusion detection based on deep learning model optimized with rule-based hybrid feature selection. Inf Secur J: A Glob Perspective 29:267–283
4. Ibor AE, Oladeji FA, Okunoye OB, Ekabua OO (2020) Conceptualisation of cyberattack prediction with deep learning. Cybersecurity 3
5. Khraisat A, Alazab A (2021) A critical review of intrusion detection systems in the internet of things: techniques, deployment strategy, validation strategy, attacks, public datasets and challenges. Cybersecurity 4
6. Shu D, Leslie NO, Kamhoua CA, Tucker CS (2020) Generative adversarial attacks against intrusion detection systems using active learning. In: Proceedings of the 2nd ACM workshop on wireless security and machine learning
7. Sornsuwit P, Jaiyen S (2019) A new hybrid machine learning for cybersecurity threat detection based on adaptive boosting. Appl Artif Intell 33:462–482
8. Keserwani PK, Govil MC, Pilli SE (2020) An optimal intrusion detection system using GWO-CSA-DSAE model. Cyber-Phys Syst 1–24
9. Hammad M, El-Medany W, Ismail Y (2020) Intrusion detection system using feature selection with clustering and classification machine learning algorithms on the UNSW-NB15 dataset. In: 2020 international conference on innovation and intelligence for informatics, computing and technologies (3ICT)
10. Al-Daweri MS, Zainol Ariffin KA, Abdullah S, Md Senan MF (2020) An analysis of the KDD99 and UNSW-NB15 datasets for the intrusion detection system. Symmetry 12:1666
11. Gandhi R, Introduction to machine learning algorithms: linear regression. https://towardsdatascience.com/introduction-to-machine-learning-algorithms-linear-regression-14c4e325882a
12. Brownlee J, Support Vector Machines for machine learning. https://machinelearningmastery.com/support-vector-machines-for-machine-learning/
13. The UNSW-NB15 Dataset, https://research.unsw.edu.au/projects/unsw-nb15-dataset
14. Moustafa N, Slay J (2015) UNSW-NB15: a comprehensive data set for network intrusion detection systems (UNSW-NB15 Network Data Set). In: 2015 military communications and information systems conference (MilCIS)

Energy Efficient VM Consolidation Technique in Cloud Computing Using Cat Swarm Optimization

Sudheer Mangalampalli, Kiran Sree Pokkuluri, Pothuraju Raju, P. J. R. Shalem Raju, S. S. S. N. Usha Devi N, and Vamsi Krishna Mangalampalli

Abstract Effective consolidation of VMS is necessary in cloud computing to pack and keep the VMs in the corresponding physical machines. It is a challenging task as incoming tasks are variable and changes with respect to time. To handle this scenario, a consolidation mechanism is needed to bundle the VMs in physical machines automatically without human intervention. Earlier authors proposed various consolidation mechanisms by using different heuristics and addressed metrics such as makespan, throughput, and execution time but authors have not addressed metrics named as energy consumption as it is one of the important metric in cloud computing. Minimization of energy consumption in cloud computing helpful in perspectives of both cloud user and provider. In the proposed approach, we have used cat swarm optimization algorithm to address metrics named as makespan and energy consumption and it is simulated on Cloudsim. The simulation results were compared against algorithms named as PSO and CS, and results revealed that proposed approach is showing significant impact over existing algorithms for the mentioned metrics named as makespan and energy consumption.

Keywords VM consolidation · Cloud computing · Makespan · Energy consumption · Cat swarm optimization · Particle swarm optimization (PSO) · Cuckoo search (CS)

S. Mangalampalli (✉)
School of Computer Science and Engineering, VIT-AP University, Amaravathi, AP, India
e-mail: mssudheer2015@gmail.com

K. S. Pokkuluri · P. Raju · P. J. R. Shalem Raju
Department of CSE, Shri Vishnu Engineering College for Women, Bhimavaram, AP, India

S. S. S. N. Usha Devi N
Department of CSE, University College of Engineering, JNTUK Kakinada, Kakinada, AP, India

V. K. Mangalampalli
Department of CSE, Aditya Engineering College, Surampalem, AP, India

V. Skala et al. (eds.), *Machine Intelligence and Data Science Applications*, Lecture Notes on Data Engineering and Communications Technologies 132,
https://doi.org/10.1007/978-981-19-2347-0_36

1 Introduction

Cloud computing is a distributed paradigm, which renders services to the users, based on the customer's needs. The entire cloud paradigm is a service oriented model in which every component will be given as a service, i.e., virtual infrastructure for storage, computations, application development, and deployment. All these services were given to the users based on agreement, which was to be made between user and provider. Users will pay for only the services for which they were subscribed and we can consider the cloud model as a subscription based model, i.e., based on the usage of the service user have to pay the price. There are many number of users will be there as cloud providers have their footprints around the world. With huge number of users, accessing cloud platforms there should be a proper scheduler through which incoming tasks from different users can be easily scheduled. All these services are provided to the users virtually by using VMs, which will be spins up from the physical machine. There is every chance that if we get high traffic onto VMs and if tasks cannot be accommodated on VMs automatically scheduler or task manager requests load balancer to spin a new VM or in the other case, if the requests were less onto cloud console, then we need an automated mechanism which can automatically consolidate VMs based on the number of requests with the help of load balancer. Consolidation is one of the main challenge as tasks onto cloud console is highly dynamic, and if proper consolidation mechanism, we can minimize makespan and energy consumption. Most of the authors addressed parameter named as makespan but makespan is not only the parameter but we also need to address the parameter named as energy consumption. In this paper, we have used status of VMs based on the usage of CPUs through which we have identified whether VMs are overloaded, under loaded, and balanced. For this approach, we have used cat swarm optimization algorithm to consolidate VMs and addressed metrics such as energy consumption and makespan. The rest of the paper is with the below sections. Section 2 discusses the related works; Sect. 3 discusses about problem formulation and proposed system architecture; Sect. 4 discusses about proposed algorithm; Sect. 5 discusses about simulation and result analysis, and finally, Sect. 6 discusses about conclusion and future works.

The main contributions of this paper are as follows:

- Consolidation technique for VMs proposed by using cat swarm optimization.
- Status of VMs were calculated based on CPU utilization.
- Simulation carried out on Cloudsim tool.
- Metrics addressed were makespan and energy consumption, and results of proposed approach is compared with PSO and CS algorithms.

2 Related Works

Sharma et al. [1] proposed a consolidation strategy, which addressed metrics, named as failure prediction and energy consumption. Authors focused on minimization of failure prone resources whenever VMs were migrating onto a physical machine, which is prone to failure. It predicts failure rate and addresses energy consumption. Tool used to implement this algorithm is MATLAB and checked the results with and without consolidation of VMs, and energy consumption is significantly decreased, and reliability of physical machines were increased as the technique is significantly predicting the failure rate of physical machines [2]. Proposed a VM consolidation mechanism, which focuses on addressing energy consumption and costs. Genetic algorithm-based methodology was used to design this consolidation technique. Cloudsim tool was used for performing experiments, and it was compared with first fit, permutation pack algorithms, and results revealed that energy consumption and costs were greatly minimized when compared with the above algorithms. Malekloo et al. [3] designed a VM consolidation and VM placement technique in which they were focused on energy consumption, QOS and SLA. They have used ant colony optimization as the design strategy to develop the algorithm. Cloudsim tool is used for simulation, and they have compared with existing metaheuristic algorithms, and it was outperformed by minimizing number of migrations, energy consumption, and SLA violations [4]. Authors focused on power consumption and reduction of active servers in datacenters. They have proposed a consolidation technique based on the workload and necessary migrations and consolidations take place based on the incoming workload onto the VMs. This is known as workload aware model. MATLAB was used for experiments and compared their technique with SA and GA, and they got significant improvement over compared approaches for energy consumption and number of migrations [5]. Uses cuckoo search for modeling their algorithm and the focus of this approach is to use group technology for consolidation of VMs. In this view, they are addressing parameters named as energy consumption and penalty cost. They have fine-tuned the cuckoo search according to the necessary requirements of cloud computing problems. It was compared over existing approaches named as FF and RR algorithms, and this approach was more superior than existing algorithms for the considered metrics. Moghaddam et al. [6] proposed a dynamic VM consolidation techniques which focuses on minimization of energy consumption and SLA violations. BPSO technique was used as a strategy for this approach. Cloudsim toolkit was used as a platform to conduct experiments, and workload is taken from the planet lab in cloudsim, and it was evaluated with Pearson coefficient, which maintains relation between different components like CPU, RAM, and bandwidth. When it was evaluated with Pearson coefficient, it greatly minimizes energy consumption and SLA violations [7]. Contains energy aware authors proposed VM consolidation mechanism which focuses on minimization of energy consumption and SLAV. Authors divided their model in three stages, i.e., detection of utilization of hosts and based on the utilization, then select the hosts which were ready for migration, and then give the place to the migrated VM to choose a best host. At first, ML model was used to predict utilization and to know how many number of hosts were about to migrate and then based on best fit decreasing model VMs are to be kept at appropriate hosts. It was compared with different prediction algorithms and results revealed that proposed technique is showing good results for the above-mentioned metrics [8]. Proposed a VM consolidation strategy focused on minimization of energy consumption, SLAV, and execution time. They have used ant colony algorithm and fine-tuned according to the cloud computing problem with double-level thresholds to keep migrated VMs at the correct hosts. It was compared with existing heuristics and simulation results revealed that it shows a great impact over these approaches for the above-mentioned parameters. Li et al. [9] aims at VM consolidation in datacenters by minimizing energy consumption and SLAV. A prediction model, i.e., linear regression was used to detect the utilization of VMs and simulation was carried out on cloudsim. Experiments were conducted with real-time traces of workloads and workload in planet lab in cloudsim. It was evaluated with existing heuristics, and this approach improves in terms of energy consumption and SLAV [10]. Focuses on aims at minimization of energy consumption due to over provisioning of resources in cloud datacenters. They proposed DVMC technique by using space aware best fit decreasing approach to place VMs in an efficient manner. Experiments were conducted on cloudsim tool with real-time traces and addressed the parameter named as energy consumption, and it was compared with existing heuristics, and it greatly minimizes energy consumption.

From the above existing works, many authors proposed their consolidation techniques by using various algorithms but still there is a huge trade-off for energy consumption while consolidation of VMs due to variable incoming traffic onto cloud console. From this point, we are proposing an energy efficient consolidation mechanism, which consolidates VMs, based on the utilization of CPU, and thereby we will set status index for normal utilization, underutilization, and over utilization of VMs. For this approach, we have used a nature-inspired algorithm named as cat swarm optimization. The below section discusses about the proposed system architecture.

3 Problem Formulation and Proposed System Architecture

We have defined the problem in a concise way. Assume that we have n number of tasks named as $T_k = \{T_1, T_2, T_3 \ldots T_k\}$, VMs are considered as n in number and mentioned here, as $V_n = \{V_1, V_2, V_3, \ldots . V_n\}$, we have hosts indicated as $H_m = \{H_1, H_2, H_3 \ldots\ldots H_m\}$, and finally, we have i number of datacenters $d_i = \{d_1, d_2, d_3 \ldots d_i\}$. Now we need to map n number of VMs onto m number of hosts based on utilization of CPU, which is known as status in this case, and it will be given by resource manager.

To consolidate VMs onto appropriate host, we have formed some set of rules by keeping a threshold value based on that system identifies whether VM is under loaded, overloaded, or in the balanced mode.

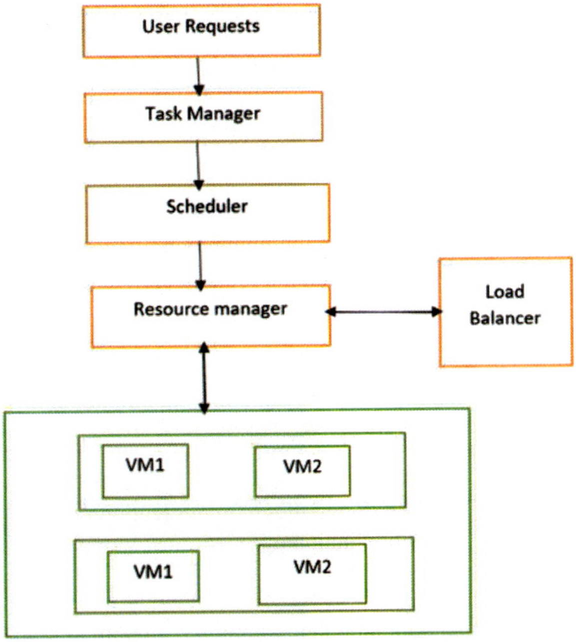

Fig. 1 Proposed system architecture

The below were the rules to be followed to detect a VM whether it is under loaded, overloaded, or balanced.

1. if cpu utilization $< 25\%$ then it is said to be in under loaded.
2. $if 60\% <$ cpu utilization $< 25\%$ then it is said to be balanced.
3. if cpu utilization $> 60\%$ then it is said to be overloaded.

Figure 1 represents proposed system architecture in which initially user requests will be submitted on cloud console, thereby task manager will collects these requests and forward these requests to scheduler which need to schedule tasks onto VMs. In the meanwhile, scheduler have to be interacted with resource manager which takes care about the availability of virtual resources in the datacenters and resource manager in turn interact with load balancer to track the status of VMs, i.e., under loaded, overloaded, or balanced. In this case, we have kept a threshold at load balancer to track the utilization of CPU based on the above-mentioned conditions. From these conditions, VMs can be consolidated into an appropriate host. The focus of this work is to consolidation of VMs onto appropriate hosts while addressing parameters named as makespan and energy consumption. The entire proposed system architecture was presented in Fig. 1.

In this paper, we are addressing the below mentioned parameters.

3.1 Makespan

Makespan is one of the important parameter need to be addressed in cloud computing as it is considered as one of the primary metric. It can be defined as "total execution time of tasks over the number of VMs available in the host". It is denoted as

$$\text{makespan}' = \text{av}^n + et^k \tag{1}$$

where av^n represents availability of a VM and et^k represents execution time of a task over a VM.

3.2 Energy Consumption

Energy consumption is also considered as one of the metric in this paper, and it is necessary and needed to minimize consumption of energy in datacenters. Energy consumption in cloud computing based on two components, i.e., consumption of energy in active and idle times of CPU. It can be denoted as

$$e^{\text{con}}(V^n) = \int_0^{m_t} e_{\text{con}}^{cs}(V^n, t) + e_{\text{idle}}^{\text{con}}(V_n, t)\text{d}t \tag{2}$$

The total energy consumption can be represented as

$$e^{\text{con}} = \sum e^{\text{con}}(V_n) \tag{3}$$

We can calculate energy consumption from Eqs. 2 and 3.

3.3 Fitness Function

We have used cat swarm optimization algorithm to model the consolidation technique. Proposed approach evaluates metrics named as makespan and energy consumption by knowing status index and thereby consolidating VMs based on utilization of VMs. The below equation represents fitness function which aims to minimize energy consumption and makespan.

$$f(x) = \min \sum_x \text{makespan}'(x), e^{\text{con}}(x) \tag{4}$$

3.4 Cat Swarm Optimization

Cat swarm optimization is one of the nature-inspired algorithm which is based on the behavior of cats [11]. Generally, cats were very alert and strive to chase targets. There are two modes were presented for cats, namely seeking and tracing mode. In the first mentioned mode, they are at the rest but they were alert. For the second mode, they were chasing targets. In this algorithm, each cat is considered as solution and they are trying to achieve targets. This mechanism will be continued until all iterations will be completed. Initially, cat population is generated randomly and the cats were presented either in seeking mode or in the tracing mode. In tracing mode, cats will be trying for targets based on the following equations. It is to be represented as follows.

$$\mathrm{Ve}_d^i(k+1) = p * \mathrm{Ve}_d^i(k) + c * r * \left(x_{\mathrm{best}}^d - x_d^i\right) \tag{5}$$

$Ve_d^i(k)$ represents velocity of a ith cat at kth iteration, c is represented as a constant, and r is represented as random number presented in between 0 and 1. x_{best}^d is a best position of a cat at that iteration. After the change of cat's position, the next position will be calculated as follows

$$x_d^i(k+1) = x_d^i + \mathrm{Ve}_d^i(k+1) \tag{6}$$

All the solutions will be calculated until all iterations will be completed and reaches an optimized value.

4 Proposed Energy Efficient Consolidation Mechanism by Using Cat Swarm Optimization Algorithm

Input: tasks $T_k = \{T_1, T_2, T_3 \ldots T_k\}$, VMs $V_n = \{V_1, V_2, V_3, \ldots . V_n\}$, hosts $H_m = \{H_1, H_2, H_3 \ldots \ldots H_m\}$ and Datacenters $d_i = \{d_1, d_2, d_3 \ldots d_i\}$.
Output: VMs allocation onto hosts by minimization of energy consumption and makespan.

```
Start

Initialize cat population randomly

Evaluate Status index of VMs based on CPU utilization

Evaluate fitness function by using eqn.4

For every task calculate solutions using eqns. 5 and 6

Assign VMs onto hosts by using status index of VMs

Replace solution with current solution

Identify VMs status, which were migrated or mapped onto
hosts

Evaluate makespan and energy consumption and if these so-
lutions were best, keep them

Update best solutions

End
```

5 Simulation and Results

Experiments were conducted by using a tool named as cloudsim [12]. For this simulation, we have considered ten datacenters, 550 hosts, 100 VMs. VM capacity is of 2048 MB, and Xen hypervisor is used in the simulation. We have used 1000 tasks to conduct experiments.

Table 1 Calculation of makespan

Tasks	PSO	CS	CSO
100	1369.7	1323.9	1274.9
500	1748.8	1776.5	1705.92
1000	2548.5	2238.7	2023.98

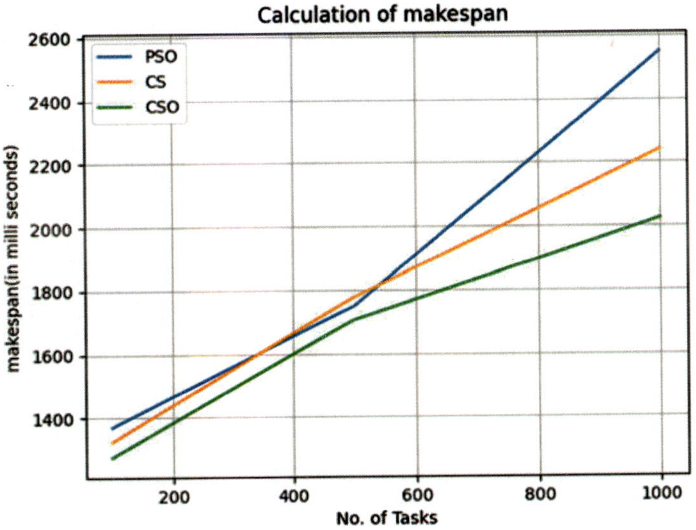

Fig. 2 Evaluation of makespan

5.1 *Calculation of Makespan*

Makespan calculation of PSO for 100, 500, and 1000 tasks are 1369.7, 1748.8, and 2548.5, respectively. Makespan calculation of CS for 100,500, and 1000 tasks are 1323.9, 1776.5, and 2238.7, respectively. Makespan calculation of CSO for 100, 500, and 1000 tasks are 1274.9, 1705.92, and 2023.98, respectively. In Table 1 and Fig. 2, we can clearly observe that proposed CSO used consolidation mechanism has achieved less makespan when compared with two other algorithms, namely PSO and CS.

5.2 *Calculation of Energy Consumption*

Energy consumption of PSO for 100, 500, and 1000 tasks are 1432, 2578, and 3276, respectively. Energy consumption of CS for 100, 500, and 1000 tasks are 1048.9, 1989.6, and 2378.9, respectively. Energy consumption of CSO for 100, 500, and 1000 tasks are 921, 1972.9, and 2107, respectively. In Table 2 and Fig. 3, we can clearly

Table 2 Calculation of energy consumption

Tasks	PSO	CS	CSO
100	1432	1048.9	921
500	2578	1989.6	1972.9
1000	3276	2378.9	2107

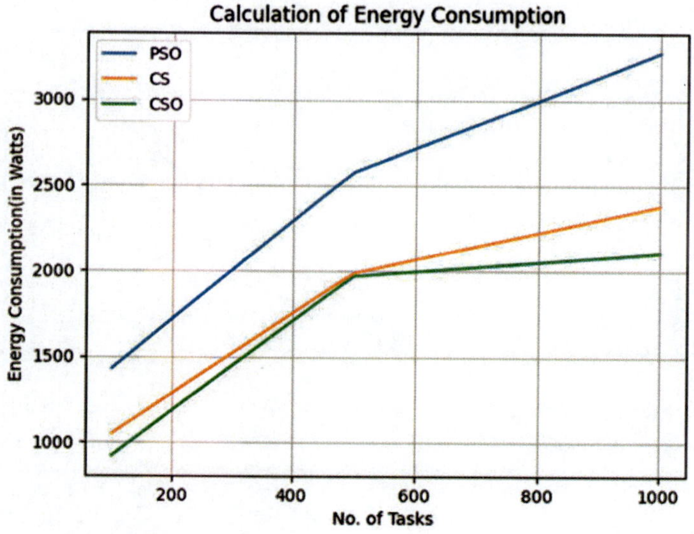

Fig. 3 Calculation of energy consumption

observe that proposed CSO used consolidation mechanism has greatly minimizes power consumption over the algorithms, namely PSO and CS.

6 Conclusion and Future Works

Effective VM consolidation is necessary in cloud computing as the incoming requests were variable, and if the VMs were overloaded, then there is a chance to migrate VMs but if it is under loaded, then it needs to be consolidated, and this have to be done automatically based on utilization of CPU. So, we have proposed an effective VM consolidation mechanism where we need to identify utilization of CPU based on rules we have mentioned in proposed architecture, and cat swarm optimization is used to model proposed algorithm, and we have addressed metrics named as energy consumption and makespan as these are the primary metrics needed for any cloud paradigm, and we have compared our proposed approach with PSO and CS algorithms, and finally, our consolidation technique greatly minimizes makespan and energy consumption. In the future, we want to evaluate other metrics named as number of migrations and SLAV to evaluate efficacy of our algorithm.

References

1. Sharma Y, Si W, Sun D, Javadi B (2019) Failure-aware energy-efficient VM consolidation in cloud computing systems. Futur Gener Comput Syst 94:620–633
2. Yousefipour A, Rahmani AM, Jahanshahi M (2018) Energy and cost-aware virtual machine consolidation in cloud computing. Softw Pract Exper 48(10):1758–1774
3. Malekloo M-H, Kara N, El Barachi M (2018) An energy efficient and SLA compliant approach for resource allocation and consolidation in cloud computing environments. Sustain Comput Inf Syst 17:9–24
4. Mohiuddin I, Almogren A (2019) Workload aware VM consolidation method in edge/cloud computing for IoT applications. J Parallel Distrib Comput 123:204–214
5. Tavana M, Shahdi-Pashaki S, Teymourian E, Santos-Arteaga FJ, Komaki M (2018) A discrete cuckoo optimization algorithm for consolidation in cloud computing. Comput Ind Eng 115:495–511
6. Mapetu JP, Buanga LK, Chen Z (2021) A dynamic VM consolidation approach based on load balancing using Pearson correlation in cloud computing. J Supercomput 77(6):5840–5881
7. Moghaddam SM, O'Sullivan M, Walker C, Piraghaj SF, Unsworth CP (2020) Embedding individualized machine learning prediction models for energy efficient VM consolidation within Cloud data centers. Fut Generat Comput Syst 106: 221–233
8. Xiao H, Zhigang H, Li K (2019) Multi-objective VM consolidation based on thresholds and ant colony system in cloud computing. IEEE Access 7:53441–53453
9. Li L, Dong J, Zuo D, Jin W (2019) SLA-aware and energy-efficient VM consolidation in cloud data centers using robust linear regression prediction model. IEEE Access 7:9490–9500
10. Wang H, Tianfield H (2018) Energy-aware dynamic virtual machine consolidation for cloud datacenters. IEEE Access 6:15259–15273
11. Chu S-C, Tsai P-W, Pan J-S (2009) Cat swarm optimization. In: Pacific rim international conference on artificial intelligence. Springer, Berlin, Heidelberg
12. Calheiros RN et al (2011) CloudSim: a toolkit for modeling and simulation of cloud computing environments and evaluation of resource provisioning algorithms. Softw: Pract Exp 41(1):23–50

A Forecast of Geohazard and Factors Influencing Geohazard Using Transfer Learning

S. Visalaxi, T. Sudalaimuthu, Tanupriya Choudhury, and A. Rohini

Abstract Geohazard is an ecological destruction problem that exists in various parts of the universe. Geohazard destroys the complete ecosystem. Geohazard results in both human and economic loss. In India, these Geohazards create an impact of 2% of loss in domestic products and 12% of economic loss. In the advancement of technology at various eras, various methodologies were implemented to predict the Geohazard. The approaches start include (a) Steel sheets technique (b) Installation of sensors (Fiber optic and Electrical Sensor) in expected place (c) Machine learning models (d) Time series analysis (e) Basic neural network structure, etc. The problem faced by conventional approaches are (a) large volume of data, (b) Satellite image-based data (c) Radar covers a small area, etc. Deep learning is the cutting-edge technology that addresses the problems faced by the traditional approaches in effectively. The usage of architectures in deep learning provides the solution for Geohazard. The proposed work implements a novel approach "Transfer learning approaches for effective prediction along with factors influencing Geohazard". VGG16 is a state-of-art technique for predicting the images more precisely with an accuracy of 80% in recognizing the occurrence of Geohazard. The various factors that influence the Geohazard are identified using Correlation mapping.

Keywords Geohazard · Deep learning transfer learning · VGG16 · Correlation

S. Visalaxi (✉) · T. Sudalaimuthu
Hindustan Institute of Technology and Science, Chennai, India
e-mail: sakthi6visa@gmail.com

T. Sudalaimuthu
e-mail: tsmuthu@hindustanuniv.ac.in

T. Choudhury (✉)
School of Computer Science, University of Petroleum and Energy Studies (UPES), Dehradun, Uttarakhand 248007, India
e-mail: tanupriya@ddn.upes.ac.in

A. Rohini
Francis Xavier Engineering College, Tirunelveli, India

V. Skala et al. (eds.), *Machine Intelligence and Data Science Applications*, Lecture Notes on Data Engineering and Communications Technologies 132,
https://doi.org/10.1007/978-981-19-2347-0_37

1 Introduction

Geohazard is unexpected damage that takes place to the environment that results in small/medium/larger risks for the ecosystem. The major Geohazard includes (a) Landslides (b) Earthquakes (c) Tsunami (d) Volcanic Eruption, etc. The most influential Geohazard is the Earthquake that happened across the globe. The cause of the Earthquake is a breakdown of rocks that happened abruptly. "Elastic strain" energy in the form of seismic waves which disseminate through the earth and results in the surface of the ground shaking. Landslides are geological hazards that occur in various parts of the world. The most important reason for the landslide is the loss of gravity in the lands. The Sand-like textures are loose particles and they are held with each other due to friction. On an inclined surface when something messes up the friction it leads to landslides. There are several reasons for landslides includes (a) Earthquakes (b) Volcanoes (c) Wildfires. The traditional scientific method of predicting landslides is done using sensors includes (a) Fiber optic sensors (b) Electrical Sensors (d) Slope geometry (e) Chemical Agents, etc. But the difficulty faced by using the legacy method is that sensors and other equipment need to replace frequently. Volcanic Eruption is yet another geological hazard that takes place in the world. The major cause of the volcanic eruption is the melting of the earth's surface which results in the formation of magma. Volcanoes emit ashes, fragments of glass, and crystals into the atmosphere at different heights. Volcanic eruptions have the possibility of mixing with groundwater surfaces. These volcanic eruptions in some cases lead to other Geohazard namely, Tsunami and earthquakes, etc. The scientific method used for predicting volcanic eruptions is still a challengeable one. Volcanoes are monitored by seismographic detection, tremor measurements of ground deformation. Although the prognostic approach of volcanic eruption is a difficult task. Geohazard is one of the major losses of the global economy. All the traditional methods for approaching Geohazard are not up to the level. Direct involvement of human beings and/or using the devices to monitor the Geohazard is a challenging task. The state-of-art for predicting Geohazard can be done effectively using deep learning techniques. This cutting-edge technology resolves the various problems in an efficient way related to the current scenario. The proposed system analyses various factors influencing Geohazard. Also, the possible prediction has been done for preventing the severity of Geohazard.

2 Related Works

Had observed that machine learning and deep learning approaches are effective techniques for handling various Geohazard. Deep learning implements classification and prediction for Geohazard. The problem facing Geohazard is dealing with a large volume of data. This can be resolved by interpreting the model with the help of AI [1]. Had analyzed the implementation of IoT for Geohazard. IoT application helps in monitoring seven types of Geohazard [2]. Had predicted the slope deformation by implementing a temporal graph convolution neural network. The proposed study is based on global information by using time series data from multiple locations [3]. Had predicted Geohazard using neural network. "Seismic images" was considered as input for training the model. The conventional approach takes time to predict the hazards [4]. Had addressed the earth fissure hazard problem. The proposed study develops a classification, regression, and prediction model for handling earth fissures. Various models were trained and compared. Random forest performs well in classification [5]. Had performed a test on stress associated with mechanical conditions like geologic and land cover patterns. Along with geomechanical models, deep learning was used for predicting the surface [6]. Had proposed a model for addressing various Geohazard that occurs in the Tibet plateau. Support vector machine, Gradient boosting, and Random Forest were used for predicting landslides, debris flow, etc. Gradient boosting achieves a higher AUC of 0.84 [7]. Had analyzed the factors that influenced landslides in Iran. A contextual neural algorithm (CNG) was used for analyzing the cluster associated with landslides. The landslides driving forces were identified and ranked using the Random Forest (RF) algorithm [7]. Had assessed the Geohazard along the China-Pakistan corridor. The proposed study focused on four machine learning approaches namely, "Logistic Regression (LR), Shannon Entropy (SE), Weights-of-Evidence (WoE), and Frequency Ratio (FR)" for identifying the factors of debris flow. Totally 13 factors were identified as attributes. The success rate was analyzed using AUC curves where logistic regression yields higher accuracy and precision [8]. Had predicted the factors for landslides using machine learning algorithms. Nearly 14 attributes were considered for training the model and classified into four groups. Three algorithms namely, Random Forest, Naïve Bayes, and Logic Boost were trained on each dataset. Random forest performs well with an AUC of 0.940 [9]. Had proposed hybrid machine learning techniques for predicting landslides. Ensembling and Reduced Error pruning trees were implemented to train the model. Root mean square values are calculated for validating the model performance [10]. Had designed a system for assessing Geohazard in underground mining. "Center for Excellence in Mining Innovation (CEMI): "Smart Underground Monitoring and Integrated Technologies" (SUMIT), "Mining Observatory Data Control Center" (MODCC), and "Ultra-Deep Mining Network" (UDMN)" were worked together in identifying the automatic assessment techniques [18]. Had implemented clustering techniques and a convolution neural network in identifying the risk of earthquakes in Indonesia [12]. Had estimated the bearing capacity of cohesion-less soil using machine learning techniques. Techniques [15] decision tree, k-nearest neighbor, multilayer perceptron artificial neural network, random forest, support vector regression, and extremely gradient boosting were employed to train the model [13]. R-squared values are used for evaluating the performance of the model [17]. Had identified that Geohazard assessment can be done using digital mapping techniques. Digital photogrammetry, laser scanning, and LiDAR were used along with GIS and BIM software for predicting Geohazard [14].

3 Transfer Learning Approach for Predicting Geohazard

Geohazard [11, 16] is one of the most significant problems for loss of economy and human values across the globe. Prediction of Geohazard is done in several ways. The recent technique known as deep learning [19] is implemented for effective prediction of Geohazard [16].

The novelty of this work is the implementation of "Transfer learning techniques along with time series analysis" for predicting the occurrence of Geohazard.

There exist various types of Geohazard across the world. They are (a) Landslides (b) Debris flow (c) Tsunami (d) Volcanic Eruption (e) Earthquakes, etc. The dataset of these Geohazard includes (a) Seismic images (b) Satellite images (c) Radar images, etc. Since deep learning performs well in handling images, these images are considered as input for predicting the Geohazard. As a result, features are listed and a correlation matrix is drawn to identify the features. Various forms of images are obtained as input from various sources. The obtained images are pre-processed, trained, tested, validated, and predicted. Figure 1 illustrates the satellite image of the geographical location causing Geohazard.

The phases of predicting the Geohazard are as follows.

1. Collection of Input Dataset (Images):
   Geohazard images are obtained from various sources. It includes Satellite images, radar images, Seismic images, etc.
2. Data preprocessing and Data Augmentation:

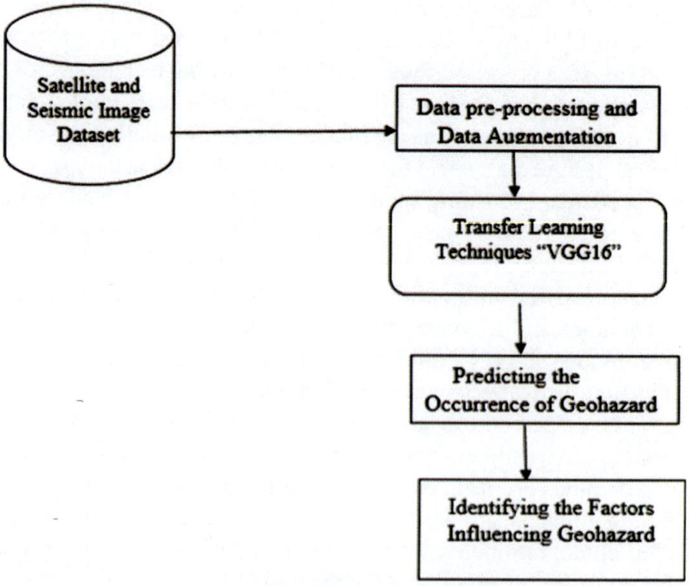

Fig. 1 System architecture for prediction of Geohazard using deep learning

3. Splitting of Images
4. Training the model
5. Model prediction

4 Result and Discussion

The obtained satellite images and Seismic images are used as input for training the model. The images are fed into the transfer learning neural network environment. VGG16 transfer learning techniques were implemented for predicting the disaster.

VGG16 transfer learning techniques train the model in an effective way for predicting the occurrence of Geohazard. The model was evaluated through their metrics includes accuracy, precision, recall, and F1-score, respectively, represented in Eqs. 1, 2, and 3.

$$\text{Accuracy} = \frac{\varphi P + \varphi N}{\varphi P + \varphi N + \omega P + \omega N} \tag{1}$$

$$\text{Precision} = \frac{\varphi P}{\varphi P + \omega P} \tag{2}$$

$$\text{Recall} = \frac{\varphi P}{\varphi P + \omega N} \tag{3}$$

The model was trained for 40 epochs with an accuracy of 80% and validation accuracy of 83.5%, precision of 80%, specificity of 73%, andsSensitivity of 71%, respectively. The performance of the proposed work was compared with ResNet50 and VGG19 architectures, respectively, where the proposed model was found to perform well in VGG16 transfer learning architectures was summarized in Table 1 as follows and illustrated in Fig. 2.

A correlation matrix was used to identify the percentage of factors influencing the Geohazard. There exist various factors which are directly associated with the occurrence of Geohazard. The percentage of factors varies for each attribute. Several attributes are not associated with the given factors. The influencing factors are represented by using Correlation mapping in Table 2. They are as follows (a) Changes in water level, (b) Earthquakes, (c) Landslides, (d) Heavy Rainfall, (e) Elevation

Table 1 Performance comparison across various transfer learning techniques

Transfer learning techniques	Accuracy (%)	Precision (%)	Recall (%)	Specificity (%)	Sensitivity (%)
VGG16	83.50	80	78	78.50	76
ResNet50	78	75	76	74.50	73
VGG19	79	77.45	76	75.50	73.65

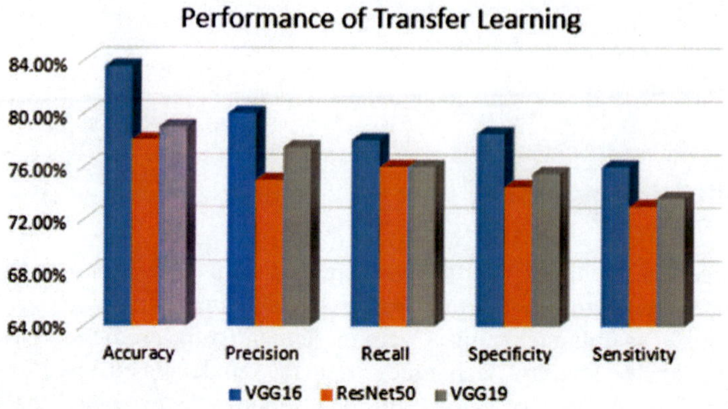

Fig. 2 Performance of various transfer learning techniques

Table 2 Correlation mapping

Factors influencing Geohazard	Landslides (%)	Earthquake (%)	Volcanoes (%)	Tsunami (%)
Changes in water level	10	50	5	
Earthquakes	50		30	50
Landslides		15		
Heavy rainfall	15			
Elevation difference		10		10
Soil quality		5		
Tectonic movements		10	15	30
Man-made activities	10			
Temperature			40	
Viscosity			10	
Ground pressure	15	10		
Extra-terrestrial collision				5
Lava entering sea				5

Difference, (f) Soil Quality, (g) Tectonic Movements, (h) Man-made Activities (i) Temperature, (j) Viscosity, (k) Ground Pressure (l) Extra-Terrestrial Collision, and (j) Lava Entering Sea.

The formula for calculating Geohazard is as follows.

$$C_{\text{coeff}} = \frac{\sum(id - id1)(d - d1)}{\sqrt{\sum(id - id1)2(d - d1)2}} \tag{4}$$

In Eq. 4 id1 represents the independent variable and d1 represents the dependent variable. From the correlation obtained the factors influencing Landslides are Changes in water level, Earthquakes, Heavy rainfall, Man-made activities, and Ground Pressure. Earthquake is the factor that major influences the occurrence of Landslide is illustrated in Fig. 3.

From the correlation obtained the factors influencing Earthquakes are Changes in water level, Landslides, Elevation Difference, Soil Quality, Tectonic Movements,

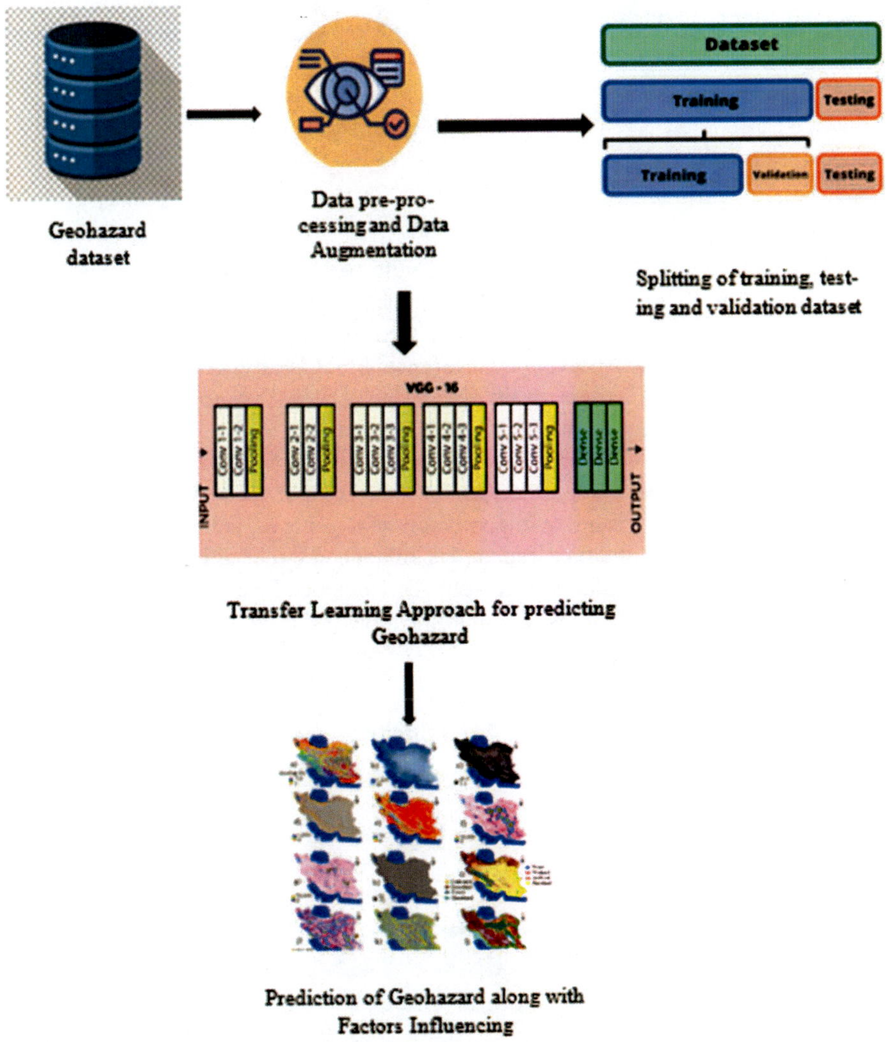

Fig. 3 Workflow of Geohazard prediction and identifying the factors influencing Geohazard

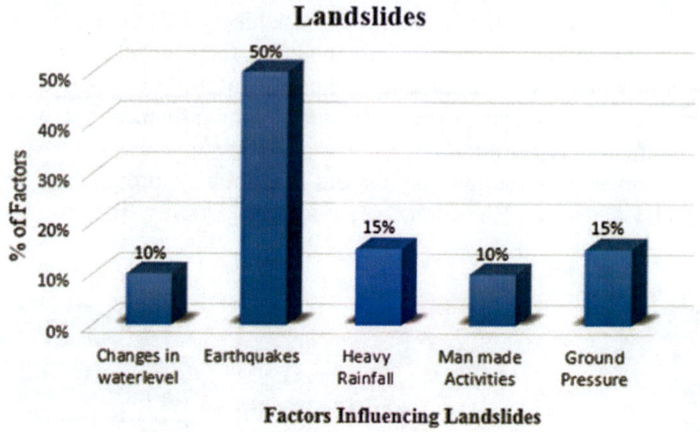

Fig. 4 Percentage of factors influencing landslides

and Ground Pressure. Changes in the Groundwater level is the factor that major influences the occurrence of Earthquake is illustrated in Fig. 4.

From the correlation obtained the factors influencing Earthquakes are Changes in water level, Landslides, Elevation Difference, Soil Quality, Tectonic Movements, and Ground Pressure. Changes in the Ground water level are the factor that major influences the occurrence of the Earthquake is illustrated in Fig. 5.

From the correlation obtained the factors influencing Volcanoes are Changes in water level, Earthquakes, Tectonic Movements, Temperature, and Viscosity. The major factor influencing is a rise in Temperature at sea level is illustrated in Fig. 6.

From the correlation obtained Tsunami is influenced by certain factors that include Earthquakes, Elevation Difference Tectonic Movements, Extra-terrestrial Collision, and Lava Entering Sea. The major factor influencing is the occurrence of Earthquakes at sea bed is illustrated in Fig. 7.

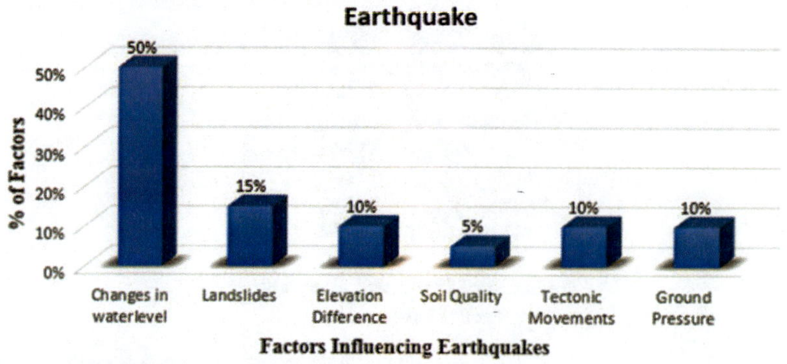

Fig. 5 Percentage of factors influencing earthquake

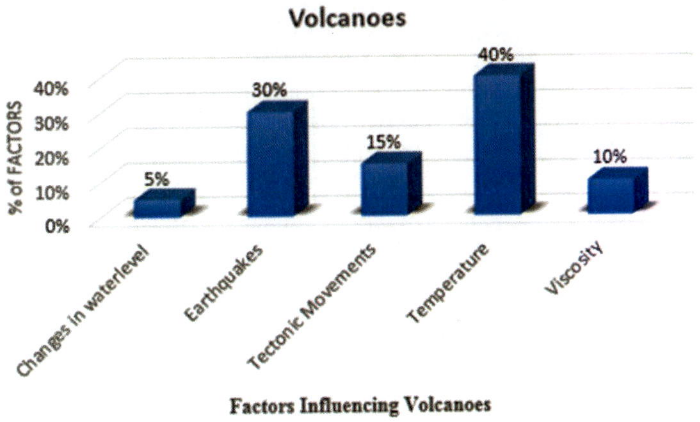

Fig. 6 Percentage of factors influencing volcano

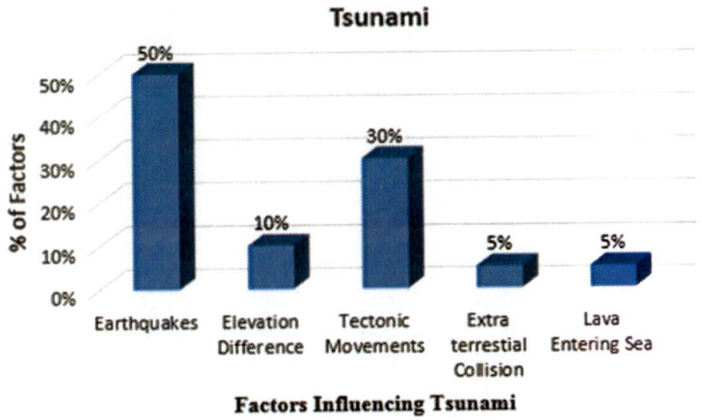

Fig. 7 Percentage of factors influencing tsunami

5 Conclusion

The proposed approach works effectively in predicting the occurrence of Geohazard. Geohazard was predicted through the various traditional approaches. That conventional way of identifying Geohazard was not so effective. The proposed work invokes a novelty-based approach through Transfer learning techniques known as VGG16. The neural network architecture VGG16 predicts the happening of Geohazard more precisely with an accuracy rate of 80%. The performance of the proposed architecture was compared with other transfer learning architectures includes VGG19 and ResNet50. The proposed model performs effectively in VGG16 architecture. Along with prediction through correlation coefficients, the factors influencing the cause of various Geohazard includes Earthquakes, Volcanoes, Tsunami, and Landslides were identified. The major influencing factor was identified as a result of correlation comparisons.

The further enhancements can be done by analyzing the time series of Geohazard prediction and implementing an effective prognostic approach. The limitation of the proposed system is the forecasting of Geohazard using the satellite images alone.

References

1. Dikshit A, Pradhan B, Alamri AM (2020) Pathways and challenges of the application of artificial intelligence to geohazards modelling. Gondwana Res, ISSN, pp 1342–1937
2. Mei G, Xu N, Qin J, Wang B, Qi P (2020) A survey of internet of things (IoT) for geohazard prevention: applications, technologies, and challenges. IEEE Internet Things J 7(5):4371–4386
3. Ma Z, Mei G, Prezioso E et al (2021) A deep learning approach using graph convolutional networks for slope deformation prediction based on time-series displacement data. Neural Comput Applic
4. Arogunmati A, Moocarme M (2019) Automatic geohazard detection using neural networks. In: Paper presented at the Offshore Technology Conference, Houston, Texas
5. Choubin B, Mosavi A, Heydari Alamdarloo E, Hosseini FS, Shamshirband S, Dashtekian K, Ghamisi P (2019) Earth fissure hazard prediction using machine learning models. Environ Res 179(Part A):108770. ISSN 0013-9351
6. Roy SG, Koons PO, Tucker GE, Upton P (2019) Advancing geo-mechanical analyses with deep learning to predict landslide susceptibility from spatially explicit strength and stress states
7. Cao J, Zhang Z, Du J et al (2020) Multi-Geohazard susceptibility mapping based on machine learning—a case study in Jiuzhaigou China. Nat Hazards 102:851–871
8. Shafizadeh-Moghadam H, Minaei M, Shahabi H et al (2019) Big data in Geohazard; pattern mining and large scale analysis of landslides in Iran. Earth Sci Inform 12:1–17
9. Ahmad H, Ningsheng C, Rahman M, Islam MM, Pourghasemi HR, Hussain SF, Habumugisha JM, Liu E, Zheng H, Ni H, Dewan A (2021) Geohazards susceptibility assessment along the upper indus basin using four machine learning and statistical models. ISPRS Int J Geo-Information 10(5):315
10. Husam A, Al-Najjar H Kalantar B, Pradhan B, Saeidi V (2019) Proceedings Volume 11156, earth resources and environmental remote sensing/GIS Applications X; 111560K
11. Pham BT, Prakash I, Singh SK, Shirzadi A, Shahabi H, Tran T-T-T, Bui DT (2019) Landslide susceptibility modeling using reduced error pruning trees and different ensemble techniques: hybrid machine learning approaches. CATENA 175:203–218. ISSN 0341-8162
12. Shukla A, Adwani N, Choudhury T et al (2021) Geospatial analysis for natural disaster estimation through arduino and node MCU approach. GeoJournal. https://doi.org/10.1007/s10708-021-10496-1
13. Jena R, Pradhan B, Beydoun G, Alamri AM, Nizamuddin A, Sofyan H (2020) Earthquake hazard and risk assessment using machine learning approaches at Palu, Indonesia. Sci Total Environ 749:141582. ISSN 0048-9697
14. Tomar R, Sastry HG, Prateek M (2020) A V2I based approach to multicast in vehicular networks. Malaysian J Comput Sci 93–107. Retrieved from https://jupidi.um.edu.my/index.php/MJCS/article/view/27337.
15. Millis1 SW (2018) Digital advancements and tools for geohazard assessment. The IEM-CIE-HKIE Tripartite Seminar, Putrajaya, Malaysia, 4
16. Jain S, Sharma S, Tomar R (2019) Integration of Wit API with python coded terminal bot. In: Abraham A, Dutta P, Mandal J, Bhattacharya A, Dutta S (eds) Emerging technologies in data mining and information security. Advances in intelligent systems and computing, vol 814. Springer, Singapore. https://doi.org/10.1007/978-981-13-1501-5_34

17. Joshi D, Patidar AK, Mishra A et al (2021) Prediction of sonic log and correlation of lithology by comparing geophysical well log data using machine learning principles. GeoJournal. https://doi.org/10.1007/s10708-021-10502-6
18. Kardani N, Zhou A, Nazem M et al (2020) Estimation of bearing capacity of piles in cohesionless soil using optimised machine learning approaches. Geotech Geol Eng 38:2271–2291
19. McGaughey WJ, Laflèche V, Howlett C, Sydor JL, Campos D, Purchase J, Huynh S (2017) Automated, real-time Geohazard assessment in deep underground mines. In: Wesseloo J (ed) Proceedings of the eighth international conference on deep and high stress mining, Australian Centre for Geomechanics, Perth, pp 521–528
20. Visalaxi S, MuthuT, Automated prediction of endometriosis using deep learning. Int J Nonlinear Anal Appl 12(2):2403–2416

Application and Uses of Big Data Analytics in Different Domain

Abhineet Anand, Naresh Kumar Trivedi, Md Abdul Wassay, Yousef AlSaud, and Shikha Maheshwari

Abstract The process of massive data screens with various data kinds—Big data—is to detect hidden patterns, unknown correlations, market trends, customer preferences and other important information. Big data analytics are an analysis technique that examines enormous data sets. The analytical results can contribute to more efficient marketing, additional income opportunities, enhanced service to the client and increased operational effectiveness. The software tools typically used in advanced analytics disciplines including predictive analysis, data mining, text analytics and statistical analysis are the forms of large data analysis. In the analytical process, mainstream BI software and data visualisation instruments can also be used.

Keywords Big data analytics · Industry 4.0 · Predictive analysis · Customer satisfaction · Business outcome

1 Introduction

Today's world generates enormous data daily. Experts said this scenario might lead to a large wave of data or possibly a tsunami. Today this enormous volume of data is

A. Anand (✉) · N. K. Trivedi · M. Abdul Wassay · S. Maheshwari
Chitkara University Institute of Engineering and Technology, Chitkara University, Punjab, India
e-mail: abhineet.anand@chitkara.edu.in

N. K. Trivedi
e-mail: nareshk.trivedi@chitkara.edu.in

M. Abdul Wassay
e-mail: abdul.wassay@chitkara.edu.in

S. Maheshwari
e-mail: shikha.maheshwari@chitkara.edu.in

Y. AlSaud
Department of Electrical Engineering and Computer Science, Howard University, Washington, D.C, USA
e-mail: yousef.alsaud@bison.howard.edu

V. Skala et al. (eds.), *Machine Intelligence and Data Science Applications*, Lecture Notes on Data Engineering and Communications Technologies 132,
https://doi.org/10.1007/978-981-19-2347-0_38

called big data. As the tsunami statistics are more or less real, we believe that we must now have a tool to systematically have these data for use in numerous domains as government, science, industry, etc. This helps to correctly analyse, store and process it.

Big data is an unrefined term for huge and complex data. The processing utilising typical processing methods is complicated and also time-consuming [1, 2].

Big data may be described as:

Volume—very significant is the amounts of data generated.

Variety—Variety is also a very important fact which must be known to analyse data in this category.

Velocity—In the context 'velocity' refers to the speed at which the data is generated and processed or how quickly it is processed.

Variability—this is the incoherence that can sometimes be demonstrated by the data. This can interfere with the process of managing and managing the data effectively.

Truthfulness—The quality of data collected can vary greatly and hence the accuracy.

Complexity—Data management may be a complex process, particularly when huge amounts of information originate from many sources [3, 4].

Big data tools are employed for the processing of this data, which examine the data and process it as necessary. Industry, academics and other key actors agree that Big Data has in recent years become a major change for most, if not all, of modern industries. As the big data is still pervading our daily life, the focus from the hoopla around it to genuine value in its usage has changed significantly [5, 6].

Whilst understand edge of the value of Big Data continues to be a challenge, other practical challenges, including funding and return on investment and skills, remain at the forefront of various large-scale industries[7]. In 2017, according to research and market reports, the worldwide big data industry reached $32 billion and by 2026 is expected to reach $156 billion [8].

Most firms generally have numerous objectives to implement Big Data projects. For most companies, the major purpose is to improve customer experience, but other aims include cost reduction, improved marketing and efficiency improvement for existing business. Data violations have also recently made increased security a major priority for big data projects [9, 10].

More significantly, however, where do you stand when it comes to big data? Probably you will also find that:

- Try to decide if or not big data are true.
- Evaluate market opportunity size.
- Development of new Big Data services and solutions.
- Big Data solutions are already used. Reposition existing services and products with the purpose of using big data or,
- Big data solutions already used.

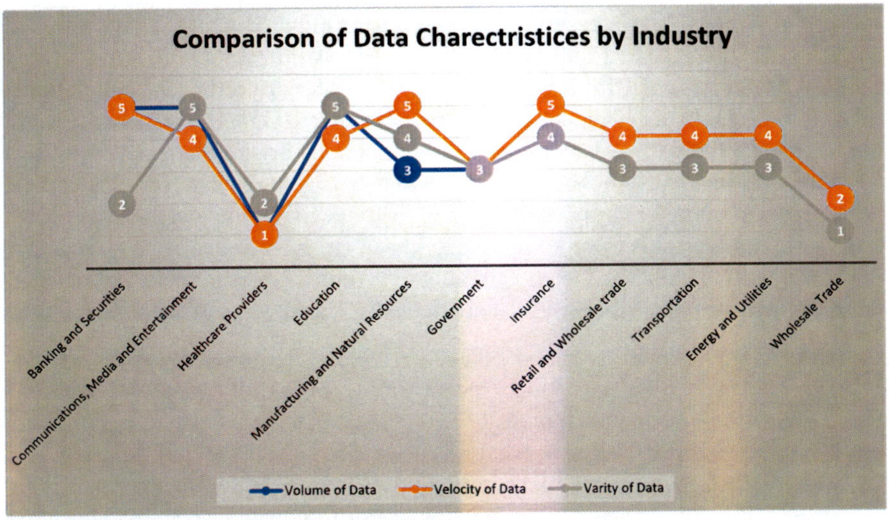

Fig. 1 Comparison of data characteristics by industry

In this respect you will understand the function of birds in large-scale data and their application in another sector, or in your industry and in several industries in the future (Fig. 1).

2 Big Data Analytics in Banking

The banking sector shows well how technology has affected consumer experience. It is time for customers to stay on their Saturday morning paycheck. Now clients can check your account balance, paychecks, pay bills and transfer money with their mobile phones—they do not need to even leave their house [11–13].

These self-service capabilities are good for customers, but are one of the key reasons why traditional banks only compete online with similar firms and financial institutions. Given that client activity is now predominantly online, several human services that brick and mortar banks provide no longer meet client requirements [14].

This is important in adopting Big Data strategies and technologies for the banking industry. Using personal or transaction information, banks can provide a 360 degree perspective of their users to:

- Track client expenditure patterns
- Customer groupings based on their profiles
- Implement processes for risk management
- Customising product range
- Incorporating retention techniques
- Gather, evaluate and answer customer comments.

Enterprises capable of measuring their profit from big data analyses have reported an average 8% increase in revenue and a 10% drop in overall costs in 2015, according to the BARC survey in 2015. Reports from BARC. We will follow Dana, a fictional client who has just opened an American One primary checking account to highlight how big data and big data analysis are used by financial organisations.

After years of unhappiness with her previous bank Dana went to America One at the advice of some of her friends. Dana is really thrilled to be with America One since her individualised customer service has been amazing to her. As America One is now officially a client, the team is willing to utilise large-scale data and banking analytics to ensure the greatest Dana experience [15].

1. Get a full view with profiling of our customers Customer segmentation has increased in the financial services industry as it allows banking and credit unions to divide their customers into neat demographic categories. But the granularity that these institutions seek is not the basic category, because they have a proper awareness of their clients' demands and aspirations. These institutions should instead utilise large data in the banking sector to construct detailed consumer profiles to the next level [16, 17].
2. Tailor each person's customer experience

   Almost one-third of customers want their employer to know their details; 33% of customers actually departed the company last year due of a lack of individualisation. As regards relationship banking, financial services are not very well known for its high level of customised service. A shift in attitudes and customer experience based on the analysis of banks is needed for banks and credit associations that not only aspire to survive but to grow [15, 18].
3. Understanding of Customers Buying pattern

   Customers generate almost all of the large data in the banking sector, whether via contacts or transactions with sales and service providers. Although two customer data sources are enormously important, transactions provide banks with a clear view of their customers' cost habits and widespread activity [19].
4. Identify up sales and cross-sales opportunities

   Businesses have a 60-70% higher chance than the prospects of selling to existing clients, which makes it simple for banks to sell across and upsell—the potential made BID analytics in banking even easier [20].
5. Reduce the risk of fraudulent conduct Identity fraud, which in 2016 reached record levels, is one of the most quickly rising forms of fraud with 16.7 million casualties alone. In order to ensure banks can use vast data to avoid fraud or keep consumers more safe, customised spending habits and unexpected behaviour can be watched.

3 Big Data Analytics Impact on Telecom Industry

Smartphones and other connected mobile devices have increased rapidly, resulting in the volume of data that passes via telecom carriers' networks. Operators need to process, store and extract information from the data supplied. By contributing to optimising network usage and services, large-scale data analysis can help them raise profitability, increase intelligence and improve security. Research has demonstrated that Big Data Analytics provides significant possibilities for telecoms companies [5, 21, 22].

However, the potential of large data creates a challenge: How can an enterprise use data to enhance revenues and profits in the value chain, encompassing networking, product development, marketing, sales and customer support?

For instance, big data analysis allows organisations to estimate how highly congestion-reduction strategies can be used by peak networks. It can also help identify customers with the greatest risk of trouble paying bills and changing operators, improving their sales [23, 24].

The standard top-down Big Data Analysis technique is generally supported by operators that identify the problem to be solved and search for data to help resolve it. Instead, the operators should focus on the data and use it for correlations and links. The data can provide insights that can be used as a basis for simpler operations when done appropriately [25].

4 Healthcare Big Data Analytics

Big data in the field of healthcare is used to describe the huge quantities of information created through digital technologies, which collect records of patients and help manage the operation of hospitals otherwise excessive and complex for traditional technology [26].

Big data analysis has several good and life-saving results in the healthcare industry. Big data refers to the large amounts of information created by digitisation, which is consolidated and analysed using special technology. It uses precise population health data (or of a specific individual) and may help prevent epidemics, cure sickness, decrease expenditures, etc. Applied in the field of health care [27].

Now that we are living longer, treatment paradigms have altered, many of which are driven by data. Doctors aim to understand a patient as well as possible and to receive warning indications as soon as possible of serious illness—the early treatment of any disease is significantly easier and much more costly. Prevention is better than cure, and the use of key performance indicators in health care and health data analyses allows the insurance company to deliver a customised package [18, 28]. This is an industry attempt to resolve patient data's silo problems: bits and bytes are collected and archived everywhere and are not adequately communicated in hospitals, clinics, surgeries, etc.

It has been costly and time-consuming for years to collect an enormous amount of data for medical purpose. With today's ever-improving technologies, the collection of such data is easier, but also comprehensive health care reports are created and converted into useful critical views, which may be used to deliver better care. Health data analyses aim to predict and resolve the problem before it is too late to employ data-driven research. Methods and treatments are also evaluated more quickly, inventories are tracked better, patients are more involved in their health and they have the tools to accomplish that [29].

5 Big Data in Education

The data was long before computers, but the amount created daily by technology has undoubtedly speeded up. We generate at least 2.5 quintillion bytes of data daily using mobile devices, the Internet of Things (IoT) and social media and other information sources. This unbelievable data volume is too complex to acquire, store and manage standard technologies. That's where a programme for data management centres. All relevant data sets are added by the correct software, which is easy to use and easily understood by compiling on a dashboard. In this article, we can learn how to find our school's proper data analytics platform. Educators, policymakers and stakeholders utilise data analytic programmes, to detect institutional flaws and identify positive change prospects. Software applications provide analysis and interpretation of a wide variety of population data and allow you to design new tactics from there to advance your organisation [30].

5.1 *It Helps You Find Answers to Hard Questions*

The greatest strategy to develop answers for the challenging difficulties facing the education profession is to evaluate your present data. The more you are aware of your history, the more you may learn. For instance, you can see enrolments fall if you are in higher education. Big data gives you the required contextual indicators to exactly find out where, when and how your registration changes. Most importantly, statistics can help to solve registration concerns such as [31, 32]:

- Does one department or the entire institution see declines?
- Is there a lecturer with low enrolment in classes?
- Can portions be combined in order to boost productivity?

5.2 *It is Accessible*

It is messy and time-consuming to search by way of a file cabinet setup. Because big data relies on technology infrastructure for information collection, storage and management, finding what you are looking for is far easier. Besides infrastructure, it can also be difficult for institutional silos to share information. Leadership may be privy to data instructors have no access to which growth and knowledge are barriers. Data analytics and the correct software help you build a more collaborative atmosphere. Since data are available at a central location, Internet access is all you need to find what you are looking for. Many software applications, such as Google Chrome and Safari, are available through browsers so you do not even need to install a plug-in or app [33].

5.3 It Can Save Costs

In higher education, proper allocation of resources is vital and your data is the key to efficiency. Firstly, your data can provide insight into the registration numbers for various classes. If ENG 102 has five sections and only two are complete, the remaining three can be merged to protect resources, such as classroom space, time for teachers and energy use. With this in mind, cloud-based systems might potentially reduce data storage expenses and lower the strain from your IT staff. The data have typically been manually sorted and transcribed, taking time and may take weeks or months. And it can take just as much time when you require specialised reports created regularly. The time your staff spend will probably be more helpful. An analytical programme, which automates a lot of this tedious effort, makes it quick and easy to retrieve data and thus saves you long-term money.

The higher education recruitment drive is another cost-saving advantage. If you look at previous school performances you can see which future pupils will be successful at your school and which will probably drop out or fail. This might allow you to design more successful processes for each student to maximise [24, 34].

5.4 It is Quick

We have briefly discussed that in terms of cost savings, but it is worth repeating. You save unbelievable weaving time with data merely to discover a particular report or information on a certain kid with all the information your school can provide in one consolidated area.

Big data is also available in real-time so that you may decide faster than previously. This is especially beneficial during registration times when instructors monitor the figures to prepare for the next six months. In order to assess how much enrolment is being paid at the same time in the previous year, your year over year enrolment reports can be automated. You can then decide to increase or maintain the stability of your inscriptions on the same day [35].

5.5 It Helps You Adapt

You can build new classes, teaching methodologies and other ways by identifying trends, to give students what they desire. Take the example of community colleges. The majority of the schools are made up of people who balance jobs, families and education, so a flexible schedule is a need for schools. As online learning expands, many of these adult students find online programmes more convenient than late classes at night or on the weekends. Big data will disclose the exact numbers behind it since online learning is preferable for specific classes—such as programmes that do not require laboratories or practical learning. Perhaps on-campus biology class registrations were consistent as English courses declined on-campus. You may customise your courses so that you always provide your pupils with the greatest options. Precision Campus makes it fast and easy to save and manage your institutional data. You may obtain the context you need to better understand your institutional performance through interactive dashboards and visual interpretations of your data [36, 37].

6 Manufacturing and Natural Resources in Big Data

The industry is using a range of production software, but frequently it is not a convenient way to link solutions together to obtain an insight into how a factory floor operates. The objective of Industry 4.0 is to resolve this. There are numerous possibilities for manufacturers to think about the many forms of software for production—ERP, MES, CMMS, manufacturing analyses. When these systems are interconnected via large data in the production sector, patterns can be recognised and problems treated in an efficient and sophisticated manner [38].

The notion underlying vast numbers of data provides an overview of all data gathered so that an overview of many equipment, lines, processes and systems may be obtained. Manufacturing, machine sensors, quality, maintenance and design data can be linked together to find patterns and collect information for advanced judgments. Big manufacturing data may include productivity data on the amount of the product you make in all metrics you need to collect for quality control. The amount of energy consumed by a machine or how much water or air the machine needs to run. Big data is a piece of data from any part of the company [39, 40].

Other software equipment like sensors, pumps, motors, Compressors or conveyors generates large-scale production data. It is also produced by external partners, suppliers and customers. It is extremely important to remember that huge data is there. If the data is generated, it may flow into the broader concept of big data.

Big data generates a lot of data in manufacture, of course. The data does not mean anything without analysis. Data visualisation, data science or data analysis should be exploited and insights should be developed to enhance the utility of the obtained data [41]. The method of data analysis is useful not only for taking decisions, but also for the final analysis of your company. Your organisation can use data analysis to:

- Enhance manufacturing
- Customise product design
- Ensure higher quality assurance
- Manage the supply chain management
- Asses any potential risk.

All of these are instances of manufacturers using Big Data analytics: oil and gas, refineries, chemical producers, auto-makers, plastics, metal shaping, food and drink manufacturing. The list continues and even goes beyond production. To make data-driven decisions, everyone and everyone can use analytics. Those who realise the most advantage efficiently employ large data by linking all their systems to acquire a broad overview of plant efficiency [42].

Think of big data analytics, which connects all. You will not discover great insights if you only look at the broader picture, but you may decide with analytics how and why your processes work as they are. It is vital to analyse how firms, especially your company, grow [43].

7 Big Data in Government Sector

In order to offer real results, large-scale data and analytics can be applied to almost every public sector activity. Analytics are utilised to detect health problems, organise thousands of displaced people and avert water scarcity problems in response to large natural disasters like Typhoon Haiyan. Recently, analytics were applied to identify regions of need and more efficient distribution of resources following hurricane Maria [44].

- Laundering anti-money. Cash-laundering and financial crimes are prevented using analytics that affects terrorist organisations directly or unfriendly foreign governments that employ illicit financial activity to support their actions.
- Threats to insiders. With analytics to detect irregular activity and abnormalities, organisations can considerably reduce the amount of data released or stolen. This helps avoid fraud and cybercrime that drains money and resources from programmes to help citizens otherwise.
- Efficiency of the workforce. Organisations can better comprehend the employee gaps which can occur as employees retire or leave the private sector. Agencies can continue to operate effectively by ensuring that new staff can meet the gaps and providing means of retaining personnel [45].

The public sector benefits enormously from large-scale data and analysis. In addition, analytics improve results that directly affect citizens. The analytical insights you can obtain from your big data store can be the difference, whether it is fighting the nationwide drug problems, answering a local tragedy, protecting against loss of sensitive information or intellectual property or simply increasing the efficient government [46, 47].

The rapid increase in customer-facing portals, cloud-related technologies and smart sensors and devices is driving government data generation and digital archiving rates. As digital information increases and gets more complicated, information can be managed, processed, stored, secured and disposed of. They can gain insight from unstructured information via new technologies, collect, search, identify and analysis companies. The State is in the process of tilting, understanding that information is a strategic asset. In order to meet mission requirements in order to secure, use and analyse, government needs both organised and unstructured information. In trying to develop data-driven agencies, government officials create the basis for the correlation between events, persons, processes and information. Governments will produce high-value solutions based on a range of the most disruptive technology.

- Cloud services
- Social business and networking technology
- Mobile devices and applications
- Big data and analytics.

The Government continues to increase the usage of laptops, smartphones and tablets. Mobile computing provides successful telework and promotes operations continuity and disaster recovery productivity for workers. Social media communication technologies allow citizens to play a proactive part in government. The usefulness of this technology, particularly if it is used to strengthen open, transparent governance and public service provision, is further enhanced with the capacity of social media [10, 16, 48].

8 Big Data in Insurance

Most insurance firms now recognise that big information should be at the heart of many of their activity, but only a few really grasp how it is processed and used effectively in their enterprise.

Big data can actually be applied to nearly everything in the insurance industry, from underwriting to claims management and customer support. The most significant role Big Data play today in pricing and contracting in insurance was revealed by a report by the European Insurance and Occupational Pension Authority (EIOPA).

Auto-mobile insurance provides a good example of this, as brokers may analyse drive behaviours with big data to anticipate risks and tailored plans for each motorist precisely.

Insurers can utilise large data in claims management to evaluate losses or harm to segment or to automate claims in particular situations. This makes it much easier for providers to decide on claims, even if a claim is paid [21, 49].

Perhaps one of the most exciting uses of big data is to forecast or even modify the behaviour of customers. This is linked to the IoT; insurers, which can analyse client behaviour accurately with data from a broad variety of devices, may be able to step in prior to making a claim to even warn policyholders to change high-risk behaviour, for example driving too quickly or forgetting to alarm.

Big data also plays a key role in the detection of fraud. Each day, 1,300 insurance scams are detected and large data may be used to screen anomaly data, analyse information from the social networks and model fraud risks. Several Big Data solutions for insurance [50].

9 Customer Acquisition

Each individual generates vast volumes of information through social media, e mail and feedback that provide significantly more accurate information than any poll or questionnaire about his/her preferences. By providing customised marketing companies for helping new clients, examination of this unstructured data could improve productivity.

9.1 Customer Retention

Algorithms can discover early indicators of unhappiness from customers based on client activity, enabling you to immediately respond and enhance services. Insurers can concentrate on solving problems for customers by using collected insight, offering reduced prices, and even modify the price model to enhance customer loyalty.

9.2 Risk Assessment

Insurers have long been focussed on verifying information for clients whilst assessing risks, and big data technology help make this process more efficient. A predictive model can be used before the final decision by an insurance firm to assess potential problems on the basis of customer data and to establish precisely its risk class.

9.3 Preventing and Detecting Fraud

The US insurance companies lose roughly $80 billion every year in fraud according to Coalition Against Insurance Fraud. Insurers can compare previously fraudulent profile information with predictive modelling and detect cases requiring further inquiry.

9.4 Cost Reductions

Many manual operations can be automated using Big data technology to make them more efficient and to reduce the cost to handle claims and administration. This will lead to lower premiums in a competitive atmosphere, attracting new customers.

9.5 Personalised Services and Pricing

Unstructured data analysis can contribute to the delivery of services that meet consumer needs. Big data life insurance, for example can be more customised by looking not only at the consumer's medical history but also at his business trackers patterns. It can also be used to identify pricing models both for profit and for customer budgets [51, 52].

10 Big Data in Retail

For merchants to remain competitive, they need to do better purchase decisions, give relevant discounts, urge customers to take advantage of new trends and remember the days of their customers—all behind scenes. How are they maintaining? Big data in retail is critical to consumer targeting and retention, operations simplified, the supply chain optimised, business choices improved and finally saved money. Prior to the cloud, firms were restricted to tracking what and when a person had purchased it. More advanced technology enables corporations to record an abundance of client information, such as their age, location, gender, preferred restaurants, the other books or news they purchase in—the lists go on. Retailers are now turning to Big Data cloud solutions to gather and manage the data [53, 54].

Big data analysis can identify trends, target the correct customer, lower marketing expenses and improve customer service quality in a timely manner [55]. The common advantages of retail use of big data include:

- Keeping a view of each customer of 360°—Create the kind of personal commitment which customers can expect, knowing every individual on an individual basis.
- Optimising price–Make the most of future trends and know if the off-trend price of products may be reduced and how much.
- Smooth back-office—Imaging that keeps the stock levels ideal all year long and collects data in real-time from registered products.
- Customer service improved—Download data hidden by registered calls, security images and commentary on social media for customer service.

10.1 360-Degree View of the Customer

The word '360° vision' is often used, but what does it mean? All of it ends up in an entirely accurate image of a customer. Distributors need to know the likes and dislikes of a client, its likelihood, its sex, its location, social media presence, etc. The combination of only a few of these data elements can lead to advanced marketing techniques. For instance, fashion businesses usually recruit costly ambassadors for famous brands. But fashion firms can be more economical and successful for representing their businesses on Instagram, paying attention to customers' sex, likes and social media activity.

10.2 Price Optimization

Big data provides companies with an advantage in product pricing. Consistent search monitoring can allow firms to predict trends before they materialise. Retailers can develop new items and anticipate a successful approach for dynamic pricing. Prices can also benefit from the customer's 360-degree view. This is because pricing is based largely on the geography and procurement habits of a customer. Beta testing for portions of their customers can be performed by companies to determine the price best. Understanding the expectations of a customer can tell retailers how to oppose their competitors.

10.3 Streamlined Back-Office Operations

Anyone who has worked at retail has had this feeling of sinking when their stocks are depleted. This manager deals with unhappy customers throughout the remainder of his shift. This issue would ideally be eliminated by companies. Big data may assist organisations to manage both supply chain and product distribution, however that may never be always achievable. Product logs and server data can provide retailers with indications about how upstream they operate. Bugs can even be exposed to the products themselves. The product's performance can be shown over time by customers who register their wearables.

10.4 Enhanced Quality of Service

Consider the last time a toll-free number was called. There is usually a notification that your call will be 'quality recorded.' Big data research can illuminate key problems with recorded calls, and then analyse the success of quality modifications driven by the organisation over time. Some retailers scan video and motion sensors in stores to enhance their client experience. Retailers evaluate how often customers gravitate to a shop and put things they wish to sell first in a strategic way. This is not a new idea—spa store designs its layout purposely, leading more food to come out.

In client reviews and comments, insights are ready to be revealed. Analysing those evaluations can enable shops to inform shoppers that certain clothing can be large or tiny. 'Sentiment analysis can also be performed in order to determine if customers talk about certain items and firms in general positively or negatively'.

11 Big Data in the Industry of Transport

Big data are used for large-scale data processing by companies across numerous transport and travel areas such as airlines, airports, freight logistics, hotel, rail and others. Every industry together with the transport sector currently collects amazing data in this networked and instrumented environment [21, 45]. Big data and analytical advantages also enable transport companies to increase model capacity, demand, turnover, pricing, client feelings, costs and many more exactly.

- Knowledge enhanced: Big data and IoT can make customers or users aware at any given time of the most effective form of transit. There are numerous train companies which have begun employing large numbers to analyse seat data obtainability in real-time and also report on platforms to passengers who are waiting for carriages having the most seats available. This benefit from big data not only enhances client experience, but also increases its knowledge. Request a demo for more information on the advantages of big data in the transport industries.
- Enhanced customer service: Big Data has increased customer experience by refining its expertise, which is one of the most essential advantages. The efficient analysis of repeated complaints made numerous times by a single client using Big Data can lead to a more successful reaction. It helps hence to provide creative solutions for some difficulties, such as smartphone technology [34, 56].
- Operating efficiently: Big data may be utilised to eliminate mistakes and to decrease wasteful expenditure. It can be used to identify problems related to transport delays and downtimes. A good illustration of how the benefits of big data may make operation efficient is the structural railway in the Netherlands. Talk to an analyst to learn more about the advantages of big data in the transport industries [57].

12 Big Data Analytics in Energy and Utilities

The data were always a crucial part of the operational procedures of E and U. However, Big Data Analytics assume prominence by introducing new sources of information and subsequently increasing the number of data created.

However, Big Data Analytics assume prominence by introducing new sources of information and subsequently increasing the number of data created.

Technology is currently being used for data transmission, storage, and correlation, such as phases or measurement units (PMUs), Advanced Metering Infrastructures (AMI), smart metres and geographic information systems (GIS).

E and U is the lifelines for every other industry since it meets the demands for power and energy for the other industries. Big data and intelligent analytics are therefore critically needed to increase service efficiency [51].

Big data analytics case in E and U practical use.

Big data analytics in energy and utilities are mostly used in the following:

12.1 Prediction, Detection and Prevention of Power Outages

A power failure can stop a whole country, such as the 2013 Northeastern blackout which hit more than 45 million people in America. One of the main causes of such outages is unfavourable weather. Nevertheless, E and U firms build intelligent facilities and sensors to improve predictability and prevent such failure scenarios. Modern power failure systems use real-time solutions based on live data and intelligent algorithms to forecast and avoid these events. These systems can estimate the influence on the network grid of any near time asset values, possible interruptions generated by smart metre events, region-specific interruptions and more.

12.2 Smart Load Management

In order to manage energy loads properly, E and U companies need to combine energy demand strategically and smartly with optimum power supplies in a specific period of time. The intelligent load management system permits them, using distributed energy sources, modern control systems and end-user devices, to cover end-to-end system management needs, including demand and energy. All management system components produce data. By implementing large-scale data analysis, companies may make correct decisions about their power planning and generation, energy consumption and performance estimates.

12.3 Preventive Asset Management

E and U is an asset-intensive business that relies largely on its network infrastructure and equipment's optimum performance. The failure of these assets could pose major difficulties for power delivery and hence diminished consumer confidence. Preventing such events is therefore one of the industry's key goals. Big data analytics are rescued for preventive equipment maintenance. The assets are incorporated into intelligent sensors, trackers and data systems that transmit information in real-time to the centre. The collected data can then be processed and analysed so that possible difficulties with equipment maintenance can be identified and proactively handled.

12.4 Greater Operational Efficiency

Leveraging real-time asset data related to business rates, business conditions, time, supply and demand analyses, and further aiding E and U organisations in optimising energy efficiency and asset performance. By continued cost/performance monitoring, Big Data Analytics apps boost their dependability, capacity and accessibility to their network assets.

13 Big Data in Wholesale

Big data has been available for a very long time on the market of the company. Big data offers different advantages which may be used to identify market trends, client requests and consumer wants, etc. Big Data Analysis could be of value to any firm using suitable technology and technology.

13.1 Comprehending Macro and Micro Trends Influencing the Market

A wholesale person's main task is to be the distributor of goods and services. The wholesaler must perform this easy and basic activity. However, it becomes a huge difficulty for a larger distributor. The continuing patterns in the market can be understood using Big Data analysis. Big data analysis will help to reduce the complexity of data and will eventually help to make more accurate business choices.

13.2 Anticipating Future Demand

Predictive data analysis was available over years, these analyses were carried out on the data, and only a few happy recipients were able to get the results. The industry as a whole did not have these outcomes as an information scientist is needed to extract the information and this was costly more than the budget of the already low margin B2B sector. But now with time, things have changed and the predictive analysis for the wholesale market has been attractive and the primary modifications are as follows: First of all, an easy real-time database available which has facilitated quick access to the results and, secondly, the predictive model can now be deployed at the cost of a data scientist. In return, both retailers and wholesalers benefit by forecasting future demand, which only by using Big Data efficiently as possible.

13.3 Improved Marketing Campaigns

The usage of large data combined with a good sales representative's skills and expertise showed that predictive analysis increased profits to the sky level. For instance, if a sales official visits a client, he/she usually spend 20–25 min at a meeting. And the salesperson can showcase and introduce a particular customer to at least three to four new products. And if the product developed is above 10, the sales agent does not know which specific product is most beneficial if shown to a particular client.

Big data has so helped businesses better address their customers' demands by providing helpful and comprehensive content.

13.4 Increasing Customer Satisfaction

Most wholesale marketers have collected a lot of information about consumer preferences and patterns of consumption. The information acquired by wholesalers covers data on various items, their demands in different locations throughout time, prices and volumes of products sold at different prices, as well as consumer and demographic data in different locations and marketplaces. The use of that data to better their procedures, solve difficulties that face them and thus raise their income makes sense for them. With this volume of data available.

If demand for different products is analysed, wholesalers can reduce costs by retaining less stock of commodities that are less demanding and greater inventory of high demand products. The strong demand for stock products results in higher customer satisfaction, therefore boosting the number of wholesale traders' repeat clients. By expressing consumer preferences to producers, they may optimise products in their inventory. The provision of products to the clients will enable wholesalers to establish relationships with these clients. Big data investment is necessary for wholesalers to detect historical sales patterns and correlate them to the population in a market, and forecast future trends. Precise demand forecasts lead to increased long-term benefits for distributors. In terms of sales volumes and growth rates, it is thought that companies who remain ahead of their competitors make use of Big Data technology most efficiently and efficiently.

References

1. Palanisamy V, Thirunavukarasu R (2019) Implications of big data analytics in developing healthcare frameworks—A review. J King Saud Univ - Comput Inf Sci 31(4):415–425. https://doi.org/10.1016/j.jksuci.2017.12.007
2. Zerdoumi S et al (2018) Image pattern recognition in big data: taxonomy and open challenges: survey. Multimed Tools Appl 77(8):10091–10121. https://doi.org/10.1007/s11042-017-5045-7
3. Pramanik MI, Lau RYK, Yue WT, Ye Y, Li C (2017) Big data analytics for security and criminal investigations. WIREs Data Min Knowl Discov 7(4):e1208. https://doi.org/10.1002/widm.1208
4. Sagiroglu S, Sinanc D (2013) Big data: a review. In: 2013 International conference on collaboration technologies and systems (CTS), pp 42–47. https://doi.org/10.1109/CTS.2013.6567202
5. Elgendy N, Elragal A (2014) Big data analytics: a literature review paper. In: Advances in data mining. applications and theoretical aspects, pp 214–227
6. Hilbert M (2016) Big data for development: a review of promises and challenges. Dev Policy Rev 34(1):135–174. https://doi.org/10.1111/dpr.12142
7. Al-Jarrah OY, Yoo PD, Muhaidat S, Karagiannidis GK, Taha K (2015) Efficient machine learning for big data: a review
8. Drosou M, Jagadish HV, Pitoura E, Stoyanovich J (2017) Diversity in big data: a review. Big Data 5(2):73–84. https://doi.org/10.1089/big.2016.0054
9. Kamilaris A, Kartakoullis A, Prenafeta-Boldú FX (2017) A review on the practice of big data analysis in agriculture. Comput Electron Agric 143:23–37. https://doi.org/10.1016/j.compag.2017.09.037
10. Hashem IAT, Yaqoob I, Anuar NB, Mokhtar S, Gani A, Ullah Khan S (2015) The rise of 'big data' on cloud computing: review and open research issues. Inf Syst 47:98–115. https://doi.org/10.1016/j.is.2014.07.006
11. Srivastava U, Gopalkrishnan S (2015) Impact of big data analytics on banking sector: learning for Indian Banks. Procedia Comput Sci 50:643–652. https://doi.org/10.1016/j.procs.2015.04.098
12. Hassani H, Huang X, Silva E (2018) Digitalisation and big data mining in banking. Big Data Cogn Comput 2(3). https://doi.org/10.3390/bdcc2030018
13. Hassani H, Huang X, Silva E (2018) Banking with block chained big data. J Manag Anal 5(4):256–275. https://doi.org/10.1080/23270012.2018.1528900
14. Sun N, Morris JG, Xu J, Zhu X, Xie M (2014) iCARE: a framework for big data-based banking customer analytics. IBM J Res Dev 58(5/6):4:1–4:9. https://doi.org/10.1147/JRD.2014.2337118
15. Chong D, Shi H (2015) Big data analytics: a literature review. J Manag Anal 2(3):175–201. https://doi.org/10.1080/23270012.2015.1082449
16. Munar A, Chiner E, Sales I (2014) A big data financial information management architecture for global banking. In: 2014 international conference on future internet of things and cloud, pp 385–388. https://doi.org/10.1109/FiCloud.2014.68
17. Kumar R, Anand A (2017) Internet banking system and security analysis. Int J Eng Comput Sci 6(6):2319–7242. https://doi.org/10.18535/ijecs/v6i4.43
18. Martin-Sanchez F, Verspoor K (2014) Big data in medicine is driving big changes. Yearb Med Inform 9(1):14–20. https://doi.org/10.15265/IY-2014-0020
19. Hale G, Lopez JA (2019) Monitoring banking system connectedness with big data. J. Econom 212(1):203–220. https://doi.org/10.1016/j.jeconom.2019.04.027
20. Srivastava A, Singh SK, Tanwar S, Tyagi S (2017) Suitability of big data analytics in Indian banking sector to increase revenue and profitability. In: 2017 3rd International conference on advances in computing, communication automation (ICACCA) (Fall), pp 1–6. https://doi.org/10.1109/ICACCAF.2017.8344732
21. Li J, Xu L, Tang L, Wang S, Li L (2018) Big data in tourism research: a literature review. Tour Manag 68:301–323. https://doi.org/10.1016/j.tourman.2018.03.009

22. Xu L et al (2019) Research on telecom big data platform of LTE/5G mobile networks. In :2019 IEEE International conferences on ubiquitous computing communications (IUCC) and data science and computational intelligence (DSCI) and Smart computing, networking and services (SmartCNS), pp 756–761. https://doi.org/10.1109/IUCC/DSCI/SmartCNS.2019.00155
23. Nwanga ME, Onwuka EN, Aibinu AM, Ubadike OC (2015) Impact of Big Data Analytics to Nigerian mobile phone industry. In: International conference on industrial engineering and operations management (IEOM), pp 1–6.https://doi.org/10.1109/IEOM.2015.7093810
24. Forouzanfar MH et al (2016) Global, regional, and national comparative risk assessment of 79 behavioural, environmental and occupational, and metabolic risks or clusters of risks, 1990–2015: a systematic analysis for the Global Burden of Disease Study 2015. Lancet 388(10053):1659–1724. https://doi.org/10.1016/S0140-6736(16)31679-8
25. Jony RI, Habib A, Mohammed N, Rony RI (2015) Big data use case domains for telecom operators. In: 2015 IEEE international conference on smart City/SocialCom/SustainCom (SmartCity), pp 850–855. https://doi.org/10.1109/SmartCity.2015.174
26. Mehta N, Pandit A (2018) Concurrence of big data analytics and healthcare: a systematic review. Int J Med Inform 114:57–65. https://doi.org/10.1016/j.ijmedinf.2018.03.013
27. Wang Y, Kung L, Byrd TA (2018) Big data analytics: understanding its capabilities and potential benefits for healthcare organizations. Technol Forecast Soc Change 126:3–13. https://doi.org/10.1016/j.techfore.2015.12.019
28. Sun J, Reddy CK (2013) Big data analytics for healthcare. In: Proceedings of the 19th ACM SIGKDD international conference on knowledge discovery and data mining, p. 1525. https://doi.org/10.1145/2487575.2506178
29. Iyengar KP, Jain VK, Vaish A, Vaishya R, Maini L, Lal H (2020) Post COVID-19: planning strategies to resume orthopaedic surgery—challenges and considerations. J Clin Orthop Trauma 11:S291–S295. https://doi.org/10.1016/j.jcot.2020.04.028
30. Khalifehsoltani SN, Gerami MR (2010) E-health challenges, opportunities and experiences of developing countries. In: 2010 International conference on e-Education, e-Business, e-Management and e-Learning, pp 264–268
31. Wang Y (2016) Big opportunities and big concerns of big data in education. TechTrends 60(4):381–384. https://doi.org/10.1007/s11528-016-0072-1
32. Pardos ZA (2017) Big data in education and the models that love them. Curr Opin Behav Sci 18:107–113. https://doi.org/10.1016/j.cobeha.2017.11.006
33. Zeide E (2017) The structural consequences of big data-driven education. Big Data 5(2):164–172. https://doi.org/10.1089/big.2016.0061
34. Fischer C et al (2020) Mining big data in education: affordances and challenges. Rev Res Educ 44(1):130–160. https://doi.org/10.3102/0091732X20903304
35. Yu X, Wu S (2015) Typical applications of big data in education. In: 2015 International Conference of Educational Innovation through Technology (EITT), pp 103–106. https://doi.org/10.1109/EITT.2015.29
36. Singh G, Dwivedi R, Anand A (2019) Attendance monitoring and management using QR code based sensing with cloud based Processing. Int J Sci Res Comput Sci Appl Manag Stud IJSRCSAMS 8(5). https://doi.org/10.21276/sjet.2018.6.2.1
37. Marín-Marín J-A, López-Belmonte J, Fernández-Campoy J-M, Romero-Rodríguez J-M (2019) Big data in education. A bibliometric review. Soc Sci 8(8). https://doi.org/10.3390/socsci8080223
38. Zhong RY, Newman ST, Huang GQ, Lan S (2016) Big data for supply chain management in the service and manufacturing sectors: challenges, opportunities, and future perspectives. Comput Ind Eng 101:572–591. https://doi.org/10.1016/j.cie.2016.07.013
39. Belhadi A, Zkik K, Cherrafi A, Yusof SM, El fezazi S (2019) Understanding big data analytics for manufacturing processes: insights from literature review and multiple case studies. Comput Ind Eng 137:106099. https://doi.org/10.1016/j.cie.2019.106099
40. Belhadi A, Kamble SS, Zkik K, Cherrafi A, Touriki FE (2020) The integrated effect of big data analytics, lean six sigma and green manufacturing on the environmental performance of manufacturing companies: the case of North Africa. J Clean Prod 252:119903. https://doi.org/10.1016/j.jclepro.2019.119903

41. Zhang Y, Ma S, Yang H, Lv J, Liu Y (2018) A big data driven analytical framework for energy-intensive manufacturing industries. J Clean Prod 197:57–72. https://doi.org/10.1016/j.jclepro.2018.06.170
42. Naik K, Joshi A (2017) Role of big data in various sectors. In: 2017 International conference on I-SMAC (IoT in Social, Mobile, Analytics and Cloud) (I-SMAC), pp 117–122. https://doi.org/10.1109/I-SMAC.2017.8058321
43. Qi Q, Tao F (2018) Digital twin and big data towards smart manufacturing and industry 4.0: 360 degree comparison. IEEE Access 6:3585–3593. https://doi.org/10.1109/ACCESS.2018.2793265
44. Jee K, Kim G-H (2013) Potentiality of big data in the medical sector: focus on how to reshape the healthcare system. Healthc Inform Res 19(2):79–85. https://doi.org/10.4258/hir.2013.19.2.79
45. Shirkhorshidi AS, Aghabozorgi S, Wah TY, Herawan T (2014) Big data clustering: a review. In: Computational science and its applications—ICCSA 2014, pp 707–720
46. Al-Sai ZA, Abualigah LM (2017) Big data and E-government: a review. In: 2017 8th international conference on information technology (ICIT), pp 580–587. https://doi.org/10.1109/ICITECH.2017.8080062
47. Gaetani E, Aniello L, Baldoni R, Lombardi F, Margheri A, Sassone V (2017) Blockchain-based database to ensure data integrity in cloud computing environments [Online]. Available: https://eprints.soton.ac.uk/411996/
48. Sharma N, An A, Husain A (2020) Cloud based healthcare services for telemedicine practices using internet of things. J Crit Rev 7(14):2605–2611. https://doi.org/10.31838/jcr.07.14.510
49. Liu J, Li J, Li W, Wu J (2016) Rethinking big data: a review on the data quality and usage issues. ISPRS J Photogramm Remote Sens 115:134–142. https://doi.org/10.1016/j.isprsjprs.2015.11.006
50. Fang X et al (2016) Evaluation of the microbial diversity in amyotrophic lateral sclerosis using high-throughput sequencing. Front Microbiol 7:1479. https://doi.org/10.3389/fmicb.2016.01479
51. Lin W, Wu Z, Lin L, Wen A, Li J (2017) An ensemble random forest algorithm for insurance big data analysis. IEEE Access 5:16568–16575. https://doi.org/10.1109/ACCESS.2017.2738069
52. Fang K, Jiang Y, Song M (2016) Customer profitability forecasting using big data analytics: a case study of the insurance industry. Comput Ind Eng 101:554–564. https://doi.org/10.1016/j.cie.2016.09.011
53. Aktas E, Meng Y (2017) An exploration of big data practices in retail sector. Logistics 1(2). https://doi.org/10.3390/logistics1020012
54. Mneney J, Van Belle J-P (2016) Big data capabilities and readiness of South African retail organisations. In: 2016 6th international conference—cloud system and big data engineering (Confluence), pp 279–286. https://doi.org/10.1109/CONFLUENCE.2016.7508129
55. Zhou Y, Wilkinson D, Schreiber R, Pan R (2008) Large-scale parallel collaborative filtering for the netflix prize. In: Algorithmic aspects in information and management, pp 337–348
56. Yu M, Yang C, Li Y (2018) Big data in natural disaster management: a review. Geosciences 8(5). https://doi.org/10.3390/geosciences8050165
57. Bilal M et al (2016) Big data in the construction industry: a review of present status, opportunities, and future trends. Adv Eng Inf 30(3):500–521. https://doi.org/10.1016/j.aei.2016.07.001

Pedestrian Detection with Anchor-Free and FPN Enhanced Deep Learning Approach

J. Sangeetha, P. Rajendiran, and Hariraj Venkatesan

Abstract Rapid development in artificial intelligence has been a huge benefit for optimizing autonomous automobiles. Although safety equipment like airbags, ABS and pre-tensioned seat belts have widely been used, pedestrian detection algorithms are an undeniably important feature to be incorporated in vehicles as driver assistance systems to avoid accidents that may injure or risk people's lives. Though sliding window and anchor-based detectors have vastly been applied in the domain of object detection, these require several configurations for the bounding boxes and are moreover sensitive to background features around the object of interest in the image. To overcome these limitations, we use Center and Scale Prediction (CSP), an anchor-free object detection technique to overcome the limitations of bounding box approaches. We extract features with a ResNet-101, a deeper convolutional neural network and use Feature Pyramid Networks to build semantically richer feature maps. Experiments performed on the CityPersons and Caltech pedestrian datasets show that our model provides promising performance compared to existing state of the art models and is robust to various occlusion levels.

Keywords Autonomous automobiles · Pedestrian detection · Anchor-Free · Residual networks

J. Sangeetha (✉)
Department of Computer Science and Engineering, Srinivasa Ramanujan Centre, SASTRA Deemed University, Kumabakaonam 612001, India
e-mail: sangita.sudhakar@gmail.com

P. Rajendiran
School of Computing, SASTRA Deemed University, Thanjavur 613401, India
e-mail: rajendranap@it.sastra.edu

H. Venkatesan
Freshworks, Chennai 600096, India
e-mail: hariraj.venkatesan@freshworks.com

V. Skala et al. (eds.), *Machine Intelligence and Data Science Applications*, Lecture Notes on Data Engineering and Communications Technologies 132,
https://doi.org/10.1007/978-981-19-2347-0_39

1 Introduction

An important area of concern for researchers and automobile makers is the safety of passengers [1, 3, 9]. The 2019 edition of the annual report by the Ministry of Road Transport and Highways states that more than 1.5 lakh people died and more than 4.5 lakh faced serious injuries from road accidents [14]. Airbags, ABS, pre-tensioned seat belts, and much more safety systems have been incorporated into automobiles. On the same note, detecting and tracking passers-by or pedestrians crossing a street is a necessary addition for all vehicles. A wide range of sensors based on using ultrasound, laser, piezoelectric, and microwave have been experimented with. Of the several methods, computer vision techniques for object detection are found to be more efficient because they are close to how humans perceive objects through sight [17, 19, 21, 28]. Fish-eye cameras provide precise information of scenes for pedestrian tracking along with zero harmful radiation.

Although vision-based pedestrian detection systems are efficient, computation time to detect pedestrians is also a bottleneck. Failing to detect pedestrians within a minimal time span may risk lives. Tracking moving pedestrians with high accuracy is also a major challenge and is essential for many applications, namely, as collision detectors, trip planners, and navigation systems. Experimental evidence suggests that using even larger datasets can significantly improve object detection efficiency [1]. The availability of open annotated datasets such as CIFAR-10, ImageNet, CityScapes, and MS-COCO have further boosted the efforts and lead to the application of object detection for various purposes such as character recognition, surveillance tracking, face recognition, and pedestrian detection [3, 5, 20] (Fig. 1).

Traditionally object detection was performed using Sliding windows and Anchor-based detection methods combined with deep learning technologies. However, these

Fig. 1 Sample labeled images of the Common Objects in Context (COCO) dataset

are very difficult to implement owing to several configurations required for the bounding box [2]. To overcome these limitations, we will adopt Center and Scale Prediction (CSP) detector in our pedestrian detector architecture.

CSP introduced by Wei Liu, Irtiza Hasan, and Shengcai Liao is an anchor-free method for pedestrian detection [13]. Instead of using bounding boxes, a convolution operation is performed to predict the center point of the pedestrian as a representative point for the object of interest. The scale of the center point is also predicted through another convolution and along with the center it aids in predicting the overall height and width of the pedestrian in the form of a heat map. To make the model more robust to losses occurring from downsampling the heat maps, we additionally predict the offset with another convolution operation. We use ResNet-101 as the backbone because deeper neural networks will be able to capture features more precisely. Compared to other deep convolutional neural networks, residual networks provide better performance with lesser complexity of floating point operations. We make use of Feature Pyramid Network (FPN) to construct the feature pyramid from successive convolutions. It follows a top-down pathway to combine the semantically weaker higher resolution features with the semantically stronger lower resolution features [12].

To summarize, the main contributions of this paper are: (i) We use CSP to overcome the configuration related limitations of anchor-based detectors. (ii) Deeper backbone of ResNet-101 is used along with Feature Pyramid networks to retain semantic information when merging multi-layer feature maps. (iii) The proposed pedestrian detector achieves higher performance compared to state-of-the-art anchor-based models on CityPersons and Caltech pedestrian datasets.

2 Related Work

Computer scientists have been working on object detection for many decades. Recent advances in computational power with GPUs empowered deep learning and led to major breakthroughs in object detection. In 2014, Dollár et al. devised an approximation technique to compute features across multiple scales directly by extrapolation instead of explicit calculation using the correlation between scales and channel features to speed up the feature pyramid computation [6]. Nguyen developed pedestrian tracking with stereo images and identification tracking [15]. The mask RCNN detector is combined with a stereo camera for computing 3-D space distances through consecutive frames to arrive at appropriate detections. Kalman filtering was used to find concrete positions of pedestrians using their previous states. Girshick introduced a faster version of RCNN for object detection [10].

Occluded individuals are more difficult to spot owing to the patterns contributed by inter-class and intra-class occlusion. In 2018, Zhang et al. combined FasterRCNN with an attention mechanism to capture the activations received from multiple channels for various parts of the body [24, 25]. Pang worked with attention-based techniques on channels to handle occlusion and achieved progress with FasterRCNN baseline on heavy subset [16]. Around the same time, Chunluan Zhou proposed using discriminative feature transformation to distinguish pedestrian and non-pedestrian for detection with partially occluded pedestrians [27]. Dalal and Triggs experimentally proved that HOG descriptor feature grids are more suited than other feature sets for human detection problems [4].

Felzenszwalb et al. used latent information of positive instances to convert the training phase as a convex problem and achieved state-of-the-art results on the PASCAL dataset [8]. Real-time detection of moving pedestrians is important in video-based surveillance systems. Stefan and Alexei combined temporal differencing and optical flow with a double background filtering method, along with morphological features to provide for faster learning of variations without any information of size and scale [18].

3 Proposed Work

3.1 Center and Scale Prediction

Similar to any object detection problem center and scale prediction works in two steps: feature extraction and detection. The feature extraction phase is performed using a convolutional neural network having multiple layers to extract semantically valuable information from the input image. Unlike traditional detectors that attempt to predict low-level features like edges and corners, CSP attempts to predict high-level semantic features, namely, center, scale, and offset. Instead of using a blob, we use the convolutions to extract scale as a feature in a single pass. Thus the detection phase will use the activation of the convolutions to predict the center, scale, and offset. This reformulation simplifies pedestrian detection to simply predicting the center and scale (Fig. 2).

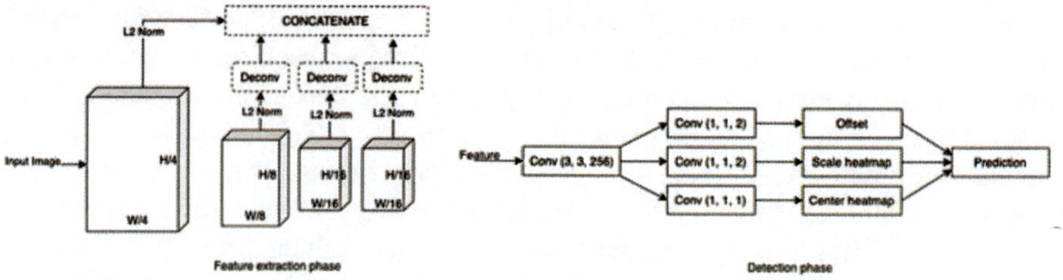

Fig. 2 Center and scale prediction using convolutions [13]

3.2 Residual Neural Networks

Deep convolutional neural networks have proven effective for visual recognition tasks. Among the popular deeper nets, residual networks have provided state-of-the-art results on ImageNet. When deeper networks start converging, degradation occurs. With increasing network depth, the accuracy plateaus and then drops rapidly. Residual neural networks (RNN) introduced by He et al., are efficiently able to overcome this degradation problem with the use of skip connections.

The decrease in training accuracy shows that deeper networks are more difficult to optimize. To understand this, let us build a deeper network formed by simply stacking additional identity layers to a shallower network. The upper bound to the training error of this constructed deeper model should be the same as the shallower network. But experimental results, shown in Fig. 4, are contradictory. The solvers face difficulties in optimizing the identity mappings for multiple non-linear layers [11].

Let $F(x)$ denote the expected mapping. We instead fit another mapping

$$F(x) = H(x) - x \tag{1}$$

He et al. hypothesize that optimizing the residual mapping is easier than for the original mapping. This reformulation helps precondition the problem because the residual can be pushed to zero rather than fitting an identity mapping by stacking non-linear layers. Thus the original mapping becomes

$$H(x) = F(x) + x \tag{2}$$

The mapping in Eq. 2 can be constructed with shortcut (skip) connections. These merely calculate identity mapping, which is combined with the stacked layers' outputs. They also do not create any extra parameters and thus do not increase complexity. Plain networks can be converted to residual networks by directly adding shortcut connections. The identity shortcuts are included when input and output are of equal dimensions. When the dimensions increase, there are two options: (A) Zero padding to increase dimensions without adding any new parameters, or (B) projection. In both options, a stride of two is performed.

3.3 Bottleneck Architecture of ResNets

To reduce training duration, the design is modified to a bottleneck design. Each residual function F is created using a stack of 3 layers-1×1, 3×3, and 1×1 convolution. The 1×1 layers decrease and restore the dimensions of the output. The 3×3 layer is the bottleneck with smaller input and output dimensions (Fig. 3).

50-layer ResNet: In place of the 2-layer blocks in Fig. 5, we use the 3-layer bottleneck block in the 34-layer network resulting in a 50-layer ResNet.

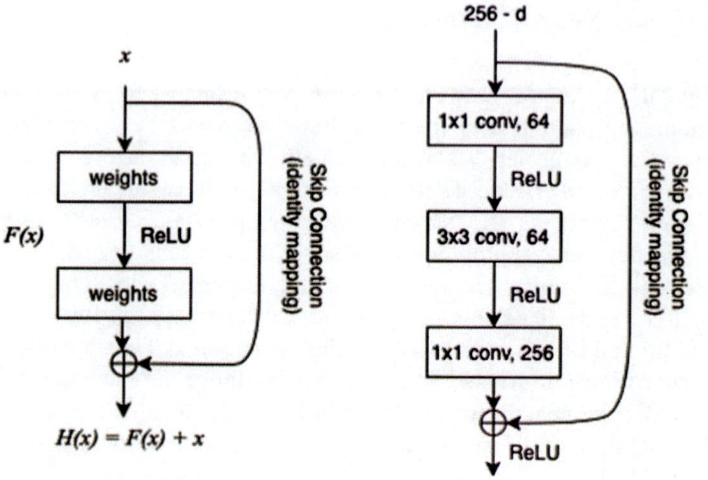

Fig. 3 Left: Building block of ResNets. Right: Bottleneck architecture-based building block used to form 34, 50, 101, and 152 layer nets

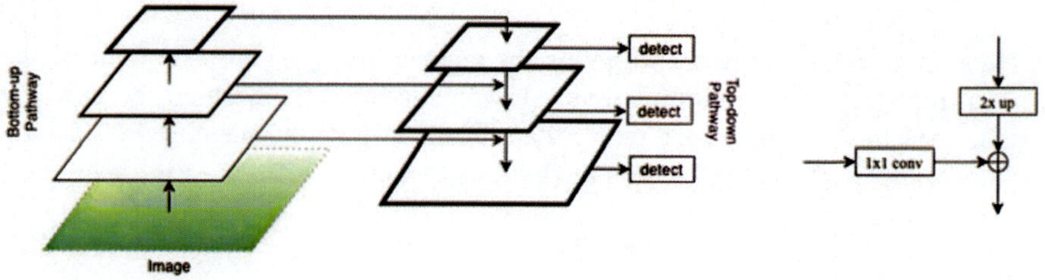

Fig. 4 Left: Construction of feature maps with FPN. Right: Lateral connection that combines bottom-up and top-down pathway outputs

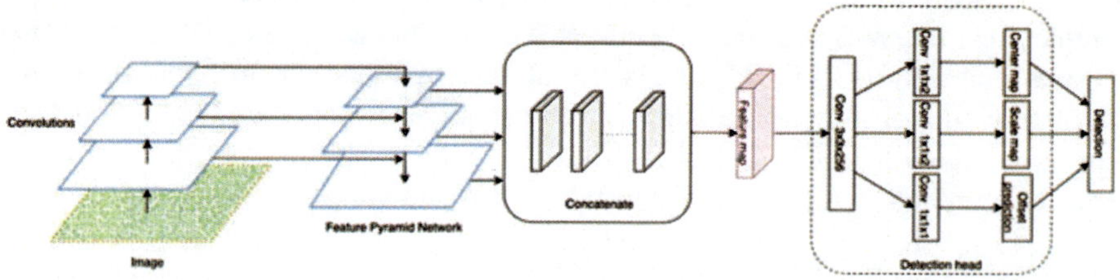

Fig. 5 Proposed architecture for pedestrian detector: features are extracted with ResNet backbone and semantically richer feature maps are constructed with FPN. Resulting single feature map is used to detect center and scale heat map and offset value of pedestrians

101 and 152-layer ResNets: Adding additional 3-layer blocks will extend the 50-layer residual network to form deeper variants like101 and 152-layer ResNets.

Deeper neural networks possess the ability to extract both location information stored in shallower feature maps and high-level semantic information in deeper ones. The nature of CSP problems involves joining different levels of feature maps. ResNets are much easier to optimize for higher depths when compared to conventional deeper networks such as VGG-Net and thus results in higher accuracy. Hence considering these reasons, our choice of ResNet-101 as the backbone for pedestrian detector is justified.

3.4 Feature Pyramid Network (FPN)

Pyramidal representations of features have long been avoided in object detectors because they were considered memory and compute intensive. However, with advances in GPUs and reducing prices of memory hardware, we can easily augment our pipelines with feature pyramids with a minimal additional overhead. Feature Pyramid Network (FPN) developed together by Facebook AI and Cornell University, when included in a FasterRCNN detector achieved better results compared to the winning model of the COCO 2016 challenge. They are useful to enrich feature maps with semantic information from multi-scale input images and can be easily fit into most deep learning architectures for object detection (Fig. 4).

The forward convolution performed with the backbone is perceived as a bottom-up pathway operation that generates pyramid shaped hierarchy of features. Each pyramid layer represents a stage and the output from the final layer of every stage becomes input to the top-down pathway to enriching features. The top-down pathway upsamples the spatially coarser and semantically stronger features from the higher levels of the pyramid by factor 2. A lateral connection is used to combine spatially similar maps of the bottom-up and top-down pathways. It consists of performing a 1×1 convolution on the bottom-up pathway maps to decrease channel dimension and is combined with the up-sampled map using element-wise addition. The process is repeated to generate higher resolution map and a 3×3 convolution is performed on the generated maps to produce a single feature map. The convolution also mitigates aliasing induced by upsampling. The addition of FPN to our pedestrian detector helps in the extraction of both high-level features that contain semantic information useful to locate the object of interest, and low-level features rich in localized information obtained from deeper layers.

3.5 Proposed Pedestrian Detector Architecture

The input image is used in its original shape of 1024 × 2048 without resizing during training. ResNet-101 network is used in the bottom-up pathway. We choose the final layer of each stage because the deepest layer will yield richer feature map. Every convolution will contain normalization, non-linearity and max-pooling operations, similar to a typical convolution network (Fig. 5).

Lateral connections will combine the outputs from the bottom-up path with those of top-down path resulting in maps of strongest representation. Downsampling by 4, 8, 16, 16 is performed on the output of each stage and the feature maps are merged to a single block of size (768, 256, 512). A 3 × 3 layer convolution is applied on this feature map resulting in a layer with 256 channels. Following this, 1 × 1 convolutions are performed to generate heat maps representing center and scale. An important observation here is the predicted center points will be slightly moved from the actual points [16]. This is caused by the strong correlation present with the height and width parameters. So, an additional 1 × 1 convolution is performed to generate the offset value. Pedestrian locations are calculated from the width and height of the heat maps and adjusted with offset.

4 Experiment

4.1 Dataset

We evaluate our model on two standard benchmarks—CityPersons and Caltech pedestrian dataset. The CityPersons dataset was introduced by Zhang et al. and derived from the CityScapes dataset [24]. The 5000 images were collected from 18 different cities in Germany containing more than 35 thousand persons consistently spread across the training, validation, and test sets. The Caltech datasets contains video captured from driving through traffic in a typical urban area and has a total duration of approximately 10 h [7]. The video of almost 250,000 frames contains 2300 pedestrians labeled with 350,000 bounding boxes. The provided test set contains 4024 frames. The training set is generated from every 30th frame of the video to yield a total of 4250 frames containing approximately 1600 pedestrians. We choose log-average miss rate as the evaluation metric for both datasets.

4.2 Loss Function

Although, L2 loss is preferred for most machine learning tasks, it is not suitable when the data contains outliers. Due to the squaring operation being performed, the effect of outliers will be profound with large updates performed in every iteration of

Table 1 Comparison with state-of-the-art on CityPersons dataset

Method	Reasonable	Heavy	Partial	Bare
TLL	15.5	53.6	17.2	10.0
RepLoss	13.2	56.9	16.8	7.6
OR-CNN	12.8	55.7	15.3	6.7
ALFNet	12.0	51.9	11.4	8.2
MGAN	10.5	47.2		
Ours	12.57	49.2	11.22	9.1

gradient descent which leads to the exploding gradient problem. On the other hand, if we opt for L1 loss function, derivatives obtained in the gradient calculation step will result in constant values. So the loss value will keep oscillating about a constant which will lead to lower stability compared to L2 loss. Therefore to obtain the best of both worlds, we use the smooth L1 loss function. Smooth L1 returns the absolute value when the input is above a threshold and the square of the value otherwise. It behaves as an L1 loss when the input parameter is above the threshold and as an L2 loss for other values. Thus our loss function includes both the steady gradient advantage of the L1 loss and the stability of L2 loss. In the context of the center and scale prediction problem for pedestrian detection, we need to factor in regression losses from the center, scale, and offset predictions. Thus the loss function will take the form of a weighted sum of the three prediction errors. For best performance the weights are set to 0.01, 1, and 0.1 for the three features, respectively [22, 23, 26].

4.3 Configuration Details

The proposed model is implemented using Pytorch's ResNet-101 as the backbone with FPN included. It is pre-trained with ImageNet to speed up convergence. Training is performed with two Nvidia Tesla K80 GPUs to accelerate convergence with two images fed to each GPU. A learning rate of 2×10^{-4} is used and training is performed for 150 epochs. Normalization is performed on a batch size of 2. Adam optimizer is used, due to its low memory footprint. It is combined with the mean teacher weight optimization to improve accuracy (Tables 1 and 2).

4.4 Comparison with State-of-the-Art

The proposed model has achieved better performance when compared to many existing state-of-the-art models. On the CityPersons dataset, it achieves better performance compared to many existing anchor-based detection methods such as RepLoss and OR-CNN. On the Caltech dataset, the model provides superior performance on

Table 2 Comparison with state-of-the-art on Caltech dataset

Method	Reasonable	Heavy
FasterRCNN	8.7	53.1
RPN + BF	7.3	54.6
ALFNet	6.1	51.0
RepLoss	5.0	47.9
OR-CNN + CityPersons	4.1	
Ours	4.5	49.5

the reasonable subset compared to existing state of the art methods such as ALFNet and RepLoss and better performance than FasterRCNN and RPN on the heavy subset. The performance of the model is attributed to our architecture consisting of CSP, ResNet, and FPN.

CSP detection provides two benefits when compared to anchor-based detection strategies [26, 27]. First, there is no requirement for complex configurations of the anchor boxes for each dataset. Second, position of the bounding box makes the network sensitive to noise from other objects in the vicinity. Detecting the center point aids in learning the entire body's features and improves the generalization ability of the network. Another major contributing factor for improved performance is the choice of ResNet as backbone. First, as mentioned in Sect. 3.2, ResNets obtain higher accuracy with lesser operations when compared to popular networks such as VGG-Net which results in better performance compared to methods such as FRCNN, OR-CNN, and HBAN. Second, the network depth of 101 layers yields better performance compared to methods such as TLL and RepLoss that rely on ResNet-50.

5 Conclusion

In this paper, we used Center and Scale prediction (CSP) to detect pedestrians from panoramic images captured from cameras mounted on a moving car. We implemented the model using the ResNet-101 backbone enriched with FPN and performed experiments on the CityPersons and Caltech datasets. The trained network was able to detect pedestrians successfully and achieved performance comparable to many state-of-the-art anchor-based models for different occlusion subsets. In the future, we wish to explore different normalization strategies and the representative point approach of anchor-free detection to increase accuracy and robustness.

References

1. Braun M, Krebs S, Flohr F, Gavrila DM (2019) EuroCity persons: a novel benchmark for person detection in traffic scenes. IEEE Trans Pattern Anal Mach Intell 41:1844–1861

2. Chen H, Zheng H (2020) Object detection based on center point proposals. Electronics 9:2075
3. Cordts M, Omran M, Ramos S, Rehfeld T, Enzweiler M, Benenson R, Franke U, Roth S, Schiele B (2016) The cityscapes dataset for semantic Urban scene understanding. In: 2016 IEEE conference on computer vision and pattern recognition (CVPR). IEEE
4. Dalal N, Triggs B (2005) Histograms of oriented gradients for human detection. In: 2005 IEEE computer society conference on computer vision and pattern recognition. IEEE, San Diego, CA, USA
5. Deng J, Dong W, Socher R, Li L-J, Li K, Fei-Fei L (2009) ImageNet: a large-scale hierarchical image database. In: 2009 IEEE conference on computer vision and pattern recognition. IEEE
6. Dollar P, Appel R, Belongie S, Perona P (2014) Fast feature pyramids for object detection. IEEE Trans Pattern Anal Mach Intell 36:1532–1545
7. Dollar P, Wojek C, Schiele B, Perona P (2009) Pedestrian detection: a benchmark. In: 2009 IEEE conference on computer vision and pattern recognition. IEEE
8. Felzenszwalb PF, Girshick RB, McAllester D, Ramanan D (2010) Object detection with discriminatively trained part-based models. IEEE Trans Pattern Anal Mach Intell 32:1627–1645
9. Geiger A, Lenz P, Urtasun R (2012) Are we ready for autonomous driving? The KITTI vision benchmark suite. In: 2012 IEEE conference on computer vision and pattern recognition. IEEE
10. Girshick R (2015) Fast R-CNN. In: 2015 IEEE international conference on computer vision (ICCV). IEEE
11. He K, Zhang X, Ren S, Sun J (2016) Deep residual learning for image recognition. In: 2016 IEEE conference on computer vision and pattern recognition (CVPR). IEEE, Las Vegas, NV, USA
12. Lin T-Y, Dollar P, Girshick R, He K, Hariharan B, Belongie S (2017) Feature pyramid networks for object detection. In: 2017 IEEE conference on computer vision and pattern recognition (CVPR). IEEE
13. Liu W, Liao S, Ren W, Hu W, Yu Y (2019) High-level semantic feature detection: a new perspective for pedestrian detection. In: 2019 IEEE/CVF conference on computer vision and pattern recognition (CVPR). IEEE
14. Ministry of Road Transport and Highways, https://morth.nic.in/road-accident-in-india. Last accessed 21 July 2021
15. Nguyen U, Rottensteiner F, Heipke C (2019) Confidence-aware pedestrian tracking using a stereo camera. ISPRS Ann Photogrammetry, Rem Sens Spat Inf Sci IV-2/W5:53–60
16. Pang Y, Xie J, Khan MH, Anwer RM, Khan FS, Shao L (2019) Mask-guided attention network for occluded pedestrian detection. In: 2019 IEEE/CVF International conference on computer vision (ICCV). IEEE
17. Sergi CM (2015) On the use of convolutional neural networks for pedestrian detection
18. Stefan Z, Alexei E (2007) Detection of multiple deformable objects using PCA-SIFT. In: Proceedings of the 22nd national conference on artificial intelligence. AAAI, Vancouver, British Columbia, Canada, pp 1127–1132
19. Szegedy C, Liu W, Jia Y, Sermanet P, Reed S, Anguelov D, Erhan D, Vanhoucke V, Rabinovich A (2015) Going deeper with convolutions. In: 2015 IEEE conference on computer vision and pattern recognition (CVPR). IEEE, Boston, MA, USA
20. The CIFAR-10 dataset, https://www.cs.toronto.edu/~kriz/cifar.html. Last accessed 21 July 2021
21. Vahab A, Naik MS, Raikar PG, Prasad SR (2019) Applications of object detection system. Int Res J Eng Technol 6(4):4186–4192
22. Wang B (2019) Research on pedestrian detection algorithm based on image. J Phys: Conf Ser 1345(6)
23. Wang W (2020) Detection of panoramic vision pedestrian based on deep learning. Image Vis Comput 103
24. Zhang S, Benenson R, Schiele B (2017) CityPersons: a diverse dataset for pedestrian detection. Preprint at https://arxiv.org/pdf/1702.05693
25. Zhang S, Yang J, Schiele B (2018) Occluded pedestrian detection through guided attention in CNNs. In: 2018 IEEE/CVF conference on computer vision and pattern recognition. IEEE, Salt Lake City, UT, USA, pp 6995–7003

26. Zhao Z-Q, Zheng P, Xu S-T, Wu X (2019) Object detection with deep learning: a review. IEEE Trans Neural Netw Learn Syst 30:3212–3232
27. Zhou C, Yang M, Yuan J (2019) Discriminative feature transformation for occluded pedestrian detection. In: 2019 IEEE/CVF International conference on computer vision (ICCV). IEEE
28. Zou Z, Shi Z, Guo Y, Ye J (2019) Object detection in 20 years: a survey. Preprint at https://arxiv.org/pdf/1905.05055

Efficient Machine Learning Approaches to Detect Fake News of Covid-19

Shagoto Rahman, M. Raihan, Kamrul Hasan Talukder, Sabia Khatun Mithila, Md. Mehedi Hassan, Laboni Akter, and Md. Mohsin Sarker Raihan

Abstract In recent years, the growth of fake news has been significantly high. Advancement in the field of technology is one of the reasons that lie behind this phenomenon. Fake news are presented in such a way that it is quite hard to identify as fake on various social platforms these days and that has a huge impact on people or communities. Such fake news is most destructive when it plays with life. COVID-19 has changed and shaken the entire universe, and fake news that are related to COVID-19 make the destruction deadlier. So, an effort regarding COVID-19-related fake news detection will guard a lot of people or communities against bogus news and can make lives better with proper news in a pandemic. For our research in this paper, a methodology has been espoused to detect COVID-19-allied fake news. Our methodology consists of two different approaches. One approach deals with machine learning models (Logistic regression, support vector machine, decision tree, random forest) using the term frequency-inverse document frequency (TF-IDF) attributes of textual documents, and the other approach involves an association of convolutional neural networks (CNNs) and bidirectional long short-term memory (BiLSTM) using the sequence of vectors of tokens or words in documents. Logistic regression using the TF-IDF is the best performer among all these models having 95% accuracy and an F1-score of 0.94 on test data with Cohen's kappa coefficient of 0.89 and Mathews correlation coefficient of 0.89.

S. Rahman · K. H. Talukder
Khulna University, Khulna, Bangladesh
e-mail: shagoto1710@cseku.ac.bd

K. H. Talukder
e-mail: khtalukder@ku.ac.bd

M. Raihan (✉) · S. K. Mithila · Md. Mehedi Hassan
North Western University, Khulna, Bangladesh
e-mail: rianku11@gmail.com; raihan1146@cseku.ac.bd

L. Akter · Md. M. S. Raihan
Khulna University of Engineering & Technology, Khulna, Bangladesh

V. Skala et al. (eds.), *Machine Intelligence and Data Science Applications*, Lecture Notes on Data Engineering and Communications Technologies 132,
https://doi.org/10.1007/978-981-19-2347-0_40

Keywords COVID-19 · Fake · TF-IDF · Tokenization · Logistic regression · Support vector machine · Random forest · BiLSTM

1 Introduction

Advancement in the field of technology has brought a revolutionary change in terms of sharing information. Today, information about anything is very easy to access on the Internet. The procedure of sharing information has brought new milestones in every aspect of life. Apart from the advantages, the thing that grabs our eyeballs these days is the sharing of fake and bogus news. Misinformation or fake news affects things in a very negative way, and it breaks all boundaries when it is related to our existence. Such an alarming thing these days is the COVID-19. In 2019, the world has seen the decade's deadliest disease that is the COVID-19. The havoc of which has advocated 75 million confirmed cases signifying the death of 1.7 million across the globe till December 1st, 2020 [1]. But, the disease can be deadlier if fake news regarding this gets viral. Every day, there is a lot of news regarding COVID-19 either how to get rid of the virus or what are the threats regarding it. So, slight misinformation can play a significant role in people's lives and can be a threat to both physical and mental health. And most of the time people share these fake news without verifying them which creates panic attacks among people. In the Munich Security Council on February 15th, 2020, the Director-General of World Health Organization (WHO), Tedros Adhanom Ghebreyesus hinted that the war is not only about fighting a pandemic but also an infodemic [2]. That is why in this research, we have addressed this problem to detect COVID-19 fake news detection.

In this paper, we have gone for the detection of fake news regarding COVID-19 realizing the havoc fake news can create in the pandemic. Two approaches have been used for that purpose. These two approaches are introduced to extract the comparison between the performance of machine learning and deep learning approaches. For the first approach, we have used the TF-IDF features to get the numeric interpretation of the text features. Advancement in natural language processing (NLP) has made classifications via text features more attractive these days. TF-IDF can play a big part in a classification model for textual data as it signifies the importance of words in a text. Then, we have applied some machine learning algorithms to the features. For the second approach, we have tokenized the data and organized the sequence of the tokens as vectors to use them as features. Then, we have used convolutional neural networks + bidirectional long short-term memory (CNN + BiLSTM) model on the features. Finally, we compared all of these models based on various statistical matrices. Unlike existing methodologies, the proposed method recognizes the importance of TF-IDF and word tokens as features and also exhibits a significant comparison among both the features with respective machine learning and deep learning algorithms.

The other part of the research study is ornamented as follows: In Sects. 2 and 3, the related work and methodology have been advocated with a distinguishing destination to the justness of the classifier algorithms, respectively. In Sect. 4, the experimented analysis has been clarified with the intend to justify the novelty of this work. Finally, this research paper is terminated with Sect. 5.

2 Related Works

The viral nature of fake news has done devastating effects in the past which has made people think and also created chances for making possible solutions regarding it. That is why many types of researches regarding this topic have been observed in recent years even with COVID-19 recently. To detect COVID-19 fake news, a method has been proposed using random forest and decision tree classifiers [3]. As for features, the authors used the TF-IDF, count vectorizer, and bag of words. Classifying results showed random forest (RF) performing better than the decision tree (DT) as the former achieved 94.49% of accuracy where the latter attained 92.07%. Elhadad et al. in their research showed features like word embedding could achieve great performance in terms of fake news detection [4]. A bunch of classifying algorithms, namely decision tree, logistic regression (LR), k-nearest neighbor (KNN), multinomial Naive Bayes (MNB), Bernoulli Naive Bayes (BNB), linear support vector machines (LSVMs), perceptron, ensemble random forest (ERF), neural network (NN), and extreme gradient boosting classifiers (XGBoost) were used in their study on various features. Neural networks, decision tree, and logistic regression were the best performers among the classifiers used. A transformer-based COVID-19 fake detection system was introduced in this paper [5]. Three types of features, namely TF-IDF, word embedding of English GloVe of dimension 300, and GloVe embedding with TF-IDF averaging were used by the authors. For classification support vector machine (SVM), probably approximately correct (PAC), multilayer perceptron (MLP), long short-term memory (LSTM) were utilized. The LSTM model performed best in their methodology. Another method was discussed by the authors on COVID-19 fake news detection with social media data in their research [6]. They used TF-IDF for the classical machine learning models. An ensemble of SVM, LR, Naive Bayes (NB), LR + NB, and bidirectional long short-term memory (BiLSTM) were used achieving an F1-score of 0.94. Raha et al. proposed a method to identify fake news corresponding to COVID-19 in social media [7]. Features like TF-IDF and Word2Vec were used with classifiers like NB, LR, RF, and XGBoost. SVM achieved the highest 90 and 93% accuracy using the two features, respectively. A symmetric feature-based fake news classification was addressed by authors in their research [8]. With the aim of reducing symmetric features, the method propagates with four evolutionary classifiers of KNN on six different datasets. The method signifies an average accuracy of 75% with TF-IDF and Word2Vec features.

3 Methodology

Our research methodology is divided into several steps. These steps include collection of data, data preprocessing, feature extraction, training of data, and tools and simulation environment. Figure 1 symbolizes the overall workflow of our methodology.

3.1 Collection of Data

For our research, we have collected COVID-19 fake news data from two sources on the Internet. From [9], we have collected a dataset that has a total of 10,201 news data of which 9727 news are labeled as **0** which means fake and 474 news are labeled as **1** which means real. So, this dataset lacks real news that is why some real news

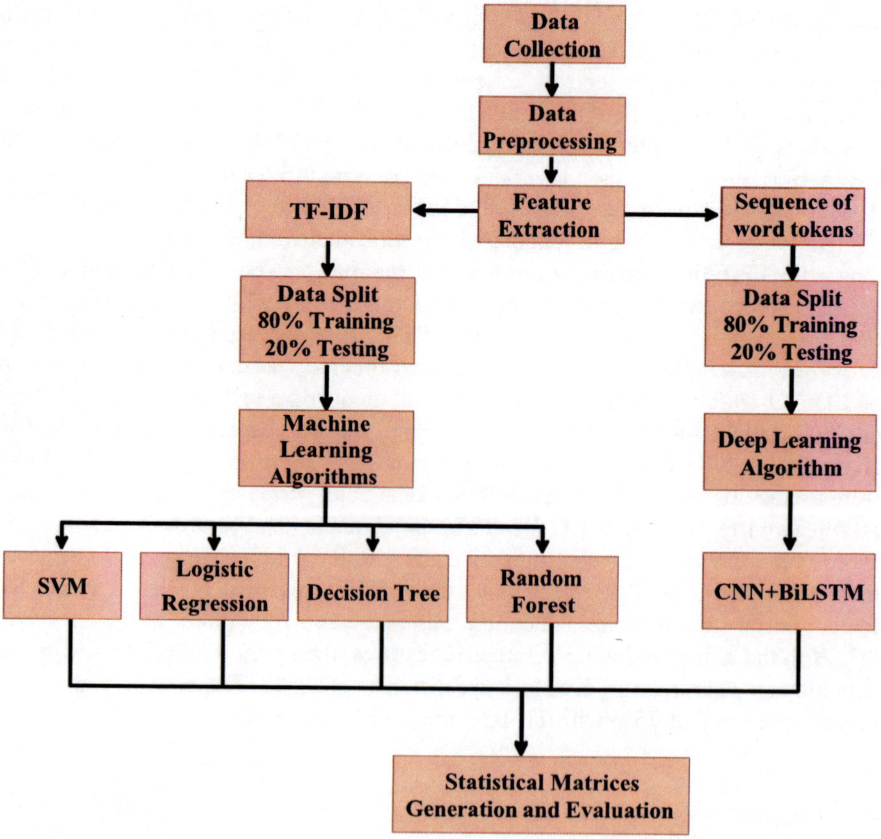

Fig. 1 Workflow of our system

from [10] has been collected. In this dataset, there is 6788 real news from **Canadian Broadcasting Corporation** (**CBC**). So, merging both datasets from two sources the final dataset has **16,989** news data of which **9727** news are fake and **7262** are real.

3.2 Data Preprocessing

For preprocessing of our data, we have performed the following steps:

Punctuation Removal: First of all, we have ensured the proper cleaning of each document by cleaning the punctuation marks. Again, many unicodes were not detected, so these unicodes were needed to be cleaned as well. Again, all of the words that had capital letters are also converted into small letters.

Tags and Duplicate Sentence Removal: Various hyperlinks and tags regarding them have been removed in this regard. Again, by creating a unique list for sentences per document, duplicate sentences have been removed.

Stemming: Stemming is a kind of process that is performed to find the root or base word origin of any word. Words like ***runs*** and ***running*** are derived from the root word ***run***. Stemming is done using Porter Stemmer from NLTK [11].

Stop words Removal: There are many stop words like ***and***, ***but***. These stop words are identified by computing frequency-based unigrams and NLTK stop words library. We have removed all the stop words from our dataset.

3.3 Feature Extraction

We have introduced two types of features here. One is for machine learning models, and the other is for the CNN + BiLSTM model. For machine learning models, we have extracted the TF-IDF features from the preprocessed dataset and represented them in vectors. For the CNN + BiLSTM model, we have tokenized the dataset and used the sequence of tokens as the input of the model.

TF-IDF: TF-IDF is a weighting process defining the importance of a word or term in a corpus. Two metrics, the term frequency and the inverse document frequency, are multiplied in this regard. Term frequency symbolizes how frequent a word is in a document. It is defined by the number of times the term occurs in the document cleaved by the total number of words in the document. Inverse document frequency (IDF) determines how important a term or word is inside a corpus or collection of documents. Logarithm of total no. of documents cleaved by the number of documents where the term is found indicates the IDF value.

Tokenization and Token Sequence: Tokenization is splitting the documents into words. Next, the words are sequenced with the document. The sequences of tokens or word vectors are forwarded into an embedding layer. This layer converts each word into a fixed length of defined size. The output of the embedding layer has been used for the CNN + BiLSTM model as for feature in this research.

3.4 Training the Data

The percentage splitting method has been used to split the dataset. We have split the dataset as 80% training data and 20% test data.

3.5 Tools and Simulation Environment

- Python 3.8.5
- Jupyter Notebook 6.0.3.

3.6 Python Packages

Pandas: For various operations regarding the dataset, we have used the pandas library of version 1.0.1.

Numpy: For working with arrays, we have used the NumPy library of version 1.18.5.

NLTK: We have done various transformations of our data using NLTK library like getting the stop words and stemming. NLTK version 3.4.5 has been used here.

Sklearn: Version 0.24.1 has been used in this research. This library helps us to import different kinds of machine learning models. Again, various evaluation matrices functions were used from this library as well.

TensorFlow: TensorFlow of version 2.4.1 has been espoused to advocate the CNN + BiLSTM model.

Matplotlib and Seaborn: For visualization of results in the evaluations section, Matplotlib library of version 3.1.3 and Seaborn of version 0.10.0 have been used in our study.

3.7 Classification Models

Support Vector Machine (SVM): The linear kernel of support vector machine (SVM) has been used in this research. Linear SVC from sklearn of version 0.24.1 has been used.

Logistic Regression (LR): Logistic regression has been used in this study from sklearn.linear model module of version 0.24.1. Tolerance for the model is 0.0001, and the inverse of regularization strength is 1.0.

Decision Tree (DT): For the decision tree, we have used the criterion parameter is **gini** for the Gini impurity and random state as 1234. We have used sklearn.tree module of version 0.24.1 here.

Table 1 Architecture of CNN + BiLSTM model

Layer type	Output shape	Parameters
Embedding	(None, 300, 128)	640,000
Convolutional	(None, 296, 128)	82,048
Max Pooling	(None, 59, 128)	0
Bidirectional	(None, 59, 128)	98,816
Bidirectional	(None, 59, 128)	98,816
Dense	(None, 59, 28)	3612
Dense	(None, 59, 14)	406
Flatten	(None, 826)	0
Dense	(None, 2)	1652

Random Forest (RF): For random forest, no. of estimators as 100, criterion as **gini,** and random state = 1234 were used. **Sklearn.ensemble** module is used here.

CNN + BiLSTM: CNN + BiLSTM is the conglomeration of CNN and BiLSTM model. The prominent fact of the model is that the CNN part brings out the lion share amount of features from the input vector. The yield of CNN goes to the input of BiLSTM, where the BiLSTM allows the chronological order between data. We have used an embedding layer before the data being passed to CNN + BiLSTM model. For the embedding layer, we have used a vocabulary size of 50,000 with an embedding dimension of 128 and input length 300. For CNN + BiLSTM configuration, we have used the CNN layer with a kernel size of 5 and filter size of 128. The CNN layer is followed by a layer of max-pooling of size 5. Two BiLSTM layers with a dropout of 0.2 have been used as well. Two dense layers with activation function **ReLU**, one flatten and one additional dense layer with softmax activation have also been used in this configuration. Table 1 illustrates the configuration of the CNN + BiLSTM model.

4 Experimental Analysis and Discussions

Figures 2, 3, and 4 show the word cloud of fake-news data, real-news data, and entire dataset, respectively.

These figures give a significant hint about the words that are most frequent in the fake, real, and entire news dataset. Words like **covid**, **19**, and **coronavirus** appear in large numbers which are shown by the sizes of the words in the figures. Figure 5 shows the ten most frequent words in the dataset after removing stop words. It symbolizes that **covid**, **19,** and **coronavirus** are most frequent in the dataset having the frequency of 6488, 6366 and 4785, respectively, objectifying that most words lie between the margin frequency of 1308 to 915 in terms of top 10 unigrams. Figure 6 portrays the TF-IDF spread of the dataset colored by pink and blue indicating fake and real class, respectively. TF-IDF of the features has been plotted with the albeit

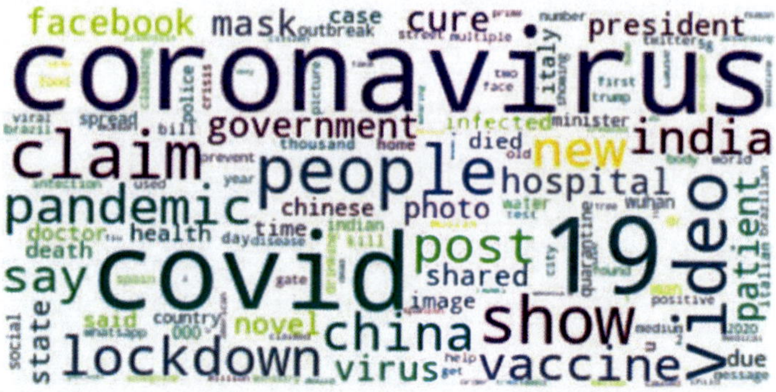

Fig. 2 Word cloud of fake-news data

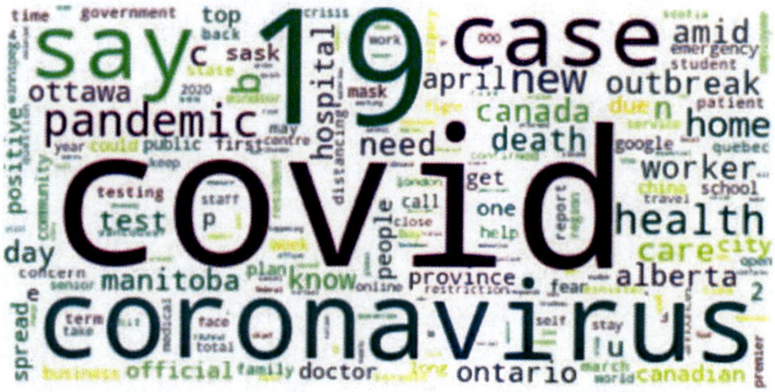

Fig. 3 Word cloud of real-news data

Fig. 4 Word cloud of entire dataset

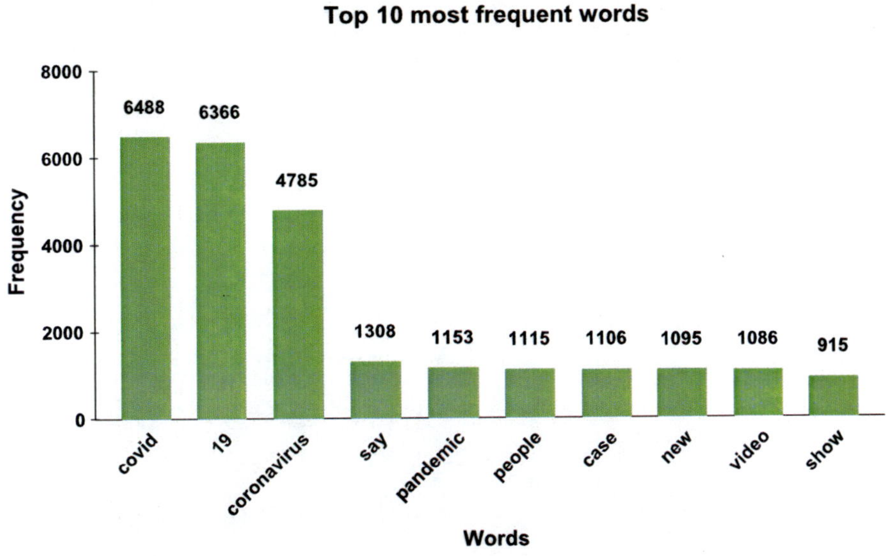

Fig. 5 Ten most frequent words in the dataset

Fig. 6 TF-IDF spread of the dataset

of Matplotlib library. The figure hints at the two clusters signifying the TF-IDF differences between the real class and the fake class. Again, the intermingling of the TF-IDF values suggests that there are words that are cross-domain. Figure 7 shows the most important word features for the TF-IDF using chi-square. Words like **video**, **claim,** and **show** have a greater influence in decision-making in terms of chi-square

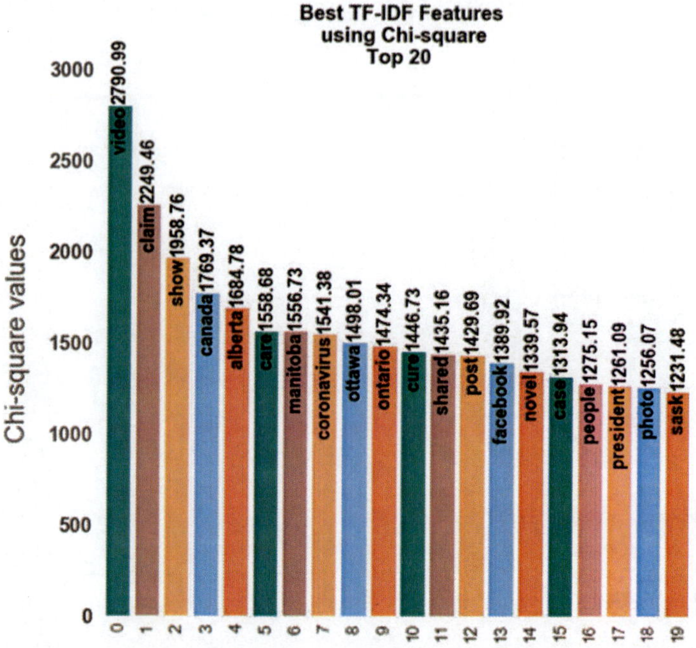

Fig. 7 Important features of TF-IDF using chi-square

of TF-IDF of words. The merged dataset is splitted 80% to 20% ratio for training and testing, respectively, and the dataset is slight imbalanced as there are 9727 fake news and 7262 real. Since it is not enormously imbalanced, the classification has not been over-fitted in this scenario as the results show the same. Figure 8 objectifies logistic regression as the best classifier comparing with other models on various

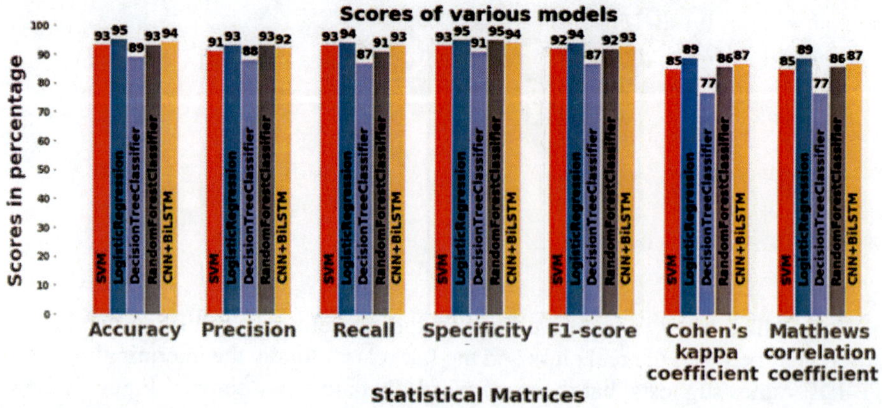

Fig. 8 Statistical matrices of various models

statistical matrices such as accuracy, precision, recall, specificity, F1-score, Cohen's kappa coefficient, and Matthews correlation coefficient. In all the criteria, logistic regression has higher values than other models. Logistic regression delivered the best accuracy of 95%, and decision tree has the least achieving 89%. Logistic regression and random forest have the best precision value of 93%, and decision tree the least with 88%. The highest recall value of 94% is attained by logistic regression, while decision tree is at the bottom with 87%. Specificity shows logistic regression and random forest have the acme value of 95%, and decision tree has the bottom value of 91%. According to F1-score, the best score is achieved by logistic regression of 94%, and decision tree has got the lowest score of 87%. Cohen's kappa coefficient and Matthews correlation coefficient score objectify that the best value for both of the statistical matrices is 89% achieved by logistic regression, and the least value for both is 77% achieved by decision tree. Figure 9 shows the area under the curve (AUC) receiver operating characteristics (ROC) curve. Logistic regression, random forest, CNN + BiLSTM have the highest AUC value of 0.98, and decision tree has the lowest with 0.89.

Table 2 exhibits the difference among various previous methods and our new proposed system on the features used, sample size, algorithms used, and accuracy. Our proposed method gives the best accuracy among the systems and that is 95%. We have used the same dataset that was used in [6, 7]. We have also added some additional data with the dataset, and with logistic regression, we have achieved an accuracy of 95% surpassing the accuracy of both [6, 7] which were 94% and 93%, respectively. Logistic regression was also used in [6] attaining accuracy of 92% where

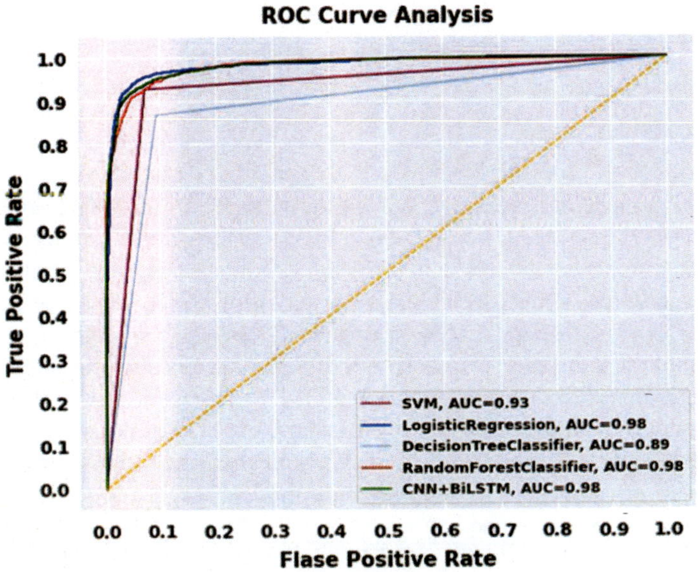

Fig. 9 AUC and ROC Curve

Table 2 Comparison with existing works

References	Features	Sample	Algorithms	Accuracy (%)
[3]	TF-IDF, count vectorizer, bag of a word	563	DT	92.07
			RF	94.49
[4]	Word embedding	1498	Neural network	92
[6]	TF-IDF	2140	LR	92
			SVM	91
			ensemble learning	94
[7]	TF-IDF	2140	SVM	93
	Word2Vec		SVM	90
Proposed system	TF-IDF	3398	LR	95
			SVM	93
			DT	89
			RF	93
	Sequence of tokens with embedding		CNN + BiLSTM	93

in our methodology, it has achieved 95%. In [4], neural network was used with a sample of 1498 data and achieved 92% accuracy and our method using 3398 samples with the CNN + BiLSTM has achieved 93% of accuracy. SVM was used in [6, 7] as well where the accuracy in [6] was 91%, and in [7], it was 93% and 90% in two different methods, respectively. In our model, the accuracy of SVM is 93% which is also the best accuracy in terms of SVM.

5 Conclusion

Fake news is a curse of modern technology these days. Effects of fake news become more deadly when it spreads in a pandemic like COVID-19. It can affect more death, raise physical, and mental stress [12]. If COVID-19-related fake news can be detected, then all of these problems can be lessened, and a lot of people can be served too. Classification of text using TF-IDF is very useful as it has significant information about the importance of every word in the corpus. Again, machine learning models are quite efficient in terms of classification using TF-IDF values and also computationally lighter than deep learning models. Keeping that thought in mind, we have proposed a system where we have got more accuracy from a machine learning model, logistic regression using TF-IDF than deep learning models using a sequence of tokens. Logistic regression has attained an accuracy of 95% and an F1-score of 0.94 with Cohen's kappa coefficient of 0.89 and Matthews coefficient of 0.89 as well. One challenge that we have faced is the limited number of data. With more data, model becomes more efficient and trustworthy. In near future, we will build a system where we will use more data and build a Web application programming interface) (API) that will detect COVID-19-related fake and real news online with the information of our research.

References

1. COVID Live Update: 216,943,564 Cases and 4,511,731 Deaths from the Coronavirus—Worldometer. Worldometers.info, 2021. [Online]. Available: https://www.worldometers.info/coronavirus/. Accessed: 28 Aug 2021
2. Munich Security Conference *Who.int*, 2021. [Online]. Available: https://www.who.int/dg/speeches/detail/munich-security-conference. Accessed: 28 Aug 2021
3. Amer AYA, Siddiqui T (2021) Detection of Covid-19 fake news text data using random forest and decision tree classifiers. Zenodo, 2021. [Online]. Available: https://zenodo.org/record/4427205#.YD5h7WgzbIU
4. Elhadad MK, Li KF, Gebali F (2020) Detecting misleading information on COVID-19. IEEE Access 8:165201–165215. https://doi.org/10.1109/AC-CESS.2020.3022867
5. Gundapu S, Mamidi R (2021) Transformer based automatic COVID-19fake news detection system, pp 1–12. [Online]. Available: https://arxiv.org/abs/2101.00180
6. Shushkevich E, Cardiff J (2021) TUDublin team at Constraint@AAAI2021 – COVID19 Fake News Detection, pp 1–8, 2021. [Online]. Available: https://arxiv.org/abs/2101.05701
7. Raha T, Indurthi V, Upadhyaya A, Kataria J, Bommakanti P, Keswani V, Varma V (2021) Identifying COVID-19 fake news in social media, pp 1–7 [Online]. Available: https://arxiv.org/abs/2101.11954
8. Al-Ahmad B, Al-Zoubi A, Abu Khurma R, Aljarah I (2021) An evolutionary fake news detection method for COVID-19 pandemic information. Symmetry 13(6):1091 Available: https://doi.org/10.3390/sym13061091
9. Banik, "COVID Fake News Dataset", *Zenodo*, 2021. [Online]. Available: https://zenodo.org/record/4282522#.YD5j-WgzbIU. Accessed: 28 Aug 2021
10. "COVID-19 Real News Data of CBC News", Zenodo, 2021. [Online]. Available: https://zenodo.org/record/4722470. Accessed: 28 Aug 2021
11. "Stemmers", *Nltk.org*, 2021. [Online]. Available: https://www.nltk.org/howto/stem.html. Accessed: 28 Aug 2021
12. Raihan M, Islam MT, Ghosh P, Hassan MM, Angon JH, Kabiraj S (2020), Human behavior analysis using association rule mining techniques. In: 2020 11th International conference on computing, communication and networking technologies (ICCCNT), pp. 1–5. Available: https://doi.org/10.1109/ICCCNT49239.2020.9225662

Application of Digital Technology in Accounting Profession for Achieving Business Goals and Sustainable Development

V. Sukthankar Sitaram, Sukanta Kumar Baral, Ramesh Chandra Rath, and Richa Goel

Abstract Digital technology has gained growing importance in professional accounting as it serves as a uniquely situated place where departments and disciplines interact around both official and non-official operations. Application of digital technology has the ability for performing a set of standards of work, develop models and generate best quality of digital information and inform to the higher authority. It noticed to various informations related with the organizational function into a standard line of execution with the principles of sustainable development. Digital technology in professional accounting research is increasingly focusing on procedures created by professional accountants with the goal of ensuring a company's long-term success. The integration of social and environmental data into financial reports has been characterized by some organizations as a way to adapt and enhance international reporting requirements. The study's primary objective is to increase public awareness of the accounting profession's contribution to national sustainable development. Because of this, we will depend on a documentary study of national and international laws for attaining organizational business goals in the matter of how digital technology influences the accounting profession for this proposed article. Thus, in order to achieve our organizational business goals, it is also referred different studies of national and international laws pertaining to the use of digital technology in the accounting profession.

V. Sukthankar Sitaram
Government College of Arts, Science and Commerce, Khandola, Marcela, Goa, India

S. Kumar Baral (✉)
Department of Commerce, Faculty of Commerce and Management, Indira Gandhi National Tribal University (A Central University), Amarkantak, Madhya Pradesh, India
e-mail: sukanta.baral@igntu.ac.in

R. Chandra Rath
Guru Gobind Singh Educational Society's Technical Campus, Chas, Bokaro Approved By AICTE Govt. of India New Delhi, and affiliated to Jharkhand University of Technology, Ranchi, Jharkhand, India

R. Goel
International Business Department, Amity University, Uttar Pradesh, Noida, India
e-mail: rgoel@amity.edu

V. Skala et al. (eds.), *Machine Intelligence and Data Science Applications*, Lecture Notes on Data Engineering and Communications Technologies 132,
https://doi.org/10.1007/978-981-19-2347-0_41

Keywords Digital technology · Development durable · Accounting profession · Business sustainability

1 Introduction

Sustainability is a term which has been prominent in the last two decades and which has to be integrated into all aspects of our societal, economic, and environmental existence. It means to live in the present and to honor the history and to preserve the future. The United Nations and other organizations created 17 sustainable development goals in 2015, using digital technologies, replacing the 8 Millennium Goals set in 2000. Sustainability economics, society, and environment may work together in three aspects to accomplish sustainable development goals. Achieving these goals needs skilled accountants and accountants [1]. Fayez Choudhary, International Accountant Federation Executive Director, has emphasized the significance of the accounting profession to help achieve these objectives directly or indirectly. Thus, accountants have a variety of options to address the problems of sustainable development [2] via their talents and expertise. Recently, the accounting profession was emphasized for its critical role in attaining these goals. "Accountants may use their skills and knowledge to help solve issues of sustainable development" [2]. With their unique position at the intersection of all sectors and departments, professional accountants are more and more important for developing sustainable development ideas in an organization's continuous, coordinated execution. Professional accountants play an increasingly significant role in organizations, since they are uniquely positioned to establish standards, build models, and provide reporting information that determines the company's consistency and participation in sustainable growth.

2 Literature Review

In the section of literature review, the researchers have trying finding out the genuine ness with its justification of the aforesaid research title "Application of Digital Technology in Accounting Profession for Achieving Business Goals and Sustainable Development". In connection with, we have taken both the method of data collection such as primary and secondary mode of data collection followed with a good questionnaire method which was asked to 10 Companies of National and International repute with 400 employees of various companies from Odisha and abroad. Respondents given their feedback and response through goggle form and questionnaire which is directly asked to them by the researcher. Further some extent, we have also collected various information from the published sources.

Gray et al. (1987) viewed environmental accounting as an extension of corporate responsibility [3]; Gray and Babbington (2000) emphasized the importance of accountancy in sustainable company development [4]; Gray and Collinson (2002)

examine the function of accountant businesses. Achieving 7 of the 17 sustainable development goals requires the assistance of the IFAC's Conference on Trade and Development International Accounting and Reporting Standards (CTD-IAS). Professional accountants, he believes, will assist accomplish goal 8—decent employment and economic growth innovate and build infrastructure—goal 9, goal 12: responsible consumption, goal 16: peace, justice, and strong institutions goal partnership (#17) [1]. In this context, our research aims to benefit both the academic and business worlds by highlighting the role of accounting in attaining sustainable development objectives.

In connection with researcher has taken two hypothesis such as null hypothesis (H_0) and alternative hypothesis (H_e) in anticipation of variable I and variable II, in variable one refers as the role of e-technology has not increased the business development and sustainability in organization. Null hypothesis (H_0) whereas the second variable refers as the execution of e-technology in business management and accounting practice has a tremendous impact for not only increases the ratio of production in different classes such as employees and youngsters in respect of male and female, but also it facilitates the accounting profession system of application of digital technology in an advanced way which enhances the sustainability (H_e). After the testing of hypothesis and the result proves that role of e-technology has a positive impact to enhance the digital accounting system and makes its sustainability.

3 Aim and Objectives of the Study

Our goal was to boost employee engagement to a new and exciting level by encouraging more people to contribute ideas, problem-solve, and contribute new ideas. In this classroom, students are encouraged to be active learners while also providing equitable access to Durham University's main researcher, Liz Burdon (2012) [3].

It is recognizing the role of diverse social, cultural, environmental, and political contexts of development and the cost of an interdisciplinary method, the route will enable contributors to painting more successfully with others to promote the develop ment and alertness of new technologies for sustainable development. The followings are the objectives of the study.

- To develop in response to the need for a more prospective of importance of accounting profession and its practice.
- To generate for sustainable improvement which offers a diverse mix professional engaged with the job.
- To study the current application of technology and identify its ways which may provide maximum assistance for sustainable development.
- To identify the importance and preference of the technical data, recent technology and setting that influence innovation, creativity, and professional sustainability.

4 Method of the Study

In connection with, the researchers have been taken both the method of data collection as the previous researchers have followed previously such as primary and secondary mode of data collection. But we have followed with a good questionnaire method which was asked to respondents through goggle form. Further some extent, we have also collected various information from the published sources, and all the data were presented, classified, analyzed, and interpreted by the help of requisite statistical tools and techniques in order to justified the aforesaid research work.

5 Formulation of Hypothesis

In this section, the researchers have taken two hypothesis such as null hypothesis (H_o) and alternative hypothesis (H_e) in anticipation of variable I and variable II, in variable one refers as the role of e-technology has not increased the business development and sustainability in organization.

5.1 *Null Hypothesis: (H_o)*

This hypothesis expected the effect and role of e-technology has not increased the business development and sustainability in organization.

5.2 *Alternative Hypothesis (H_e)*

This hypothesis also focused on application of digital technology how enhancing the performance of accounting profession system in an advanced way which enhances the economic development and its sustainability in to a great extent.

6 What is Digital Technology in Accounts?

Digital technology in accounting is described as an array of equipment used for assisting workers in improving their skills, as well as measuring their reasons for doing something. Digital technology is the use of various digital tools and resources to help you accomplish your educational goals. Virtual generation in accounting depends on a wide meaning of the term era, which includes big machinery and resources.

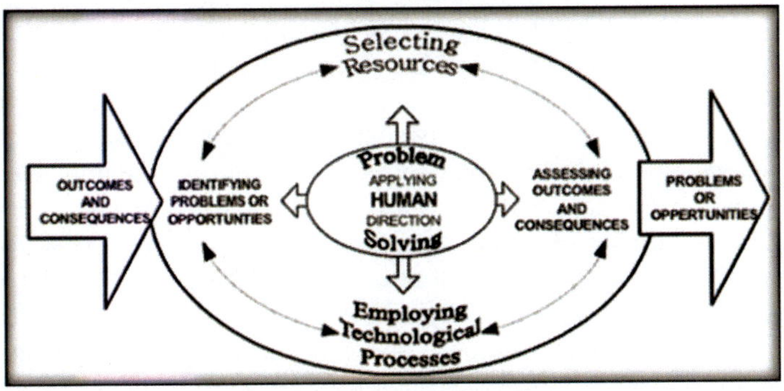

Fig. 1 Model-I of digital technology of accounting profession for problem solving. *Source* Goel R, Pattanayak M, Rath RC and Baral SK, "An Impact of Digital Technology on Education Industry for Innovative Practices of Teaching and Learning Process: Emerging Needs and Challenges", 2021 9th International Conference on Reliability, Infocom Technologies and Optimization (Trends and Future Directions) (ICRITO), 2021, pp. 1–5. https://doi.org/10.1109/ICRITO51393.2021.9596202

7 Uses of Technology in Accounting and Education

Education is necessary for all to learn in order to distinguish ourselves from the dark to the light, i.e., "ignorance to light" enchanting the related Sanskrit Sloke of Indian Vedic (Aryan) Culture "Saa Vidya Jaa Bimuktate Tamosa Maa Jyotri Gamayae" because it assists all to become a well cultured and civilized person. "However, its history is most often associated with the introduction of instructional films in the 1900s or Sidney Presser's mechanical teaching machines in the 1920s. V79.2 7". Here, a model of technology in accounting digital technology and education has been present for the purpose of solving human problems in the fields of accounting education from the following model (Fig. 1).

8 Accountants' Role in Sustainable Development

Now, instructors assisted students in comprehending the function of professional accountants in promoting sustainable development. Accountants and their training in the UK: a long-term development perspective, Gray and Collinson [5], Lovell and Mackenzie [6], Gray and Babington [4]. "IFAC President Warren Allen sees the accounting profession as a crucial collaborator in achieving seven of the 17 SDGs" (2011) (ISAR UNCTAD). Goals 4 and 8 emphasize excellent education, while industry, innovation, and infrastructure are covered in objective 9. "Responsible consumption and production" 17 is for "partnership for Goa" and 13 is for "climate action". The accounting profession may help achieve sustainable development goals by offering key methods to act in this study. "An accountant's duty is to serve the public while staying accountable to the financial actors: shareholders/friends, employees, customers, suppliers, the country, banks, and investors". Assisting businesses in generating value over time and reacting to risks, enhancing decision-making processes, fig. "An accountant's position inside a company may be a lever for the business's long-term success". Professional accountants possess the information, attitudes, judgements, and techniques necessary to facilitate the implementation of a long-term company plan. Accountants provide company-friendly reporting that helps create long-term development and economic equilibrium.

Accountants' Methods for Sustainable Development: "Accountants have various methods for carrying out sustainable operations in line with SDGs. That is why, in Chap. 1 accountant support tools for sustainable development, we view corporate laws, supply chain stress, stakeholder engagement, taxation, and government subsidies as a means of incentivizing businesses to adopt a long-term creative and prophetic approach to sustainability".

Accountants' Role in Sustainable Development: Now, instructors assisted students in comprehending the function of professional accountants in promoting sustainable development. Accountants and their training in the UK: a long-term development perspective, Gray and Collinson [5], Lovell and Mackenzie [6], Gray and Babington [4]. "IFAC President Warren Allen sees the accounting profession as a critical partner in achieving seven of the seventeen sustainable development goals" (2011) (UNCTAD ISAR). Excellent education is addressed in goals 4 and 8, while industry, innovation, and infrastructure are addressed in goal 9. Goal 12 is devoted to "responsible consumption and production", goal 13 is devoted to "climate action", and goal 17 is devoted to "partnership for Goa". The accounting profession may help achieve sustainable development goals by offering key methods to act in this study. "An accountant's duty is to serve the public while staying accountable to the financial actors: shareholders/friends, employees, customers, suppliers, the country, banks, and investors". Assisting businesses in generating value over time and reacting to risks, enhancing decision-making processes, fig. "An accountant's position inside a company may be a lever for the business's long-term success". Professional accountants possess the information, attitudes, judgements, and techniques necessary to facilitate the implementation of a long-term company plan. Accountants provide company-friendly reporting that helps create long-term development and economic equilibrium.

9 Accountants' Methods for Sustainable Development

Accountants have various methods for carrying out sustainable operations in line with SDGs. That is why, in Chap. 1 accountant support tools for sustainable development, we view corporate laws, supply chain stress, stakeholder engagement, taxation, and government subsidies as a means of incentivizing businesses to adopt a long-term creative and prophetic approach to sustainability (Fig. 2).

Those methods must be based on trustworthy in formation and reporting in order for them to be applicable to accountants and to appropriately achieve the goals

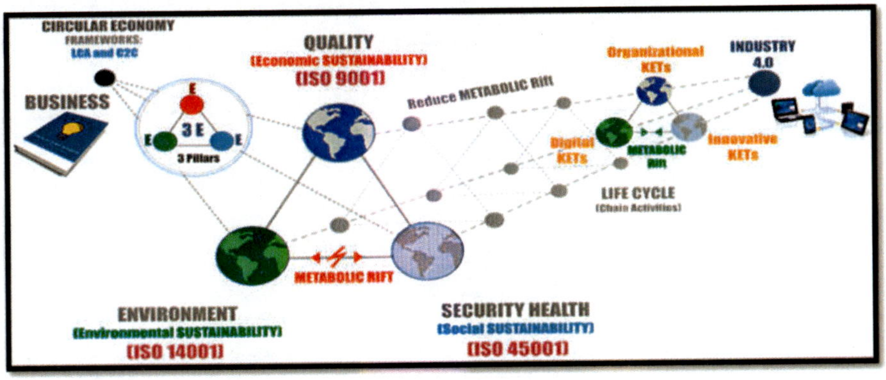

Fig. 2 Model-II digital accounting profession mechanism. *Source* María Jesús Ávila-Gutiérrez et. al, Standardization Framework for Sustainability from Circular Economy 4.0, Sustainability 2019, 11, 6490, Sustainability, An Open Access Journal from MDPI. http://dx.doi.org/10.3390/su11226490

of long-term improvement. The accounting profession's contribution to economic sustainability is characterized by strong international and domestic professional organizations that collaborate with governments to create and execute accounting standards.

10 Sustainable Reporting of Environmental Information

Standards for reporting on environmental and social data were established and made accessible to companies so that they may achieve not only sustainable but also regulated outcomes. Other expert groups have also addressed the problem of harmonizing social and environmental data reporting.

Laws, financial reporting requirements, shareholder obligations, and transparency are all included. It recognized three corporate sustainability phases: strategic, operational, and reporting. Financial and accounting records must be audited for materiality, comparability, and completeness by independent national and international experts. Accounting companies offer different compliance services, highlighting the profession's involvement in achieving the SDGs. In addition to performing environmental reporting research and providing helpful sustainability reporting standards, accounting firms and other non-profit organizations help and encourage sustainable development. It is still not possible to reconcile the gaps between conventional accounting and reporting for sustainable development and reporting. Due to the accounting profession's critical role in a business's long-term economic performance, differentiation 2 focuses on how the professional accountant works to achieve long-term development objectives.

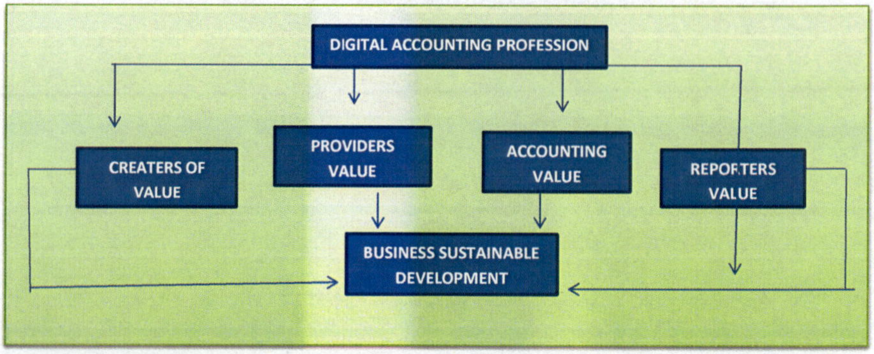

Fig. 3 Processing function of the professional accountants in sustainable development by the researchers. *Source* Compiled by the authors

Accounting and accounting contribute to eight of the 17 sustainable development goals of the UN Agenda 2030 (SDGs). Accounting, according to a recent IFAC study, helps accomplish these goals: Recent IFAC research demonstrates how accounting is critical to achieving these objectives. Accounting is critical to the profitability and sustainability of businesses, financial markets, and economies, according to Fayez Choudhury, CEO of IFAC. "We need to be mindful of our contribution as a profession." Accountants give both strategic and operational value. They provide leadership in long-term value planning. We can encourage high-quality corporate reporting by including accountants in the SDGs. Assuring a company's long-term viability usually falls on the accountant. According to the International Federation of Accountants (IFAC), accountants may participate as business partners and contribute to a more sustainable strategy by focusing on performance and business model (Fig. 3).

"The sustainability of an organization must be reviewed, analyzed, and reported, and the accountant is usually responsible for these duties". The International Federation of Accountants (IFAC) recommends two practical ways for accountants to participate as business partners and contribute to a more sustainable approach [7]: performance and business model.

It identifies and incorporates essential natural and social capital concerns into the decision-making system; income generation; control; great information inspires user trust and is critical to business sustainability.

11 Data Table: I

In this section, the researcher asked to 200 respondents (employees/staff) from 10 number of National and International Companies like

11.1 *Users of Digital Technology in Accounting Profession*

Company	No. of users/respondent's			Favor/against		% of positive response
	T	M	F	Y	N	
Nalco Ltd	40	20	20	23	17	34(40)
Hindalco	40	20	20	26	14	37(40)
Jindal	40	20	20	28	12	34(40)
Nilachal Ispat	40	20	20	19	11	35(40)
TATA	40	20	20	22	18	33(40)
ESSAR	40	20	20	16	24	27(40)
Mahindra	40	20	20	18	22	31(40)
Birla Tire	40	20	20	24	16	37(40)
IFAC	40	20	20	28	12	31 (40)
Total	400	200	200	282/400 = 70.5	118/400 = 29.5	299/400 = 74.75

NB: *T—Total *M—Male, *F—Female, *Y—Yes. *N—No and *AM = Average mean

From this aforesaid data table, we may able to know that all companies are favor of application of digital technology in the accounting profession for smooth operation of accounting work and the percentage of positive response is.

12 Final Result Table

Companies	10
Respondents	400
Mean difference Y-70.05 N-29.5	41% (difference)
Sex ratio	50% (equal)
Positive response (yes)	282
Negative response (no)	118

12.1 Semantic Differential Model in Pie Chart of Response

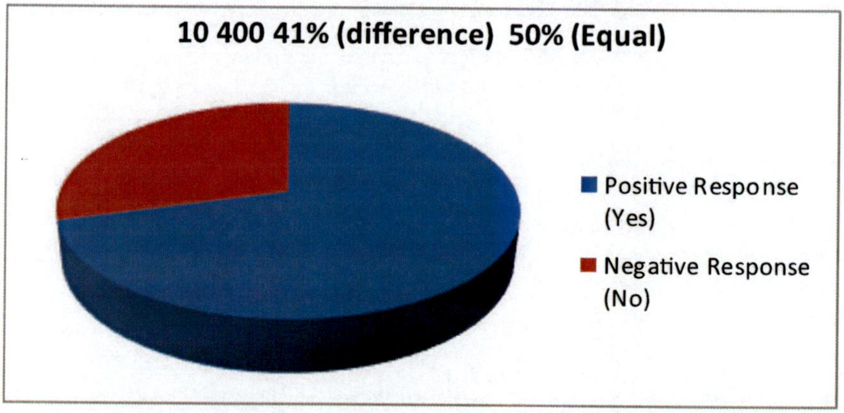

12.2 Semantic Differential Model in bar Graph of Response

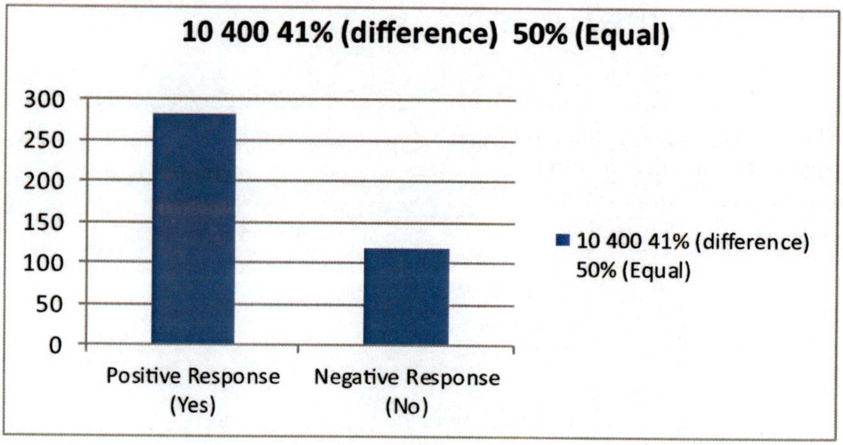

13 Hypothesis Testing

After analyzing the evidence, we reject the null hypothesis (H_O) and adopt the alternate hypothesis (H_e) since it is more important. Similarly, the difference in sex ratio is nil (Zero) in National Aluminum Company (NALCO) and in Nilachal Ispat Nigam Limited (NINL), so Nilachal Ispat Nigam Limited (NINL) has been able to maintain a sex ratio of female employees as equal to male employees of TATA Company by utilizing e-technology. Same also maintained by other companies also but due to more positive response (70.50) regarding the application of digital technology in accounting profession for enhancing the business goal and its sustainability.

14 Discussion and Conclusion

To sum up, digital technology has revolutionized the way we live. Professional accountants are actively involved in the business administration of a company where corporate sustainability efforts are promoted. As a consequence, the accounting profession should rethink its role in attaining the SDGs and work to include sustainable development ideas into decision-making at all levels, including commercial operations (management, budgetary projection, evaluation, and reporting). Therefore, accountants are strategic value builders, operational value suppliers, and reporting value maintainers and reporters. This will need a re-evaluation of the long-term development role of professional accountants. Accountants are responsible for sustainable reporting that is essential in order to collect information and promote sustainable development. Sustainability problems will be measured, recorded, and interpreted by accountants from organizations with substantial environmental or social effects. Environmental and social data will be required in future information and control management systems.

Once again, professional accountants are closely linked to the promotion of business sustainability efforts. As a result, the accounting profession should reconsider its role in achieving the SDGs and participate in integrating sustainable development criteria at all levels—strategic, operational, and tactical (management, budgetary projection, evaluation, and reporting). Accountants are value creators at the strategic level, operational value providers at the operational level, and report keepers and reporters at the reporting level. As a result, the role of professional accountants in sustainable development must evolve. The mechanisms of competence of professional accountants need more analyzes since accountants are accountable for sustainable reporting, a vital information, and sustainability assistance. The measuring, recording, and interpretation of sustainable problems will include accountants from organizations with major environmental or social effects. Environmental and social data will be required in future information and control systems.

References

1. https://www.ifac.org/news-events/201511/accountancy-profession-and-sustainable-development-goals. Accessed on 13 Mar 2018
2. ACCA (2014) Sustainability and business: the next 10 years. http://www.accaglobal.com/content/dam/acca/global/PDFtechnical/sustainabilityreporting/presentationsustainability-and-business-thenext-10-years.pdf
3. Gray R, Owen D, Maunders K (1987) Corporate social reporting: accounting and accountability. Prentice-Hall, London, p 4
4. Gray R, Bebbington J (2000) Environmental accounting, managerialism and sustainability: is the planet safe in the hands of business and accounting? In: Advances in environmental accounting and management, vol 1. Emerald Group Publishing Limited, p 1–44
5. Gray R, Collison D (2002) Can't see the wood for the trees, can't see the trees for the numbers? Accounting education, accounting and financial control sustainability and the public interest. Crit Perspect Acc 13(5–6):797–836
6. Lovell H, MacKenzie (2011) Accounting for carbon: the role of accounting professional organisations in governing climate change. Antipode 43(3):704–730
7. Jen KC, Jen FC (1961) The investment performance I: an empirical investment of timing selectivity and market efficiency. J Bus 52:263–290

Application of Robotics in the Healthcare Industry

Vishesh Jindal, Shailendra Narayan Singh, and Soumya Suvra Khan

Abstract Robots have made their presence everywhere because of their uncounted advantages. Health care is a field that faced a lot of challenges in reducing the huge gap between those in need of services and those who receive services. With the technological advancement, healthcare and automobile sectors are benefitting the most by adopting advanced robotics. In health care, robotic surgery brings precision with minimal invasion and the use of miniaturized surgical instruments on patients. Having this technology, surgeons can perform various complex surgeries like cardiac, urological, kidney, and many more with ease by assisting robots. This paper provides research and analysis on surgical and non-surgical robots by highlighting their evolution, design along with its uses, and limitations of robots in medicine, and health care. This term paper also focuses on various scopes and development of robotics in the healthcare sector with an immense number of outcomes and conclusions.

Keywords Robots · Humanoid robots · Robotic surgery · Healthcare sector

1 Introduction

Health care deals with the well-being of humans. It majorly focuses on prevention, diagnosis, treatment, etc., for the ailments. With the advancement in medical science, the role and application of "technology" calls for the need for discussions on benefits in a long way. This term paper will be discussing Robotic Applications in the field of health care [1].

V. Jindal · S. Narayan Singh
Department of Computer Science, Amity School of Engineering and Technology, Amity University, Noida, Uttar Pradesh 201313, India
e-mail: snsingh36@amity.edu

S. Suvra Khan (✉)
Department of Computer Science and Engineering, Meghnad Saha Institute of Technology, Kolkata, West Bengal 700150, India
e-mail: soumyasuvra@msit.edu.in

V. Skala et al. (eds.), *Machine Intelligence and Data Science Applications*, Lecture Notes on Data Engineering and Communications Technologies 132,
https://doi.org/10.1007/978-981-19-2347-0_42

The use of robotics in the surgery which was specifically tested had enhanced the emergence of robots in the whole medical sector. Holding of this technology or methodology [2, 3] has benefited the surgeons in performing complex operations/surgeries like urological, cardiothoracic, etc., with ease as the robotic technology uses the sensor technology which enhances the precision in performing various tasks [4].

4 major robots constitute the pillars of health care, and they are in the following:

- Surgical Robots
- Exoskeletons
- Care Robots
- Hospital Robots

These robots are not only part of science fiction movies but also are the pillars or we can say the backbone for the healthcare industry.

Robotics is a vast field which majorly constitutes under the field of artificial intelligence; to be more precise, it is the reduction of human effort by replacing them with machines [5, 6] which function on programming and set of instructions that have been faded by humans to function that particular task with zero causalities and zero errors [7]. Figure 1 discusses the **R4H** innovation themes in the healthcare industry.

The most important [8] factor that links them is various innovative themes that cannot be neglected at any price [9, 10]. Themes are classified as follows:

- The first and foremost theme discusses giving the right guidance and assistance by the robots to the nurses in the hospital as well as outside the hospital and taking care of patients.

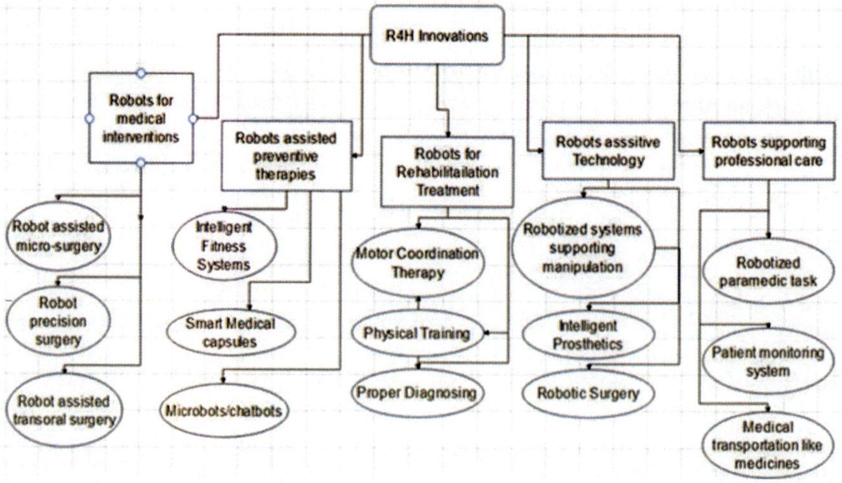

Fig. 1 R4H innovation themes

- The second theme focuses on the facilities provided by the robot like the assistance and robots in surgery.
- The third theme puts pressure on discussing the efficiency increase by the use of robotic system and prevention of the ailment as well as diagnosing at a proper time helps in curing it much faster.
- The fourth theme focuses on the rehabilitation of patients, i.e., treating the patients very carefully after the treatment by the hospital and assisting them in making their situation from better to the best [11].

2 Literature Review

Table 1 will discuss different types of robots which are used in the healthcare sector from the years 1985 to 2020 along with mentioning some applications also.

3 Evolution of Robotics in the Healthcare Industry

In the earliest period of the twentieth century, robots were not yet coming into existence in real or even science fiction. By the year 1921, when the renowned novelist wrote the play Rossum's Universal Robots, then the concept of robotics came to the market. After the discovery, it remained a matter of debate in the Czech literary world about the coining of the term "robota" which signifies worker or laborer [4].

After the introduction of robotics in movies, production lines in factories, war, etc., led to rising in the industry of health care which was the breakthrough in the technology sector as it involves the development of robots doing operations on humans [19]. The first hands-on experience by the robots and guidance by the surgeons took place in 1985 when PUMA 560 robotic surgical arm was used in neurosurgical/brain biopsy which is used to guide the needle in CT guidance. This system was used in very few patients approximately in numbers of 10 [8].

Later on, in 1992, ROBODOC was developed by Integrates Surgical Systems which worked for hip replacements or orthopedic surgeries; at a point, it was approved by FDA because of the high precision of the orthopedic operation but later this system failed because of various side effects on the patients and lead to various complications which therefore lead to stoppage of availability in the market [5].

4 Designing of Medical Robots

Various tools are required for the development of robots. The architecture/design presented by Kimming et al. [20] reflect different tasks for robot control; primarily

Table 1 A brief survey on the healthcare robotic application

Known as	Proposed in	Ability
PUMA560 robotic arm	1985	Neurosurgical biopsy
The Elektrohand 2000	1988	EMG signal processing
The cyber knife	1990	Radiation therapy to tumors
ROBODOC	1992	Orthopedic surgery/Hip replacement
Arm guide system	1999	Diagnosing options and therapy in 2-D motion
ASIMO	2000	Humanoid robot for walking, running, climbing stairs, pushing carts
DA Vinci robot system	2000	Minimal invasive approach for performing complex
ZEUS system	2001	Assistance in minimal invasive surgery
PARO therapeutic robot	2003	Helpful in improving quality of life during recovery, i.e., from surgery to treatment
ARMAR-III	2006	Support tasks in human-centered environments
ILimb	2007	Avoiding crushing of objects
RIBA	2009	Lifting and transferring patients with high load capacity [12]
CODY	2009	Helping in patient hygiene like bed baths
PR2	2010	Mobile humanoid robots supporting in human environments
Xenex germ zapping system	2011	Disinfect hospitals with a large spectrum of UV light
The TUG	2015	Transferring supplies, meals, and other materials in hospitals
UVD robot	2016	Used for cleaning rooms by emitting UV light
Tele-healthcare humanoid robots [13]	2017	Tele-healthcare with humanoid robots
Surgical robotics 4.0 [14]	2018	Agile extension of the human eyes, hands, and skillful and smart partner of their human counterpart
Healthcare 5.0 [15]	2019	Machine-to-machine (M2M) or device-to-device (D2D) communication in remote surgeries
Visual-MIMO [12]	2019	Measure heart rate (HR) and blood oxygenation (BO)
Concentric tube robot [2]	2020	Minimally invasive surgery
Homecare robotic systems [16]	2020	CPS-based HRS are proposed
Hello, Dr. [17]	2021	Self-diagnose, emergency care
Healthcare robotics [3]	2021	Provides disinfection and logistics services that support patients and healthcare professionals

(continued)

Table 1 (continued)

Known as	Proposed in	Ability
2-DoF meso-scale continuum robotic tool [18]	2021	Helps in pediatric neurosurgery

motor control, and supervisory control, which in turn calculate the position and projectile.

Figure 2 discusses the designing of robots which mainly hold 4 major requirements to have robots in proper design for good functioning. These are classified as follows:

- Majorly for the simple requirements which include tasks for the positioning of robot require COTS motion controllers and talks or connects to them by ISA or PCI or having a strong network like Ethernet, etc. These controllers are required for providing motor control and coordinated motion which comes under low-level functionality. These are mainly done on either Windows or Linux Operating systems [21].
- The second point discusses the application part which is required for the regulation of design in hospitals. The main aim is to develop a safe system by providing high-quality software and using a good amount of documentation for the initialization process. By separating the core software from the specific software, this can be achieved and will be helpful in learning [6].

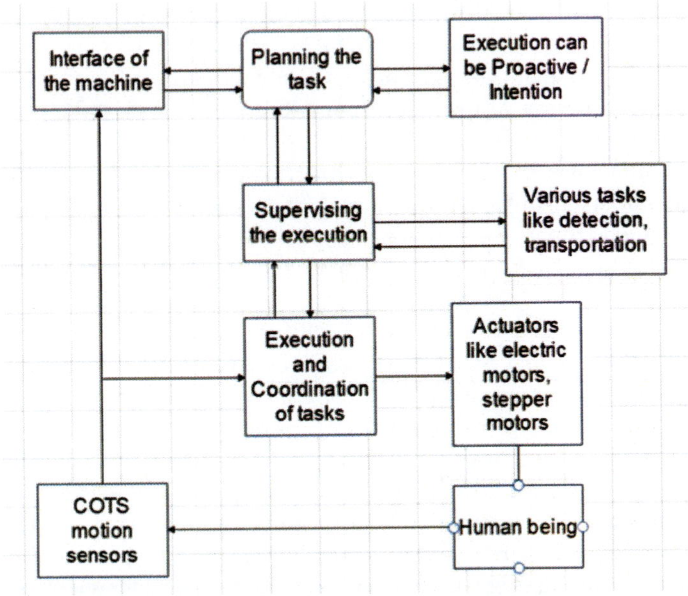

Fig. 2 Robot controller architecture

- This point focuses on maximum usage of robotics in the healthcare sector, for this development robot needs to have appropriate design and functioning, it needs to be applied in the larger application domain. It must be consistent with the software-based devices. Take an example of managing the position of the instrument by various sensors and tracking systems [19, 22].
- The last requirement deals with the complexity of the devices which should not be limited to the chronological order, i.e., [20] the robot system consists of various sensors like navigation sensors, velocity sensors, force sensors, position sensors, etc.

5 Uses of Robotics in the Healthcare Sector

Robotics in the healthcare sector has provided immense benefits and a good push in the development of the technology which was never thought of before. As a result, robots are being developed to perform various tasks within the medical scenario [23].

Various uses are classified as follows:

- Usage of telepresence robots which connects the robots with the physicians via the call or the Bluetooth devices for which they need not be physically present in remote or backward locations. This kind of system has benefitted various sectors including health care, education, manufacturing, and many more. Telepresence robots are also used for organizing rounds for patients in hospitals.
- Sanitation and Disinfectant robots are much more in use as these robots are resistant to bacteria and deadly diseases like SARS, Ebola, etc. Nowadays, these robots are being used by healthcare facilities to disinfect rooms and surfaces. Moreover, these robots can clean or disinfect them within minutes by using UV light and hydrogen peroxide vapors [24].
- The next main use of robots in the healthcare industry is assisting remote-controlled robots in surgical procedures. The delicate arm of the robot is manipulated and guided by surgeons through remote control for treating various surgeries like Minimal Invasive Surgery and complex surgeries being done in the operation theater.
- With this advancement of robotics in health care, nowadays dispensaries being operated by robots that provide prescriptions are in great demand. As we are all aware that robots are meant for high speed and accuracy, they are being programmed in handling viscous materials, powder, liquid medicines, etc. [19].
- Then comes the robots which are the support system of the patients, i.e., they help the patients in improving their balance, help them in walking, and help in various other motor functions who are suffering from various disabilities, strokes, trauma, and spinal injuries. These robots give the patients new life by helping them in improving strength, mobility, coordination, and better quality of life [25].

6 Challenges Faced by Robotics in Health care

Robotics in the healthcare industry is growing tremendously fast in terms of the services and the technology used. Parallel at some points, it is also lagging and various issues are being raised. Some of the challenges or issues faced are classified as follows:

- The first issue is the adaptability and adjustments of the robots with the humans. It tells that those humans of any age group work differently according to their capabilities, strength, mobility, etc. [26]. Patient wants to go to the washroom in between the destination and need to stop.
- The next is the communication between the robots and humans which needs to be a two-way process. HRI studies are trying to build social intelligence algorithms and multiple perspectives to enable robots to engage in different modes of communication that are intuitive and natural to the users.
- Then, the next challenge is lack of social intelligence, this requirement focuses on effective communication and the interaction between the robots and the users/humans and the level [19] of understanding between them, and their main aim is not to replace the nurses or doctors with robots but to help robots in guiding and assisting them properly for better treatment.
- The next issue is errors in the decision-making capabilities of robots which need to be minimized to a major extent as robots are engaged in tasks of longer duration and of high complexity without any technical help [27].

7 Humanoid Robot in Healthcare Service

Overview

Various types of robots are added to service in the healthcare industry. One of them is the Humanoid Robot, as the name suggests, this robot understands all the human interactions and is endowed with all the human capabilities to carry out tasks and in assisting the surgeons particularly in contaminated areas. These robots are controlled by remotes and increase the capability of human–robot interaction [25] contaminated areas.

A mobile humanoid robot is in great demand as this model is designed to perform various tasks in the environment which include opening or closing the doors and emptying dishwashers. Figure 3 mentioned below depicts a humanoid robot having special features and applications.

Features

There are various features of a humanoid robot that helps them in functioning, and some of them enlisted here:

Fig. 3 Humanoid robot

- The first feature is the sensing behavior of the robots, and there is *n* number of sensors are there in robots that help them in functioning better and understanding. Specifically, for lifting multiple objects and performing tasks arm sensors are used [8]. Then, the sense of touch is the most important sensing behavior which is used by the robots.
- The next is the manipulation-related tasks, as we are aware that humans have 30° of freedom and but the humanoid robot is built with the design of a light cable-driven manipulator which then converted to form a large and advanced SCARA type manipulator which is used for palpation.
- The next feature deals with the mobile-based platform where robots need to be flexible in performing various tasks for easy manipulation and providing better service to the patients and in caregiving, based on that there are two types of the mobile-based platform which are short distance-based platform and long distance-based platform [11].

Applications

Humanoid robots in the healthcare industry are growing very rapidly, and more interest in these fields is now being shown due to the success in the healthcare sector. There are three main applications or uses of humanoid robots in the healthcare industry which are as follows:

- The first is the controlling of the robot from a particular distance in uncontrolled environments that are remotely controlled by humans, i.e., Teleoperated Service Robot, this type of robot enables the operator in copying the manipulations of the operator at a certain distance [22].
- The next is carrying out tasks for the aging population or helping them in making their life and surroundings better. For this purpose, there is a strong demand for humanoid robots to solve and understand the problems faced by elderly people or handicapped people.

- Last is the pain relief from the medical procedure with the application of techno-psychological distraction for the children, as children feel the most pain and distraction needs to be there in avoiding pain and giving them a smile on the face without knowing them that they are being operated on parallel.

8 Robotic Surgery

Overview

With the due course of time, robotic surgery has evolved in the healthcare industry and it showed various benefits too. Robotic surgeries are being carried out by Surgical Robots which use very small tools which are fixed with the robot's arm to carry out the operations and tasks [26].

When the robotic surgery takes place ,robots most popularly Da Vinci Robot seated in front of patients for carrying out the surgery on the command of the surgeon and makes [27] small incision in the body to operate the particular part or to insert various instruments. Robots' arm copies the task of human arms for doing the operations and surgeries which are controlled by surgeons in a computerized manner. Figure 4 mentioned below depicts the performing of surgery with the robot assistance specifying its various features and applications.

Characteristics

Robotic Surgery is a known method of performing various surgeries with the help of robots which are operated by doctors through computers that give a certain number of instructions using small tools attached to the robotic arm [8]. This system involvers and deals with various characteristics and features which are enlisted below:

- The first is the reduction in pain and approximately no amount of discomfort after surgery which means that having the robotized surgery involves a lot of benefits

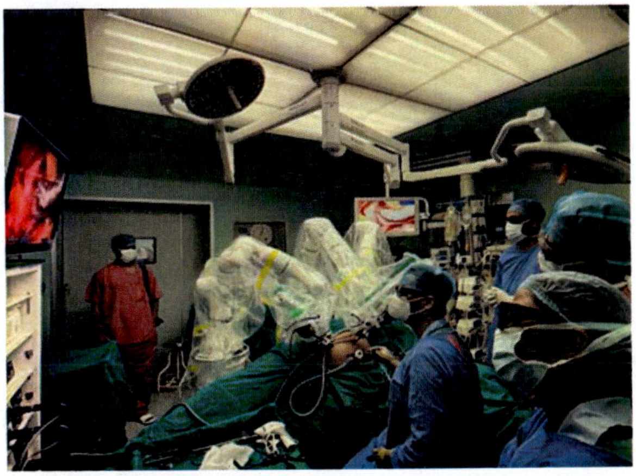

Fig. 4 Robotic surgery

as compared to normal surgery done by the surgeons. This surgery is done with the robotic arms which are controlled by the system.

- Next is the shorter amount of hospitalization, and the patient is discharged or can go home the same day. There is no need to stay for a longer duration in hospitals when having robotic surgery.
- Next is the faster recovery of the patient and coming back to normal activities that any person performs in day-to-day life. There are some complications after robotic surgery, but the patient recovers from it very fast. In a normal surgery, it takes approximately a month or 2 to take normal activities but in robotized surgery, it will take 2–3 weeks [21].

Application

As we are aware that the Food and Drug Association (FDA) approved the first robotic surgery a way back in the year 2000, since then the robotic surgery has evolved and provided various benefits [17]. It has various applications in terms of surgical procedure which are listed below:

- The first is kidney surgery where robot-assisted laparoscopic surgery is emerging for kidney patients because they are minimally invasive which leads to a reduction in pain and less blood loss and faster recovery due to the small incisions [11].
- Then comes the Gallbladder surgery in which cholecystectomy is used for the removal of the gallbladder, but after the coming of robotic devices introduced minimal invasive procedures that involve only a small incision in the navel [8].
- Next is orthopedic surgery in which includes a CAD system for precise activity and surface preparation for knee replacement, this all comes under a robotic surgical system for orthopedic surgery. For that 3-D workstation is of utmost importance for pre-operation surgery planning.

9 Future Prospects and Conclusion

Talking about future scopes, there is an immense amount of revolution which is taking place in the healthcare sector which is very much crucial for medical staff shortages, aging population, and austerity. Future surgeries will now be accompanied by robots that are going to assist surgeons and will make operations, surgeries, tasks, or even the complex situation is much more fast, accurate, safe, and hygienic. Shortly with the help of surgical robots and other robotic devices like Telepresence, Disinfectant, Transportation, robots will help in cutting down the cost of health care.

As the term paper concludes, we can say that the concept of robotics started its journey from science fiction movies and now robots play an integral role in modern life as robots are the creation of humans for helping them in various aspects of life.

However, in the future there are a great number of advances in robotics and scientists are optimistic about finishing the disadvantages by lowering down the cost.

References

1. Butter M, Rensma A, van Boxsel J, Kalisingh S, Schoone M et al (2008) Robotics for healthcare—final report. European Commission, DG Information Society, Brussels, p 12
2. Alfalahi H, Renda F, Stefanini C (2020) Concentric tube robots for minimally invasive surgery: current applications and future opportunities. IEEE Trans Med Robot Bionics 2(3):410–424. https://doi.org/10.1109/TMRB.2020.3000899
3. Jovanovic K et al (2021) Digital innovation hubs in health-care robotics fighting COVID-19: novel support for patients and health-care workers across Europe. IEEE Robot Autom Mag 28(1):40–47. https://doi.org/10.1109/MRA.2020.3044965
4. Kwoh YS, Hou J, Jonckheere EA, Hayati S (1988) A robot with improved absolute positioning accuracy for CT guided stereotactic brain surgery. IEEE Trans Biomed Eng 35:153–160
5. Khan ZH, Siddique A, Lee CW (2020) Robotics utilization for healthcare digital in covid management. Int J Environ Res Public Health 17
6. Scutti S (2015) Medical robots are not just the future of healthcare, but part of the present. 2pp
7. Tuffield P, Elias H (2003) The shadow robot mimics human actions. Ind Robot 30:56–60
8. Azeta J, Bolu C, Abioye AA, Oyawale FA (2017) A review on humanoid robotics in healthcare 33:88–96
9. Narang S, Nalwa T, Choudhury T, Kashyap N (2018) An efficient method for security measurement in internet of things. In: International conference on communication, computing and Internet of Things (IC3IoT), pp 319–323. https://doi.org/10.1109/IC3IoT.2018.8668159
10. Tomar R, Sastry HG, Prateek M (2020) A V2I based approach to multicast in vehicular networks. Malays J Comput Sci 93–107. Retrieved from https://jupidi.um.edu.my/index.php/MJCS/article/view/27337
11. Yang GZ et al (2020) Combating COVID-19—the role of robotics in managing public health and infectious diseases. Sci Robot 5:eabb5589
12. Kwon T, An K, Banik PP, Kim K (2019) Wearable visual-MIMO for healthcare applications. In: 1st International conference on advances in science, engineering and robotics technology (ICASERT), pp 1–4. https://doi.org/10.1109/ICASERT.2019.8934600
13. Panzirsch M, Weber B, Rubio L, Coloma S, Ferre M, Artigas J (2017)Tele-healthcare with humanoid robots: a user study on the evaluation of force feedback effects. In: IEEE world haptics conference (WHC), pp 245–250.https://doi.org/10.1109/WHC.2017.7989909
14. Haidegger T (2018) The next generation of robots in the operating room—surgical robotics 4.0. In: IEEE 16th world symposium on applied machine intelligence and informatics (SAMI), pp 000013–000013. https://doi.org/10.1109/SAMI.2018.8324841
15. Mohanta B, Das P, Patnaik S (2019) Healthcare 5.0: a paradigm shift in digital healthcare system using artificial Intelligence, IOT and 5G Communication. In: International conference on applied machine learning (ICAML), pp 191–196. https://doi.org/10.1109/ICAML48257.2019.00044
16. Yang G et al (2020) Homecare robotic systems for healthcare 4.0: visions and enabling technologies. IEEE J Biomed Health Inform 24(9):2535–2549. https://doi.org/10.1109/JBHI.2020.2990529
17. Mohd F, Elanie Mustafah NI (2021) "Hello, Dr.": A healthcare mobile application. In: 4th International symposium on agents, multi-agent systems and robotics (ISAMSR), pp 20–23.https://doi.org/10.1109/ISAMSR53229.2021.9567764
18. Chitalia Y, Jeong S, Yamamoto KK, Chern JJ, Desai JP (2021) Modeling and control of a 2-DoF meso-scale continuum robotic tool for pediatric neurosurgery. IEEE Trans Rob 37(2):520–531. https://doi.org/10.1109/TRO.2020.3031270
19. Dhamija J, Choudhury T, Kumar P, Rathore YS (2017) An advancement towards efficient face recognition using live video feed: "for the future". In: 3rd International conference on computational intelligence and networks (CINE), pp 53–56. https://doi.org/10.1109/CINE.2017.21

20. Kimmig R, Verheijen RHM, Rudnicki M (2020) Robot assisted surgery during the COVID-19 pandemic, especially for gynecological cancer: a statement of the Society of European Robotic Gynecological Surgery (SERGS). J Gynecol Oncol 31:e59 (2020). Chang C, Murphy RR (2007) Towards robot-assisted mass-casualty triage. In: IEEE international conference networking, sensing and control. IEEE
21. Kraft K, Chu T, Hansen P, Smart WD (2016) Real-time contamination modeling for robotic health care support. In: IEEE International conference on intelligent robots and systems. IEEE
22. Jain S, Sharma S, Tomar R (2019) Integration of wit API with python coded terminal bot. In: Abraham A, Dutta P, Mandal J, Bhattacharya A, Dutta S (eds) Emerging technologies in data mining and information security. Advances in intelligent systems and computing, vol 814. Springer, Singapore. https://doi.org/10.1007/978-981-13-1501-5_34
23. der Loos V, Machiel HF, Ullrich N, Kobayashi H (2003) Development of sensate and robotic bed technologies for vital signs monitoring and sleep quality improvement. Auton Robots 15:67–79
24. Anderson J (1999) Image guided robotic assisted percutaneous interventions. J Vasc Interv Radiol 10(suppl):198–201
25. Wall J, Chandra V, Krummel T (2008) Robotics in general surgery, medical robotics. Tech Publications, pp 491–506
26. Taylor RH (2006) A perspective on medical robotics. IEEE Proc 94:1652–1664
27. Chen AI, Balter ML, Maguire TJ, Yarmush ML (2020) Deep learning robotic guidance for autonomous vascular access. Nat Mach Intell 2:104–115

Context-Driven Method for Smarter and Connected Traffic Lights Using Machine Learning with the Edge Servers

Parminder Singh Sethi, Atishay Jain, Sandeep Kumar, and Ravi Tomar

Abstract In today's world, traffic congestion has become a major issue in nearly all of the major cities around the globe. Due to an increase in the number of vehicles on the road, the intersections (or junctions) are major hotspots of congestion, so to maintain a proper flow, traffic intersections are installed with traffic lights that control the flow of the traffic moving in a particular direction. Traffic lights were first used nearly a century ago, but there have not been any major advancements in the functioning of traffic lights that can cope up with the requirements of today's traffic. Nowadays, due to a lack of optimization in the timings of traffic lights, they can further cause more delays and congestions which causes environmental issues (like increased CO_2 emissions and noise pollution), road rage, mental frustration, etc. This paper addresses the problem of traffic lights not having the optimized timings of signal for each direction due to the unavailability of robust, pro-active and real-time analysis of traffic and then analyzing and predicting the best timings of traffic lights based upon the size (small, medium, or large) and class (emergency and normal) of vehicles.

Keywords Machine learning · Traffic lights optimization · Edge server · Image processing · OpenCV · Classification

1 Introduction

Nearly 1.2 billion active vehicles are running on the roads of the world and are estimated to increase to two billion by 2035 [1]. Due to augmentation in the number of

P. S. Sethi · A. Jain
Dell EMC, Bangalore, India

S. Kumar
IIMT Group of Colleges, Greater Noida, India

R. Tomar (✉)
School of Computer Science, University of Petroleum and Energy Studies, Dehradun, India
e-mail: ravitomar7@gmail.com

V. Skala et al. (eds.), *Machine Intelligence and Data Science Applications*, Lecture Notes on Data Engineering and Communications Technologies 132,
https://doi.org/10.1007/978-981-19-2347-0_43

vehicles, the congestion on roads is one of the biggest problems faced by humanity. The increased number of vehicles causes more congestion which increases the waiting time of vehicles that leads to colossal environmental issues like greenhouse gas emission and global warming. The few existing methods used to manage congestion on roads are traffic lights at junctions, increased awareness to use public transport instead of personal vehicles, etc. currently most widely adopted and dependable method is the use of traffic lights. Traffic lights were first used in 1868 [2], and they were designed in context to the traffic requirements of that time. There have been few amendments to optimize the use of traffic lights to cope up with the requirements of today, but still, there is a requirement of much more efficient, rapid, and robust transformation which can help traffic lights better manage the traffic as per the upcoming traffic requirements and by taking various factors in consideration.

Traffic lights are a set of electrically operated signal lights with one of a set of the color red, yellow, and green lights controlling the traffic by signaling when vehicles have to stop and when they can go at road intersections, pedestrian crossings, railway crossing, and other locations. The timings are determined by the amount of traffic that goes through each day, and they may vary at different times of the day (the busiest road gets a longer green light). Modern traffic lights contain sensors that identify the incoming vehicles, which is especially beneficial for motorists traveling at night, as the sensor will turn their light green if alternative roads are clear. Traffic lights if not timed efficiently can cause huge traffic congestions on busy roadways. Traffic lights only depend on the vehicles approaching it directly and based upon this data the traffic controller (or controlling system) times the traffic light duration. This process is only dependent upon one set of data that is traffic approaching directly to traffic lights due to which the traffic controller is not able to estimate the correct time of traffic lights efficiently.

The above process can be highly optimized if the traffic controller is provided with data of vehicles coming to it from its previous traffic lights, that is, by the use of an interconnected system the upcoming traffic light is provided by data such that it is already aware of the amount of traffic released by the previous traffic light, this will help time the traffic in a much more optimized manner. Moreover, if any emergency vehicle approaches the traffic light, it would never have to stop using this approach. The work discussed in this paper proposes a similar approach.

The work discussed uses machine learning and image recognition using OpenCV technology to determine the type (two-wheeler, four-wheeler, etc.), the number of vehicles, and class (hatchback, SUV, and sedan) of vehicles at a given traffic light. Machine learning is an application of artificial intelligence (AI) that provides systems the ability to automatically learn and improve from experience without being explicitly programmed [3]. It is the process of parsing data, learning from it, and then making a decision or prediction about anything in the real world using algorithms (like linear regression, logistic regression, decision tree, SVM, naive Bayes, kNN, etc.). The recent advancements in the machine learning domain have aided the development of computer vision and image recognition technologies. Image recognition is the ability of a system or program to recognize objects, people, places, and activities in photographs. Image recognition is mostly done using OpenCV. Open source computer vision library (OpenCV) is a free, and open source computer vision library that comprises hundreds of algorithms [4]. Using OpenCV once these vehicles are identified, this data is fed to the edge server for processing. Edge servers are powerful computing devices placed at the network's "edge" where data computing is required. They are physically near to the systems or applications that generate the data that the server stores or uses.

2 Literature Review

Traffic congestion on roads is one of the biggest contributors to various phenomena like road rage, increase CO_2 emissions, and mental frustration in motorists. Over time there have been various approaches that are discussed to reduce traffic congestion using efficient timing of traffic lights [5] uses video and image processing to analyze the traffic coming at a junction from all four directions and then processing it to time the traffic lights. This approach provides efficiency over the old traffic lighting system which uses hardcoded light timings. On similar grounds, [6] uses YOLO and simple online and real-time tracking algorithm (SORT) detection techniques to track vehicles. This method uses deep learning, to analyze the environment around the traffic light to time the light efficiently, this model learns about the environment and analyzes it to make better judgments and predictions in the future, which helps in better configuration of green lights timings. This [7] method also analyzes the upcoming traffic using an adaptive traffic light control system by using the image processing technique and uses the Canny edge detection technique. This method having similar benefits as above also reduces the hardware requirements and enhanced image detection in various lights. Another method [8] uses double-phase traffic light fuzzy logic for analyzing the upcoming traffic and priced similar benefits as above. This paper [9] a multi-modal smart traffic control system (STSC) for smart city infrastructure, which can be widely used in smart city applications for intelligent transportation systems by timing the traffic lights efficiently. It proposes a design for EVSP scenario, it can alert all drivers at the junction when an EV is coming, improving traffic flow, and increasing safety without incurring additional hardware expenditures. Traffic lights if timed properly can be used to clear out traffic for emergency vehicles efficiently without disturbing the flow of traffic, this [10] proposes a novel approach for timing the traffic lights using various sensors such that emergency vehicles can pass as soon as possible. In [11], author presents a machine learning model to optimize and regulate the flow of traffic in Hong Kong city, it uses an intelligent traffic light system (ITLS)-based machine learning model to analyze a pre-recorded video and then tries to generate a simulation that predicts the numbers of vehicles and pedestrians on the road. It then compares the ITLS model to the fixed cycle traffic light (FCTL) system and provides a result depicting optimized waiting time for vehicles and pedestrians. In [12], author discusses the rising traffic problems in Lebanon and proposes an adaptive traffic light system implemented using reinforcement learning, tested using data from Lebanese traffic. The proposed model showed a significant reduction in traffic length queues. After discussing the above describes methods, we can concur that their methods work on the upcoming traffic to the traffic light and do not take into consideration the amount of traffic released by the previous traffic light. The approach discussed in this paper discusses a method to use the data from previous traffic lights to time the traffic light efficiently for the coming traffic to optimize the stopping time of vehicles based on various factors and providing privilege to emergency vehicles along with over speeding detection [13, 14].

3 Methodology

The approach we present in the paper addresses one of the major limitations of existing models, which is analyzing the data beforehand even before the vehicle has approached the traffic light so that the signal can be timed in an appropriate manner and in a sufficient amount of time.

The above image (Fig. 1) describes a set of four junctions where at each junction four roads meet. Each junction has its traffic light which controls the movement to

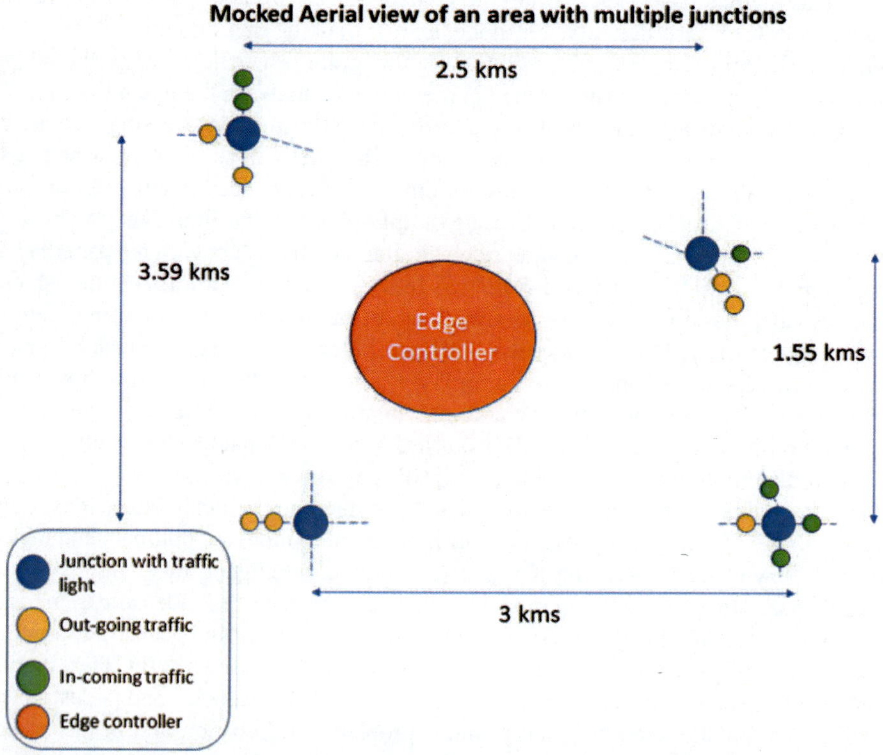

Fig. 1 Aerial view of four junctions of traffic light connected to edge controller

and fro from the junction. All the junctions are connected to the edge controller [15] which is the processing unit of the entire system. Each of the junction traffic lights will have a camera that will take real-time images and send them back to the edge server for processing.

Let's suppose a certain number of cars are passing through traffic light A, then traffic light A will capture images of the cars and will send the images back to the edge server for processing. Edge server upon receiving the images uses image processing [16] using OpenCV [4] to find out various factors like the number of cars, type of cars, and class of cars. Once these factors are mapped edge server pulls various other parameters like average time taken by a car to reach the next traffic light, the distance between the traffic lights, data received from other directions of the same traffic light, etc. All this information is fed to the machine learning algorithm to determine the best signal timing for the next upcoming traffic light on that route. Then, this information is sent to the next upcoming traffic light on the route so that even before the cars have reached the next traffic light, the traffic light is having the data about the best timing of signal lights. When cars reach the next traffic light, the signal is ready with the optimized signal. Upon arriving at any junction, the above-discussed process restarts.

Let's suppose an emergency, vehicle is traveling on a route and it has crossed traffic light A at a junction, then this image will be captured by the traffic light at junction A, and this information will be sent back to the edge server. Edge server upon receiving the image will use image processing to determine that is an emergency vehicle, once it is known by the edge server that it is an emergency vehicle, the edge server calculates the best signal timings for the next upcoming signal by supplying various factors to machine learning model which determines the best signal timing. Then, edge server sends the information of best signal timings to the next upcoming traffic light to let the emergency vehicle pass as early as possible [17, 18].

4 Implementation

The implementation of the above-discussed approach uses four stages to complete the flow, where based on the data from traffic light A the optimized signal timing to traffic light B is sent. The stages are discussed below:

4.1 Image Captured by the Traffic Light at a Junction

As shown in Fig. 2, the process starts when vehicles approach the traffic light at a junction, where the camera installed at the traffic light continuously clicks pictures of the traffic passing through and sends them to the edge server. The picture is clicked at the traffic light whenever the color of the traffic light changes to green because at that time vehicles can start moving in their particular direction, and in this way, based on the images OpenCV can easily judge in which direction vehicles are moving.

Fig. 2 Illustration of connection between traffic light at a junction and edge controller

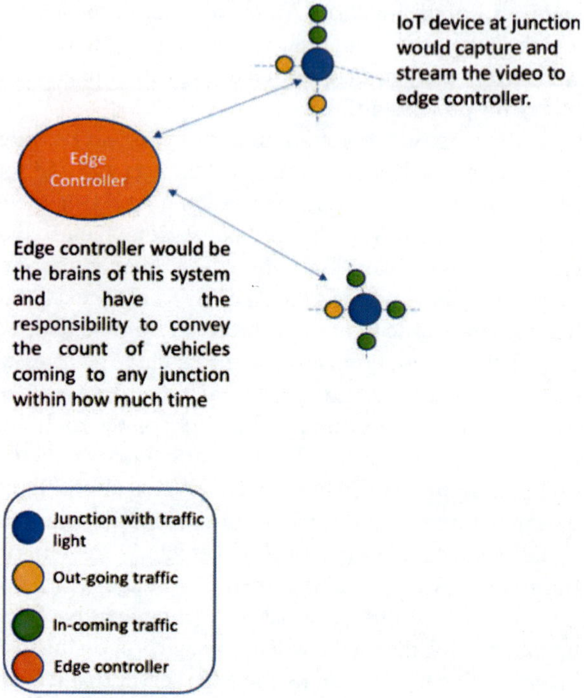

4.2 The Image Received at Edge Controller

An edge controller is a centralized edge server that is connected to a group of junction traffic lights. Edge controller uses image processing using OpenCV library to classify the image and identify various factors like:

(a) Type of vehicle—two-wheeler, four-wheeler, trucks, lorries, etc.
(b) Class of vehicle—SUV, sedan, mini-truck, hatchback, etc.
(c) Number of vehicles—number of vehicles crossing the junction moving in the outgoing or incoming direction.

The model broadly classifies three main types of information, first, the number of vehicles of each class (like big vehicles, medium-sized vehicles, and small-sized vehicles in Table 1), second, the direction of the vehicles they are moving toward, and third will be how many emergency vehicles are moving toward the next traffic

Table 1 Classification of vehicles using OpenCV

Vehicle type	Vehicles
Big	Trucks, bus, SUV's, pick up, tractors
Medium	Hatchback, mini SUV's, sedan
Small	two-wheelers, auto-rickshaw, three-wheelers

Table 2 List of classified emergency vehicles using OpenCV

S. No.	Emergency vehicles
1	Police vehicles
2	Ambulance
3	Embassy vehicles
4	Bank vehicles
5	Fire trucks
6	Press vehicles
7	Administrative vehicles

lights illustrated in Table 2. This information helps in the collection of data required by the machine learning model to determine the optimized time for signal light.

Once the above factors are determined the edge controller pulls out the static information stored in its database like:

(a) Average time taken to travel from traffic light A to traffic light B.
(b) Distance between traffic light A and traffic light B.
(c) The number of exits between the traffic lights.
(d) Average time required for light to be kept green in a particular direction at the junction.
(e) The sequence in which each direction of the traffic light is turned green.
(f) The number of vehicles that can pass through the traffic light in the maximum value of time the light can stay green.

After collecting this information, the data is fed to a pre-trained machine learning model which analyzes the parameters and gives an optimized time for signal light. The machine learning model provides the best optimized time for each traffic light based upon the sequence in which they turn green. The time allotted to each direction of a traffic light is based on how much traffic is at each direction of the traffic light, if let's suppose in direction A there are ten cars, in direction B there are 100 cars, direction C there are 20 cars, and direction D there are five cars, so the model will take into consideration the number of vehicles in each direction of the traffic light and will allocate the time required by each direction accordingly in the sequence that is fixed, clearly for above case vehicles coming from the direction B will have the highest light time as green because if more vehicles are waiting to pass and are stopped at a traffic light for more time it will lead to more environmental issues like noise pollution, CO_2 emissions, extra fuel consumption, etc., as compared to fewer vehicles waiting for little longer time. The model will also consider the number of vehicles that can pass through the traffic light in the maximum allowed value of light that it can remain green in one direction. The entire process restarts as soon as vehicles from the direction of traffic lights crosses and the time required for each traffic light is calculated again based on the number of vehicles. The sequence in which the lights turn green will not change until there is an emergency vehicle approaching from any direction of the light. In that scenario, the edge server will calculate the approximate time when the emergency vehicle will arrive at the next light so that the light can stay green at that time and the emergency vehicle does not have to wait.

As soon as any camera detects an emergency vehicle using OpenCV, its direction is determined, that is, in which direction vehicles in moving. Once the direction of the emergency vehicle is determined at a junction, edge controlled identifies the type of emergency vehicles (listed in Table 2) and pulls out the average speed at which that type of emergency vehicle moves, from the database. Then, the edge controller calculates an estimated time (using the average speed of the vehicle and distance between two junctions, i.e., Time = Distance/Speed) at which the emergency vehicle will reach the next junction in the direction in which it is moving. Once the time is calculated, the edge controller times the next junction such that at the calculated time the light of the next junction will be green such that the emergency vehicle can pass through without stopping. This helps to ensure that emergency vehicles are slowed to a minimum and can pass through as fast as possible.

4.3 Sending the Optimized Time to a Traffic Light

Figure 3 demonstrates the type of data structure that will be used for communication between the edge server and traffic signal junctions. It will be a JSON object which will tell the traffic light which time to display on which side of the intersection on the traffic light signal. There will be two separate JSON for outgoing and incoming signals, which will have the count of each vehicle type and coming from which direction.

4.4 Displaying the Signal of the Traffic Light

Upon receiving the JSON, the small IoT device will read the JSON and display the signals on each side of the intersection for provided time.

5 Conclusion

The approach described in this paper solves a major setback of existing models available for traffic lights, that is, pro-actively determining the timing for traffic lights even before the vehicles have reached the traffic light using the robust and real-time analytical engine deployed at an edge server which uses machine learning and image processing using OpenCV. The analytical engine uses the data of images captured by previous traffic lights, by the use of this data the image processing at edge server can classify vehicles of different sizes (as different sizes of vehicles take different times to cover a distance) and accordingly determine the time required by

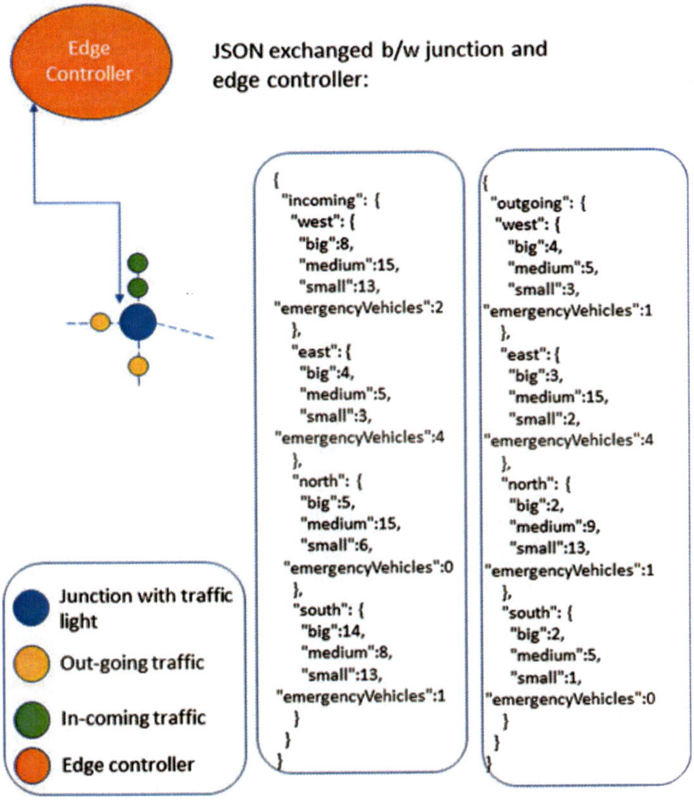

Fig. 3 JSON data structure exchanged between traffic junction and edge controller

each type of vehicle to reach the next traffic light so that best optimized time can be used by the traffic light. The model also classifies emergency vehicles and always prioritizes that in case of an emergency vehicle the timings are analyzed and modeled in such a way the emergency vehicle does not have to wait at any traffic signal and passes through as early as possible. The lesser vehicles have to wait at traffic lights the lesser are the issues like increased CO_2 emissions, consumption of fuel, noise pollution, road rage, etc.

References

1. 1.2 billion vehicles on world's roads now, 2 billion by 2035: report. https://www.greencarreports.com/news/1093560_1-2-billion-vehicles-on-worlds-roads-now-2-billion-by-2035-report. Accessed 24 Jul 2021
2. The history of traffic lights—lighting equipment sales. http://lightingequipmentsales.com/the-history-of-traffic-lights.html. Accessed 24 Jul 2021

3. What is the definition of maequal rights chine learning?|Expert.ai|Expert.ai. https://www.expert.ai/blog/machine-learning-definition/. Accessed 24 Jul 2021
4. Home—OpenCV. https://opencv.org/. Accessed 24 Jul 2021
5. Kanungo A, Sharma A, Singla C (2014) Smart traffic lights switching and traffic density calculation using video processing. Recent Adv Eng Comput Sci RAECS 2014. https://doi.org/10.1109/RAECS.2014.6799542
6. Sharma M, Bansal A, Kashyap V, Goyal P, Sheikh TH (2021) Intelligent traffic light control system based on traffic environment using deep learning. IOP Conf Ser Mater Sci Eng 1022(1):012122. https://doi.org/10.1088/1757-899X/1022/1/012122
7. Meng BCC, Damanhuri NS, Othman NA (2021) Smart traffic light control system using image processing. IOP Conf Ser Mater Sci Eng 1088(1):012021. https://doi.org/10.1088/1757-899X/1088/1/012021
8. Bagheri E, Bagheri E, Faizy M (2007) A novel fuzzy control model of traffic light timing at an urban intersection
9. Lee W-H, Chiu C-Y (2020) Design and implementation of a smart traffic signal control system for smart city applications. Sensors 20(2):508. https://doi.org/10.3390/S20020508
10. Srinath S, Bhandari SS (2014) A novel approach for emergency vehicle detection and traffic light control system. Sci Technol Eng 3:400–405. Accessed: 24 Jul 2021. [Online]. Available: www.erpublications.com
11. Ng S-C, Kwok C-P (2020) An intelligent traffic light system using object detection and evolutionary algorithm for alleviating traffic congestion in Hong Kong. Int J Comput Intell Syst 13(1):802–809. https://doi.org/10.2991/IJCIS.D.200522.001
12. Natafgi MB, Osman M, Haidar AS, Hamandi L (2018) Smart traffic light system using machine learning. In: IEEE international multidisciplinary conference on engineering technology IMCET. https://doi.org/10.1109/IMCET.2018.8603041
13. Joshi T, Badoni P, Choudhury T, Aggarwal A (2019) Modification of Weiler-Atherton algorithm to address loose polygons. J Sci Ind Res 78:771–774
14. Goel S, Sai Sabitha A, Choudhury T, Mehta IS (2019) Analytical analysis of learners' dropout rate with data mining techniques. In: Rathore V, Worring M, Mishra D, Joshi A, Maheshwari S (eds) Emerging trends in expert applications and security. Advances in intelligent systems and computing, vol 841. Springer, Singapore. https://doi.org/10.1007/978-981-13-2285-3_69
15. What are edge servers and how are they used?|OnLogic|Edge Servers. https://www.onlogic.com/company/io-hub/what-are-edge-servers/. Accessed 13 Jul 2021
16. 1. Introduction to image processing|Digital Image Processing. https://sisu.ut.ee/imageprocessing/book/1. Accessed 24 Jul 2021
17. Tomar R, Sastry HG, Prateek M (2020) A V2I based approach to multicast in vehicular networks. Malays J Comput Sci 93–107. Retrieved from https://jupidi.um.edu.my/index.php/MJCS/article/view/27337
18. Jain S, Sharma S, Tomar R (2019) Integration of wit API with python coded terminal bot. In: Abraham A, Dutta P, Mandal J, Bhattacharya A, Dutta S (eds) Emerging technologies in data mining and information security. Advances in intelligent systems and computing, vol 814. Springer, Singapore. https://doi.org/10.1007/978-981-13-1501-5_34

Forecasting of COVID-19 Trends in Bangladesh Using Machine Learning Approaches

Chayti Saha, Fozilatunnesa Masuma, Nayan Banik, and Partha Chakraborty

Abstract Bangladesh, a low-to-middle income economy country with one of the world's densest populations, is ranked 26th worldwide having a positive case rate of 25%-30% of COVID-19 confirmed instances as of July 28, 2021. The recent researches related to COVID-19 focus on addressing mental health problems caused by it, and fewer works have been performed to forecast its trends using machine learning (ML), especially in Bangladesh. Therefore, this research attempts to predict the infected, death, and recovery cases for COVID-19 in Bangladesh using four ML techniques FB Prophet, ARIMA, SARIMAX, LSTM and com pare their forecasting performance to find out the best prediction model. The experimental results showed that for 'Detected' and 'Death' case, LSTM and SARIMAX performed better than other models with (RMSE = 1836.79, MAE = 1056.36) and (RMSE = 24.70, MAE = 15.54), respectively. In the 'Recovery' case, the best result was found in the ARIMA model with RMSE = 558.87, MAE = 299.64. According to the analysis, this research work can help predict the trends of COVID-19 cases in the future and help policymakers taking necessary precautions to control the detection and death rate in the country.

Keywords Forecasting · COVID-19 · Machine learning · Trend analysis · Bangladesh

1 Introduction

Coronavirus disease 2019 (COVID-19), a contagious disease that can be passed from person to person, is the result of a new coronavirus. The number of infected patients around the world is increasing daily at an alarming exponential growth rate.

C. Saha (✉) · F. Masuma · N. Banik · P. Chakraborty
Department of Computer Science and Engineering, Comilla University, Cumilla 3506, Bangladesh
e-mail: chaytisaha98@gmail.com

P. Chakraborty
e-mail: partha.chak@cou.ac.bd

V. Skala et al. (eds.), *Machine Intelligence and Data Science Applications*, Lecture Notes on Data Engineering and Communications Technologies 132,
https://doi.org/10.1007/978-981-19-2347-0_44

In Bangladesh, infected patients were first found on March 08, 2020, and since then, it is engulfing the whole country.

Situation data and other insights are crucial in COVID-19 pandemic situations because by analyzing those data, policymakers can understand the trends of this virus and take necessary preventive measures like which area to put under strict lockdown, which area is less risky, or which area has a sudden growth of infected cases, etc. As a result, researchers are continuously trying to forecast the trends like detection of infected patients, death rate, and recovery rate using different ML approaches on collected data. Advancing that knowledge, four ML-based forecasting models, Facebook Prophet (FB Prophet), autoregressive integrated moving average (ARIMA), seasonal autoregressive integrated moving average with eXogenous factors (SARIMAX), and long short-term memory (LSTM) were investigated, and their comparative analyses were performed in this work.

The delineation of the paper starts with the existing works done by researchers in Bangladesh perspectives in Sect. 2. Proposed methodologies with dataset description are provided in Sect. 3. The trend analysis using models is given in Sect. 4. Obtained results and analysis on them are covered in Sect. 5, and the paper culminates in Sect. 6 with future references.

2 Literature Review

The increasing consequences of the COVID-19 global pandemic draw the attention of the research community due to the prompt update from country-level data [5]. This growing data from Bangladesh's perspective also requires some works in forecasting the trends of this disease, but the existing literature shows descriptive analysis with limited applications of ML-based approaches. In [17], the authors discussed the financial situation regarding the pandemic and proposed developing SMEs and their management. The public health challenges and the government's steps to manage the unprecedented condition were analyzed in [8, 13]. Similar to that, authors in [12] analyzed the divisional gender and age vulnerability of COVID-19 till June 2020 using ML techniques. Authors in [10] used adaptive neuro-fuzzy inference system (ANFIS) and LSTM to predict the newly infected cases in Bangladesh, and after comparing the results, they claimed that the LSTM performed better than ANFIS. Additionally, authors in [16] showed that FB Prophet model outperformed support vector regression (SVR), Holt's linear regression (HLR), Holt's winter additive model (HWAM), etc. in forecasting the pandemic situations. Authors in [19] developed a ML-based interactive forecasting web portal for Bangladesh that provides a present indication, trend analysis and predicts the possible number of detecting and death cases of COVID-19 in the future. They applied ML-based regression models such as linear regression, polynomial regression, SVR, multilayer perception, polynomial multilayer perception, ridge regression, Holt linear, Holt winter, ARIMA, and FB Prophet models to inspect and predict COVID-19 cases. The authors in [7] used various forecasting models, such as the regression model, the nonlinear exponential regression model, and some time series models based on daily positive, deceased, recovered cases, and some environmentally important variables, such as average daily temperature and daily humidity, to forecast daily active COVID-19 cases for Bangladesh. According to their findings, time series forecasting models are superior at predicting daily active cases. Depending on data gathered via online surveys, the authors in [14] developed a binary logistic regression model to examine the influence of the COVID-19 concern in Bangladesh. By using the ARIMA model, they were able to estimate the next 40 days of COVID-19 confirmed deaths. The authors of paper [15] proposed combining the unscented Kalman filter (UKF) with the traditional SIRD model to explain epidemic dynamics in individual districts across the country. They showed that the UKF-SIRD model can predict transmission dynamics for 1–4 months with a high degree of accuracy. They used the robust UKF-SIRD model to forecast the epidemic's progress in various parts of Bangladesh. The authors of paper [18] have given a review and brief analysis of the most relevant machine learning forecasting models against COVID-19 which were used in many countries like South Africa, Hubei Province, India. The vibrant activities of the aforementioned research domain indicate that further investigation is required in terms of the forecasting of COVID-19 trends in Bangladesh.

3 Methodology

The detailed workflow of the experiment is outlined in this section with the description of data collection first. Minimal preprocessing was performed as the data were free from any noise and on a similar scale. The four experimental models were described with their working procedures, and the employed evaluation metrics were defined. The high-level overview of the process is depicted in Fig. 1.

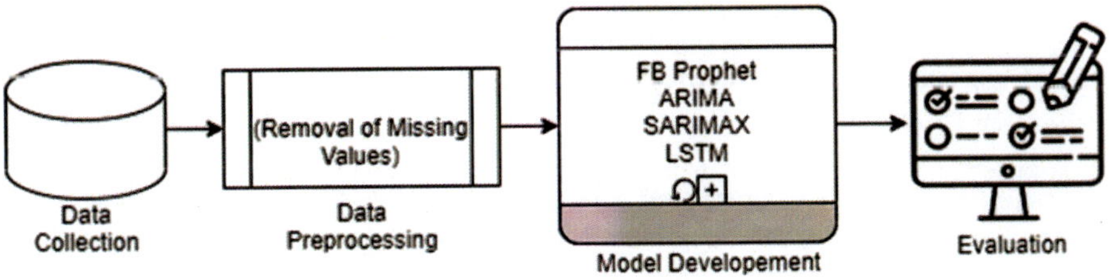

Fig. 1 Overview of the proposed system workflow

Table 1 Dataset statistics

Property	Count
No. of samples (days)	512
No. of attributes	7
Minimum value in 'Detected per day'	0
Average value in 'Detected per day'	1902
Maximum value in 'Detected per day'	8822
Minimum value in 'Death per day'	0
Average value in 'Death per day'	30
Maximum value in 'Death per day'	119
Minimum value in 'Recovery per day'	0
Average value in 'Recovery per day'	1678
Maximum value in 'Recovery per day'	7266

3.1 Data Collection

The data were collected from several online resources including COVID-19 Information Bangladesh [3], Worldometer Bangladesh COVID Live Update [1], Coronavirus COVID-19 Dashboard, 2020 by Directorate General of Health Services (DGHS) [2], and live Facebook situation briefing by Institute of Epidemiol ogy Disease Control and Research (IEDCR) [4]. The compiled dataset contained datewise COVID-19-related data from March 8, 2020 (the day when the first COVID-19 patient was detected in Bangladesh) to July 31, 2021 and was further cross-checked from the aforementioned sources. Summary statistics of the dataset are provided in Table 1.

3.2 Data Preparation

The raw data containing missing values might degrade system performance and result in unpredictable problems. So, it is required to analyze the data to remove them. Few missing values were found only in the field 'Recovery per day'. All of those missing values were filled with an average of seven days with the missing values in middle in the same column. Since all the values of the dataset were in discrete range without any outlier, scaling was not performed. In total, around 70% and 30% of the data have been used for training and testing purpose.

3.3 Model Descriptions

In this work, four ML-based forecasting models FB Prophet, ARIMA, SARIMAX, and LSTM were used to predict the three cases (Detected case, Death case, Recovery case) of the dataset. The working principles and properties of each model were described accordingly.

FB Prophet FB Prophet makes it much more straightforward to create a reasonably accurate forecast that includes many different forecasting methods such as ARIMA, smoothing with an exponential function. It flexibly fits the trend component, accurately models seasonality, and results in a robust forecast. Moreover, this model handles missing data more gracefully.

Despite the benefits, FB Prophet cannot perform well in highly irregular time series with exogenous variables. Furthermore, the algorithm requires data to be fed in a predefined format, and it does not work with multiplicative models [20].

FB Prophet is used to forecast data in time series depending on an additive model where seasonality has holiday impacts on nonlinear patterns, which are fit on an annual, weekly, and daily basis. The additive regression model uses a decomposing time series model containing three key components named trend, seasonality, and holidays. These can be calculated through the equation below:

$$y(t) = g(t) + s(t) + h(t) + \epsilon t \ (1) \tag{1}$$

where $g(t)$ denotes the modeling non-periodic changes in time series with a logis tic growth curve, $s(t)$ denotes periodic changes (e.g., weekly/annual seasonality), $h(t)$ denotes the effects of holidays (user provided) with irregular schedules, and ϵt adjusts for any unexpected changes that the model does not account for.

ARIMA and SARIMAX ARIMA describes a given time series depending on its prior values, such as its own lags and lagged prediction mistakes, in order to anticipate future values with the equation. It is applicable to any time series that is not seasonal which exhibits patterns and is not just random white noise. An ARIMA model is distinguished by three terms called p, d, and q, where p is the sequence of the AR term, q is the sequence of the MA term, and d denotes the number of deviations necessary to render the time series stationary. If seasonal trends exist in a time series with exogenous factors, the addition of seasonal terms makes it SARIMAX [9].

A pure AR paradigm is one in which Y_t is solely reliant on its own lags.

Specifically, Y_t is defined by the 'lags of Y_t:

$$Y_t = \alpha + \beta_1 Y_{(t-1)} + \beta_2 Y_{(t-2)} + \ldots + \beta_p Y_{(t-p)} + \epsilon_1 \tag{2}$$

where $Y_{(t-1)}$ is the series' first lag, β_1 denotes the coefficient as per the estimation of the model and α denotes intercept term, likewise calculated by the model. As well, a pure moving average (MA) model is that one where Y_t is solely determined by the lag prediction errors.

LSTM The LSTM is a type of recurrent neural network that may learn order dependency in prediction of sequence issues. It is trained using backpropagation through time. We can use it to process, predict, and classify based on time series data. Memory blocks, rather than neurons, are used in LSTM networks, which are linked by layers. Each block has components and recent sequences, stored in memory. A block additionally has gates that control its current condition and output. Every gate within a block uses the sigmoid activation units to regulate the input sequences. It causes a state change and the addition of input information to flow via the conditional block [11].

All of these gates make LSTM models much better than typical convolution feedforward networks or RNNs at sequence predictions because they are able to store past information. This is important in this work as the previous cases of Detected, Recovery, and Death are crucial in predicting future cases of them. Because memory-bandwidth-bound computation is required, and solutions based on neural networks are limited in their applicability, LSTMs are challenging to train.

3.4 Evaluation Metrics

To evaluate the proposed system's dependability, three metrics of evaluation, root mean squared error (RMSE), mean squared error (MSE), and mean absolute error (MAE), were utilized for each of the models. RMSE follows a normal distribution on a hypothesis that errors are unbiased. This metric measures the differences between the actual X_i and the predicted $(X^{\hat{}})$ numbers of COVID-19 confirmations, recoveries, and deaths. The main advantage of RMSE is that it penalizes large prediction errors.

$$\text{RMSE} = \sqrt{\sum_{i=1}^{N} \left(x_i - \hat{x}_i\right)^2} \tag{3}$$

MSE firstly measures the square of the forecast error, then finds the average of it. That is why for the same data, MSE is larger. It can be defined generally as follows:

$$\text{MSE} = \frac{\sum_{i=1}^{N} \left(X_i - \hat{X}_i\right)^2}{N} \tag{4}$$

Though RMSE and MSE are good at performance measuring, outliers may control the resulting error of both RMSE and MSE, as the square of the error is a part of the main calculation.

$$\text{MAE} = \frac{\sum_{i=1}^{N} \left|\left(X_i - \hat{X}_i\right)\right|}{N} \tag{5}$$

4 Trend Analysis

For FB Prophet, cross-validation was done for the cases (Detected, Death, Re covery) considering the parameters (initial = '358 days', period = ''20 days', horizon = '40 days'). Thus, 6 forecasts with cutoffs between 2021–03-13 and 2021–06-21 were obtained. Then future date frames for 150 days were created, respectively. Based on the future date frames, Detected, Death, and Recovery cases were predicted by different components for the cases separately. Prediction plots (from March 08, 2020, to July 31, 2021) are given in Fig. 2. In Fig. 2, (a)–(c) depicted the predicted projection of Detected, Death, and Recovery cases. In those plots, the black dots represented the actual Detected, Death, and Recovery cases, respectively. Blue dots represented the prediction for the respective cases (yhat), and the light blue color plot represented the range of prediction (yhat-upper, yhat-lower). It was clear from the blue plot that the future detected case was upward.

For ARIMA/SARIMAX, the data were checked whether it was stationary or not. To check this, both rolling statistics (plotting the rolling mean and rolling standard deviation) and augmented Dickey-Fuller (ADF) test were carried out. The outcome of the ADF test is given in Table 2.

From Table 2, as weak evidence against the null hypothesis was obtained, and the time series had a unit root, it was indicating the dataset to be non-stationary.

The non-stationary data were converted into stationary using difference. The 'auto arima' function from the 'pmdarima' library then assisted in determining the most

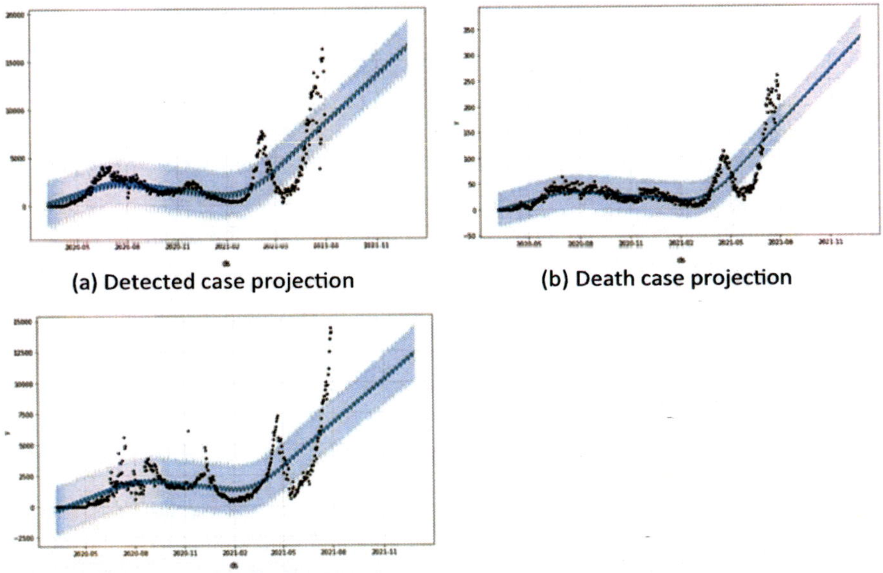

(a) Detected case projection

(b) Death case projection

(c) Recovery case projection

Fig. 2 Predicted projection and components of cases using FB Prophet

Table 2 Results of ADF tests for three cases

Case	ADF test statistics	*P*-value	Lags used	No. of observations
Detected	−0.506009702269	0.8908383982	16	494
Death	0.525575943359	0.9856300717	18	495
Recovery	0.590183936245	0.9873711644	15	492

optimum ARIMA, and SARIMAX models for all three cases and returned the fitted ARIMA models and SARIMAX models. The best model for Detected case ARIMA (0,0,0) and SARIMAX (0,0,0)(2,1,1)[30], the best model for Death case ARIMA (0,0,0) and SARIMAX (0,0,1)(2,1,0)[30], the best model for Recovery case ARIMA (3,0,0) and SARIMAX (3,0,0)(2,1,0)[30] were recorded, and based on these models, the results for desired cases were predicted. Actual forecast and predicted forecast using ARIMA model are shown in Fig. 3. From Fig. 3, in (a), (c), and (e), the actual data from March 08, 2020, to July 31, 2021 were shown in blue color plots, whereas the orange color plots denoted the forecasted data. The trends showed that the ARIMA model generalized well on past data. In (b), (d), and (f), the orange color plot represented the forecasted cases from July 2021 to the next 5 months. The rigid shape of orange lines showed that the model performed poorly on prediction. Actual forecast and predicted forecast using SARIMAX model are shown in Fig. 4 wherein (a), (c), and (e), the actual data from March 08, 2020, to July 31, 2021 were shown in blue color plots, whereas the orange color plots denoted the forecasted data. In (b), (d), and (f), the orange color plot represented the forecasted cases from July 2021 to the next 5 months. The trends showed that the SARIMAX model generalized well both on past data and in predicting the future. For LSTM network, there were 100 epochs, batch size of 1. The prediction plot for three cases using LSTM model is shown in Fig. 5.

In Fig. 5, the blue color plot represented the actual detected case using the LSTM model from March 08, 2020, to July 31, 2021, the orange color plot represented the predicted detected case for the training dataset, and the green color plot represented the predicted detected case for testing data, set. The plot for prediction on the testing data set is mixed with downward and upward shapes.

5 Result and Discussion

The evaluative results of four models FB Prophet, ARIMA, SARIMAX, and LSTM are represented in Table 3.

From Table 3, it is clear to see that for the Detected case, LSTM performed better than other models with RMSE and MAE values of 1836.79 and 1056.36, respectively. The other models, in this case, are SARIMAX, ARIMA, and FB Prophet, respectively. According to paper [6], the ability of this network to ac quire long-term dependencies is its most notable characteristic, and the forget gate allows the network

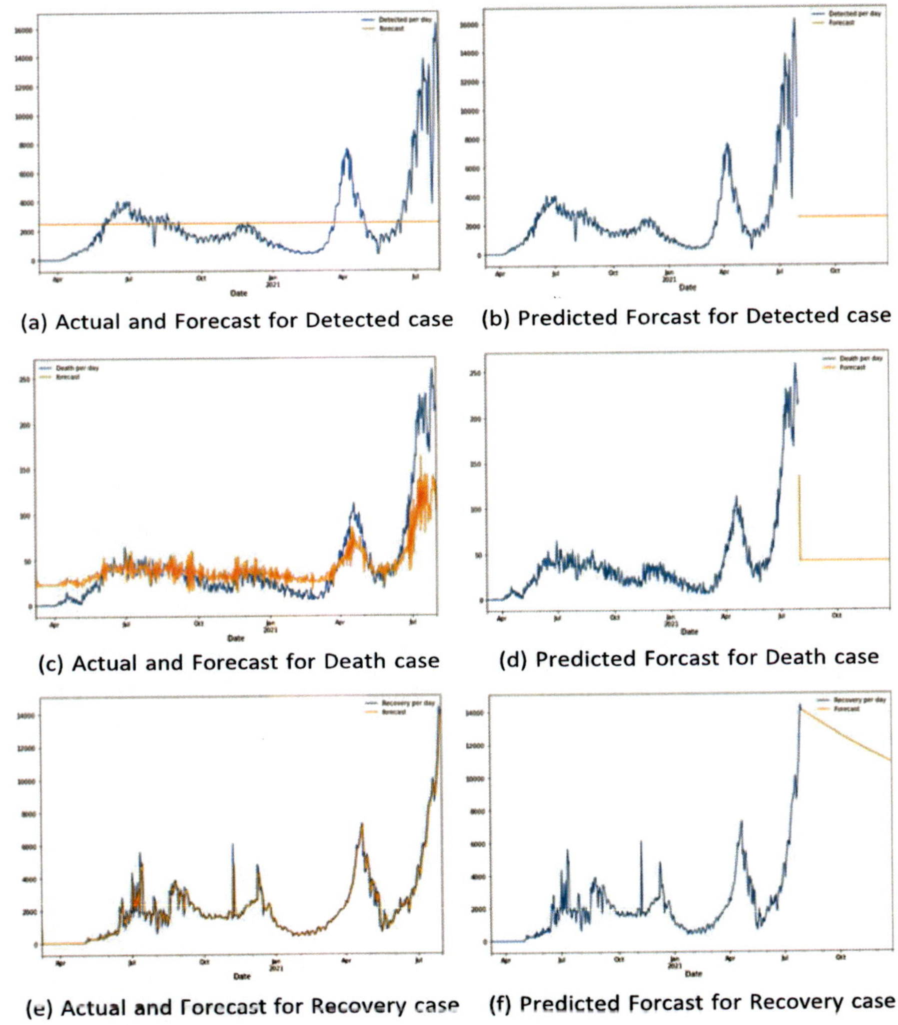

(a) Actual and Forecast for Detected case (b) Predicted Forcast for Detected case

(c) Actual and Forecast for Death case (d) Predicted Forcast for Death case

(e) Actual and Forecast for Recovery case (f) Predicted Forcast for Recovery case

Fig. 3 Actual and predicted forecasting using ARIMA

to maintain or forget the necessary amount of past information, and as a result, it aids in the improvement of the modeling. For the Death case, SARIMAX performed best with RMSE and MAE of 24.70 and 15.54 respectively. The rest in the ranking were ARIMA, LSTM, and FB Prophet.

For the Recovery case, ARIMA performs better with RMSE and MAE values of 558.87 and 299.64, respectively, and the rest are SARIMAX, LSTM, FB Prophet. ARIMA is one of the most acceptable models, despite the fact that the transmission of the virus is impacted by various factors that may regulate the forecasts [21].

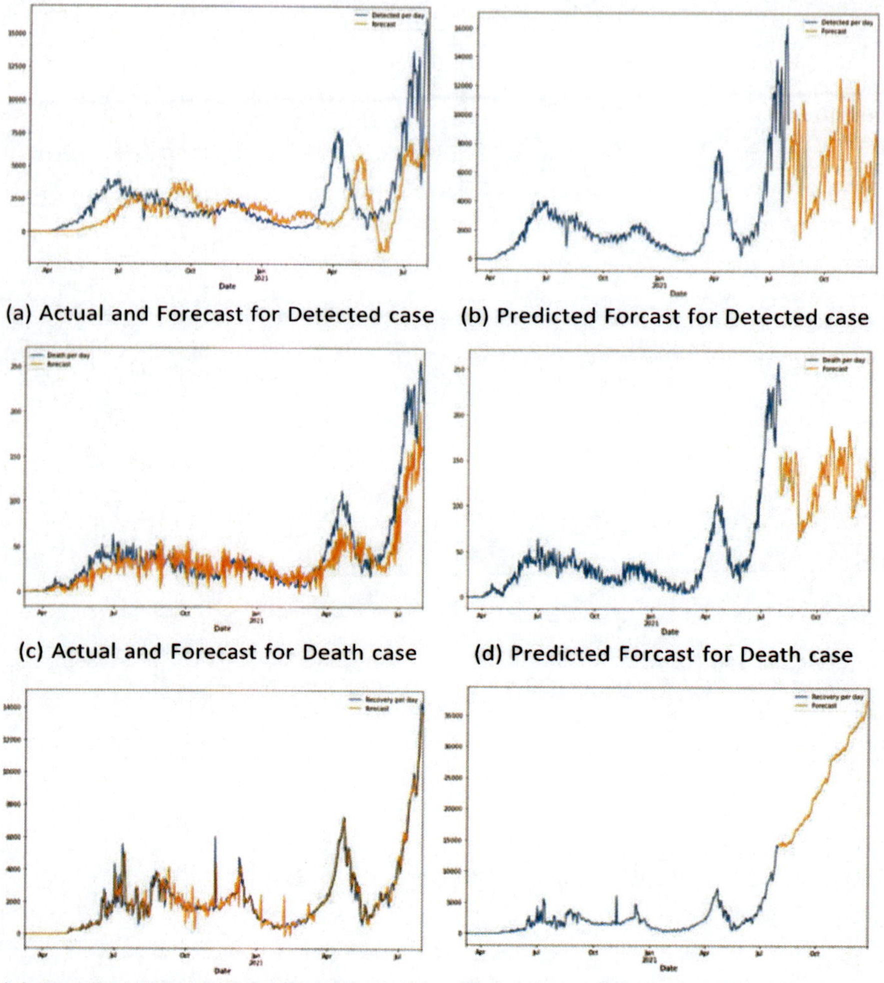

(a) Actual and Forecast for Detected case (b) Predicted Forcast for Detected case

(c) Actual and Forecast for Death case (d) Predicted Forcast for Death case

(e) Actual and Forecast for Recovery case (f) Predicted Forcast for Recovery case

Fig. 4 Actual and predicted forecasting using SARIMAX

6 Conclusion

Combating COVID-19, the global pandemic, requires insight from historical data. Forecasting trends using machine learning approaches provide action- oriented insights that help policymakers take prompt decisions to handle critical situations. Among many ML-based forecasting models, four widely used algorithms, FB Prophet, ARIMA, SARIMAX, and LSTM, were investigated in this work. The comparative analysis showed that depending on different cases like Detected, Death, and Recovery, certain algorithms performed better than others. From Bangladesh's

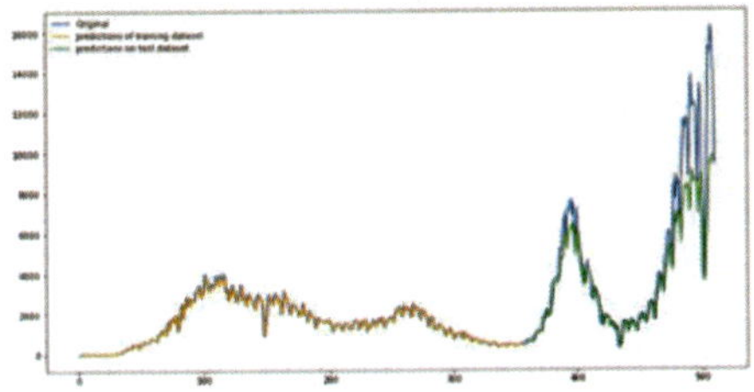

(a) Detected case

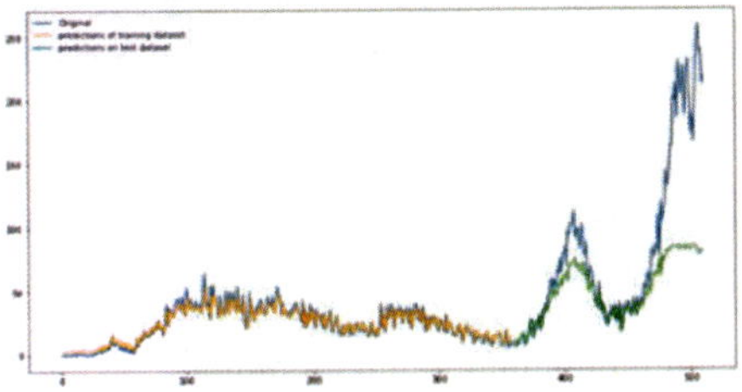

(b) Death case

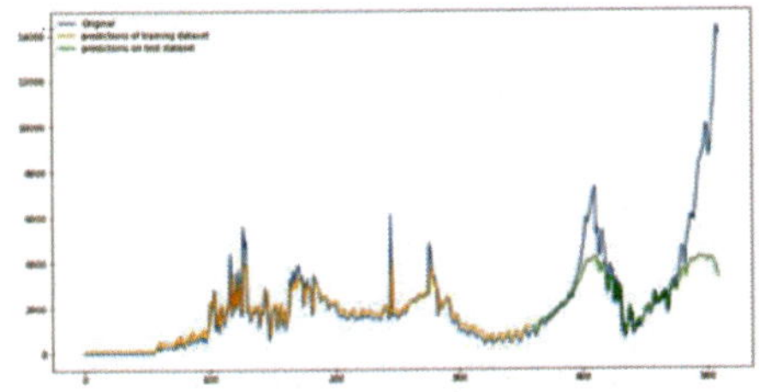

(c) Recovery case

Fig. 5 Prediction for three cases using LSTM

Table 3 Performance of forecasting models

FB Prophet				ARIMA		
	MSE	RMSE	MAE	MSE	RMSE	MAE
Detected	20910024.98	4450.32	3760.20	7269790.40	2696.25	1736.59
Death	5042.39	68.08	57.65	801.01	28.30	18.05
Recovery	11627030.41	3240.48	2760.96	312337.72	558.87	299.64
SARIMAX				LSTM		
	MSE	RMSE	MAE	MSE	RMSE	MAE
Detected	5174991.93	2274.86	1498.04	3373783.37	1836.79	1056.36
Death	610.01	24.70	15.54	3138.88	56.03	32.56
Recovery	383317.26	619.13	365.50	5910474.73	2431.15	1280.24

perspective, the trends are better captured by the LSTM and the ARIMA/SARIMAX models. The detailed analysis in this work will help the research in this domain to step forward.

References

1. Bangladesh covid live update-worldometer. https://www.worldometers.info/coronavirus/country/bangladesh/. Accessed on 08 Mar 2021
2. Coronavirus covid-19 dashboard, 2020 by dghs, bd. http://dashboard.dghs.gov.bd/webportal/pages/covid19.php. Accessed on 08 Mar 2021

3. Coronavirus disease 2019 (covid-19) information Bangladesh—corona.gov.bd. https://corona.gov.bd/press-release. Accessed on 08 Mar 2021
4. Iedcr, covid-19 control room—facebook. https://www.facebook.com/IedcrCOVID-19-Control-Room-104339737849330/. Accessed on 08 Mar 2021
5. Ahmed I, Bhuiyan MEM, Helal MSA, Banik N (2020) Hybrid instruction: Post covid-19 solution for higher education in Bangladesh. Int J Modern Trends Sci Technol 6(10):20–25
6. Apaydin H, Feizi H, Sattari MT, Colak MS, Shamshirband S, Chau KW (2020) Comparative analysis of recurrent neural network architectures for reservoir inflow forecasting. Water 12(5):1500
7. Banik S, Answer M, Kibria B (2021) Prediction of daily covid-19 active cases in Bangladesh. Int Jr Infect Dis Epidemlgy 2(1):25–33
8. Begum M, Farid MS, Alam MJ, Barua S (2020) Covid-19 and Bangladesh: socio- economic analysis towards the future correspondence. Asian J Agric Extension, Econ Sociol 143–155
9. Chaurasia V, Pal S (2020) Covid-19 pandemic: ARIMA and regression model-based worldwide death cases predictions. SN Comput Sci 1(5):1–12
10. Chowdhury AA, Hasan KT, Hoque KKS (2021) Analysis and prediction of covid-19 pandemic in Bangladesh by using ANFIS and LSTM network. Cogn Comput 13(3):761–770
11. Greff K, Srivastava RK, Koutn´ık J, Steunebrink BR, Schmidhuber J (2016) LSTM: a search space odyssey. IEEE Trans Neural Netw Learn Syst 28(10):2222–2232
12. Haque A, Pranto TH (2020) Covid-19 (sars-cov-2) outbreak in Bangladesh: situation according to recent data analysis using covid-19 data set for Bangladesh. Tahmid Hasan, COVID-19 (SARS-CoV-2) outbreak in Bangladesh: situation according to recent data analysis using COVID-19 data set for Bangladesh, 7 June 2020
13. Haque A (2020) The covid-19 pandemic and the public health challenges in Bangladesh: a commentary. J Health Res
14. Hossain M, Saleheen AAS, Haq I, Zinnia MA, Hasan M, Kabir S, Methun M, Haq I, Nayan M, Hossain I et al (2021) People's concerns with the prediction of covid-19 in Bangladesh: application of autoregressive integrated moving average model. Int J Travel Med Global Health 9(2):84–93
15. Islam MS, Hoque ME, Amin MR (2021) Integration of kalman filter in the epidemiological model: a robust approach to predict covid-19 outbreak in Bangladesh. Int J Modern Phys C 2150108
16. Hossain MF, Islam S, Chakraborty P, Majumder AK (2020) Predicting daily closing prices of selected shares of Dhaka stock exchange (DSE) using support vector machines. Internet Things Cloud Comput 8(4):46
17. Qamruzzaman M (2020) Covid-19 impact on SMEs in Bangladesh: an investigation of what they are experiencing and how they are managing? Available at SSRN 3654126
18. Nagavelli U, Samanta D, Chakraborty P (2022) Machine learning technology-based heart disease detection models. J Healthc Eng
19. Sarker A, Chakraborty P, Sha SS, Khatun M, Hasan MR, Banerjee K (2020) Improvised technique for analyzing data and detecting terrorist attack using machine learning approach based on twitter data. J Comput Commun 8(7):50–62
20. Kundu LR, Islam S, Ferdous MZ, Hossain MF, Chakraborty P (2021) Forecasting economic indicators of Bangladesh using vector autoregressive (VAR) model
21. Tran T, Pham L, Ngo Q (2020) Forecasting epidemic spread of sars-cov-2 using arima model (case study: Iran). Global J Environ Sci Manage 6(Special Issue (Covid-19)):1–10

A Comparative Study of Machine Learning Algorithms to Detect Cardiovascular Disease with Feature Selection Method

Md. Jubier Ali, Badhan Chandra Das, Suman Saha, Al Amin Biswas, and Partha Chakraborty

Abstract Heart disease is considered one of the calamitous diseases which eventually leads to the death of a human, if not diagnosed earlier. Manually, detecting heart disease needs doing several tests. By analyzing the result of tests, it can be assured whether the patient got heart disease or not. It is time consuming and costly to predict heart disease in this conventional way. This paper describes different machine learning (ML) algorithms to predict heart disease incorporating a Cardiovascular Disease dataset. Although many studies have been conducted in this field, the performance of prediction still needs to be improved. In this paper, we have focused to find the best features of the dataset by feature selection method and applied six machine learning algorithms to the dataset in three steps. Among these ML algorithms, the random forest algorithm gives the highest accuracy which is 72.59%, with our best possible feature setup. The proposed system will help the medical sector to predict heart disease more accurately and quickly.

Keywords Heart disease · ML algorithms · Feature selection

1 Introduction

Also known as cardiovascular disease, heart disease is a condition that affects the heart and circulatory system. If there is an abnormality in the heart, that is considered

Md. Jubier Ali (✉) · B. Chandra Das
Bangladesh University of Business and Technology, Dhaka, Bangladesh
e-mail: shondhi1997@gmail.com

S. Saha
Bangabandhu Sheikh Mujibur Rahman Digital University, Kaliakair, Bangladesh

A. A. Biswas
Daffodil International University, Dhaka, Bangladesh

P. Chakraborty
Comilla University, Cumilla, Bangladesh
e-mail: partha.chak@cou.ac.bd

V. Skala et al. (eds.), *Machine Intelligence and Data Science Applications*, Lecture Notes on Data Engineering and Communications Technologies 132,
https://doi.org/10.1007/978-981-19-2347-0_45

heart disease. There are many reasons for heart disease. Every year a significant number of people die due to this fatal disease. If any abnormality in the heart can be detected at an early stage, a person can take the necessary steps to be safe from any devastating effects of it. Thus, the death rate can be controlled. So, it is highly needed to identify heart disease at the initial stage in order to be safe and conscious. Though many types of tests are needed to detect heart disease manually, that is the only conventional way to detect it, which is time consuming and costly. Therefore, an automatic heart disease detection system can ease some sort of difficulties when it comes to time and money.

In this paper, we will be using ML algorithms to predict the heart disease of a person by analyzing the attributes of a cardiovascular dataset. There are several works with the problem, but their accuracy is not so high to the Cardiovascular Disease dataset. In this research, by using the feature selection approach, we will find the different experimental configurations of features. We have used six ML algorithms (Naïve Bayes, Logistic Regression, K-nearest neighbor, Support Vector Machine (SVM), Random Forest, and Decision Tree) in order to detect the heart disease and find out the performance measurement of each ML algorithms. The feature selection method has been used to find out the best features of the dataset based on the scores provided by it. Then, we applied six ML algorithms on the different sets of features separately. Using the feature selection strategy, we have found out our important features. Three different feature sets have been set through the outcomes from the feature selection method.

The following are the main contributions of this paper:

- Using the feature selection approach, we have selected the different experimental configurations for features.
- We have applied six ML algorithms and measured the performance of different experimental configurations for each set of features.
- An extensive experiment has been performed on a cardiovascular dataset and compared the performance of applied ML algorithms.

The remaining parts of this paper is constructed as follows. The related works are described in Sect. 2. Section 3 gives the description of Cardiovascular Disease dataset is included. Section 4 consists of the methodology. Section 5 contains the ML algorithm's evaluation and comparison of the results. In Sect. 6, we have discussed the outcome of the result of the algorithms for all feature configurations. At last, we have concluded the paper in Sect. 7.

2 Related Works

2.1 *Heart Disease Prediction System and Application*

Gavhane et al. [1] discussed how to anticipate heart disease at the initial stage. They proposed to develop an application that can predict heart disease by taking some symptoms as their features. They found neural network has given the most accurate result. Palaniappan et al. [2] proposed an Intelligent Heart Disease Prediction system (IHDPS) by developing a Web site for the IHDPS with the best data mining technique. They used Naïve Bayes, decision tree, and neural network techniques. Shah et al. [3] proposed a system that can easily predict the early diagnosis of heart disease. A dataset from the database which is named as Cleveland database of heart disease patients repository. They considered only 14 important attributes among 73 attributes. Among ML algorithms, KNN gave the highest accuracy. Rajdhan et al. [4] have proposed a system to predict the chances of heart disease and classify the patient's risks. Among some ML algorithms, random forest yielded the highest accuracy which is 90.16%. They also compared the results of different ML algorithms on the dataset. Repaka et al. [5] proposed a system for disease prediction that uses Nave Bayesian (NB) techniques for dataset categorization and the advanced encryption standard (AES) algorithm for safe data transit. Mohan et al. [6] proposed a hybrid technique, dubbed a hybrid random forest with a linear model by (HRFLM). Kelwade et al. [7] have devised a system that uses a radial basis function neural network (RBFN) to predict eight heart arrhythmias. The linear and nonlinear aspects of each arrhythmia's heart rate time series are detected. Jabbar et al. [8] discussed how Hidden Naïve Bayes (HNB) can be used to forecast heart disease. They showed how HNB performed on the heart disease dataset. They also described the algorithm and the performance of the algorithm in different parameters.

2.2 *Comparative Analysis of Machine Learning Algorithms to Predict Heart Disease*

Bhatla et al. [9] have executed a comparative analysis of different types of data mining classifiers and various techniques. Sowmiya et al. [10] cited a comparative study on heart disease dataset to predict the heart disease diagnosis by different data mining techniques. They used some medical symptoms or attributes to predict heart disease. They compared the result of data mining techniques or algorithms. Thomas et al. [11] have shown a comparative study to predict heart cancer using different data mining techniques. They have tried to find the risk rate of heart disease by using these algorithms. Singh et al. [12] have discussed a systematic review of different types of machine learning techniques to predict heart diseases.

Table 1 Features of dataset

Features	Data type	Description
id	Numerical	Anonymous ID of patients
age	Numerical	Age of patients in days
gender	Categorical	Gender of patients. (women or men)
height	Numerical	Patient's height in cm
weight	Numerical	Patient's weight in kg
ap_hi	Numerical	Systolic blood pressure
ap_lo	Numerical	Blood pressure in the diastole
cholesterol	Categorical	Cholesterol status of a patient (normal, above normal, or well above normal)
gluc	Categorical	Glucose status of patient (normal, above normal, or well above normal)
smoke	Categorical	whether patient smokes (1) or not (0)
alco	Categorical	whether patient drinks alcohol (1) or not (0)
active	Categorical	Physically active (1) or not (0)
cardio	Categorical	The result of a candidate value 0: no heart disease value 1: heart disease

2.3 Feature Selection to Predict Heart Disease

Using data science, Bashir et al. [13] proposed a cardiac disease prognosis. Their study focused on feature selection algorithms and strategies for various heart disease prediction datasets.

3 Dataset

We collected the dataset from Kaggle[1]. The dataset called Cardiovascular Disease dataset is formed on a CSV file with 70,000 instances and 12 features. In this dataset, data attributes are the important features that can help us to predict heart disease or cardiovascular disease. The value of the target column is dependent on 12 column features which are discussed in Table 1. There are 12 features such as id, age, gender, height, weight, ap_hi, ap_lo, cholesterol, gluc, smoke, alco, and active. The target column is cardio, which represents whether a patient is affected by heart disease or not. In the dataset, there is a total of 70,000 instances.

Table 2 shows some of the instances of the dataset. The target column "cardio" has two values, they are—0 (no heart disease) and 1(heart disease). It is working as a dependent variable which depends on the other 12 independent variables at its left.

[1] https://www.kaggle.com/sulianova/cardiovascular-disease-dataset.

Table 2 Instances of dataset

ID	age	gender	height	weight	ap_hi	ap_lo	cholesterol	gluc	smoke	alco	active	cardio
1	20228	1	156	85.0	140	90	3	1	0	0	1	1
2	18857	1	165	64.0	130	70	3	1	0	0	0	1
3	17623	2	169	82.0	150	100	1	1	0	0	1	1
4	17474	1	156	56.0	100	60	1	1	0	0	0	0
8	21914	1	151	67.0	120	80	2	2	0	0	0	0
9	22113	1	157	93.0	130	80	3	1	0	0	1	0

4 Methodology

The proposed method of this task has been initiated by the collection of the Cardiovascular Disease dataset. In the next phase, data pre-processing is performed. Then, the feature selection method is used by us, which is the most significant part of our research. Then, we applied the following six ML algorithms, decision tree, logistic regression, SVM, Naive Bayes, random forest, and KNN. Finally, we compared results and discussed the outcomes of those machine learning models. Figure 1 shows the overall procedure of our work.

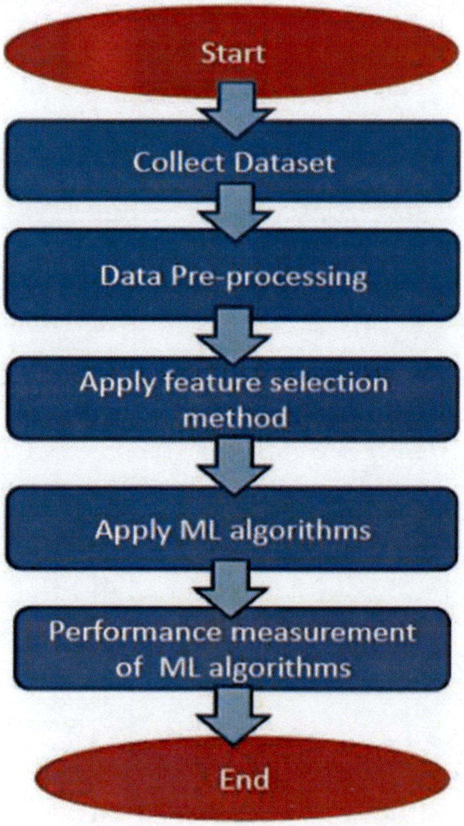

Fig. 1 Methodology of detecting cardiovascular disease with feature selection method

4.1 Dataset Collection

We collected the dataset from Kaggle. The name of our dataset is the Cardiovascular Disease dataset. In the dataset, there are 70,000 instances. These instances help us to identify whether a patient has heart disease or not.

4.2 Data Pre-processing

Data pre-processing is provided in a comprehensible manner by transforming raw data into an understandable context for a specific goal. We performed several pre-processing steps like data cleaning (null and incomplete instances removal, redundancy elimination), outlier removal, and data transformation on the dataset.

4.3 Feature Selection

Finding a subset of input features that are most correlated with the target feature can be done by the feature selection method. After the data has been pre-processed, we applied the feature selection approach to find the importance of features. There were 12 features columns and a target column. The dataset was subjected to feature selection, and the score values for each feature were obtained. The ways of selecting features is described as follows.

ANOVA F-statistic. ANOVA views an analysis of variance and parametric statistical hypothesis test which determines the means of more than two instances are belong to the same trading or not. It is also depicted as F-test and it belongs to a type of ANOVA that appraises the proportion of variance values, like whether the variance from the different samples or not and elucidated or not. An ANOVA f-test is a form of F-statistic that uses the ANOVA approach. We utilized *scikit-learn* machine library to get the feature scores of all the independent attributes. *f_classif()* function includes an implementation of ANOVA f-test which can be used in a feature selection technique, like—using *SelectKBest* class to find the top *k* most admissible features (highest values). *f_classif()* function incorporates an implementation of ANOVA f-test which can be used in a feature selection technique, like—using *SelectKBest* class to find the top *k* most admissible features (highest values).

Dividing the features into different sets. For our work, we have used the ANOVA F-statistic feature selection method because there are numerical and categorical values in the dataset. We used *f_classif()* function for our work. We used SelectKBest to find the best features. We divided the features into three parts based on the feature scores ranged in three categories and sort these features in descending order. First, we have selected three features for the first part. In the second part, we have included the first

Table 3 Features with feature selection score value

Features	Features selection score
age	4209.008
cholesterol	3599.361
weight	2388.777
gluc	562.773
ap_lo	303.629
ap_hi	208.339
active	89.091
smoke	16.79
height	8.197
gender	4.604
alco	3.761
id	1.010

three features and added more five features with them. In the third part, we have included all 12 features for the experiment. We termed these parts to feature sets. For every feature set, we have applied ML algorithms and measured the performance of ML algorithms. Table 3 shows the score value of every feature in descending order.

4.4 Machine Learning Models for Different Feature Sets

In this phase, we have applied six machine learning algorithms on the pre-processed Cardiovascular Disease dataset for all of the features sets constructed by the values of feature selection method.

Figure 2 shows our working process. First, we performed pre-processing of the data. After data pre-processing, we have applied the feature selection approach and feature importance, with the highest score value, firstly we considered three features, then eight features, and at last we considered 12 features. We divided the dataset into train and test data phases. We applied ML algorithms to the dataset. Then, we measured the performance of every ML algorithms through the widely known evaluation metrics. We do our experiment using *Scikit-learn* library of Python in order to apply ML algorithms to our dataset. Following ML algorithms are applied to our pre-processed data.

Decision Tree: It is also known as decision support, uses a tree-like graph to represent various outcomes, such as chance event outcomes, resource codes. It is a conditional control statement-based representation of an algorithm. The values of the parameters are as follows. Criterion = 'Gini', Splitter = 'Best', Minimum samples split = 2.

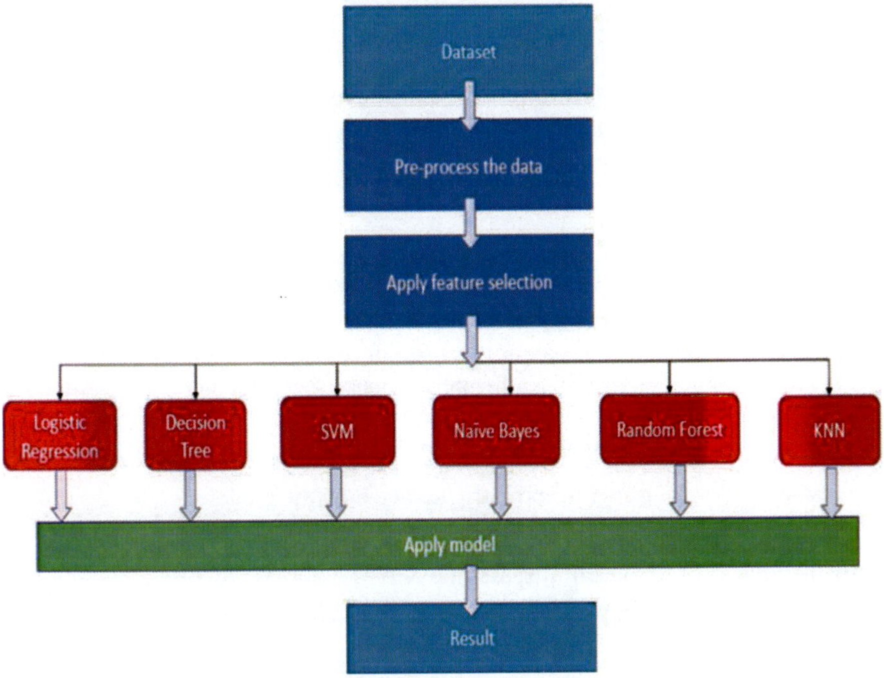

Fig. 2 The process of how ML algorithms applied to the dataset

Logistic Regression: Logistic regression evaluates the parameters of the logistic model in regression analysis. Here, we used Penalty = '12', Tolerance for stopping criteria = 0.0001, Maximum iteration = 10.

Support Vector Machine (SVM): SVM is precise and generates better results in terms of classification. Kernel type = 'rbf', Degree of the polynomial kernel function = 3, gamma = 'scale', Tolerance for stopping criteria = 0.001.

Naïve Bayes: The probability of given dataset is calculated by Nave Bayes classifiers when they do classification. Each property in a set of data is treated as though it were unrelated to the others. The output class of a particular instance is the high probability class. Prior probabilities = None, Portion of the largest variance = 1e-09.

Random Forest: It is a well-known model for classification as well as regression analysis. Multiple trees are built, and the majority voting is used to classify the records. No. of estimators = 100, Criterion = 'Gini', Splitting the smallest samples = 2, Minimum samples leaf = 1.

***K*-nearest neighbor (K-NN)**: The resemblance is assumed by KNN algorithm, between the new data and puts the new data into the category that is most similar to the available categories. No of neighbors = 5, Weights = 'Uniform', Metric = 'Minkowski'.

5 Experimental Results

The analysis of the acquired results is presented in this section. Applying the feature selection method to the dataset, we found the score value of every feature.

5.1 *Performance Measurement*

We compared the results of ML algorithms to the Cardiovascular Disease dataset and all the outcomes of all six algorithms for all of our features configurations. We constructed a confusion matrix for the performance measurement and then demonstrate the results in Sect. 5. For the confusion matrix, there are four terms described as follows as per our dataset and experiment.

True positive: Heart disease is correctly predicted by the model.

True negative: The model indicates that the patient does not have heart disease, and the patient does not have heart disease in reality.

False positive: Despite the fact that the patient does not have heart disease, the model predicts that he or she has it.

False negative: The model predicts that the patient does not have heart disease, yet the patient does.

The mathematical formulas and the notations are as follows [14].

Accuracy (Acc) describes the correct classification from all samples.

$$\text{Acc} = \frac{(\text{TP} + \text{TN})}{(\text{TP} + \text{TN} + \text{FP} + \text{FN})} \tag{1}$$

Sensitivity (Sen) refers to a test's capacity to appropriately identify patients who have cardiac disease. It is also called recall.

$$\text{Sen} = \frac{\text{TP}}{(\text{TP} + \text{FN})} \tag{2}$$

Specificity (Spec) describes the ability of a test to correctly identify people without the heart disease.

$$\text{Spe} = \frac{\text{TN}}{(\text{TN} + \text{FP})} \tag{3}$$

Precision (Prec) describes the ratio of truly positive samples among all the samples that model predicted positive.

Table 4 Performance measurement of ML algorithms for top three features

Algorithms	Sensitivity (%)	Specificity (%)	Precision (%)	Accuracy (%)
Logistic regression	*84.8*	14.9	50.61	50.34
Support vector machine (SVM)	59.5	60.52	60.52	*60.01*
Random forest	57.34	57.2	57.95	57.27
Decision tree	53.86	58.69	57.29	56.24
Naïve Bayes	45.58	*69.03*	*69.03*	54.9
K-nearest neighbor (KNN)	55.7	59.6	58.65	57.62

$$\text{Prec} = \frac{\text{TP}}{(\text{TP} + \text{FP})} \tag{4}$$

5.2 *Performance of ML Algorithms for Top Three Features*

There is a total of 70,000 instances in the dataset. We put 49,000 instances for training and 21,000 for testing. Using the feature selection method, first, we considered the top three features: age, cholesterol, and weight. Then applied all six ML algorithms for this feature set. The performance of ML algorithms through the mentioned evaluation metrics has shown in Table 4.

Table 4 shows the performance measurement of ML algorithms to the three features of the dataset. These three features give a higher score value of feature selection than other features. For sensitivity, logistic regression gives the highest value, which is 84.8%. Naive Bayes gives the highest specificity value, which is 69.03%. In terms of precision, Naive Bayes gives the highest precision score value which is 69.03%. Again, SVM gives the highest accuracy, which is 60.01%.

5.3 *Performance of ML Algorithms for Top Eight Features*

After taking the top three features, we included have more features with the previous set. Thus, we got top eight features in total (age, weight, ap_hi, ap_lo, cholesterol, gluc, smoke, and active). Then again, we applied six ML algorithms to the new feature set. The performance measurement of ML algorithms has shown in Table 5 for the top eight features.

Table 5 shows the performance measurement of ML algorithms to the eight features of the dataset. For sensitivity, random forest gives the highest value, which is 69.49%. Naive Bayes gives the highest specificity value, which is 90.23%. For precision, Naive Bayes gives the highest precision score value which is 73.92%. In terms of accuracy, random forest gives the highest accuracy which is 69.41%.

Table 5 Performance measurement of ML algorithms for top eight features

Algorithms	Sensitivity (%)	Specificity (%)	Precision (%)	Accuracy (%)
Logistic regression	62.28	48.52	55.45	55.5
Support vector machine (SVM)	54.98	66.11	62.63	60.47
Random forest	*69.49*	69.33	69.97	*69.41*
Decision tree	62.8	64.25	64.37	63.51
Naïve Bayes	26.92	*90.23*	*73.92*	58.13
K-nearest neighbor (KNN)	66.24	70.4	69.71	68.29

Table 6 Performance Measurement of ML algorithms for 12 features

Algorithms	Sensitivity (%)	Specificity (%)	Precision (%)	Accuracy (%)
Logistic regression	65.49	73.82	72.01	69.6
Support vector machine (SVM)	65.42	54.29	59.55	59.93
Random forest	*69.48*	75.99	74.85	*72.69*
Decision tree	63.62	63.42	64.14	63.52
Naïve Bayes	11.74	*96.66*	*78.32*	53.6
K-nearest neighbor (KNN)	54.74	56.38	56.35	55.5

5.4 Performance of ML Algorithms for 12 Features

After taking the top eight features, we considered all 12 features. Then applied six ML algorithms to the top 12 features. The performance measurement of ML algorithms has shown in Table 6 for the top 12 features.

Table 6 shows the performance measurement of ML algorithms to the 12 features of the dataset. For sensitivity, random forest gives the highest value, which is 69.48%. Naive Bayes gives the highest specificity value, which is 96.66%. For precision parameter, Naive Bayes gives the highest precision score value which is 78.32%. In terms of accuracy parameter, random forest gives the highest accuracy which is 72.69%.

Figure 3 depicts the graphical representation of the accuracy of all six Machine learning algorithms for different experimental configurations of features. From this figure, it can be easily diagnosed that SVM gives the maximum accuracy for three features. Random forest gives the highest accuracy for 8 and 12 features, respectively.

6 Discussion

Using feature selection score value, we considered 3, 8, and 12 feature sets respectively for this work and then applied ML algorithms. We discussed all performance measurements of ML algorithms for, respectively, 3, 8, and 12 features. It is clear

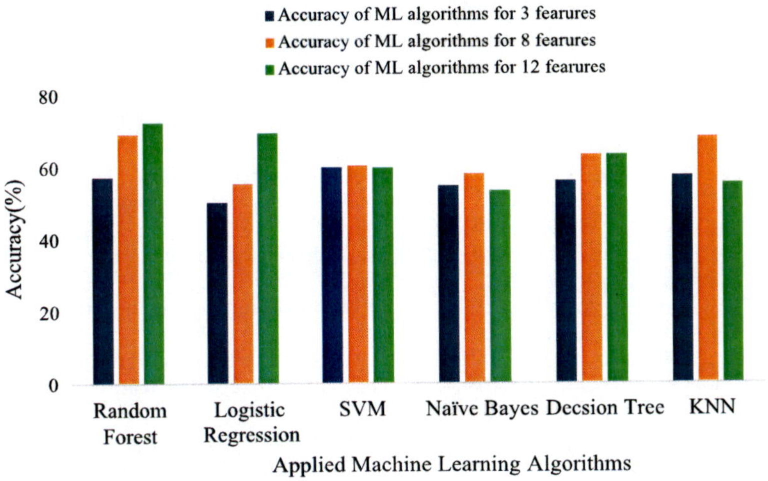

Fig. 3 Accuracy of ML algorithms to 3, 8 and 12 features

from the resulting numbers in terms of the evaluation measurements that when we considered three features, the performance of ML algorithms is quite low. The accuracy is not that much good. For eight features, ML algorithms gave a little better performance than the previous set of features. When we consider all the features. Now, the ML algorithms gives the highest performance. For precision and specificity parameters, Naive Bayes gives the highest score. Random forest gives the highest score in sensitivity and accuracy parameters. It can be inferred that incorporating more features yields better performance in terms of evaluation metrics.

7 Conclusion and Future Scope

The work represents the comparative study of different supervised ML algorithms to the Cardiovascular Disease dataset. The overall aim is to define various ML algorithms useful to effective heart disease prognosis. By employing the feature selection method, we took different experimental setups of features and applied ML algorithms to them. Only 12 critical features are considered in this study. We compared the result of ML algorithms, respectively, for different experimental configurations for different feature sets. In the future, we are planning to expand this research incorporating new contexts, like, how the heart condition of a person changes as he or gets older. We are about to build up a recommendation system to what extent a person should get aware of his or her heart diseases based on the deciding factors of heart disease.

References

1. Gavhane A, Kokkula G, Pandya I, Devadkar K (2018) Prediction of heart disease using machine learning. In 2018 Second international conference on electronics, communication and aerospace technology (ICECA), pp 1275–1278
2. Palaniappan S, Awang R (2008) Intelligent heart disease prediction system using data mining techniques. In 2008 IEEE/ACS international conference on computer systems and applications, pp 108–115
3. Shah D, Patel S, Bharti SK (2020) Heart disease prediction using machine learning techniques. SN Comput Sci 1(6):1–6
4. Rajdhan A, Agarwal A, Sai M, Ravi D, Ghuli P (2020) Heart disease prediction using machine learning. Int J Res Technol 9(04):659–662
5. Repaka AN, Ravikanti SD, Franklin RG (2019) Design and implementing heart disease prediction using naives Bayesian. In 2019 3rd International conference on trends in electronics and informatics (ICOEI), pp 292–297
6. Mohan S, Thirumalai C, Srivastava G (2019) Effective heart disease prediction using hybrid machine learning techniques. IEEE access 7:81542–81554
7. Kelwade JP, Salankar SS Radial basis function neural network for prediction of cardiac arrhythmias based on heart rate time series. In 2016 IEEE first international conference on control, measurement and instrumentation (CMI), pp 454–458
8. Jabbar MA, Samreen S (2016) Heart disease prediction system based on hidden naïve bayes classifier. In 2016 International conference on circuits, controls, communications and computing (I4C), pp 1–5
9. Bhatla N, Jyoti K (2012) An analysis of heart disease prediction using different data mining techniques. Int J Eng 1(8):1–4
10. Sowmiya C, Sumitra P (2017) Analytical study of heart disease diagnosis using classification techniques. In 2017 IEEE international conference on intelligent techniques in control, optimization and signal processing (INCOS), pp 1–5
11. Thomas J, Princy RT (2016) Human heart disease prediction system using data mining techniques. In 2016 International conference on circuit, power and computing technologies (ICCPCT), pp 1–5
12. Singh D, Samagh JS (2020) A comprehensive review of heart disease prediction using machine learning. J Crit Rev 7(12):281–285
13. Bashir S, Khan ZS, Khan FH, Anjum A, Bashir K (2019) Improving heart disease prediction using feature selection approaches. In 2019 16th international Bhurban conference on applied sciences and technology (IBCAST), pp 619–623
14. Zul ker MS, Kabir N, Biswas AA, Nazneen T, Uddin MS (2021) An in-depth analysis of machine learning approaches to predict depression. Curr Res Behav Sci 2:100044

A Pilot Study for Devanagari Script Character Recognition Using Deep Learning Models

Ayush Sharma, Mahendra Soni, Chandra Prakash, Gaurav Raj, Ankur Choudhary, and Arun Prakash Agrawal

Abstract Across the globe, over 300 million people understand the Devanagari Script (DS) Based languages i.e. Hindi. Despite the number of speakers, there is no accurate Optical Character Recognition (OCR) System available for the Hindi language/other DS-based languages. Various problems are faced while developing an efficient OCR for DS such as segmentation of characters (As there is no gap between characters in DS-based languages), touching characters (characters formed by the combination of pure consonants and consonants). Given the complexity of DS-OCR, it has emerged as an interesting area of research. In this paper, Deep Neural Models (VGG16, ResNet50, and MobileNet) are explored for Handwritten Hindi Characters Recognition. VGG16 model outperforms with an accuracy of 98.47%.

Keywords Deep Learning · Devanagari Script (DS) · Optical Character Recognition (OCR)

1 Introduction

OCR is an important domain of computing. It has emerged as a helping hand in the wake of digitalization. Devanagari Script (DS) is used to write various languages like Hindi, Sanskrit and is used by millions of people around the globe. Despite

A. Sharma · M. Soni · C. Prakash
Department of CSE, NIT, Delhi, India
e-mail: 171210016@nitdelhi.ac.in

M. Soni
e-mail: 171210036@nitdelhi.ac.in

C. Prakash
e-mail: cprakash@nitdelhi.ac.in

G. Raj (✉) · A. Choudhary · A. P. Agrawal
Department of Computer Science and Engineering, School of Engineering and Technology, Sharda University, Greater Noida, India
e-mail: er.gaurav.raj@gmail.com

V. Skala et al. (eds.), *Machine Intelligence and Data Science Applications*, Lecture Notes on Data Engineering and Communications Technologies 132,
https://doi.org/10.1007/978-981-19-2347-0_46

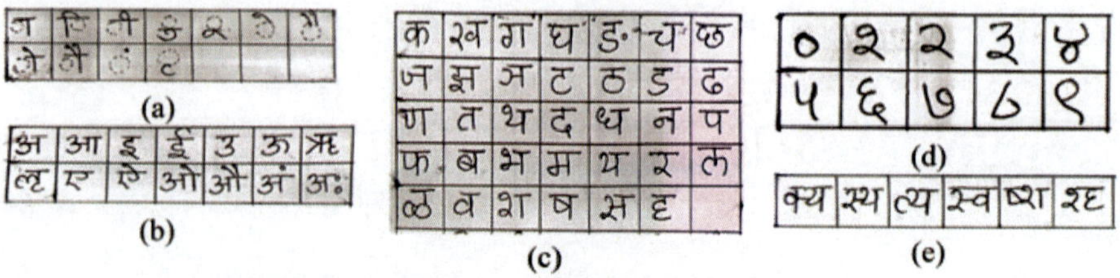

Fig. 1 Sample of Handwritten Devanagari Characters, **a** Modifiers, **b** Vowels, **c** Consonants, **d** Digits, **e** some form of conjuncts

the huge amount of speakers, there is no promising Optical Character Recognition System available for DS. It is a left-to-right script developed in ancient India from 1st to fourth century and was originally developed to write Sanskrit language but later became a foundation of various other languages like Hindi, Sanskrit, Nepali. Over 300 million people use DS for communication and documentation. In India, almost all of the Government offices, Banks, Schools are using the Hindi language for documentation. Therefore, the Optical Character Recognition(OCR) system for DS is a domain of interest.

DS has 34 consonants and 14 vowels. Apart from vowels and consonants, some conjuncts are formed by the combination of pure consonants with consonants. Figure 1 shows the modifier, vowels, digits, consonants, and some conjuncts in DS.

In the literature various segmentation techniques for Handwritten Hindi Text to improve the Segmentation accuracy for lines and words [1–3]. Segmentation is the most challenging task for developing an Optical Character Recognition System for DS as all the characters written in DS are connected with a horizontal line (called Shirorekha) to form a word. Also, there are problems faced with the segmentation of touching characters (or conjuncts). Other challenges include the classification of Handwritten Characters and the limited availability of Dataset.

After segmentation of a Handwritten page into characters, we have to classify them from different classes. For this purpose, we need to extract features from the text image to recognize the true class of the image character. The performance of an OCR system is greatly affected by the feature extraction technique used. Kumar [4] recognizes the performance of 5 different feature extraction methods using MLP and SVM classifiers. Narang et al. [5] improved the results of recognition used in ancient characters of Devanagari using Scale Invariant Feature Transform (SIFT) and Gabor filters for extracting features. SVM is used for the classification task. We need to define features implicitly beforehand for these traditional techniques. In contrast, Deep Neural Networks work on the input pixel data to produce the best features for the input image, therefore it is not indispensable to define features implicitly. In [6] features are extracted by underlying convolution layers by applying 3 × 3 filters across images to convert them into feature maps. The performance of Deep Neural Networks is better for Recognition of Handwritten Devanagari Characters.

Class	क	ख	ग	घ	ङ	च	छ	ज	झ	ञ	ट	ठ
#	2,000	2,000	2000	2,000	2,000	2,000	2,000	2000	2,000	2,000	2,000	2,000
Class	ड	ढ	ण	त	थ	द	ध	न	प	फ	ब	भ
#	2,000	2,000	2000	2,000	2,000	2,000	2,000	2000	2,000	2,000	2,000	2,000
Class	म	य	र	ल	व	श	ष	स	ह	क्ष	त्र	ज्ञ
#	2,000	2,000	2000	2,000	2,000	2,000	2,000	2000	2,000	2,000	2,000	2,000
Total	72,000											

Fig. 2 Consonant character dataset [7]

Class	०	१	२	३	४	५	६	७	८	९
#	2,000	2,000	2000	2,000	2,000	2,000	2,000	2000	2,000	2,000
Total	20,000									

Fig. 3 Vowel character dataset [7]

In this paper, different Deep Neural Networks model has been explored and compared over DHCD (Devanagari Handwritten Characters Dataset) and propose a model with the best performance for OCR of Handwritten Hindi Characters.

2 DHCD Dataset

Devanagari Handwritten Character Dataset (DHCD) is an image dataset of Handwritten Devanagari Characters. It was introduced in [7]. It has 92,000 images of Handwritten Devanagari Characters with 46 different classes, 36 classes for constants, and 10 classes for digits. Each image is of size 32×32 pixels in .png format. The actual character is centered within 28×28 pixels, padding of 2 pixels is added on four sides of the actual character. Each class has 2000 images which is shown in Figs. 2 and 3.

3 Methodology

In this section the proposed methodology has been discussed in Fig. 4. Input Image is a handwritten character image. The input image is passed for feature extraction. After extracting features they are detected by the hidden layer and the result is passed for classification. Once the image is classified the final output has been discussed.

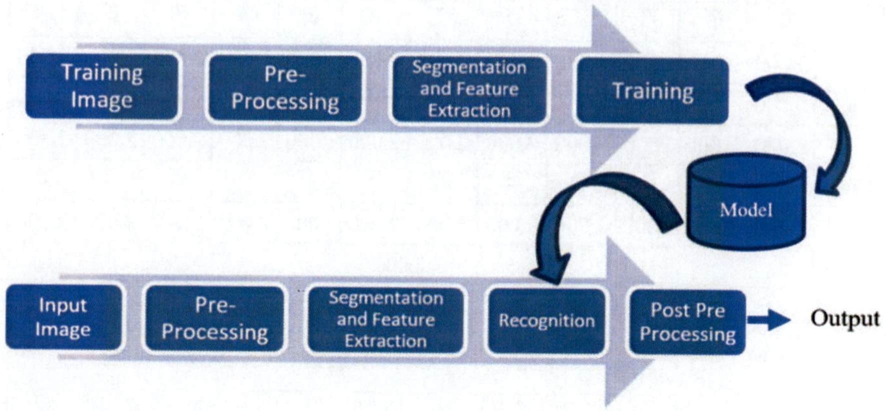

Fig. 4 Proposed methodology for the DS Character Recognition

3.1 Dataset Preparation

The Dataset (DHCD) was randomly split to 85% training set and 15% testing set. Training set has 78,200 images and Testing set has 13,800 images. After splitting, the models were trained using this dataset.

3.2 Deep Learning Model Used for the DS Character Recognition

VGG16, ResNet50 and MobileNet Model have been used in this study to explore the DHCD dataset.

VGG 16 Model. VGG stands for Visual Geometric Group where VGG-16 is containing 16 layers having the parameters which may be trained in a shorter time and it includes other layers further like the Max pool layer. The flattened architecture of VGG16 is additionally shown in Fig. 5.

The architecture is incredibly simple as it has two contiguous blocks of two convolution layers which is followed by a max-pooling layer. It has three contiguous blocks of convolution layers followed by the max-pool layer. Finally, it has 3 dense

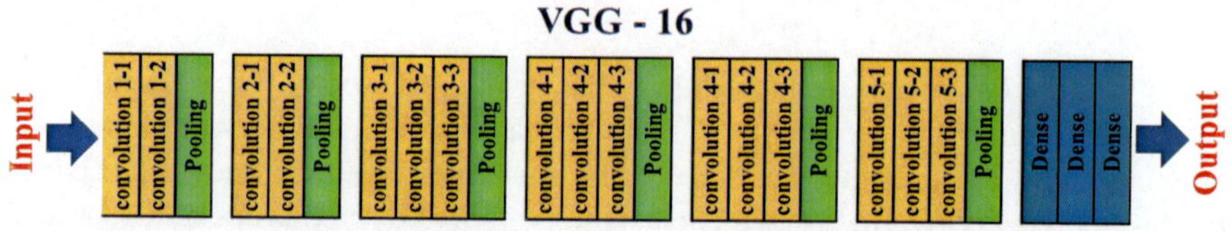

Fig. 5 Flattened VGG-16 architecture [8]

layers, have different depths in several architectures. The point of consideration is that after each max-pool layer, the dimension is reduced by half.

ResNet50 Model. ResNet50 could be a variation of the ResNet model [9] that has 48 Convolution layers together with 1 Average-Pool and 1 Max-Pool layer to perform floating points operations. It is the general ResNet model and Due to the framework that ResNets presented, it is feasible to train ultra-deep-NN where the network may contain hundreds/thousands of layers to attain good performance. Initially, the ResNets were used in image recognition but later on, the framework was also utilized in other projects and provided better accuracy too [9].

The ResNet50 Model architecture is shown in Fig. 6. This ResNet50 Architecture has a convolution with 64 different kernels with 7 × 7 kernel size with the tread of size gives us a layer.

In the next subsequent convolution, there are 64 kernels of 1 × 1 following this 64 kernel of 3 × 3, and finally 256 kernels of 1 × 1. Using the following sequence three times gives us 9 layers. Following is 128 kernel of 1 × 1, then again 128 kernel of 3 × 3, and ultimately 512 kernel of 1 × 1. This step was repeated 4 times so gave us 12 layers. After that, there's a kernel of 1 × 1,256 and two more kernels with 3 × 3,256 and 1 × 1,1024 and this could be recurrent 6 times giving us a total of 18

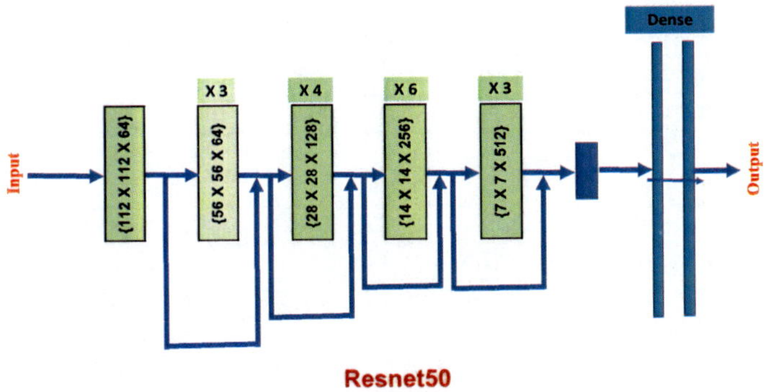

layer name	output size	18-layer	34-layer	50-layer	101-layer	152-layer
conv1	112×112	7×7, 64, stride 2				
conv2_x	56×56	3×3 max pool, stride 2				
		[3×3, 64; 3×3, 64] ×2	[3×3, 64; 3×3, 64] ×3	[1×1, 64; 3×3, 64; 1×1, 256] ×3	[1×1, 64; 3×3, 64; 1×1, 256] ×3	[1×1, 64; 3×3, 64; 1×1, 256] ×3
conv3_x	28×28	[3×3, 128; 3×3, 128] ×2	[3×3, 128; 3×3, 128] ×4	[1×1, 128; 3×3, 128; 1×1, 512] ×4	[1×1, 128; 3×3, 128; 1×1, 512] ×4	[1×1, 128; 3×3, 128; 1×1, 512] ×8
conv4_x	14×14	[3×3, 256; 3×3, 256] ×2	[3×3, 256; 3×3, 256] ×6	[1×1, 256; 3×3, 256; 1×1, 1024] ×6	[1×1, 256; 3×3, 256; 1×1, 1024] ×23	[1×1, 256; 3×3, 256; 1×1, 1024] ×36
conv5_x	7×7	[3×3, 512; 3×3, 512] ×2	[3×3, 512; 3×3, 512] ×3	[1×1, 512; 3×3, 512; 1×1, 2048] ×3	[1×1, 512; 3×3, 512; 1×1, 2048] ×3	[1×1, 512; 3×3, 512; 1×1, 2048] ×3
	1×1	average pool, 1000-d fc, softmax				
FLOPs		1.8×10^9	3.6×10^9	3.8×10^9	7.6×10^9	11.3×10^9

Fig. 6 ResNet50 model architecture [9]

layers. And nevertheless, a 1 × 1,512 kernel with two more of 3 × 3, 512, and 1 × 1, 2048, and this were repeated 3 times giving us a complete of 9 layers. In the next step, the median pool has been applied and finished by a completely linked layer comprising of 1000 nodes and at the end of layers, the soft-max function introduces one layer at the end.

MobileNet Model. MobileNet is the efficient architecture that adopts depth-wise separable convolutions to construct lightweight Deep CNN and provides a well-organized model [10].

Depthwise separable convolution filters are composed of depth-wise and point-wise convolution filters. The depth-wise convolution filter implements the single convolution on an independent input channel, whereas the point-wise convolution filter linearly joins the depth-wise convolution and 1 × 1 convolutions.

The architecture MobileNet Model is shown in Fig. 7. For MobileNets the depth-wise convolution implement one filter to all input channel. The point-wise convolution then applies one 1 × 1 convolution to combine all outputs with the depth-wise convolution.

Type / Stride	Filter Shape	Input Size
Conv / s2	3 × 3 × 3 × 32	224 × 224 × 3
Conv dw / s1	3 × 3 × 32 dw	112 × 112 × 32
Conv / s1	1 × 1 × 32 × 64	112 × 112 × 32
Conv dw / s2	3 × 3 × 64 dw	112 × 112 × 64
Conv / s1	1 × 1 × 64 × 128	56 × 56 × 64
Conv dw / s1	3 × 3 × 128 dw	56 × 56 × 128
Conv / s1	1 × 1 × 128 × 128	56 × 56 × 128
Conv dw / s2	3 × 3 × 128 dw	56 × 56 × 128
Conv / s1	1 × 1 × 128 × 256	28 × 28 × 128
Conv dw / s1	3 × 3 × 256 dw	28 × 28 × 256
Conv / s1	1 × 1 × 256 × 256	28 × 28 × 256
Conv dw / s2	3 × 3 × 256 dw	28 × 28 × 256
Conv / s1	1 × 1 × 256 × 512	14 × 14 × 256
5× Conv dw / s1	3 × 3 × 512 dw	14 × 14 × 512
5× Conv / s1	1 × 1 × 512 × 512	14 × 14 × 512
Conv dw / s2	3 × 3 × 512 dw	14 × 14 × 512
Conv / s1	1 × 1 × 512 × 1024	7 × 7 × 512
Conv dw / s2	3 × 3 × 1024 dw	7 × 7 × 1024
Conv / s1	1 × 1 × 1024 × 1024	7 × 7 × 1024
Avg Pool / s1	Pool 7 × 7	7 × 7 × 1024
FC / s1	1024 × 1000	1 × 1 × 1024
Softmax / s1	Classifier	1 × 1 × 1000

Fig. 7 MobileNet architecture [10]

Table 1 Results of CNN architecture

S.No	*CNN architecture*	*Train score*	*Test score*
1	VGG16	99.49	98.47
2	ResNet50	99.12	98.89
3	MobileNet	99.06	97.64

4 Model Results and Analysis

The performance of any model can be measured by its validation accuracy and model loss. The plot shown in Fig. 6 suggests that MobileNet has the highest model loss value whereas VGG16 Model computes a varial loss with ResNet 50 Model. After training all the models, Train Score and Test score were calculated for each model. The results are shown in Table 1.

Although the accuracy of the RestNet50 Model is the highest, it is not an appropriate model to use. Because it is evident from Fig. 5, it has a higher model-loss value than VGG16 Model. Also, its model-loss value increases exponentially with each epoch. MobileNet model has lesser accuracy than VGG16 Model and higher model-loss value which increases with each epoch as well. So it can be concluded that the VGG16 model is the finest among all tested models because it was able to find more features as compared to any other model. Hence, VGG16 Model is the proposed model with an accuracy of 98.47%.

5 Conclusion and Future Work

In the last two decades, a lot of research work has been done for optical character recognition of various languages. The segmentation and recognition of the Devnagari language script have great importance as many of the great vades have been written. In this paper three effective models VGG16, ResNet50 and MobileNet have been applied to DHCD dataset. The results of applied models show that the testing accuracy of ResNet50 is better than VGG16 and MobileNet with 98.89%. In future, hybrid deep learning can be applied to address selected DS recognition issues.

References

1. Pattnaik I, Patnaik T (2020) Optical segmentation of DS: a scientific analysis. academia.edu
2. Iman B, Haque MA (2020) A slice-based character recognition technique for handwritten DS. In: ICSES transaction on image processing and pattern recognition, vol 6, no 1, Apr 2020
3. Palakollu S, Dhir R, Rani R (2012) Handwritten hindi text segmentation techniques for lines and characters. WCECS 2012, San Francisco, USA, Oct 24–26 2012
4. Kumar S (2009) Performance comparison of features on Devanagari hand-printed dataset. Int J Recent Trends Eng 1(2)

5. Narang SR, Jindal MK, Ahuja S, Kumar M (2020) On the recognition of Devanagari ancient handwritten characters using SIFT and gabor features. Soft Comput
6. Shetty S, Shetty S (2020) Handwritten Devanagari character recognition using convolutional neural networks. J Xi'an Univ Archit Technol XII(II)
7. Acharya S, Pant AK, Gyawali PK (2015) Deep learning based large scale handwritten Devanagari character recognition. In: Proceedings of the 9th international conference on software, knowledge, information management and applications (SKIMA), pp 121–126
8. Simonyan K, Zisserman A (2015) Very deep convolutional networks for large-scale image recognition. ICLR
9. He K, Zhang X, Ren S, Sun J (2015) Deep residual learning for image recognition. Cornell University
10. Howard AG, Zhu M, Chen B, Kalenichenko D, Wang W, Weyand T, Andreetto M, Adam H (2017) MobileNets: efficient convolutional neural networks for mobile vision applications. Cornell University

COVID-19 in Bangladesh: An Exploratory Data Analysis and Prediction of Neurological Syndrome Using Machine Learning Algorithms Based on Comorbidity

Shuvo Chandra Das, Aditi Sarker, Sourav Saha, and Partha Chakraborty

Abstract COVID-19 is caused by the SARS-CoV-2 virus, which has infected millions of people worldwide and claimed many lives. This highly contagious virus can infect people of all ages, but the symptoms and fatality are higher in elderly and comorbid patients. Many COVID-19 survivors have experienced a number of clinical consequences following their recovery. In order to have better knowledge about the long-COVID effects, we focused on the immediate and post-COVID-19 consequences in healthy and comorbid individuals and developed a statistical model based on comorbidity in Bangladesh. The dataset was gathered through a phone conversation with patients who had been infected with COVID-19 and had recovered. The results demonstrated that out of 705 patients, 66.3% were comorbid individuals prior to COVID-19 infection. Exploratory data analysis showed that the clinical complications are higher in the comorbid patients following COVID-19 recovery. Comorbidity-based analysis of long-COVID neurological consequences was investigated and risk of mental confusion was predicted using a variety of machine learning algorithms. On the basis of the accuracy evaluation metrics, decision trees provide the most accurate prediction. The findings of the study revealed that individuals with comorbidity have a greater likelihood of experiencing mental confusion after COVID-19 recovery. Furthermore, this study is likely to assist individuals dealing with immediate and post-COVID-19 complications and its management.

Keywords COVID-19 · Exploratory data analysis · Comorbidity · Mental confusion · Long-COVID

S. Chandra Das
Department of Biotechnology and Genetic Engineering, Noakhali Science and Technology University, Noakhali, Bangladesh

A. Sarker · P. Chakraborty (✉)
Department of Computer Science and Engineering, Comilla University, Cumilla, Bangladesh
e-mail: partha.chak@cou.ac.bd

S. Saha
Department of Computer Science and Engineering, United International University, Dhaka, Bangladesh
e-mail: ssaha152058@bscse.uiu.ac.bd

V. Skala et al. (eds.), *Machine Intelligence and Data Science Applications*, Lecture Notes on Data Engineering and Communications Technologies 132,
https://doi.org/10.1007/978-981-19-2347-0_47

1 Introduction

The coronavirus disease 2019 (COVID-19), which is caused by the severe acute respiratory syndrome coronavirus-2 (SARS-CoV-2), was discovered for the first time in Wuhan, China, in 2019 [1]. SARS-CoV-2 is a naturally occurring positive-sense RNA virus that belongs to the Coronaviridae family [2]. Its genome is highly homologous to that of SARS-CoV, which was responsible for the pandemic in 2002–2003 [3]. This bactocoronavirus is highly contagious and spreading rapidly over the world and was declared a pandemic on March 11, 2020. This virus is transmitted from human to human through droplets and aerosols. For viral entrance into the host cell, SARS-CoV-2 uses two key host proteins: angiotensin-converting enzyme 2 (ACE2) and the cell surface transmembrane protease serine 2 (TMPRSS2) [4]. The virus's primary target is the alveolar epithelium, where expression of the ACE-2 receptor is high [5]. The average incubation period of COVID-19 is 5 days, but it can last up to 14 days [6]. COVID-19 patients experience a variety of clinical symptoms, ranging from asymptomatic infection to mild, moderate, or severe illness. The most common symptoms are fever, headache, cough, and loss of taste and smell [7]. Some of the less common COVID-19 symptoms include aches and pains, diarrhea, conjunctivitis, and skin rashes. Breathing difficulties or shortness of breath, chest discomfort or pressure, and a loss of speech or movement are all significant warning symptoms [8]. Viral pneumonia and inflammation of the lower respiratory tract are serious symptoms that can lead to death in rare situations. Most COVID-19 patients, on the other hand, may not exhibit any symptoms.

Despite the fact that the majority of COVID-19 patients recover within two weeks of their illness, many survivors experience a wide range of post-COVID-19 clinical symptoms. Patients may have a range of clinical problems after healing from COVID-19. Patients with COVID-19 have a high likelihood of hospitalization and mortality due to hypertension, cardiovascular disease (CVD), diabetes mellitus, and chronic obstructive pulmonary disease (COPD) [9]. Long-term consequences of COVID-19 (long COVID) may include symptoms or abnormal clinical parameters that remain for two weeks or more after the beginning of the disease and do not return to a healthy baseline. In Bangladesh, there is a significant study gap on the impact of post-COVID complications and comorbidities in bigger cohorts. This nationwide retrospective study is focused on the exploratory data analysis (EDA) of immediate and post-COVID-19 complications of COVID-19-affected comorbid patients. EDA analysis is a process by which the given dataset is analyzed to generate useful information. In this research work, EDA depicts the data in a visual form, enabling a better understanding of the immediate and post-COVID-19 complications in Bangladesh. Additionally, this study also predicts the risk of mental confusion after COVID-19 recovery employing various machine learning strategies.

2 Related Works

Several studies have reported that the presence of comorbidity in COVID-19 patients has been associated with higher and longer hospitalizations and death in China, the USA, Canada, the UK, Spain, Portugal, and Hong Kong [10–16]. The incidence and influence of comorbidities on illness prognosis among COVID-19 patients in Bangladesh were reported in a publication [17]. COVID-19 was the subject of an exploratory data study by Indian researchers [18]. The findings of the study indicate the impact of COVID-19 in India on a weekly and regular basis, as well as the response of India's healthcare sector to the epidemic. Dsouza et al. used exploratory data analysis to make conclusions regarding the correlation of COVID-19 instances [19]. The dataset is then subjected to a data visualization approach, which is then used to create patterns in order to better comprehend the pandemic's effects in connection to the features in the dataset. A recently published article demonstrates that [20] the epidemiological outbreak of COVID-19 was investigated using a visual exploratory data analysis approach.

COVID-19 symptoms and diseases have also been studied using statistical machine learning regression models and different machine learning methods. The machine learning regression model Li-MuLi-Poly was used to predict COVID-19 statistics and reported in an article [21]. This model anticipates COVID-19 fatalities in the USA. The number of COVID-19 instances is predicted using a support vector regression technique in paper [22]. The researchers in another study [23] used "Linear Regression" to estimate the "case fatality rate (CFR)" by examining worldwide regular information such as confirmed, mortality, and healed cases. In article [24], researchers utilize "Bailey's model" to investigate and evaluate COVID-19 viral transmission statistics from various countries. To assess the effectiveness of comorbidities on COVID-19 patients, we employed both statistical analysis and machine learning models.

3 Methodology

The initial phase in the data analysis procedure was to receive the raw data from the survey (Fig. 1). The information gathered was accurate. Each text category was transformed to numbers during preprocessing for further analysis. After that, the dataset was cleaned and made ready for analysis. During the data cleaning stage, irrelevant data was removed. The exploratory data analysis (EDA) was then performed on the clean data. The result of EDA might be an improved form of data as well as data visualizations. Suitable features are extracted from clean data for prediction. These features include comorbidities (diabetics, cardiovascular diseases, hypertension, and asthma) prior to COVID-19. Post-COVID neurological effects were also predicted using machine learning methods. Following that, the most accurate models were selected depending on the accuracy evaluation metrics.

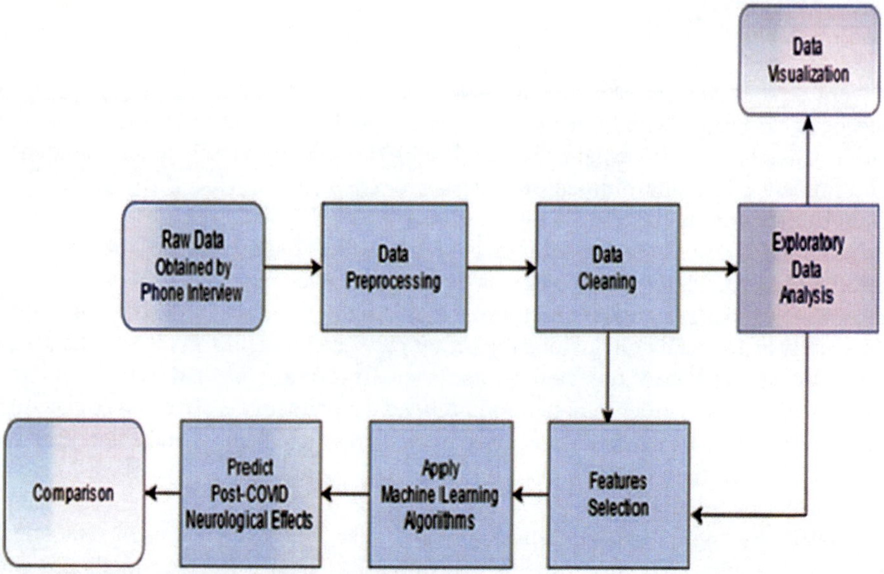

Fig. 1 Flow diagram of overall work process

Different visualization techniques based on the Python data analysis packages NumPy, Pandas, Matplotlib, and Seaborn were used for EDA analysis and data visualization. Age, gender, clinical symptoms, pre-COVID comorbidities, post-COVID complications and so on were all taken into account in the analysis. For predicting the post-COVID-19 mental confusion, regression methods such as linear regression, random forest, support vector machine, and decision tress algorithms were used. The overall process of our research is depicted in Fig. 1.

3.1 Data Collection

The patient's diagnosed with COVID-19 infection was selected by reverse transcriptase-polymerase chain reaction (RT-PCR) assay "between" January 2021 to May 2021. Patients' information was enrolled for this research if they tested negative after two consecutive RT-PCR tests at least 24 h and 28 days previous to the interview date. Telephone interviews were used to collect all of the data for the retrospective study, which was then recorded in a well-structured questionnaire. Out of 3000 patients, 705 (male 480 and female 225) patient's data were selected. Collected data were saved in Microsoft Excel. The data were gone through normalization and filtering to select critical columns, deriving new columns, and presenting the data in a graphical way. Table 1 depicts the demographic information of comorbid and non-comorbid COVID-19 positive patient.

Table 1 Demographic information of symptomatic and asymptomaticCOVID-19 patients

Characteristics	Comorbidities patients $N = 469$	Non-comorbidities patients $N = 236$
Sex		
Male	324	145
Female	145	91
Age range	2–100	5–66

Prior to COVID-19 infection, 469 patients had comorbidities such as diabetes mellitus, cardiovascular disease, hypertension, and asthma while 236 patients had none, EDA analysis was performed to determine the percentage of each comorbidity and its impact on post-COVID complications.

Immediate COVID-19 symptoms (fever, headache, cough, breathlessness, weakness, tastelessness, sore throat, loss of smell, vomiting, and diarrhea) were asked to COVID positive patients. We also gathered data on patient's post-COVID-19 complications like asthma, hypertension, diabetes, vision problems, allergies, weight loss, hair loss, mental confusion, weakness, tiredness, and other consequences are just some of the issues that can arise. We conducted an EDA analysis on post-COVID-19 complications based on comorbidity. Age, gender, symptoms, and comorbidity-wise EDA analysis were performed in this paper.

4 Experimental Results and Discussion

4.1 Exploratory Data Analysis of COVID-19 in Bangladesh

The current COVID-19 pandemic has motivated an EDA analysis on the datasets obtained through a phone interview. To process and extract information from the dataset, this article employed "Python" for "data processing" and the "pandas" package. The Python libraries "Matplotlib" and "Seaborn" were used to produce appropriate graphs for better visualization.

4.1.1 Age and Gender-Wise Analysis of COVID-19 in Bangladesh

Both symptomatic and asymptomatic patients are considered in this age- and gender-based analysis. The bar chart evaluates COVID-19 patients distribution in Bangladesh to see which age groups and genders are mostly affected. Male patients in the age bracket 41–50, which is considered middle age, are the most affected, with 15.4%, 0.5.0% are between the ages of 71 and 100, 8.7% are between the ages of 61 and 70, 12.3% are between the ages of 51 and 60, 14.1% are between the ages of 31 and 40, 9.5% are between the ages of 21 and 30, 1.9% are between the ages of 11

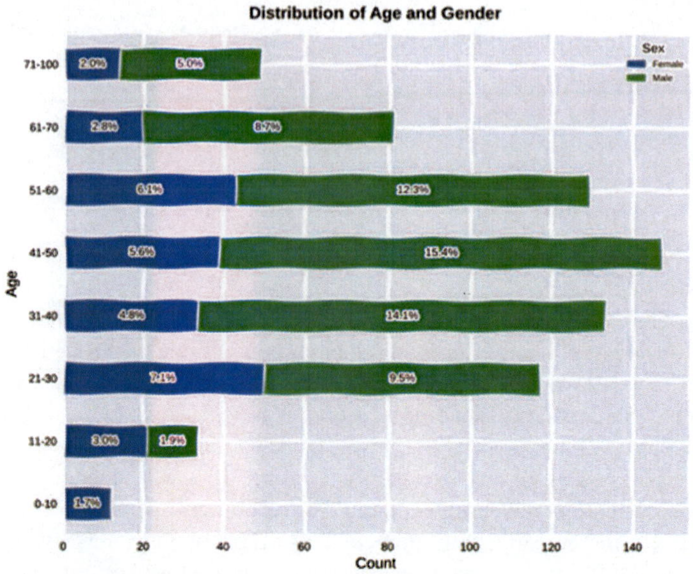

Fig. 2 Age and gender-wise analysis of COVID-19 patients in Bangladesh

and 20, and no male patients are found in our survey data between the ages of 0 and 10. Patients between the ages of 21 and 30 are the most impacted, accounting for 7.1% of all female patients. Between the ages of 51 and 60, 6.1% of the female patients is between those ages. 5.6% are aged 41–50, 4.8% are aged 31–40, 3.0% are aged 11–20, 2.8% are aged 61–70, 2.0% are aged 71–100, and 1.7% are aged 0–10 (Fig. 2).

4.1.2 Symptoms and Gender-Wise Analysis of COVID-19 Patients in Bangladesh

The *X*-axis of the bar graph in Fig. 3 depicts the percentage of male and female patients, while the *Y*-axis represents the names of symptoms that were evaluated from the persons who were tested. This is the observation to retain the common symptoms checklist that was established to keep track of new patients and to assist identify them as positive or negative. Fever is the most common symptom among male patients, accounting for 11.6% of all cases. Other commonly observed symptoms include loss of smell (9.0%), weakness (8.2%), headache (7.8%), cough (7.7%), breathlessness (5.9%), sore throat (5.8%), tastelessness (4.9%), diarrhea (3.9%), and vomiting (2.8%). As a result of the investigation, it was shown that the most common symptom for patients is fever and vomiting is a symptom that is rarely found in male patients. The most prevalent symptom among female patients is weakness, which accounts for 5.0% of all cases. Fever (4.8%), loss of smell (4.4%), headache (4.1%),

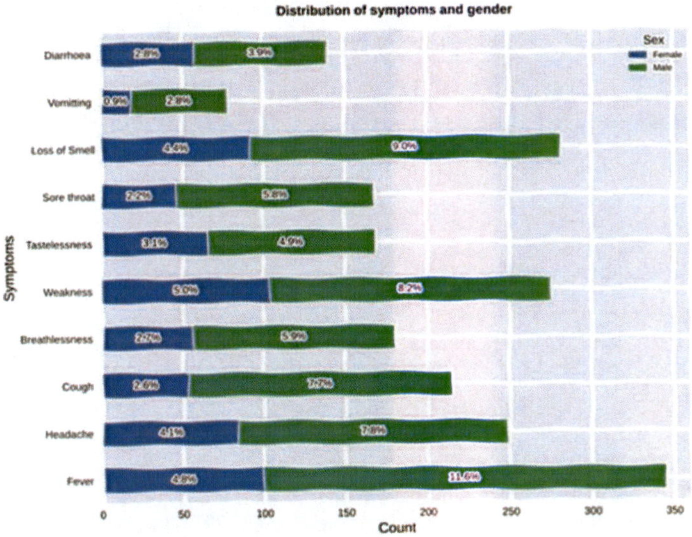

Fig. 3 Symptoms and gender-wise analysis of COVID-19 patients in Bangladesh

tastelessness (3.1%), diarrhea (2.8%), breathlessness (2.7%), cough (2.6%), sore throat (2.2%), and vomiting are some of the most prevalent symptoms (0.9%) among the female patients. As a result of the inquiry, it was reviled that fever, weakness, headache, loss of smell are the most common symptoms among female patients. Vomiting is a symptom that occurs in a small percentage of male and female patients.

4.1.3 Comorbidities Diseases Analysis in Bangladesh

The proportion of patients without comorbidities in our dataset is 33.7%, while the number of patients with comorbidities is 66.3%. The comorbidities diseases analysis of COVID-19 in Fig. 4 shows that the number of patients with comorbidities is higher than the number of patients without comorbidities. Patients had comorbidities such as asthma, hypertension, diabetes mellitus (DM), and cardiovascular disease. According to the analysis of comorbidities diseases, 41.7% of patients had diabetes, 20.7% of patients had asthma, 20.0% of patients had hypertension, and 17.6% of patients had cardiovascular diseases before being impacted by COVID-19. People with diabetes, in particular, are more affected by COVID-19 than patients with other concomitant conditions. This is due to the presence of enhanced basal level of inflammatory cytokines in diabetic patients, allowing a cytokine storm in response to SARS-CoV-2 infections [25]. Furthermore, several studies also revealed that patients with other comorbid conditions such as hypertension cardiovascular disease, asthma are more likely to experience quick and more worsen outcomes than those without [26].

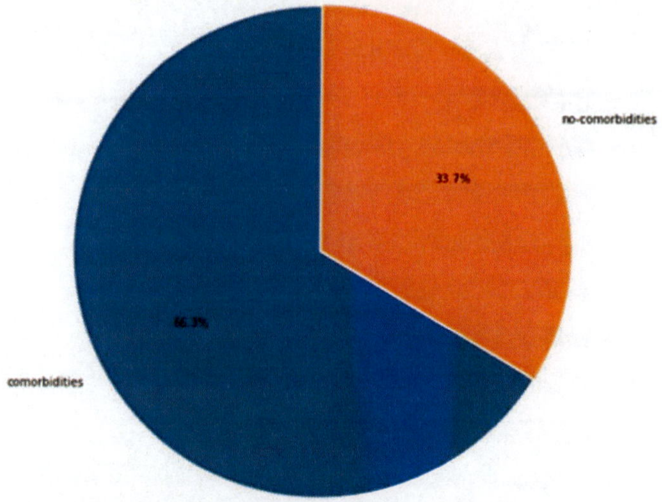

(a) Percentage of comorbid and healthy COVID-19 patients

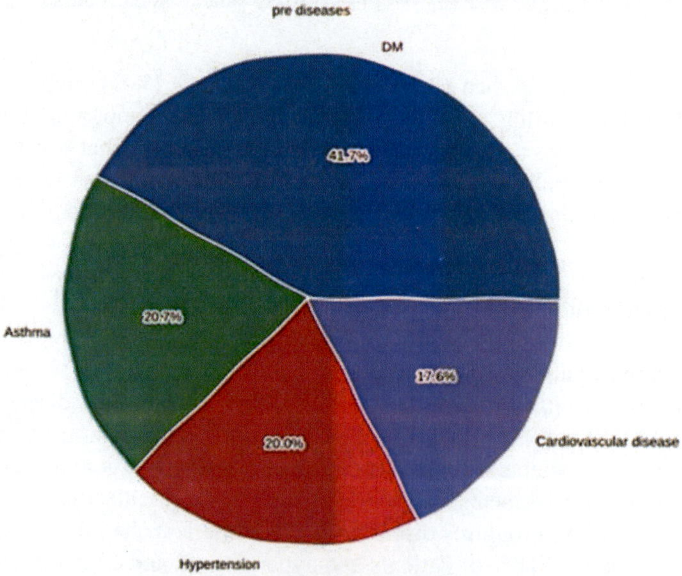

(b) Percentage of COVID-19 patients with various comorbid conditions

Fig. 4 Comorbidity percentage of patients before COVID-19 infection

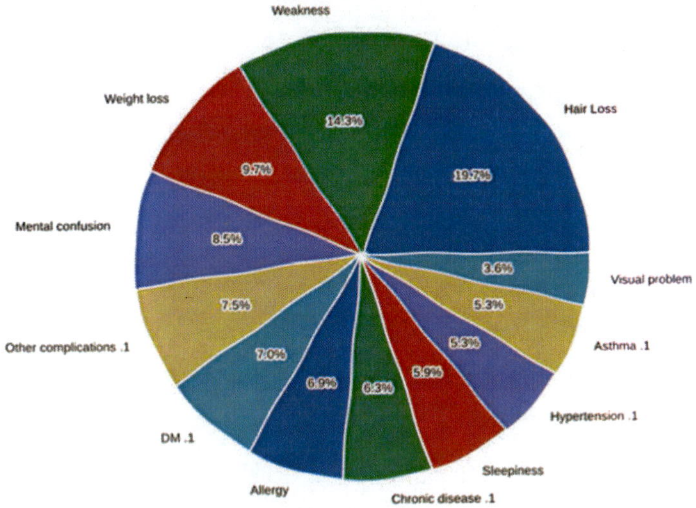

Fig. 5 Post-COVID-19 complications analysis

4.1.4 Post-COVID-19 Complications Analysis

According to Fig. 5, recovered patients showed different clinical complications like hair loss in 19.7% patients, weight loss in 9.7% patients, while mental confusion arise in 8.5% patients, diabetics in 7.0%, allergy in 6.9% patients, tiredness in 5.9% patients, vision difficulties in 3.6% patients, and other consequences in 7.5% patients. Newly developed comorbidities like asthma (5.3%), hypertension (5.3%) and chronic diseases (6.3%) were also arisen in COVID-recovered patients. Hair loss, weakness, weight loss, and mental confusion were the most common post-COVID complications.

4.1.5 Comorbidity (Before COVID) and Symptom-Wise Analysis of COVID-19

Figure 6 illustrates a symptom-by-symptom examination of COVID-19-based comorbidities in a bar chart. The number of patients with comorbidities is represented by the green color, while the number of patients without comorbidities is shown by the blue color of the bar chart. Fever symptoms are seen by 10.6% of comorbid patients and 5.6% of non-comorbid patients, and these symptoms are seen by the largest number of patients 10.9% of the patients with comorbid conditions. Loss of smell is the most common symptom across comorbid patients (10.9%) and non-comorbid individuals (3.7%). Accordingly, 8.4% of the patients have weakness, 8.2% have a headache, 7.6% have a cough, 6.8% have a sore throat, 5.8% have dyspnea, 4.6% have tastelessness, 3.6% have diarrhea, and 1.8% have vomiting due

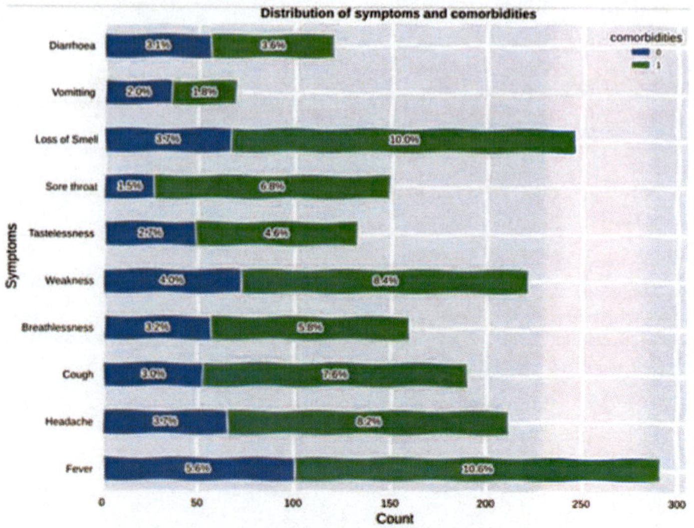

Fig. 6 Comorbidities and symptom-wise diseases analysis in Bangladesh

to various comorbidities. Meanwhile, weakness is shown to 4.0%, headache is shown to 3.8%, breathlessness is shown to 3.2%, diarrhea is shown to 3.1%, cough is shown to 3.0%, tastelessness is shown to 2.7%, vomiting is shown to 2.0%, and sore throat are shown to 1.5% patients who had not any comorbidities. According to the above analysis, symptoms are mostly viewed in the patients with comorbidities.

4.2 *Post-COVID Neurological Effect Analysis*

4.2.1 Statistical Analysis

It is already well established that SARS-CoV-2 can infect a wide variety of tissues and organs in the presence of adequate angiotensin-converting enzyme-2 (ACE2) receptors [27]. In particular, SARS-CoV-2 infect nerve cells specifically neurons and can damage nervous system, resulting in encephalopathy ranging from mild headache to mental confusion, dementia, etc. [28]. In this study, the post-COVID-19 neurological complications (mental confusion, visual problems, and sleepiness) were analyzed based on the comorbid conditions of the patients. The result is depicted in Table 2.

Here, 0 denotes the absence of comorbidity prior to COVID-19, while 1 denotes the presence comorbidities prior to COVID-19. Asthma, hypertension, cardiovascular disease, and diabetes mellitus are the four comorbidities taken for the analysis of post-COVID neurological effects. Based on comorbid conditions neurological syndromes (mental confusion, visual problem, and sleepiness) were analyzed.

Table 2 Post-COVID-19 neurological effect analysis

Category	Probability	Neurological disorder level	Number of patients
0	0.00	Low	143
	0.33	Middle	83
	0.66	Above middle	10
1	0.00	Low	274
	0.33	Middle	169
	0.66	Above middle	17
	1.00	High	5

The risk of mental confusion as a post-COVID-19 complication is divided into four categories. Low levels suggest a 0.00 probability of mental disorientation, and there are no neurological effects after COVID. After COVID, any one neurological effect is detected at the mid-level, which indicates a 0.33 chance of having mental confusion. Above the middle level, there is a 0.66 chance of mental disorientation, and after COVID-19, any two neurological effects are visible. And a high level implies a 1.00 chance of mental disorientation, and all three neurological consequences are visible after COVID-19. After recovering from COVID, 143 non-comorbidities patients have a low risk of mental disorientation, 83 have a middle risk, and 10 have an above-medium risk. Furthermore, no patient is at serious danger of becoming mentally confused. However, 274 people with comorbidities have a low risk of mental confusion, 169 people have a middle risk, 17 people have an above middle risk, and 5 people have a high risk of mental confusion.

Following the study of Table 2, we found that patients with comorbidities have a higher risk of mental confusion as a post-COVID-19 consequence than patients without comorbidities.

4.2.2 Prediction of Mental Confusion in COVID Survivors Using Machine Learning

According to the EDA analysis, the number of patients with comorbidities, both male and female, is significantly larger than the number of patients without comorbidities.

And post-COVID-19 problems occur more frequently in patients with comorbidities than in people without comorbidities. And, according to our statistical research of post-COVID-19 neurological effects, out of 465 comorbidities patient's 191 patients have a greater risk of developing mental confusion after COVID. Additionally, we identified patients with comorbidities between the ages of 10 and 100. As a result, we realize that we should apply different machine learning approach to predict the risk of mental confusion after COVID based on comorbidities.

SVM, DT, RF supervised learning methods [29–32], and a linear recreation [33] model are used in this work to analyze and predict the post-COVID neurological effects regarding the participant's comorbidities. To predict the risk of post-COVID

Table 3 Machine learning model evaluation

Models	MSE	MAE	RMSE
Decision tree	0.0384	0.0742	0.1960
Random forest (RF)	0.0565	0.1631	0.2377
Linear regression	0.0674	0.2005	0.2597
SVM	0.0674	0.2005	0.2597

mental confusion age, gender, and comorbidities diseases are taken as input variables. All of the models mentioned above are trained on 80% of the data and tested on 20% of the data.

The most accurate models were selected on the basis of the accuracy evaluation metrics root mean square error (RMSE), mean square error (MSA), and mean absolute error (MAE). A good model has the qualities of minimizing the root mean square error and the mean absolute error. Decision tree (DT) and random forest (RT) models are presented in Table 3 as models that have lower error than other models. So, DT and RF provide the most accurate prediction of post-COVID-19 mental disorientation.

5 Conclusion

This study provides EDA analysis of COVID-19 patients based on age, gender, comorbidity, clinical symptoms, and post-COVID complications. The major outcome of this study is that comorbidities have a significant impact on post-COVID-19 complications in COVID-19 patients. Additionally, machine learning algorithms used to predict neurological complexities that will provide a valuable insight into the treatment and management of COVID-19 survivors in Bangladesh. The EDA analysis is based on the retrospective investigations in which clinical data from COVID-19 patients were collected and examined. More information about COVID-19 patients will help us to improve the findings of our research.

References

1. Kong WH, Li Y, Peng MW, Kong DG, Yang XB, Wang L, Liu MQ (2020) SARS-CoV-2 detection in patients with influenza-like illness. Nat Microbiol 5(5):675–678
2. Lippi A, Domingues R, Setz C, Outeiro TF, Krisko A (2020) SARS-CoV-2: at the crossroad between aging and neurodegeneration. Movement Disorders 35(5)
3. Yuki K, Fujiogi M, Koutsogiannaki S (2020) COVID-19 pathophysiology: a review. Clin Immunol 215:108427
4. Pillay TS (2020) Gene of the month: the 2019-nCoV/SARS-CoV-2 novel coronavirus spike protein. J Clin Pathol 73(7):366–369
5. Li W, Moore MJ, Vasilieva N, et al (2003) Angiotensin-converting enzyme 2 is a functional receptor for the SARS coronavirus. Nature 4:426:450

6. Parasher A (2021) COVID-19: Current understanding of its pathophysiology, clinical presentation and treatment. Postgrad Med J 97(1147):312–320
7. Hossain M, Das SC, Raza MT, Ahmed IU, Eva IJ, Karim T, Chakraborty P, Gupta SD (2021) Immediate and post-COVID complications of symptomatic and asymptomatic COVID-19 patients in Bangladesh: a cross-sectional retrospective study. Asian J Med Biol Res 7(2):191–201
8. Salman FM, Abu-Naser SS (2020) Expert system for COVID-19 diagnosis
9. Emami A, Javanmardi F, Pirbonyeh N, Akbari A (2020) Prevalence of underlying diseases in hospitalized patients with COVID-19: a systematic review and meta-analysis. Archives Acad Emerg Med 8(1)
10. Guan WJ, Liang WH, Zhao Y, Liang HR, Chen ZS, Li YM, et al (2020) Comorbidity and its impact on 1590 patients with COVID-19 in China: a nationwide analysis. Eur Respir J 55(5)
11. Nisha KA, Kulsum U, Rahman S, Hossain MF, Chakraborty P, Choudhury T (2022) A comparative analysis of machine learning approaches in personality prediction using MBTI. In: Das AK, Nayak J, Naik B, Dutta S, Pelusi D (eds) Computational intelligence in pattern recognition. Adv Intell Syst Comput Vol 1349. Springer, Singapore. https://doi.org/10.1007/978-981-16-2543-5_2
12. Chakraborty P, Yousuf MA, Rahman S (2021) Predicting Level of Visual Focus of Human's Attention Using Machine Learning Approaches. In: Kaiser MS, Bandyopadhyay A, Mahmud M, Ray K (eds) Proceedings of international conference on trends in computational and cognitive engineering. Adv Intell Syst Comput vol 1309. Springer, Singapore. https://doi.org/10.1007/978-981-33-4673-4_56
13. Chakraborty P, Ahmed S, Yousuf MA, Azad A, Alyami SA, Moni MA (2021) A human-robot interaction system calculating visual focus of human's attention level. In: IEEE Access. 9:93409–93421. https://doi.org/10.1109/ACCESS.2021.3091642
14. Nagavelli U, Samanta D, Chakraborty P (2022) Machine learning technology-based heart disease detection models. J Healthc Eng 2022(9) Article ID 7351061. https://doi.org/10.1155/2022/7351061
15. Chakraborty P, Sultana S (2022) IoT-based smart home security and automation system. In: Sharma DK, Peng SL, Sharma R, Zaitsev DA (eds) Micro-electronics and telecommunication engineering. Lecture notes in networks and systems vol 373. Springer, Singapore. https://doi.org/10.1007/978-981-16-8721-1_48
16. Nogueira PJ, de Araújo Nobre M, Costa A, Ribeiro RM, Furtado C, Bacelar Nicolau L, et al The role of health preconditions on COVID-19 deaths in Portugal: evidence from survey
17. Nadim S, Opu RR, Ahmed SN, Sarker MK, Jaheen R, Daullah MU, Khan S, Mubin M, Rahman H, Islam F, Khan FB, Haque N, Ayman U, Shohael AM, Dey SK, Talukder AA (2021) Prevalence and impact of comorbidities on disease prognosis among patients with COVID-19 in Bangladesh: a nationwide study amid the second wave. Diab Metabol Syndr: Clin Res Rev, Elsevier 15
18. Mittal S (2020) An exploratory data analysis of COVID-19 in India. Int J Eng Res Technol (IJERT) 09(04):28
19. Dsouza J, Senthil S (2020) Using exploratory data analysis for generating inferences on the correlation of COVID-19 cases. In: 11th International conference on computing, communication and networking technologies (ICCCNT), India
20. Samrat k, Mahbubur R, Umme R, Arpita H (2020) Analyzing the epidemiological outbreak of COVID-19: a visual exploratory data analysis approach. J Med Virol 92:632–638
21. Singh H, Bawa S (2021) Predicting COVID-19 statistics using machine learning regression model: Li-MuLi-Poly. Springer, Multimedia Systems. https://doi.org/10.1007/s00530-021-00798-2
22. Peng Y, Nagata MH (2020) An empirical overview of nonlinearity and overfitting in machine learning using COVID-19 data. Chaos Solitons Fractals, Elsevier, 139
23. Hoseinpour Dehkordi A, Alizadeh M, Derakhshan P, Babazadeh P, Jahandideh A (2020) Understanding epidemic data and statistics: a case study of COVID-19. 92(7)

24. Gondauri D, Mikautadze F, Batiashvili M (2020) Research on covid-19 virus spreading statistics based on the examples of the cases from different countries. Electron J General Med 17(4)
25. Azar WS, Njeim R, Fares AH, Azar NS, Azar ST, El Sayed M, Eid AA (2020) COVID-19 and diabetes mellitus: how one pandemic worsens the other. Rev Endocr Metab Disord 21(4):451–463
26. Sanyaolu A, Okorie C, Marinkovic A, Patidar R, Younis K, Desai P, Hosein Z, Padda I, Mangat J, Altaf M (2020) Comorbidity and its impact on patients with COVID-19. In: SN comprehensive clinical medicine, pp 1–8
27. Robba C, Battaglini D, Pelosi P, Rocco PR (2020) Multiple organ dysfunction in SARS-CoV-2: MODS-CoV-2. Expert Rev Respir Med 14(9):865–868
28. Liotta EM, Batra A, Clark JR, Shlobin NA, Hoffman SC, Orban ZS, Koralnik IJ (2020) Frequent neurologic manifestations and encephalopathy-associated morbidity in Covid-19 patients. Ann Clin Transl Neurol 7(11):2221–2230
29. Rohini M, Naveena K, Jothipriya G, Kameshwaran S, Jagadeeswari M (2021) A comparative approach to predict corona virus using machine learning. In: International conference on artificial intelligence and smart systems (ICAIS)
30. Hossain MF, Islam S, Chakraborty P, Majumder AK (2020) Predicting daily closing prices of selected shares of Dhaka stock exchange (DSE) using support vector machines. Internet Things Cloud Comput 8(4):46
31. Faruque MA, Rahman S, Chakraborty P, Choudhury T, Um JS, Singh TP (2021) Ascertaining polarity of public opinions on Bangladesh cricket using machine learning techniques. Spatial Inform Res 1–8
32. Sarker A, Chakraborty P, Sha SS, Khatun M, Hasan MR, Banerjee K (2020) Improvised technique for analyzing data and detecting terrorist attack using machine learning approach based on twitter data. J Comput Commun 8(7):50–62
33. Argiawu A (2020) Linear regression model for predictions of COVID-19 new cases and new deaths based on may/june data in Ethiopia. https://doi.org/10.21203/rs.3.rs-61667/v1

Ransomware Family Classification with Ensemble Model Based on Behavior Analysis

Nowshin Tasnim, Khandaker Tayef Shahriar, Hamed Alqahtani, and Iqbal H. Sarker

Abstract Ransomware is one of the most dangerous types of malware, which is frequently intended to spread through a network to damage the designated client by encrypting the client's vulnerable data. Conventional signature-based ransomware detection technique falls behind because it can only detect known anomalies. When it comes to new and non-familiar ransomware traditional system unveils huge shortcomings. For detecting unknown patterns and sorts of new ransomware families, *behavior-based anomaly detection* approaches are likely to be the most efficient approach. In the wake of this alarming condition, this paper presents an *ensemble classification model* consisting of three widely used machine learning techniques that include decision tree (DT), random forest (RF), and K-nearest neighbor (KNN). To achieve the best outcome, ensemble *soft voting* and *hard voting* techniques are used while classifying ransomware families based on attack attributes. Performance analysis is done by comparing our proposed ensemble models with standalone models on behavioral attributes-based ransomware dataset.

Keywords Ransomware · Behavior analysis · Cybersecurity · Machine learning · Ensemble model · Supervised classification

1 Introduction

The Internet usage pattern has changed dramatically over the years. With the advent of web services, cyber threats are also increasing gradually [14]. A cyber threat is a form of malicious activity that attempts to disrupt client data which includes malware, data breaches, zero-day attacks, identity theft, and other harmful practices [19]. In

N. Tasnim · K. T. Shahriar · I. H. Sarker (✉)
Department of Computer Science and Engineering, Chittagong University of Engineering and Technology, Chittagong 4349, Bangladesh
e-mail: iqbal@cuet.ac.bd

H. Alqahtani
College of Computer Science, King Khalid University, Abha, Saudi Arabia

V. Skala et al. (eds.), *Machine Intelligence and Data Science Applications*, Lecture Notes on Data Engineering and Communications Technologies 132,
https://doi.org/10.1007/978-981-19-2347-0_48

recent years, ransomware has become one of the most serious digital crimes which affects organizations pitifully. The Federal Bureau of Investigation (FBI) reports that the losses incurred in 2016 due to ransomware are approximately 1 billion US dollars [4]. Brewer et al. [4] informed that in 1989, Dr. Joseph Popp circulated a trojan called PC Cyborg where a malware program would cover all the folders and clash with records on Computer's C: drive. Then the attacker sent a ransom demanding 189 US dollars to recover the data from the affected computer and manage the malicious programs. In short, ransomware is a self-spreading malicious program that uses encryption and locking mechanisms to capture victims' information and demand a ransom for recovery. Thus, it is a great challenge to ensure the security and safety of digital documents from ransomware. To meet the challenges of the Fourth Industrial Revolution (Industry 4.0), Sarker et al. [17] provided a direction with artificial neural network (ANN) and deep learning (DL) methods that can also be used to protect computer networks. However, in this paper, we focus on machine learning techniques that can detect the cyber-anomalies effectively [16]. Bendovschi et al. [3] reported that ransomware has exceeded 33% by 2020 than 2019. Hence, with the increase of ransomware attacks, it is essential to detect attacks effectively and minimize financial loss.

With the advancement of information technology, cybercriminals develop new types of attacks and tactics to make the computer system vulnerable and remain untraceable. Signature-based anomaly detection is not useful to detect ransomware because attackers could change and increase the malicious program to bypass the detection mechanisms of anti-virus softwares [12]. However, Kruegel et al. [10] presented the complexity of storing a large number of signatures of known anomalies in the signature-based ransomware detection system. The authors also suggested a strategy that evaluates each web application and compares it with standard log files to detect anomalies. Based on the concept of AI-based cybersecurity, Sarker et al. [19] presented a summary by addressing the challenges of traditional methods and focusing on data-driven intelligent decision support to protect the system from cyber-attacks on the perspective of machine learning.

In this paper, we propose a behavior-based ransomware family classification system. We collect a ransomware dataset having 85 behavioral features for the classification analysis. We select the 20 best features from 85 features by using an effective feature selection technique. We consider the correlation value less than 0.95 to avoid multicollinearity problems. The main advantage of this method is that it does not consider any traditional static methods but focuses on the dynamic prediction method. The primary contributions of this paper are given below:

- Our proposed approach performs a behavior-based analysis by using the machine learning approach and acquires the ideal precision of ransomware classification.
- The proposed approach effectively detects ransomware families by applying an ensemble voting classifier with the implementation of three machine learning techniques.

- The range of experiments presents a comparison of standalone models with the ensemble voting-based models with an evaluation of standard deviation and means of accuracy to show the effectiveness of our proposed approach.

The rest of the paper is organized as follows. Section 2 reviews the related works, and Sect. 3 precisely describes the working procedure of the proposed method. Section 4 contains the result and performance analysis. And finally, in Sect. 5 we conclude the paper by summarizing the work.

2 Related Work

Ransomware is a detrimental kind of malware that can lock the victims' screen or illegally encrypt their confidential documents for ransom, resulting in significant damage to clients. Zhang et al. [20] separated ransomware by families which helps to distinguish the variation of the ransomware test. They achieved an accuracy rate of 91.43 percent by using an opcode sequence for each sample of ransomware and converting it to an N-gram sequence. The classification method of Pirscoveanu et al. [14] implemented a cognitive combination of features that achieves a high degree of accuracy with a typical AUC value of 0.98 for random forest classifier. Pekta et al. [13] particularly addressed that in runtime analysis of malware, file system, network, registry activities, and API calls are the most important behavioral attributes. They also used N-gram display over API calls to separate malware families. Daku et al. [7] proposed primarily two approaches: a repetitive approach to recognize behaviors for high-level classification performance, and a collective approach for highly related behaviors. Alaeiyan et al. [1] recommended another order of trigger-based malware classification by following evasive and elicited practices. Both of these practices address the specification of environmental conditions. However, evasive practices focus on self-defense while elicited practices show the benefits of malware for malignant demonstrations. Chen et al. [6] presented how to measure ransomware behavior from a secure log called a cuckoo sandbox. They considered all logs from contaminated hosts as individual records and looked at the features from the infection report by using the TF-IDF process. Galal et al. [9] discussed statistical-based, graph-based, polymorphic, and metamorphic malware structure. To avoid problems with signature-based detection Canfora et al. and Bazrafshan et al. [2, 5] provided some alternative detection methods such as obfuscation strategy and heuristic technique.

By considering the above works, in this paper, we focus on the behavior of anomaly to handle the rise of cybercrime and the problems of the traditional signature-based detection system. Moreover, our feature selection technique provides better classification accuracy of more than 97 percent in ten different ransomware families. The proposed approach is based on an ensemble model which is incorporated with three basic machine learning classifiers.

3 Methodology

In this section, we present the methodology of our proposed approach that performs the classification process on ten different ransomware families. We dynamically select the best 20 features out of 85 features by using the correlation matrix and the value of feature importance without any user involvement. We develop an ensemble model by implementing three popular machine learning models: K-neighbor, decision tree, and random forest classifier [18]. Our approach automatically selects the optimized K value for the K-neighbor classifier. Finally, we present a comparison of standalone models with the ensemble models. We illustrate the whole process of the proposed ensemble model-based ransomware classification approach in Algorithm 1.

Algorithm 1: Ransomware family classification with ensemble model

Input: Dataframe df containing ransomware instances of n different families.
Result: Predict family class label of n instances.

ObjList = df.select_data_type(object);
for *feature in ObjList* **do**
  | df[feature] = convert to numeric
end
Correlation_matrix = df.corr();
Drop upper[columns], where correlation_value> 0.95 ;
Feed df into mutual_classif_info to find importance with respect to target label;
sort importance in descending order;
if *importance <threshold* **then**
  | drop less important features;
end
for *i in range (1, 40)* **do**
  | error_rate = KNeighborClassifier(k = (i + 1);
end
Select k value for minimum error_rate;
Pass optimized KNN, DT and RF to Ensemble model;
Feed dataset into Ensemble model;
Compare standalone model with ensemble model by mean value and standard deviation of classification accuracy;

At first, the algorithm takes the dataset as input that contains ransomware instances. Each instance has 85 behavioral attributes. In the next step, we perform the data preprocessing for the feature selection. Then the processed data access to the feature selection phase by implementing a twofold method. In the first fold, the less important features are removed using the information gain method. The threshold value is automatically generated according to the data pattern. The second fold of the feature selection method considers the correlation value to remove the highly correlated features. In the final phase of the proposed algorithm, the ensemble model is implemented to classify the ransomware family.

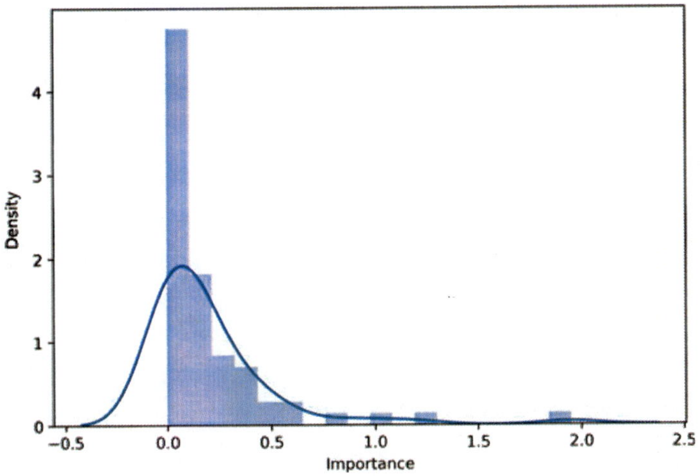

Fig. 1 Density of feature importance

3.1 Data Preprocessing

We collect ten different ransomware family datasets with similar attributes from the Canadian Institute for Cybersecurity [8, 16] and combine them into a single dataset. The final dataset contains 107,700 instances with 85 features and 10 different class labels. Then the system observes data to check whether it fits for the next procedure. Few feature selection methods do not accept non-numerical attributes. Hence, data preprocessing plays a vital role in suiting the data into the adopted feature selection method. The proposed system analyzes the features and converts non-numerical features to numerical ones. We apply the information gain method and correlation matrix [11, 15] to select features for further processing.

3.2 Feature Selection

The value of the attribute can be determined significantly if the dataset is managed to contain a large number of attributes. This type of dataset is often referred to as a high-dimensional dataset. This high dimensionality comes with a number of problems; for example, this will typically create overhead to train the machine learning model and increase the complexity of the model. Figure 1 shows that the density of the feature value is high in a portion of the feature section. Thus, it is important to extract important features and remove unnecessary features to get a balanced feature set to train the model. Figure 1 reflects that feature importance values are ordered in descending manner that helps to perform simple and appropriate analysis.

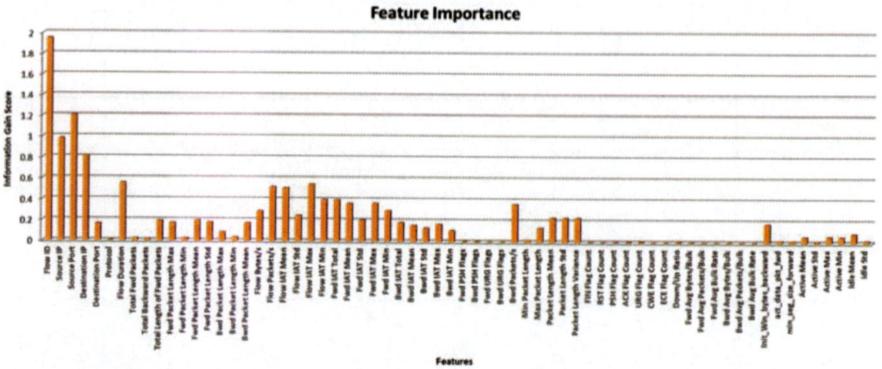

Fig. 2 Feature importance with respect to target feature

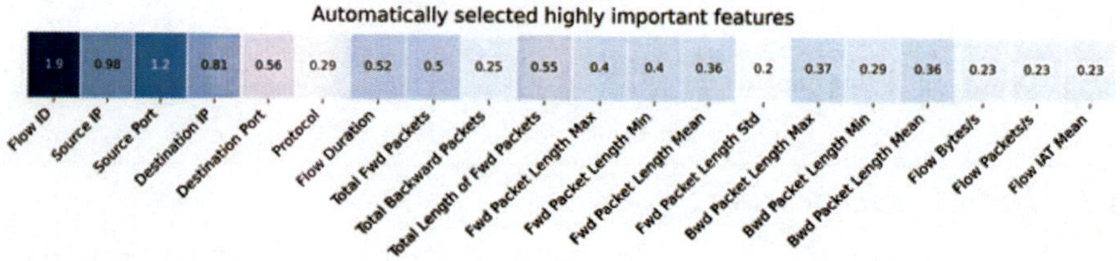

Fig. 3 Automatically selected top 20 important features

The proposed method contains two completely automated major steps to extract features dynamically. The first step is accomplished by implementing a correlation matrix. Correlation measures the internal dependency of features. The high value of the correlation is better in the sense that one attribute can be predicted from another. But often the problem arises in the case of a perfect positive or a perfect negative correlation between two attributes. Hence, it is required to manage multicollinearity problems; otherwise, it can lead to wrong predictions. Thus, the proposed system drops the attributes which have a correlation value of more than 0.95. By applying this process, 17 attributes are removed and 68 attributes are extracted. In the second step of feature selection, we measure each feature's importance by concerning the target feature. Some of the features possess almost zero importance density. Figure 2 shows the importance range with respect to the target feature. After examining all features, the information gain method exhibits that there are a few less important features.

To extract the best features, we follow an automatic selection process by generating a standard threshold and selecting the most important features. The information gain method helps to identify the key attributes to classify the training data accurately. Hence, no user interpretation is required to select the number of attributes. The proposed system automatically generates the necessary parameters to select features using the following equations.

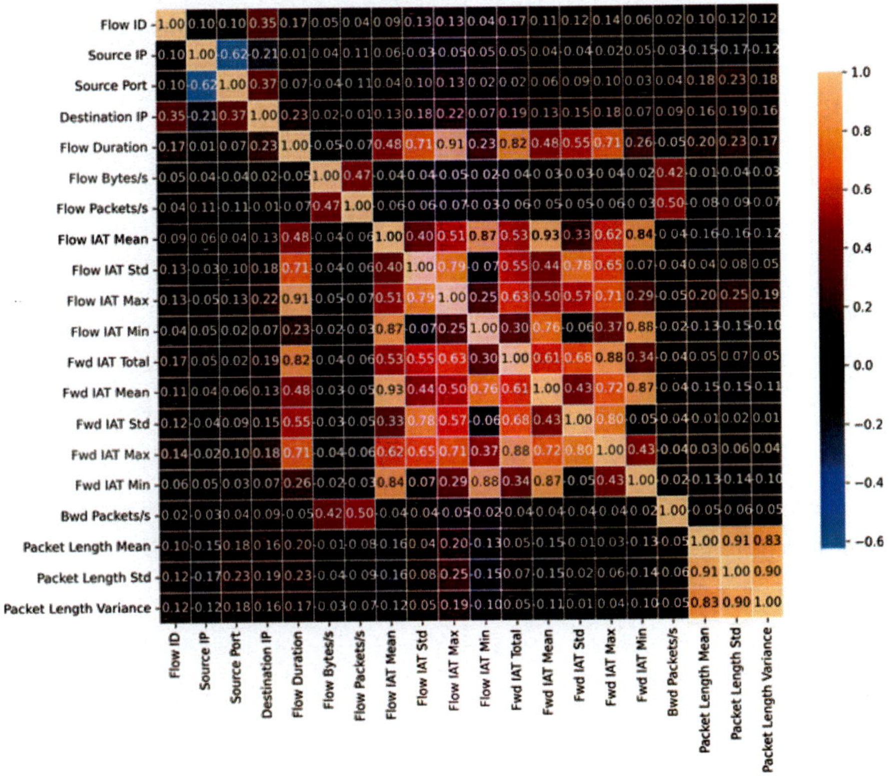

Fig. 4 Correlation matrix of finally selected features

$$a = \text{pd.Series.mean(importance)} \tag{1}$$

$$\text{new\_series} = \text{importance[(importance} > \text{a).any(1)]} \tag{2}$$

$$\text{threshold} = \text{len(new\_series)} \tag{3}$$

$$p = (\text{threshold/len(importance)}) * 100 \tag{4}$$

Here p is the percentile parameter to pass into the feature selection method. Our adopted technique selects $p\%$ features from the total number of attributes. However, in this process, only key attributes are selected that have a value that exceeds the mean importance value and produces a percentage of the total length of the series. The dynamic feature selection is one of the main contributions. Figure 3 presents automatically selected features by the proposed system. Eventually, 20 features are selected. Figure 4 shows the correlation matrix of the finally selected features, which has value less than 0.95.

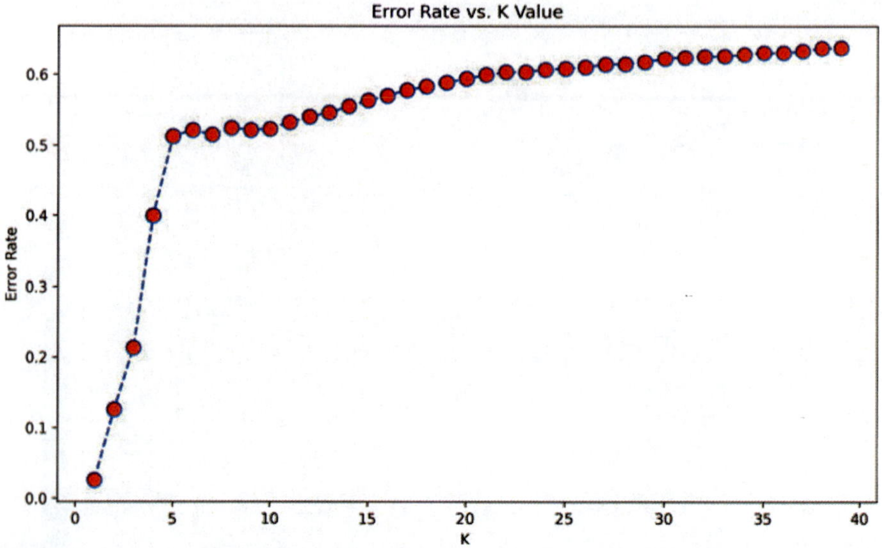

Fig. 5 Error rate w.r.t *K* value in KNN classifier

3.3 Ensemble Model

We propose an ensemble model to detect ransomware families effectively. The ensemble model measures the performance of multiple models with a variety of voting techniques. We analyze both hard and soft voting with a standard deviation of accuracy score. We incorporate the decision tree, random forest, and K-neighbour algorithm in the ensemble model. However, the regression models perform like predictors and show little less efficiency because the features contain high correlation. Thus, in this case, multicollinearity and high correlations between predictors often mislead the performance. Moreover, the classifiers that use the entropy measure of attributes while taking decisions get the priority because we extract features based on the information gain method. The final execution of the ensemble model is followed by selecting decision tree classifier, random forest classifier, and the optimum version of the K-neighbor classifier. In Fig. 5, error rate with respect to each k value for K-neighbor classifier is plotted. Here, the k value is 1 with a minimum error rate of −0.025. Finally, the automatically generated optimized k value is transferred to the ensemble model.

4 Experiment and Result Analysis

This section includes results analysis with a confusion matrix, precision, recall, f1-score for each ransomware family available in the dataset. The proposed ensemble model is prepared by the three machine learning classifiers. We use the mean and

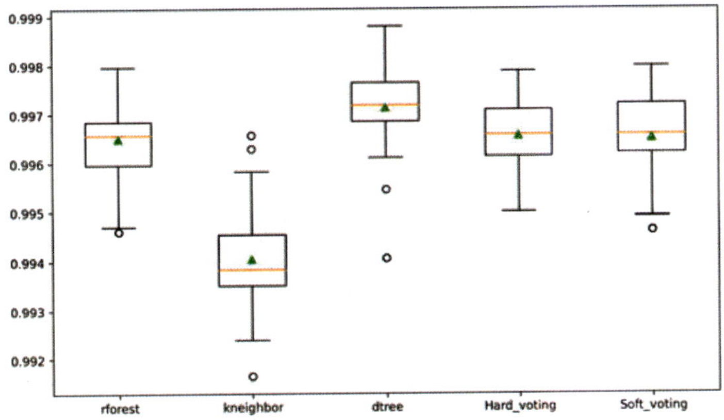

Fig. 6 Accuracy mean and std. score of standalone models and ensemble voting models

Table 1 Comparison of standalone models and ensemble model

Model	Accuracy mean	Accuracy Std.
DecisionTree	0.97716	0.00185
RandomForest	0.97211	0.00168
KNearestNeighbor	0.95214	0.00205
Ensemble hard voting	0.97299	0.00175
Ensemble soft voting	*0.97568*	*0.00155*

standard deviation methods to compare the standalone models with the ensemble models. Standard deviation is a factual estimation of the sum of numbers that fluctuate from the normal numbers in a series. A low standard deviation implies that the information is firmly associated with the normal. High standard deviation implies that there is a significant difference between the data and the mean value. Table 1 presents that soft voting of the ensemble generates high accuracy with minimum standard deviation.

Though the accuracy mean of the decision tree is higher than the soft voting of the ensemble model, it has a greater standard deviation compared to others.

The performance of the standalone models and ensemble models with two types of voting techniques is estimated and plotted in Fig. 6 also shows the standard deviation and mean score for each model. The accuracy of each model is almost similar, but according to standard deviation, soft voting for the ensemble model provides the most reliable performance by being closely related to the mean value. A large value of standard deviations is considered less reliable because it varies greatly with mean values. We present a classification report analysis of the ensemble soft voting model in Fig. 7 as it is evaluated as the best model. We use the tenfold cross-validation method to improve the reliability of the performance of the model.

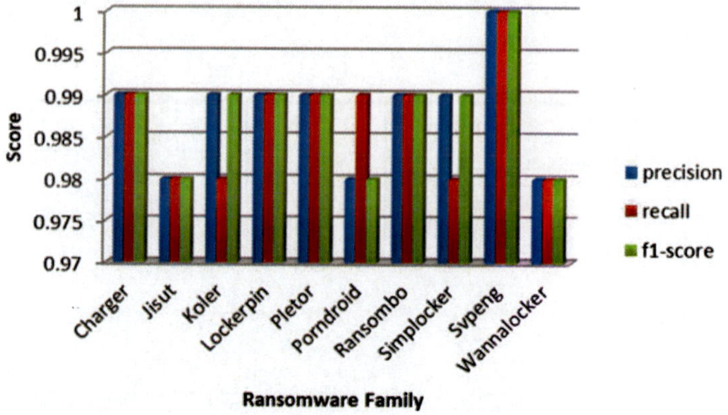

Fig. 7 Classification report of ensemble soft voting model

5 Conclusion

In this paper, we propose a method of classifying the ransomware family by using a unique ensemble method based on behavioral analysis. The behavior-centric detection system enables the most significant results as the attackers also work continuously to avoid security measures. Selecting the appropriate attribute to achieve the best accuracy is a great challenge. To overcome this challenge, in this paper, we compile two levels of the feature selection process by implementing the information gain method and correlation value. We also develop formulas to generate the importance threshold value automatically depending on the dataset pattern. Our proposed ensemble technique is based on decision tree, random forest, and the K-neighbor classifier where the best K-neighbor classifier version is selected with the minimum error rate. We analyze and evaluate the performance of the model by executing two types of voting classifiers: soft voting and hard voting. Choosing the ideal standard deviation and mean value of accuracy are the important factors for the classification purpose. Classification accuracy with minimum standard deviation is considered the most reliable one. The experimental results show that the ensemble model with the soft voting classifier performs better resulting in an accuracy of 97% with a minimum standard deviation of 0.00155.

References

1. Alaeiyan M, Parsa S, Conti M (2019) Analysis and classification of context-based malware behavior. Comput Commun 136:76–90
2. Bazrafshan Z, Hashemi H, Fard SMH, Hamzeh A (2013) A survey on heuristic malware detection techniques. In: The 5th conference on information and knowledge technology. IEEE, pp 113–120

3. Bendovschi A (2015) Cyber-attacks-trends, patterns and security countermeasures. Proced Econom Finance 28:24–31
4. Brewer R (2016) Ransomware attacks: detection, prevention and cure. Netw Secur 2016(9):5–9
5. Canfora G, Di Sorbo A, Mercaldo F, Visaggio CA (2015) Obfuscation techniques against signature-based detection: a case study. In: 2015 Mobile systems technologies workshop (MST). IEEE, pp 21–26
6. Chen Q, Bridges RA (2017) Automated behavioral analysis of malware: a case study of wannacry ransomware. In: 2017 16th IEEE international conference on machine learning and applications (ICMLA). IEEE, pp 454–460
7. Daku H, Zavarsky P, Malik Y (2018) Behavioral-based classification and identification of ransomware variants using machine learning. In: 2018 17th IEEE international conference on trust, security and privacy in computing and communications/12th IEEE international conference on big data science and engineering (TrustCom/BigDataSE). IEEE, pp 1560–1564
8. Ferrag MA, Maglaras L, Moschoyiannis S, Janicke H (2020) Deep learning for cyber security intrusion detection: approaches, datasets, and comparative study. J Inf Secur Appl 50:102419
9. Galal HS, Mahdy YB, Atiea MA (2016) Behavior-based features model for malware detection. J Comput Virology Hack Tech 12(2):59–67
10. Kruegel C, Vigna G (2003) Anomaly detection of web-based attacks. In: Proceedings of the 10th ACM conference on computer and communications security, pp 251–261
11. Lee C, Lee GG (2006) Information gain and divergence-based feature selection for machine learning-based text categorization. Inf Proc Manage 42(1):155–165
12. Patcha A, Park JM (2007) An overview of anomaly detection techniques: existing solutions and latest technological trends. Comput Netw 51(12):3448–3470
13. Pektaş A, Acarman T (2017) Classification of malware families based on runtime behaviors. J Inf Secur Appl 37:91–100
14. Pirscoveanu RS, Hansen SS, Larsen TM, Stevanovic M, Pedersen JM, Czech A (2015) Analysis of malware behavior: type classification using machine learning. In: 2015 International conference on cyber situational awareness, data analytics and assessment (CyberSA), IEEE, pp 1–7
15. Roobaert D, Karakoulas G, Chawla NV (2006) Information gain, correlation and support vector machines. In: Feature extraction. Springer, pp 463–470
16. Sarker IH (2021) Cyberlearning: effectiveness analysis of machine learning security modeling to detect cyber-anomalies and multi-attacks. Internet of Things 14:100393
17. Sarker IH (2021) Deep learning: a comprehensive overview on techniques, taxonomy, applications and research directions. SN Comput Sci 2(6):1–20
18. Sarker IH (2021) Machine learning: algorithms, real-world applications and research directions. SN Comput Sci 2(3):1–21
19. Sarker IH, Furhad MH, Nowrozy R (2021) Ai-driven cybersecurity: an overview, security intelligence modeling and research directions. SN Comput Sci 2(3):1–18
20. Zhang H, Xiao X, Mercaldo F, Ni S, Martinelli F, Sangaiah AK (2019) Classification of ransomware families with machine learning based on n-gram of opcodes. Future Generat Comput Syst 90:211–221

UCSP: A Framework to Tackle the Challenge of Dependency Chain in Cloud Forensics

Prajwal Bhardwaj, Kaustubh Lohani, Navtej Singh, Vivudh Fore, and Ravi Tomar

Abstract Coined around 2006, the term 'cloud computing' has risen since and has revolutionized how we work. Parallelly, the traditional forensics practices have been modified to include cloud computing services as a part of the investigation. However, the journey has not been easy. With the enormous advantages that cloud computing entails, it has been quite a challenge to conduct digital forensics over the cloud. It is essential to address the challenges as well as to innovate the technological field. This paper tries to put forth several challenges faced by cloud forensic investigators and summarize them. One of them is the increasing dependency of cloud providers on other cloud providers, and how it challenges forensic practices. Many vendors collaborate and provide services in conjunction with other vendors, and this teaming results in many hindrances while going for a forensic investigation. This paper will try to see the challenges from the confidentiality, integrity, and availability (CIA's) lens. Moreover, a logical model, explained with an example, is proposed to help cope with this challenge.

Keywords Cloud forensics · Cloud computing · CSP · SLA · CIA · Cloud forensics challenges · Cloud forensics model

1 Introduction

With the ever-increasing flush of cloud computing services in the market, industries were already reaping the benefits of the cloud. However, the worldwide lockdown

P. Bhardwaj · K. Lohani (✉) · N. Singh · R. Tomar (✉)
School of Computer Science, University of Petroleum and Energy Studies (UPES), Dehradun, Uttarakhand 248007, India
e-mail: kaustubhlohani25@gmail.com

R. Tomar
e-mail: rtomar@ddn.upes.ac.in

V. Fore
Gurukul Kangri University, Haridwar, India

V. Skala et al. (eds.), *Machine Intelligence and Data Science Applications*, Lecture Notes on Data Engineering and Communications Technologies 132,
https://doi.org/10.1007/978-981-19-2347-0_49

due to coronavirus provided the impetus needed to move toward the online world, thus bringing cloud computing to the center stage. Connected online over the cloud, the world learned how to adjust to the ever-changing economies.

Although cloud computing has numerous advantages, cloud services also led to new methods of conducting or organizing crimes [1]. Due to the combat security features available, it is often challenging for forensic investigators to get hold of data in the cloud. Privacy concerns when a cloud environment is hosted for many tenants are relatively common [2]. Due to the increasing complexities of cloud infrastructures and service models, forensics in the cloud is becoming increasingly difficult. Legal challenges often remain because of the current complexities of information security laws inherent in most countries. Technical challenges are often presented when different services are clubbed to form a comprehensive solution.

Challenges with forensic investigations have been increasing ever since. The volume of data in an average investigation case is becoming increasingly large, and the advanced features of the cloud make the investigations even more complex and tricky. Since more and more CSPs are collaborating to provide more comprehensive cloud solutions, increasing inter-dependency, there is a need for more in-depth investigation methodologies. This paper explores the challenges and proposes a logical model for solving one of the open challenges of dependency in cloud providers. We have also included the essential introduction to the cloud and its concepts and an introduction to other challenges faced.

This paper starts with Sect. 2 that sheds light on the background information about cloud computing, information security, and digital forensics; after that, Sect. 3 goes on to discuss the main discussion of the challenge at hand and the proposed model; and Sect. 4 finally lists out the conclusion and where future research activities can be focused.

2 Background

This section presents precursor information about cloud forensics, its challenges, and information about the confidentiality, integrity, and availability (CIA). Moreover, cloud computing, alongside its benefits, is also introduced in this section.

2.1 Information Security

This section provides context on basic building blocks of information security: confidentiality, integrity, and availability introduces cloud forensics and gives an overview of challenges faced in an investigation process and finally explains challenges related to forensics in a cloud environment.

CIA: Confidentiality, integrity, and availability, often termed as CIA, is regarded as the cornerstone of information security. Any challenge in the information space can be defined along with these aspects.

Confidentiality: According to NIST, confidentiality ensures that access to data, or any type of information, is available or provided to legitimized personnel only [3].

Integrity: According to NIST, integrity corroborates that the state of data remains unaltered throughout the transmission of information. It helps to ensure that the expected and precise state of information is maintained [4].

Availability: According to NIST, availability makes it possible that information or data can be promptly accessed by authorized personnel. It helps to ensure reliable access whenever required [5].

CIA is a very versatile lexicon and can be applied to any field. For example, a hospital, which has to deal with all the patient's information such as admission records, test reports, or even personal information, must ensure that confidentiality, integrity, and availability is preserved for all the processes. In a scenario, if confidentiality of information is not maintained, the patient's prescriptions, personal information, medical test reports, or even vaccinations can be altered by someone, resulting in lethal ramifications. If the integrity of information is not maintained, the test reports could be tampered with as they flow from laboratory to doctors and eventually to patients. In this case, alteration in reports could result in a mismatch between prognosis and the required diagnosis, resulting in disturbing repercussions in terms of resources or survival. If the availability of information is not maintained, and constant access is not necessitated, implications might include prescribing wrong dosages, ill-use of resources such as expensive or rare medications, or unanticipated medical reactions.

2.2 Cloud Computing Features and Models

Cloud computing is innovating exponentially and providing companies with the resources, and they require most without compromising on the quality or other functional aspects. Different cloud models have emerged, and the ever-changing technology space is creating even more hybrids, but at the base, cloud computing can be divided into three types: infrastructure as a service (IaaS), platform as a service (PaaS), and software as a service (SaaS).

Infrastructure as a Service (IaaS): A service that offers the basic blueprint of computing resources. It offers the lowest level of abstraction among the three. It may offer the necessary resources such as network frameworks, storage services, or compute services through virtual or hybrid models.

Platform as a Service (PaaS): A service that provides resources that enable a development engineer to deploy or build applications. The level of abstraction that PaaS resources offer is somewhere in the middle between IaaS and SaaS. PaaS services often include the basic frameworks such as the whole technology stack to build Web applications.

Software as a Service (SaaS): A cloud service that offers the highest level of abstraction. It offers ready-made software directly for use for front-end users. Most users have to subscribe to use the services. They will be very much dependent on the cloud vendors for the services.

Below we explain the three models through an example:

Someone wants to go from P1 to P2, covering a distance of 2600 miles. He wants to go by air. Now, he can either lease an airplane, rent an airplane, buy a ticket, and utilize the services of an airline.

Leasing will mean he will be in charge of fuel, maintenance, and every cost associated with it. He will be the part-owner for the time being. This type of mode can be termed IaaS. Renting will mean he just uses the plane for going from P1 to P2 for one-time usage. He is satisfied by the services as long as he gets the job done. This type of model can be seen as PaaS. If he decides to buy a ticket, he just needs to sit on his designated seat and pay for the ticket while everything is being taken care of; this model can be seen as SaaS. The responsibilities for the ownership get reduced while we go from IaaS to SaaS or from leasing to buying a ticket.

There are several features of cloud computing, as depicted in Fig. 1:

- Resource Pooling: Cloud vendors can pool multiple resources and host services for a mix of clients. For example, the same server could be used by multiple vendors, running multiple instances.
- On-demand Service: A host of self-services available for the end users ensures that they have sustained access to different visuals to monitor the system capabilities.
- Easy Maintenance: Servers available in a cloud environment are increasingly becoming more efficient, with close to 100% availability in some cases.
- Large Network Access: The resources in cloud computing can be accessed from multiple locations and are independent of the geographies, and the resources had been deployed.

Fig. 1 Features of cloud computing

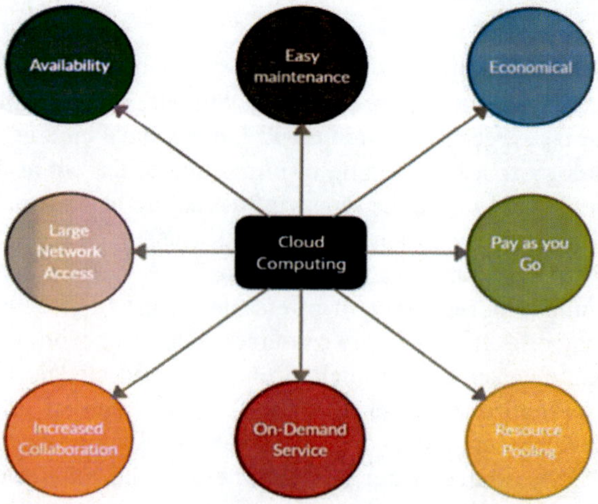

- Availability: With most of the services available through Web-based consoles or secured logins, there is no physical dependency, and end users can readily access the services just from anywhere in the world.
- Economical: Due to the reduced infrastructure requirement and reusable instances supporting a multi-tenant environment, cloud computing turns out to be economical for both customers and vendors.
- Pay-as-you-go Subscription Model: A user pays only for the services that he uses. There are no overhead costs involved, and exciting offers from vendors often make it a much lucrative choice.

Increased Collaboration: With cloud computing, collaboration efforts have taken center stage, and it is essential than ever. With multiple teams and different technologies, cloud computing makes collaboration an easy and much more effective process.

2.3 *Cloud Forensics*

Cloud Forensics: Taking keywords, 'cloud' from cloud computing and 'forensics' from digital forensics, cloud forensics is a subdivision in digital forensics that can be used in investigations involving a cloud environment. Cloud computing is supporting more and more features nowadays, making forensic investigations increasingly complex. There is a need to integrate all the processes involved in forensic investigations to have a smooth experience in an investigation. The primary purpose of any forensic investigation is to find evidence that could then be presented in a court of law.

Digital forensics encompasses multiple designated fields from computer science to forensics and beyond [6].

Some of the primary principles that every digital forensic investigation has to follow [7], as depicted in Fig. 2, are

- Preservation: includes preserving the state of the system or environment so that information cannot be altered, further meaning there should be restricted access to data, and all the changes should be noted appropriately.

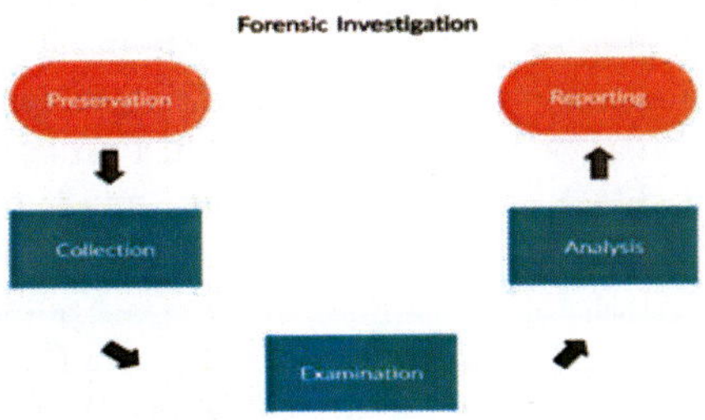

Fig. 2 Different stages in a forensic investigation

- Collection: revolves around searching the devices or systems for information and collecting all the data that might be relevant for the investigation. This involves freezing the system's state, and in some cases, documenting the whole organizational structure for data.
- Examination: includes searching for the relevant information that can be eventually represented in the court of law. This might require going through all the data collected in the previous steps, sometimes resulting in an abundance of information, while sometimes may result in recollection of data.
- Analysis: After the examination, we need to see the strength of the case, and we might have to analyze what all evidence has been collected and how it can relate to the case at hand.
- Reporting: This includes writing a detailed report of all the processes followed, giving all the pertinent information related to the case at hand.

2.4 Challenges in Cloud Forensics

Challenges faced during the forensic investigation of cloud are described below.

- Jurisdiction: Different jurisdictions have different data holding and access requirements, posing a significant challenge for forensics investigations. Lack of coordination or too many requirements to access data within other jurisdictions takes up a lot of time and effort. Furthermore, sometimes, this prolonged wait can affect the investigation. Many cloud forensic research works such as Zawod et al. [8], Ruan et al. [9], Lopez et al. [10], and Alqahtany et al. [11] have addressed the issue.
- Lack of International Collaboration: The challenge of different jurisdictions is often exacerbated by the lack of international collaboration, further impending delays in investigations. [9]
- Data Correlation Issues: Even if we can get the data from different sources about the activities from one account, there is still the issue of correlation as everyone could be using different formats and systems, consequently syncing all data to get some helpful information would still be a herculean task. [10]
- Chain of Custody: While performing a digital forensic investigation, we need to follow the rules of forensic investigations, which state the relevance of preserving systems and paths. However, since in the cloud, everything is virtual, and it is impossible to follow the trail or maintain a chain of custody [8, 10, 12].
- Access to Deleted Data: With the proliferation of cloud solutions, deleting data has become a cakewalk. Just order the deletion and within seconds, the whole block would be unmounted and may be deleted further. Sure, some companies might be willing to retain the data for the very purpose of compliance, but some notorious criminals may choose not to go with this option, and this feature could act as a roadblock while doing a forensic investigation [9].
- Data on the Cloud: One of the most intrinsic steps involved in any digital investigation is preservation. However, with all the data hosted on the cloud, pinpointing

the exact place is unrealistic. Investigators have to trust the CSPs often as they are the only point of contact. This entails both data and trust issues [13–15].

- Dependence on Cloud Providers: Due to the relative flexibility and ever-changing dynamic nature, the cloud can offer many services without much agony. Making the host of services often more lucrative, sometimes, more than one vendor collaborates and offers some service in conjunction. If each connection can be represented as one link, the whole chain of dependency, from the first vendor to the last dependent, is known as the chain of dependency. Moreover, it sprung up many new challenges to investigate the whole chain [8, 13, 15–17].

3 Challenges Related to Dependence on Cloud Providers

Nowadays, several CSPs often combine with other CSPs to provide a comprehensive solution to a consumer. To depict our framework, we will take an example of four CSPs, such as A1, A2, A3, and A4, and the related services provided by them as L1, L2, L3, L4, and L5. A1 is our primary vendor who provides service L1 and L5, but for the L2 service, A1 needs to team up with A2 to provide a comprehensive solution by loaning service L2 from A2. Now, for providing service L2, A2 has further teamed up with A3 and A4 to get the prerequisite services L3 and L4, without which L2 cannot be provided.

The consumer is getting a solution with services L1 and L2, but he has no idea what goes behind, as depicted in Fig. 3. Forensic investigations focus on traversing each of the links and face some challenges while doing the same.

Most of the challenges related to this domain can be compiled under coordination or correlation.

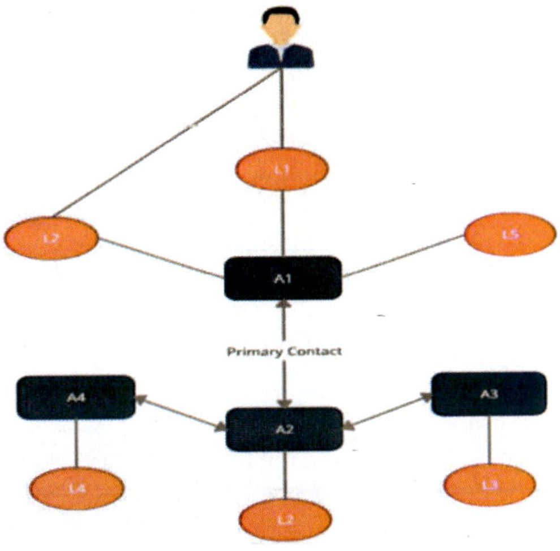

Fig. 3 Default connections across vendors

3.1 Challenges Related to Coordination

A1 might not know how a service L3 or L4 is aiding in its L2 service offering, and it might be storing some sensitive and vital information. Forensics investigation has to compile all the data and correlate the pertinent records to make sense. However, if the two vendors in the same chain are not having the pertinent information about how they are linked, it might be challenging to coordinate and even form the initial contact. Sometimes, this could also obscure essential information that might make or break the case.

3.2 Challenges Related to Correlation

The data might be segregated. Even if we can successfully get the data about a person P1 from A1, we might not get the concomitant records from the vendors A2, A3, and A4 providing services L2 and L3 or L4. In some cases, we might end up with many host data without relevant keys or pertinent records to correlate. So, we indeed need a better correlation mechanism across the vendors.

3.3 Other Challenges Related to Information Security

Confidentiality: If A2 has a shared model, IaaS, SaaS, or PaaS, to provide services, a forensics investigator cannot just subpoena all of the records. If vendor A2 refuses to hand out records, there would be *jurisdictions*, local or international, involved to get the records. A2 cannot present some records containing sensitive information because then the *confidentiality* clause for the data of other consumers would be invoked. If the investigator gets access to all the records, there might be an issue with data privacy for other consumers. Moreover, there might be downtime involved, in case, the issue comes into a deadlock, invoking the *availability* clause. Since the type of data can also be varied, different standards for different types of data would be in place, making the correlation of data a strenuous activity.

4 Proposed Model

We propose a concept of ultimate CSP (UCSP). A UCSP should support all forensic investigation requirements (collection, preservation, examination, analysis, and reporting) without forgoing the existing features/requirements for particular standards or certifications. Moreover, a UCSP should also be bound by service level agreement (SLA) and only engage with or loan services from vendors who are also UCSP's.

4.1 Why is UCSP Effective?

Let us take an example of arbitrary vendors who collaborate or pool services to provide a comprehensive solution. They might offer or support all the services that a customer requires, but they might not have all the controls or processes in place that are required from a forensic standpoint.

For example, to fully support a forensic investigation, a vendor must follow all the stages—preservation, collection, examination, analysis, and reporting. A vendor might only be loaning the service of database storage through a collaborative effort, but in order to support an investigation, he should have processes in place to readily support collection and examination as well.

When we try to enforce a UCSP, we ensure that a vendor is already well versed with all the requirements of supporting investigation activities. Thus, he would already have what an investigation might need. Moreover, the SLA in place would ensure that all other associated parties have the same background, thus aiding in a successful forensic investigation.

From an economic standpoint, getting a UCSP tag could motivate a vendor to maintain the status quo and create one additional plus point to market its services to customers.

4.2 How Does This Model Solve the Correlation and Coordination Challenges Already Posited?

As discussed in the earlier example, a client buys a solution for services L1 and L2 from A1. Moreover, A1 and A2 have allied to offer the services L1 and L2, and for providing the service L2, A2 has further teamed up with A3 and A4 to get the prerequisite services L3 and L4.

How will this scenario change if all vendors are UCSPs? In this case, if all vendors are UCSPs, then even if A1 does not have a formal contract between A3 or A4, these vendors would be well versed to follow the processes and standards for a forensic investigation. So, when an investigation is followed from A1, the pertinent records from A3 and A4 would be similar in format, easy to correlate and form clear connections with the original data. Moreover, because an SLA would now bound the vendors with similar vendors, coordinating activities related to an investigation could be done much more quickly. The contracting terms might not be alike in every contract, but they contain all the required information about forensic investigations. As a result of this model, an investigation would be easy to follow and traverse through the entire chain from A1 to A4. Even if there is no formal contract, UCSP binds them together in a transitive chain of confidence extending from A1 to A4. This model will help form a transitive chain that will extend from the first vendor involved in the process through the last vendor, helping in the forensic investigations and complying with the industry standards, as depicted in Fig. 4.

Correlation: Now, the data correlation challenge would be minimized as the vendor would be able to support the forensic investigation with relevant data in an efficient manner. Since the dependency chain will consist of all trusted vendors, each vendor will have more responsibility for data correlation activities.

Coordination: As all the chains consist of UCSPs, bound by SLAs will be the core part of every connection. Even though there might be no connection between A1 and A3 or A4, one can still get all the relevant records for entries in A2, A3, and A4 from A1.

With this enhanced model, all services: L1, L2, L3, L4 will be more intricately connected, as depicted in Fig. 5.

Fig. 4 UCSP vendors forming a transitive chain

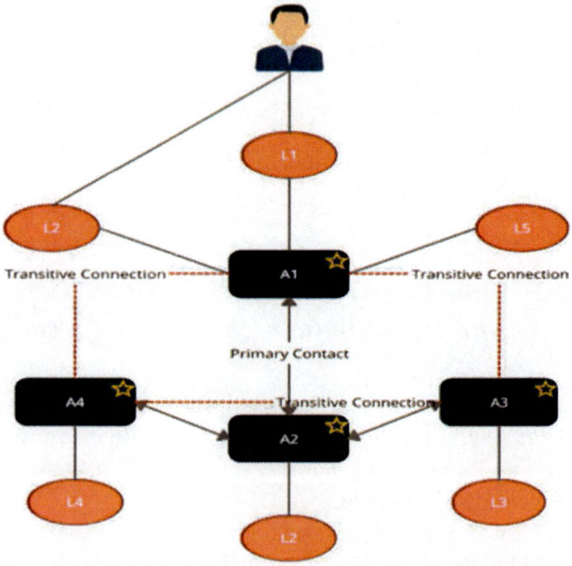

Fig. 5 Vendor services connected more intricately

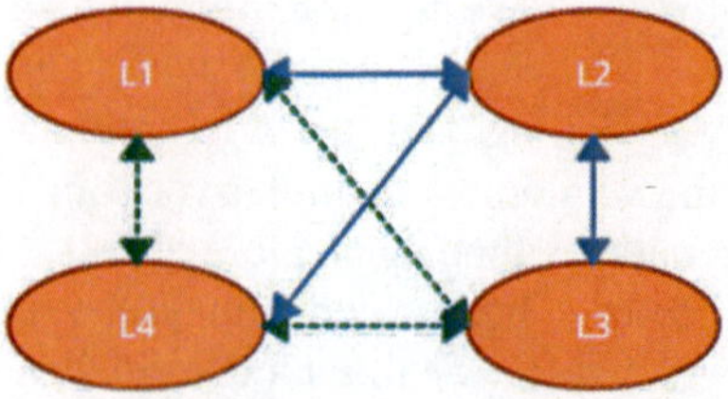

A UCSP may vary; for example, the UCSP for the finance industry might differ from the UCSP for the healthcare industry. However, if they are instead certified in their domain, they can be assured that the process will be followed. For example, let us say A1 is not PCI DSS certified and cannot hold the financial data. At the same time, A2 might be PCI DSS. So, A1 can ask for the services of A2, provided A2 is a UCSP, and an SLA is in place to provide a comprehensive solution to the customers. Since A2 would be a UCSP, it would help aid in forensic investigations as well, without requiring additional work.

Conjugating the Taylor et al. [18] suggestion of a standardized SLA, this model will ensure that CSPs specifically support the requirements of a forensic investigation and indulge in partnerships that also do the same.

4.3 Extending the Proposed Model to Multi-variety Vendors

In a practical scenario, customers or organizations might need to loan services from CSPs and different vendors to ensure all their business or personal requirements are met. Let us consider a scenario that involves infrastructure vendor (Iv), cloud vendor (Cv), and solution vendor (Sv). Although in such cases, an Iv can easily comply with the investigation when compared with a Cv because of the inherent nature of the cloud structure.

The proposed model can aid in scenarios involving different vendors such as Iv, Cv, or Sv. If all Iv, Cv, and Sv support requirements of UCSP, they will already have control or processes in place for supporting the applicable requirements of forensic activities. Thus, coordination and correlation of data or information will already be addressed with no additional efforts.

From a consumer standpoint, if one consumer has loaned or purchased a subscription to services from all Iv, Cv, and Sv, and all these meet the conditions of UCSP, they can easily support the forensic activities. Even from the vendor standpoint, issues related to coordination and correlation of information for the support of investigations will be resolved as these vendors will already be following the best practices for established standards and meeting the requirements of UCSP, thus having controls in place to ensure that data can be extracted whenever and however, needed by following the defined steps.

4.4 Addressing the Already Existing Challenges

Our model also helps address some of the challenges pointed out by Taylor et al. [18], Shetty [19], and Sree et al. [20]. First, as Taylor et al. [18] pointed out, due to the lack of a standardized process, it has been very problematic to write the programs for data extraction in a forensic investigation because there are many different platforms and their proper usage formats. If this model is implemented, the usage formats can be fixed according to the requirements posed by forensic investigation processes, and then, the ready algorithm can be scaled to different platforms. Second, according to Shetty [19] and Sree et al. [20], there has been many dependencies on the CSPs to comply with the investigations and pitch-in additional data. A UCSP would be required to comply with the investigation, and the resulting relationship would be already bounded by an SLA. Lastly, as M Taylor et al. [18] pointed out, the tools for forensic investigations may not support the virtual environment of the cloud. However, if the CSP is bound to support a forensic investigation, that deadlock might be mitigated as there would have to be some common ground to be in a binding agreement successfully.

Simou et al. [21] summarized various challenges and their solutions in their comprehensive study and often quoted the works in which SLAs or binding contracts can be included based on customer and vendor requirements for addressing the dependency and agreement challenges.

Damshenas et al. [22], Ruan et al. [9], Baset [23], Thorpe et al. [24], Serrano et al. [25], Biggs et al. [26], and Zargari et al. [27] voiced the need for a comprehensive SLA to be included between the vendors and customers to address the dependency and trust issues while dealing with cloud environments. However, the situation complicates when multiple vendors are included in this CSP-customer agreement.

Birik et al. [28] and Haeberlen et al. [29] presented the requirement of an external or a third-party auditor in their papers. However, the primary reliance will still be on the CSP to comply. Still, it will be their word against the outside auditors. This might help pacify some cases, but this does not work every time with every other case. We need a more robust and easily scalable model.

Dykstra et al. [30], KO et al. [31], Nurmi et al. [32], and Bouchenak et al. [33] manifested an enhanced trust model in cloud computing that could be used to address the over-reliance on CSP making accountability the center of the model. However, from the forensic investigation view, the onus remains on a CSP at the end of the day to provide the information to move the investigation further.

Patel et al. [34], Busalim et al. [35], and Keller et al. [36] also revealed a new SLA framework for Web-based and e-commerce services, listing out the specific requirements that would be the pertinent cases but that may miss out certain cases when the vendors are from other domains. Alhamad et al. [37] put forward a conceptual framework that can be used to negotiate a reliable SLA with CSP, but its applicability to a forensic investigation domain is not explored conclusively.

These models present some great solutions to the ever-increasing problems in business at the intersection of forensics and cloud computing. However, some of the services required as a precursor are yet to be explored fully when multiple vendors are involved. Moreover, by building on that situation, our model tries to give a framework that can be scaled according to the forensic requirements and help solve dependency in a forensic investigation [38, 39] (Table 1).

Table 1 Summarizing the existing challenges and highlighting how the proposed framework solves them

S. No.	Existing literature	The issue addressed or idea proposed in the existing literature	Challenges	How the proposed framework addresses the issue
1	[29]	Acquisition and analysis of digital evidence in the cloud	Data extraction hindered due to multiple platforms	Platforms and usages can be fixed in SLA
2	[30]	Trust model to enhance trust for a CSP	Inadequate binding agreements with CSP	CSPs would be bound by comprehensive SLAs beforehand
3	[31]	Data collection techniques within cloud forensics		
4	[33]	Comprehensive SLA to enhance forensic practices	Comprehensive SLAs did not address the various deadlocks introduced when multiple vendors were included	USCP would form a dependency chain; wherein multiple vendors would be bound by the existing model of SLA
5	[35]	Challenges in digital forensics by technical, organizational, and legal dimensions		
6	[36]	Redefined and comprehensive SLA for cloud forensics		
7	[36]	Service-based framework for auditing logs		
8	[32]	Enhancing the performance and dependability for online cloud services		
9	[24]	Comprehensive SLAs for fighting cybercrime		
10	[27]	Overview of cloud forensics, including the issues and the existing challenges		
11	[34]	Digital forensics from a technical POV	For a forensic investigation, the onus remains on the CSP	USCP makes the CSP a part of the dependence chain, so accountability is inherently implied
12	[34]	Accountable cloud for customers and providers		
13	[33]	Remote data acquisition methods for cloud forensics		
14	[35]	Enhanced framework for accountability via technical and policy-based approaches		

(continued)

Table 1 (continued)

S. No.	Existing literature	The issue addressed or idea proposed in the existing literature	Challenges	How the proposed framework addresses the issue
15	[36]	An open-source framework for cloud computing		
16	[23]	An enhanced cloud model for fully autonomic cloud services		
17	[30]	SLA framework for monitoring and enforcement tasks	The case of multiple vendors from different industries was scarcely addressed	USCP behaves as an industry-agnostic model
18	[31]	SLA framework for e-commerce cloud end user		
19	[22]	Web SLA for cloud and Web services		
20	[25]	Conceptual SLA framework for cloud	Application of the conceptual framework to a forensic investigation is not explored	UCSP brings in a new model to address issues in the dependency chain

5 Future Work

The cloud computing field is ever-evolving. There cannot be one solution to all problems, but we can help refine the processes to find the optimal solution. Certain domains within this model might need more work to improve and serve the value proposition. Some of them are

- Cost factor: The trusted vendors might be few, and getting every vendor to be classified as a UCSP would be a far-fetched effort that would take support from all the industries. This can become a de facto standard if the start is good. No vendor will want to spend a considerable sum of money if it is not getting something in return for them. So, there could be an incentive involved in the process as well.
- Regulatory Body: There are a lot of standards available for compliance with various security and forensics measures. If there is one regulatory body that can help the vendors get certified on the forensic front, most of the issues related to investigations can be addressed.
- Standards Modification: There is one more alternate route where the existing standards can add additional clauses regarding forensic practices in their general compliance rule-set. This way, most vendors might not need additional certifications as most of the requirements will already be addressed. This requirement has

been suggested many times in the already cited works, but there is a need for a more robust rule-set explicitly tailored to the cloud forensic part.

- Standards: Either there could be industry-specific standards in place, or there could be a more general approach. Either way, there should be someone to regulate the standards.
- This type of model has to be widespread for the whole cloud community to gain an advantage.

6 Conclusion

Since the last decade, much work has been done in cloud forensics, and new challenges have never stopped seeping into the online world. Cloud forensics poses peculiar challenges that might need a tailored approach. Through this paper, we tried to discuss a few challenges that are often faced by forensic investigators while keeping the focus of our discussion more toward dependency chains. We also proposed a model based on enhancing the transitive relationships that could help in addressing this challenge. Research on this model might need further adaptations and thoughts, and we would expect researchers to join hands in innovating this model to be readily adapted by the industry.

References

1. Alghamdi MI (2020) Digital forensics in cyber security-recent trends, threats, and opportunities. Periodicals Eng Nat Sci (PEN) 8(3):1321–1330
2. Dezfoli FN, Dehghantanha A, Mahmoud R, Sani NFBM, Daryabar F (2013) Digital forensic trends and future. Int J Cyber-Secur Dig Forensics 2(2):48–77
3. Hash J, Bowen P, Johnson A, Smith CD, Steinberg DI (2005) An introductory resource guide for implementing the health insurance portability and accountability act (HIPAA) security rule. US Department of Commerce, Technology Administration, National Institute of Standards and Technology
4. Kuhn DR, Hu VC, Polk WT, Chang SJ (2001) Introduction to public key technology and the federal PKI infrastructure. National Inst of Standards and Technology Gaithersburg MD
5. CNSSI 4009-2015 from NSA/CSS Manual Number 3–16 (COMSEC)
6. Edward JD, Memon ND (2009) "Digital Forensics", Article in IEEE Signal Processing Magazine · April 2009 https://doi.org/10.1109/MSP.2008.931089 · Source: IEEE Xplore
7. Kaur R, Kaur A (2012) Digital forensics. Int J Comput Appl 50(5)
8. Zawod S, Hasan R (2013) Cloud forensics: a meta-study of challenges, approaches, and open problems. arXiv:1302.6312v1 [cs.DC] 26 Feb 2013
9. Ruan K, Carthy J, Kechadi T, Crosbie M (2011) Cloud forensics. In: IFIP international conference on digital forensics. Springer, Berlin, Heidelberg, pp 35–46
10. Miranda Lopez E, Moon SY, Park JH (2016) Scenario-based digital forensics challenges in cloud computing. Symmetry 8(10):107
11. Giova G (2011) Improving chain of custody in forensic investigation of electronic digital systems. Int J Comput Sci Netw Secur 11(1):1–9
12. Prayudi Y, Azhari SN (2015) Digital chain of custody: state of the art. Int J Comput Appl (0975–8887) 114(5)

13. Alqahtany S, Clarke N, Furnell S, Reich C (2015) Cloud forensics: a review of challenges, solutions and open problems. In: 2015 international conference on cloud computing (ICCC). IEEE, pp. 1–9
14. Daryabar F, Dehghantanha A, Choo KKR (2017) Cloud storage forensics: MEGA as a case study. Aust J Forensic Sci 49(3):344–357
15. Quick D, Martini B, Choo R (2013) Cloud storage forensics. Syngress
16. Jain KN, Kumar V, Kumar P, Choudhury T (2018) Movie recommendation system: hybrid information filtering system. In: Bhalla S, Bhateja V, Chandavale A, Hiwale A, Satapathy S (eds) Intelligent computing and information and communication. advances in intelligent systems and computing, vol 673. Springer, Singapore. https://doi.org/10.1007/978-981-10-7245-1_66
17. Srivastava R, Tomar R, Sharma A, Dhiman G, Chilamkurti N et al (2021) Real-time multimodal biometric authentication of human using face feature analysis. Comput Mater Cont 69(1):1–19
18. Taylor M, Haggerty J, Gresty D, Lamb D (2011) Forensic investigation of cloud computing systems. Netw Secur 2011(3):4–10
19. Shetty J, Anala MR, Shobha G (2014) A study on cloud forensics: challenges, tools and CSP features. Biom Bioinform 6(6):149–153
20. Sree TR, Bhanu SMS (2020) Data collection techniques for forensic investigation in cloud. In: Digital Forensic Science. IntechOpen.
21. Simou S, Kalloniatis C, Gritzalis S, Mouratidis H (2016) A survey on cloud forensics challenges and solutions. Secur Commun Netw 9(18):6285–6314
22. Damshenas M, Dehghantanha A, Mahmoud R, Bin Shamsuddin S (2012) Forensics investigation challenges in cloud computing environments. In: Proceedings title: 2012 international conference on cyber security, cyber warfare and digital forensic (CyberSec) pp 190–194. IEEE
23. Baset SA (2012) Cloud SLAs: present and future. ACM SIGOPS Operat Syst Rev 46(2):57–66
24. Thorpe S, Grandison T, Campbell A, Williams J, Burrell K, Ray, I (2013) Towards a forensic-based service oriented architecture framework for auditing of cloud logs. In: 2013 IEEE ninth world congress on services. IEEE, pp 75–83
25. Serrano D, Bouchenak S, Kouki Y, Ledoux T, Lejeune J, Sopena J, Arantes L, Sens P (2013) Towards qos-oriented sla guarantees for online cloud services. In: 2013 13th IEEE/ACM international symposium on cluster, cloud, and grid computing. IEEE, pp 50–57
26. Biggs S, Vidalis S (2009) Cloud computing: the impact on digital forensic investigations. In: 2009 international conference for internet technology and secured transactions (ICITST). IEEE, pp 1–6
27. Zargari S, Benford D (2012) Cloud forensics: concepts, issues, and challenges. In: 2012 third international conference on emerging intelligent data and web technologies. IEEE, pp 236–243
28. Birk D, Wegener C (2011) Technical issues of forensic investigations in cloud computing environments. In: 2011 sixth IEEE international workshop on systematic approaches to digital forensic engineering, IEEE, pp 1–10
29. Haeberlen A (2010) A case for the accountable cloud. ACM SIGOPS Operat Syst Rev 44(2):52–57
30. Dykstra J, Sherman AT (2012) Acquiring forensic evidence from infrastructure-as-a-service cloud computing: exploring and evaluating tools, trust, and techniques. Digit Investig 9:S90–S98
31. Ko RK, Jagadpramana P, Mowbray M, Pearson S, Kirchberg M., Liang Q, Lee BS (2011) TrustCloud: a framework for accountability and trust in cloud computing. In: 2011 IEEE world congress on services. IEEE, pp 584–588
32. Nurmi D, Wolski R, Grzegorczyk C, Obertelli G, Soman S, Youseff L, Zagorodnov D (2009) The eucalyptus open-source cloud-computing system. In: 2009 9th IEEE/ACM international symposium on cluster computing and the grid. IEEE, pp 124–131
33. Bouchenak S, Chockler G, Chockler H, Gheorghe G, Santos N, Shraer A (2013) Verifying cloud services: present and future. ACM SIGOPS Operat Syst Rev 47(2):6–19
34. Patel P, Ranabahu AH, Sheth AP (2009) Service level agreement in cloud computing

35. Busalim AH, Ibrahim A (2013) Service level agreement framework for e-commerce cloud end-user perspective. In: 2013 international conference on research and innovation in information systems (ICRIIS). IEEE, pp 576–581
36. Keller A, Ludwig H (2003) The WSLA framework: specifying and monitoring service level agreements for web services. J Netw Syst Manage 11(1):57–81
37. Alhamad M, Dillon T, Chang E (2010 . Conceptual SLA framework for cloud computing. In: 4th IEEE international conference on digital ecosystems and technologies. IEEE, pp 606–610
38. Sarishma D, Sangwan S, Tomar R, Srivastava R (2022) A review on cognitive computational neuroscience: overview, models, and applications. In: Tomar R, Hina MD, Zitouni R, Ramdane-Cherif A (eds) Innovative trends in computational intelligence. EAI/Springer Innovations in Communication and Computing. Springer, Cham. https://doi.org/10.1007/978-3-030-78284-9_10
39. Gupta PT, Choudhury T, Shamoon M (2019) Sentiment analysis using support vector machine. In: 2019 International conference on contemporary computing and informatics (IC3I), pp 49–53. https://doi.org/10.1109/IC3I46837.2019.9055645

DeshiFoodBD: Development of a Bangladeshi Traditional Food Image Dataset and Recognition Model Using *Inception V3*

Samrat Kumar Dey, Lubana Akter, Dola Saha, Mshura Akter, and Md. Mahbubur Rahman

Abstract For a multitude of reasons, including restaurant selection, travel location selection, dietary calorie intake, and cultural awareness, traditional Bangladeshi cuisine picture classification has become more important. However, creating an efficient and usable traditional labeled (English and Bengali) food dataset for study in Bangladesh is quite difficult. In this article, the 'DeshiFoodBD' dataset is given in both Bengali and English for traditional Bangladeshi food classification. Web scraping and camera photos (digital, smartphone) are used to create food images. The dataset includes 5425-labeled pictures of 19 popular Bangladeshi dishes, including Biriyani, Kalavuna, Roshgolla, Hilsha fish, Nehari, and others. There are a number of convolutional neural network (CNN) architectures that may be utilized with this dataset including ImageNet, ResNet50, VGG-16, R-CNN, YOLO, DPM, and so on. It is currently available at Mendeley data repository for research purposes and further use. Our proposed Inception v3-based food recognition model, DeshiFoodBD-Net, had higher test accuracy of ~97% with the DeshiFoodBD dataset.

Keywords Bangla food · Food classification · Convolutional neural network · Inception v3 · Food image dataset · Computer vision

S. Kumar Dey (✉)
School of Science and Technology (SST), Bangladesh Open University (BOU), Gazipur 1705, Bangladesh
e-mail: samrat.sst@bou.ac.bd

L. Akter · D. Saha · M. Akter
Department of Computer Science and Engineering (CSE), Dhaka International University (DIU), Dhaka 1205, Bangladesh

Md. Mahbubur Rahman
Department of Computer Science and Engineering (CSE), Military Institute of Science and Technology (MIST), Dhaka 1216, Bangladesh

V. Skala et al. (eds.), *Machine Intelligence and Data Science Applications*, Lecture Notes on Data Engineering and Communications Technologies 132,
https://doi.org/10.1007/978-981-19-2347-0_50

1 Introduction

Nowadays, people are more conscious of their dietary habits. A significant portion of the population of any country is interested in food culture. A comparison between beliefs, castes, and ethnic groups reveals that traditional food in every nation differs. All foods that have been popular for centuries are referred to as traditional foods. Bangladesh has a variety of popular foods that have become well-known around the world. At the current time, Bangladesh is also very famous around the world for its different types of food diversity. It is, however, a more difficult challenge to create a complete Bangladeshi traditional cuisine dataset and categorize it using the latest deep learning architecture. Food has its own distinct color, flavor, shape, and size, all of which will vary if the ingredients are changed due to the vast variety of foods available. Previously, no dataset has been designed dedicatedly focusing on enlisting available famous foods of Bangladeshi culture. This research examined the creation of the first food picture datasets in both Bengali and English that may be employed to train and test different deep learning models for food identification, classification, and successful segmentation. Apart from that, the proposed *DeshiFoodBD* dataset [1] will be useful to deep learning and machine learning researchers working on conventional food image recognition, detection, and segmentation. More precisely, this dataset will aid researchers in the development of individual food habit analysis applications. The development of large labeled databases [2] and profoundly trained representations [3] has revolutionized object recognition and scene classification. In recent times, deep learning (DL) and computer vision (CV)-based techniques gaining much popularity in recognizing specific foods from images. Among several DL approaches, CNN is seen as the modern technology for the classification of images because of its ability to automatically remove the features from the image that are needed for image classification. Therefore, in this article, Google's Inception v3 model of CNN architecture is employed on the proposed *DeshiFoodBD* dataset of 19 food classes. The Inception v3 model has been established to attain more than 78.1 percent accuracy on the ImageNet dataset. During the development of this article, the significant contributions of this study are the development of the *DeshiFoodBD* dataset for public use and the approach of the Inception v3-based food category recognition algorithm. The following is a breakdown of the rest of the article: Numerous models utilized for food recognition and categorization are discussed in Sect. 2 of the associated study. The proposed methodology is discussed in Sect. 3, which is aimed to discuss the dataset development procedure and the proposed model. Section 4 highlights the available results and accuracy improvement procedure. Finally, the paper ends with the conclusion in Sect. 5.

2 Related Work

This section will briefly highlight some of the popular food image benchmark datasets of different countries along with their characteristics. In addition, food recognition approaches dependent on CNN will also be addressed. For food image classification and labeling, researchers have released several datasets such as Recipe1M + [4], KenyanFood13 [5], FoodX-251, and ISIA Food-500 [6], ChineseFoodNet [7], ETH Food-101 [8], EC FOOD 100, and UEC FOOD 256 [9], FoodAI [10]. DeshiFoodBD, on the contrary, is the first detailed dataset focused on Bangladesh's traditional and most common foods. Previously, no dataset dedicated to the Bangladeshi community's food culture has been developed. The authors identified the final recipe and its item in Recipe1M+, a recent extensive, arranged corpus of over 1 million culinary food items and 13 million pictures of food, using a combined neural embedding model with semantic regularization technique. By using the methodology of scrape-by-location, authors have created a food image dataset from Instagram posts and published two different datasets of 104,000 and 8174 image/caption pairs for traditional Kenyan food datasets, namely KeneyanFood13. It used the ResNeXt101 architecture, pre-trained based on ImageNet in recognizing foods based on images and captions. For a fine-grained type food classification, P. Kaur et al. have proposed FoodX-251, a dataset of 251 fine-grained food categories with 158 k images collected from the Web and implemented a naive baseline using a pre-trained ResNet-101 network using ADAM optimizer. W. Min et al., on the contrary, argue that the ISIA Food-500 is a more comprehensive food dataset that, with 399,726 pictures of 500 categories, outperforms existing popular benchmark datasets in terms of category coverage and data volume. In another exploration, P. Pandey et al. have suggested an automated food classification method based on deep CNN that recognizes the contents of a meal from food images using ETH Food-101 dataset and a new Indian food image database. Subsequent research on food image classification that was solely based on the use of CNN's [11] built a five-layer CNN to identify a subset of ImageNet data that consisted of ten food groups. The performance of the CNN model after using the data augmentation approach provided more than 90% accuracy on the Food-101 dataset. In recent time, Liu et al. recently introduced DeepFood [12], a CNN-based solution inspired by LeNet-5 [13], AlexNet, and GoogleNet [14] that uses Inception modules to maximize total network depth. After 300,000 epochs, DeepFood achieved 77.4 percent top-1 accuracy on the Food-101 dataset. Tasnim et al. [15] in another work proposed an Inception v3-based food image classification approach of only 7 traditional Bangladeshi food and obtained an accuracy of 95.2% approximately. However, for recognizing food items authors only collected 700 images of different foods and utilized them as a dataset. In the category of dataset development and model utilization, Park et al. [16] have developed their own Korean food-based K-foodNet dataset that contained 92,000 images into 23 groups, and by using the DCNN approach, it achieved an accuracy of 91.3% with 0.4 ms faster recognition time.

3 Methodology

The implementation procedure of recognition of traditional Bangladeshi food is described here in Fig. 1. Data Description, Model Development, and Training are the three subsections in this section.

3.1 Data Description

DeshiFoodBD is a comprehensive image dataset consisting of 19 traditional food classes of Bangladesh. The foods are as follows: (i) Hilsha Fish, (ii) Biriyani, (iii) Khichuri, (iv) Morog Polao, (v) Dohi, (vi) Roshgolla, (vii) Porota, (viii) Fuchka, (ix) Roshmalai, (x) Kachagolla, (xi) Kalavuna, (xii) Haleem, (xiii) Nehari, (xiv) Kebab, (xv) Egg omelet, (xvi) Beguni, (xvii) Mashed Potato, (xviii) Chick Peas, and (xix) Bakorkhani. Moreover, we also developed class file in Bengali language. For Bengali language, the classes are as follows: (i) *ইলিশ মাছ* (ii) *বিরিয়ানি* (iii) *খিচুড়ি* (iv) *মোরগ পোলাও* (v) *দই* (vi) *রসগোল্লা* (vii) *পরোটা* (viii) *ফুচকা* (ix) *রসমালাই* (x) *কাচা গোল্লা* (xi) *কালা ভুনা* (xii) *হালিম* (xiii) *নেহারি* (xiv) *কাবাব* (xv) *ডিম ভাজি* (xvi) *বেগুনি*, (xvii) *আলু ভর্তা* (xviii) *ছোলা ভুনা* and (xix) *বাকড়খানি*. The dataset contains five hundred twenty-five (5425) images with a high diversity of poses, angles, lighting conditions, and backgrounds. Four (4) different parent categories have chosen to identify the 19 classes of traditional food items of Bangladesh. The categories are vegetables, non-vegetables, sweets, and snacks. Table 1 shows the alignment rubrics of food classes with the parent food categories.

All of the images are in digital image format in the dataset. Sample images from each food class of the dataset are presented in Fig. 2.

The sample image data files are divided into 19 different folders for both *food_data_bengali* and *food_data_english*. Each folder contains images of one specific food class. The 'Train' and 'Test' folders contain images for training and testing data, respectively. For both the English and Bengali labeling, DeshiFoodBD dataset contains four folders including Images, Meta, Test, and Train. Images folder

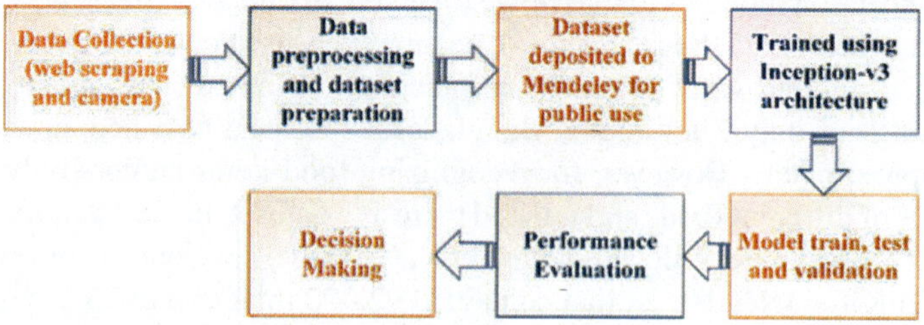

Fig. 1 The implementation procedure of recognition of traditional Bangladeshi food based on DeshiFoodBD dataset using Inception v3

Table 1 Alignment of food classes with the parent food categories

Food classes	Vegetables	Non-vegetables	Sweets	Snacks
Hilsha Fish		×		
Biriyani		×		
Khichuri		×		
Morog Polao		×		
Dohi			×	
Roshgolla			×	
Porota				×
Fuchka				×
Roshmalai			×	
Kachagolla			×	
Kalavuna		×		
Haleem		×		
Nehari		×		
Kebab		×		
Egg omelet		×		
Beguni	×			
Mashed Potato	×			
Chick Peas				×
Bakorkhani				×

contains all the food images of the dataset while Meta folder contains two text files with labels and classes. Table 2 shows the number of images per class as well as other details.

3.2 *Model Development*

This section briefly discusses the potential deep learning architecture in which the proposed dataset can be deployed. In order to check the appropriateness of Deshi-FoodBD dataset, this research used the Inception v3 architecture to recognize the different foods and classify them according to their class. The Inception v3 of Google's Image Recognition Inception Architecture was released in 2015, and it is a pre-trained model based on ImageNet. The basic architecture of the Inception v3 model is given in Table 3. Our proposed Inception v3 model is depicted in Fig. 3.

Fig. 2 Sample images of *DeshiFoodBD* dataset with respective food class representation

3.3 *Training*

To train the models, this exploration has used Google Colab Platform which is based on Python 3 Google Compute Engine backend (GPU), with 12 GB of memory for training and testing the model. With a batch size of 16, the data have been trained in 30 epochs. Our training goal is to make the cross-entropy as minor as possible by focusing on the trend of loss curves. However, by feeding the cache value into the layer for each image, the training process runs smoothly. After the successful training phases, we have enlisted the available results of validation accuracy, training accuracy, and cross-entropy.

Table 2 Data description of 'DeshiFoodBD' dataset

S. No.	Classes	Cameras	Web scrapping	Augmented data	Total appearance
1	Hilsha Fish	159	211	N/A	370
2	Biriyani	90	105	N/A	195
3	Khichuri	123	211	N/A	334
4	Morog Polao	98	84	N/A	182
5	Dohi	76	123	N/A	199
6	Roshgolla	145	252	N/A	397
7	Porota	234	331	N/A	565
8	Fuchka	59	106	N/A	165
9	Roshmalai	73	93	N/A	166
10	Kachagolla	59	51	N/A	110
11	Kalavuna	145	66	N/A	211
12	Haleem	122	266	N/A	388
13	Nehari	134	234	N/A	368
14	Kebab	224	195	N/A	419
15	Egg omelet	214	371	N/A	585
16	Beguni	197	97	N/A	294
17	Mashed Potato	35	52	N/A	87
18	Chick Peas	128	96	N/A	224
19	Bakorkhani	87	79	N/A	166
	Total	2402	3023	–	5425

Table 3 The architecture of *Inception v3*

Type/stride	Patch size	Input size
Conv/s2	3 × 3	299 × 299 × 3
Conv/s1	3 × 3	149 × 149 × 32
Conv padded/s1	3 × 3	147 × 147 × 32
Pool/s2	3 × 3	147 × 147 × 64
Conv/s1	3 × 3	73 × 73 × 64
Conv/s2	3 × 3	71 × 71 × 80
Conv/s1	3 × 3	35 × 35 × 192
3 × Inception	Inception module	35 × 35 × 288
5 × Inception	Inception module	17 × 17 × 768
2 × Inception	Inception module	8 × 8 × 1280
Avg pool	Pool 8 × 8	8 × 8 × 2048
FC	2048 × 1000	1 × 1 × 2048
Softmax	Classifier	1 × 1 × 1000

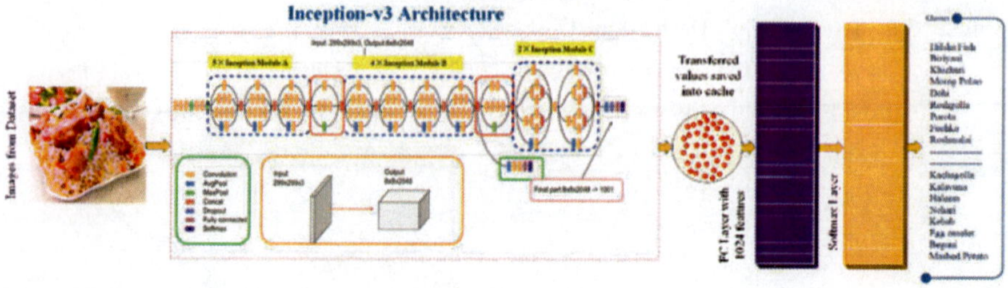

Fig. 3 Proposed Inception v3 model

4 Results

This research focuses on developing a dedicated food image dataset based on traditional foods of Bangladesh. Apart from that, it also employed Inception v3 model on the dataset to find the accuracy in recognizing and labeling specific food categories. The ratio of correctly recognized images to the number of total images based on the Inception v3 recognition model resulted in an overall accuracy is shown in Fig. 4 for training and test images, respectively. Figure 4a illustrates the accuracy curves with training accuracy of 97% and validation accuracy of 91%, respectively. On the contrary, Fig. 4b contains the train and validation loss of the proposed model. With the increase of each epoch from 0 to 30, train and validation loss decrease gradually from around 1.2 and 1.1 to 0.2, respectively. Overall, the suggested and the developed deep model produced an accuracy of 97% in recognizing food class based on the *DeshiFoodBD* food dataset.

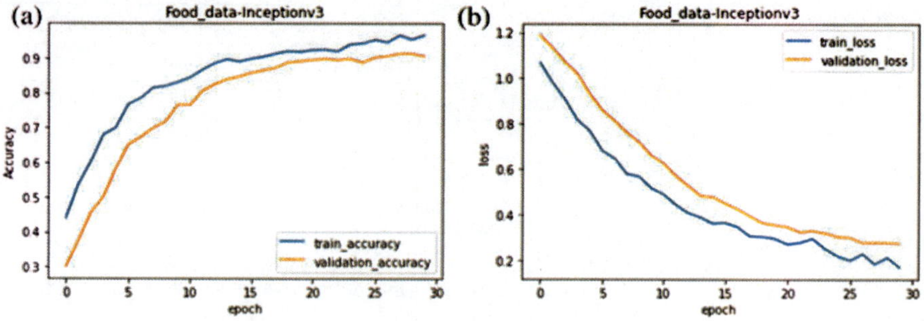

Fig. 4 Accuracy and loss curve for the Inception v3 model

5 Conclusion

In this research, we developed and deposited a novel food dataset of traditional Bangladeshi cuisine pictures, which included 19 classes and over 5 K photos. DeshiFoodBD may be improved in better way by increasing the number of photos per class and adding new food categories, as well as the quantity of food images in each class. This dataset may be used to track food intake and could be useful for patients who require assistance during their regular meals (e.g., elderly individuals with obesity and/or diabetes, people with food allergies). When compared to alternative CNN-based techniques, the baseline findings utilizing the state-of-the-art Inception v3 classifier demonstrate acceptable results with a low error rate. In the future, different deep learning models can enhance and increase classification accuracy as this dataset is expanded.

References

1. Dey SK, Shibly KH, Akhter L, Saha D, Akter M, Rahman DMM (2021) DeshiFoodBD 1. https://doi.org/10.17632/tczzndbprx.1
2. Zhou B, Lapedriza A, Xiao J, Torralba A, Oliva A (2014) Learning deep features for scene recognition using places database. In: Proceedings of the 27th international conference on neural information processing systems, vol 1. MIT Press, Cambridge, MA, USA, pp 487–495
3. Krizhevsky A, Sutskever I, Hinton GE (2017) ImageNet classification with deep convolutional neural networks. Commun ACM 60:84–90
4. Marín J, Biswas A, Ofli F, Hynes N, Salvador A, Aytar Y, Weber I, Torralba A (2021) Recipe1M+: a dataset for learning cross-modal embeddings for cooking recipes and food images. IEEE Trans Pattern Anal Mach Intell 43:187–203
5. Jalal M, Wang K, Jefferson S, Zheng Y, Nsoesie EO, Betke M (2019) Scraping social media photos posted in kenya and elsewhere to detect and analyze food types. In: Proceedings of the 5th international workshop on multimedia assisted dietary management Association for Computing Machinery, New York, NY, USA, pp 50–59
6. Min W, Liu L, Wang Z, Luo Z, Wei X, Wei X, Jiang S (2020) ISIA Food-500: a dataset for large-scale food recognition via stacked Global-local attention network. arXiv:2008.05655 [cs]
7. Chen X, Zhu Y, Zhou H, Diao L, Wang D (2017) ChineseFoodNet: a large-scale image dataset for chinese food recognition. arXiv:1705.02743 [cs]
8. Pandey P, Deepthi A, Mandal B, Puhan NB (2017) FoodNet: recognizing foods using ensemble of deep networks. IEEE Sig Proc Lett 24:1758–1762
9. Hassannejad H, Matrella G, Ciampolini P, De Munari I, Mordonini M, Cagnoni S (2016) Food image recognition using very deep convolutional networks. In: Proceedings of the 2nd international workshop on multimedia assisted dietary management. Association for Computing Machinery, New York, NY, USA, pp 41–49
10. Sahoo D, Hao W, Ke S, Xiongwei W, Le H, Achananuparp P, Lim E-P, Hoi SCH (2019) FoodAI: food image recognition via deep learning for smart food logging. In: Proceedings of the 25th ACM SIGKDD international conference on knowledge discovery and data mining, pp 2260–2268
11. Zhang W, Zhao D, Gong W, Li Z, Lu Q, Yang S (2015) Food image recognition with convolutional neural networks. In: 2015 IEEE 12th International conference on ubiquitous intelligence and computing and 2015 12th IEEE international conference on autonomic and

trusted computing and 2015 IEEE 15th international conference on scalable computing and communications and its associated workshops (UIC-ATC-ScalCom), pp 690–693
12. Liu C, Cao Y, Luo Y, Chen G, Vokkarane V, Ma Y (2016) DeepFood: deep learning-based food image recognition for computer-aided dietary assessment. In: Chang CK, Chiari L, Cao Y, Jin H, Mokhtari M, Aloulou H (eds) Inclusive smart cities and digital health. Springer International Publishing, Cham, pp 37–48
13. Lecun Y, Bottou L, Bengio Y, Haffner P (1998) Gradient-based learning applied to document recognition. Proc IEEE 86:2278–2324
14. Szegedy C, Liu W, Jia Y, Sermanet P, Reed S, Anguelov D, Erhan D, Vanhoucke V, Rabinovich A (2015) Going deeper with convolutions. In: 2015 IEEE conference on computer vision and pattern recognition (CVPR), pp 1–9
15. Tasnim N, Romyull Islam Md, Shuvo SB (2020) A convolution neural network based classification approach for recognizing traditional foods of bangladesh from food images. In: Abraham A, Cherukuri AK, Melin P, Gandhi N (eds) Intelligent systems design and applications. Springer International Publishing, Cham, pp 844–852
16. Park S-J, Palvanov A, Lee C-H, Jeong N, Cho Y-I, Lee H-J (2019) The development of food image detection and recognition model of Korean food for mobile dietary management. Nurs Res Pract 13:521–528

Sentiment Analysis of E-commerce Consumer Based on Product Delivery Time Using Machine Learning

Hasnur Jahan, Abu Kowshir Bitto, Md. Shohel Arman, Imran Mahmud, Shah Fahad Hossain, Rakhi Moni Saha, and Md. Mahfuj Hasan Shohug

Abstract In this modern era, e-commerce sites, online selling, and purchasing are at the top of the list. Product quality and delivery time usually divert people's sentiments about e-commerce. We conducted a sentiment analysis of consumer comments on Daraz and Evaly's Facebook pages, and data were gathered from these two pages comments of Facebook. We evaluated the mood of client comments in which they expressed their opinions and experience regarding e-commerce pages services. With diverse models such as logistics regression, decision tree, random forest, multinomial naive Bayes, K-neighbors, and linear support vector machine in n-grams, we employ unigram, bigram, and trigram features. With 90.65 and 89.93% accuracy in unigram and trigram, random forest is the most accurate. With an accuracy of 88.49% in bigram, decision tree is the most accurate. Among the finest fits are the unigram feature and random forest.

Keyword Digital Bangladesh · E-commerce · Consumer sentiment · N-grams · Machine learning

1 Introduction

Electronically buying and selling services online is said to the e-commerce. Electronic commerce is the abstract form of e-commerce. With the development of technology in communication and information systems in recent years, people are moving toward e-commerce-based services worldwide day by day. In an e-commerce system,

H. Jahan · A. Kowshir Bitto · Md. Shohel Arman (✉) · I. Mahmud · S. Fahad Hossain · R. Moni Saha · Md. M. H. Shohug
Department of Software Engineering, Daffodil International University, Dhaka, Bangladesh
e-mail: sshuvo27@gmail.com

I. Mahmud
e-mail: imranmahmud@daffodilvarsity.edu.bd

R. Moni Saha
e-mail: rakhi35-2416@diu.edu.bd

V. Skala et al. (eds.), *Machine Intelligence and Data Science Applications*, Lecture Notes on Data Engineering and Communications Technologies 132,
https://doi.org/10.1007/978-981-19-2347-0_51

any company or any seller uses the Internet to communicate with customers, give them service with the Internet, and confirm the delivery with the Internet also [1]. Internet helps them to make more good communication.

But the idea of e-commerce was introduced about 40 years ago [2]. The world's first e-commerce company was established in 1982 named Boston Computer Exchange. At first, it was a site where people can sell their used computers. Then in these several decades, there have established a lot of these types of websites worldwide. Its beginning in Bangladesh was in the 90 s. But at that time, it was used to send a gift to non-residual people of abroad. But all that sites were from abroad, and in Bangladesh, they had only some branches. But now there has a lot of e-commerce services there. In 2009, Bangladesh bank first time permit online payment. The trade body of Bangladesh e-commerce is the e-commerce association. On basis of this association, information about 8000 Facebook pages of e-commerce are there only in Bangladesh. In 2016, Bangladesh e-commerce have been opened for every district by the government. Taxes on e-commerce were removed from that year by recommendation of FBCCI. On basis of reports of 2017, every year transactions of about 10 billion takas are done with e-commerce.

Any e-commerce site or website's popularity mostly depends on the customer's satisfaction. In e-commerce, customer satisfaction is dependent on any type of thing. They are a well-designed interface, information customization, proper product information, pricing, the behavior of staff sellers on the site, delivery system, and many more things [3]. It needs to feel by the customers that the service will keep their information confidential [1]. By improving all these service quality, online retailers will be able to make their customer satisfaction. And customer satisfaction will help to increase profit, the number of customers, sell and also favorable forget good words from them. Therefore, it is necessary to understand customers' needs and need to identify all factors which are affecting the business. They can understand the customer's need and their services problem by analyzing sentiment. It is a process where from text pieces we can categorize the opinion of the writers [4]. It finds out if the text is positive, negative, or neutral. On a page of e-commerce, we can find out about their service is bad or good by manually seeing the comment. But when the company analyzes the comment, they need to analyze a thousand comments it is not an easy way. For that reason, companies are using sentiment analysis to understand if their services are good or bad. So, we implemented that sentiment analysis on two companies in Bangladesh.

In Bangladesh at present, there have a lot of e-commerce websites and pages on Facebook. Every, Daraz, Chaldal, Bikroy.com and Rokomari.com are some popular business to customer (B2C) e-commerce websites of Bangladesh. According to an SEO audit according to user's behavior and source of traffic top websites are Evaly, Daraz, Chaldal.com, Bikroy.com, Rokomari.com, etc. Each is an e-commerce website that is built by a developer sharp-minded. They supply every type of product, and they look over and take the opinion of their customer very seriously. The application is easy to use and understand. And this site is popular for its payment range. Among 7 to 45 days, the delivery is made by them. According to the SEO audit, they have 97.29% of users are from BD and 10.53 million people visited the site in July 2021.

In Bangladesh, another site Daraz has recently become one of the most online shopping. They have a quick delivery system. They grunted their product as 100% authentic and warranty. They offer many types of coupons and discount facilities [5]. And they have any type of payment options. According to an SEO audit in July 2021, total visits were 7.35 million. And among all the users 95.72% were BD users.

In our work, we will make a sentiment analysis of comments of consumers on Daraz and Evaly's Facebook page. We have collected the data of these two websites from Facebook. Customers give their opinion on basis of their services. We have analyzed the sentiment of the comments where customer have given their opinion about their services.

2 Literature Review

There have previously been numerous papers, studies, and research articles produced by several writers that are similar to our work. To go along with our work, we've included some work reviews below:

Yang [2] used a sentiment lexicon with a convolutional neural network (CNN) and an attention-based bidirectional gated recurrent unit (BiGRU) to crawl and sanitize the real book assessment of dangdang.com, a well-known Chinese e-commerce website. There are 100,000 reviews in the dataset, 50,000 of which are favorable, and 50,000 of which are negative. The main purpose was to evaluate user evaluations that can assist businesses on platforms of e-commerce in obtaining user input promptly, allowing them to enhance facility standards and get more consumers. The next stage will be to investigate text sentiment fineness categorization.

With four objectives, Giao [6] have taken interviews of 200 customers utilizing a convenient sampling method at the E-commerce platform named Tiki.vn. The four objectives are: The factors that affect the satisfaction of customers online of Tiki.vn, factors impact level measuring, among different groups of customer test the satisfaction level, for improving the quality of services on online proposing some new implications. This study finds out the scale of reliability by Alfa coefficient of Cronbach's, analysis of linear regression, analysis of exploratory factor. This analysis result shows that customer service and reliability influence the satisfaction of customers mostly.

Gajewska et al. [7] measured customer satisfaction using the serval method. The study compared two groups. One is the e-commerce facility standard before purchase and the other is after purchase e-commerce service quality. Also used Kaiser–Meyer–Olkin test for testing relations of the variable is enough for carrying the analysis of factor or not. This study finds out that the customer mostly values the guarantee.

Identifying cubic kernel provides the best accuracy on a given dataset, and Noor et al. [3] utilized the vector space model, a.k.a bag of words model, for feature extraction, which was then passed to support vector machines (SVMs). In the data, there are a total of 20,285 reviews from the e-commerce website of Pakistan–Daraz.pk.

This is a review-based study on PConline.com e-commerce. To characterize the review of online here used probability multivalued neutrosophic linguistic numbers (PMVNLNs). The model's fuzzy characteristics find the differences and similarities among negative and positive reviews. In PConline.com, there are already four existing models. Ji et al. [8] compared the model they developed and the existed four models by their accuracy. By the good accuracy of the model, the study proposed that for e-commerce customers the built model is a promising option.

Jabbar et al. [9] utilized the data that are product reviews online obtained from Amazon.com to offer a natural-time emotional examination on e-commerce application product reviews so that the user experience may be improved using a machine learning approach called support vector machine (SVM). Advanced machine learning and deep learning methods can be used to improve sentiment analysis applications. E-commerce applications will also benefit from this transformation.

Akter et al. [10] used some machine learning models to analyze the e-commerce Bengali sentiment. They have taken data from Daraz only. Five machine learning models are used there. The models are KNN, SVM, XGBoost, logistic regression, random forest classifier. They got an accuracy of 96.25 and precision, recall, F1-score of 0.96.

Jha et al. [11] analyzed e-commerce sentiment by utilizing natural language processing. They collect structural and nonstructural data for sentiment analysis. Resilient distributed datasets (RDD) and PySpark were used for sentiment analysis. For collecting data with web scrapping, they used Scrapy and Restful APIs. Their finding is that natural language processing for e-commerce has increased efficiency in analyzing the sentiment for products.

On an e-commerce platform, sentiment analysis and classification of product reviews are done by Munna et al. [12]. In the study, they have proposed a deep learning network, and natural language processing (NLP) was applied to solve the problem of sentiment analysis. They focused on Bangla data. They got an accuracy of 0.84 and 0.69 for sentiment analysis. This will help the customer for choosing the correct products.

3 Methodology

We started by collecting and preprocessing data. After that, we optimized the parameters for our training algorithm. Then we trained and tested it with our dataset. Then we put our output to the test to see if our algorithms were correct and finally determine algorithms accuracy.

Fig. 1 Evaly and Daraz customer sentiment data

	Comment	Tag
0	আদৌও কি আমার রিফান্ড পাবো, প্রিয় ই- ভ্যালি?????...	Bad
1	আপনাদের প্রায়োরিটি স্টোর/ T10 এর আগের পণ্য গু...	Bad
2	Order করেছি কিন্তু এখনো নোটিফিকেশন এর মাধ্যমে ...	Bad
3	পেমেন্ট করতে চাইলে নিচের মেসেজ টা শো করছে।আমার...	Neutral
4	msung M01s) এটা ৭ দিনের গ্যারান্টেড ডেলিভারির ...	Bad
...	...	...
944	শুধু চীন নয়, আরো দেশ থেকে আরো ক্রস বর্ডার পণ্...	Neutral
945	আমি দারাজ মূলকে ভালোবাসি কারণ 10% ক্যাশব্যাক +...	Good
946	দারাজমল 100% প্রামাণিক পণ্য ডেলিভারি দে .. আইখ...	Good
947	আজ অবধি, আল্লাহর রহমতে আমি দারাজ থেকে যতটা সম্...	Good
948	আগামি 10 বছর একটি দারাজ পৃথীবির পোটাক্তা কোনায...	Neutral

3.1 Data Collection and Preprocessing

The data we used for our study was gathered from the Evaly and Daraz e-commerce sites' Facebook pages' comment sections. We utilized the web scraping method to obtain reviews. The dataset contains two features: One is client feedback, and the other is data kind. The total number of reviews collected is 949. We carefully identified and ensured that all negative reviews are labeled as 'Negative,' positive reviews as 'Positive,' and neutral reviews as 'Neutral.' There are 364 good reviews, 401 bad reviews, and 183 neutral reviews. We wrote all of our comments in Bangla, and some of them were a mix of Bangla and English (Fig. 1).

For implementing a machine learning model into a dataset preprocessing of collected data are the most important process. If data are preprocessed, then it will help to get a good result.

Here we import re-library to remove unnecessary punctuations from the collected comment. Then we used the lambda function to remove low length comments which contained less the 2 words. With that process, we removed 97 small reviews, and then our total comment number became 852.

3.2 Logistic Regression

It is one kind of regression analysis to predict. It is used only when there has a binary target variable. So for binary classification, it is used. There will have two outcomes only in the target variable (0, 1). Logit function is used there to predict binary outcome's probability. For this reason, it is a special kind of linear regression. To transfer outcome into categorical value activation functions are used there. An example of an activation function is sigmoid (σ). This is used to keep the value between 0 and 1. Normally if the value is less than 0.5, then it will be counted as 0, and if the value is greater than 9.5, then the value will be counted as (1). Equation of logistic regression equation is: σ (WTX) = Sigmoid Function (1):

$$W^T X => \sigma\left(W^T X\right) => \sigma\left(\frac{1}{e^{-w^T x}}\right) \tag{1}$$

3.3 *Decision Tree*

The decision tree model is used to get the decision of the target value which one gets by seeing an item. It is one kind of prediction model used in computer science in the field of machine learning, data mining, and statistics. Another name of it is classification trees. Here target variable is normally taken as a discrete value's set. Trees nodes are the representation of class labels, and conjunction properties which maintain the class labels are represented by the branches. If the target variable has a continuous value, then it will be called a regression tree. For making the decision, the decision tree is implemented. For describing the data in data mining, it is used. Selecting node randomly may occur problems. For this reason, there some approaches are followed. Like entropy. In entropy information, randomness which is being processed by the tree is measured. Equation of the randomness is (2):

$$E(S) = \sum_{i=1}^{C} -P_i \log_2 p_i \tag{2}$$

3.4 *Random Forest*

For classification, random forest is an ensemble learning. In different subsamples, a random forest set decision trees classifier. It is a meta estimator that used an average of subsamples for improving the accuracy in prediction and for controlling overfitting. It is normally trained with the method of bagging. Bagging is the method of combining learning models. Both classifier and regression are used. It adds randomness to the model for growing trees. It finds the best feature from the random splitting subset. For splitting nodes, the feature's random subset is taken into consideration. For missing data imputation, clustering, selection of feature random forest is used.

3.5 Multinomial Naïve Bayes

It is normally used in natural language processing. It is a probabilistic learning method. The probability of each tag is predicted by it from the given samples. The probability of each tag is counted by it. Then it shows the output from the highest probability. It is the combination of some algorithms where there has the common principle of all the algorithms. One feature's presence does not affect the other features. We can calculate class A's probability when B predictors are already there (3):

$$P(A|B) = P(A) \times \frac{P(B|A)}{P(B)} \tag{3}$$

Here *P(B)* = *B*'s prior probability.
P(A) = *A*'s prior probability.
P(B|A) = incident of predictor *B* for provided class *A* probability.

3.6 K-Neighbors

KNN's full form is *K*-nearest neighbors. It is one kind of supervised learning. The data which we think to make classify is the *K* value. It finds out the distance from data points which is unknown from all the data. From ascending to descending orders, data points are sorted on basis of the value of the distance. From sorted data points, the first K numbers will be taken. The class which has the most data points will have used to find out the data points which is unknown:

3.7 Linear Support Vector Machine

Linear support vector machine is supervised very popular learning. The hat is utilized for both classification and regression problems. Normally that is used for the problem of classification. Its main goal is to make the best line between data to make category prediction of new data correctly. The boundary of that best decision is called a hyperplane. This SVM finds the highest point to make the hyperplane. The cases of extreme point are called support vector (SV), and algorithms are called support vector machine (SVM):

$$f(x) = w_t x + b = \sum_{k=0}^{n} w_j y_j + b = 0 \tag{4}$$

Here w is one of M-dimensional vector (4).

b is scalar and used for defining the hyperplane (4).

3.8 Performance Measuring

On the basis of its prediction error, the estimated models are assessed and contrasted. We evaluate the accuracy, precision, recall, and the $f1$-score in our job.

Accuracy is the numbers of comments which is predicted correctly. Equation is (5):

$$\text{Accuracy} = \frac{\text{predictions of correct comment}}{\text{Predictions of allcomments}} \tag{5}$$

Precision is the proportion of all positive comments which is predicted and rightly predicted all positive comments. Equation is (6):

$$\text{Precision} = \frac{\text{True positive comments}}{\text{True Positive comments} + \text{False Positive comments}} \tag{6}$$

Recall is the proportion of all the comments and just the positive comments which is predicted correctly. Equation is (7):

$$\text{Recall} = \frac{\text{True positive comments}}{\text{True Positive comments} + \text{False Negative comments}} \tag{7}$$

$F1$ is the precision and recall's and weighted average. Equation is (8):

$$\frac{\text{True Positive comments}}{\text{True Positive comments} + \frac{\text{False Negative comments} \times \text{False Positive comments}}{2}} \tag{8}$$

4 Result and Discussion

By using the python function on the comment, we were able to remove superfluous punctuation. Then 97 short reviews with a very short length were deleted. So, after cleaning the data, we have a total of 853 records. We create a summary of negative, positive, and unique data after cleaning the data (Fig. 2). There the total number of sentences in the negative class is 391. Where the number of words is 8227, the number of unique words is 1732, and the top ten words are shown. In the positive remark class, the total number of sentences is 303, the number of words is 3276, the number of unique terms is 648, and the top ten most often used words are also shown. The total number of sentences in the neutral remark class is 157, the number

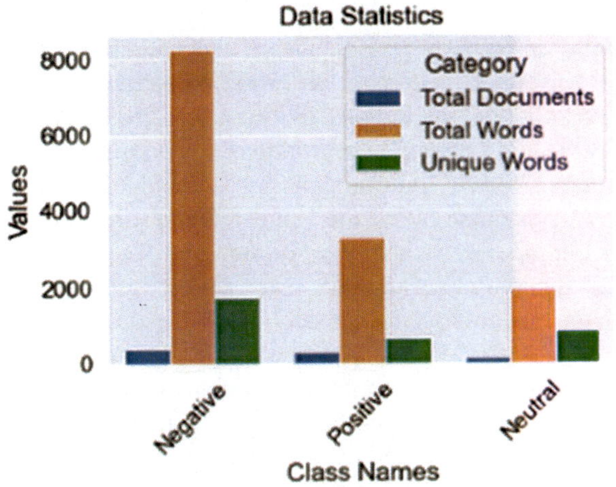

Fig. 2 Data statistics visualization

of words is 1937, the number of unique terms is 841, and the top ten most frequently used words are displayed. The comment length-frequency distribution is shown in Fig. 3.

Then, using only the positive and negative tag comments, we create a new dataset for building the model. In the dataset, there have 391 negative data points and 303 positive data points. Label encoding was used for this data collection. In that process datasets, sentiment labels were transformed into a Numpy array and then defined the class name again. Then for encoded labels, we returned the labels with the condition. After that, we divided our dataset into training and testing data. We utilized the holdout method of cross-validation to divide our dataset, where the training data to testing data ratio is 80:20.

Then we used the TF-IDF method to extract our feature, and it changed the way we looked at things. For converting every single word of dataset into numeric form

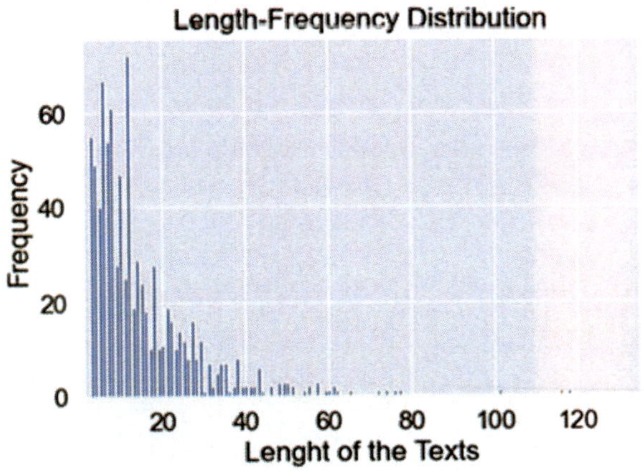

Fig. 3 Length-frequency distribution

TF-IDF Vectorizer was used. For retaining the context of conversation converted with vectorizer n-gram helps and as a tokenizer, Lambda is used there.

On our spitted data, we used different models. Because this is a classification problem, we employ the majority of the classification algorithm. We utilized SVM, SGD classifier, and Naïve Bayes, logistic regression model, decision tree, random forest, multinomial Naive Bayes, K-neighbors classifier, linear support vector machine models to achieve the best results. We just edited some parameters of our models, without that we didn't edit any model. For linear regression, we used 123 random states; in the decision tree, we used 0 random state and entropy criterion; in the random forest, we used 0 random state and 100 estimators, multinomial Naïve Bayes used 0.15 alpha parameter, K-neighbors classifier used five neighbors, linear, and kernel SVM both used 0 random states. We utilized n-grams features in each model. This program is widely used for text mining and natural language processing. Unigram, bigram, and trigram were utilized.

We calculated the performance of each model into unigram, bigram, and trigram after fitting our data into the model. There are differences between unigram, bigram, and trigram models. With 90.65% accuracy, random forest plays the most important function in the unigram feature (Table1). In bigram, the decision tree has the highest accuracy, at 88.49 percent (Table 2). Random forest had the most accuracy in the trigram feature, with 89.93 percent accuracy (Table 3).

We also showed a comparison table of all model's accuracy and F1-value for unigram feature (Fig. 4), bigram feature (Fig. 5), trigram feature (Fig. 6).

Table 1 Performance in unigram feature

	Accuracy	Precision	Recall	F1 score	Model name
1	87.05	95.92	74.60	83.93	LR
2	87.05	83.58	88.89	86.15	DT
3	90.65	94.64	84.13	89.08	RF
4	88.49	82.19	95.24	88.24	MNB
5	87.05	82.61	90.48	86.36	KNN
6	84.17	95.56	68.25	79.63	Linear SVM

Table 2 Performance in bigram feature

	Accuracy	Precision	Recall	F1 score	Model name
1	87.05	95.92	74.60	83.93	LR
2	88.49	85.07	90.48	87.69	DT
3	87.77	94.23	77.78	85.22	RF
4	84.89	76.25	96.83	85.31	MNB
5	87.05	82.61	90.48	86.36	KNN
6	82.73	95.35	65.08	77.36	Linear SVM

Table 3 Performance in trigram feature

	Accuracy	Precision	Recall	F1 score	Model name
1	87.05	95.92	74.60	83.93	LR
2	87.77	83.82	90.48	87.02	DT
3	89.93	100.00	77.78	87.50	RF
4	81.29	71.26	98.41	82.67	MNB
5	86.33	81.43	90.48	85.71	KNN
6	80.58	95.00	60.32	73.79	Linear SVM

Fig. 4 Accuracy and F1 value comparison for unigram feature

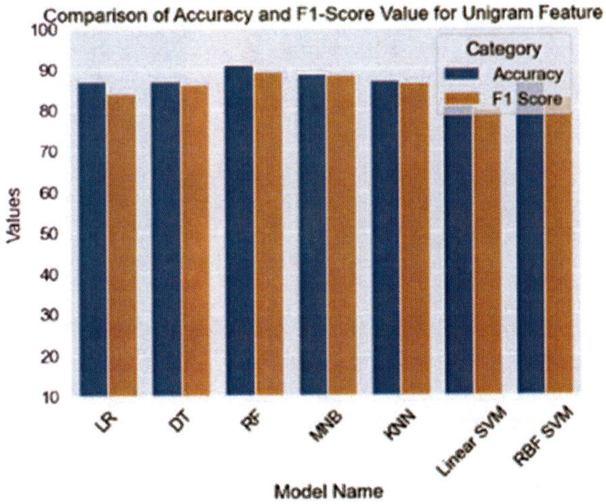

Fig. 5 Accuracy and F1 value comparison for bigram feature

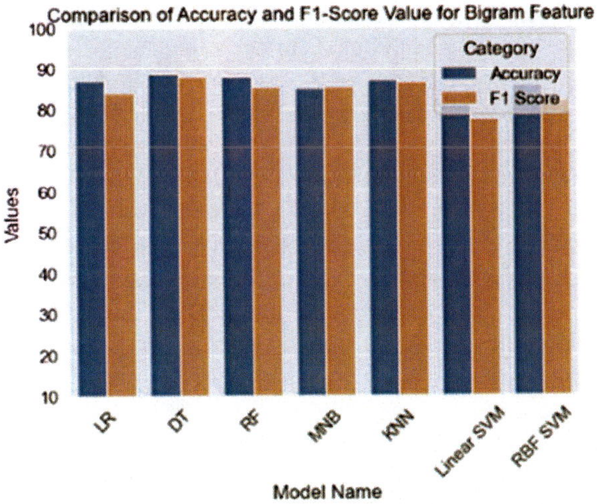

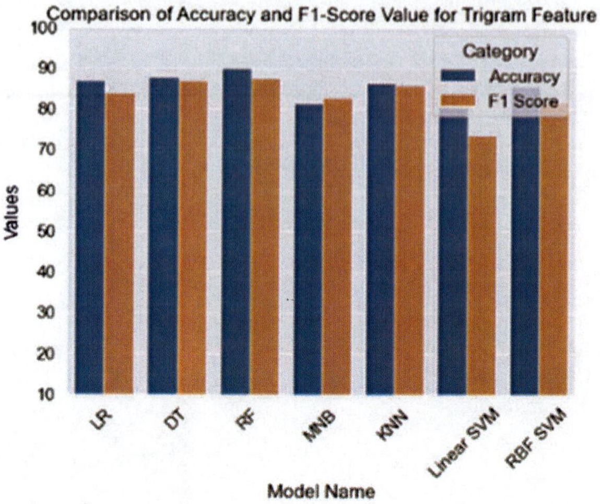

Fig. 6 Accuracy and F1 value comparison for trigram feature

5 Conclusion

E-commerce is the top trending topic in the world nowadays. E-commerce means electronic media. These e-commerce sites are online-based. Businesses man and normal people are participating more and more in these sites. They are making their decision through online sites and buying or selling products. Evaly and Daraz are Bangladesh's most popular e-commerce sites. Sentiment analysis is used to find out the sentiment type of any word. By applying sentiment analysis on customer's feedback, it is easy to understand the type of their feedback. So, this sentiment analysis will help an e-commerce site to analyze sentiment and to understand the type of sentiment from customer's feedback. By understanding them, the company will be able to increase its service quality. As a result, their number of customers will increase along with their reputations. We scraped the Facebook pages of these e-commerce companies and gathered client feedback. Most of the information was in the Bengali language, and some of it has been combined into both English and Bengali languages. With diverse models such as logistics regression, decision tree, random forest, multinomial Naive Bayes, K-neighbors, and linear support vector machine in n-grams, we employ unigram, bigram, and trigram features. Random forest is the most accurate with 90.65 and 89.93% accuracy in unigram and trigram. Among the finest fits are the unigram feature and random forest. In this paper, we have worked with positive and negative reviews of the customer. So, the e-commerce company Daraz and Evaly will be able to analyze the sentiment of customers from the comments of customers and understand the service quality. As it is a Bengali comment sentiment analysis process and it will be very helpful for any Bangladeshi company. So, as Bangladeshi companies, Daraz and Evaly will be able to analyze sentiment effectively and will be able to understand their service quality easily. Then they can increase their service and increase their company quality. In the future, we have planned to work with neutral reviews and their impact on the e-commerce business.

References

1. Wilson N, Christella R (2019) An empirical research of factors affecting customer satisfaction: a case of the Indonesian e-commerce industry. DeReMa J Manage 14(1):21–44
2. Yang L, Li Y, Wang J, Sherratt RS (2020) Sentiment analysis for E-commerce product reviews in Chinese based on sentiment lexicon and deep learning. IEEE Access 8:23522–23530
3. Noor F, Bakhtyar M, Baber J (2019) Sentiment analysis in e-commerce using svm on roman urdu text. In: International conference for emerging technologies in computing, pp 213–222 Springer, Cham
4. Kumar S, Yadava M, Roy PP (2019) Fusion of EEG response and sentiment analysis of products review to predict customer satisfaction. Inf Fus 52:41–52
5. Zhao H, Liu Z, Yao X, Yang Q (2021) A machine learning-based sentiment analysis of online product reviews with a novel term weighting and feature selection approach. Inf Process Manage 58(5):10265
6. Giao HNK (2020) Customer satisfaction at Tiki. vn E-Commerce platform. J Asian Finance, Econom Bus 7(4):173–183
7. Gajewska T, Zimon D, Kaczor G, Madzík P (2019) The impact of the level of customer satisfaction on the quality of e-commerce services. Int J Product Perform Manage
8. Ji P, Zhang HY, Wang JQ (2018) A fuzzy decision support model with sentiment analysis for items comparison in e-commerce: the case study of http://PConline.com. IEEE Trans Syst Man Cybernet Syst 49(10):1993–2004
9. Jabbar J, Urooj I, JunSheng W, Azeem N (2019) Real-time sentiment analysis on E-commerce application. In: 2019 IEEE 16th international conference on networking, sensing and control (ICNSC), pp 391–396. IEEE
10. Akter MT, Begum M, Mustafa R (2021) Bengali sentiment analysis of E-commerce product reviews using K-Nearest neighbors. In 2021 International conference on information and communication technology for sustainable development (ICICT4SD), pp 40–44. IEEE
11. Jha BK, Sivasankari GG, Venugopal KR (2021) Sentiment analysis for E-commerce products using natural language processing. Ann Rom Soc Cell Biol 166–175
12. Munna MH, Rifat MRI, Badrudduza ASM (2020) Sentiment analysis and product review classification in E-commerce platform. In: 2020 23rd International conference on computer and information technology (ICCIT). IEEE, pp 1–6
13. Karthik RV, Ganapathy S (2021) A fuzzy recommendation system for predicting the customers interests using sentiment analysis and ontology in e-commerce. Appl Soft Comput 108:10739
14. Ktaviani V, Warsito B, Yasin H, Santoso R (2021) Sentiment analysis of e-commerce application in Traveloka data review on Google Play site using Naïve Bayes classifier and association method. In: Journal of Physics: Conference Series. IOP Publishing, vol 1943, no 1, p 012147
15. Parveen N, Santhi MVBT, Burra LR, Pellakuri V, Pellakuri H (2021) Women's e-commerce clothing sentiment analysis by probabilistic model LDA using R-SPARK. Mater Today: Proceed
16. Zhang S (2021) Sentiment analysis based on food e-commerce reviews. In: IOP Conference series: earth and environmental science. IOP Publishing, vol 792, no 1, p 012023
17. Latha SS (2021) An experimental analysis on E-commerce reviews, with sentiment classification using opinion mining on web. Int J Eng Appl Sci Technol 5(11):2455–2143
18. Zhang Y, Sun J, Meng L, Liu Y (2020) Sentiment analysis of e-commerce text reviews based on sentiment dictionary. In: 2020 IEEE international conference on artificial intelligence and computer applications (ICAICA), IEEE, pp 1346–1350

Applications of Artificial Intelligence in IT Disaster Recovery

Kaustubh Lohani, Prajwal Bhardwaj, Aryaman Atrey, Sandeep Kumar, and Ravi Tomar

Abstract Artificial intelligence (AI) changed several industries and processes over the last decade. One such process is disaster recovery (DR) planning to ensure business continuity in an adverse event. Over the past decade, computing power has grown tremendously. At the same time, AI has also grown, making it possible to have complex real-time analytics, which was previously difficult, making time-sensitive applications such as DR management possible. AI can automate the disaster recovery processes and ensure a quick initiation of the disaster recovery plan in case of an adverse event by shifting the critical operations to a secondary site, issuing automatic alerts to the corresponding administrator, and providing insights to deal with the situation. Consequently, using AI to implement a DR plan can be a practical move. Utilizing AI instead of manual processes to initiate the DR plan ensures faster initiation, reliability, and availability. This paper discusses the possible use cases for utilizing AI in DR workflow to ensure business continuity. Moreover, possible use cases for three phases of disaster recovery workflow, pre-disaster, implementation, and the aftermath, will be discussed. Furthermore, potential benefits and challenges of utilizing AI in the disaster management workflow will be highlighted.

Keywords Disaster recovery · Business continuity · Backup · Artificial intelligence · Downtime

---

K. Lohani (✉) · P. Bhardwaj · A. Atrey · R. Tomar (✉)
School of Computer Science, University of Petroleum and Energy Studies (UPES), Dehradun, Uttarakhand 248007, India
e-mail: kaustubhlohani25@gmail.com

R. Tomar
e-mail: rtomar@ddn.upes.ac.in

S. Kumar
IIMT Group of Colleges, Greater Noida, India

V. Skala et al. (eds.), *Machine Intelligence and Data Science Applications*, Lecture Notes on Data Engineering and Communications Technologies 132,
https://doi.org/10.1007/978-981-19-2347-0_52

1 Introduction

Artificial intelligence is the ability to program machines intelligently that mimic the natural intelligence found in humans, meaning a well-trained AI model can mimic the accurate real-time decision making of humans as required in a particular situation.

Disaster recovery (DR) is the process of setting up policies and procedures to ensure business continuity in case of a natural or a human-induced adverse event such as a flood, fire, or cyber-attack. Ensuring business continuity in a disaster is a complex multi-phase process that requires quick and accurate decision making.

The first phase encompasses planning, which generally includes identifying the assets and potential risks to those assets. Artificial intelligence can be utilized to provide insights and recommendations in this stage to maximize the effectiveness of the DR plan.

The second phase of DR and business continuity is activated when a disaster strikes. A secondary site or backup site is created by replicating the tools, technologies, and data present at the primary site to offset the risk posed to an asset class. Under DR procedure, all critical operations are transferred to the secondary site if a disaster hits the primary site to ensure business continuity. The time elapsed until the switch to the secondary site happens is called downtime. The downtime is one of many factors that should be minimized to ensure business continuity and achieve recovery. At this stage, AI can be primarily utilized to make the switch faster and more reliable, leading to a minimum downtime and aiding business continuity.

The third phase deals with controlling the aftermath of the disaster, including monitoring the loss of reputation to the organization and investigating the cause of the disaster. AI systems can aid the objectives of this stage by monitoring social media and aiding in sifting through colossal amounts of metadata where the potential cause for the disaster resides.

Incorporating AI into the DR workflow can be very much beneficial. However, some challenges come along with it.

The subsequent sections will discuss the potential use cases of AI in the disaster management workflow with its benefits and challenges. First, some basic terminologies that characterize an ideal DR plan will be discussed. Furthermore, the phases of DR workflow will be discussed. Moreover, the potential use cases of utilizing AI in the DR workflow to ensure business continuity will be discussed. Finally, the benefits and challenges of incorporating AI in the DR workflow will be outlined.

2 Background

2.1 Business Continuity

Business continuity is the capability of an organization to ensure product or service delivery during a disruption that is identical to or acceptable in a usual operation scenario. The disruption can take many forms. They can be natural such as floods or fire, human induced such as cyber-attacks, or they can be simple such as a critical employee leaving the organization.

Such disruptions or disasters are countered by devising a business continuity plan (BCP) that outlines the policies and procedures to deal with potential disruptions. A good BCP better prepares the organization to deal with exceptions or disruptions.

2.2 Failover

Failover is the process of switching business-critical operations from a primary site to a secondary site in case of a disaster at the primary site. Initiating a failover procedure activates the replicated modules in the backup site and shifts critical operations there [1].

2.3 Failback

Failback is the procedure initiated to shift the operations back to the primary site once recovered from the disaster. Failback is necessary as operations on the secondary site can incur additional costs for the organization [1].

2.4 Recovery Time Objective or RTO

In a DR plan, RTO signifies the maximum amount of accepted time for which a system can be down without causing significant damage to the business operations. Essentially, RTO is the maximum allowed downtime for operations to resume. RTO is one of the critical parameters to measure the effectiveness of the DR plan. Ideally, the lower the RTO higher is the effectiveness of the DR plan. The measurement of the RTO starts the moment the disaster strikes a site and ends when the operations of that particular site are resumed on the backup site or in other words, RTO is the time elapsed until a failover is completed.

Minimizing the RTO can be challenging as organizations often use manual or semi-manual disaster detection with manual site switching procedures [1].

2.5 Recovery Point Objective or RPO

RPO is a parameter to denote the maximum acceptable amount of data lost after a switch to a secondary site is done. Furthermore, RPO also signifies the frequency of the backup on the secondary site. For example, if RPO is 24 h, the data should be backed up every 24 h, and if a disaster strikes before the backup is done, then the maximum amount of data loss will be data generated in the last 24 h. Ideally, critical systems in the primary site should be in sync with the secondary site, meaning at any given point in time, both the sites should duplicate each other [1].

2.6 Consistency and Performance Impact

Consistency and performance impact are other parameters to measure the effectiveness of the DR workflow.

Consistency means that the DR implementation should ensure the accuracy of business operations even after switching to the backup server. A low level of consistency between the primary and backup sites could mean more disruption in business processes, thereby failing the goal of business continuity.

Moreover, DR implementation should have the least amount of performance impact on the organization's day-to-day operations. If a DR implementation starts interfering in carrying out regular tasks, the DR implementation or the DR plan is ineffective [1].

3 Disaster Recovery Phases

The disaster recovery lifecycle can be broadly divided into three phases. Preparing for each of these phases is extremely important to achieve the goals of business continuity.

The three phases of disaster recovery described in detail below are before planning, implementation, and the aftermath.

3.1 Planning or Pre-disaster Phase

This phase concerns with planning the actions to ensure business continuity in a disaster. Assets and their potential risks are listed in this phase to assess the overall threat to business continuity. A detailed plan consisting of policies and procedures is created in this phase that sets the maximum acceptable RTO, RPO, and downtime. Moreover, in this phase, recovery strategies are chosen out of two broad categories, traditional and cloud based. Choosing the correct strategy is vital as the site location is also determined based on the chosen strategy. There are several disaster recovery site alternatives such as on-site, colocation, or cloud [2]. Moreover, possible implementation techniques and suggested intelligence integration differ based on the chosen strategy [3].

To begin the planning phase, organizations define the scope of the plan and then create a list of hardware devices (servers, laptop computers, desktop computers, and other devices), software, and data deployed at the primary site that might need protection in case of a disaster to achieve the goal of business continuity. The critical data identified in the previous step is then prepared for deployment to a secondary site as per the chosen strategy (traditional or cloud based).

Furthermore, the planning phase will also include the steps that need to be taken to comply with the local law and regulations. Furthermore, DR plan testing and dry-run procedures to ensure preparedness and quick response are also detailed in this phase.

3.2 *Implementation Phase*

This phase is initiated to prepare the secondary site as a backup to the primary site in case a disaster strikes the primary site. This phase encompasses the processes to enable switching to a secondary site if a disaster strikes at the primary location. Furthermore, organizing and managing the backups to the secondary site and keeping the operational components at the secondary site up to date with the primary site are also taken care of at this stage. Processes at this phase are designed keeping in mind the RPO and RTO agreed in the planning phase.

Essentially once the disaster strikes, this phase is initiated for recovery and restoration of the business processes. Once the disaster strikes the primary site, the business operation recovery is initiated by implementing the failover followed by a list of procedures detailed in the planning phase. Moreover, after the switch happens restoration process begins by carrying out forensics at the primary site to determine the cause and the type of the disaster. The restoration process ends after the primary site is up and fit to execute the business operations, after which failback is executed.

3.3 *Aftermath Phase*

This phase starts as the failback is executed. The organization has to report their incidents to their clients, consumers, and other organizations to comply with the local laws and standards.

Furthermore, the cause of the disaster is addressed in this phase, and the planning phase is executed again to ensure that the disaster will not be repeated. Moreover, a team of personnel from the organization is deputed to monitor the loss to the reputation, start actions to repair it and regain the lost trust of clients and consumer (Fig. 1).

4 Artificial Intelligence in Disaster Recovery

AI can be utilized on all the phases of the disaster recovery mentioned above to streamline the execution and planning of the DR plan.

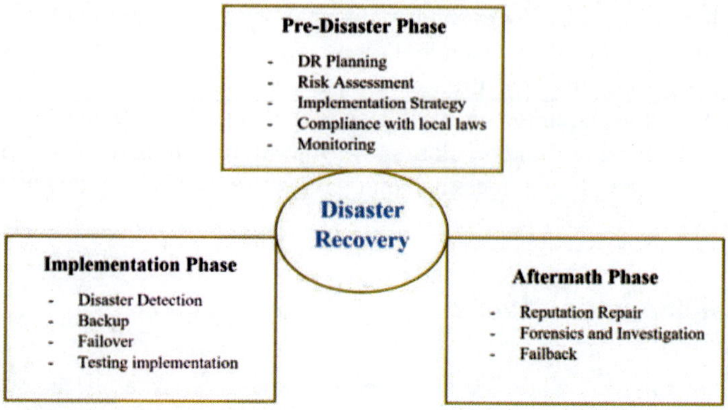

Fig. 1 Phases of disaster recovery

In the pre-disaster phase, AI can provide insights and recommendations by analyzing previous disasters and help planners devise a more effective plan. Furthermore, in the implementation phase, AI can help automate the backups to the secondary site and keep it ready for operation in case of an adverse event at the primary site. Finally, in the aftermath phase, AI can help the organization monitor the reputation loss after the disaster and help the team devise pragmatic data-driven strategies to repair it.

Possible use cases for AI in the DR workflow are mentioned below.

4.1 Pre-disaster Phase

4.1.1 Assist in Disaster Recovery Planning

AI can analyze data from previous disasters in the same business setup to spot common patterns and mistakes. These patterns can be further utilized to create an effective DR plan that can ensure that previous mistakes are not repeated. Furthermore, AI can be integrated with the planning phase to provide recommendations and examine the plan for interdependencies in assets, which might be missed in a disaster scenario. Moreover, AI can conduct risk assessments and calculate the impact on the business if a particular asset gets affected.

Authors of [4] have proposed an intelligent system based on fuzzy cognitive maps (FCM) to analyze and monitor disaster recovery plans. The proposed model based on FCM can analyze the "what-if" scenarios to provide the team managing the DR plan with insights and improve their understanding of the consequences of the plan. However, here, the FCM model requires weight assignment that the subject matter experts should undertake. Moreover, the weight assignment ensures that the FCM is customized to the need of that particular organization.

Similarly, the authors of [5] presented an AI model designed using feed-forward neural networks to aid in business continuity planning. The proposed model considers business continuity budget, allocated staff, support, geographic area of operations, and possible hazards, among other factors. The model is trained on data from 283 organizations operating in Toronto, Canada.

Utilizing AI models to aid in DR planning would result in an effective and efficient plan that could reduce downtime in case of a disaster if implemented in the manner specified.

4.1.2 Continuous Monitoring

AI models can be trained to recognize characteristics of regular network traffic. This model can then be deployed over the network to scan for anomalies and trigger an alert if abnormal values are registered. These predictions can then be used to trigger the failover procedure or sent to the appropriate personnel for further investigation, triggering the failover mechanism manually if found valid.

Furthermore, AI models can be deployed to monitor weather data, temperature sensor data continuously, and other physical security sensor data to scan for anomalies and predict the possibility of a disaster before it is manually detected.

Results from the continuous monitoring by the AI can be further tied to the disaster management team who can take appropriate actions based on the information provided by the AI system.

The authors of [6] utilized deep learning methodologies to propose an artificial intelligence-based intrusion detection system called AI-IDS. The proposed system monitors the web traffic in real time and flags any abnormal traffic using the intelligent modeling techniques employed.

Deploying AI systems to continuously monitor the environment for any disaster threats would result in disaster prevention or give the organization a prior notification to switch operations to the secondary site, thereby reducing the potential downtime.

4.1.3 Running Simulations

Running prior simulations can significantly benefit the planning process of disaster recovery. AI models can simulate threats and worst-case scenarios and identify the infrastructure's vulnerabilities and points of failure. The vulnerabilities and weak points identified can be fortified by allocating extra protection to address them in the planning phase.

The authors of [7] used AI to develop a novel framework called scenario planning adviser (SPA) that takes information from media and domain knowledge from experts to generate scenarios that explain key risks and depict possible futures.

The previous simulation of disaster scenarios would enable better planning and preparation, resulting in low downtime if a disaster strikes [8, 9].

4.2 Implementation Phase

4.2.1 Automate Backups

Organizations need to keep the primary and secondary sites in sync to achieve business continuity in a disaster. Backup operations can be performed in two ways, instruction-based backup and intelligent backup.

Generally, organizations use an instruction-based system where a backup instruction is programmed to perform a backup to the secondary site after a specific time interval. For example, every fortnight, a backup will be performed at 10:00 a.m. However, an instruction-based backup system poses certain risks to business continuity in a disaster. First, if a device misses the backup window due to unavailability or maintenance issues, then the data will be out of sync in the secondary site. Moreover, if the data is mission-critical, it could render the whole system unusable at the secondary site during the disaster. Secondly, if a fortnight backup is scheduled for tomorrow and a disaster strikes before the backup is performed, the data for the past fifteen days would be absent at the secondary site. Thus, manual data migration procedures would be required to update the data at the secondary site before shifting business operations, increasing downtime.

Another backup system is intelligent backup, wherein AI models can be deployed to carry out backup tasks. AI models can monitor the network traffic and intelligently schedule backups when the workload for the system is lower. Utilizing AI to backup primary site has several advantages. The backup operation can be stopped if the workload increases and resumed when the system workload decreases. Moreover, if a device misses a backup window, AI can schedule a separate backup window for that particular device. Thus, data could be backed up to the secondary site multiple times by fixing a window, thereby reducing the chance of the secondary site going out of sync.

Ideally, organizations should implement a hybrid strategy, using an instruction-based backup model according to mission criticality and implementing an AI-based intelligent system to backup the primary site on the secondary site to ensure no failure occurs to keep both the sites in sync.

Deploying AI-based backup systems would enable multiple backup windows for the organization instead of one fixed backup window. This would reduce downtime, RTO, and RPO. Furthermore, deploying AI for backup would reduce the performance impact as AI systems would initiate backup only when the workload is low and pause when it increases. Moreover, as AI creates multiple backup windows rather than a fixed backup window, data and operation consistency would be maintained when the switch happens to the secondary site.

The authors of [10] have discussed how implementing machine learning (ML)-based security schemes often are not enough to provide reliable IoT services. Therefore, it is necessary to have backup solutions integrated with ML models to have reliable backups in a disaster situation. However, specific studies are needed to develop and evaluate automated backup solutions that function across platform.

4.2.2 Automate Failover

Automating the failover procedure is complex and risky as false disaster alarms can unnecessarily trigger the failover procedure, causing unnecessary complexities to the business operations.

Data captured at the monitoring module of the AI deployed for disaster management can automatically initiate the failover procedure. AI models can be deployed to interpret the captured network parameters, which can then be analyzed to predict the likelihood of disaster, which, if higher than the set threshold, can trigger the failover procedure and shift operations to the secondary site.

The authors of [11] proposed a novel solution called ITRIX, a web application that enables consultants to convert textual DR procedures and plans to executable actions. The solution uses natural language processing and machine learning techniques to suggest actions from the text plans. These actions can significantly reduce the time taken to complete the failover procedure. However, the proposed solution needs additional methods to judge the magnitude of the disaster and suggest the ideal failover strategy. For example, if a disruption occurs at only one server, the system should recommend switching the affected server rather than suggesting shifting the operations of the whole data center.

Utilizing AI to perform failover sequences in the event of a disaster would ensure quick decision making and result in reduced downtime, RTO, and RPO.

4.2.3 Rapid Recommendations and Insights

Human experience and decision making are incredibly crucial for the organization in case of a disaster. While AI cannot eliminate the much-required human experience, it can efficiently provide timely recommendations to deal with the situation. For example, AI systems can be trained to carry out operations to identify the nodes not affected by the disaster. Moreover, AI systems can be deployed to automatically deal with blanket situations such as power cuts or system-wide failure.

Prior studies on the development of recommendation systems for recovery in case of an adverse event is minimal. However, DR vendors have modern tools, wherein these types of recommendation systems are baked in. The primary cause for this discrepancy is that there is no one size fits all recipe for success in recommendation engines. Moreover, recommendations systems are designed with keeping in mind the whole process of a DR plan. Since not every organization has the same plan, it becomes challenging to have one recommendation system designed for all. In contrast, modern DR vendors offer their own customized DR solutions with a fixed process to follow. Therefore, it becomes easy for them to design a recommendation engine that revolves around their offering and suggests steps that are already baked in their offering.

4.2.4 Automated Detection

Data from the continuous monitoring module can be fed to an AI system trained to analyze and interpret different network parameters to determine the onset of the disaster. The results from this module can be used in a variety of ways. They can be used to automatically sound the disaster alarm and issue notifications to appropriate departments and administrators. Moreover, the results can begin the failover procedure and execute the other parts of the DRP.

Furthermore, AI can be used to detect contaminated or corrupted data which is causing operational issues. Since AI is custom programmed as per the business requirement, it can detect anomalies in the data before it escalates into a disaster situation.

The authors of [12] utilized deep learning techniques to detect advanced persistent threats (APT). The authors applied an intrusion detection system based on recurrent neural networks to classify threats to the system.

Similarly, authors of [13] have proposed an intelligent system for flood detection, which could cause substantial damage to the primary data center applications, thus hampering business operations. The proposed model uses data collected through a wireless sensor network and feeds it to the support vector machines (SVMs) to classify them as flood causing nor no-flood causing.

4.3 Aftermath Phase

4.3.1 Social Media Monitoring

Disaster can severely impact the reputation of the organization. The organization has to answer to the media, clients, and the local governing bodies while quashing any incorrect information before it gains the attention of other stakeholders.

AI models can be deployed to monitor social media and gauge the people's opinions on the disaster, and a relevant strategy can be formed to address the concerns. AI systems can filter out any mention of the disaster or the affected organization and further, perform a sentiment analysis and present the disaster management team with clear statistics and information necessary to formulate a plan for reputation repair.

Moreover, AI systems can monitor information relating to the disaster on social media platforms, enabling the organization to quash any incorrect information before it causes severe damage.

Authors of [14] proposed a novel framework called social intelligence advisor capable of automatically analyzing raw data from social media and presenting the marketing teams with actionable conclusions. However, the model does not explicitly focus on reputation repair after a disaster but can be modified to add additional weight to the sentiment analysis phase.

4.3.2 Forensics and Investigation

AI models can be deployed to sift through colossal amounts of metadata and logs containing information about the disaster's cause. An AI system can be trained to identify correct parameters and identify the anomalies that can be beneficial to determine the cause of the disaster from the metadata files.

Authors of [15] have proposed a novel AI-based multi-agent system to aid in digital investigations. The system is comprised of multiple agents, each with domain knowledge and expertise of a human domain expert. The proposed system utilizes an intelligent content-based reasoning model to select the agent appropriate for a particular investigation. Moreover, the authors compared the findings of the proposed model with a human expert and found significant positive results amounting to reduced turnaround time. The results were encouraging primarily due to the involvement of colossal amounts of data involved in a digital forensics investigation which AI can help co-relate and present in a structured manner to the human investigator.

Moreover, the authors of [16] have highlighted the aspects of shifting computer forensics to intelligent forensics. Moreover, knowledge-based AI systems can be developed to capture the knowledge of a domain expert and further, apply the domain knowledge to spot anomalies in the data [17, 18] (Table 1; Fig. 2).

4.4 Benefits of Utilizing AI in DR Workflow

- As opposed to a traditional algorithm, AI has the potential for continuous improvement. More data can be fed to the AI algorithms to improve them further. Additionally, data from past disasters can also be used to improve the AI models. Thus, AI can continuously improve and learn from its mistakes, thereby getting better at predicting outcomes and detecting disasters.
- In contrast with the human workforce, an AI is available 24 × 7 without any holidays, ensuring quick response in a disaster. Moreover, AI does not have human emotions, making them incorruptible. Thus, if AI is doing its job well, there is no reason for it to perform inaccurately unless tweaked which can be prevented by limiting access to the core of AI.
- Incorporating AI in the DR workflow ensures a quick and decisive response if a disaster strikes the business. AI can quickly activate the DR plan and initiate the failover procedure in a disaster, reducing downtime. Moreover, AI enables automated backup to the secondary site making it up to date with the latest data from the primary site. Thus, requiring minimal manual intervention to shift processes to the secondary site and reducing the downtime.
- Finally, AI enables automation of several processes in the DR workflow, such as disaster detection, backup, and failover. Automation in the DR workflow minimizes the chances of human error, increases the availability, and reliability of the overall system.

Table 1 Summarizing the applications of AI in the DR workflow with potential impact over DR workflow

S. No	Application of AI	Disaster recovery phase	Potential impact	Related works
1	Assist in disaster recovery planning	Pre-disaster phase	Better plans lead to low downtime	[4, 5]
2	Continuous monitoring	Pre-disaster phase	Reduced downtime due to early disaster detection	[6]
3	Running simulations	Pre-disaster phase	Better preparation leads to low downtime	[7]
4	Automate backups	Implementation phase	Reduced RPO, low-performance impact and improved consistency	[10]
5	Automate failover	Implementation phase	Reduced RTO and downtime	[11]
6	Rapid recommendations and insights	Implementation phase	Reduced downtime	–
7	Automated detection	Implementation phase	Reduced downtime	[12, 13]
8	Social media monitoring	Aftermath phase	Aids in reputation repair in the aftermath of the disaster	[14]
9	Forensics and investigation	Aftermath phase	Faster cause detection leads to quick failback, thereby saving operational costs	[15, 16]

4.5 Challenges of Utilizing AI in DR Workflow

- Training an AI model requires colossal amounts of data pertaining to that particular use case. For example, training the AI model to detect a disaster automatically requires network traffic data that depicts normal and anomalous behaviors. Sometimes, data required to develop an AI does not exist for a particular use case, meaning it needs to be collected by the organization itself, acquired from a third party or a survey.
- AI models learn from the data that is being fed in intelligent algorithms. If that data is biased or does not represent the statistics about a use case, then the AI model will suffer from a data bias, and the outputs or predictions will not be necessarily accurate.
- Furthermore, organizations still do not trust machines to entirely take decisions that directly impact their business finances or reputation. This trust deficit will most likely improve as more organizations start utilizing AI for disaster recovery.

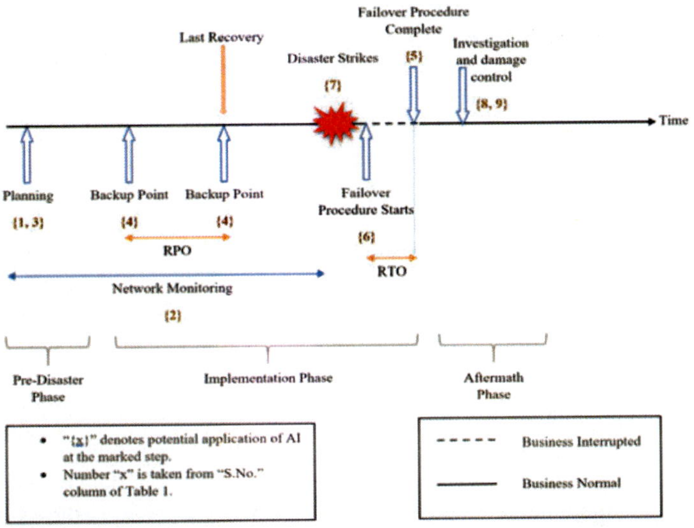

Fig. 2 Highlighting the utilization of AI in the lifecycle of disaster recovery planning

- Finally, a significant hurdle in AI playing a significant role in the DR workflow is the lack of contextual awareness. AI models cannot sometimes understand the context of a situation leading to AI making incorrect or partially correct assumptions. For example, if a node or a component in a network is down for maintenance, the network parameters for that node may appear anomalous to the AI monitoring the network, prompting the AI to raise a false disaster alert resulting in unnecessary complexities (Table 2).

Table 2 Summarizing the benefits and challenges of utilizing AI in the disaster recovery workflow

S. No.	Benefits of utilizing AI in IT disaster recovery workflow	Challenges of utilizing AI in IT disaster recovery workflow
1	Potential for continuous improvement	Training an AI model requires vast amounts of data that is not always readily available
2	Continuous availability	Possibility of incorrect predictions through a model trained on biased data
3	Limited human involvement resulting in the quick response in disaster scenarios	Organizations still do not trust a machine to take business-critical decisions
4	Potential for automation of several aspects of disaster recovery workflow	Lack of contextual awareness in AI models leads to unintended situations

5 Conclusion

Disaster recovery is an integral part of an organization's plan to deal with interruptions and ensure business continuity even in the worst cases.

In this paper, first, the phases of the disaster recovery workflow are depicted. Next, the role of artificial intelligence in the DR workflow is discussed. Furthermore, the possible use cases of AI in all three phases mentioned above are detailed. Moreover, the benefits and challenges of incorporating AI in the DR workflow are also highlighted. Furthermore, the above sections discuss how AI can automate several aspects of the DR workflow, eventually reducing downtime and improving RTO and RPO.

AI has several potential use cases in disaster recovery; however, the technology is not yet matured or entirely developed due to the lack of available data, among other reasons. Moreover, organizations do not trust the machines to independently make decisions without human intervention in some cases. The challenges mentioned in the above section should be overcome to benefit from incorporating AI in the DR workflow.

References

1. Wood T, Cecchet E, Ramakrishnan KK, Shenoy PJ, van der Merwe JE, Ven-kataramani A (2010) Disaster recovery as a cloud service: economic benefits & deploy-ment challenges. HotCloud 10:8–15
2. Alhazmi OH, Malaiya YK (2012) Assessing disaster recovery alternatives: on-site, colocation or cloud. In: 2012 IEEE 23rd international symposium on software reliability engineering workshops. IEEE, pp 19–20
3. Ju H (2014) Intelligent disaster recovery structure and mechanism for cloud computing network. Int J Sens Netw 16(2):70–76
4. Mohammadian M, Yamin M (2017) Intelligent decision making and analysis using fuzzy cognitive maps for disaster recovery planning. Int J Inf Technol 9(3):225–238
5. Asgary A, Naini AS (2011) Modelling the adaptation of business continuity planning by businesses using neural networks. Intell Syst Account Finance Manage 18(2–3):89–104
6. Kim A, Park M, Lee DH (2020) AI-IDS: application of deep learning to real-time Web intrusion detection. IEEE Access 8:70245–70261
7. Sohrabi S, Riabov, AV, Katz M, Udrea O (2018) An AI planning solution to scenario generation for enterprise risk management. In: Thirty-second AAAI conference on artificial intelligence
8. Parashar N, Soni R, Manchanda Y, Choudhury T (2018)3D Modelling of human hand with motion constraints. In: 2018 International conference on computational techniques, electronics and mechanical systems (CTEMS), pp 124–128. https://doi.org/10.1109/CTEMS.2018.8769229
9. Khullar R, Sharma T, Choudhury T, Mittal R (2018) Addressing challenges of hadoop for big data analysis. In: 2018 International conference on communication, computing and internet of things (IC3IoT), pp 304–307. https://doi.org/10.1109/IC3IoT.2018.8668136
10. Xiao L, Wan X, Lu X, Zhang Y, Wu D (2018) IoT security techniques based on machine learning: how do IoT devices use AI to enhance security? IEEE Sig Proc Mag 35(5):41–49
11. Sherchan W, Vaughn G, Pervin S, Barone B (2021) ITRIX-an AI enabled solution for orchestration of recovery instructions. In: Proceedings of the AAAI conference on artificial intelligence, vol 35, no. 18, pp 16106–16107

12. Eke HN, Petrovski A, Ahriz H (2019) The use of machine learning algorithms for detecting advanced persistent threats. In: Proceedings of the 12th international conference on security of information and networks, pp 1–8
13. Al Qundus J, Dabbour K, Gupta S, Meissonier R, Paschke A (2020) Wireless sensor network for AI-based flood disaster detection. Ann Operat Res 1–23
14. Perakakis E, Mastorakis G, Kopanakis I (2019) Social media monitoring: an innovative intelligent approach. Designs 3(2):24
15. Hoelz BW, Ralha CG, Geeverghese R (2009) Artificial intelligence applied to computer forensics. In: Proceedings of the 2009 ACM symposium on applied computing, pp 883–888
16. Irons A, Lallie HS (2014) Digital forensics to intelligent forensics. Fut Internet 6(3):584–596
17. Tomar R, Tiwari R, Sarishma (2019) Information delivery system for early forest fire detection using internet of things. In: Singh M, Gupta P, Tyagi V, Flusser J, Ören T, Kashyap R (eds) Advances in computing and data sciences. ICACDS 2019. Communications in computer and information science, vol 1045. Springer, Singapore. https://doi.org/10.1007/978-981-13-9939-8_42
18. Tomar R, Patni JC, Dumka A, Anand A (2015) Blind watermarking technique for grey scale image using block level discrete cosine transform (DCT). In: Satapathy S, Govardhan A, Raju K, Mandal J (eds) Emerging ICT for bridging the future—Proceedings of the 49th annual convention of the computer society of India CSI volume 2. Advances in intelligent systems and computing, vol 338. Springer, Cham. https://doi.org/10.1007/978-3-319-13731-5_10

Mushroom Classification Using MI, PCA, and MIPCA Techniques

Sunil Kaushik and Tanupriya Choudhury

Abstract Consumption of mushroom is seen to be increasing across the world since last decade, because mushroom is a vital source of protein and has huge medicinal constituents which can even cure cancer. Mushrooms are fungi and hence can be cultivated anywhere. Mushrooms have huge varieties and not all varieties are edible for humans. Hence, it is important to classify mushrooms in edible or poisonous categories. Recently, machine learning techniques are used to help in this categorization. Current research presents a novel method to classify mushrooms quickly and accurately using feature selection and machine learning techniques.

Keywords Mushroom · Mutual information · Principal component analysis · DT · Machine learning

1 Introduction

Mushroom, a multi-cellular fungi, is a nutrient rich food which contains the essential minerals such as calcium, phosphorus, vitamins, and minerals for human beings [1–3]. Based on the richness of the nutrition, ancient romans called mushroom as super food or food for the gods. Mushroom is able to help cell to absorb these nutrients [4]. These properties have made mushroom a super food that rejuvenates the cells and increase the human immune system [3, 4]. Studies have shown that mushroom is capable of killing various viruses and cancer cells [5]. However, not all the mushrooms are edible and few of them are poisonous [1, 5]. Few mushrooms contain mycotoxins which makes them poisonous [6, 7]. Thus, it has become necessary to identify the mushroom and classify them for human consumption. Luckily, few of

S. Kaushik (✉)
University of Petroleum and Energy Studies, Dehradun, UK, India
e-mail: Sunil.kaushik2011@gmail.com

T. Choudhury (✉)
School of Computer Science, University of Petroleum and Energy Studies (UPES), Dehradun, Uttarakhand 248007, India
e-mail: tanupriya@ddn.upes.ac.in

V. Skala et al. (eds.), *Machine Intelligence and Data Science Applications*, Lecture Notes on Data Engineering and Communications Technologies 132,
https://doi.org/10.1007/978-981-19-2347-0_53

the structural features of the mushroom can tell us about existence of the mycotoxins in the mushroom. Mushroom processing industries usually use automated systems to visually identify the edibility of the mushroom [8, 9]. The visual techniques are inflicted with the noise on the picture arising from various sources such as light, dirt on surface, and angle of the camera. Thus, it has become necessary to classify the mushroom [10, 11] based on the morphological feature such as shape of umbrella, gill size, gill shape, and position of the rings [12].

The previous studies have listed a number of classifiers [9, 13] resulting the high accuracy and precisions. These methods range from autoencoders to random forest. This paper proposes a novel method using the mutual information for feature selection with PCA for feature extraction and compare the accuracy and time taken by the neural networks and tree-based classifiers.

2 Literature Review

2.1 Related Work

Verma and Dutta [14] used the ANN and adaptive neuro-fuzzy inference system technique for classifying the mushroom dataset in edible or poisonous. They tried using different percentage of training set and found that ANN achieved maximum classification accuracy at 70% test train split. This test train split also showed the low MAE and minimum incorrectly classified instances. On the other hand, adaptive neuro-fuzzy inference system was able to correctly classify the dataset at test train split of 80%. This like ANN also showed the least MAE and minimum incorrectly specified instances. ANN was able to achieve an accuracy of 96%, and ANFIS was able to achieve accuracy of 99%.

Khan et al. [3] used the clustering techniques to correctly classify the mushrooms based on edibility. They used the UCI machine learning dataset and found out that K-means achieved highest accuracy of the 62% followed by the farthest east at 60% and distantly followed by expectation maximization algorithm at 42%. Authors also observed that EM algo took approximate 1000 secs for training while K-means took only 0.1 secs.

Alkronz et al. [15] used the ANN with backpropagation of 22 input neurons, 3 hidden layers with $(2 \times 1 \times 3)$ neurons, and an output layer with 1 neuron. Authors were able to achieve the accuracy of 99%. The accuracy was comparable to that of feed forward NN.

Yildrim and Bingol [16] have used the ensemble algorithms for classifying the poisonous mushrooms correctly. Authors were able to see the accuracy of the 99% with subspace KNN with a training time of 33.2% seconds. While other ensemble methods fairly agreeable at accuracy 89% at training time of 27–30 secs. These models, however, outperformed the simple machine learning algorithm such as SVM.

Wagner et al. [12] carried out experiments using Naive Bayes, logistic regression, linear discriminant analysis (LDA), and random forests (RFs). They were able to

Table 1 Feature selection techniques in last 4 years

Technique	Count
SVM, NB, DT	2
Neural networks	4
PCA-DT	1
KNN/	4
Ensemble	2

achieve 100% accuracy with random forest, but the accuracy for others accuracy was in range of 75% to 80%. They also created a new dataset for the mushroom classification for 173 species. However, they have not published any dataset.

Hamonangan et al. [17] using the UCI mushroom dataset applied the Naïve Bayes and KNN classifiers. They were able to achieve accuracy of apx 90% and 99% using the KNN algorithm.

Ismail et al. [18] used the PCA technique on the UCI machine learning mushroom dataset. They used the ML techniques such as ID3, J48, and C4.5 and were able to classify the edible and poisonous mushroom, but the accuracy of the experiments is not mentioned.

Zahan et al. [19] used the deep learning approaches like InceptionV3, VGG16, and ResNet50 on the dataset of 8190 records and were able to achieve an accuracy of 88.4%.

Aljojo et al. [20] used the multi-layer feed forward ANN with 22 input neurons, 2 hidden layers with (3×1) neurons, and an output layer with 1 neuron with JNN tool on the UCI machine learning dataset and were able to achieve an accuracy of 99%.

Chumuang et al. [21] used 200 sample of toxic and nontoxic mushroom to create a balanced sample set and used the classifiers Naive Bayes updateable, Naive Bayes, SGD text, LWL, and *K*-nearest neighbor (KNN). They were able to achieve an accuracy of 100% with KNN as given in Table 1.

2.2 Research Gap

Careful analysis [2, 22–25] of the literature suggests that there are very less studies done to identify the edible mushroom using feature selection or dimension reduction techniques for pre-processing.

This paper presents a novel method using MI and PCA for feature selection to enhance the accuracy of the classifiers and reduce the time taken to classify.

2.3 Dataset

The study uses the mushroom dataset provided at the University of California at Irvine. The dataset contains the data of edible and poisonous mushrooms of 23

Table 2 Feature of the dataset [26]

Feature	Values
Classes	Edible, poisonous
Cap-shape	Conical, convex, knobbed, bell, flat, sunken
Cap-surface	Grooves, smooth, fibrous, scaly,
Cap-color	Buff, gray, brown, green, pink, cinnamon, purple, white
Bruises	Bruises, no
Odor	Anise, creosote, almond, foul, musty, fishy, none, spicy
Gill attachment	Free, descending, attached, notched
Gill spacing	Crowded, distant, close
Gill size	Narrow, broad
Gill color	Brown, buff, black, gray, green, purple, orange, red
Stalk-shape	Tapering, enlarging
Stalk-root	Rhizomorphs, cup, bulbous, equal, rooted, club
Stalk-surface-above-ring	Silky, scaly, smooth, fibrous
Stalk-surface-below-ring	Silky, scaly, smooth, fibrous
Stalk-color-above-ring	Pink, buff, gray, orange, red, yellow, cinnamon, white, brown
Stalk-color-below-ring	Pink, buff, gray, orange, red, yellow, cinnamon, white, brown
Veil-type	Universal, partial
Veil-color	Yellow, orange, brown, white
Ring-number	Two, one, none
Ring-type	Sheathing, flaring, large, none, evanescent, pendant, zone, cobwebby
Spore-print-color	Orange, black, yellow, brown, buff, purple, green, white, chocolate
Population	Abundant, clustered, numerous, scattered, several
Habitat	Waste, meadows, paths, grasses, urban, woods, leaves

species in the Agaricus and Lepiota family. The features of the mushroom dataset are given in Table 2 [26] (Fig. 1).

2.4 Machine Learning Classifiers

Gaussian Naïve-Bayes Algorithm (NB)—This algorithm is based on the classical Bayes theorem with its foundation on the probability theory. This classification algorithm is Naïve in that it considers that all the features in a dataset are independent of each other. This algorithm calculates the posterior probability as a product of prior information and likelihood for given to evidence.

$$P(E_i|A) = \frac{P(E_i).P(A|E_i)}{\sum_{i=1}^{n} P(E_i).P(A|E_i)} \tag{1}$$

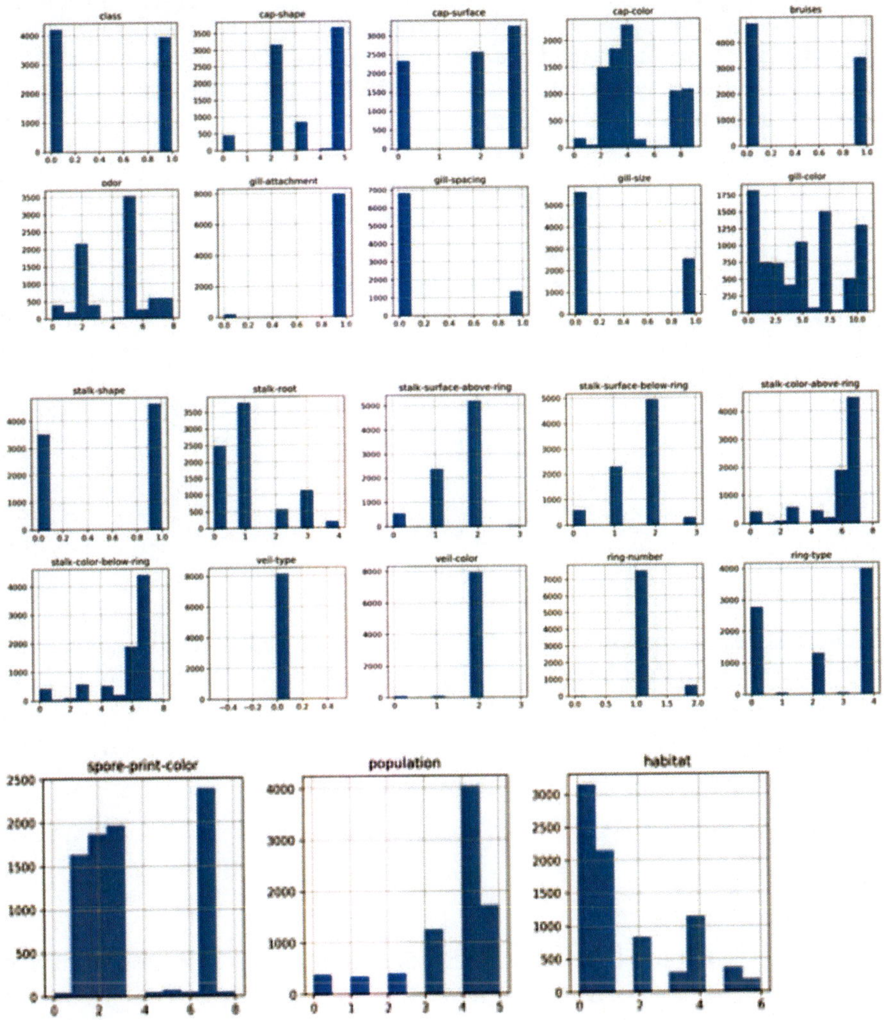

Fig. 1 Correlation of features in dataset

Here, $E_1, E_2, E_3 \ldots E_n$ are events in any random experiment, and A is any random event in tandem with event E.

Support Vector Machine Algorithm (SVM)—This algorithm is non-probability-based model and tries to classify the events in a space separable by hyperplanes. The equation gives the hyperplanes.

$$f(x) = (\beta_0 + \beta_t x) \quad (2)$$

Here, β_0 is the bias, and β_t is the weight.

Logistic Regression Algorithm (LR)—The algorithm is used to classify the dataset into two classes, i.e., it is used for binary classification. The regression is a few of the oldest algorithms used in classification problems and uses the sigmoid function (given in Eq. 3) as an activation function.

$$f(x) = \frac{1}{1 + e^{-x}} \tag{3}$$

Decision Tree (DT)—This classification algorithm uses the information gain at each step to classify the tuple in a dataset (T) concerning a given feature (F) to a target class ©. The information gain (IG) can be easily said as a difference of entropy E. The information gain and entropy are given in Eqs. 4 and 5.

$$\text{IG}(T, F) = E(T) - E(T, F) \tag{4}$$

$$E(S) = \sum_{i=1}^{n} -\text{pi} \log \text{pi} \tag{5}$$

Random Forest Algorithm (RF)—This algorithm is based on the DT, and it regulates the DT by creating the height of the only N where N is a number of features for better accuracy.

AdaBoost Algorithm (AD): This algorithm combines multiple weak classifiers and provides a strong classifier. It can be represented by Eq. 6.

$$H(x) = \sin \sum_{t=}^{T} \alpha_{1'} h_t(x) \tag{6}$$

Here, $h(x)$ is the output of the weak classifier, and α is the weight associated with the classifier.

2.5 *Principal Component Analysis*

PCA is a feature extraction technique which is used to generate the new features from initial features. The new features are linear combination of the initial features. PCA maps each instance of the given dataset present in a d dimensional space to a k dimensional subspace such that $k < d$. The set of k new dimensions generated is called the principal components (PC), and each principal component is directed toward maximum variance. The principal components can be represented as the following:

$$PC_i = a_1X_1 + a_2X_2 + \ldots + a_dX_d \tag{7}$$

where PC_i—principal component 'i'.
X_j—original feature 'j'.
a_j—numerical coefficient for X_j.

2.6 *Mutual Information*

Mutual information (MI) is used to measure the dependency between two random variables. The higher MI score means the higher the dependency, and low score means the low dependency. It is given by the Eq. 8.

$$I(X; Y) = H(X) - H(X|Y) - H(Y|X) \tag{8}$$

2.7 *Performance Measures*

- **Accuracy**—Accuracy can be thought of as the ratio of correctly classified to the total number of observations.
- **Precision**—Precision can be considered the ratio of actual observations to the observations predicted by the classifier.
- **Recall**—This can be regarded as the ratio of observations classified as true in the sample.
- **F1 Score**—This can be regarded as the harmonic mean of precision and recall.
- **Time**—Execution time of the classifier to classify the mushrooms.

3 Methodology

In this paper, a novel methodology for accurately and precisely classifying edible and poisonous mushrooms is discussed. The new methodology considers the MI and PCA for feature selection and reduction and DT classifier for the classifications. The accuracy of the model is compared with the ML algorithm such as LR, SVM, NB, LDA, and ensemble algorithms such as random forest and AdaBoost with the MI and PCA feature selection techniques.

The study considers the mushroom classification dataset provided by UCI. The dataset has 23 feature including the target feature and has 8124 tuples distributing over the edible and poisonous labels. The dataset contains the morphological sample of the 23 species of the gilled mushrooms. Morphology of a gilled mushroom is given in Fig. 2 (Figs. 3 and 4).

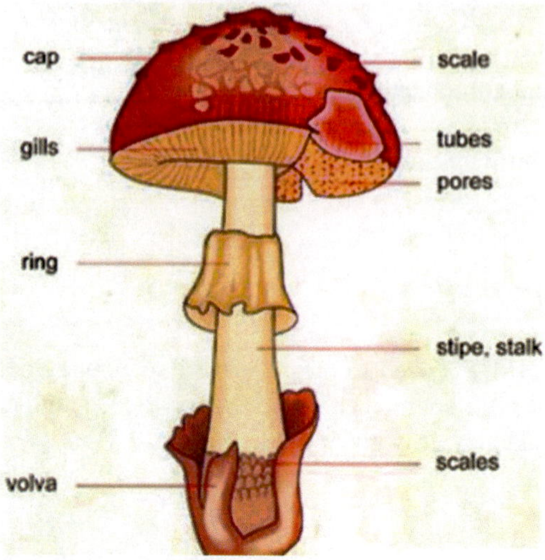

Fig. 2 Morphology of mushroom [26]

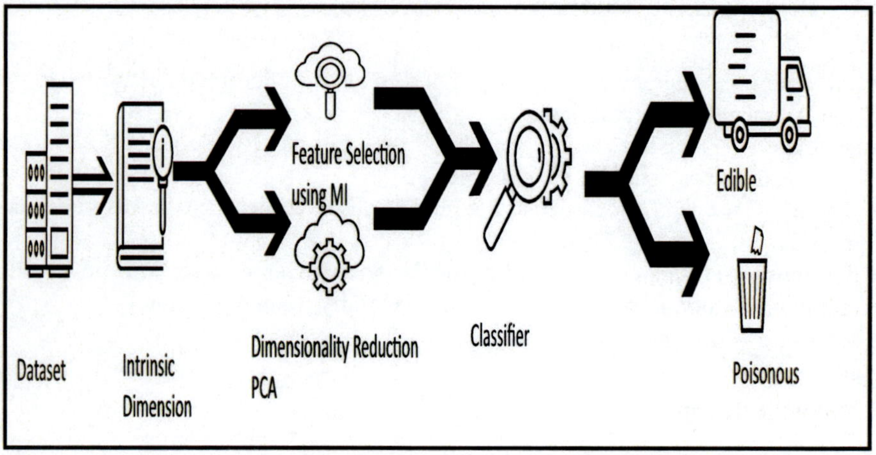

Fig. 3 Framework of MIPCA model

Algorithm 1: Mushroom classification using MI and PCA.

Algorithm 1

```
INPUT
        : D - UCI Mushroom Data
         ML = {LR,LDA, NB, SVM, AD, DT,RF}
         P = {FULL SET, MI, PCA, MI-PCA}
OUTPUT
     : T – Execution time of classifier
       A – Accuracy of classifier
PROCESS :
Read the UCI Mushroom Data D
Find the Intrinsic Dimension to find the number of optimum features I : I
→D
Standardize the data
LOOP For P
   Create a copy of data
   F : P→D
   LOOP For ML
     Get T for ML→D
        Get A for ML→D
   END LOOP
END LOOP
Find Best P and ML for argmax( T,A)
```

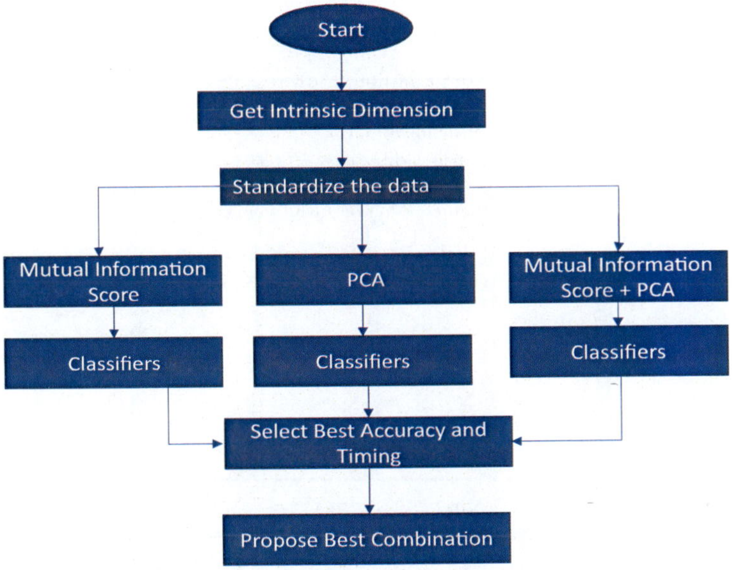

Fig. 4 Research methodology

4 Results and Discussion

Publicly available UCI mushroom dataset was applied to the ML models (LR, LDA, NB, DT, and SVM) and ensemble methods (AdaBoost and random forest), and first experiment was carried out. This is referred as full set in Table 3 and Fig. 5. The first set of experiments was set as a baseline. The intrinsic dimension for the dataset was found to be 16 which means that at least 16 features are needed to make the correction with justifiable accuracy. The MI was calculated between the features, the target was carried out, and 16 best features were selected. This set of 16 features was fed to ML and ensemble models. The output is given in Table 3 and Fig. 5 and is referred as MI. Similarly, the PCA for the original dataset was calculated, and the newly extracted features were applied to ML and ensemble models. The output is referred as PCA in Table 3 and Fig. 5. The final experiment was carried out with data

Table 3 Time taken by various algorithms using feature selection and extraction techniques

Classifier	Full set	MI	PCA	MIPCA
LR	99,358	683	752	1123
LDA	729	568	696	556
NB	1191	574	885	510
SVM	36,238	26,073	30,940	21,919
AB	42,364	43,087	47,933	37,698
DT	511	447	450	466
RF	4253	4238	3195	3166

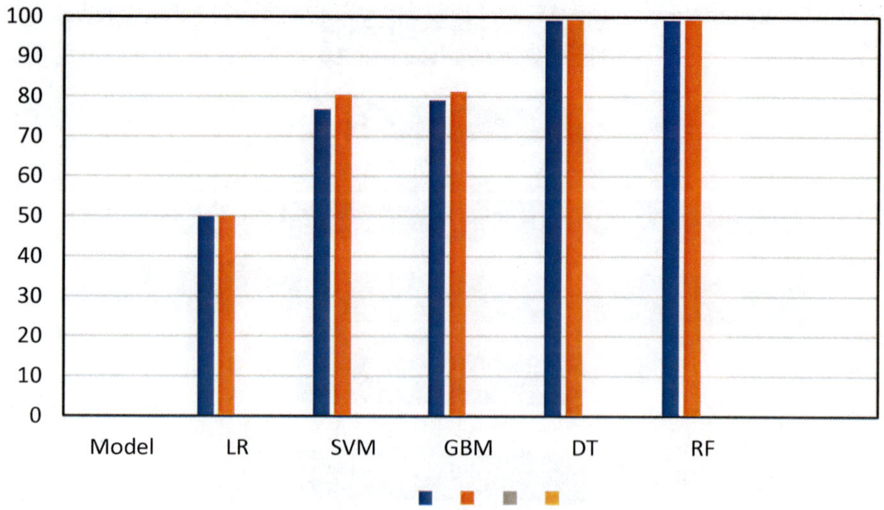

Fig. 5 Accuracy of classifiers

created using PCA applied on the 16 best features selected based on MI. The output is referred as MIPCA in Table 3 and Fig. 5.

LR algorithm took the highest time while the DT took the least time for the full set data. ML algorithms achieved an accuracy of 92 to 95% approximately with an exception to the SVM which achieved an accuracy of the 100% which means all the instances were correctly classified by SVM using the unprocessed data. However, this accuracy had a trade-off time which made the SVM second highest time consumer. LR and SVM, respectively, took 99,000 microsecs and 36,000 microsecs approximately, while the other ML classifiers took lesser than 1200 microseconds to classify the mushrooms with a good accuracy ranging from 92–95%. On the other hand, ensemble classifier with full dataset classified the mushrooms with 100% accuracy and took 42,000 microsecs and 4200 microsecs for AD and RF classifiers, respectively. Looking at the accuracy achieved (Fig. 5) and time consumed (Table 3) by each classifier, it can be concluded that DT performed best in the class.

All the ML classifiers took very less time with increase or same accuracy as that of the full dataset. However, NB classifier showed diminished accuracy with the dataset having feature selected using MI. This dip can be explained by the understanding that MI is calculated after removing the conditional entropy jointly caused by two variables in system, and conditional entropy is calculated using the conditional probability of two variables. However, all the ML classifiers monotonically showed huge decrease in classification time. This is in line with the understanding that less features cause less time. On the other hand, ensemble classifiers showed the same accuracy with very little dip in classification time. The data in Table 3 and Fig. 5 for MI lead us to conclude that DT classifier works better with data with features selected through MI.

PCA is the dimensional reduction technique. All the ensemble algorithms (AD and RF), SVM, and NB showed similar or an increase in the accuracy when fed with data dimensionally reduced using PCA. The accuracy of NB classifier increased by 1% with saving of apx 300 microseconds. However, the classification time was more than 50% of the time taken by same classifier with data selected using the MI. On the

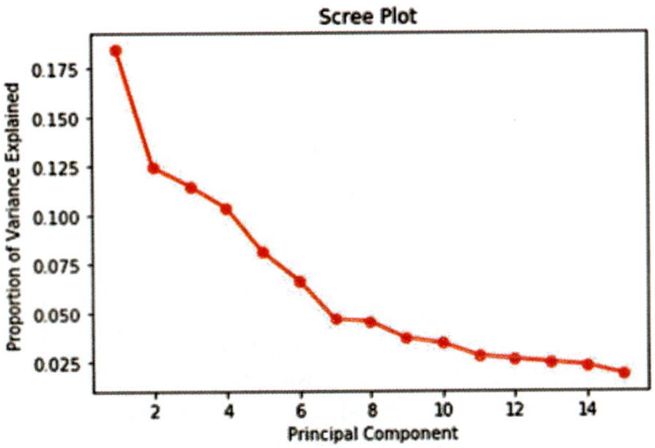

Fig. 6 Scree plots on principal components

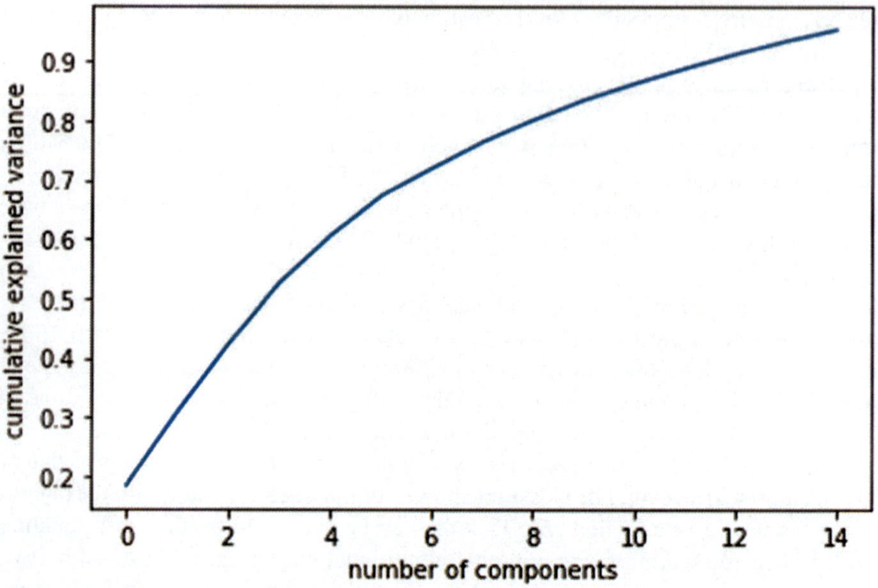

Fig. 7 Cumulative variance of eigen vectors

other hand, LR and LDA took lesser time but the accuracy to classify was reduced by 2–3%. LR and LDA, inherently, are good for classifying the linear or near linear data but in case of PCA data does not remain linear but becomes aligned to the direction of eigen vectors. All the classifiers took lesser time than the classifiers with full dataset except the AdaBoost classifier. PCA usually does well with data which has higher variance along with the first two principal components, but in case of our data, it was found that 95% of the variance was distributed across 13 principal components (shown in Fig. 6) and proportion of the explained variance looked to be good and finally setting after principal component 7 (shown in Fig. 7) but the dataset with 7 principal component captured only 70% of the variance of total dataset. Hence, the dataset with 13 principal components was considered. PCA analysis could have taken lesser time and have shown a better accuracy then the current one if dataset had most of the data distributed along with first 2–3 components.

Finally, the PCA was applied to the feature vector with16 best features selected using MI. The newly extracted features using PCA were applied to the ML classifier and ensemble classifier. ML classifiers—LR, LDA, and NB, showed a miserable accuracy, i.e., fall of 4–5% in accuracy with respect to the full set with reduction in time. Unlike other ML classifiers, SVM reacted very positively to the dataset curated with MIPCA and showed no decrease in accuracy but showed a dip of 40% in time consumed by the classifiers. Like SVM, both ensemble algorithms showed no change in the accuracy but showed a dip of 10–20% timing while compared to time consumed by same classifiers with full dataset. Both ensemble algorithm showed lowest time with respect to all the experimental datasets.

The proposed technique to use MI, PCA, or MIPCA shows the accuracy as 91.2, 93, and 92%, respectively, for NB classifier, while the Hamonangan et al. [17] were able to achieve only 90% accuracy using the NB classifier. Also, they were able to achieve 100% accuracy using KNN on same data, and proposed technique using MI-DT is able to reach the 100% accuracy in 450 microseconds.

Similarly, [16] were able to achieve an accuracy of 89% with the boosted ensemble techniques in 28–30 secs, but all the boosted ensemble classifiers using MI, PCA, and MIPCA showed the accuracy of 100% in maximum 1 s. Further, they were able to achieve 100% accuracy with training time of 28–30 s, and proposed system was able to achieve 100% accuracy using MI-DT with training time of 0.5 secs.

Hence, it can be easily concluded that proposed system is able to achieve the higher or equal accuracy in the lesser time than the previously proposed frameworks and techniques.

5 Conclusion

In the previous section, we carried out the experiments with various classifiers and the UCI mushroom data with no treatment, feature selected through MI, dimensionally reduced using PCA, and treated with combination of MI and PCA techniques. The experiments results were given in table and figure, and this leads us to conclude that MI is a good feature selection technique for the UCI mushroom dataset followed by the PCA. Classifiers using the data selected using the MI showed very less time and showed a reduction of 20–50%. The reduction was more than 99% in case of the LR. DT classifier showed a consistent accuracy irrespective of the feature selection, extraction, or full dataset, and it also showed the least time among all the classifier in all the experiments. However, the execution time of DT was drastically reduced with MI or PCA. Hence, authors recommend that framework containing MI and DT could be used in industry, where high volume of data is to be classified with good accuracy in less time.

The UCI mushroom dataset used in experiments dataset was an unbalanced dataset. Future research can be done using MI and PCA coupled up with balanced dataset to study the performance.

References

1. Badalyan SM, Barkhudaryan A, Rapior S (2019) Recent progress in research on the pharmacological potential of mushrooms and prospects for their clinical application. Med Mushrooms 1–70
2. Sarishma D, Sangwan S, Tomar R, Srivastava R. (2022) A review on cognitive computational neuroscience: overview, models and applications. In: Tomar R, Hina M.D, Zitouni R,

Ramdane-Cherif A (eds) Innovative trends in computational intelligence. EAI/Springer innovations in communication and computing. Springer, Cham. https://doi.org/10.1007/978-3-030-78284-9_10
3. Thu ZM, Myo KK, Aung HT, Clericuzio M, Armijos C, Vidari G (2020) Bioactive phytochemical constituents of wild edible mushrooms from Southeast Asia. Molecules 25(8):1972
4. Klančnik A, Šimunović K, Sterniša M, Ramić D, Možina SS, Bucar F (2021) Anti-adhesion activity of phytochemicals to prevent Campylobacter jejuni biofilm formation on abiotic surfaces. Phytochem Rev 20(1):55–84
5. Tung YT, Pan CH, Chien YW, Huang HY (2020) Edible mushrooms: novel medicinal agents to combat metabolic syndrome and associated diseases. Curr Pharm Des 26(39):4970–4981
6. Li Y, You L, Dong F, Yao W, Chen J (2020) Structural characterization, antiproliferative and immunoregulatory activities of a polysaccharide from Boletus Leccinum rugosiceps. Int J Biol Macromol 157:106–118
7. Pinky NJ, Islam SM, Rafia SA (2019) Edibility detection of mushroom using ensemble methods. Int J Image, Graph Sig Proc 11(4):55
8. Saxena V, Aggarwal A (2020) Comparative study of select non parametric and ensemble machine learning classification techniques. In: 2020 2nd international conference on advances in computing, communication control and networking (ICACCCN). IEEE, pp 110–115
9. Wibowo A, Rahayu Y, Riyanto A, Hidayatulloh T (2018) Classification algorithm for edible mushroom identification. In: 2018 international conference on information and communications technology (ICOIACT). IEEE, pp 250–253
10. Alkronz ES, Moghayer KA, Meimeh M, Gazzaz M, Abu-Nasser BS, Abu-Naser SS (2019) Prediction of whether mushroom is edible or poisonous using back-propagation neural network
11. Dawood KJ, Zaqout MH, Salem RM, Abu-Naser SS (2020). Artificial neural network for mushroom prediction. Int J Acad Inf Syst Res (IJAISR) 4(10)
12. Wagner D, Heider D, Hattab G (2021) Mushroom data creation, curation, and simulation to support classification tasks. Sci Rep 11(1):1–12
13. Husaini M (2018) A data mining based on ensemble classifier classification approach for edible mushroom identification July:1962–1966
14. Verma SK, Dutta M (2018) Mushroom classification using ANN and ANFIS algorithm. IOSR J Eng (IOSRJEN) 8(01):94–100
15. Khan AAR, Nisha SS, Sathik MM (2018) Clustering techniques for mushroom dataset (June):1121–1125
16. Yildirim Ş (2020) Classification of mushroom data set by ensemble methods. Recent Innov Mechatron 7(1):1–4
17. Hamonangan R, Saputro MB, Atmaja CBSDK (2021) Accuracy of classification poisonous or edible of mushroom using naïve bayes and k-nearest neighbors. J Soft Comput Explor 2(1):53–60
18. Ismail S, Zainal AR, Mustapha A (2018) Behavioural features for mushroom classification. In: 2018 IEEE symposium on computer applications & industrial electronics (ISCAIE). IEEE, pp 412–415
19. Zahan N, Hasan MZ, Malek MA, Reya SS (2021) A deep learning-based approach for edible, inedible and poisonous mushroom classification. In: 2021 international conference on information and communication technology for sustainable development (ICICT4SD). IEEE, pp 440–444
20. Aljojo MS, Dawood KJ, Zaqout MH, Salem RM (2021) ANN for mushroom prediction
21. Chumuang N, Sukkanchana K, Ketcham M, Yimyam W, Chalermdit J, Wittayakhom N, Pramkeaw P (2020) Mushroom classification by physical characteristics by technique of k-nearest neighbor. In: 2020 15th international joint symposium on artificial intelligence and natural language processing (iSAI-NLP). IEEE, pp 1–6
22. Jain KN, Kumar V, Kumar P, Choudhury T (2018) Movie recommendation system: hybrid information filtering system. In: Bhalla S, Bhateja V, Chandavale A, Hiwale A, Satapathy S. (eds) Intelligent computing and information and communication. Advances in intelligent systems and computing, vol 673. Springer, Singapore. https://doi.org/10.1007/978-981-10-7245-1_66

23. Srivastava R, Tomar R, Sharma A, Dhiman G, Chilamkurti N et al (2021) Real-time multimodal biometric authentication of human using face feature analysis. Comput Mater Continua 69(1):1–19
24. Jain A, Choudhury T, Mor P, Sabitha AS (2017) Intellectual performance analysis of students by comparing various data mining techniques. In: 2017 3rd international conference on applied and theoretical computing and communication technology (iCATccT), pp 57–63. https://doi.org/10.1109/ICATCCT.2017.8389106
25. Goel S, Sai Sabitha A, Choudhury T, Mehta IS (2019) Analytical analysis of learners' dropout rate with data mining techniques. In: Rathore V, Worring M, Mishra D, Joshi A, Maheshwari S (eds) Emerging trends in expert applications and security. Advances in intelligent systems and computing, vol 841. Springer, Singapore. https://doi.org/10.1007/978-981-13-2285-3_69
26. UCI Mushroom Dataset at https://archive.ics.uci.edu/ml/datasets/mushroom. Accessed on July 2021

Factors Influencing University Students' E-Learning Adoption in Bangladesh During COVID-19: An Empirical Study with Machine Learning

Rakib Ahmed Saleh, Md. Tariqul Islam, and Rozi Nor Haizan Nor

Abstract The present study investigates the factors that influence Bangladeshi university students' behavior toward adopting the e-learning system during the COVID-19 emergency. A conceptual framework was developed by adopting the variables from several previously published studies to meet the aims of the study. The current study was conducted in the quantitative approach by performing a survey on 393 university students of Bangladesh. All the obtained data were analyzed by using SPSS, AMOS, and machine learning algorithms. The findings of the study indicate that facilitating condition, effort expectancy, performance expectancy, and social influence affect the behavioral intention of the students to adopt e-learning, and the current study provides both practical and theoretical contributions. Theoretically, it provides a research framework and the literature to analyze the factors that trigger the behavioral intention of Bangladeshi university students. Practically, the findings will assist the policymakers of the education industry in Bangladesh.

Keywords E-learning adoption · COVID-19 · Behavioral intention · Machine learning

1 Introduction

The development and application of information technology (IT) have transformed and given a new shape to every sector of today's world [1]. Technological development has stimulated the educational intuitions to adopt recent trends in teaching–learning methods, and e-learning is one of its outcomes. E-learning is considered a potential tool that has transformed the conventional process of teaching and learning as it provides a practical and convenient platform to the educational institutions for

R. A. Saleh (✉) · R. N. H. Nor
Department of Information System and Software Engineering, Faculty of Computer Science and Information Technology, Universiti Putra Malaysia, 43400 Serdang, Selangor, Malaysia
e-mail: rasaleh19@gmail.com

Md. T. Islam
School of Business and Economics, Universiti Putra Malaysia, 43400 Serdang, Selangor, Malaysia

V. Skala et al. (eds.), *Machine Intelligence and Data Science Applications*, Lecture Notes on Data Engineering and Communications Technologies 132,
https://doi.org/10.1007/978-981-19-2347-0_54

conducting teaching and learning, as well as distributing effective knowledge among students [1, 2]. E-learning can be defined as the assessment of Internet-based technological tools used both in the classroom and outside the classroom, where the entire teaching–learning process is thoroughly conducted online. The learners can assess the learning resources by using information from anywhere [3]. The principal motive of e-learning is to transform the learning pattern to the student-centered from the teacher-centered and increase learning and problem-solving ability [1]. The educational sectors all over the world are affected by pandemics as the government of almost every country declared the closure of educational institutions to minimize the spread of viruses that disrupted the learning process of the students [3]. In addition, all the universities in Bangladesh have resumed their teaching–learning activities through online platforms due to the closure of educational institutions. Apart from a few institutions, most of the institutions have adopted e-learning technology. That situation influences a rapid transformation to Internet-based modalities such as video conferencing through Zoom meetings, Google meet, and Google Classroom. Universities have become more interested in an asynchronous and synchronous teaching method to deliver quality course content to their students. Numerous factors influence the students' intention toward attending online classes and adopting e-learning, but the previous studies revealed that those factors differ from region to region [1, 3]. Congruent with this effort, the current study aims to analyze the factors that influence the behavioral intention of university students in Bangladesh toward adopting e-learning. The present study will provide both theoretical and practical contributions; as the academic contribution, a framework has been developed to examine the factors that influence students' e-learning adoption, and practically, the policymakers of the education sectors in Bangladesh will be benefitted by findings of the current study.

2 Literature Review

This section depicts the factors that stimulate the adoption of e-learning among the university students of Bangladesh, and doing so, six variables; self-efficacy, effort expectancy, facilitating condition, performance expectancy, social influence, hedonic motivation as an independent variable and one variable; Behavioral intention dependent variable has been adopted from the previously published literature.

Behavioral Intention (BI): Behavioral intentions are the motivational factors that affect a specific behavior. The intensity of the intention to do the behavior directly influences the propensity of the activity to be performed. Behavioral intention toward adopting e-learning can be described as the learners' willingness, attitude, and decision to use e-learning technology [4].

Self-efficacy (SE): Self-efficacy is the individuals' internal factor which can be defined as the individuals' perception and judgment about their competencies

of organizing and performing the course of action required to achieve specific performance [2]. Self-efficacy has been found in many past studies as an essential factor that influences students' behavioral intentions [1]. Hence, this study predicts that;
H1: Self-efficacy influences the behavioral intention of students toward adopting e-learning.

Facilitating Condition (FC): Facilitating condition is an environmental factor that refers to the availability of the external facilities, resources, support, and assistance needed to use the e-learning system, and it affects the students' perception toward the ease or difficulties of carrying out any task related to e-learning [2]. The facilitating condition has been found in past studies as a significant factor influencing students' behavior intentions [4]. Hence, this study predicts that;
H2: Facilitating condition influences the behavioral intention of students toward adopting e-learning.

Effort Expectancy (EE): Effort expectancy in e-learning is the degree of ease connected with the usage of the e-learning system, and the extent to which a student considers that using the e-learning technology will be convenient and effort-free [2]. Prior studies have found that effort expectancy influences the students' behavioral intentions [1]. Hence, this study predicts that;
H3: Effort expectancy influences the behavioral intention of students toward adopting e-learning.

Performance Expectancy (PE): In the perspective of e-learning, performance expectancy can be described as the degree to which the students think that their performance will be uplifted by utilizing e-learning technology [1]. Performance expectancy is found in past studies as an important element that influences students' behavioral intentions [5]. Therefore, this study predicts that;
H4: Performance expectancy influences the behavioral intention of students toward adopting e-learning.

Social Influence (SI): Social influence in the context of e-learning is considered as the degree to which a learner's behavioral intention of e-learning adoption is triggered by others' opinions, and it represents the pressure of subjective norm [1, 2]. Social influence has been proven as a factor that influences students' behavioral intentions [5]. Therefore, this study predicts that;
H5: Social influence influences the behavioral intention of students toward adopting e-learning.

Hedonic Motivation (HM): Hedonic motivation in the context of e-learning is connected to the students' perceived playfulness, enjoyment, and entertainment with the online learning system and strategy [4]. Prior studies have found hedonic motivation as a factor that influences students' e-learning adoption [1]. Therefore, this study predicts that;
H6: Hedonic motivation influences the behavioral intention of students toward adopting e-learning (Fig. 1).

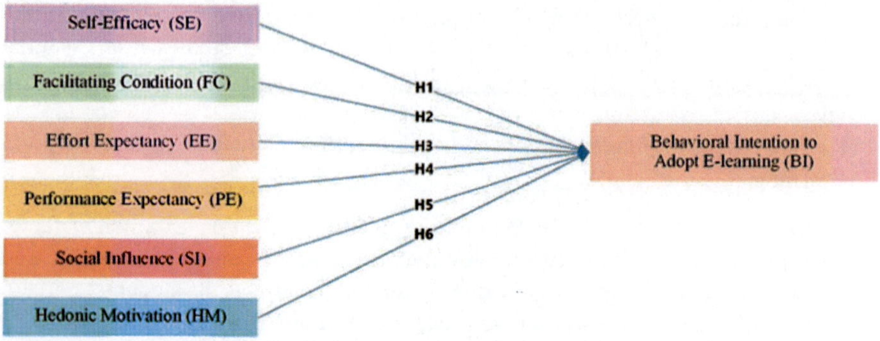

Fig. 1 Conceptual research framework

3 Research Methodology

The current study exerts the positivist philosophical paradigm along with a quantitative data approach by conducting an online questionnaire-based survey. The quantitative research method assists the researchers in examining the relationship between the independent and dependent variables, which will provide the evidence to test the proposed hypotheses whether these are accepted or rejected. Six hypotheses have been developed in the current study, which makes it explanatory research, and the flowchart of research methodology has been depicted in Fig. 2.

3.1 Sample and Data Collection

The current study followed the purposive sampling technique, as the target populations are Bangladeshi university students who attended online classes during that COVID-19 emergency. A structured online questionnaire was used to collect the data from primary sources. Several academic articles and books were used as prime sources for secondary data collection to identify the research gaps and develop a

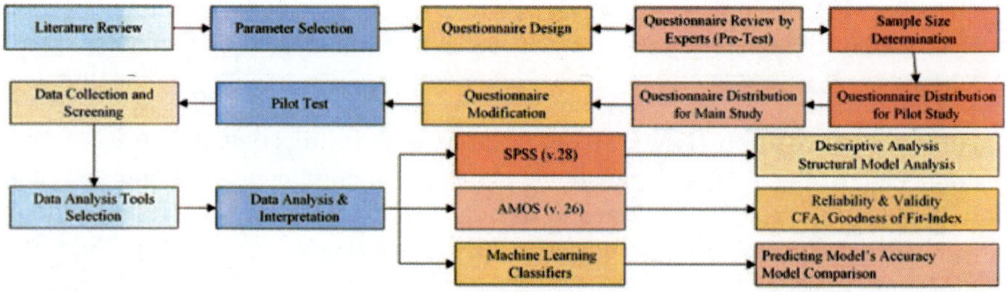

Fig. 2 Flowchart of research methodology

Table 1 Measurement items of the variables

Variables	Item number	Source
Self-efficacy (SE)	04	[2]
Effort expectancy (EE)	03	[1]
Facilitating condition (FC)	03	[2]
Performance expectancy (PE)	03	[4]
Social influence (SI)	03	[1]
Hedonic motivation (HM)	03	[2]
Behavioral intention (BI)	03	[4]

framework for the study. A total number of 410 was collected, with 393 data being used following a data screening to carry out this study. The measurement items of the variables were adopted to prepare the questionnaire from the previously published studies. Prior to the distribution, a pre-testing of the questionnaire was accomplished to identify and correct questionnaire-related issues like terminology, layout, design, and appropriateness, etc., [6] with the assistance of three academics with significant expertise in e-learning operations. The questionnaire was sent to the 30 respondents for the pilot study to assess the feasibility of the study, to examine the adequacy of the research instruments, and to test the reliability and validity of the measurement items of the variables [6]. The questionnaire was modified and finalized according to the result of the pilot test and feedback from the respondents (Table 1).

The questionnaire contained two parts; the demographic questions of the respondents and the perceptions of the students' regarding e-learning were asked in the two parts, respectively. A 5-point Likert scale was utilized to quantify the perception of the students, where 1 represents 'strongly disagree' and 5 illustrates 'strongly agree.' All the participants were informed in the self-administered online survey that their responses will be confidential, conforming to which the dataset is available at https://github.com/rasaleh19/dataforelearning.git.

3.2 Data Analysis Tools and Techniques

The present study has two phases. In the first phase, the descriptive statistical analysis, variables' reliability and validity, and the hypotheses testing of the model have been carried out by SPSS (version 28.0) and structural equation modeling (AMOS v.26.0) software, respectively. In the second phase, the research predicts the dependent variables entailing the conceptual model with the help of five machine learning (ML) learners: OneR, artificial neural network (ANN), logistic, decision tree (J48), and Naïve Bayes. As the traditional statistical analysis method like structural equation modeling (SEM) oversimplifies the relationship between dependent and independent variables, ML classifiers were concurrently used to investigate complex nonlinear relationships of variables, thereby harnessing the result's precision [7]. Weka (ver. 3.8.5) is employed for modeling, training, and evaluating the model's accuracy.

4 Data Analysis and Findings

4.1 Descriptive Statistics

The samples' demographics are illustrated in Table 2, which include 49.4% male and 50.6% female respondents. The bulk of responders (45.8%) belongs to the ages of 18 to 24 (45.8%) and are undergraduate students (51.9%). Concerning the academic discipline, the majority (40.5%) have a science background. The respondents are almost equally distributed to all universities, namely public, private, and national. Most of them use Zoom (51.7%) and Google Meet (40.2%) as their e-learning platforms.

Table 3 shows that the standard deviation is relatively minimal, with values ranging between 1.1. Skewness and kurtosis are used to assess the symmetry, and the combined size of data distribution tails, skewness's absolute values fluctuate between

Table 2 Respondents' demographic profile

Demographic variable	Classification	Frequency	Percentage
Age	18–24	180	45.8
	25–34	153	38.9
	35–44	60	15.3
Gender	Male	194	49.4
	Female	199	50.6
Level of education	Bachelor	204	51.9
	Masters	147	37.4
	PhD	42	10.7
Discipline	Business studies	98	24.9
	Humanities	136	34.6
	Science	159	40.5
Category of university	National	130	33.1
	Private	127	32.3
	Public	136	34.6
E-learning platform	Google meet	158	40.2
	Moodle	3	0.8
	Others	9	2.3
	University owned	20	5.1
	Zoom	203	51.7

Table 3 Descriptive statistics of the constructs

Items	Mean	SD	Skewness	Kurtosis
SE	3.70	1	−0.378	−0.360
FC	3.78	1.03	−0.386	−0.728
HM	3.68	1.06	−0.364	−0.692
PE	3.85	1	−0.618	−0.155
SI	3.78	1.04	−0.462	−0.678
EE	3.84	1	−0.429	−0.694
BI	3.86	1.03	−0.569	−0.544

1, which are both within the acceptable absolute values of 2 and 3 for Skewness and Kurtosis, respectively [8]. This implies that the samples of the current study have a normal univariate distribution [9].

4.2 *Reliability and Validity*

AMOS 26 was used to perform confirmatory factor analysis to test how well the items represent the constructs. As $p < 0.001$ (Table 4), the KMO and Bartlett's test seemed to have statistical significance, which confirm the effectiveness of factor analysis of the correlation [9]. Figure 2 illustrates the measurement model with all factors loading. The SEM analysis of the path model formulated in this empirical study complied to an ideal fit for the structural equation model ($\chi 2/\text{df} = 2.627$, $P < 0.001$, IFI = 0.933, TLI = 0.917, CFI = 0.932, RMSEA = 0.064, RMR = 0.0415) [10].

The constructs' reliability was measured by conducting Cronbach's alpha and compound reliability (CR) (Table 5). CFA results suggest that all factor loadings (FL) have scored above 0.5. For every variable, the value of Cronbach's alpha (CA) and composite reliabilities (CRs) has been found more than 0.7, thus confirming the measurement items' reliability [11]. Although the recommended score of average variance extracted (AVE) is anything above 0.5 [12], a score of 0.4 is acceptable if AVE is smaller than 0.5, but CR is more than 0.6 for the same variables, yielding in convergent validity of the constructs [13]. Thus, strong evidence of CR greater than 7 and AVE around 5 demonstrates satisfactory validity.

Table 4 KMO and Bartlett's test

KMO		0.956
Bartlett's test of sphericity	Approximate chi-square	4633.582
	Degree of freedom	231.000
	Significance	0.000

Table 5 Measurement model reliability and validity

Variables	FL	CA	CR	AVE	KMO
SE	0.687	0.79	0.789	0.485	0.778
FC	0.7	0.762	0.762	0.516	0.689
EE	0.698	0.73	0.744	0.492	0.683
PE	0.697	0.726	0.726	0.469	0.681
SI	0.734	0.745	0.748	0.498	0.682
HM	0.672	0.74	0.74	0.487	0.647
BI	0.69	0.73	0.73	0.474	0.673

4.3 Structural Model Results

The next stage is to conduct a structural test after the research model has been confirmed to pass the measurement test. In this part, hypothesis testing is done by evaluating the path coefficient score, as illustrated in Table 6. For a significant relationship between the constructs, the t-statistic score must be greater than the absolute value of 1.96 [14]. Based on this, hypotheses H2, H3, H4, and H5 should be accepted. Therefore, the study can be concluded that the facilitating condition (FC), effort expectancy (EE), performance expectancy (PE), and social influence (SI) are significant for the behavioral intention of Bangladeshi university students to adopt e-learning during the COVID-19 emergency. Other hypotheses H1 and H6 should be rejected.

This result suggests that self-efficacy (SE) and hedonic motivation (HM) are not significant for the behavioral intention of Bangladeshi university students to adopt e-learning during the pandemic situation (Fig. 3).

Table 6 Path coefficients and t-statistics result

Model	Unstandardized coefficients		Standardized coefficients	*t*	Sig	Hypotheses result
	B	Std. Error	Beta			
(Constant)	0.339	0.138		2.452	0.015	
H1: SE → BI	0.017	0.053	0.016	0.322	0.748	Not supported
H2: FC → BI	0.146	0.055	0.146	2.645	0.009	Supported
H3: EE → BI	0.202	0.054	0.198	3.707	0.000	Supported
H4: PE → BI	0.347	0.054	0.333	6.446	0.000	Supported
H5: SI → BI	0.125	0.049	0.127	2.570	0.011	Supported
H6: HM → BI	0.087	0.047	0.091	1.872	0.062	Not supported

a. Dependent variable: BI

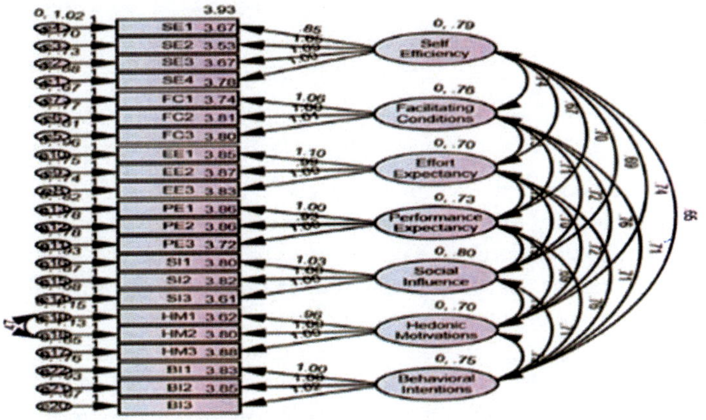

Fig. 3 Confirmatory factor analysis

4.4 Model Accuracy Testing Using Machine Learning Algorithms

From the values of the items, the corresponding mean values of the six independent variables are calculated and considered as the input of the classification models. Likewise, the mean value of the dependent variable, namely behavioral intention, has been converted into categorical values by five distinct classes (1–5) with an interval value of 0.8, and these classes are deemed to be the output of classification models. To prevent overfitting of the model, the researcher designed *K*-fold cross-validation with a splitting of 80% of the training set and 20% as the testing set. Table 7 describes the tenfold cross-validation results attained from the five different ML models using Weka (ver. 3.8.5) to predict the proposed research model [15]. As to accuracy, decision tree (J48) is uppermost with values of 0.70, while Naïve Bayes is the least accurate model with only 0.65. Artificial neural network (ANN) and OneR come up with the same accuracy of 0.68, while logistic and Naïve Bayes provide almost the same accuracy of 0.66 and 0.65, respectively. In respect of the F-measure, precision, and recall, it is salient that the J48 model surpasses other ML

Table 7 Comparison of machine learning models

Algorithm	TP[1] rate	FP[2] rate	Precision	Recall	*F*1-score	AUC[3]	Accuracy
J48	0.709	0.114	0.708	0.709	0.706	0.807	**0.70**
Logistic	0.658	0.186	0.565	0.658	0.615	0.853	0.66
ANN	0.684	0.132	0.734	0.684	0.674	0.894	0.68
OneR	0.684	0.120	0.677	0.684	0.678	0.782	0.68
Naïve Bayes	0.646	0.124	0.651	0.646	0.647	0.864	0.65

[1]TP: True positive, [2]FP: False positive, [3]AUC: Area under the curve

Table 8 Predicting BI by FC, EE, PE, and SI

Algorithm	TP rate	FP rate	Precision	Recall	Accuracy
J48	0.722	0.117	0.709	0.722	0.72 ↑

models, while logistic has the minimum precision value of 0.565. Concerning the AUC score, ANN outruns the other ML models with the values of 0.894, followed by the Naïve Bayes and logistic. As J48 outperforms other ML classifiers, J48 is chosen to investigate further the association between the input constructs and the output of the revised research model that defines BI for e-learning adoption.

Table 8 illustrates that the accuracy, precision, and recall of the revised research model increases, thus strengthening the correctness of the revised research model (SE and HM being dropped) as supported earlier by hypotheses testing with path coefficient score.

5 Discussion

The study has designed the research model based on an extensive literature review, considering the demographic variables. For testing the proposed model, the research undertakes a complementary multi-analytical approach by combining SEM, statistical-based multi-variate modeling, and supervised machine learning algorithms, convenient for exploring nonlinear relationships between constructs. The J48 decision tree outperformed all other classifiers. Thus, the significance of influencing factors of e-learning adoption was prioritized and defined precisely by choosing the optimal ML classifier model and hypothesis testing by SEM.0000. Effort expectancy and performance expectancy are found to be the most significant factors of e-learning adoption, followed by facilitating condition and social influence, as aligned with the previous studies [1]. This is probably because university students of Bangladesh prefer a user-friendly e-learning platform and want to keep up their studies without any disruption by staying safe at home during the COVID-19 emergency. Faculty members should therefore ensure that the e-learning contents of the system are considered valid and up to date, which better addresses students' demands, and the e-learning platform should not need additional skills to operate. Besides this, they want support staffs or facilities always available to assist. Hence, the decision-makers should make an effort to provide learner–content interaction and flexible learning process control. On the contrary, self-efficacy (SE) and hedonic motivation (HM) are not recognized as significant factors for e-learning adoption amid COVID-19 in Bangladesh. Finally, an association between social influence (SI) and behavioral intention (BI) strengthens the significance of positive enforcement from classmates, family members, supervisors, and faculty members. As a result, to hasten e-learning adoption, policymakers may persuade members of reference groups to spread positive words regarding e-learning acceptance.

6 Conclusion

The advancement of technology has revolutionized education, transforming it from a conventional offline system to an online one. During this pandemic, online learning has been adopted by almost every university around the world. Still, the achievement of e-learning adoption, especially in developing countries, depends on the users' engagement. However, the current study's findings provide a brief concept about the antecedents that influence the students' behavioral intention to adopt the e-learning system. The study has found that among the six proposed hypotheses, four are accepted, indicating that university students of Bangladesh are affected by facilitating conditions, effort expectancy, performance expectancy, and social influence to adopt e-learning. In contrast, self-efficacy and hedonic motivation did not have a significant impact on the behavioral intention of the students. The study's findings indicate that Bangladeshi authorities concentrate on these issues in order to encourage university students' favorable attitudes and spontaneous involvement in the e-learning system.

Implication and Limitation of the Study:

The current study has both practical and theoretical implications. Theoretically, it provides a conceptual research model containing the constructs that leverage students' behavioral intention, and practically, it gives the data and information to the education stakeholders. Moreover, the present study is a cross-sectional study, and the human behavioral intention may change with time; hence, the longitudinal research can be conducted in the future with a larger sample size. In addition, further investigation can be conducted by adopting more variables to evaluate the barriers to e-learning adoption and the spontaneous participation of the stakeholders with the e-learning platform.

References

1. Samsudeen SN, Mohamed R (2019) University students' intention to use e-learning systems: a study of higher educational institutions in Sri Lanka. Interact Technol Smart Educ 16(3):219–238. https://doi.org/10.1108/ITSE-11-2018-0092
2. Tarhini A, Deh RM, Al-Busaidi KA, Mohammed AB, Maqableh M (2017) Factors influencing students' adoption of e-learning: a structural equation modeling approach. J Int Educ Bus 10(2):164–182. https://doi.org/10.1108/JIEB-09-2016-0032
3. Maheshwari G (2021) Factors affecting students' intentions to undertake online learning: an empirical study in Vietnam. Educ Inf Technol 2021:1–21. https://doi.org/10.1007/S10639-021-10465-8
4. Mehta A, Morris NP, Swinnerton B, Homer M (2019) The influence of values on E-learning Adoption. Comput Educ 141. https://doi.org/10.1016/J.COMPEDU.2019.103617
5. Ali M, Raza SA, Qazi W, Puah CH (2018) Assessing e-learning system in higher education institutes: evidence from structural equation modelling. Interact Technol Smart Educ 15(1):59–78. https://doi.org/10.1108/ITSE-02-2017-0012/FULL/HTML

6. Sekaran U (2016) Research methods for business: a skill-building approach, 4th edn. John Wiley & Sons, New York
7. Gerlein EA, McGinnity M, Belatreche A, Coleman S (2016) Evaluating machine learning classification for financial trading: an empirical approach. Exp Syst Appl 54:193–207. https://doi.org/10.1016/J.ESWA.2016.01.018
8. Kline R (2015) Principles and practice of structural equation modeling. Guilford Publications, Fourth Editon
9. Treiblmaier H, Filzmoser P (2010) Exploratory factor analysis revisited: How robust methods support the detection of hidden multivariate data structures in IS research. Inf Manag 47(4):197–207. https://doi.org/10.1016/J.IM.2010.02.002
10. Awang Z, Hui LS, Zainudin NFS (2018) Pendekatan mudah SEM-structural equation modelling. MPWS Rich Publication
11. Khan GF, Sarstedt M, Shiau WL, Hair JF, Ringle CM, Fritze MP (2019) Methodological research on partial least squares structural equation modeling (PLS-SEM): an analysis based on social network approaches. Internet Res 29(3):407–429. https://doi.org/10.1108/INTR-12-2017-0509/FULL/
12. Hair JF, Tatham RL, Anderson RE, Black W (1998) Multivariate data analysis, 5th edn. Prentice Hall
13. Fornell C, Larcker DF (1981) Evaluating structural equation models with unobservable variables and measurement error. J Mark Res 18(1):39. https://doi.org/10.2307/3151312
14. Berger JO, Delampady M (1987) Testing precise hypotheses. Statist Sci 2(3):317–335. https://doi.org/10.1214/SS/1177013238
15. Arpaci I (2019) A hybrid modeling approach for predicting the educational use of mobile cloud computing services in higher education. Comput Hum Behav 90:181–187. https://doi.org/10.1016/J.CHB.2018.09.005

Nano Rover: A Multi-sensory Full-Functional Surveillance Robot with Modified Inception-Net

Sheekar Banerjee, Aminun Nahar Jhumur, and Md. Ezharul Islam

Abstract In this rigorous world of civilized progression, terrorism and other subsidiary security threats are undoubtedly intimidating to us. In order to fight back these mischievous activities, we focus to the technical development of our surveillance and reconnaissance systems. Military reconnaissance might be the ancient most solution for preventing the violent outcomes of terrorism but still the unfortunate chance of human life loss exists there. For this why, the solution appears to be using the unmanned surveillance vehicle or machine which would be able to transmit visual data of a certain dangerous area to the controlling headquarters of the particular law enforcement agency. Nano Rover is a significant approach of cost-efficient surveillance and reconnaissance robot which is fully functional and cost-efficient at the same time. It features the service of active reconnaissance mode with LIDAR sensor, location tracking with GPS Neo 6 M module, visual information collection, person detection, weapons detection and identification, gender and age prediction of the hostile and other artificial threat detection, etc. Remote navigation plays as the core controlling system of the robot which is also modifiable through replacement with Internet and satellite navigation system. We modified the conventional Inception-Net architecture with a better hyper-parameter tuning for the successful execution of image processing tasks with a better level of accuracy so far.

Keywords Surveillance robot · Weapon detection · Deep learning · Inception-Net · LIDAR sensor

---

S. Banerjee (✉)
Cisscom LLC, Irvine, CA, USA
e-mail: sbanerjee@thetitantech.com; 16203084@iubat.edu

A. N. Jhumur
IUBAT—International University of Business Agriculture and Technology, Dhaka 1230, Bangladesh
e-mail: aminun.nahar@iubat.edu

Md. E. Islam
Jahangirnagar University, Savar, Dhaka 1342, Bangladesh
e-mail: ezharul.islam@juniv.edu

V. Skala et al. (eds.), *Machine Intelligence and Data Science Applications*, Lecture Notes on Data Engineering and Communications Technologies 132,
https://doi.org/10.1007/978-981-19-2347-0_55

1 Introduction

The use of modern technology in the sector of security for human and natural resources is visible from the very dawn of civilization. Following the sequence of progress for the modern technology, robotic inventions became imminent. It is not very hard to notice that nowadays robots are being developed rapidly for the execution of very risky and dangerous tasks.

Previously, human beings executed quite dangerous tasks by themselves physically which sometimes brought the fatal outcomes with intolerable losses of lives and casualties. Scientists and researchers are striving very hard to pave the way of developing the unmanned vehicles or machines, namely 'robots,' those can execute the dangerous tasks of information gathering, transmission of camera visuals and data, wanted personnel detection from any specific hostile district or even behind the enemy lines.

Nano Rover is the cost-efficient outcome of such kind of state of the art practice of modern robotics in the field of surveillance, reconnaissance, face object and threat detection, and unmanned execution of tasks into the dangerous areas. We introduced an improvised implementation of Inception-Net architecture with 20 different layers inside it. Primarily, we approached the Bluetooth navigation system but it was not limited to this while the WI-FI, intranet, Internet and even the satellite navigation have also become a part of the Nano Rover. For the surveillance of the 360° surroundings of the bot, LIDAR sensor has been utilized while Neo 6 M module has been used for the location tracking. Nano Rover shows full functionality of its continuous object and person detection and tracking activity with Inception-Net and single shot detection (SSD).

However, our proposed research work is focusing to contribute upon the field of novel amalgamation of robotic surveillance, high precision image processing, and the precise interfacing of cutting-edge smart sensors simultaneously.

2 Related Works

Military robot has always been a very important keyword in modern technological development. Previously, there have already been several developments of military robots which were also deployed with explosive disposal tasks in the real combat and severely hostile war zones like Iraq, Afghanistan, and others countries in the Middle East. But most of those war robots were used for the purpose of visual data collection of land mines and RDX explosives. In the very recent years, we came across with the cutting-edge implementation of multiple neural nets in field of computer vision which are mainly based on software platform [1–5]. Those implementations are not yet tested for the moving robotic mechanism or machinery in the real time [6–11]. Facial recognition, weapons' detection, and terrorist personnel's identification features were reasonably absent on those implementations.

Tank-based military robot (TBM) was introduced several years earlier. But the Indonesian research effort gave a new dimension where the new features of object detection and target locking systems were included in it. It was normally initiated with the help of OpenCV [12]. The main body structure was specially based on the genuine architectural shape of a legitimate tank, but unfortunately, the entire issue of security was not flawless enough to reach the satisfactory point. Despite the flaws, it can be noticed that it is inspired the other tank- related military robot research work [13].

Semi-autonomous gun and movement detection robot (SAG) included a portable command station which showed adequate work for further development. The researchers invested particular concentration in the main structure of the robot very precisely [14]. For this why, the prototype ended up being the all-terrain executable military operational robot. The features of object tracking and gun detection are also noticeable but the real-time implementation does not seem to be up to the mark [15–17]. On the other side, the various similar approaches for gender and age classification humanoid features with linear regression method can be notified but the combination of everything was nonexistent.

Weapon detection robot (WDR) with AI and DL research seems to be very rich with information and real-time implementation. It particularly concerns about the implementation of several computer vision algorithm approach for the various types of weapons recognition and detection at the same time [18]. It can easily draw someone's attention due to the almost correct model training accuracy for the weapons image datasets [19]. Though it showed very exquisite improvement in the genre of military computer vision, the prototype does not concern about the development and security of the structural body of the robot [20–22].

There are some significant key features of functionality in Nano Rover that keeps it ahead from the other related research works happened previously. The most important features are: gazing capability, object detection, continuous object tracking, use of GPS for tracking bot location, LIDAR sensor for 360° surveillance. Presence of all these features could not be found in other related works. The comparative feature analysis of Nano Rover with other previously executed related works are represented in Table 1.

Table 1 Comparison between Nano Rover and other research works with feature analysis

Features	Nano Rover	TBM	SAG	WDR
Gazing capability	Yes	Yes	Yes	No
Object detection	Yes	Yes	Yes	Yes
Continuous object tracking	Yes	Yes	No	Yes
GPS for bot location	Yes	No	Yes	No
LIDAR for 360° surveillance	Yes	No	No	No

3 Proposed Method

The particular methodology of Nano Rover are detailed within four main components of the entire machinery and they are: nano rover navigation, sensor data transmission (LIDAR, GPS) and real-time visual data transmission with image processing. The robot navigation was related to the precise manipulation of the data collection structure of the body frame through the subjective hostile environment. The precise assembly of electronic equipment, connection, and responsible act of rechecking were the prime factors in this section. The HC-05 module Bluetooth connection exhibited the utilization of close ranging reconnaissance activity (10 m) whether the Internet and satellite navigation explored the wider ranging capability of that very Nano Rover. At that time, HC-05 module could be replaced with ESP-8266 or SIM-808 module for wide ranging of navigation. The schematic diagram, representing the top level view of navigation of Nano Rover is articulated in Fig. 1. Nano Rover navigation is eventually dependent upon the transmission and receiving protocol (TX-RX), multiple DC motors connection and manipulation with the utmost profound efficiency and frequency. It was a major concern that the error of motor movement, rotation, and navigation could cause the fatal destruction of the robot in such hostile area concerning what it was designed for. Single packet signal tracing was implemented to avoid the signal jamming and direction error.

Namely four frequency coded DC motors attached wheels were used for locomotion, and L298N bridge motor driver was used for the regular transmission of electrical impulses through the controlled frequency of the DC motors with better pulse-width modulation (PWM). Concerning the long-term power supply challenge of the bot, we initiated rechargeable Tiger LIPO 3S battery having 4500 mAh, 11.1 voltage, continuous discharge rate 30C, maximum charging rate 5C. The uninterrupted three hours of navigation could be possible for these powerful dry cells within such a tiny prototype. Arduino ATMega 2560 was used for the micro-controlling

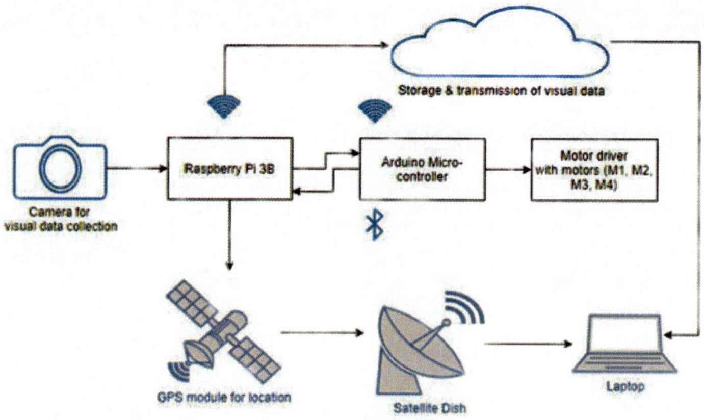

Fig. 1 Schematic diagram for navigation of Nano Rover

device between the abovementioned actuators and GPS Neo 6 M module sensor. The virtual navigation circuitry is articulated in Fig. 2.

The inclusion of Raspberry Pi3B initiated the installation Raspbian OS so that the visual data examination could be executed over that platform. TF lite was installed with CLI commands for better index of image processing accuracy in the Raspberry hardware. The usage of light detection and ranging (LIDAR) sensor escalated the 3D representation of the Rover surroundings for the protection of external security threats. Adafruit VL6180X library was installed in Arduino IDE for LIDAR activation. The circuitry between RPi 3B and LIDAR sensor is represented in Fig. 3. The real-time assembly of the compact Bot is showcased in Fig. 4.

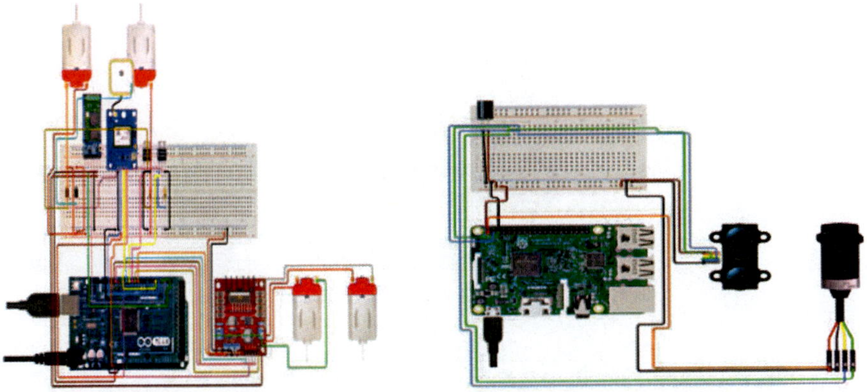

(a) The navigation circuitry of the Nano Rover.

(b) LIDAR sensor circuitry of the Nano Rover with Raspberry Pi3B.

Fig. 2 The navigation circuitry with Arduino ATMega (**a**) and LIDAR sensor circuitry with Raspberry Pi3B (**b**)

Fig. 3 Real-time assembly of Nano Rover

Fig. 4 The architecture of modified Inception-Net with 20 layers for image processing task of Nano Rover

The pin configurations between motor driver and Arduino, LIDAR sensor and Arduino, and Bluetooth module and Arduino are articulated in Tables 2, 3, and 4, respectively.

Table 2 Pin configuration between motor driver and Arduino

L298N bridge	Arduino	Description
In1	4	Streaming output data to the left motor (+ side)
In2	5	Streaming output data to the left motor
In3	7	Streaming output data to the right motor (+ side)
In4	8	Streaming output data to the left motor (− side)
ENA	3	Activation of pulse-width modulation for left motor
ENB	6	Activation of pulse-width modulation for right motor
12 V	Vin	External power source with higher voltage up to 12 V
5 V	5 V	Connection to positive of power source
GND	GND	Connection to negative of power source

Table 3 Pin configuration between LIDAR and Arduino

LIDAR	Arduino	Description
SCL	A5 (analog side)	Serial clock pin
SDA	A4 (analog side)	Serial data pin
5 V	5 V	Connection to positive of power source
GND	GND	Connection to negative of power source

Table 4 Pin configuration between Bluetooth module and Arduino

HC-05	Arduino	Description
TX	RX	Transmission data pin
RX	TX	Receiving data pin
Vcc	5 V	Connection to positive of power source
GND	GND	Connection to negative of power source

The algorithms for LIDAR sensor processing and OpenCV image processing are represented, respectively, in Algorithm 1.

Algorithm 1 LIDAR sensor processing algorithm.

```
begin
import serial processing library
 SETUP Process
procedure SETUP()
    defining port-name as COMx
    creating variable myPort with serial monitor library
    initializing LIDAR interface background
end procedure
DRAW Process
procedure DRAW(a, b)
    if (myPort.available() > 0) then
        val = myPort.readStringUntil ('\n')
        if (val! = null) then
            String[]nums = split (val, "p")
            if (nums.length=3) then
                set LIDAR range
                initialize LIDAR position
                if (i=180) then
                    set LIDAR background
                end if
            end if
        end if
    else
        interpret the line drawing procedure for LIDAR
    end if
end procedure
end
```

For the genuine purpose of weapon and human age–gender detection, we had to go through the creation of real-time visual dataset with the real imagery of male and female of different ages. The creation of different types of weapons' image dataset was also a great challenge. At the developmental phase of our research, there were no such dataset or library containing the pixel-defined image dataset for person and weapon detection simultaneously. We had to build our own dataset with efficient data labeling with CVAT. We trained our ML model both locally (Pycharm IDE) and in cloud (Google AutoML). In local run, we saved our model as h5 file and also converted it in TensorFlow JS (TFJS) format for the Web engineering compatibility.

For image processing implementation, the components we used were: OpenCV–Python with latest version 4.5.3.56, Python interpreter 3.9 with global variables and packages, COCO-SSD image-labeled dataset from Microsoft and self-made custom image dataset. Keras and TensorFlow with version 2.6.0 were used for importing the core Inception-Net architecture. On the other hand, the utilized hardware components were: Intel Core i5—10th Generation CPU with 2.11 GHz of clock speed. CUDA 10.1 and CuDNN (CUDA deep neural network) libraries were installed for the compatible GPU support for NVIDIA Geforce MX110 2 GB Graphics Card.

In Fig. 4, we demonstrated the implementation of conventional Inception-Net with more improvisation and modification with 20 different layers. Though the input images toward the convolution (Conv2D) layer of 7 × 7 filter size of pixel entity, followed its way to Max pooling (MaxPool2D) layer of 3 × 3 filter size. The hidden layers of Conv2D initialized 3 × 3 filter size for the ongoing preprocessing. Conv2D filter numbers were changed in different layers as 64 and 192 filters, respectively, where padding, strides, and activation functions were kept same in these parameters. Rectified linear unit (ReLU) function was called for triggering multiple classification efficiently. At first, pretrained Inception-Net was imported from Keras library, then the hyper-parameters were tuned efficiently with Conv2D, MaxPool2D, Average Pooling 2D, Dropout and Flatten layers functions. Six classes were defined for six different types of weapons' detection while gender and age classification were implemented from open-source COCO-SSD dataset. Images were resized and preprocessed from a 225 × 255 pixel perception. Afterward, inception (3a, 3b), (4a, 4b, 4c, 4d, 4e), (5a, 5b) modules were defined, separated by Max Pooling 2D (3 × 3 filter size) layers. Then Average Pooling 2D (7 × 7 filter size) was assigned with strides number 3. The 2D tensor values were converted into 1D image data through Flatten layer. To overcome the overfitting, we had to add a dropout layer (45%). 'Softmax' function was utilized for multi-class output whether binary classification output for gender detection was executed through 'Sigmoid'.

For large size of image data, parameter tuning, training, and testing, a large number of epoch were also needed. Device-installed GPU took care of faster large model training where TensorFlow-GPU was also installed for the backup support. Epoch number was defined 400, batch size was 42, and initial learning rate was 0.01. Stochastic gradient descent (SGD) was activated for defining vanishing gradient properties. We trained on 50,000 image samples and left 10,000 sample for validation. We trained and saved the improvised Inception-Net model as .h5 file and .pkl artifact for further training and testing. We also utilized the rich functions of computer vision algorithms from OpenCV and improvised our own image processing algorithm which is articulated in the section Algorithm 2.

Algorithm 2 OpenCV Image Processing function algorithm.

```
begin
import cv2, math and argument parsing libraries
FACE HIGHLIGHT Process
procedure HIGHLIGHTFACE (NET, FRAME)
defining frame of opencv deep neural net
defining frame height
    defining frame width
    defining threshold as 0.7
    calculate: blob = cv2.dnn.blobFromImage()
    getting input from setInput(blob) function
    get detection value from net.forward() function
    for i do in range (detections.shape[2])
        confidence = detections [0, 0, i, 2]
        if confidence > conf_theshold then
            x1 = int (detections [0, 0, i, 3] * frameWidth)
            y1 = int (detections [0, 0, i, 4] * frameHeight)
            x2 = int (detections [0, 0, i, 5] * frameWidth)
            y2 = int (detections [0, 0, i, 6] * frameHeight)
            faceBoxes.append ([x1, y1, x2, y2])
            frame buffering with cv2.rectangle function
            give frameOpencvDnn, faceBoxes
            delay the streaming of bit every 1 millisecond
        end if
    end for
end procedure
end
```

4 Experimental Analysis

Entering the main segments of real-time testing, we partitioned the entire mechanism into two significant stages. They were namely expected to be like: ground navigation and real-time visual data transmission, LIDAR sensor and GPS module testing, and person and weapons detection testing. We vigorously tested the robot continuously for 62 days and ended up getting some negligible error upon our Rover ground navigation mechanism. However, we figured out the bug and corrected that with precise manner. We demonstrated a very noble nature of iterative testing into various kinds of scenarios and different types of hostile situations during the navigation of the Recon robot, namely Nano Rover. The precise navigation can be seen in the figures. Here, Fig. 5a represents the real-time visual data transmission of Nano Rover while navigating on a hard soil ground. On the other hand, Fig. 5b explains the robot camera facial detection preciseness.

(a) Real time visual data transmission of Nano Rover.

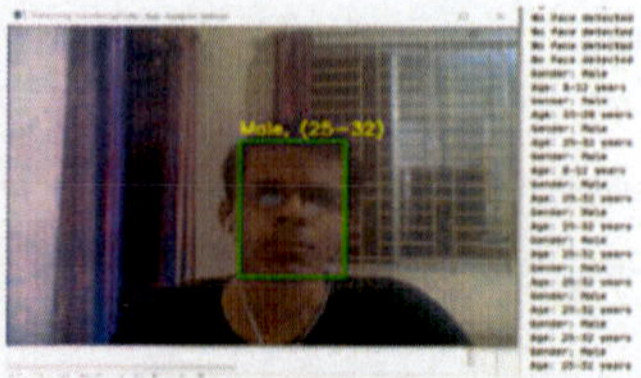

(b) Robot camera facial detection preciseness testing with high accuracy gender and age specification.

Fig. 5 Real-time visual data transmission (**a**) and facial detection preciseness of Nano Rover (**b**)

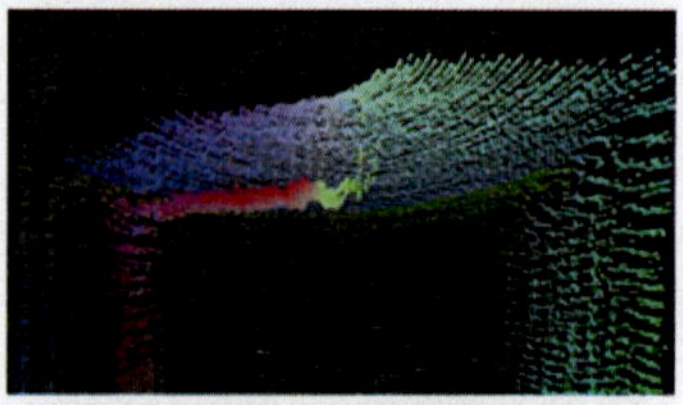

(a) LIDAR frame buffer showing the 360° observation of the surroundings of the rover with 3D representation.

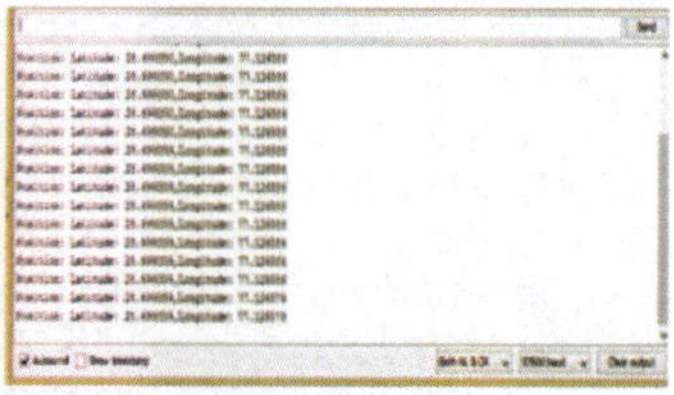

(b) The live location streaming from the GPS module of the Nano Rover.

Fig. 6 Real-time 3D observation of LIDAR Sensor (**a**) and live GPS location streaming of Nano Rover (**b**)

We activated LIDAR sensor's frame buffer console monitor of Arduino IDE from Raspbian OS. The real-time 3D surrounding observation of LIDAR can be noticed from Fig. 6a while the live coordinates (longitude, latitude) streaming at Arduino COM-serial monitor can be observed from Fig. 6b. At first, the person detection was showing only 71.47% of average testing accuracy while training accuracy was more than 87%. We reinstated several image data augmentation functions with parameters like re-scale = 1./255, rotation range = 360, horizontal-vertical flip, height-width shift, and zoom-brightness range. It regenerated the image data and made the dataset richer. When the model was retrained with the same epoch iterative process and retested, the testing accuracy jumped up to be 85.59% on average. We increased batch size to 50 and epoch number to 550 and saw even more accuracy. Overall, we could undertake the integrated process of robot navigation, person and weapons' detection with an average accuracy rate of 89.65% of the preciseness of the CNN-algorithmic improvised Inception-Net-based visual data model training and testing.

The live real-time person and weapons detection visual data streaming can be noticed at Fig. 7. The snapshots were collected from the streamed live video data which are converted automatically as .avi file due to Inception-Net model training.

After testing for more than 30 days, we tried the person and weapons detection mechanism with the proper usage of wide angle optical lenses (24, 35, 56 mm) of the dedicated gyroscopic action camera. In Table 5,—the average testing accuracy and fps

(a) AK-47 detection.

(b) Benelli M4 detection.

(c) Glock-19 detection.

(d) HK-MP5 detection.

(e) Sig Sauer-226 detection.

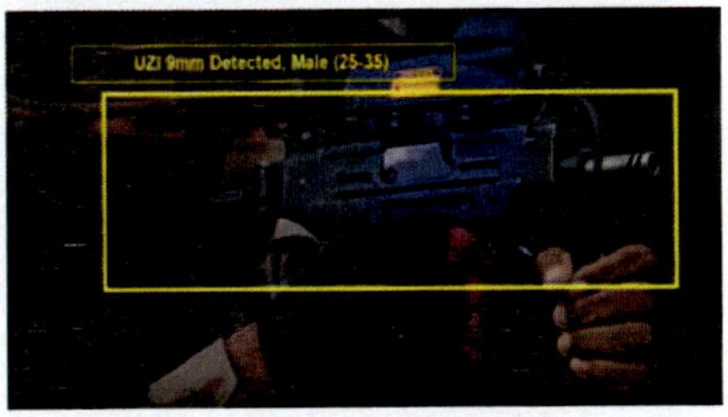

(f) UZI 9mm detection.

Fig. 7 Real-time detection of Ak-47, Benelli M4, Glock-19, HK-MP5, Sig Sauer-226 semi-automatic pistol, and UZI 9 mm submachine gun with hostile's gender and age specification from the static image dataset

Table 5 Performance analysis for modified inception-net algorithm

Gun model	Speed (seconds)	Average detection accuracy (%)	Detection happened	Correct classification
AK-47	0.62	93.63	Yes	Yes
Benelli M4	0.78	92.28	Yes	Yes
Glock-19	1.06	84.32	Yes	Yes
HK-MP5	0.73	89.85	Yes	Yes
Sig Sauer-26	1.09	88.78	Yes	Yes
UZI 9 mm	0.98	91.23	Yes	Yes

Table 6 Performance analysis of modified Inception-net algorithm for real -time person, gender and age detection

Parameters	Speed (seconds)	Average detection accuracy (%)	Detection happened
Person	0.36	98.53	Yes
Gender	0.89	91.78	Yes
Age	1.12	84.16	Yes

Table 7 Performance analysis of LIDAR sensor and GPS Neo 6 M module

Sensors	Stream buffer (bits/second)	Average sensory data time interval (milliseconds)	Activity	Average accuracy (%)	Probable threat detection success
LIDAR	19,200	500	Surrounding reconnaissance	94.53	Yes
GPS Neo 6 M	9600	500	Location tracking (longitude, latitude)	96.82	No

detection speed—we found for AK-47, Benelli M4, Glock-19, HK-MP5, Sig Sauer-226, and Uzi 9 mm are 93.63, 92.28, 84.32, 89.85, 88.78, and 91.23%, respectively. The performance analysis of the same modified Inception-Net for person, gender and age detection are represented in Table 6 while in Table 7 shows the performance analysis of LIDAR and GPS Neo 6 M module in five different parameters. We implemented the GPS location tracking, LIDAR sensor and navigation procedures in a fully functional way. For the person and weapons detection feature, we uniquely modified Inception-Net architecture with better hyper-parameter tuning and testing accuracy. In the arena of robot vision, this kind of unified accomplishment is very rare which we successfully executed. We studied the previously executed related works while our research work stays quite ahead of everything.

5 Conclusion and Future Works

The principal objective of this research work was to expand the boundaries of robotics approach into the field of engineering for surveillance, reconnaissance, and security of life. It is quite haunting to believe the truth that we are entering into such kind of era where the extremist people are constantly threatening the humanity with their power of violence as well as the usage of modern technology into a wrong way. So, becoming concerned citizens of the world of scientific thinking, it was our responsibility to keep our vision always one step ahead of the extremist activists. We executed our unique research work that way. Still there were some limitations in this research work concerning the run-time and memory allocation complexities and configurations between the modified Inception-Net programs and recent TensorFlow packages. We successfully improvised Inception-Net for more efficient image processing for robot vision with good accuracy. However, we are focusing to take our research at further level of implementation of new cutting-edge neural network architectures (Res-Net, Mobile-Net, Efficient-Net, etc.) which will sharpen our visionary perception in the field of artificial intelligence, sensor fusion, and robotics.

References

1. Yang R, Singh SK, Tavakkoli M, Amiri N, Yang Y, Karami MA, Rai R (2020) Cnn-lstm deep learning architecture for computer vision-based modal frequency detection. Mech Syst Signal Process 144:106885
2. Goh G, Cammarata N, Voss C, Carter S, Petrov M, Schubert L, Radford Olah C (2021) Multimodal neurons in artificial neural networks Distill 6(3):e30
3. Wu C, Yu H, Lee S, Peng R, Takeuchi I, Li M (2021) Programmable phase-change metasurfaces on waveguides for multimode photonic convolutional neural network. Nat Commun 12(1):1–8
4. Malik J, Kiranyaz S, Gabbouj M (2021) Self-organized operational neural networks for severe image restoration problems. Neural Netw 135:201–211
5. Kang S, Iwana BK, Uchida S (2021) Complex image processing with less data—document image binarization by integrating multiple pre-trained u-net modules. Pattern Recogn 109:107577
6. Boulila W, Sellami M, Driss M, Al-Sarem M, Safaei M, Ghaleb FA (2021) Rsdcnn: A novel distributed convolutional-neural-networks based-approach for big remote-sensing image classification. Comput Electron Agric 182:106014
7. Zhu C, Chan E, Wang Y, Peng W, Guo R, Zhang B, Soci C, Chong Y (2021) Image reconstruction through a multimode fiber with a simple neural network architecture. Sci Rep 11(1):1–10
8. Sirichotedumrong W, Kiya H (2021) A gan-based image transformation scheme for privacy-preserving deep neural networks. In: European signal processing conference (EUSIPCO). IEEE, Netherlands, pp 745–749
9. Lin J, Li Y, Yang G (2021) Fpgan: Face de-identification method with generative adversarial networks for social robots. Neural Netw 133:132–147
10. Wan S, Goudos S (2020) Faster R-CNN for multi-class fruit detection using a robotic vision system. Comput Netw 168:107036
11. Jia W, Tian Y, Luo R, Zhang Z, Lian J, Zheng Y (2020) Detection and segmentation of overlapped fruits based on optimized mask R-CNN application in apple harvesting robot. Comput Electron Agric 172:105380
12. Budiharto W, Andreas V, Suroso JS, Gunawan AAS, Irwansyah E (2019) Development of tank-based military robot and object tracker. In: Asia-Pacific conference on intelligent robot systems (ACIRS). IEEE, Japan, pp 221–224
13. Budiharto W, Gunawan AA, Irwansyah E, Suroso J (2019) Android-based wireless controller for military robot using bluetooth technology. In: 2nd world symposium on communication engineering (WSCE). IEEE, Japan, pp 215–219
14. Islam MZ, Ahsan A, Acharjee R (2019) A semi-autonomous tracked robot detection of gun and human movement using haar cascade classifier for military application. In: International conference on nascent technologies in engineering (ICNTE). IEEE, India, pp 1–6
15. Eidinger E, Enbar R, Hassner T (2014) Age and gender estimation of unfiltered faces. IEEE Trans Inf Forensics Secur 9(12):2170–2179

16. Salihbašic·A, Orehovački T (2019) Development of android application for gender, age and face recognition using opencv. In: 42nd international convention on information and communication technology, electronics and microelectronics (MIPRO). IEEE, Croatia. pp 1635–1640
17. Meena M, Thilagavathi P (2012) Automatic docking system with recharging and battery replacement for surveillance robot. Int J Electron Comput Sci Eng 1148–1154
18. Jain H, Vikram A, Kashyap A, Jain A, et al. (2020) Weapon detection using artificial intelligence and deep learning for security applications. In: International conference on electronics and sustainable communication systems (ICESC). IEEE, pp. 193–198
19. Shaikh Z, Gaikwad P, Kare N, Kapade S, Korade M (2017) An implementation on-surveillance robot using raspberry-pi technology. Int Res J Eng Technol 4(4):1910–1913
20. Chemel B, Mutschler E, Schempf H (1999) Cyclops:.miniature robotic reconnaissance system. In: IEEE international conference on robotics and automation (Cat. No. 99CH36288C), vol 3. IEEE, South Korea, pp 2298–2302
21. Liu G-H, Wen S-F, Chen P-C, Shih W-P (2011) Real-time humanoid visual system based on region segmentation. In: 9th world congress on intelligent control and automation. IEEE, China, pp 347–352
22. Bekios-Calfa J, Buenaposada JM, Baumela L (2010) Revisiting linear discriminant techniques in gender recognition. IEEE Trans Pattern Anal Mach Intell 33(4):858–864

Bangla Handwritten Character Recognition Using Convolutional Neural Network

Partha Chakraborty, Afroza Islam, Mohammad Abu Yousuf, Ritu Agarwal, and Tanupriya Choudhury

Abstract Recognizing handwritten characters is more challenging than recognizing printed characters. The size and shape of a handwritten character written by different people are not the same. The function of character recognition is complicated by the numerous variants in writing styles. Bangla handwritten character recognition had already been the focus of many researchers. Due to its specific characteristics of feature extraction and classification, the convolutional neural network (CNN) has recently shown notable progress in the fields of image-based recognition, video analytics, and natural language processing. As a result, this research provides a deep convolutional neural network (DCNN)-based mechanism for recognizing Bengali handwritten characters. In the field of pattern recognition, one of the most efficient ways to achieve higher accuracy or a lower error rate is to use a deep, optimized architecture that can process a large amount of data. Therefore, this paper has used DCNN for feature extraction and classification. This paper compares their accuracies by applying the DCNN model over three datasets: BanglaLekha-Isolated dataset, CMATERdb dataset, and our created dataset. By applying DCNN in researchers created dataset, accuracy has achieved up to 93.07%. This paper has also worked on simple CNN, DCNN, more advanced VGG-16 models for classification. And finally, the accuracies obtained from all the datasets have been compared.

Keywords Handwritten character recognition · Deep convolutional neural network · Visual geometry group

---

P. Chakraborty (✉) · A. Islam
Department of Computer Science and Engineering, Comilla University, Cumilla 3506, Bangladesh
e-mail: partha.chak@cou.ac.bd

M. Abu Yousuf
Institute of Information Technology, Jahangirnagar University, Savar, Dhaka 1342, Bangladesh
e-mail: yousuf@juniv.edu

R. Agarwal
Raj Kumar Goel Institute of Technology, Ghaziabad, Uttar Pradesh 201003, India

T. Choudhury
Department of Informatics, School of Computer Science, University of Petroleum and Energy Studies, Dehradun, India

V. Skala et al. (eds.), *Machine Intelligence and Data Science Applications*, Lecture Notes on Data Engineering and Communications Technologies 132,
https://doi.org/10.1007/978-981-19-2347-0_56

1 Introduction

This study focuses on offline unconstrained Bangla handwritten character recognition, which is still considered a challenging topic in the research community.

The main challenge in handwritten character identification is dealing with the large variety of handwriting styles used by numerous authors. In certain cases, characters are written separately from each other, depending on the language (e.g., Laos and Japanese). In some of these cases, they are written in cursive, and the characters are joined to each other. (e.g., English, Bangladeshi, and Arabic). Many experts in the field of natural language processing have already recognized this challenge (NLP) [1–3]

Bangla is one of the most frequently spoken languages in the world, with more than 200 million speakers [4, 5]. It is the official language of the country of Bangladesh as well as India's second most commonly spoken language. So, scholars from around the world are working to computerize the Bangla language. Because of the following factors, recognizing handwritten characters in Bangla is more difficult than recognizing printed characters: (1) Handwritten characters by multiple writers are not only nonidentical but also variety in shape and size; Character identification is difficult due to the numerous variances in writing styles among the characters; (3) The character recognition challenge is made more difficult by the similarities in shape of various characters and the overlaps of surrounding characters.

In this paper, Bengali handwritten character recognition technique has accomplished by using simple CNN architecture, DCNN architecture, and VGG-16 model architecture which have been described in chapter four.

The entire manuscript is organized in such a way that the literature review is done in Sect. 2, the overview of the proposed method is done in Sect. 3, the experimental details in Sect. 4, the result analysis is in Sect. 5, and the final Sect. 6 is in conclusion.

2 Literature Review

Halima Begum et al. "recognition of handwritten Bangla characters using Gabor filter and artificial neural network" [6] used a dataset they produced that included 95 participants, and their suggested model achieved recognition rates of roughly 68.9% without feature extraction and 79.4% with feature extraction, respectively.

The author proposed document classification model using machine learning approaches [7–10].

Chakraborty and Jahan API [11] suggested NumtaDB dataset, in which they used CNN to understand Bangla handwritten numerals. Their model had a 98% accuracy rate.

A Bangla document classification using deep recurrent neural network with BiLSTM by Saifur Rahman and Partha Chakraborty et al. for text classification. This approach achieved 98.33% accuracy and an F1-score of 0.98 [12].

Chatterjee et al. [13] presented a solution which used state-of-the-art techniques in deep learning to tackle the problem of Bengali handwritten character recognition (HCR). In comparison with most other comparable approaches, their approach needed fewer iterations to train. The BanglaLekha-Isolated dataset was used to test their methodology.

Das et al. [14] discovered a new variation of feature set which significantly outperformed on handwritten Bangla alphabet from the previously used feature set. About 132 number of features in all, viz. modified shadow features, octant and centroid features, distance-based features, and quad tree-based longest run features were used their paper. On the same dataset, using that feature set, recognition performance improved dramatically from 75.05% in their existing works [15] to 85.40% on 50 character classes using an MLP-based classifier.

To this end, the previous research has achieved promising results using several methods. So, with the help of the above papers researchers have mainly worked on Bangla character recognition. This research has been done with simple CNN architecture, DCNN with dropout, and VGG-16 model architecture. It has been possible to apply the models to researcher's created dataset by studying the different models from the above papers and also worked with other datasets like CMATERdb dataset, BanglaLekha-Isolated dataset.

3 Proposed Method

Figure 1 shows the block diagram of the proposed handwritten Bengali character recognition system which involves three main stages: preprocessing, feature extraction, and classification.

The description of the whole diagram is given in the following sections.

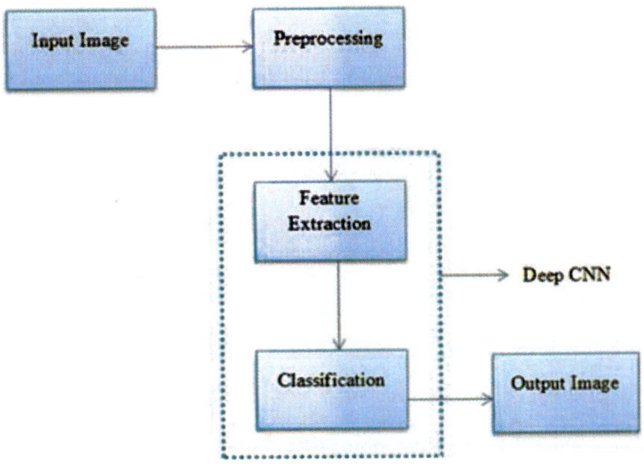

Fig. 1 Block diagram of showing the steps of proposed Bangla handwritten character recognition system

3.1 Dataset Collection and Preparation

The proposed method utilized use of three datasets: researcher's created dataset, BanglaLekha-Isolated dataset and CMATERdb dataset. A large publicly accessible dataset free of biases arising from geographical location, gender, and age is needed to benchmark Bengali character recognition algorithms. With this aim in mind, BanglaLekha-Isolated, a dataset consisting of more than 10,0000 images of handwritten Bengali characters and digits, has been assembled. This dataset contains Bangla handwritten numerals, basic characters, and compound characters. A total of 81,794 characters have been used for processing from this dataset. The CMATERdb dataset contains 15,000 character pictures in total. The image was a noise-free 32 × 32 pixel image. For this research researchers also used their created dataset which consists of 2600 handwritten characters. Figure 2 showing an example of different datasets.

In this thesis work, at first, CMATERdb dataset has been used, and these images have been 32 × 32 pixel noise-free images. The BanglaLekha-Isolated dataset contains with an inverted foreground and background noise removal with a median filter, edge thickening filter, and resizing into a square shape with appropriate paddings. And the images of our own created dataset named BanglaCharacter dataset have different pixel size. So, all the datasets have been preprocessed to improve the quality of the CNN algorithm in the following way:

- Split the training data to have a validation set similar in size and distribution to the test set. There are 13,205 and 2400 validation alphabet images with uniform distribution for BanglaLekha-Isolated and CMATERdb datasets. Further, there are 416 validation alphabet images with uniform distribution for researcher's created dataset.
- All the images have been rescaled by dividing every pixel in every image by 255. As a result, the scale is now 0–1 rather than 0–255.
- The input images were cropped at random into 64 × 64 gray-scale images.
- One-hot encoding for the categories to compare in the classification stage. For instance, label = 3 is converted to a vector as [0 0 0 1 0 0 0 0 0 0 …. 0] (labels starts from 0 to 49 for the 50 classes)

Fig. 2 **a** Researcher's created dataset. **b** BanglaLekha-Isolated dataset. **c** CMATERdb dataset

3.2 Model Architecture

Handwritten character recognition is a difficult task that usually requires the manipulation of two-dimensional images. And CNN architecture is well suited for image processing. Because other conventional models need a separate process for feature extraction, which CNN does automatically. So, in this thesis, DCNN architecture has been used for the recognition of Bengali characters. The CNN architecture for handwritten character recognition has been shown in Fig. 3.

Step 1: The preprocessed gray-scale image data in 64×64 size is the input to the convolution layer. Convolution is applied to an input image using feature detectors (also known as filters or kernels) in a neural network, resulting in a feature map. Researchers usually create a lot of feature maps by using a lot of feature detectors. After training with back propagated gradient descent, the feature detectors are automatically set as part of the supervised learning algorithm.

Step 2: For implementing nonlinearity to the output, an activation layer will be needed like ReLU activation.

Step 3: Pooling guarantees that the neural network has spatial invariance, or that it accounts for any distortions, since it captures the function inside a given matrix. In this case, MaxPooling has been used.

Step 4: The final pooled feature maps are flattened into a column vector at the end of the convolution layers so that they can be interpreted by a traditional artificial neural network (ANN) with one or more densely connected hidden layers and a softmax classifier as the final output layer.

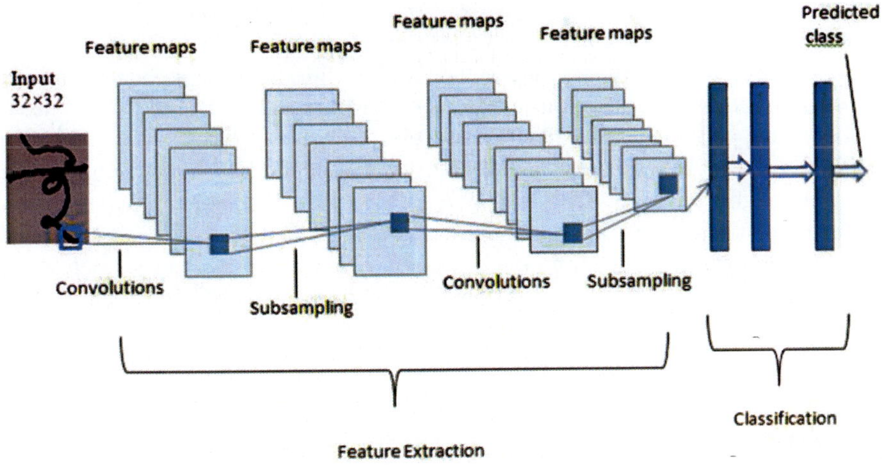

Fig. 3 Basic CNN architecture of Bengali handwritten character recognition

Step 5: Finally, a fully connected (dense) ANN is added to the convoluted layer output. And the column vector is passed through a traditional tightly connected ANN for final classification.

4 Experimental Details

4.1 Train, Test and Validation Sets

For researcher's created dataset, 2080 images used as train image and 520 for the test image. For CMATERdb, the dataset has total 15,000 images, where we use 9600 as training, 2400 as validation and 3000 as test images. For BanglaLekha-Isolated 81,794 images, 66,023 (80%) images were taken for training and 15,771 (20%) image as a test set

4.2 Model Evaluation

At first, simple CNN model has been set up with 2 convolutional layers using ReLU activation and 1 fully connected layer terminating in 50 softmax classifiers. We got an accuracy of just over 09% for my created dataset on the training data after training the model for 100 epochs and got accuracy of 36 and 26% for BanglaLekha-Isolated dataset and CMAREdb dataset which suggests a very high bias.

So more layers have been added to the network, settling on 5 convolution layers and 2 dense layers in the end. The model summary is given in Table 1.

With this model, it has been possible to get 91.90% accuracy for researcher's created dataset, while the test accuracy was 93.07% with same batch size and number of epochs. And for BanglaLekha-Isolated and CMATERdb datasets, the test accuracies were 84.53% and 93.43%, respectively. This model has been trained with augmented data, and train speed has also been increased. Finally, it has been possible to get the highest test accuracy for CMATERdb dataset rather than two datasets. In this experiment, researcher's dataset has also been trained with more advanced VGG-16 through which maximum accuracy of 98% has been achieved. Below are the graphs of training and validation obtained by applying VGG-16 model on this dataset (Fig. 4).

5 Result Analysis

In this section, the training and testing accuracy for these datasets have been compared and tried to find out the highest accuracy from Table 2.

Table 1 Model summary of DCNN architecture

Layer	Output shape
Conv2D 2	(None, 63, 63, 16)
MaxPooling2D 2	(None, 31, 31, 16)
Dropout	(None, 31, 31, 16)
Conv2D 3	(None, 30, 30, 32)
MaxPooling2D 3	(None, 15, 15, 32)
Dropout 1	(None, 15, 15, 32)
Conv2D 4	(None, 14, 14, 64)
MaxPooling2D 4	(None, 7, 7, 64)
Dropout 2	(None, 7, 7, 64)
Conv2D 5	(None, 6, 6, 128)
MaxPooling2D 5	(None, 3, 3, 128)
Dropout 3	(None, 3, 3, 128)
Conv2D 6	(None, 2, 2, 256)
MaxPooling2D 6	(None, 1, 1, 256)
Dropout 4	(None, 1, 1, 256)
Flatten	(None, 256)
Dense 1	(None, 256)
Dense 2	(None, 512)
Dense 5	(None, 512)
Dense 3	(None, 50)

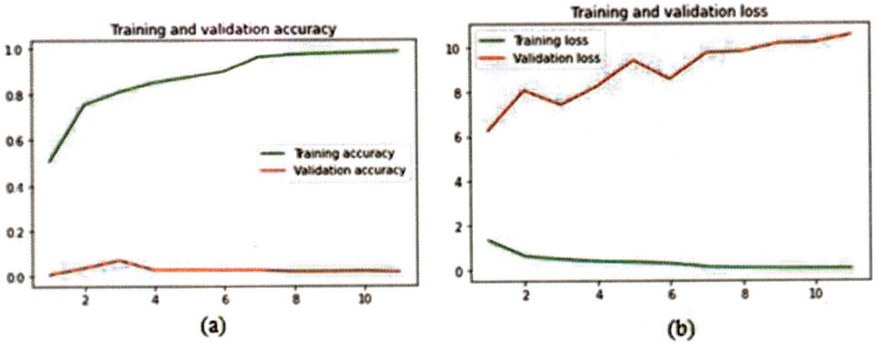

Fig. 4 Graph for researcher's dataset **a** Training and validation accuracy. **b** Training and validation loss

Analysis of Tables 2 and 3 shows that up to 93.07% accuracy was achieved by applying the DCNN model to the dataset researchers created. And the highest accuracy was found in the CMATERdb dataset from all the datasets which is 93.43%. This accuracy has been achieved by applying deep CNN (Fig. 5).

Table 2 Table with training accuracy

Datasets	Simple CNN model (%)	DCNN model (%)	DCNN with augmentation (%)
Researcher's created dataset	09.98	91.90	78.05
CMATERdb dataset	27.92	95.53	81.75
BanglaLekha-isolated	36.91	70.83	69.40

Table 3 Table with test accuracy

Datasets	Simple CNN model (%)	DCNN model (%)	DCNN with augmentation (%)
Researcher's created dataset	14.62	93.07	66.35
CMATERdb dataset	28.66	93.43	96.17
BanglaLekha-Isolated	35.99	84.53	92.89

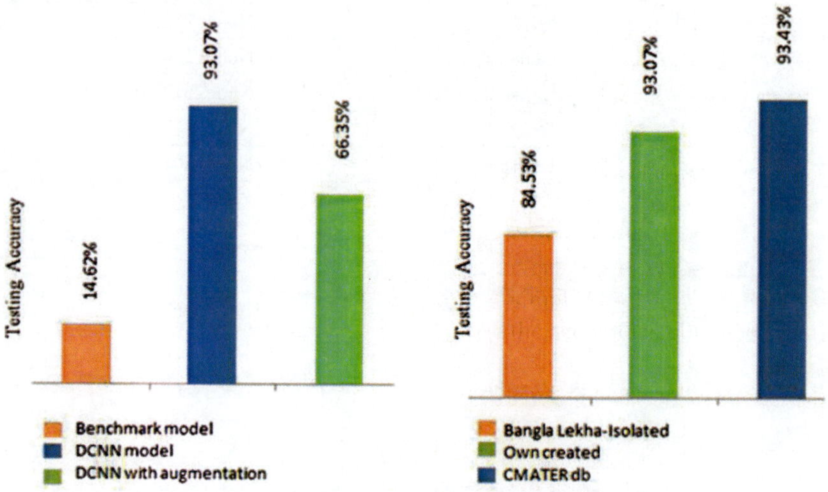

Fig. 5 Dance charts were made from the above tables **a** Testing accuracy for our created dataset. **b** Testing accuracy of all the datasets by applying the DCNN model

5.1 Error Analysis

The misclassified images have been interpreted toward error analysis in order to understand the root cause of the classification issues. This was expected because of the similarities between these characters, and handwritten characters are very difficult to distinguish (Fig. 6).

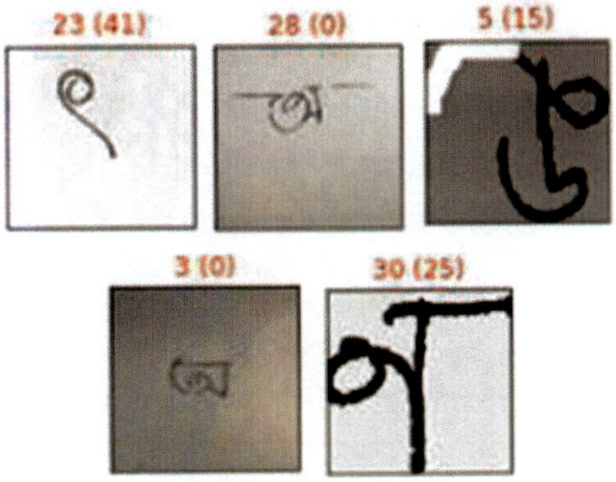

Fig. 6 Some misclassified alphabet of the created dataset

5.2 Comparison with Previous Research

In Table 4, the accuracy obtained from this research paper is compared with the accuracy obtained from some of the previous research papers. And in this research, it has been possible to get more than 90% accuracy for researcher's created dataset which is 93.07%.

Table 4 Comparison of result with some previous works

The work reference	Datasets	Total samples	Accuracy (%)
Recognition of handwritten Bangla characters using Gabor filter and artificial neural network [6]	Their created dataset	Data was collected from 95 volunteers	79.04
An improved feature descriptor for recognition of handwritten Bangla alphabet, 2015 [14]	Their created dataset	10,000 samples	85.40
Bengali handwritten character classification using transfer learning on deep convolutional neural network, 2020 [13]	BanglaLekha-isolated	166,105 samples	96.12
Bengali handwritten character classification using deep convolutional neural network	**Researcher's created dataset**	**2600 samples**	**93.07**

Definition for the significance of bold is comparison of the acquired result with some previous works

6 Conclusion

All the models have been applied on three datasets to find out the better accuracy. Specifically, recognition rate of 93.07% has achieved from researcher's created dataset by applying DCNN model. And by applying the deep CNN model, CMATERdb dataset has achieved the highest accuracy compared to other datasets. The research objective of this thesis is basically to work with own dataset and comparing the results obtained with the results of other datasets.

There have not any compound characters or digits in our dataset so it has become easy to process the data which is one of the limitation. So in the future, researchers will try to work with digits and compound characters together. Some fusion-based DCNN models, such as inception recurrent convolutional neural network (IRCNN) [16–18] will be explored and created in the future for handwritten Bangla character recognition.

References

1. Cires¸an DC, Meier U, Gambardella LM, Schmidhuber J (2010) Deep, big, simple neural nets for handwritten digit recognition. Neural Comput 22(12):3207–3220
2. Meier U, Ciresan DC, Gambardella LM, Schmidhuber J (2011) Better digit recognition with a committee of simple neural nets. In: Proceedings of international conference on document analysis and recognition (ICDAR), Beijing, China, September 2011, pp 1250–1254
3. Song W, Uchida S, Liwicki M (2011) Comparative study of part-based handwritten character recognition methods. In: Proceedings of international conference on document analysis and recognition (ICDAR), Beijing, China, September 2011, pp 814–818
4. Khan HA, Al Helal A, Ahmed KI (2014) Handwritten Bangla digit recogni-tion using sparse representation classifier. In: Proceedings of 2014 international conference on informatics, electronics and vision (ICIEV), IEEE, Dhaka, Bangladesh, May 2014, pp 1–6
5. Pal U, Chaudhuri B (2000) Automatic recognition of unconstrained off-line Bangla handwritten numerals. In: Proceedings of advances in multimodal interfaces–ICMI 2000, Springer, Beijing, China, October 2000, pp 371–378
6. Begum H et al Recognition of handwritten Bangla characters using Gabor filter and artificial neural network. Int J Comput Technol Appl 8(5):618–621. ISSN:2229-6093
7. Rahman MM, Pramanik MA, Sadik R, Roy M, Chakraborty P (2020) Bangla documents classification using transformer based deep learning models. In: 2020 2nd International Conference on Sustainable Technologies for Industry 4.0 (STI), pp 1–5. IEEE
8. Ahammad K, Shawon JA, Chakraborty P, Islam MJ, Islam S (2021) Recognizing Bengali sign language gestures for digits in real time using convolutional neural network. Int J Comput Sci Inf Secur (IJCSIS) 19(1)
9. Sultana M, Chakraborty P, Choudhury T (2022) Bengali abstractive news summarization using Seq2Seq learning with attention. In: Cyber intelligence and information retrieval. Springer, Singapore, pp 279–289
10. Ahmed M, Chakraborty P, ChoudhuryT (2022) Bangla document categorization using deep RNN model with attention mechanism. In: Cyber intelligence and information retrieval. Springer, Singapore, pp 137–147
11. Chakraborty P, Jahan Api SS, Choudhury T Bangla handwritten digit recognition

12. Rahman S, Chakraborty P (2021) Bangla document classification using deep recurrent neural network with BiLSTM. In: Proceedings of international conference on machine intelligence and data science applications. Springer, Singapore, pp 507–519
13. Chatterjee S, Dutta RK, Ganguly D, Chatterjee K (2020) Handwritten character classification using transfer learning on deep convolutional neural network. In: International conference on intelligent human computer interaction, April 2020
14. Das N, Basu S, Sarkar R, Kundu M, Nasipuri M, Kumar Basu D (2015) An improved feature descriptor for recognition of handwritten Bangla alphabet. arXiv:1501.05497 [cs], Jan 2015 [Online]. Available: arXiv:1501.05497. Accessed 16 Nov 2018
15. Basu S, Das N, Sarkar R, Kundu M, Nasipuri M, Basu DK (2009) A hierarchical approach to recognition of handwritten Bangla characters. Pattern Recogn 42:1467–1484
16. Chakraborty P, Nawar F, Chowdhury HA (2022) Sentiment analysis of Bengali facebook data using classical and deep learning approaches. In: Mishra M, Sharma R, Kumar Rathore A, Nayak J, Naik B (eds) Innovation in electrical power engineering, communication, and computing technology. Lecture notes in electrical engineering, vol 814. Springer, Singapore. https://doi.org/10.1007/978-981-16-7076-3_19
17. Chakraborty P, Nawar F, Chowdhury HA (2022) A ternary sentiment classification of Bangla text data using support vector machine and random forest classifier. In: Mandal JK, Hsiung PA, Sankar Dhar R (eds) Topical drifts in intelligent computing. ICCTA 2021. Lecture notes in networks and systems, vol 426. Springer, Singapore. https://doi.org/10.1007/978-981-19-0745-6_8
18. Sarker A, Chakraborty P, Shaheen Sha SM, Khatun M, Hasan MR, Banerjee K (2020) Improvised technique for analyzing data and detecting terrorist attack using machine learning approach based on twitter data. J Comp Commun 8(7):50–62

Music Genre Classification Using Light Gradient Boosting Machine: A Pilot Study

Akhil Sibi, Rahul Singh, Kumar Anurag, Ankur Choudhary, Arun Prakash Agrawal, and Gaurav Raj

Abstract The world is changing at a rapid pace with the advent of mind blowing, sophisticated yet elegant technologies like artificial intelligence, robotics, and cloud computing. Music streaming industry is not untouched and is also evolving to adapt to the ever changing needs of their customer base. Addition of large number of songs to these streaming platforms every day sometimes suffers from the problem of correct classification of genre of songs and sometimes recommends wrong song to the listener. This usually happens because of the inefficient classification by the algorithm to correct genre of the song and to the listener's interest. Till date a lot of approaches have been proposed neither of them is perfect and hence, leave room for improvement. This paper seeks the application of light gradient boosting machine (LightGBM) and proposes an approach to address this issue. Proposed approach was also compared with existing state-of-the-art approaches, viz. convolutional neural networks (CNN), artificial neural networks (ANN), LSTM (RNN) and gated recurrent unit (GRU-RNN), gradient boosting machine (GBM), LightGBM, CatBoost, XGBoost, and AdaBoost. All the approaches under study were applied on GTZAN dataset retrieved from Kaggle. The results indicate that LightGBM is the best model for the genre classification, which achieves the highest accuracy of 92%.

Keywords Music genre classification · GTZAN · RNN · LTSM · CNN · ANN · GBM

A. Sibi · R. Singh · K. Anurag · A. Choudhary · A. Prakash Agrawal · G. Raj (✉)
Department of Computer Science and Engineering, School of Engineering and Technology, Sharda University, Greater Noida, India
e-mail: er.gaurav.raj@gmail.com

R. Singh
e-mail: grayhathacks10@gmail.com

V. Skala et al. (eds.), *Machine Intelligence and Data Science Applications*, Lecture Notes on Data Engineering and Communications Technologies 132,
https://doi.org/10.1007/978-981-19-2347-0_57

1 Introduction

A lot of music streaming platforms are in existence today, viz. Spotify, Hungama Music, Gaana.com, Amazon Music, etc. Millions of subscribers are getting attracted to these platforms as thousands of songs are being added to these platforms on daily basis. Because of large volume of songs these platforms suffer from the problem of correct classification of songs according to their genre like Jazz, Pop, Opera, Rock, Country Music, etc.

The problem of music genre classification dates back to 2002 when Tzanetakis developed the widely used GTZAN dataset [1] and classified the music in 10 different genres: Blues, Classical, Country, Disco, Hip-hop, Jazz, Metal, Pop, Reggae and Rock. His work was later on followed by several other researchers and contributed to the music genre classification problem.

From 2007 onwards, researchers started using machine learning to achieve the better accuracy on the dataset, and then from 2010 onwards, they started to use deep learning approaches. Application of so many approaches for music genre classification motivated us to conduct this pilot study. Authors in this paper conducted a pilot study on ten such widely used approaches for music genre classification. All the approaches under study were implemented in Python and were applied to GTZAN dataset.

Audio samples taken from the dataset in waveforms were first converted into FFT spectrograms for further processing. FFT spectrograms were later on converted into Log spectrograms which are in MFCC form. 70% of the samples from the dataset were used for the training purpose, and rest 30% were utilized for the tuning and testing of the approaches under study. All the approaches under study were then trained on the training samples from the GTZAN dataset. Approach using LightGBM achieved the best accuracy among all the compared ones.

2 Literature Review

Steep rise in application of machine learning (ML) and deep learning (DL) toward genre classification has revolutionized the music information retrieval (MIR) systems. In this revolution, many sophisticated but fascinating applications like Spotify, Resso, YouTube Music, and many others have emerged and are redefining the music streaming. Let us have a look at the history of MIR systems which dates back to early 2000s.

Tzanetakis et al. [2] in 2002 proposed the automatic classification of audio signals for different music genre using three different feature sets of audio signals which were timbral texture, rhythmic content, and pitch content. These features were trained using statistical pattern classifiers, which on testing gave an accuracy of 61% for ten genres. This research work proved to be a milestone and paved the way for other researchers to work in this direction. Further researchers used other feature extraction techniques and used machine learning and deep learning models for training.

Li et al. [3] in 2003 proposed a novel and unique feature extraction method DWCHs for music genre classification which capture the local and global information of music signals simultaneously by computing histograms on their Daubechies wavelet coefficients at various frequency bands to improve the classification accuracy. They also provided a comparative study of various classification methods on different features sets and DWCHs. Cataltepe et al. [4] later on worked on the audio features from MIDI files separately and combined together for MIDI music genre classification. They used normal compression distance (NCD) to compute distances for feature extraction and used 12 different classifiers like K-NN, LDC, etc., and combined them by majority voting to get the best results.

Elbir et al. [5] used the acoustic features from different genres of music by using digital signal processing techniques, and these features were trained using different classification algorithms in which it was found that support vector machine (SVM) achieved the best results for classification. Silla et al. [6] emphasized the use of the middle part of the music for feature extraction, which shows better results in comparison with the first and end parts, and then different ML algorithms were used to train the features. Lidy and Rauber [7] studied the rhythmic histogram features and statistical spectrogram descriptors of the audio files and evaluated their efficiency compared to other feature sets and trained them on SVM. Vishwanathan and Sundaraj [8] analyzed the song as a wave for feature extraction and then used frequency and tempo features to train the machine learning models. However, a smaller dataset was used which was not effective for large-scale purposes.

Li and Ogihara [9] presented a unique classification technique: hierarchical taxonomy based on linear discriminant projection to classify music genres. They conducted empirical experiments on two datasets and the taxonomy achieved an accuracy of 72.7 and 70.12% on the two datasets. They also highlighted the relationship of dependence between different genres and provided valuable sources of information for genre classification.

Liu et al. [10] proposed a convolutional neural network (CNN)-based architecture which could take full advantage of low-level information of Mel spectrogram for making the classification decisions on three benchmarked dataset of GTZAN, Ballroom and Extended Ballroom. They achieved accuracies of 93.9%, 96.7%, and 97.2%, respectively, and the trainable model was so lightweight that it could be applied in mobile phones, etc. This work was notable for achieving the best results on the mentioned datasets so far. Pelchat and Gelowitz [11] research used images of spectrograms generated from time slices of songs as the input into a neural network to classify the songs into their respective music genres.

Ghildiyal et al. [12] did the comparative analysis of various ML and DL models by analyzing the Mel spectrogram of audio files from GTZAN dataset, and it was found that CNN achieved the higher accuracy of 91%. In a similar approach of comparative analysis, Upadhyay et al. [13] compared the accuracy of K-NN and SVM algorithms.

All these works motivated us to find new ways for music genre classification with better results, accuracy, and in an optimal way.

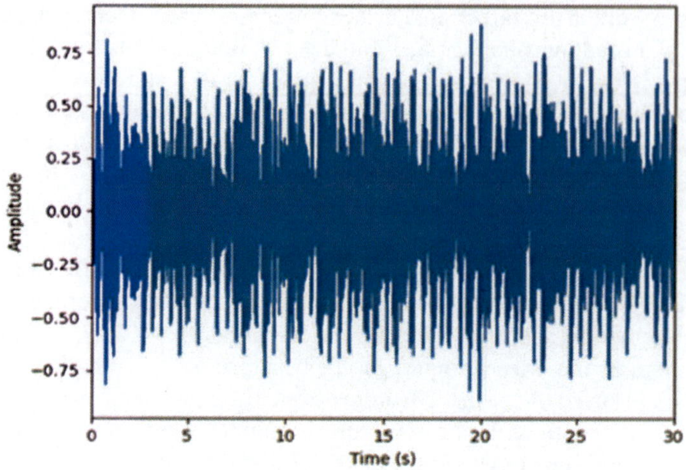

Fig. 1 Conversion of audio clips from waveform to waves

3 Methodology

3.1 Data Preparation and Analysis

GTZAN dataset is used for feature selection and extraction, and this dataset has ten classes—Blues, Classical, Country, Disco, Hip-hop, Jazz, Pop, Metal, Reggae, and Rock. Each class contains 100 audio clips of 30 s each in .wav format and sampled at 22,050 Hz in 16 bits. We divided each .wav file into 5 segments to augment training data [2]. Figure 1 presents the conversion of these audio clips into waves.

3.2 Data Preprocessing

We convert these waveforms into the FFT spectrum as shown below.

Figure 2 shows the FFT spectrum's rate of change in magnitude with frequency. Then we convert these FFT spectrums to Log spectrograms. Here, third axis color values co-relate to decibel values.

Figure 3 shows the rate of change of frequency with time in Log spectrogram. At last, these Log spectrograms are converted to MFCC [final dataset form]. MFCC is preferred over FFT because they have perceptually motivated features to match how humans perceive pitch. Figure 4 shows the rate of change of MFCC with time. At last, we stored the complete dataset in this format and feed this data to different neural networks:

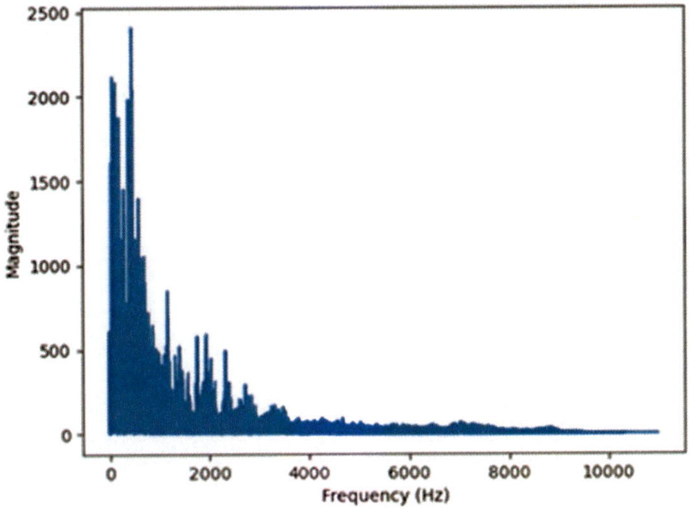

Fig. 2 Conversion of waveform into FFT spectrum

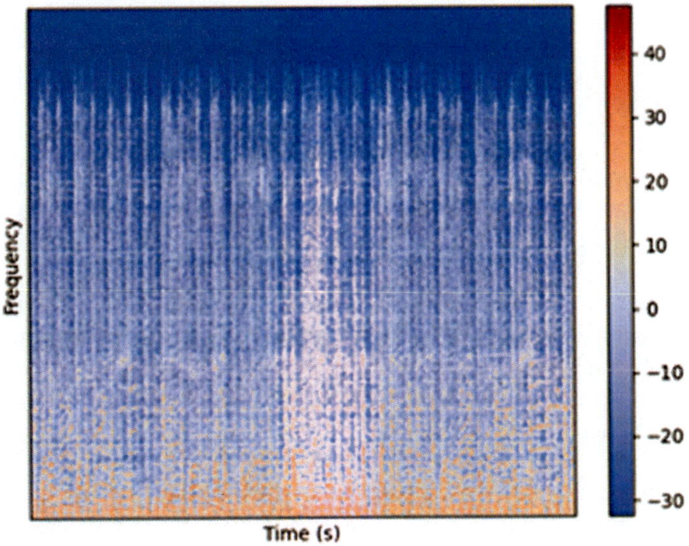

Fig. 3 Log spectrogram


```
{
"mapping" : ["classical","blues",...],
"mfcc": [[[...],[...],...,[...]],...],
"labels": [0,2,...]
}
```

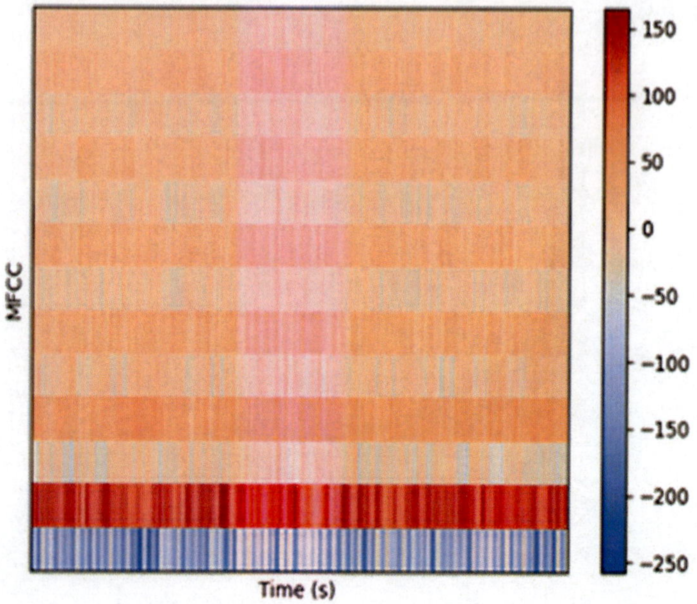

Fig. 4 MFCC

3.3 Modeling

We used the MFCC feature of the music files and started training the models one by one.

ANN

It is the simplest basic neural network with the input, hidden and output layers and we do perform regularization on it. Figure 5 shows the detailed architecture with the layers in ANN and processing of inputs and outputs. Over-fitting happens in ANN during training time, which is Kernel regularized with dropout.

CNN

Convolutional neural networks are used for image processing tasks, and they easily process and train on MFCCs. CNN is always trained in 3D data music but since GTZAN was 2D data, one more dimension was added (e.g., 130, 13, 1). Figure 6 shows the layers of CNN architecture with the inputs and outputs.

LSTM (RNN)

Figure 7 shows the layers of LSTM architecture with the inputs and outputs.

GRU (RNN)

Figure 8 shows the architecture of GRU with the layers and inputs and outputs. We have also successively used intuitive and high performing machine learning models

Fig. 5 Architecture of ANN

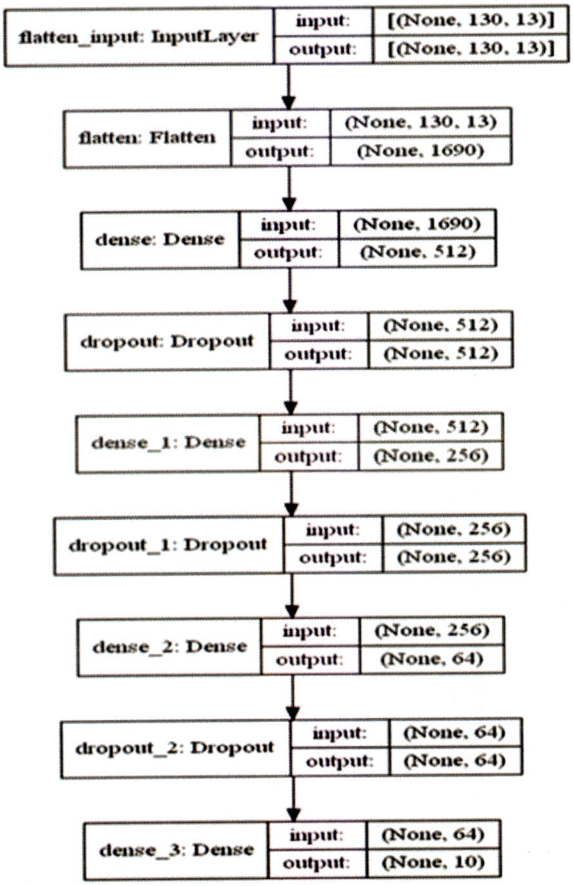

on GTZAN dataset like CatBoost, XGBoost, GBM, LightGBM, and AdaBoost. and compared their accuracies with the deep learning network.

4 Results

Figure 9 shows the accuracy of ANN on the testing data and from the graph we can see that the accuracy is 58%. Figure 10 shows the accuracy of CNN on the testing data and from the graph, we can see that CNN achieves the accuracy of 70% which means that CNN performs well. CNN is also fast to train.

Figure 11 shows the accuracy of LSTM on the testing data and it achieves an accuracy of 64%; however, it is slower to train.

Figure 12 shows the accuracy of GRU on the testing data and it achieves an accuracy of 66%; however, it is as slow as LSTM to train.

Fig. 6 Architecture of CNN

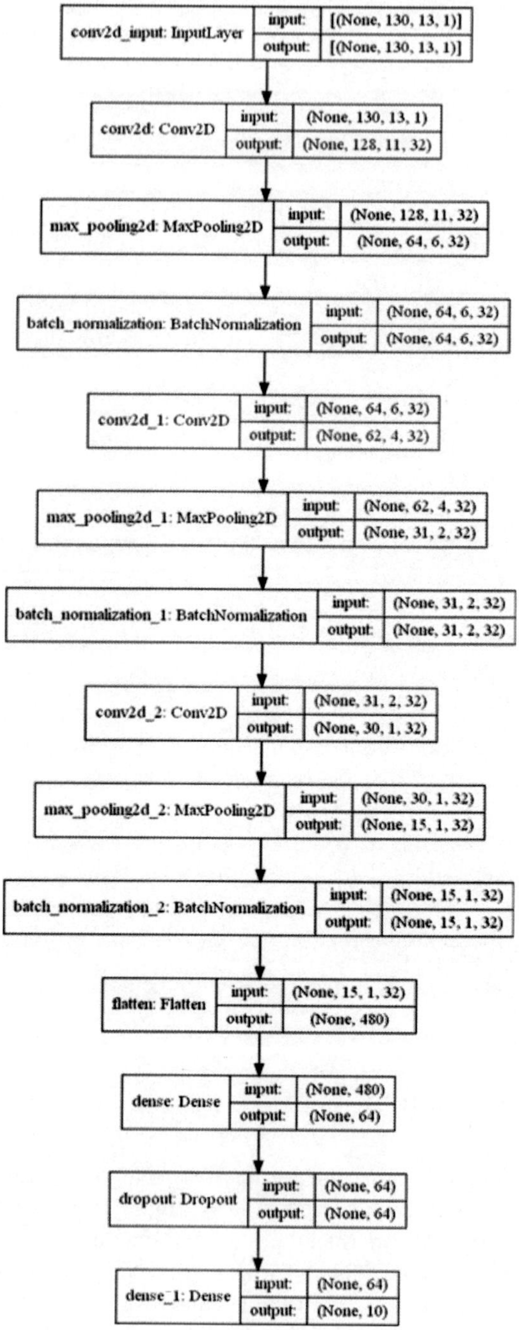

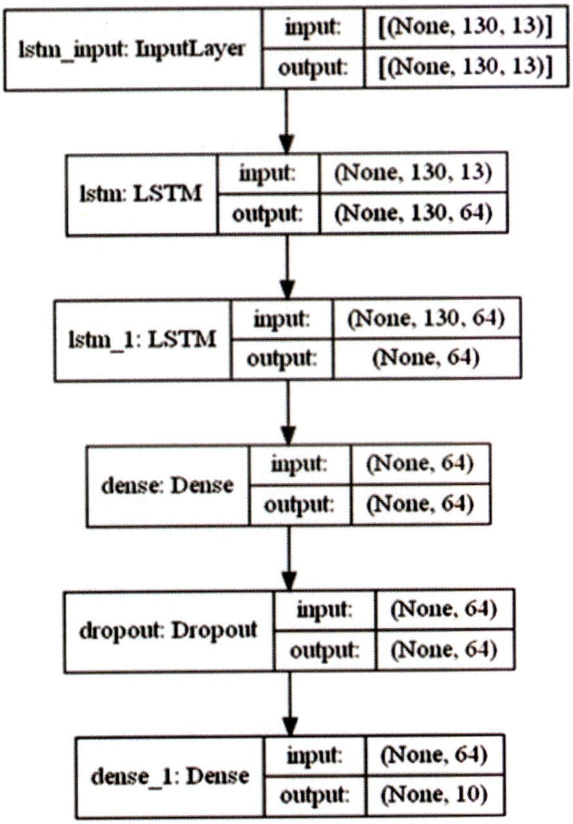

Fig. 7 Architecture of LSTM

These machine learning models from Figs. 13, 14, 15, 16, 17, 18 and 19 were executed on Kaggle notebooks and achieved their respective accuracy on the testing data. Table 1 shows the accuracy of models from the testing and by having done the comparative analysis, we can say that the LightGBM is the best model in terms of speed and accuracy and is best suited for real-time classification for music files.

5 Conclusion and Future Work

In our research work, we have extracted the MFCCs from music files of the GTZAN dataset through conversion from waveforms to FFT spectrum, Log spectrogram, and then the final form: MFCC. We have successfully trained and tested varying machine learning models and deep learning neural networks on MFCCs, and we have observed from the accuracies that LightGBM is the best performing model with the highest accuracy of 92% followed by CatBoost (89.5%), XGBoost (87.8%), and RF (86.8%). During our observations, we also noted that machine learning algorithms perform far

Fig. 8 GRU architecture

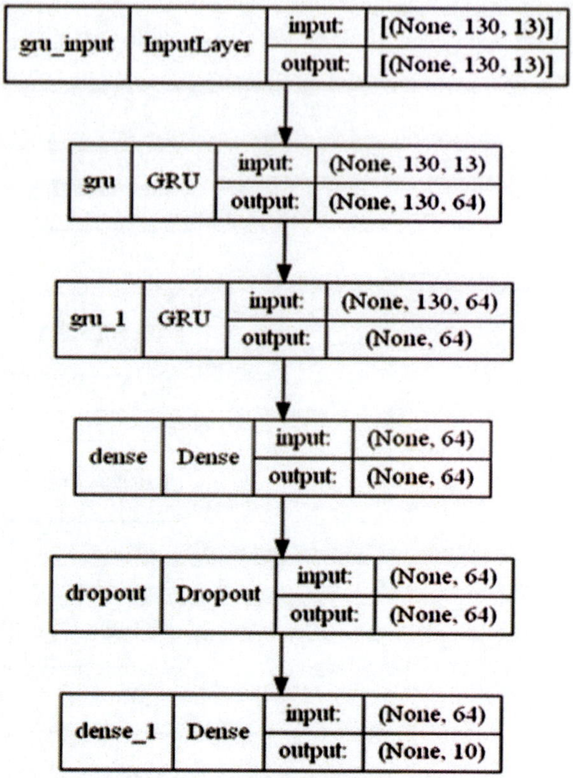

Fig. 9 Accuracy of ANN

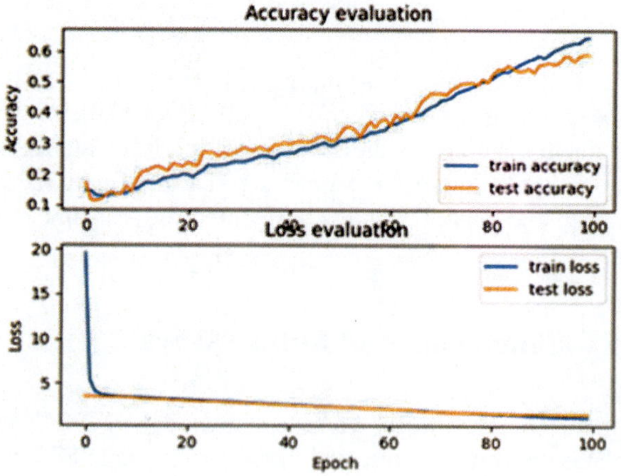

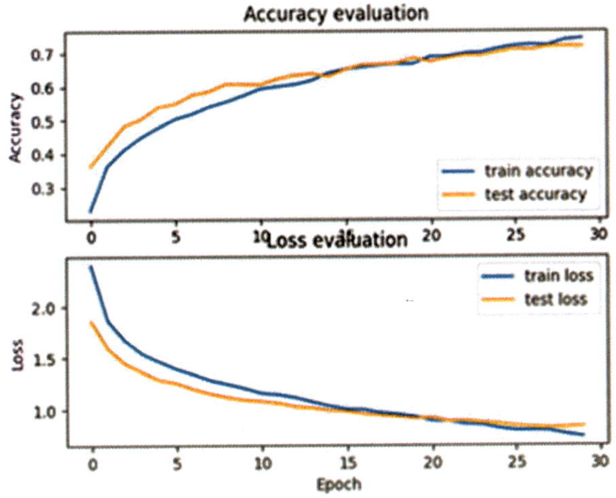

Fig. 10 Accuracy of CNN

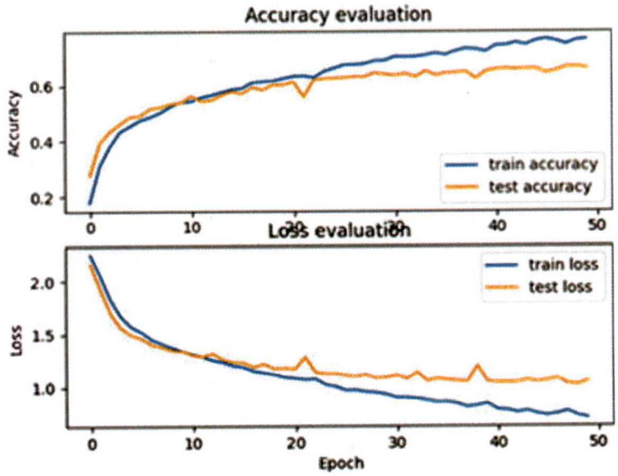

Fig. 11 LSTM (RNN)

better than deep learning neural networks in terms of accuracy and performance on GTZAN dataset like LightGBM, CatBoost, random forest, etc. and so we can further improve their accuracy by hyper-parameter tuning. These tried and tested models with higher accuracies can be deployed in real-time classification of music genre on streaming platforms and can perform with good results.

Future work in our research may include the usage of new improved architectures and modifications to the algorithms. Other considerations may include addition of more data to the dataset for improved training, performance and accuracy of the models and adding more classifiers to genre classification so that the quality of work will be enriched with new possibilities and observations.

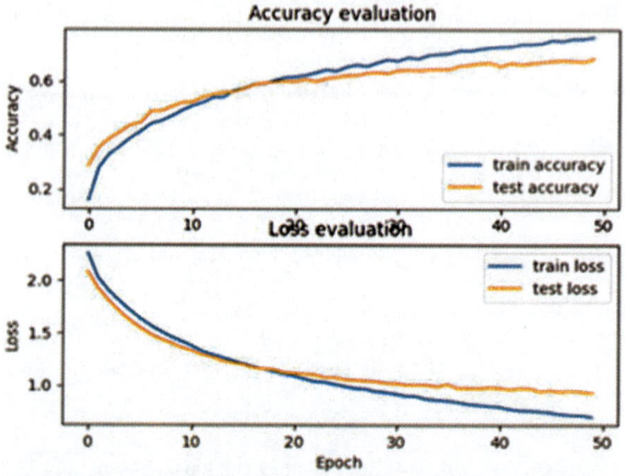

Fig. 12 GRU (RNN)

Fig. 13 CatBoost accuracy

	precision	recall	f1-score	support
0	0.915	0.874	0.894	198
1	0.960	0.960	0.960	198
2	0.833	0.888	0.860	197
3	0.895	0.859	0.876	198
4	0.918	0.904	0.910	197
5	0.867	0.919	0.892	198
6	0.930	0.944	0.937	198
7	0.936	0.884	0.909	198
8	0.857	0.909	0.882	198
9	0.852	0.813	0.832	198
accuracy			0.895	1978
macro avg	0.896	0.895	0.895	1978
weighted avg	0.896	0.895	0.895	1978

Fig. 14 XGBoost accuracy

	precision	recall	f1-score	support
0	0.901	0.823	0.860	198
1	0.939	0.939	0.939	198
2	0.784	0.883	0.831	197
3	0.872	0.823	0.847	198
4	0.908	0.898	0.903	197
5	0.882	0.904	0.893	198
6	0.917	0.949	0.933	198
7	0.891	0.864	0.877	198
8	0.886	0.899	0.892	198
9	0.814	0.798	0.806	198
accuracy			0.878	1978
macro avg	0.879	0.878	0.878	1978
weighted avg	0.879	0.878	0.878	1978

Fig. 15 LightGBM accuracy

	precision	recall	f1-score	support
0	0.940	0.936	0.938	700
1	0.940	0.986	0.962	699
2	0.877	0.920	0.898	698
3	0.908	0.888	0.898	699
4	0.920	0.894	0.907	699
5	0.916	0.930	0.923	700
6	0.941	0.954	0.948	700
7	0.956	0.960	0.958	700
8	0.892	0.886	0.889	700
9	0.915	0.851	0.882	698
accuracy			0.920	6993
macro avg	0.920	0.920	0.920	6993
weighted avg	0.920	0.920	0.920	6993

Fig. 16 GBM accuracy

	precision	recall	f1-score	support
0	0.813	0.768	0.790	198
1	0.948	0.929	0.939	198
2	0.690	0.746	0.717	197
3	0.718	0.707	0.712	198
4	0.779	0.716	0.746	197
5	0.805	0.854	0.828	198
6	0.871	0.889	0.880	198
7	0.843	0.843	0.843	198
8	0.722	0.788	0.754	198
9	0.731	0.672	0.700	198
accuracy			0.791	1978
macro avg	0.792	0.791	0.791	1978
weighted avg	0.792	0.791	0.791	1978

Fig. 17 AdaBoost accuracy

	precision	recall	f1-score	support
0	0.338	0.475	0.395	198
1	0.595	0.934	0.727	198
2	0.295	0.132	0.182	197
3	0.361	0.414	0.386	198
4	0.344	0.228	0.274	197
5	0.578	0.338	0.427	198
6	0.575	0.874	0.693	198
7	0.453	0.788	0.576	198
8	0.317	0.167	0.219	198
9	0.385	0.152	0.217	198
accuracy			0.450	1978
macro avg	0.424	0.450	0.410	1978
weighted avg	0.424	0.450	0.410	1978

Fig. 18 Random forest accuracy

	precision	recall	f1-score	support
0	0.869	0.838	0.853	198
1	0.936	0.965	0.950	198
2	0.769	0.843	0.804	197
3	0.865	0.838	0.851	198
4	0.928	0.853	0.889	197
5	0.852	0.904	0.877	198
6	0.893	0.929	0.911	198
7	0.891	0.864	0.877	198
8	0.841	0.884	0.862	198
9	0.843	0.758	0.798	198
accuracy			0.868	1978
macro avg	0.869	0.868	0.867	1978
weighted avg	0.869	0.868	0.867	1978

Fig. 19 Logistic regression accuracy

	precision	recall	f1-score	support
0	0.645	0.742	0.690	198
1	0.944	0.929	0.936	198
2	0.599	0.538	0.567	197
3	0.591	0.621	0.606	198
4	0.669	0.543	0.599	197
5	0.723	0.843	0.779	198
6	0.774	0.848	0.810	198
7	0.818	0.793	0.805	198
8	0.608	0.641	0.624	198
9	0.491	0.399	0.440	198
accuracy			0.690	1978
macro avg	0.686	0.690	0.686	1978
weighted avg	0.686	0.690	0.686	1978

Table 1 Comparative analysis

Model	Accuracy
ANN	58%
CNN	70%
LSTM(RNN)	64%
CatBoost	89.50%
GRU(RNN)	66%
XGBoost	87.8
LightGBM	92%
GBM	79.1
AdaBoost	45%
Random forest	86.8
Logistic regression	69

References

1. https://www.kaggle.com/andradaolteanu/gtzan-dataset-music-genre-classification
2. Tzanetakis G, Cook P (2002) Musical genre classification of audio signals. IEEE Trans Speech Audio Process 10:293–302. https://doi.org/10.1109/TSA.2002.800560
3. Li T, Ogihara M, Li Q (2003) A comparative study on content-based music genre classification, pp 282–289. https://doi.org/10.1145/860435.860487
4. Cataltepe Z, Yaslan Y, Sonmez A (2007) Music genre classification using MIDI and audio features. EURASIP J Adv Signal Process. https://doi.org/10.1155/2007/36409
5. Elbir A, Bilal Çam H, Emre Iyican M, Öztürk B, Aydin N (2018) Music genre classification and recommendation by using machine learning techniques. In: 2018 innovations in intelligent systems and applications conference (ASYU), 2018, pp 1–5. https://doi.org/10.1109/ASYU.2018.8554016.

6. Silla C, Koerich A, Kaestner C (2008) A machine learning approach to automatic music genre classification. J Braz Comp Soc 14:7–18. https://doi.org/10.1007/BF03192561
7. Lidy T, Rauber A (2005) Evaluation of feature extractors and psycho-acoustic transformations for music genre classification, pp 34–41
8. Viswanathan A, Sundaraj S (2015) Music genre classification. Int J Eng Comput Sci. https://doi.org/10.18535/IJECS/v4i10.38
9. Li T, Ogihara M (2005) Music genre classification with taxonomy. In: Proceedings. (ICASSP '05) IEEE international conference on acoustics, speech, and signal processing, 2005, vol 5, pp v/197–v/200. https://doi.org/10.1109/ICASSP.2005.1416274
10. Liu C, Feng L, Liu G, Wang H, Liu S (2021) Bottom-up broadcast neural network for music genre classification. Multimedia Tools Appl 80:1–19. https://doi.org/10.1007/s11042-020-09643-6
11. Pelchat N, Gelowitz CM (2020) Neural network music genre classification. Can J Electr Comput Eng 43(3):170–173, Summer 2020. https://doi.org/10.1109/CJECE.2020.2970144
12. Ghildiyal A, Singh K, Sharma S (2020) Music genre classification using machine learning. In: 2020 4th international conference on electronics, communication and aerospace technology (ICECA)
13. Upadhyay U, Kumar S, Dubey P, Singh A, Geetanjali M (2021) Music genre classification. Int J Innov Res Technol 8(1):725–728

HRPro: A Machine Learning Approach for Recruitment Process Automation

Atri Biswas, Shourjya Chakraborty, Debarshi Deb, and Rajdeep Chatterjee

Abstract The interview and hiring process for an average IT applicant clocks in at about 33 days. Currently using the outdated man-driven SOP, companies, and candidates lose out on precious time and resources which could be utilized for critical operations and tasks. A contemporary application for the automation of base-level interviews has been developed using open-source machine learning (ML) libraries that intensively investigate the candidate's speaking patterns and facial orientation to generalize a report of the candidate. A general score and rating have been generated that accurately reflects the level of proficiency of the interviewee in terms of soft skills and body language. This application is named HRPro. The application can successfully conduct a short 5 min interview and generate a post-processing report. The report is ideally custom mailed manually to the applicant at the end of the processing time. Information consists of emotion analysis, interview points over time graph, and facial orientation over time graph. Interview points are assigned against a log of 100 separate good and objectively bad interviews used to train the model using pseudo ML algorithms.

Keywords Automation · Computer vision · Machine learning · Interview · Recruitment process · Speech recognition

1 Introduction

Machine learning is commonly used to describe a branch of computer sciences that enables a computer to learn patterns in data without explicit instructions. The modeling of learning multiple processes in their multiple manifestations is the primary subject matter of ML. ML application ranges in this age from credit card processing to video-game AI and everything in between. The basis behind the project is the extended time lag between interviewee application and processing by the

A. Biswas · S. Chakraborty · D. Deb · R. Chatterjee (✉)
School of Computer Engineering, KIIT Deemed to be University, Bhubaneshwar, Odisha 751024, India
e-mail: rajdeepfcs@kiit.ac.in

V. Skala et al. (eds.), *Machine Intelligence and Data Science Applications*, Lecture Notes on Data Engineering and Communications Technologies 132,
https://doi.org/10.1007/978-981-19-2347-0_58

company resources of the same. It is a known statistic that on average, around 250 applications are accepted for a corporate job, and of those, only 4–6 get called for an interview [1], and one gets selected from that bunch recruiters often take less than 6 s, to scan a resume, which might give an unfair or inefficient understanding of the applicant's skill-set or experience. The average time to finish the hiring process for a company is 39 days [2], which is outrageously slow and painstakingly inadequate for keeping up with the ever fast-growing economy [3]. The project aims to look for ways to automate the initial interview process in which the most number of applicants appear, thus cutting down on time required to scan and screen the applicants substantially. It is also a sure-fire way to set objective requirements for the position by obtaining the basic minimum verbiage of the applicant, gaining the analytic regarding the emotional state of the applicant, and the orientation and posture of the applicant. Using HRPro, the company benefits time-wise from an array of painstakingly long processes and gains objective statistics regarding the skill-set of the interviewee. He/She might use the information presented in bullet format for further questioning, reducing the time for future in-person interviews and gauging the applicant's proficiency in the English language. The average amount of time for a job-seeker to find an appropriate job is 12 weeks [4, 5]. For an applicant, this project is the perfect place to train their skill-set from the point of view of English verbiage and basic interview etiquette. This module trains the applicant to sharpen their choice of words during an interview and corrects bad posture. It also is the perfect application for increasing their neutrality when it comes to emotional speech, pointing out peaks, and valleys in their choice of emotional words to a high amount of accuracy [3, 6, 7].

Here, the novelty of the work is a multi-modal engineering solution to automate the recruitment process. The current prototype has been developed to analyze and report only based on the HR round of interviews (non-technical). The proposed bot asks relevant questions related to the HR interview, such as "What are your strengths and weaknesses?," "how do you define success?," "Are you willing to relocate?," "Why are your grades not higher?" so on and so forth. In return, the subject (interviewee) replied within a predefined time range and continue this conversation with the HRPro bot. During this interaction, machine learning modules such as speech to text, grammar checker, facial expression, and emotions are computed to generate a consolidated report.

The paper has been divided into six Sections. The related work has been discussed in Sect. 2. The background concepts, including different open-source libraries, are described in 3. The subsequent section tells us about the proposed framework. The results and analysis are explained in 5. Finally, the paper is concluded in 6.

2 Related Works

Related works in the field include a study of bias in multi-modal AI [8, 9]. Multi-modal refers to the process of obtaining data from multiple heterogeneous sources of information. The study speaks of the significant shifts into a society where everything from social media giants to corporate businesses uses raw data and processes them using machine learning algorithms to create more thoughtful options for users and more excellent opportunities for companies to grow faster and faster. Inherently, a structured algorithm built on societal conditioning will inherit some bias, as society itself is a tiny creature. To be more precise, artificial intelligence has been a growing force in auto recruitment drives, where candidates are screened automatically by an AI model based on previous recruitment data. In this model, computer vision and natural language processing boost algorithmic capabilities beyond the usual standards [10–13]. Further, a new model called FairCV is trained and tested.

In this particular field, there is a vacuum in the sense that there has not been an attempt to automate soft skill interviews directly with a working user friendly interface. Our project fills this void aptly with an application that is working with high accuracy.

3 Background Concepts

The very basis of the project is artificial intelligence and machine learning. The basic model of the project is based on Artificial intelligence and pseudo ML. We have used a number of machine learning libraries such as gTTs [14], NLTK [11, 15], OpenCV [16], and more to help us achieve this goal (Figs. 1 and 2).

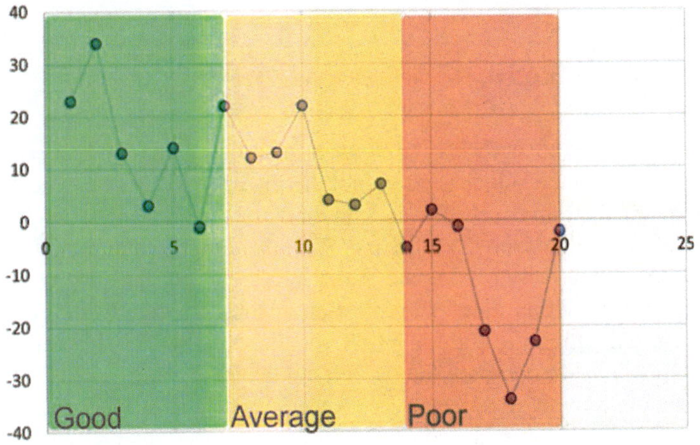

Fig. 1 Validation testing of the application

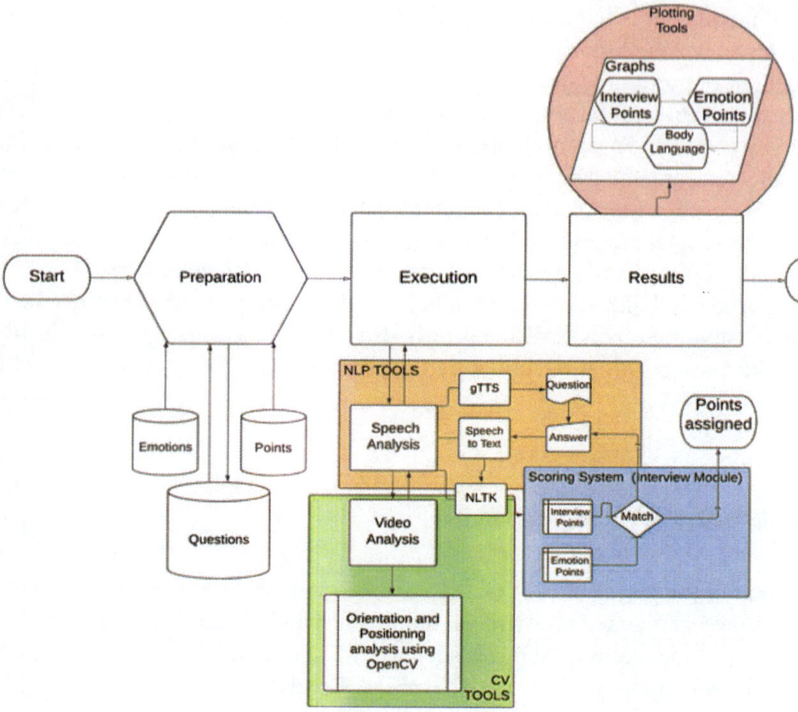

Fig. 2 Workflow of the HRPro application

4 Proposed Method

The first objective of the module is to create a functioning chatbot that can smoothly "communicate" with the subject to conduct the interview. It has been achieved with the help of the gTTS framework. The second fundamental element is the video capture achieved with the help of the CV2 library. With the assistance of the gTTS module, we are able to interactively participate in the interview process, further blurring the line between machine and man. The questions which are core to the interview are hard coded into the module, and the gTTS vocalized them. The third element is the data frame to which we compare the speech made by the interviewee. Extraction of the data frame is done by iterating over a set of good (and bad) speeches that have been imported into a .txt file; the source of these speeches is an internet article on the college-grad Web site, where tough interview questions are provided along with an amateurish right answer, experienced correct answer and blunderous wrong answer. All these are being fed to the algorithm that counts each word used in each speech and assigns positive points for every time they appear on the good speeches and negative points for every time they appear on the bad speeches. One hundred fifty such speeches are, thus being given to the code in a .txt file. The word point list is saves them locally as a .csv file. This .csv file is, hence loaded into the program during

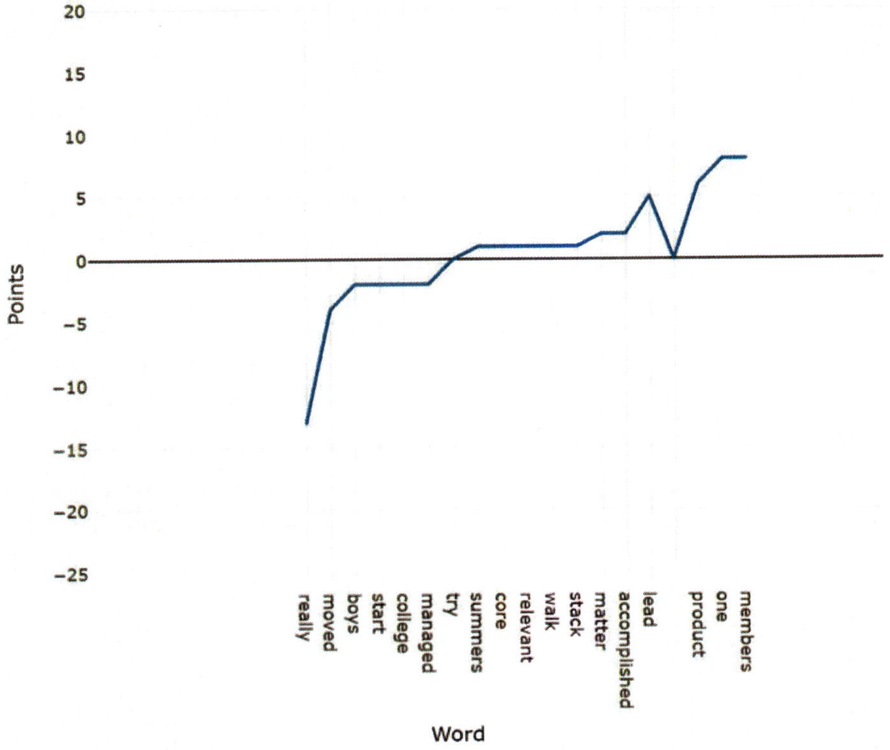

Fig. 3 Word to point mapping of select ten words

run-time as a data frame and acted upon. Similarly, a data-set of emotion-based word classification is downloaded for free from Kaggle, and a similar algorithm is applied to find words and classify them into the six main emotional landscapes (Fig. 3).

Moreover, a facial recognition module is used to study and locate the local coordinates of the face and its facial features. The orientation part of the module is achieved by calculating a mean of the facial features, namely eyes, nose, and two points of the mouth, with relation to the average location of the face, which in turn produces a cuboid structure that can display the relative orientation of the face. We can use this information to estimate the posture of the applicant correctly and use it to assign points to the same to show an average of his/her body posture points.

5 Results and Analysis

In the ongoing project that is HRPro, we are trying to include several other modules that include facial recognition and CV analysis. The video orientation mapping module, with the help of traditional image processing, has a confidence level of

92%, the emotion graphing module currently under development has a confidence level of 75%, and the interview point mapping, when tested with a variety of test cases achieved a level of 82% accuracy.

Firstly, the future application asks the user to upload their resume, which undergoes thorough checking for requirements and relevant skills that fall under the domain of the recruiter. After the validation and scoring of the resume, the workflow goes to the interview module of the application, where the current working model comes into play. The interviewee is scored and graded with the help of natural language processing and speech to text libraries. A custom-built grading system has been developed to assign a score to each interviewee. The detailed mechanism to compute the score is explained in the subsequent sub-section. After the chatbot asks a question, the applicant is asked to answer as precisely as possible, then after scoring and video ID validation, the results are either shown to the applicant or forwarded to the hiring organization for further review (Fig. 4).

Coming back to the current working model, the test mentioned above via the HRPro application gives us a few diagnostics to look over. They include the interview face deviation, the interview points, and the interview emotion analysis (Fig. 5).

5.1 *Interview Points*

The interview point analysis is the crutch of the HRPro application. With a data frame of over 2000 words and phrases lined up to measure against the speech made by the applicant or the candidate, it is quickly deducing the efficiency of the candidate while making the speech relative to the sample space of over 50 good and 50 bad speeches. Individual words are detected and tokenized using various natural language processing libraries such as NLTK. These words are compared one by one to those uttered by the candidate, and accordingly, specific points are assigned to them throughout the whole speech. This iteration occurs over the whole speech and produces a list of numerical points ready to be plotted. The points are, then passed on to the plot() function where they are plotted against time to give the interviewee some idea of when and where the interview took off or went down, giving them some idea regarding how they should model or remodel the speech to provide more clarity or create more substance to back up their speech (Fig. 6).

5.2 *Face Deviation*

This is a section of the result which has been obtained by using OpenCV. cv2.CascadeClassifier function has been used to first create a box around the applicant's face as visible through the webcam. Then, another box is created using the estimated positioning of the eyes, nose, mouth ends, giving us two boxes and an approximate relative positioning of the various facial features with relation to the

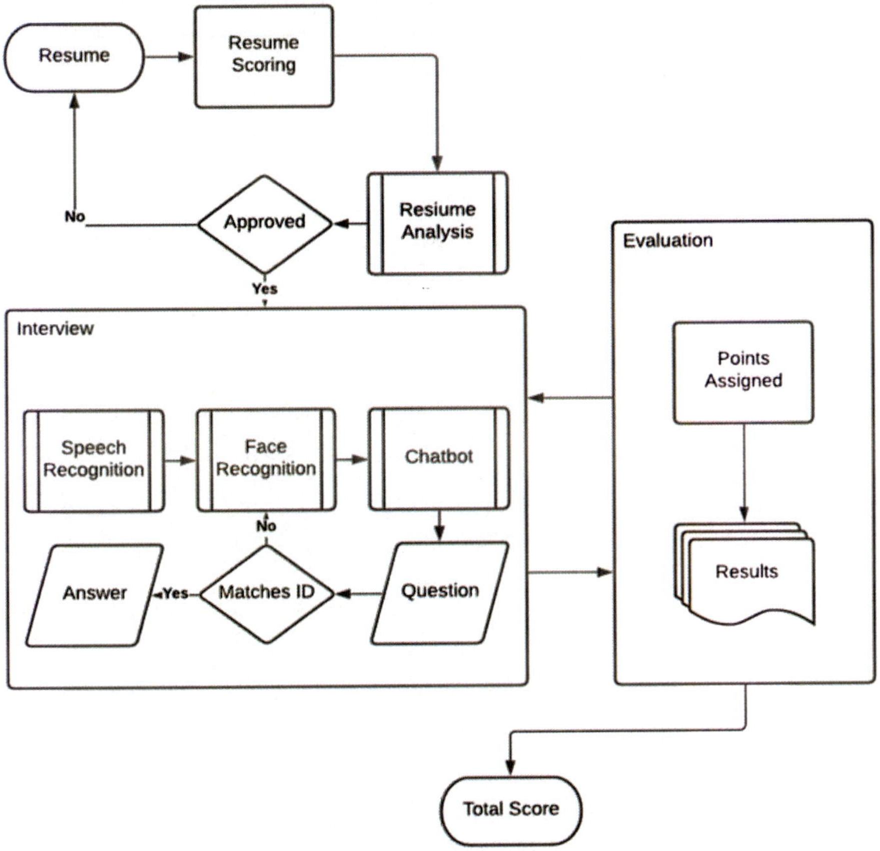

Fig. 4 Extended project prospects for HRPro. The diagram explains in detail the functioning of individual modules that are currently under development

entire face [17, 18]. This is what is used as the orientation of the face. When the applicant turns right, left, up, or down, the orientation of the face deviated from the center. Additionally, a variable has been added to measure the positioning of the face as well. These two variables have given us a good understanding of how fidgety or distracted the applicant is during the tenure of the interview. These variables have been taken and plotted them together in the same graph for the user to understand if he/she is positioned correctly and has good posture during the interview (Figs. 7 and 8).

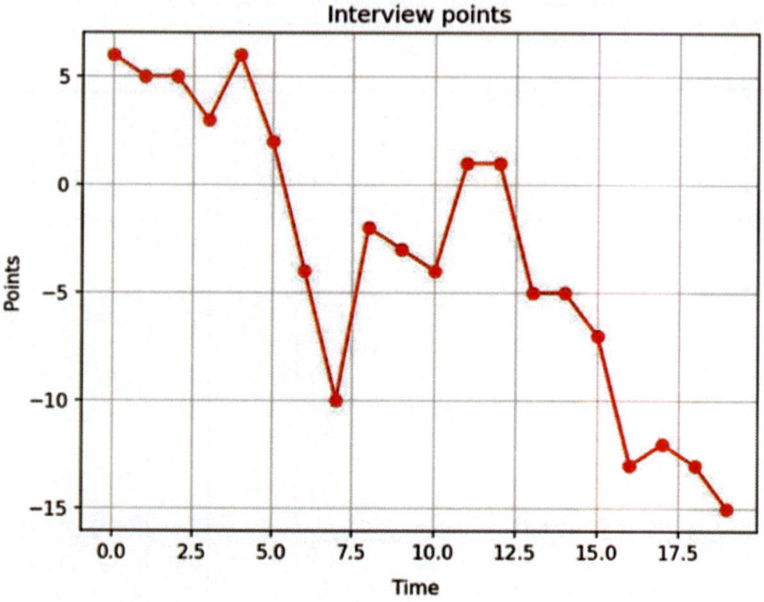

Fig. 5 Interview points over time

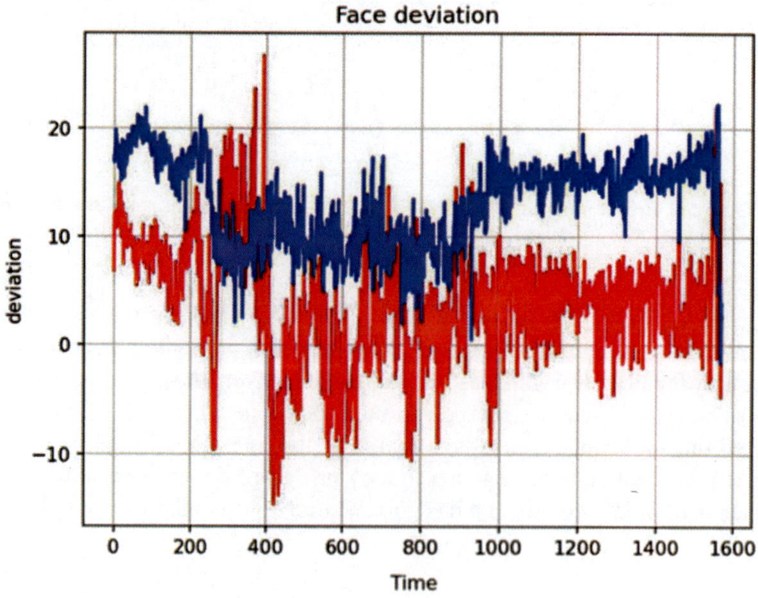

Fig. 6 Face deviation over time

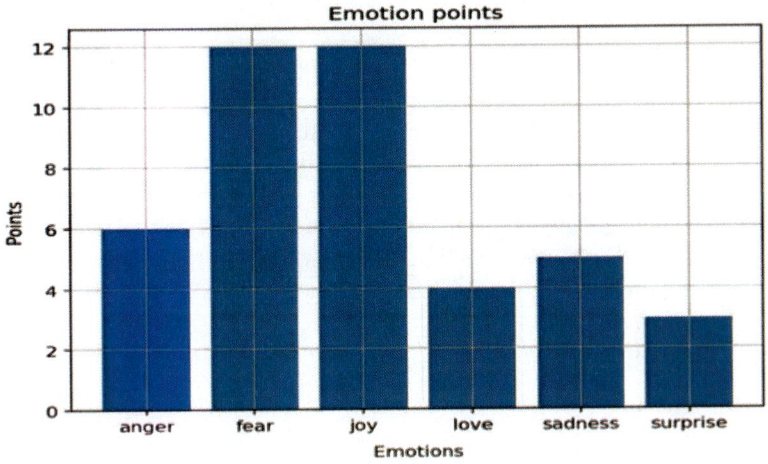

Fig. 7 Emotional analysis

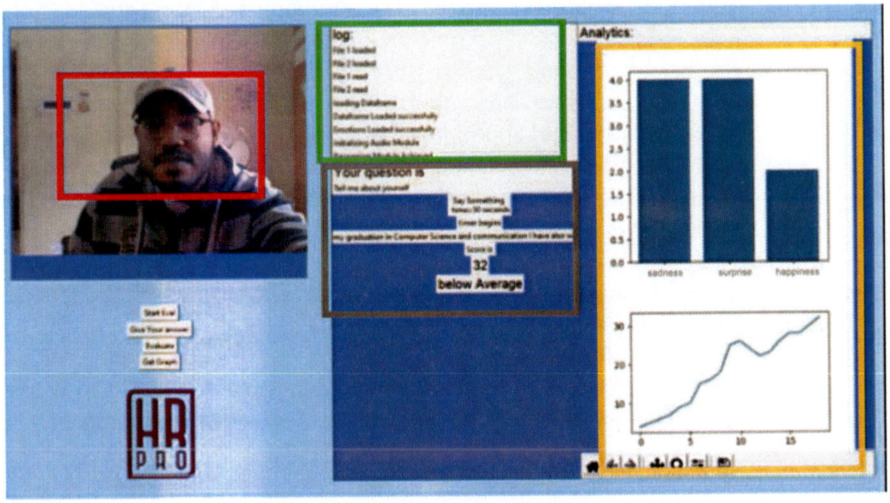

Fig. 8 GUI of the application. The red box is the video input taken by the application, after which the facial orientation function deduces the deviation of the face from the center of the screen. The green box is the log where the errors (if any) and the loading are printed. The gray box is the question, answer, and points for the respective answer is printed. We can see the various diagnostics in the yellow box, including the emotion detection, interview points, and orientation and facial positioning graphs

5.3 Emotion Points

This module is fairly simple, a simple pseudo ML technique is used to assign points to the words in the speech made by the candidate, with reference to a data frame of "words to emotion" points. Similar to interview points, the words are assigned certain points with regards to their standing in the data frame. Then, a dictionary of "emotion to points" is created with individual emotions carrying certain weight that can be represented using a bar graph. The same is done for the emotion-point graph [19, 20]. The link to github repository is provided.[1]

6 Conclusion

In conclusion, the standing model of HRPro is functioning and fairly unique. However, there are a few current limitations that we face and are working to neutralize over time. The most noticeable setbacks are the time required to process and generate the reports for analysis. The standing time frame for a 10 min interview comes close to 30 min of post-processing, primarily because of video analytic tools, which are considerably heavy and time consuming. The image processing module takes upwards of 20 min to analyze and approximate the facial deviation of the applicant.

However, we are able to understand and create a multi-modal chatbot that understands, analyzes, and reports the interviewee's abilities or the candidate's capacity to give a good interview. This model can be used as a training module or as a complete-fledged onboarding tool in the future. With the emerging need for an automated recruitment system, as companies need faster and more efficient screening methods, HRPro holds the industry standard for quick results.

As for future potential for this project, it is planned that we implement a certain resume analysis tool including social media background checker and add to the already existing module to further streamline the workflow of the application.

References

1. Turczynski B (2020) Hr statistics: job search, hiring, recruiting & interviews [electronic resource]/turczynski b–2020. https://zety.com/blog/hr-statistics. Accessed 15 Aug 2021
2. Melián-González S, Bulchand-Gidumal J (2017) Why online reviews matter for employer brand: evidence from glassdoor
3. Yakubovich V, Lup D (2006) Stages of the recruitment process and the referrer's performance effect. Organ Sci 17(6):710–723
4. Armstrong M, Taylor S (2020) Armstrong's handbook of human resource management practice
5. Storey J (2016) Human resource management. Edward Elgar Publishing Limited
6. Fry R (2018) 101 great answers to the toughest interview questions. Open Road Media

[1] github repo: https://github.com/atribiswas/HRpro-repository

7. Berman EM, Bowman JS, West JP, Van Wart MR (2019) Human resource management in public service: paradoxes, processes, and problems. CQ Press
8. Ferrer X, van Nuenen T, Such JM, Coté M, Criado N (2021) Bias and discrimination in AI: a cross-disciplinary perspective. IEEE Technol Soc Mag 40(2):72–80
9. Cukurova M, Kent C, Luckin R (2019) Artificial intelligence and multimodal data in the service of human decision-making: a case study in debate tutoring. Br J Edu Technol 50(6):3032–3046
10. Deng L, Liu Y (2018) Deep learning in natural language processing. Springer
11. Srinivasa-Desikan B (2018) Natural language processing and computational linguistics: a practical guide to text analysis with Python, Gensim, spaCy, and Keras. Packt Publishing Ltd
12. Li S, Deng W (2020) Deep facial expression recognition: a survey. IEEE Trans Affective Comput
13. Chen CH (2015) Handbook of pattern recognition and computer vision. World Sci
14. Durette PN and Contributors. gtts (google text-to-speech), a python library and cli tool to interface with google translate's text-to-speech api. https://github.com/pndurette/gTTS, May 2021
15. Nltk: Natural language toolkit. https://www.nltk.org/, May 2021
16. Kaehler A, Bradski G (2016) Learning OpenCV 3: computer vision in C++ with the OpenCV library. O'Reilly Media, Inc.
17. Chatterjee R (2020) Deep ensemble learning-based smart teaching
18. Halder R, Chatterjee R, Sanyal DK, Mallick PK (2020) Deep learning-based smart attendance monitoring system. In: Proceedings of the global AI congress 2019. Springer, pp 101–115
19. Chatterjee R, Mazumdar S, Sherratt RS, Halder R, Maitra T, Giri D (2021) Real-time speech emotion analysis for smart home assistants. IEEE Trans Consum Electron 67(1):68–76
20. Yu D, Deng L (2016) Automatic speech recognition. Springer

A Deep Convolutional Generative Adversarial Network-Based Model to Analyze Histopathological Breast Cancer Images

Tanzina Akter Tani, Mir Moynuddin Ahmed Shibly, and Shamim Ripon

Abstract Breast cancer is one of the severe cancers, and early detection is needed to remedy the severity. With the advancement of AI technology, machine learning and deep learning are performing a very important role by automatically finding the types of tumors. A machine learning model needs many data to learn the features of the dataset more precisely. Data scarcity can lead to weak accuracy and can introduce bias toward the higher class. In this study, the BreakHis dataset has been used which contains histopathological images of breast tissues of different magnification factors. The datasets have two classes, but the distribution of images is quite imbalanced. The dataset is balanced by adding generated images to the minority class by using a deep convolutional generative adversarial network (DCGAN). For the classification, four pre-trained deep convolutional neural networks (deep CNN), namely DenseNet, MobileNet, ResNetV2 and Xception, have been applied. After applying DCGAN, the performance has been improved with a maximum increase of 2.92%. As for the classification, the DenseNet model with DCGAN has given the overall top performance in this study. Moreover, the result of this study has outperformed most of the state-of-the-art works in identifying malignancy of breast tissue from histopathological images.

Keywords Breast cancer · Deep CNN · BreakHis dataset · GAN · DCGAN

1 Introduction

Breast cancer is considered as one of the severe cancers, and according to [19], it is the fifth leading cause of mortality in cancer, affecting over 2.3 million women worldwide. A medical test on breast tissue is required to determine the type of breast cancer. The biopsy test is performed where tissue is observed under the microscope and the pathologist examines that the tissue is either benign or malignant lesion [2, 10]. The benign lesion is considered as non-cancerous tumor, whereas the malignant

T. A. Tani · M. M. A. Shibly · S. Ripon (✉)
Department of Computer Science and Engineering, East West University, Dhaka, Bangladesh
e-mail: dshr@ewubd.edu

V. Skala et al. (eds.), *Machine Intelligence and Data Science Applications*, Lecture Notes on Data Engineering and Communications Technologies 132,
https://doi.org/10.1007/978-981-19-2347-0_59

can be invasive cancer that spreads uncontrollably. If the detection of tissue comes out wrong for an actual malignant patient, it can be a life-threatening situation. Therefore, accurate detection is necessary for proper treatment. Different properties like pattern and texture in the tissue images are helpful to the pathologists to perform diagnosis [1, 12]. But the manual process is very time consuming as analyzing images requires focusing and scanning the images, and these can lead to inaccurate diagnosis if the pathologists are not very expert [1, 6]. However, with the advancement of AI, machine learning and deep learning are utilized in various automatic detection processes of breast cancer which is less time-consuming and more reliable than manual expertise [22]. Deep learning can learn very hidden features and correlations of those features in images which is capable of providing accurate classification results [9, 17]. Therefore, this study aims at developing deep learning-based models to identify malignancy of breast tissue from histopathological images.

A machine requires a large number of input data to learn the features of data more precisely. In the medical field, it is difficult to obtain data because of the requirement of the patient's consent. As a result, in most cases, the breast cancer image datasets have a very limited number of samples, or the samples are mostly imbalanced [22]. But it can lead to weak accuracy and overfitting. Many systematic methods, such as data augmentation and/or the ensemble approach, can be used to deal with this problem. But producing new synthetic images by the generative adversarial network (GAN) can bring many complex modifications of images [16]. Since its inception in 2014 [7], many researchers have applied different types of GANs to improve classification performance. However, according to [25], in medical field, there are only 9% of GANs-related publications on histopathology images and there are only 8% of GANs related publications on classification study on or before January 1, 2019. As GANs offer the flexibility of data distribution on the whole sample [4], there is a scope for applying GANs in histopathology images for classification related tasks. In this study, a deep convolutional generative adversarial network (DCGAN) [14] is proposed to generate histopathological breast cancer images to improve the deep learning-based classification model outcome.

This study aims at improving the identification of malignancy from histopathological breast cancer images. A DCGAN model has been proposed to balance the classes of breast cancer images. In addition, various deep learning models have been applied for classification. Finally, a comparative analysis of the related study has been conducted with the proposed approaches.

2 Literature Review

In [13], CNN, LSTM and a combination of CNN and LSTM have been proposed on the BreakHis [18] dataset where the CNN model has outperformed the other two methods. SVM and softmax layers have been used for the classification after the features are extracted from DNN models. The best outcome of this study is 91% accuracy on the 200× dataset and 96% precision on the 40× dataset. Transfer learning and the deep feature extraction process have been used in [5] on the BreakHis dataset. AlexNet and Vgg16 networks have been utilized to extract features from a fivefold organized dataset, and then, classification is done with the aid of SVM. As a transfer learning technique, a pre-trained model AlexNet is developed for further fine-tuning. The transfer learning approach outperformed the deep feature extraction and SVM classification. The best obtained accuracy is 93.57%. Another histopathological image classification has been done in [3] by using three different machine learning classifiers, namely logistic regression (LR), SVM and KNN. To improve the performance, four pre-trained CNN models, ResNet50, InceptionV3, InceptionResNetV2 and Xception, have been applied for feature extractions. 96.24% accuracy has been obtained from the combination of ResnNet50 and LR. Alom et. al. [1] proposed an inception recurrent residual convolutional neural network model to improve the classification performance of the BreakHis and breast cancer classification challenge 2015 datasets. The proposed techniques attained a maximum accuracy of 97.95% in image-level classification. To classify pathological images, [23] has presented a deep CNN model called the BiCNN model. To increase classification accuracy, an advanced data augmentation approach and a transfer learning method are used and 97% accuracy has been achieved.

GAN as a data augmentation technique has been proposed in many biomedical experiments, such as progressive growing GAN on CT images [4], DCGAN in chest X-ray images [21] and white blood cell images [8], deep GAN on mammography dataset [17] and many more. To tackle the imbalanced class distribution problem, [15] proposed the DCGAN model as a data augmentation approach. The DCGAN is used to balance the dataset by generating synthetic benign images. Although the DCGAN failed to generate high-quality images, the proposed work has obtained the highest test accuracy of 96.5% and outperformed other state-of-the-art deep networks such as bag of visual words, CNN and few others. Man et al. [11] presented DenseNet121-AnokoGan to classify the histopathological BreakHis dataset, which comprises of two parts: screening mislabeled patches with the GAN model AnoGan and extracting multilayered features of discriminative patches with DenseNet. The proposed method has been evaluated by fivefold cross-validation. The best result of the study is 99.13% accuracy and 99.38% F_1 score for the 40× factor on the image level. Other models, such as AlexNet, VGG19, VGG16 and ResNet50, have also been explored, both with and without AnoGan. Deep learning, transfer learning and GAN have been combined in [20] to increase the classification efficiency on the histopathological BreakHis dataset. To generate synthetic images, Pix2Pix and StyleGAN have been used, but the resulting images produced a lot of noises, which affected the classification accuracy. Before applying CNN for classification, fine-tuning on VGG16 and VGG19 has been utilized for feature extraction, which greatly improved the performance. A conditional GAN model named HistoGAN has been developed by Xue et al. [24] which is evaluated by using two different datasets: cervical histopathology and a lymph node histopathology dataset. The experiment has shown a substantial increase in classification accuracy.

The literature review reveals that there is room for improvement in the DCGAN generated images. As authors in [15] were unable to produce high-quality benign samples, an approach with a new DCGAN model has been proposed in this paper to improve the realism of synthetic data which can contribute to the performance of classifiers.

3 Materials and Methods

A series of steps have been taken to analyze the histopathological breast cancer images. Figure 1 illustrates a schematic representation of these steps.

3.1 Dataset

The histopathological BreakHis dataset proposed in [18] has been used for the experiments of this study. There are 7909 images in this dataset taken from 82 patients. The dataset is split into two categories: benign and malignant, with eight sub-cancer kinds and contains four magnification factors (MF): 40×, 100×, 200× and 400×. Table 1 illustrates the distribution of categories of data of the BreakHis dataset. A sample of images from each MF is illustrated in Fig. 2.

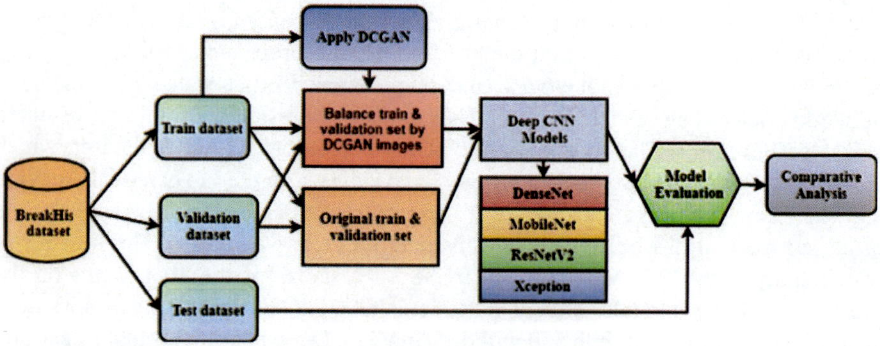

Fig. 1 Schematic diagram of the proposed methodology

Table 1 Dataset overview

Magnification factors	Benign	Malignant	Total
40×	625	1370	1995
100×	644	1437	2081
200×	623	1390	2013
400×	588	1232	1820
Total	2480	5429	7909

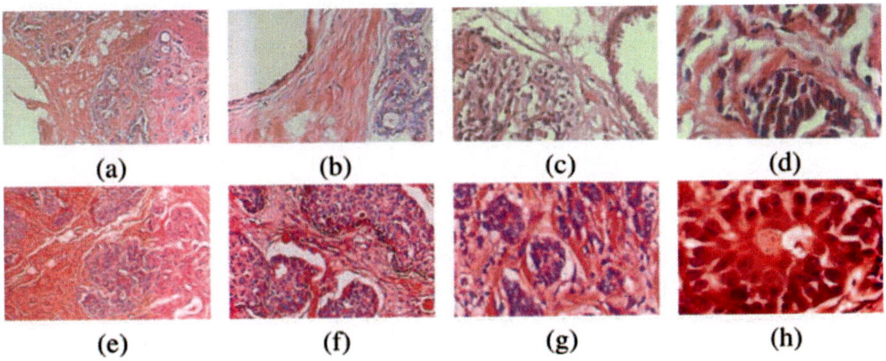

Fig. 2 Dataset overview

3.2 *Generative Adversarial Network*

The generative adversarial network (GAN), introduced in [7], is a method that consists of two models, the generator and the discriminator. The generator (G) learns the data distribution P_{data} from the sample x and creates new samples updating the input noise components z. Discriminator (D) distinguishes between real and fake images and gives a probability value. The fight between discriminator and generator continuous by following min-max strategy:

$$\min_G \max_D V(D, G) = \mathbb{E}_{x \sim p_{\text{data}}(x)}[\log D(x)] + \mathbb{E}_{z \sim p_z(z)}[\log(1 - D(G(z)))] \quad (1)$$

Later, [14] introduced a GAN version called DCGAN, which uses deep convolutional neural networks to build the discriminator and generator following the same concepts of GAN. In this study, a DCGAN approach has been adopted for creating synthetic images to balance the breast cancer dataset.

DCGAN Method

Preprocessing: Before the DCGAN model is trained, the images have been preprocessed. In order to avoid generating the same images from the original dataset, only the training data has been used in the DCGAN model to generate the benign images. The benign class has a very small dataset in each MF. DCGAN also needs more data for training so that it can produce more synthetic images. The training images are augmented three times with a maximum 60° rotation. After that, the augmented and train data are resized in 128 × 128 pixels. All the images are then normalized to the range of −1 and +1, which is the range of the Tanh activation function. The DCGAN model is run separately for each of the MFs.

Generator Architecture: The generator model takes a 100-size random noise vector from the uniform distribution as an input and passes it to the fully connected dense layer to reshape the size of 8 × 8 × 512. After that, to generate the 128 × 128 × 3 size of a benign image, four 2D convolutional layers (256, 128, 64 and 32 filters,

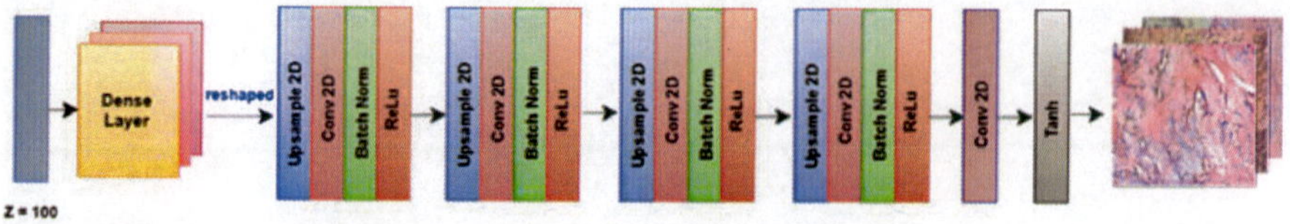

Fig. 3 Generator architecture

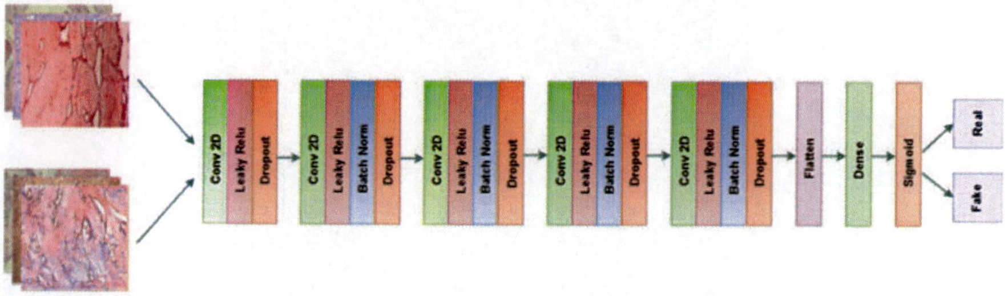

Fig. 4 Discriminator architecture

respectively) with the following upsample are given to the network. Batch normalization is used in every layer except the output layer which normalizes the input to have zero mean and unit variance in order to keep the learning process stable [14]. Also, the ReLu activation function has been used in every layer except the output layer, where the Tanh activation function has been used. Another 2D convolutional layer is used before the Tanh function. A 3×3 kernel size is used in each convolutional layer. Figure 3 shows the architecture of the model.

Discriminator Architecture: The discriminator architecture is shown in Fig. 4. The original images from the dataset and the generated images from the generator model are given as input to the discriminator. Five convolutional layers with kernel size 3×3 are used in the architecture. Thirty-two filters are given to the initial Conv2D layer. The rest factor of 512 is used in respective Conv2D layers. Leaky ReLU activation function is used in the discriminator model after each convolutional layer as it helps to perform well [14]. Also, batch normalization is used in each layer of the model except the input layer. 25% dropout is used in all layers to keep away the model from over fitting. Finally, the output layer is flattened, and 1 unit of output is used in the dense layer with the sigmoid activation function. This function gives the probability of the image between 0 and 1.

Training Process: To train the DCGAN model, the Adam optimizer with a learning rate of 1.5e−4 and momentum term $\beta_1 = 0.5$ are used in both the discriminator and generator network. As the discriminator model performs the distinction between real and fake, so the binary cross-entropy is used to find both the discriminator and generator losses. The 64-batch size and 2000 epochs are run for each MF of the benign data. The generated images are shown in Fig. 5.

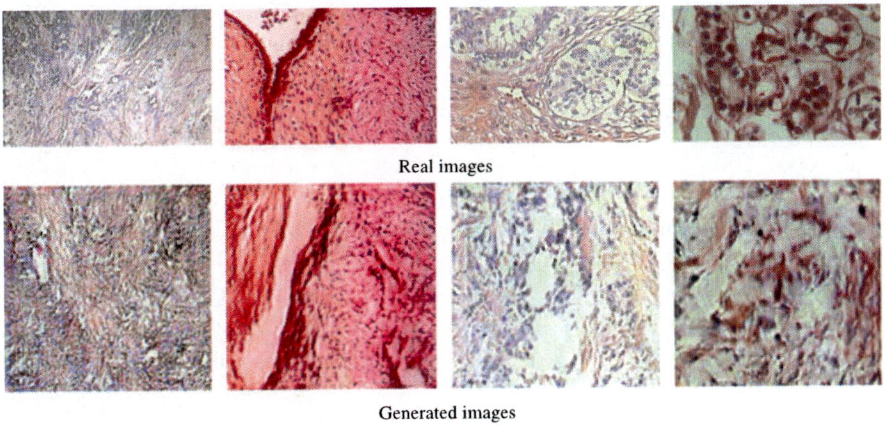

Fig. 5 Real versus generated images

3.3 Classification Models Setup

Four pre-trained deep CNN models, namely DenseNet, MobileNet, ResNetV2 and Xception, have been used for the classification. All the models are trained with the same input shape 128 × 128 × 3. In the output layer of each model, a single neuron with the sigmoid activation function has been applied such that when the neural network receives positive class input it becomes activated; otherwise, it remains deactivated. The positive class, in this case, is malignant images, whereas the negative class is benign images. Before the output layer, a dense layer of 256 has been given to each model. Before feeding the training dataset into the model, the images are rescaled by 255. Stochastic gradient descent (SGD) optimizer with a 0.001 learning rate and 0.9 momentum value has been used in all the experiments. The experiments have been conducted with a batch size of 32 along with binary cross-entropy loss function.

3.4 Train, Test and Validation

The BreakHis dataset has been split into the train, test and validation sets. The ratio of train, test and validation sets is 7:2:1. As the dataset is imbalanced, the DCGAN has been applied to balance the dataset and again performed the classification as a second approach. The DCGAN-generated synthetic images are only added to the train and validation dataset for balancing purposes. But test dataset has remained unchanged so that the model can evaluate more accurately without introducing bias. Keras and TensorFlow have been used for the pre-trained CNN models and DCGAN model. All the experiments are done with the help of Google Colab which facilitates free GPU connection.

4 Result

The test performance of the four pre-trained CNN models with and without DCGAN generated benign images has been evaluated by using the same test data for each MF 40×, 100×, 200× and 400×. All the predictions are calculated from the confusion matrix by taking 'malignant' as a positive class and 'benign' as a negative class. The performance of all the proposed approaches is shown in Table 2.

Table 2 Performances for all four magnification factors

MFs	Approach	Precision (%)	Recall (%)	F_1-Score (%)	Specificity (%)	Accuracy (%)
40×	DenseNet	95.68	97.08	96.38	90.91	95.07
	DenseNet+DCGAN	97.16	**100**	**98.56**	93.60	**97.99**
	MobileNet	98.18	98.54	98.36	96.21	97.78
	MobileNet+DCGAN	**99.26**	97.81	98.53	**98.40**	**97.99**
	ResNetV2	96.06	97.81	96.93	91.67	95.81
	ResNetV2+DCGAN	96.77	98.54	97.65	92.80	96.74
	Xception	95.10	99.27	97.14	89.39	96.06
	Xception+DCGAN	97.13	98.91	98.01	93.60	97.24
100×	**DenseNet**	**98.25**	97.91	98.08	**96.12**	97.36
	DenseNet+DCGAN	97.95	**99.65**	**98.79**	95.35	**98.32**
	MobileNet	95.29	98.61	96.92	89.15	95.67
	MobileNet+DCGAN	97.23	97.91	97.57	93.80	96.63
	ResNetV2	94.90	97.21	96.04	88.37	94.47
	ResNetV2+DCGAN	96.91	98.26	97.58	93.02	96.63
	Xception	96.54	97.21	96.87	92.25	95.67
	Xception+DCGAN	96.59	98.61	97.59	92.25	96.63
200×	**DenseNet**	98.22	**99.28**	**98.75**	96.00	**98.26**
	DenseNet+DCGAN	**99.27**	97.84	98.55	**98.40**	98.01
	MobileNet	98.55	97.84	98.19	96.80	97.52
	MobileNet+DCGAN	98.92	98.56	98.74	97.60	**98.26**
	ResNetV2	96.75	96.40	96.58	92.80	95.29
	ResNetV2+DCGAN	98.51	95.32	96.89	96.80	95.78
	Xception	97.49	97.84	97.67	94.40	96.77
	Xception+DCGAN	98.91	97.84	98.37	97.60	97.77
400×	DenseNet	97.20	98.78	97.98	94.07	97.25
	DenseNet+DCGAN	98.77	97.97	98.37	97.46	97.80
	MobileNet	98.33	95.93	97.12	96.61	96.15
	MobileNet+DCGAN	99.16	95.93	97.52	**98.31**	96.70
	ResNetV2	97.55	97.15	97.35	94.92	96.43
	ResNetV2+DCGAN	97.93	95.93	96.92	95.76	95.88
	Xception	98.37	98.37	98.37	96.61	97.80
	Xception+DCGAN	**99.19**	**100**	**99.60**	**98.31**	**99.45**

'Bold' indicates best results compared to other approaches.

All the approaches for each MF have achieved over 94% test accuracy. The DenseNet with DCGAN has achieved the best performance on average in all MFs. Besides, this approach has attained the best test accuracy for 40× and 100× datasets. MobileNet with DCGAN has also attained the best accuracy for 40× and 200× but could not overcome DenseNet on the average result. Though Xception with DCGAN has given the best accuracy of 99.45% on 400× factor which is the highest performance among all MFs but could not attain better results in other MFs. Among all the methods, ResNetV2 with and without DCGAN has given the worst performance. Although about 2.16% accuracy has improved from only ResNetV2 to ResNetV2 with DCGAN. The DCGAN method has improved most of the performance of CNN models. About 2.92% accuracy has improved after applying DCGAN on DenseNet for 40× MF which is the highest improvement of DCGAN. An improvement of 1.65% accuracy has been shown in the Xception with DCGAN on 400× factor. Precision, recall and specificity are also important to evaluate the models. DenseNet with DCGAN has achieved the highest 99.27% precision for 200× MF and 100% recall for 40× MF. Xception with DCGAN has also achieved a 100% recall score for the 400× MF. Accurately recognizing benign tumors is important which can be evaluated by specificity. The best specificity result is 98.40% in all MF-based datasets. All the specificity result has been improved after adding DCGAN produced benign images except DenseNet for 100× MF.

4.1 Models Performance

All the models have been validated with the validation set to check whether the training of the model has been done properly. The validation accuracy for all MFs is visualized in Fig. 6. DenseNet with and without DCGAN and MobileNet with DCGAN have achieved 100% validation accuracy for the 40× MF. For 100× MF, MobileNet with DCGAN has given 98.61% accuracy. DenseNet with DCGAN shows the highest accuracy of 99.61% and 99.11% for 200× and 400× respectively. All the validation accuracies have been improved with DCGAN added data except DenseNet and Xception for 100× MF. From the figure, it can be visualized that the DCGAN has helped the models to be more stable which enhances the classification performance.

5 Discussion

In this study, four deep CNN models have been used on BreakHis data to classify benign and malignant. The classes are balanced by generating new benign instances by using DCGAN, and this approach has improved most of the classification models and given better results compared to models trained by the imbalanced datasets. The same fine-tuning technique has been kept for both balanced and imbalanced datasets. The models are saved twice: once when the validation accuracy has improved and

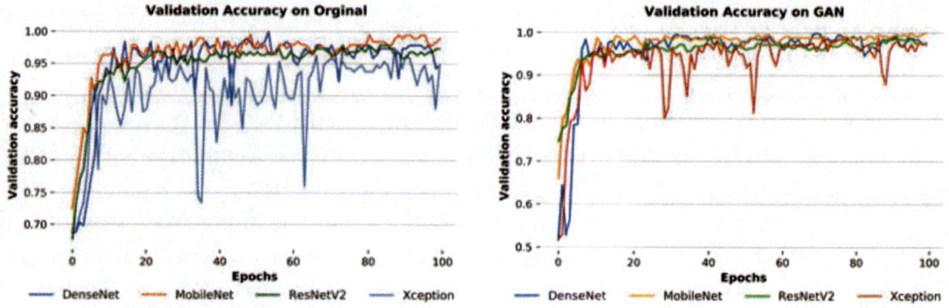

(a)

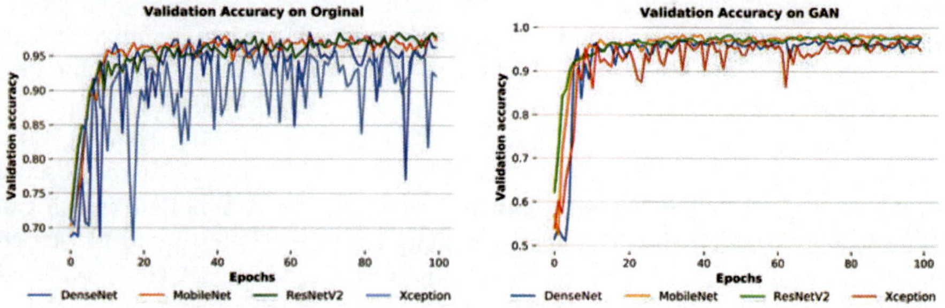

(b)

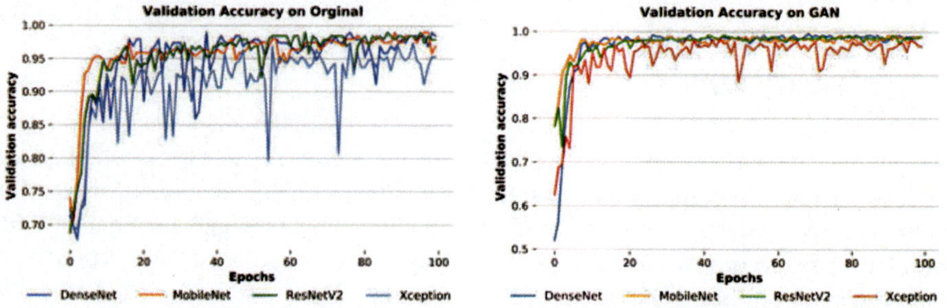

(c)

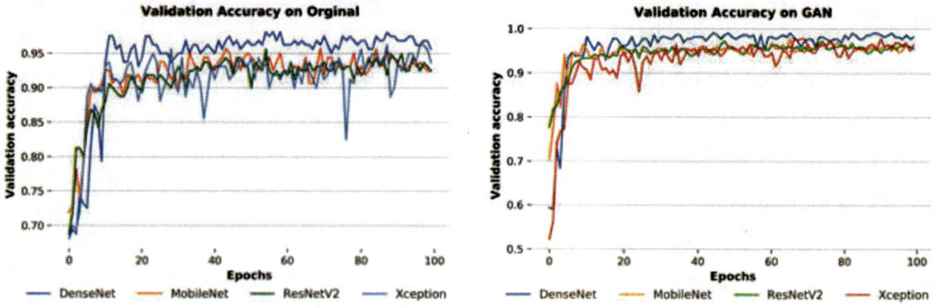

(d)

Fig. 6 Validation accuracy per epochs for **a** 40× **b** 100× **c** 200× **d** 400× MFs: without DCGAN (left) and with DCGAN (right) another after 100 epochs. Only the best model has been considered after tested both saved models on the testset. As we have not evaluated the produced images, the distribution differences between the real and produced images could not be identified. The model has been evaluated by observing the images and classification performances. But as our classification result has improved by the produced images, it can be stated that the proposed DCGAN for balancing the BreakHis dataset has a significant impact on this study of breast cancer detection.

5.1 Comparison with Previous Related Study

The performance of the proposed approaches has outperformed the other studied related work except for [11, 20]. The study of [11] has given the highest score for 40× MF, but for other MFs, our proposed methods have achieved better result. Our study has achieved almost the same result as [20] for 100× and 200× MFs, but for the 400× MF, the present work outperformed their work. Our proposed work also outperformed [15] where DCGAN produced images by them have created more noise and affected the classification performance. In our study, the DCGAN produced images have shown improved performance for most of the classifications. If only one approach is considered, DenseNet with DCGAN has given the best result compared to others except [20]. Our proposed approaches with DCGAN have outperformed the transfer learning methods as well. The performance comparison with the state of the art has been summarized in Table 3.

Table 3 Comparison of performances with the state of the art

Reference work	Methods	40× (%)	100× (%)	200× (%)	400× (%)
Nahid et al. [13]	CNN+Softmax	90	90	91	90
Deniz et al. [5]	Transfer learning AlexNet + VGG16	93.57	92.50	93.24	91.87
Bhuiyan et al. [3]	Transfer learning and supervised classifier	96.24	92.81	94.29	92.86
Wei et al. [23]	BiCNN	97.89	97.64	97.56	97.97
Alom et al. [1]	IRRCNN + Aug(image level)	97.95	97.57	97.32	97.36
Saini and Susan [15]	DCGAN + VGG16	96.5	94	95.5	93
Man et al. [11]	DenseNet121-AnoGAN	**99.13**	96.39	86.38	85.20
Thuy and Hoang [20]	VGG16 + VGG19 + CNN	98.2	**98.3**	**98.2**	97.5
Proposed approach	**DenseNet+DCGAN**	97.99	**98.32**	98.01	97.80
Proposed approach	**MobileNet +DCGAN**	97.99	96.63	**98.26**	96.70
Proposed approach	**Xception + DCGAN**	97.24	96.63	97.77	**99.45**

'Bold' indicates best results compared to other approaches.

6 Conclusion

The medical image data are difficult to train for automation systems due to some obstacles such as dataset scarcity, imbalance dataset and the complex structure of the images. In this study, the histopathological images have been explored with four pre-trained deep CNN models. Though all the models have provided better result, still there is a scope to improve the model training. Some fine-tuning steps can be applied in the future. Also, to balance the dataset, a DCGAN approach has been proposed which has improved the classification result. Most of the CNN models have performed well with DCGAN. Also, DCGAN with DenseNet classifier has provided maximum 2.92% improvement. But for some MFs, the results have been improved little. This indicates that the DCGAN model needs some improvement. Despite the limitations, the DCGAN model has facilitated most of our classifiers which is noteworthy, and such improved models will help us to automate breast cancer detection.

References

1. Alom MZ, Yakopcic C, Nasrin MS, Taha TM, Asari VK (2019) Breast cancer classification from histopathological images with inception recurrent residual convolutional neural network. J Digit Imaging 32(4):605–617. https://doi.org/10.1007/s10278-019-00182-7
2. Bardou D, Zhang K, Ahmad SM (2018) Classification of breast cancer based on histology images using convolutional neural networks. IEEE Access 6:24680–24693. https://doi.org/10.1109/ACCESS.2018.2831280
3. Bhuiyan MNQ, Shamsujjoha M, Ripon SH, Proma FH, Khan F (2019) Transfer learning and supervised classifier based prediction model for breast cancer. In: Dey N, Das H, Naik B, Behera HS (eds) Big data analytics for intelligent healthcare management. Advances in ubiquitous sensing applications for healthcare, Chap 4, pp 59–86. Academic Press. https://doi.org/10.1016/B978-0-12-818146-1.00004-0
4. Bowles C, Chen L, Guerrero R, Bentley P, Gunn RN, Hammers A, Dickie DA, del C Valdés Hernández M, Wardlaw JM, Rueckert D (2018) GAN augmentation: augmenting training data using generative adversarial networks. CoRR **abs/1810.10863**
5. Deniz E, Şengür A, Kadiroğlu Z, Guo Y, Bajaj V, Budak Ü (2018) Transfer learning based histopathologic image classification for breast cancer detection. Health Inf Sci Syst 6(1):1–7. https://doi.org/10.1007/s13755-018-0057-x
6. Elston CW, Ellis IO (1991) Pathological prognostic factors in breast cancer. I. The value of histological grade in breast cancer: experience from a large study with long-term follow-up. Histopathology 19(5):403–410. https://doi.org/10.1111/j.1365-2559.1991.tb00229.x
7. Goodfellow I, Pouget-Abadie J, Mirza M, Xu B, Warde-Farley D, Ozair S, Courville A, Bengio Y (2014) Generative adversarial nets. In: Advances in neural information processing systems, vol 27
8. Hartanto CA, Kurniawan S, Arianto D, Arymurthy AM (2021) DCGAN-generated synthetic images effect on white blood cell classification. IOP Conf Seri Mater Sci Eng 1077(1):012033. https://doi.org/10.1088/1757-899x/1077/1/012033
9. He K, Zhang X, Ren S, Sun J (2015) Delving deep into rectifiers: surpassing human-level performance on imagenet classification. In: Proceedings of the IEEE international conference on computer vision, pp 1026–1034. https://doi.org/10.1109/ICCV.2015.123

10. He L, Long LR, Antani S, Thoma GR (2012) Histology image analysis for carcinoma detection and grading. Comput Methods Programs Biomed 107(3):538–556. https://doi.org/10.1016/j.cmpb.2011.12.007
11. Man R, Yang P, Xu B (2020) Classification of breast cancer histopathological images using discriminative patches screened by generative adversarial networks. IEEE Access 8:155362–155377. https://doi.org/10.1109/ACCESS.2020.3019327
12. McCann MT, Ozolek JA, Castro CA, Parvin B, Kovacevic J (2015) Automated histology analysis: opportunities for signal processing. IEEE Signal Process Mag 32(1):78–87. https://doi.org/10.1109/MSP.2014.2346443
13. Nahid AA, Mehrabi MA, Kong Y (2018) Histopathological breast cancer image classification by deep neural network techniques guided by local clustering. BioMed Res Int. https://doi.org/10.1155/2018/2362108
14. Radford A, Metz L, Chintala S (2016) Unsupervised representation learning with deep convolutional generative adversarial networks. In: 4th International conference on learning representations, ICLR 2016—conference track proceedings
15. Saini M, Susan S (2020) Deep transfer with minority data augmentation for imbalanced breast cancer dataset. Appl Soft Comput J 97:106759. https://doi.org/10.1016/j.asoc.2020.106759
16. Sedigh P, Sadeghian R, Masouleh MT (2019) Generating synthetic medical images by using GAN to improve CNN performance in skin cancer classification. In: ICRoM 2019—7th International conference on robotics and mechatronics, pp 497–502. https://doi.org/10.1109/ICRoM48714.2019.9071823
17. Shams S, Platania R, Zhang J, Kim J, Park SJ (2018) Deep generative breast cancer screening and diagnosis. In: International conference on medical image computing and computer-assisted intervention, pp 859–867. https://doi.org/10.1007/978-3-030-00934-2_95
18. Spanhol FA, Oliveira LS, Petitjean C, Heutte L (2016) A dataset for breast cancer histopathological image classification. IEEE Trans Biomed Eng 63(7):1455–1462. https://doi.org/10.1109/TBME.2015.2496264
19. Sung H, Ferlay J, Siegel RL, Laversanne M, Soerjomataram I, Jemal A, Bray F (2021) Global cancer statistics 2020: GLOBOCAN estimates of incidence and mortality worldwide for 36 cancers in 185 countries. CA Cancer J Clin 71(3):209–249. https://doi.org/10.3322/caac.21660
20. Thuy MBH, Hoang VT (2020) Fusing of deep learning, transfer learning and GAN for breast cancer histopathological image classification. In: Advances in intelligent systems and computing. https://doi.org/10.1007/978-3-030-38364-0_23
21. Venu SK, Ravula S (2021) Evaluation of deep convolutional generative adversarial networks for data augmentation of chest x-ray images. Future Internet 13(1). https://doi.org/10.3390/fi13010008
22. Vo DM, Nguyen NQ, Lee SW (2019) Classification of breast cancer histology images using incremental boosting convolution networks. Inf Sci 482:123–138. https://doi.org/10.1016/j.ins.2018.12.089
23. Wei B, Han Z, He X, Yin Y (2017) Deep learning model based breast cancer histopathological image classification. In: 2017 2nd IEEE International conference on cloud computing and big data analysis, ICCCBDA 2017, pp 348–353. https://doi.org/10.1109/ICCCBDA.2017.7951937
24. Xue Y, Ye J, Zhou Q, Long LR, Antani S, Xue Z, Cornwell C, Zaino R, Cheng KC, Huang X (2021) Selective synthetic augmentation with HistoGAN for improved histopathology image classification. Med Image Anal 67:101816. https://doi.org/10.1016/j.media.2020.101816
25. Yi X, Walia E, Babyn P (2019) Generative adversarial network in medical imaging: a review. Med Image Anal 58:101552. https://doi.org/10.1016/j.media.2019.101552

What Drives Adoption of Cloud-Based Online Games in an Emerging Market? An Investigation Using Flow Theory

Ashok S. Malhi, Raj K. Kovid, Abhisek Dutta, and Rajeev Sijariya

Abstract Intention to adopt playing games on electronic gadgets has well been explored by researchers. However, cloud of things (CoT) enabled online games, as a new service, has yet to see its widespread acceptance and usage by consumers in emerging markets and hence, the need for research on factors determining behavioural intention to adopt it. Using flow theory, this study aims at investigating the role of some additional variables such as enjoyment and usage cost in shaping the behavioural intention to adopt playing CoT-enabled online games. The data, collected from 145 respondents through an online survey was analysed using PLS-SEM. The results show that enjoyment had a direct significant effect on behavioural intention to use cloud-based online games as well as indirect effect via flow state. The facilitating conditions and effort expectancy influenced behavioural intention through flow state, whereas social influence and usage cost had direct influence on behavioural intention. The study provides meaningful insights for managers in the online games industry and future direction for research.

Keywords Cloud of things · CoT · Flow theory · Online games · Technology adoption · Social influence · Emerging markets · Generation Y · Generation Z

1 Introduction

Video gaming is a computationally demanding task. Advanced computer graphics allow for the production of realistic and seamless gaming experiences. Thus, video gaming need advanced technology which is often beyond the means of a mobile

A. S. Malhi · R. K. Kovid (✉)
Sharda University, Uttar Pradesh, India
e-mail: kovidrk@gmail.com

A. Dutta
NIIT University, Rajasthan, India

R. Sijariya
Jawaharlal Nehru University, Delhi, India

V. Skala et al. (eds.), *Machine Intelligence and Data Science Applications*, Lecture Notes on Data Engineering and Communications Technologies 132,
https://doi.org/10.1007/978-981-19-2347-0_60

device or even a home computer. While fast mobile and broadband Internet, and mobile devices like smartphones and tablets, have become more common, playing a high-quality game is still not a common phenomenon. Rather than using the powerful but power-hungry processors that are built into mobile devices, a greener option for thin or mobile clients is to use cloud-based gaming, which leverages cloud computing for gaming. Cloud gaming offloads processing to cloud servers by using communication infrastructure. The cloud-based gaming can lead to decreased software maintenance cost, better hardware scalability, and longer equipment lifespans. In addition, it also provides a cutting-edge strategy, called layered coding, which may decrease the required bit rate of game streaming by taking advantage of the growing graphical processing power of a mobile client.

Worldwide, mobile internet is expanding at an exponential rate. The CNNIC [1] report says that mobile internet users in China, which represents 72% of the country's online population, account for 388 million, surpassing PC internet users for the first time. To cope with the evolving marketplace, carriers have introduced a number of mobile services, including mobile instant messaging, mobile news, mobile payment, and mobile gaming. Of these, certain services have been widely used by customers. As shown by a study, the number of mobile users who have utilized instant messaging through mobile devices (83%) is substantial [1]. This survey provides some of the insight that just 30% of the respondents are playing mobile games, a personal leisure application, and this is a stark contrast to the 70% who have used e-commerce solutions to answer how far the various types of mobile gaming user adoption have spread across the globe. Users are going to utilize whatever service providers choose to offer. Therefore, service providers need to understand the variables that impact their clients' actions. They may use efficient methods to make mobile games accessible and usable by users, which is essential to the success of the project. Another benefit of mobile games is that they are everywhere, independent of the users' location. Users can use their phone's internet connection to play games at any place, even while travelling. The strategy may result in a better experience for customers and encourage their behaviour. Mobile devices, such as smaller displays and subpar input methods, may lessen the user experience [2]. Mobile phones, in contrast to desktop PCs, have smaller displays and lesser resolution. This may have an impact on the way mobile games are visually presented. Even while some phones provide touch-sensitive controls, there are still many users who have to utilize tiny keys to enter information. Even worse, a bad network connection may impair user experience. Customers may experience service disruptions and downtime when using mobile games in the car. If mobile game does provide a positive user experience, users will not want to utilize the technology.

Studies on mobile usage behavioural intention have used various theories of information systems such as technology adoption model [3], innovation diffusion theory [4], task technology fit [5, 6], and unified theory of acceptance and use of technology. The task, technology, and user-belief factors of usefulness, relative advantage, fit, and performance expectation are considered to influence the users' intention to adopt. The motives, however, are primarily instrumental, and are considered extrinsic. Mobile users' behaviour, in case of emerging technologies and usage, has seldom been researched in the context of user experience. Extrinsic motivation is concerned with the results of use, while intrinsic motivation is concerned with the process of use [7]. Users' intentions are affected by both external and internal factors. Another work showed that a "flow" experience—an ideal user experience, has also been found to influence user behaviour [8]. For better flow, content quality, connection quality, and simplicity of use are believed to be important. We have included the variables of social impact and cost of use to the model as well. Some research suggest that the usage cost of the network depends on the capacity of the infrastructure to make network use easier, while social influence is defined by the impact of other influential individuals on individual users [9]. UTAUT provides that the impact of social influence and facility conditions on user behaviour exists.

This study, conducted in Indian context, build on the flow theory to investigate the role of certain factors in determining the intention to adopt playing the CoT-enabled online games. The hypothesized determinants include facilitating conditions, enjoyment, effort expectancy or ease of use, usage cost (price), and social influence. We also see whether flow mediate the impact of enjoyment on behavioural intention. The study contributes by testing the flow theory with presence of additional variables in new technology driven context. The paper is organized as follows. Section 2 analysis the existing research on the concepts of flow and CoT-enabled online game adoption and establishes the hypothesis followed by the hypothesized research model. Third section contains details on research methods, instrument development, and data gathering. The results are shown in Sect. 4, while Sect. 5 provides for discussion on the results followed by the future scope of research. Section 6 concludes the study.

2 Literature Review and Hypotheses Development

2.1 Flow

One researcher observed that the feeling of flow is based on a psychological notion, and it is defined as a complete engagement that individuals experience while they do an activity [10]. Another researcher describe flow as a condition of (a) uninterrupted machine interaction that promotes, (b) intrinsic pleasure, (c) self-consciousness, and (d) a positive feedback loop [8]. While challenges and abilities are in equilibrium, there is flow. When consumers lack the abilities to meet the demands of difficulties, they feel uneasy. When abilities are over the level of difficulty, users get bored. When either challenge or skill levels fall below the cut-off points, the user experience is characterized by indifference. Users are only able to experience flow when the skill and challenge levels are both above the defined thresholds and are compatible. Flow, a protean idea, encompasses a number of elements. Customers engaged in online shopping experience six dimensions: focus, perceived control, feeling of merging action and awareness, the change of time, the transcendence of self, and the autotelic experience [11]. A number of factors determine whether or not a person have achieved the state of flow while playing games online [8] including difficulty, attention, control, and pleasure. A researcher hypothesized that flow is characterized by control, interest, attention, and curiosity [12]. Some of the components seen in such designs include different aspects of pleasure, control, and attention concentration. It is believed that an individual's perception of a system's usefulness provides insight into how much fun a person has with the system. The level of control you feel over the project and the world around you influences how you feel about your ability to manage the situation. User attention is a reflection of how immersed and involved a person is with their information technology. For the purposes of online commerce [11], virtual worlds [13], instant messaging [14], sports team websites [15], and e-learning [16], the concept of flow has been used. It is determined that the factors of complexity, interaction, sociability, and telepresence influence the flow. More recently, flow has been used to study the behaviour of mobile users. Content quality impacts the flow of mobile TV use [17]. Researchers found that both flow and network externalities are important for mobile instant messaging users [18].

2.2 *Adoption of Cloud of Things Enabled Online Games*

Research has been done on user behaviour because of the poor uptake of mobile services. Theories such as technology acceptance model (TAM), information diffusion theory (IDT), task technology fit (TTF), and unified theory of acceptance and use of technology (UTAUT) are frequently utilized as theoretical foundations for information systems. TAM suggests that two key elements in terms of user adoption of an information technology are the usability and utility of the technology. TAM has been popular for examining the uptake of mobile health care [3], mobile payment [19], and mobile internet [3] because of its frugality and efficacy. IDT explains that five factors influence the rate at which an innovation is adopted by users: relative advantage, compatibility, perceived complexity, trialability, and observability. Research in the area of mobile banking and mobile payments has utilized IDT to investigate the intention to use these services [4, 20]. TTF says that consumers will only embrace a technology when task-based attributes align with technology-based attributes. TTF has been used to research mobile security systems [21], user acceptance of mobile work [5], and locatable information systems [22]. The UTAUT methodology asserts that four aspects, including performance and effort expectations, social influence, and supportive circumstances, have a major impact on user adoption. In studies of mobile data services [6] and mobile technologies [23], UTAUT was used to investigate user behaviour.

2.3 Hypotheses Development

Flow and Behavioural Intention to Adopt

Optimal experience is reflected in flow. It is quite a positive experience for users when they feel immersed in mobile games and can find a great deal of enjoyment while playing. They do not seem to be concerned with the environment or time passing quickly for them. A pleasurable experience like this may increase their willingness to use it. Studies in mobile TV [17], sporting team websites [15], and online shopping [24] have reported the effect of flow on user behaviour. Based on the aforementioned studies, we hypothesize:

H1: Flow will positively affect the behavioural intention.

Enjoyment

Since enjoyment is a subjective word, we need to define it in our conceptual model since we believe it affects the flow factor and also affects the individual's adoption intention while utilizing CoT-enabled games [9, 25, 26]. It provides enjoyable and exciting activities that add pleasure to one's life, whether in this world or the next. It may also help you enhance your degree of entertainment. We attempted to uncover another dimension in our model in which pleasure has control over the flow factor, and if anybody has a high degree of experience and they want to remain in that condition, while the activity's importance is of great relevance in acquiring CoT-enabled games [25, 26].

H2: Enjoyment will positively affect the behavioural intention.

H3: Enjoyment will positively affect the flow.

Facilitating Conditions and Flow

The most often noted element influencing the adoption of mobile health technology was the ability of technology to address the existing problems. This study supported earlier findings on the significance of enabling circumstances, but it also pointed out that people had all the required resources to utilize CoT-enabled games [25, 26]. Other technologies are also compatible with CoT-enabled games if the consumer has the option, and in this case, the client should have quick customer assistance to keep him motivated to use the games [9].

H4: Facilitating conditions will positively affect the flow.

Effort Expectancy

If a person uses that platform, but does not receive pleasure, then the likelihood of this kind of fringe failing is considerably higher. That's why one must have an executable ability such allows him/her to utilize that arrangement with minimum effort. The level of trust in the capability to do CoT-enabled games tasks will be increased if the system is simple to use and interaction with it is intuitive [9, 25, 26].

H5: Effort expectancy will positively affect the flow.

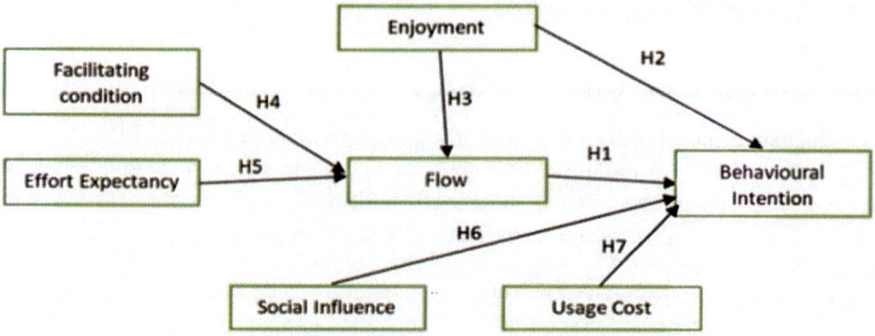

Fig. 1 Hypothesized model

Social Influence

Social influence is a reflection of the influence that other significant people have on individual users. Like the subjective norm of the theory of planned behaviour, it is subjective. A user is more likely to adopt a mobile game when his or her friends and peers encourage him or her to use it. UTAUT claims that social influence plays a key role in influencing user adoption [9]. According to existing research, the effect of social influence on behavioural intention has been uncovered in the contexts of mobile data [6] services, instant messaging [27], and multimedia messaging services [19, 28]. This view is supported by the research that we just reviewed.

H6: Social influence will positively affect the behavioural intention.

Usage Cost

The use cost of mobile games takes into account communication and transaction expenses. It is a hardship for the majority of customers to deal with these costs. They may refuse to use the games if they believe they are too costly. With regards to the issue of use cost, UTAUT incorporates enabling circumstances, which are an important factor in influencing user adoption [9]. Earlier studies have found that the cost to use mobile browsing services [29], m-commerce [30], short message services [31], and 3G mobile value-added services [32] have been affected by user adoption.

H7: Usage cost will positively affect the behavioural intention.

The hypothesized model is given in Fig. 1.

3 Methods

This study analysed the data to determine the effects of a number of hypothesized variables on cloud-based online gaming service adoption. Data was gathered through an online survey. The study was carried out during July and August of the year 2021. The sample population is composed of university students and professionals from the workplace. A convenient response-collection method was employed to conduct a survey. We used different social media channels to disseminate the online questionnaire to persons having some exposure to online gaming and associated service. All the participants were given five-point Likert-type questions to record their response. The questionnaire was sent to 352 people out of whom a total of 147 people responded, and 145 of them were found usable after cleaning the responses. People who claimed they never utilized cloud-based services were excluded from the survey. Based on the recommendation for the number of relationships in partial least square structural equation modelling (PLS-SEM), the data was gathered based on a sample size of ten times that number. The partial least square technique of structural equation modelling was used to test the hypothesized model. The analysis was conducted using WarpPLS 7.0. PLS-SEM, which may be used in place of covariance-based SEM (CB-SEM) as it allows route modelling to overcome the issue of limited sample size, and can be used to run simulations on a variety of models and systems. PLS-SEM can calculate models with a massive number of latent variables, even if they are very complex. The assumptions of PLS-SEM are less restrictive with respect to the distribution of variable and error terms, and it may be used for both reflective and formative measures [33].

4 Results

The demographic composition of the participants included 41% female and 59% male. The ages of the survey participants ranged from 14 to 50 years including those from Generation Y and Generation Z. Gen Y consists of people born between 1970 to 1986 and Gen Z are people born between 1986 and 1996. Gen Y and Z are considered to be quick in adopting new technologies since they are one who have seen the advent of personal computers and the internet in early age. Most importantly, all the respondents have played some kind of online game like massive multiplayer online role play game (MMORPG), etc. The study has used the flow theory to build upon the constructs which may influence the user's adoption of cloud-based online games. The variables enjoyment, social influence, usage cost had positive significant effect on behavioural intention to adopt cloud-based online games, whereas facilitating condition, effort expectancy had indirect effect on behavioural intention through flow state. Even, enjoyment was found to have an indirect effect via flow state on behavioural intention towards cloud-based online games. Reflective measurement model assessment is performed by determining indicator loadings and suggesting the recommended measurement when any loading of 0.751 or more is found as the construct can explain 50% or more of the variation in indicators. Average variance extracted (AVE)s are used to assess. The convergent validity was assessed using the AVE for each construct (Table 1). A composite reliability of 0.70 or more is recommended [34].

Table 1 Measurement model results

	BI	EE	FC	FL	SI	ENJ	UC
Composite reliability	0.959	0.949	0.935	0.91	0.952	0.976	0.946
Cronbach's alpha	0.946	0.932	0.918	0.867	0.932	0.971	0.924
Variance inflation factor	2.592	2.934	3.126	2.645	2.838	3.232	1.998

The internal consistency reliability is measured using the composite reliability [35]. In exploratory research, reliability levels of 0.60–0.70 are regarded as "acceptable," and reliability values of 0.70–0.90 are considered "satisfactory to good." Values of variance inflation factor (V.I.F) for all variables lies between one and five and hence, all variables are valid. Items with composite reliability of greater than 0.95 may be considered problematic since they display redundancy [36]. Table 1 details Cronbach's alpha and composite reliability values for each construct, both of which were determined using Cronbach's alpha and composite reliability (CR). The Cronbach's Alpha of all constructs exceeds 0.70. Composite reliability values for all constructs are over 0.70, which indicates that the level of reliability is high. In order to assess discriminant validity of the variables, the researchers used the heterotrait–monotrait ratio (HTMT) and found all the ratios to be less than one (see Table 2), thus showcasing that the variables which may be similar in meaning are not considered same by the respondents, thus, keeping them distinct [37].

Assessment of the structural model

The structural model's evaluation aids researchers in explaining the link between latent constructs [34]. Figure 2 shows the structural model with the newly included constructs, along with all path coefficients, *p*-values, and R^2 values. A few main metrics, such as the coefficient of determination (R^2) and the Q^2 score, may be used to evaluate the structural model. The hypothesis testing results can be viewed in Table 3. The R^2 value for behavioural intention and flow is 0.609. As a result, the structural model's components can predict 61% of the variance in behavioural intention and 61% of the variance in flow state when it comes to cloud-based online games. Past studies have reported R^2 values in determining adoption of technologies

Table 2 Heterotrait–monotrait ratio (HTMT)

	BI	EE	FC	FL	SI	ENJ	UC
BI							
EE	0.492						
FC	0.599	0.787					
FL	0.508	0.758	0.776				
SI	0.775	0.569	0.705	0.63			
ENJ	0.674	0.721	0.711	0.782	0.656		
UC	0.608	0.615	0.495	0.555	0.639	0.634	

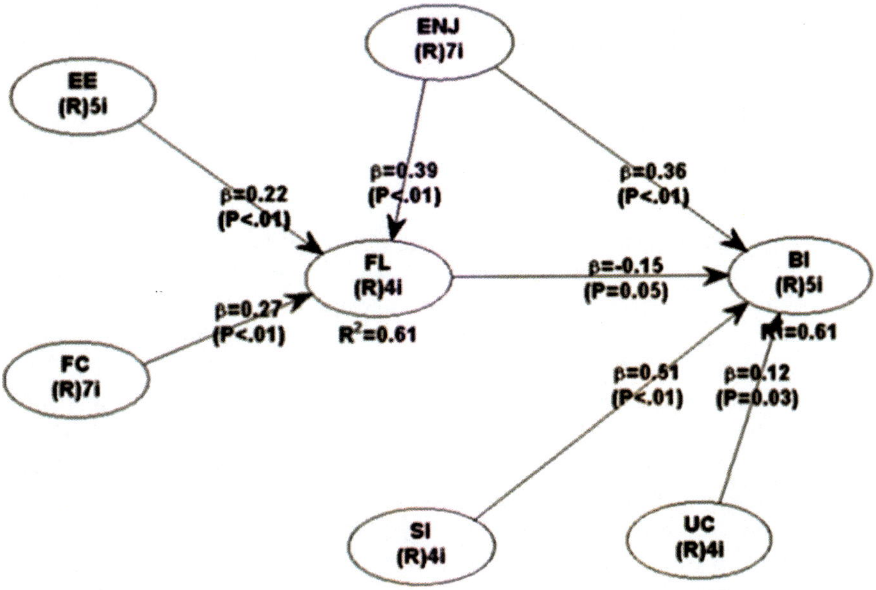

Fig. 2 PLS-SEM output model with path coefficients and p-value

Table 3 Hypothesis testing summary

#	Hypothesis	Beta	*p*-value	Decision
H1	FL → BI	−0.15	0.05	Supported
H2	ENJ → BI	0.36	<0.01	Supported
H3	ENJ → FL	0.39	<0.01	Supported
H4	EE → FL	0.22	<0.01	Supported
H5	FC → FL	0.27	<0.01	Supported
H6	SI → BI	0.51	<0.01	Supported
H7	UC → BI	0.12	0.03	Supported

such as the study of mobile banking adoption [38] which reported R^2 of 0.60. The hypothesis that flow may affect behavioural intention was supported ($b = -0.15$, $p = 0.05$) but negatively. Thus, every unit increase in flow state of the user, the behavioural intention to adopt cloud-based online games decreases. The effect of enjoyment on behavioural intention was supported ($b = 0.36$, $p < 0.01$); Enjoyment also had a positive effect on flow ($b = 0.39$, $p < 0.01$). Effort expectancy has a positive effect on flow state ($b = 0.22$, $p < 0.01$). Facilitating conditions also had a significant impact on flow state ($b = 0.27$, $p < 0.01$) towards cloud-based online games. Also, as expected, the hypothesis (H6), that is, social influence has a positive effect on behavioural intention was supported. Even usage cost considered by users in adopting cloud-based online games was also found to be significantly affecting behavioural intention. Table 3 provides a summary of results on hypothesis testing.

5 Discussion

One of the most striking outcomes of this study is the negative relation between flow state and behavioural intention towards adopting cloud-based online games. Past researches have shown both positive and negative effects of flow state on users' behaviour towards an activity or a game. Flow is defined as a state of peak performance in which one is completely immersed in a task, and this feeling of immersion in the task is supplemented by a "sense of effortless ease and fluency in one's movements and thoughts" [39]. Though the positive aspects of flow have been studied in the past and have been used in various contexts, the issues of negative effect cannot be ruled out. Csikszentmihalyi himself stated that a flow was a state where for the user nothing else seems to matter, and the enjoyment of doing the activity is so high that the individual will do it even at high cost for the sheer sake of doing it [39, 41]. Thus, it means that the person gets addicted to the activity that the individual becomes captive to the order of the activity and might not be willing to cope with realities of life. Such dangerous outcomes have been pointed out and reported by news in the recent past where young users have got engrossed in online games resulting in addiction and psychosocial consequences for the people affected [40]. Popularly known as video game addiction is a pattern of gaming behaviour whether online or offline, wherein the user finds it difficult to control and continues it with vigour even when it may rise to serious negative consequences in other areas of gamers' life [41, 42]. The pinpointing of such problems by the mainstream media and medical practitioners have led to an additional awareness towards gaming addiction. Thus, it is highly possible that the respondents might have demonstrated this self-awareness of not adopting cloud-based online games due to the perceived understanding of games leading to addiction and further psychological imbalance. The study contributes to better understand the role of various factors in shaping the intention to use CoT-enabled games, especially social influence was acknowledged as one of the most important factors leading to positive intention towards such games. Thus, game designers should integrate social interactions as a part of the game experience as well as physical and online engagement via events and social media to keep the gamers involved and invested in online games.

Limitations and Future Research Directions

The study has certain limitations too. This study has not considered the technology infrastructure such as quality of internet may have some influence on users' intention to adopt online games. Moreover, the study has not taken into consideration the credibility of technology hardware and software which may also influence the behavioural intention. So, these factors can be considered by future researchers. Also, gender may have some role in shaping the intention. So, the study of comparisons between genders in understanding how males and females respond to flow state in adoption of online gaming can be conducted in Indian context as many studies in the past have pointed out that the negative effect of video game addiction was observed to be higher among teenage and adolescent males than that among women [43].

6 Conclusion

The study investigated the influence of facilitating conditions, effort expectancy, enjoyment, mediated through flow state, and on users' behavioural intention to play CoT-enabled online games in Indian context. It also has examined the impact of social influence and usage cost or price on behavioural intention. The flow state significantly mediates the effect of enjoyment, facilitating conditions, and effort expectancy on behavioural intention towards using cloud-based online games. Social influence and usage cost had direct influence on behavioural intention. However, study found negative relation between flow state and behavioural intention towards adopting cloud-based online games. Thus, this finding, inconsistent with earlier studies, provides a strong base for further studies seeking to settle the debate. Moreover, the study has implications for managers in the online games industry and social implications for young generation.

References

1. CNNIC (2012) 30th statistical survey report on the internet development in China, China Internet Network Information Centre
2. Lee YE, Benbasat I (2004) A framework for the study of customer interface design for mobile commerce. Int J Elect Comm 8(3):79–102
3. Wu I-L, Li J-Y, Fu C-Y (2011) The adoption of mobile healthcare by hospital's professionals: an integrative perspective. Decis Support Syst 51(3):587–596
4. Lin H-F (2011) An empirical investigation of mobile banking adoption: the effect of innovation attributes and knowledge-based trust. Int J Inf Manage 31(3):252–260
5. Yuan Y, Archer N, Connelly CE, Zheng W (2010) Identifying the ideal fit between mobile work and mobile work support. Inform Manag 47(3):125–137
6. Thong JYL, Venkatesh V, Xu X, Hong S-J, Tam KY (2011) Consumer acceptance of personal information and communication technology services. IEEE Trans Eng Manage 58(4):613–625
7. Davis FD, Bagozzi RP, Warshaw PR (1992) Extrinsic and intrinsic motivation to use computers in the workplace. J Appl Soc Psychol 22(14):1111–1132
8. Hoffman DL, Novak TP (2009) Flow online: lessons learned and future prospects. J Inter Mark 23(1):23–34
9. Chuah S, Yuen C, Cheung N (2014) Cloud gaming: a green solution to massive multiplayer online games. IEEE Wirel Commun 21(4):78–87. https://doi.org/10.1109/mwc.2014.6882299
10. Venkatesh V, Morris MG, Davis GB, Davis FD (2003) User acceptance of information technology: toward a unified view. MIS Q 27(3):425–478
11. Csikszentmihalyi M, Csikszentmihalyi IS (1988) Optimal experience: psychological studies of flow in consciousness. Cambridge University Press, Cambridge

12. Guo YM, Poole MS (2009) Antecedents of flow in online shopping: a test of alternative models. Inform Syst J 19(4):369–390
13. Ho L-A, Kuo T-H (2010). How can one amplify the effect of e-learning? An examination of high-tech employees' computer attitude and flow experience. Comput Hum Behav 26(1):23–
14. Wang LC, Baker J, Wagner JA, Wakefield K (2007) Can a retail web site be social? J Mark 71:143–157
15. Zaman M, Anandarajan M, Dai Q (2010) Experiencing flow with instant messaging and its facilitating role on creative behaviors. Comput Hum Behav 26(5):1009–1018
16. O'Cass A, Carlson J (2010) Examining the effects of website induced flow in professional sporting team websites. Internet Res 20(2):115–134
17. Animesh A, Pinsonneault A, Yang SB, Oh W (2011) An odyssey into virtual worlds: exploring the impacts of technological and spatial environments on intention to purchase virtual products. MIS Q 35(3):789–810
18. Jung Y, Perez-Mira B, Wiley-Patton S (2009) Consumer adoption of mobile TV: examining psychological flow and media content. Comput Hum Behav 25(1):123–129
19. Kim C, Mirusmonov M, Lee I (2010) An empirical examination of factors influencing the intention to use mobile payment. Comput Human Behav 26(3):310–322
20. Zhou T, Lu Y (2011) Examining mobile instant messaging user loyalty from the perspectives of network externalities and flow experience. Comput Hum Behav 27(2):883–889
21. Mallat N (2007) Exploring consumer adoption of mobile payments—a qualitative study. J Strateg Inf Syst 16(4):413–432
22. Liu Z, Min Q, Ji S (2010) An empirical study of mobile securities management systems adoption: a task-technology fit perspective. Int J Mobile Commun 8(2):230–243
23. Junglas I, Abraham C, Watson RT (2008) Task-technology fit for mobile locatable information systems. Decis Support Syst 45(4):1046–1057
24. Park J, Yang S, Lehto X (2007) Adoption of mobile technologies for Chinese consumers. J Elect Comm Res 8(3):196–206
25. Hausman AV, Siekpe JS (2009) The effect of web interface features on consumer online purchase intentions. J Bus Res 62(1):5–13
26. Narwane VS, Raut RD, Gardas BB, Kavre MS, Narkhede BE (2019) Factors affecting the adoption of cloud of things: the case study of Indian small and medium enterprises. J Syst Inf Technol 21(4):397–418. https://doi.org/10.1108/JSIT-10-2018-0137
27. Kim KK, Shin HK, Kim B (2011) The role of psychological traits and social factors in using new mobile communication services. Electron Commer Res Appl 10(4):408–417
28. Narwane VS, Raut RD, Mangla SK et al (2020) Mediating role of cloud of things in improving performance of small and medium enterprises in the Indian context. Ann Oper Res. https://doi.org/10.1007/s10479-019-03502-w
29. Shin YM, Lee SC, Shin B, Lee HG (2010) Examining influencing factors of post-adoption usage of mobile internet: focus on the user perception of supplier-side attributes. Inform Syst Frontier 12(5):595–606
30. Yun H, Lee CC, Kim BG, Kettinger WJ (2011) What determines actual use of mobile web browsing services? A contextual study in Korea. Commun Assoc Inform Syst 28(1):313–328
31. Wei TT, Marthandan G, Chong AYL, Ooi KB, Arumugam S (2009) What drives Malaysian m-commerce adoption? An empirical analysis. Ind Manage Data Syst 109(3–4):370–388
32. Lu Y, Deng Z, Wang B (2010) Exploring factors affecting Chinese consumers' usage of short message service for personal communication. Inform Syst J 20(2):183–208
33. Kuo Y-F, Yen S-N (2009) Towards an understanding of the behavioral intention to use 3G mobile value-added services. Comput Hum Behav 25(1):103–110
34. Henseler J, Ringle CM, Sinkovics RR (2009) The use of partial least squares path modeling in international marketing. Adv Int Mark 20:277–319. https://doi.org/10.1108/S1474-7979(2009)0000020014
35. Hair JF, Black WC, Babin BJ, Anderson RE, Tatham RL (1997) Multivariate data analysis. New Jersey

36. Jöreskog KG (1971) Simultaneous factor analysis in several populations. Psychometrika 36(4):409–426. https://doi.org/10.1007/BF02291366
37. Teo TS, Srivastava SC, Jiang LI (2008) Trust and electronic government success: an empirical study. J Manag Inf Syst 25(3):99–132. https://doi.org/10.2753/MIS0742-1222250303
38. Diamantopoulos A et al (2012) 'Guidelines for choosing between multi-item and single-item scales for construct measurement: a predictive validity perspective. J Acad Mark Sci 40(3):434–449. https://doi.org/10.1007/s11747-011-0300-3
39. Luo X et al (2010) Examining multi-dimensional trust and multi-faceted risk in initial acceptance of emerging technologies: an empirical study of mobile banking services. Decis Support Syst 49(2):222–234. https://doi.org/10.1016/j.dss.2010.02.008
40. Psychology Today. Video game Addiction. https://www.psychologytoday.com/us/basics/video-game-addiction (30th Aug 2021)
41. Csikszentmihalyi M, Csikzentmihaly M (1990) Flow: the psychology of optimal experience, vol 1990. Harper & Row, New York
42. Esposito MR, Serra N, Guillari A, Simeone S, Sarracino F, Continisio GI, Rea T (2020) An investigation into video game addiction in pre-adolescents and adolescents: a cross-sectional study. Medicina 56(5):221
43. Griffiths M, Kuss J, King L (2012) Video game addiction: past, present and future. Curr Psychiatry Rev 8(4):308–318

Method to Recommend the Investment Portfolio with the Mix of Various Schemes and Options Using Reinforcement Learning

Parminder Singh Sethi, Vasanth Sathyanaryanan, Nalam Lakshmi, Ravi Tomar, and Vivudh Fore

Abstract In today's scenario, everybody wants to invest their money wisely to get high returns. Unfortunately, a plan that gives you high returns with low risks does not exist. However, to minimize the overall risk and to maximize the chance of high returns, it is recommended to diversify your investment portfolio in such a way that investment is done in multiple asset classes and securities. Diversification does not always help because it ensures low risk but cannot predict returns. Nowadays, machine learning is widely used in stock trading to predict and recommend actions to buy, sell, or hold. This is mostly restricted to a particular asset type such as equity. In general, we are likely to invest in different types of investment options, such as fixed deposits, mutual funds, equity, house, and so on. Therefore, it is wise to build a machine learning model which can recommend an investment portfolio across all investment options for an investor based on his/her expected returns and risk level. In this paper, we propose a machine learning model based on reinforcement learning that can continuously learn the emerging trend of each investment type and recommend reallocation of assets to increase return on investment (ROI).

Keywords Reallocate investment portfolio · Reinforcement learning · Recommend investment portfolio

P. S. Sethi · V. Sathyanaryanan (✉) · N. Lakshmi
Dell EMC Bangalore, Bangalore, India
e-mail: vasanthds@gmail.com

R. Tomar (✉)
School of Computer Science, University of Petroleum and Energy Studies (UPES), Dehradun, Uttarakhand 248007, India
e-mail: rtomar@ddn.upes.ac.in

V. Fore
Gurukul Kangri University, Haridwar, India

V. Skala et al. (eds.), *Machine Intelligence and Data Science Applications*, Lecture Notes on Data Engineering and Communications Technologies 132,
https://doi.org/10.1007/978-981-19-2347-0_61

1 Introduction

A portfolio is a collection of assets in which we would have invested our money. Generally, we use to invest in assets such as stocks, bonds, provident funds, NPS, fixed deposits, gold, and immovable assets such as plots, commercial properties, and houses. Portfolio management is the process of relocating various assets to increase the expected returns at the desired risk level. It comprises different strategies to maximize the expected return on investment (ROI), however, at the same time, the aim is to minimize the financial risk.

As one of the portfolio management strategies to minimize the overall risk and to maximize the chance of high returns, it is recommended to diversify your investment portfolio in such a way that your money is invested in different asset classes and securities. Unfortunately, portfolio management or diversification helps you minimize the risk on your principle, but there is no assurance that you will get the desired returns from the investment portfolio. In today's scenario, everybody wants to invest their money wisely to get high returns. Unfortunately, a plan that gives you high returns with low risks does not exist.

The following are some of the investment options available, categorized based on the risk level and returns. It is inevitable to expect high returns without high risk. The returns and risk go hand in hand.

- **High risk and high returns**: Direct equity, mutual funds
- **Low risk and low returns**: Mutual funds, house, plots, corporate bonds
- **Very low risk and very low returns**: Fixed deposits, provident fund, gold.

This research paper explains how we can use reinforcement learning to recommend the investment portfolio for an investor based on the existing investment portfolio, expected returns, and risk level. This paper is structured in a way that it provides an overview about how reinforcement learning can be used to recommend investment portfolios and in subsequent sections explain each component in reinforcement learning.

2 Literature Review

In today's world, as an investor, we have multiple investment options available for us to invest in. Ranging from high-risk investment options such as equity to very low-risk options such as fixed deposits. In the case of equity, we can choose a stock to invest out of thousands, and in the case of asset types such as mutual fund, fixed deposit, provident fund, pension plan, so on, we have different schemes or plans offered by different vendors. As the available options for investment increase, the complexity of finding the right portfolio for investment also increases. To handle this scenario and propose an investment portfolio, various machine learning models were developed. Yang et al. [1] use machine learning to recommend stocks from S&P 500. It uses machine learning methods such as linear regression, ridge regression, stepwise regression, random forest, and generalized boosted regression. Paiva et al. [2] use a machine learning model developed using a fusion approach of a classifier based on machine learning, with the support vector machine (SVM) method, and the mean–variance (MV) method for portfolio selection. [3] surveys recent literature in the domain of machine learning techniques and artificial intelligence used to forecast stock market movements. The publications are categorized according to the machine learning technique used, the forecasting timeframe, the input variables used, and the evaluation techniques employed. Xue et al. [4] use a bi-direction evolution recommendation system (BDE_RS) to recommend techniques for portfolio management based on investor transaction and investor profile. Pathak and Shetty [5] use both sentiment analysis and neural network techniques to provide more accurate and flexible recommendations. Xu et al. [6] use LightGBM, a novel machine learning model to predict stock trends based on empirical results. This model predicts based on the different effects of analyst attitude and crowd sentiment on stick prices.

3 Reinforcement Learning Model to Recommend Investment Portfolio

In the realm of investment, it is difficult to find out the investment portfolio that can give you high returns based on the risk level. A reinforcement model can be used to learn the current and historical trends of each investment option and recommend an investment portfolio for the investors. The reinforcement machine learning model [7] also uses passive concept drift detection to continuously learn the emerging trend of each asset in the portfolio while preserving the knowledge acquired by using the historical data [8]. The proposed reinforcement machine learning model comprises of following components:

- **Environment**: The data lake contains the historical investment data of each investor, and it is used as an environment to experiment with the actions.
- **Agent**: The agent gathers the input data such as details of the existing portfolio, expected return in years, risk level, and then applies the actions on the environment.
- **Q-Learning Algorithm**: Q-learning is an off-policy reinforcement learning algorithm that determines the best action to take provided the current state. The Q-learning function learns from actions that are outside the current policy, like taking random actions. The Q-learning seeks to learn a policy that maximizes the total reward.
- **State**: The environment shares the state (observation) of the environment with the agent each time an action is performed. That is, it shares the trend or change in the environment that occurred based on actions.

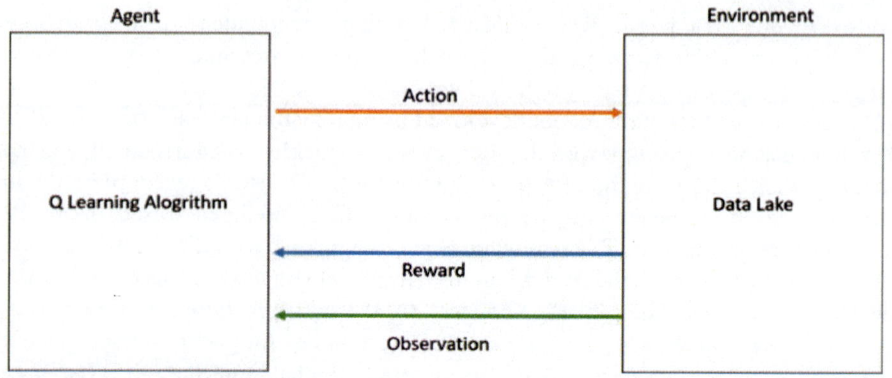

Fig. 1 Proposed reinforcement learning model

- **Reward**: A positive reward is granted if a particular action resulted in expected returns. A negative reward is granted if a particular action resulted in low returns.

The following diagram illustrates the proposed reinforcement learning model. In the subsequent sections, each component in the model is explained in detail (Fig. 1).

4 How This RL Model Works

Initially, a reward matrix is created from each state and possible actions. Then, the Q-learning algorithm is used to train the reinforcement model based on the q value obtained for each possible state and action. The q value is updated after each step or action and ends when an episode is done. Done in this case means reaching some terminal point by the agent. A terminal state for example can be a state where an investment portfolio gives high returns or completes some desired objective. The agent will not learn from a single episode. It learns eventually with enough exploring (steps and episodes) to learn the optimal q-values.

5 Agent

The goal of the agent [9–13] is to receive observation and reward from the uncertain environment and perform the action on the environment. The agent contains a Q-learning algorithm. The algorithm is used to update the Q value continuously based on the actions, state, and rewards. The learning algorithm aims to determine the policy to optimize the expected cumulative reward.

6 Q-Learning Algorithm

In this research paper, we use Q-learning because it builds the policy by taking random actions and seeks to learn a policy that maximizes the total reward in the training phase [14]. The Q-leaning algorithm includes the following steps:

1. Select the gamma parameter and environment reward matrix.
2. Initialize Q-matrix to zero.
3. Select the random initial state.
4. Set the initial state to the current state.
5. Select one among all possible actions from the current state.
6. Using this possible action, go to the next state.
7. Get maximum Q value for the next state based on all the possible actions.
8. Compute Q value again using the following formula.
   $Q(state, action) = R(state, action) + Gamma * Max[Q(next\ state, all\ actions]$
9. Repeat step 3 to 8.

6.1 Assign Gamma Parameter

Gamma parameter has a range from 0 to 1. If gamma is closer to zero, the agent will tend to consider only immediate rewards which are exploitation. If gamma is closer to one, the agent will tend to consider future rewards with greater weights which are exploration. Exploration is about exploring and capturing more information about an environment. Exploitation is about using the already known exploited information to optimize the rewards.

6.2 State and Action

In this research, each possible state in the environment is assigned with a code for reference as shown in the Table 1.

Also, each possible action in the environment is assigned with a code for reference as shown in the Table 2.

6.3 Reward Matrix Table

Based on the state and action, reward matrix table is created for each possible state and action (Table 3).

Table 1 Code assigned to each state in the environment

State	Code
Start investment processing	0
Investment scan and consolidate	1
Diversify portfolio validation	2
Increase principal validation	3
Hybrid validation	4
Goal—overall reward	5
Failed	10
Remediation	11
Centralized store	12

Table 2 Code assigned to each action in the environment

Action	Code
Start	0
Scan	1
Diversify scan	2
Post principal scan	3
Hybrid scan	4
Score check	5
Store	12

Table 3 Reward matrix table based on each possible state and action

State/action	0	1	2	3	4	5	10	11	12
0	−1	20	−1	−1	−1	−1	−10	−1	−1
1	−1	−1	30	−1	−1	−1	−10	−1	−1
2	−1	−1	−2	40	−1	−1	−10	−1	−1
3	−1	−1	−2	−2	50	−1	−10	−1	−1
4	−1	−1	−2	−2	−1	100	−10	−1	−1
10	−1	−1	−1	−2	−1	−1	−10	10	−1
11	0	20	30	40	50	100	−10	10	−1
5	−1	−1	−1	−1	−1	−1	−10	−1	200

Table 4 Initial stage of Q-matrix

State/action	0	1	2	3	4	5	10	11	12
0	0	0	0	0	0	0	0	0	0
1	0	0	0	0	0	0	0	0	0
2	0	0	0	0	0	0	0	0	0
3	0	0	0	0	0	0	0	0	0
4	0	0	0	0	0	0	0	0	0
10	0	0	0	0	0	0	0	0	0
11	0	0	0	0	0	0	0	0	0

6.4 Q-Matrix Table

After creating the reward matrix, the next step is to enable the agent to interact with the environment and make updates to the state-action pairs in a Q-matrix table Q[state, action]. The Q-matrix table acts as a memory that stores the learning of the agent. The agent uses the Q table as a reference and views all possible actions from the present state. The agent then selects the action based on the max value of those actions.

Initially, all state-action pairs are set to "0". As the agent interacts with the environment and traverses through the different states by taking actions, the Q value is updated for each state-action pair. In Q-matrix, each row represents the current state of the agent, and each column represents the possible actions leading to the next state (Table 4).

The following example shows how the Q value is determined for the state-action pair (0,1) (Table 5).

$$Q(state,\ action) = R(state,\ action) + Gamma{*}Max[Q(next\ state,\ all\ actions)]$$
$$Q(0,1) = R(0,1) + 0.8{*}Max[Q(1,10), Q(1,2)]$$
$$Q(0,1) = 20 + 0.8{*}Max[-10,\ 30]$$
$$Q(0,1) = 20 + 0.8{*}30$$
$$Q(0,1) = 44$$

Table 5 Q-matrix value updated based on a stage and action

State/action	0	1	2	3	4	5	10	11	12
0	0	44	0	0	0	0	0	0	0
1	0	0	0	0	0	0	0	0	0
2	0	0	0	0	0	0	0	0	0
3	0	0	0	0	0	0	0	0	0
4	0	0	0	0	0	0	0	0	0
10	0	0	0	0	0	0	0	0	0
11	0	0	0	0	0	0	0	0	0

The agent proceeds to the next state and action returned by environment and reach the goal in one episode. Several episodes are being executed in the training phase. The updated Q-matrix which is the result of the training phase is used by the agent to infer and recommend an investment portfolio.

7 Reward Function

The AI with the prediction model just uses the existing data to predict desired outcomes. For example, in the online banking system, the banking network decides whether to pass each transaction or not. The system tries to allow genuine transactions and decline hoax transactions. In this sort of scenario, even a well-trained algorithm can also make the wrong predictions because it is based on the data and not based on the judgment capabilities that humans possess [15].

Reinforming learning solves this problem by using a reward function. A reward is a feedback mechanism by which we measure the success or failure of an agent's actions in each state. From any given state, an agent sends output in the form of actions to the environment, and the environment returns the agent's new state as well as rewards. Rewards can be immediate or delayed. In this research paper, the reinforcement model uses a reward function to make a judgment to reallocate your investments for better returns. A positive reward is received if an action resulted in better returns. A negative reward is received if an action does not meet the expected returns. Our idea is to optimize the action based on the positive rewards and narrow down the actions that are resulted in negative rewards.

8 State

The agent needs meaningful information to take action. This meaningful information is the state. For example, a reinforcement learning model predicts and recommends the reallocation of funds from a mutual fund scheme to equity [16].

To make this decision, the reinforcement model needs meaningful information about the historical performance data of the current mutual fund scheme and the data about equity. All this information collected becomes the state.

The reinforcement model learns the environment using the Q-learning algorithm. In general, the state information is used by the model to decide on the next action to take.

9 Expected Behavior and Outcome from This Model

Consider a customer has invested in an investment portfolio, which comprises only low-risk assets such as fixed deposits, house, and mutual funds. The customer is not satisfied with the overall returns and thus uses an application that uses the proposed reinforcement learning model to optimize the portfolio for high returns. The reinforcement model uses the historical data of other customers and recommends a portfolio that has given high returns or suggests an action to diversify the portfolio of the customer.

The following diagram describes how this model works in the real-time scenario. Customer A is a medium-risk profile customer and invested in low-risk investment options such as fixed deposit, house, mutual fund, and so on. However, customer A wants to reallocate his funds for high returns and expects a return of 30% YoY. The customer A portfolio is fed into the reinforming learning model. The model based on the knowledge acquired through historical data takes an action to reallocate the customer A portfolio similar to customer Y (a medium-risk profile customer) or allocates the funds and investment options that yielded high returns in the past. After a while, the agent in RL receives a positive or negative award based on the performance of the new allocation. Thus, the RL learns and makes a judgment based on the historical data and the emerging trends (Fig. 2).

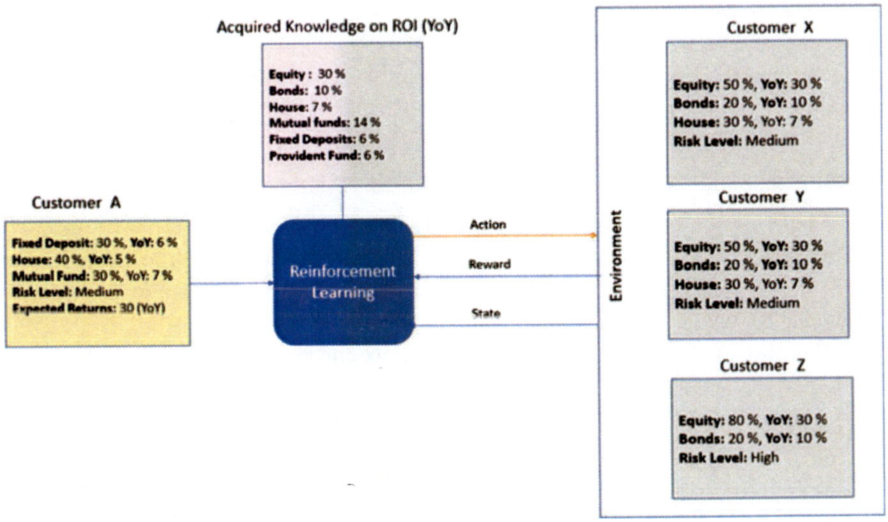

Fig. 2 Expected behavior and outcome

10 Experiment Data

The experiment data used by the reinforcement learning to learn and recommend the investment portfolio are stored in a data lake in both structured and unstructured form. Generally, a data lake with such data already exists, and financial organizations can access this data lake. Such data lake contains the investment portfolio of each investor. See the Table 6. This data lake will be used by the reinforcement learning to perform the action and learn based on the rewards.

11 Feature Selection

As we are predicting and recommending the portfolio optimization across all investment segments, such as equity, mutual fund, house, gold, and so on, the proposed model would use financial indicators that are generic across the segments. The financial indicators employed in our proposed model are

- **Initial investment**—Initial investment made by the investor.
- **Tenure**—Indicates the investment period in years.
- **Investment per month**—Investment is done by the investor every month.
- **YoY returns**—Return on investment (in percentage) per year.

12 Examples

Example 1:

A customer with identification number "a944dc86-41b3-4966-b34d-6488d4765502" has already invested in the investment portfolio listed in the Table 7. The customers want to reallocate the existing investments to grow to 40 million in 20 years.

In the case of this scenario, the agent predicts and provides the following result (Table 8):

Example 2:

A customer with identification number "a944dc86-41b3-4966-b34d-6488d4765523" has already invested in the investment portfolio listed in the Table 9. The customer wants to reallocate the existing investments to grow to 150 million in 20 years.

In the case of this scenario, the agent predicts and provides the following result (Table 10):

Table 6 Experiment data used in the proposed solution (data lake)

Customer identification number	Category	Subcategory	Plan	Initial investment	Investment in years	Investment per month	YoY return (%)	Compound interest	Rounded interest
a944dc86-41b3-4966-b34d-6488d4765502	Debt	Company X	Growth plan	25,000	20	10,000	10	7,041,187	70 Lakhs
a944dc86-41b3-4966-b34d-6488d4765502	Dept	Company Y	Growth plan	25,000	20	10,000	12	8,887,450	88.8 Lakhs
a944dc86-41b3-4966-b34d-6488d4765502	Assets	Land	NA	100,000	15	20,000	8	18,230,273	1.8 Crores
a944dc86-41b3-4966-b34d-6488d4765502	Equity	Mutual funds	Direct/indirect plan	10,000	5	5000	20	471,379	4.7 Lakhs
a944dc86-41b3-4966-b34d-6488d4765523	Dept	Company Y	Growth plan	20,000	20	20,000	10	7,007,549	1.38 Crores
a944dc86-41b3-4966-b34d-6488d4765523	Assets	Land	NA	100,000	20	30,000	8	16,940,402	1.7 Crores
a944dc86-41b3-4966-b34d-6488d4765523	Assets	House	NA	100,000	20	30,000	5	12,169,073	12 Crores
a944dc86-41b3-4966-b34d-6488d4865506	Equity	Corporate bonds	Direct plan	50,000	20	100,000	9	61,672,364	6.1 Crores
a944dc86-41b3-4966-b34d-6488d4865506	Equity	Shares	NA	200,000	10	50,000	35	36,775,406	3.7 Crores

Table 7 Example 1 of existing customer portfolio

Customer identification number	Subcategory	Plan	Initial investment	Tenure	Investment per month	YoY return (%)s
a944dc86-41b3-4966-b34d-6488d4765502	Company X	Growth plan	25,000	20	10,000	10
a944dc86-41b3-4966-b34d-6488d4765502	Company Y	Growth plan	25,000	20	10,000	12
a944dc86-41b3-4966-b34d-6488d4765502	Land		100,000	15	20,000	8
a944dc86-41b3-4966-b34d-6488d4765502	Mutual funds	Direct/indirect plan	10,000	5	5000	20

Table 8 Goal prediction by the agent (Example 1)

Approach	Result
Existing portfolio of customer	The goal is achievable. (Failed stage)
Diversify portfolio	The goal is achievable. (Will go to next stage)
Increase principal	The goal is achievable. (Will go to next stage)

13 Conclusion

This paper proposes a model for portfolio optimization based on reinforcement learning. The proposed model not only recommends portfolio reallocation for an investment type, such as trading. Instead, it considers all the available investment options and plans available in the market and recommends the reallocation of investments across all investment options. The proposed model continuously learns the trend of each asset while preserving the knowledge acquired using historical data. The model learns and evolves as the environment changes. Therefore, it can recommend the portfolio reallocation not only based on the historical data also using the recent investment and trends.

Table 9 Example 2 of existing customer portfolio

Customer identification number	Subcategory	Plan	Initial investment	Tenure	Investment per month	YoY returns (%)
a944dc86-41b3-4966-b34d-6488d4765523	Company Z	Growth plan	20,000	20	20,000	10
a944dc86-41b3-4966-b34d-6488d4765523	Land		100,000	20	30,000	8
a944dc86-41b3-4966-b34d-6488d4765523	House		100,000	20	30,000	5
a944dc86-41b3-4966-b34d-6488d4865506	Corporate bonds	Direct plan	50,000	20	100,000	9
a944dc86-41b3-4966-b34d-6488d4865506	Shares		200,000	10	50,000	

Table 10 Goal prediction by the agent (Example 2)

Approach	Result
Existing portfolio of customer	The goal is not achievable
Diversify portfolio	The goal is not an achievable goal if standalone (Failed stage)
Increase principal	The goal is achievable with a standalone approach (will go to the next stage)
Hybrid (increase principal after diversify portfolio)	The goal is achievable with a hybrid approach (will go to next stage)

References

1. Yang H, Liu XY, Wu Q (2018) A practical machine learning approach for dynamic stock recommendation. In: Proceedings—17th IEEE International Conference Trust Security Private Computing Communication. 12th IEEE International Conference Big Data Science Engineering Trust, pp 1693–1697. https://doi.org/10.1109/TRUSTCOM/BIGDATASE.2018.00253
2. Paiva FD, Cardoso RTN, Hanaoka GP, Duarte WM (2019) Decision-making for financial trading: a fusion approach of machine learning and portfolio selection. Expert Syst Appl 115:635–655. https://doi.org/10.1016/J.ESWA.2018.08.003
3. Financial time series forecasting with machine learning techniques: a survey|Paper|Microsoft Academic. https://academic.microsoft.com/paper/1575726819/reference/search?q= Financial time series forecasting with machine learning techniques: A survey. Accessed 05 Sept 2021
4. Xue J, Huang L, Liu Q, Yin J (2017) A bi-directional evolution algorithm for financial recommendation model. Commun Comput Inf Sci 768:341–354. https://doi.org/10.1007/978-981-10-6893-5_25
5. Pathak A, Shetty NP (2019) Indian stock market prediction using machine learning and sentiment analysis. Adv Intell Syst Comput 711:595–603. https://doi.org/10.1007/978-981-10-8055-5_53
6. Wu X, Ye Q, Hong H, Li Y (2020) Stock selection model based on machine learning with wisdom of experts and crowds. IEEE Intell Syst 35(2):54–64. https://doi.org/10.1109/MIS.2020.2973626
7. Reinforcement learning: what is, algorithms, applications, Example. https://www.guru99.com/reinforcement-learning-tutorial.html. Accessed 05 Sept 2021
8. Cavalcante RC, Oliveira ALI (2015) An approach to handle concept drift in financial time series based on extreme learning machines and explicit drift detection. In: Proceedings international joint conference neural networks. https://doi.org/10.1109/IJCNN.2015.7280721
9. Kaur H, Choudhury T, Singh TP, Shamoon M (2019) Crime analysis using text mining. In: 2019 international conference on contemporary computing and informatics (IC3I), pp 283–288. https://doi.org/10.1109/IC3I46837.2019.9055606
10. Chandel A, Aggarwal A, Mittal A, Choudhury T (2019) Comparative analysis of AES & RSA cryptographic techniques. Int Conf Comput Intelli Knowl Econ (ICCIKE) 2019:410–414. https://doi.org/10.1109/ICCIKE47802.2019.9004338
11. Anuj Kumar Y, Tomar R, Kumar D, Gupta H (2012) Security and privacy concerns in cloud computing. Int J Adv Res Comput Sci Softw Eng 2(5)
12. Bansal P, Aggarwal B, Tomar R (2019) Low-voltage multi-input high trans-conductance amplifier using flipped voltage follower and its application in high pass filter. In: 2019 International Conference on Automation, Computational and Technology Management (ICACTM), pp 525–529. https://doi.org/10.1109/ICACTM.2019.8776789

13. Reinforcement Learning Agents—MATLAB & Simulink. https://www.mathworks.com/help/reinforcement-learning/ug/create-agents-for-reinforcement-learning.html. Accessed 05 Sept 2021
14. Simple Reinforcement Learning: Q-learning | by Andre Violante|Towards Data Science. https://towardsdatascience.com/simple-reinforcement-learning-q-learning-fcddc4b6fe56. Accessed 16 Sept 2021
15. Reinforcement Learning in Portfolio Management by Bijula Ratheesh|Analytics Vidhya|Medium. https://medium.com/analytics-vidhya/reinforcement-learning-in-portfolio-management-4e4eed75961d. Accessed 09 Sept 2021
16. Reinforcement learning in trading. https://blog.quantinsti.com/reinforcement-learning-trading/. Accessed 09 Sept 2021

Deep Learning Approach for Electricity Load Forecasting Using Multivariate Time Series Data

Shishir Zaman, Md. Nayeem, Rifah Tatrapi, and Shamim Ripon

Abstract Electricity load forecasting plays a vital role in planning power systems in any region. A reliable forecast model is essential to implement an affordable and sustainable energy system. Several studies reveal that weather conditions in a region have a strong relation with energy generation. This paper presents deep learning models to forecast electricity demand considering both energy generation and weather factors in a region. The deep learning models consist of LSTM, stacked LSTM, and CNN-LSTM. In addition, ARIMAX forecasting is applied to compare the effectiveness of the deep learning models. With RRMSE of 1.528%, the CNN-LSTM model outperforms not only the other LSTM models but also ARIMAX confirming the usability of the proposed model.

Keywords Electricity load forecasting · LSTM · Stacked LSTM · CNN-LSTM

1 Introduction

Forecasting electricity load is crucial for ensuring a secure electricity supply [1] and an efficient energy management system. It plays an important role to mitigate pollution and boost economic activity. Recent rapid population and economic growth are accelerating the augment in electricity demand [2]. In the next 20 years, global energy demand is expected to increase by 40%, and by this process, carbon emissions will increase by 25%, which will influence climate change [3]. Therefore, there should be a fair balance between electricity demand and generation to minimize that effect. The generated electricity is hard to store and cannot be stored in a considerable quantity. So, an efficient electricity demand forecasting approach is required to ensure a stable power supply. Several statistical and deep learning-based machine learning approaches have been employed for electricity demand forecasting. It has been found that the accuracy of deep learning-based forecasting models is sufficiently well but also have some issues including overfitting, long training time, etc.

S. Zaman · Md. Nayeem · R. Tatrapi · S. Ripon (✉)
Department of Computer Science and Engineering, East West University, Dhaka, Bangladesh
e-mail: dshr@ewubd.edu

V. Skala et al. (eds.), *Machine Intelligence and Data Science Applications*, Lecture Notes on Data Engineering and Communications Technologies 132,
https://doi.org/10.1007/978-981-19-2347-0_62

Inefficient resource utilization, significant distribution losses, and high fluctuation of the electricity market are the cause of the overestimation of electricity load. Also, financial losses and unmet demands occur due to the underestimation of electricity load demand [4]. Thus, efficient methods are required to accurately forecast the electricity load. In this paper, we implement three prominent deep learning model architectures (LSTM, Stack-LSTM, CNN-LSTM) to forecast electricity load and conduct behavioral and performance analysis among them. Along with them, we employed ARIMAX models for comparative analysis. These models are implemented on a composite dataset of four years of electricity generation, consumption, and weather data for Spain, and the predictions have an hourly frequency.

The rest of this paper is structured as follows. Section 2 explains the literature review of electricity load forecasting. A brief overview of the dataset is provided in Sect. 3. Section 4 demonstrates data cleaning, feature selection, and model architecture. Analysis of experimental results is provided in Sect. 5, and finally, Sect. 6 concludes the paper by summarizing the finding and outlining future plans.

2 Related Work

In order to supply stable electricity, it is crucial to anticipate electricity load. For that purpose, a lot of studies have been conducted to find a perfect machine learning electricity demand forecasting model. To forecast electricity demand, Yukseltan E. et al. proposed a feedback-based methodology and an autoregression part to forecast hourly, daily, and yearly demand [5]. At every step of moving average feedback, the coefficient was updated, and the model extrapolates demand based on updated parameters. Biswas et al. [6] proposed a neural network-based prediction model to predict residential house electricity load. The prediction accuracy of the proposed model was compared with other forecast models and found promising. The artificial neural network approach has good forecasting ability and can handle nonlinear problems, but it contains some overfitting and long training time issues. These problems were solved by the support vector machine model [7, 8]. Hamzaçebi et al. [9] proposed a gray forecasting model SGM (1,1) to forecast electricity load in Turkey. The model was used to forecast the monthly electricity demand between 2015 and 2020 in Turkey, and the model accuracy was 5.18 in MAPE evaluation metrics. Guo et al. [10] proposed an electricity load forecasting model based on the seasonal vector error correction model (SVECM) and self-adaptive screening (SAS) model. The obtained MAPE value was 1.66, which signifies a higher accuracy than ANN, SVM, X-12-ARIMA, and SVR with X-12-ARIMA model. Li et al. [11] proposed a hybrid electricity demand forecasting method based on Adaptive Fourier Decomposition (ADF) with Support Vector Machine (SVM) and got higher accuracy than ARMA, SVM, AFDSCASVM, SACSVM, BPNN, and ELM model. Zhang et al. [12] proposed a hybrid-based SVR model with meteorological factors and electricity price where SVR was used to develop and analyze forecasting model and an improved genetic algorithm (IAGA) was employed to optimize penalty factor C

Table 1 Dataset distribution

	Energy-related data	Weather-related data
Number of features	29 (continuous)	12 (continuous) and 5 (categorical)
Total sample size	35,064	178,396

and Gaussians kernel function. In 2018, a combined forecasting model of empirical mode decomposition (EDM) and LSTM was proposed by Bedi [4] where time series signals were decomposed into intrinsic mode functions by the EDM method and the LSTM model was trained by extracted IMFs and residuals. Based on a comparison of the forecasted results, they claimed that the proposed model (EMD + LSTM) outperforms other regression models.

As mentioned above, various exploration research has been conducted to explore the novel electricity load forecasting approach and from much research, it was found that deep learning-based models outperform conventional statistical models. Also, deep learning-based models promise as alternatives to traditional statistical models [13]. From this motivation, we have selected three prominent deep learning-based forecasting models LSTM, Stacked LSTM, and CNN-LSTM to extrapolate electricity load. CNN-LSTM model performed better by a small margin among the proposed deep learning models.

3 Dataset Overview

The dataset that has been used in this paper is collected from Kaggle [14] and is public for use. This dataset consists of electricity generation, pricing and consumption for Spain, and weather data for the five largest cities in Spain. The dataset is formed with 4 years of hourly data from the year 2015 to 2018. Table 1 represents distribution of the dataset.

Each energy generation sample consists in Megawatt and there is hourly time series data, so each faction denotes Megawatt hours.

4 Method

The schematic view of the proposed workflow is depicted in Fig. 1. The composite dataset of weather and energy is driven to the feature selection process after conducting cleaning operations, where Granger Causality feature selection technique is performed to extract important features. The obtained features from the feature selection process are then employed in the deep learning and statistical-based forecasting models. Several types of evaluation metrics are employed to evaluate the

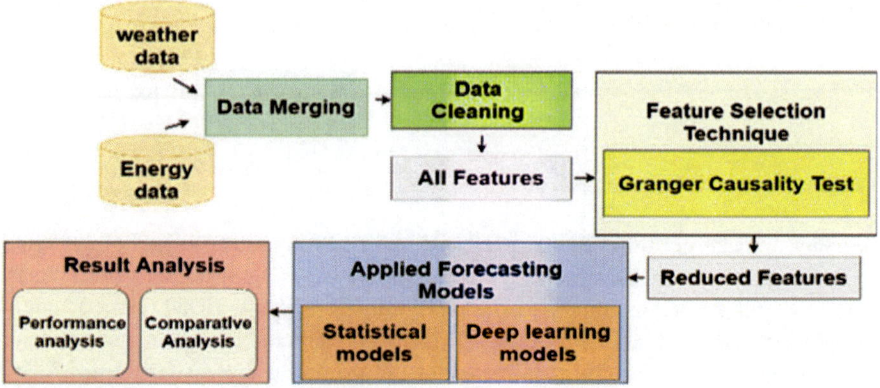

Fig. 1 Proposed workflow scheme

performance of each model, and a comparative analysis is also conducted among them.

4.1 Data Cleaning and Analysis

Data cleaning aims to identify and remove irrelevant data, e.g., missing values, duplicates, or not associated with analysis data, pertinent to create reliable data. To assert the aim, features constituted by zeros and missing values are firstly omitted from the dataset. There are some forecasted features in the dataset that is irrelevant to the analysis. Thus, all the features that are irrelevant to the analysis are omitted from the dataset. In our dataset, there is the existence of missing values in energy-related data and duplicate and outlier values in weather-related data. Since simply dropping missing values will not be a good idea in time series analysis, missing values are filled using the time interpolation method with a forward direction considering the existence of non-linearity in the target feature. Although it may cause noisy input data, the number of null values is considerably minor than the total sample size, and it may not affect the performance noticeably. Thereafter, comprehended duplicates are eliminated by keeping the first row of the duplicate values simultaneously constrain. In addition, to check data consistency, a dataset is created by keeping the last row of the duplicate values simultaneously constrain. To check the consistency of the dataset, Coefficient of Determination metric (R-squared) has been employed. Before that, categorical features from the dataset are encoded from strings to numerical labels. After applying the Coefficient of Determination metric, features with an R-squared value of 1 are kept and omit others features that are induced to data inconsistency. Then the available outliers in weather pressure are replaced compared with the highest (1051 hPa) and lowest (931 hPa) air pressure filed in the Liberian peninsula [15] and compared with the estimated maximum limit of the wind speed

(>50 ms^{-1}) of F1 (Fujita scale) tornado [16] wind speed outliers are replaced. Furthermore, the Spearman correlation test is applied concerning the target feature. Features that depict null correlation with the target feature are dumped from the dataset. After that, the weighted mean of each weather feature for the five cities is calculated. The population of the cities is used for weight terms in our analysis. The weighted mean of each weather feature is calculated by using Eq. (1).

$$\text{weighted mean weather feature} = \sum_{city=1}^{5}\left(\text{weather feature} \times \frac{\text{population}}{\sum_{city=1}^{5}\text{population}}\right) \tag{1}$$

Thereafter, only weighted mean weather features are kept in the analysis from the weather-related features. Finally, both energy-related and weather-related data are merged and there is a total of 23 features including the target feature with a total sample size of 35,064.

With that, there is a chance of the existence of seasonality and trend in the time series data. To check trend and seasonality, Augmented Dickey–Fuller (ADF) [17] and Kwiatkowski–Phillips–Schmidt–Shin (KPSS) tests [18] are applied. The ADF is a unit root test that defines the constancy of a time series by a trend. Null (H_0) and alternate (H_1) hypothesis of the ADF test are, respectively, "the time series has a unit root, thus it is non-stationary" and "there is no unit root, thus the time series is stationary." On the contrary, Null (H_0) and alternate (H_1) hypothesis of the KPSS test are, respectively, "The time series is stationary as there is a unit root in the time series" and "The time series is not stationary as there is no unit root in the time series." After applying ADF and KPSS test on the target feature, with 1% significance level the alternate hypothesis of the ADF test is accepted as critical value at 1% (−3.430537) is greater than ADF statistics (−21.420316). Conversely, with 1% significance level the null hypothesis of the KPSS test is accepted as critical value at 1% (0.347000) is less than KPSS statistics (0.936473).

So, from both of these statistical tests, it can be said that the electricity load (target feature) is stationary; hence, there is no existence of trend and seasonality or existence of constant trend and no effect of seasonality on the electricity load. After that, autocorrelation and partial autocorrelation are applied to the electricity load. The partial autocorrelation plot depicts that direct relationship of hour (t) is strongest with the previous hours $t-1$, $t-2$, and $t-25$ time-steps. Therefore, in this study the previous 25 time lags are used for feature selection and deep learning models.

4.2 Feature Selection

In a time series to observe causality between two features, the Granger causality [19] has been widely used. It is a statistical hypothesis test. If the autoregressive model of Y is outstandingly precise with the past values of X and Y compared to the only past values of Y, then X is called Granger cause of Y. The Granger causality can be employed on the stationary time series variables. For that, the stationarity of the features has been checked by applying the ADF test similar to the electricity load, and that results in the time series is stationary for all features. For instance, assume x_t and y_t are two stationary time series sequences, $x_{t\text{-}p}$ and $y_{t\text{-}p}$ are the past values p of x_t and y_t, respectively. Then, conducting two regressions shown in Eqs. (2)–(3) Granger causality is performed.

$$\widehat{y_{t_1}} = \sum_{p=1}^{m} a_p y_{t-p} + \varepsilon_t \tag{2}$$

$$\widehat{y_{t_2}} = \sum_{p=1}^{m} a_p y_{t-p} + \sum_{p=1}^{n} b_p x_{t-p} + \eta_t \tag{3}$$

Here, y_{t1} is first regression fitting value and y_{t_2} is second regression fitting value, respectively, the highest number of lag x_t and y_t are represented by m and n, and least square ascertained coefficient vectors represented by $a_p, b_p \in \mathrm{R}$, ε_t, η_t are white noises. P-value is obtained by applying the F test. Null hypothesis H_0 of the Granger causality test is that x_t Granger does not cause y_t (p-value > 0.05) whether alternate hypothesis H_1 is that x_t Granger causes y_t. In this study, Granger causality test for the target feature (electricity load) is applied with max lags 25. The outcome of the Granger causality test selects 13 features that are Granger cause of the target feature.

Subsequently, along with the target feature (electricity load), these selected features are used for multivariate data where the previous 25 h' data is used for forecasting present hour data. With that, these selected features are used as exogenous factors in the ARIMAX models.

4.3 Long Short-Term Memory Network (LSTM)

LSTM is a more efficient artificial neural network-based architecture and has a strong capability in learning long-term time dependency information. It converges the optimum minima attributed to the problem of vanishing gradient by incorporating memory cells and gating mechanisms instead of using conventional perceptron architecture. The LSTM model contains cell memory (internal state), input gate, output gate, and forget gate [20] which controls the flow of information. The basic architecture of an LSTM memory cell is depicted in Fig. 2.

The LSTM models with a single hidden LSTM layer and multiple hidden LSTM layers (stacked LSTM) are assigned to extrapolate the future electricity load. In the single hidden LSTM layer-based model, 100 LSTM units are set to learn time dependency for different time lags and output of the previous LSTM hidden layer is driven to a dense layer with 200 neurons. To improve generalization and regularization for reducing overfitting, a dropout unit is assigned before the output layer.

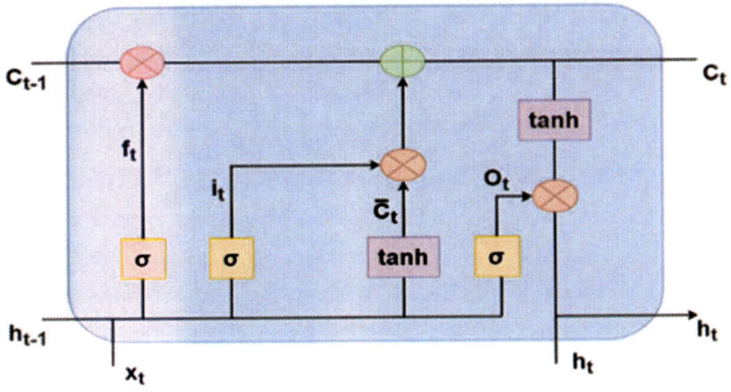

Fig. 2 LSTM memory cell

Deep neural network with multi-layer architecture has a better generalization and ability to learn specific patterns than a single-layer model [21]. Inspired by that, we have trained the LSTM model with two consecutive LSTM layers, where the first layer (input layer) contains 250 LSTM units and the second layer contains 150 LSTM units (Fig. 3). A dense layer with 150 neurons and a 10% dropout unit is incorporated in between the second LSTM layer and the output dense layer. The model training phase for the single hidden LSTM layer model and stacked LSTM model is conducted for 120 epochs with 32 batch sizes. Each batch is formed by the previous 25 lags to forecast one-hour future electricity demand.

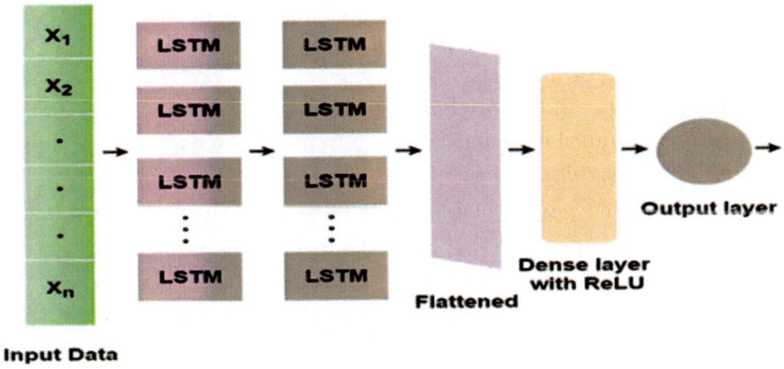

Fig. 3 Proposed stacked LSTM model

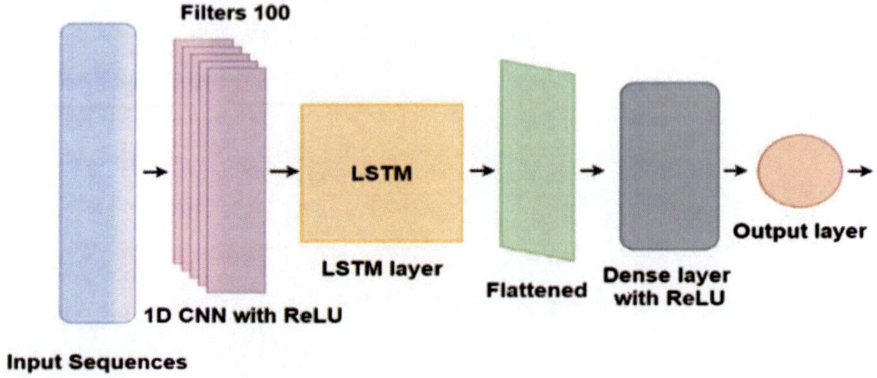

Fig. 4 Proposed CNN-LSTM model

4.4 CNN-LSTM

CNN is an artificial neural network architecture and has been widely used to eliminate data redundancy and attain important data features [22]. The CNN hidden layer is generally divided into a convolutional layer, pooling layer, and dense layer. Equation (4) refers to the l-th convolutional layer output for input vector $x_i^0 = \{x_1, x_2, x_3, \ldots, x_n\}$ where b_j^1 represent bias for j-th feature map, m is the index filter value and kernel weight is w.

$$y_{ij}^l = \sigma\left(b_j^l + \sum_{m=1}^{M} W_{m,j}^l x_{i+m-l}^0, j\right) \tag{4}$$

The CNN-LSTM model consists of a series of connections where the CNN layer can receive various input variables and output values from the previous CNN layer driven to LSTMs gate units. In the proposed model (Fig. 4), The CNN block consists of one 1D convolutional layer with filter size 100 and kernel size 2. The CNN output data is driven to LSTM through the Rectified Linear Unit (ReLU) activation function. In the LSTM section, 100 LSTM units are set to learn long-term time dependency and a dense layer is incorporated with 100 neurons before getting the final output. The model training phase is conducted for 120 epochs with 32 batch sizes, where each batch is formed by the previous 25 lags to forecast one-hour future electricity load.

4.5 ARIMAX

ARIMAX is an extension model of Auto-Regressive Integrated Moving Average (ARIMA) with exogenous variables X and can handle multivariate datasets, that is a combined model of Auto-Regressive (AR) and Moving Average (MA) model with a differencing method. The ARIMAX architecture has three factors which are autoregressive (p), integration (d), and moving average (q) [23]. ARIMAX model is compatible for analysis where there are extra features that have a possible effect on predicted value. The equation for ARIMAX shown in Eq. (5).

$$y_t = \mu + \sum_{i=1}^{k} \frac{\Theta_i(B)}{\Phi_i(B)} B^{l_i} x^{it} + \varepsilon_t \tag{5}$$

$$\varepsilon_t = \frac{\Theta(B)}{\Phi(B)} a_t$$

where $\Theta_i(B)$ and $\Phi_i(B)$ indicate average and autoregression multinomial coefficients of the i - th and I - th input time series, respectively; lag degree of i - th input variables is indicated by l_i; ε_t indicates time series regression residual; $\Theta(B)$ and $\Phi(B)$ indicate residual series' moving average and autoregression coefficients' multinomial, respectively, and a_t indicates white noise. By analyzing ACF and PACF function, we observe a significant partial autocorrelation in AR (2) and AR (25) series while correlation was not remarkable in moving average series. Therefore, the proposed ARIMAX model was trained for (2,0,0) and (25,0,0) parameter sets. The ARIMAX (7,0,0) model was also developed following the rule of minimizing AIC [24].

5 Result

The entire dataset is split into train, test, and validation sections, where train data is applied to fit the proposed model, validation data is used to appraise the model's fit on the training dataset by fine-tuning the model hyperparameters and test data is applied to estimate the model performance.

To estimate the performance of the proposed deep learning models, five statistical metrics entitled Root Mean Squared Error, Relative Root Mean Squared Error, Root Mean Squared Percentage Error, Mean Absolute Error, and Mean Absolute Percentage Error are employed. The equation of the evaluating statistical metrics is addressed in Eqs. (6–10).

$$\text{RMSE} = \sqrt{\frac{1}{n}\sum_{i=1}^{n}(\text{predicted load} - \text{actual load})^2} \tag{6}$$

$$\text{RRMSE} = 100 \times \frac{\sqrt{\frac{1}{n}\sum_{i=1}^{n}(\text{predicted load} - \text{actual load})^2}}{\text{mean of actual laod}} \tag{7}$$

$$\text{RMSPE} = 100 \times \sqrt{\frac{1}{n}\sum_{i=1}^{n}\left(\frac{\text{predicted load} - \text{actual laod}}{\text{actual load}}\right)^2} \tag{8}$$

Table 2 Performance of the models

Approach	RMSE (MW)	RMSPE (%)	RRMSE (%)	MAE (MW)	MAPE (%)
LSTM	500.117	1.810	1.723	349.464	1.244
Stacked LSTM	458.439	1.659	1.580	315.340	1.118
CNN-LSTM	**443.283**	**1.603**	**1.528**	**300.173**	**1.065**
ARIMAX(2,0,0)	2156.523	7.531	7.432	1743.476	6.114
ARIMAX(7,0,0)	2161.791	7.544	7.450	1749.009	6.132
ARIMAX(25,0,0)	4127.847	12.915	14.225	3405.856	11.075

$$\text{MAE} = \frac{1}{n}\sum_{i=1}^{n}|\text{predicted load} - \text{actual load}| \quad (9)$$

$$\text{MAPE} = 100 \times \frac{1}{n}\sum_{i=1}^{n}|\text{predicted load} - \text{actual load}| \quad (10)$$

Here, n represents the total number of predicted load (or actual load) values. Multiple evaluation metrics are utilized to evaluate every model's performance in that an evaluation matrix might not be effective to exhibit all shortcomings of the model [25]. RMSE and MAE values that are most imminent to zero are considered as a best performing model. Over a range of statistically different hydrological flows, RRMSE, RMSPE, and MAPE are also applied to describe the model's behavior [26]. More specifically, a model performance with RRMSE < 10% is considered excellent, with 10% < RRMSE < 20% is good, with 20% < RRMSE < 30% is fair, and with RRMSE > 30% is poor [27].

Table 2 depicts the forecasting performance of the proposed models (LSTM, Stacked LSTM, CNN-LSTM) and the ARIMAX models in the test period by using root mean squared error (RMSE), root mean squared percentage error (RMSPE), relative root mean squared error (RRMSE), mean absolute error (MAE), and mean absolute percentage error (MAPE).

From Table 2, it is demonstrated that CNN-LSTM model achieves maximum forecasting exactitude by a small margin compared with the other two proposed deep learning models with distinction of RMSE = 56.834 MW, RMSPE = 0.207%, RRMSE = 0.195%, MAE = 49.291 MW, MAPE = 0.179% for LSTM and RMSE = 15.156 MW, RMSPE = 0.056%, RRMSE = 0.052%, MAE = 15.167 MW, MAPE = 0.053% for Stacked LSTM. Figure 5 exhibits the forecasted electricity load of Last 48 h of the proposed deep learning models along with actual electricity load in the test period.

Here, the CNN-LSTM model achieved scantily better than the other two deep learning models, particularly in lower and higher electricity load bounds.

In contrast to the proposed models, ARIMAX (2,0,0), ARIMAX (7,0,0), and ARIMAX (25,0,0) are employed. Among ARIMAX (2,0,0), ARIMAX (7,0,0), and ARIMAX (25,0,0), ARIMAX (2,0,0) performed highest forecast accuracy difference of RMSE = 5.268 MW, RMSPE = 0.013%, RRMSE = 0.018%, MAE = 5.533 MW

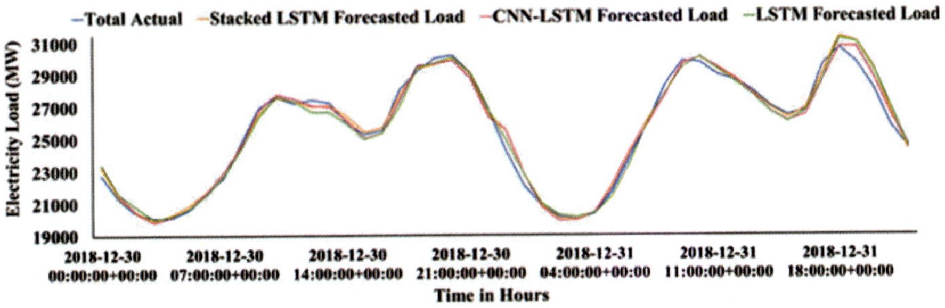

Fig. 5 Last 48 h forecasted load of the proposed models along with actual load

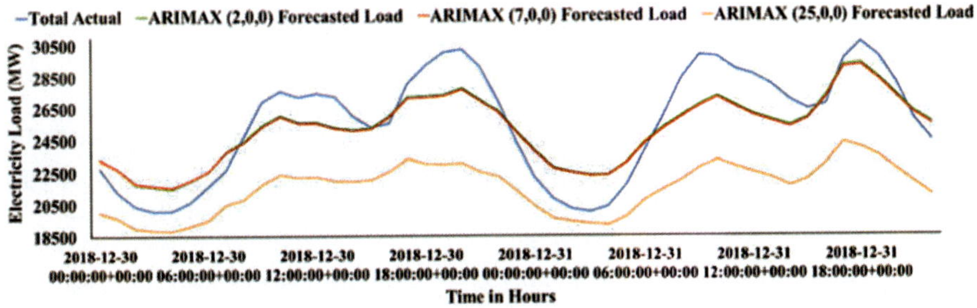

Fig. 6 Last 48 h forecasted load of the ARIMAX models along with actual load

and MAPE = 0.018% with ARIMAX (7,0,0) and RMSE = 1971.324 MW, RMSPE = 5.384%, RRMSE = 6.793%, MAE = 1662.38 MW, MAPE = 4.961% with ARIMAX (25,0,0). Figure 6 spectacles the forecasted electricity load of the last 48 h of the ARIMAX models along with the actual electricity load in the test period.

Here, ARIMAX (25,0,0) model underperformed in contradiction of the ARIMAX (2,0,0) and ARIMAX (7,0,0) where ARIMAX (2,0,0) performed slightly better than ARIMAX (7,0,0).

With that, CNN-LSTM outperformed in comparison with the ARIMAX models with RMSE = 443.283 MW, RMSPE = 1.603%, RRMSE = 1.528%, MAE = 300.173 MW, and MAPE = 1.065% where ARIMAX (2,0,0) model RMSE = 2156.523 MW, RMSPE = 7.531%, RRMSE = 7.432%, MAE = 1743.476 MW and MAPE = 6.114%. Figure 7 illustrates the forecasted electricity load of the last 48 h of the CNN-LSTM model in contradistinction to ARIMAX (2,0,0) along with the actual electricity.

Along with that, the CNN-LSTM model performed extremely well with the lowest RRMSE value. The lower and upper forecasted peaks of the electricity loads are closer to peaks of the actual electricity loads (Figs. 5 and 7). This defines the reliability and ability of the CNN-LSTM model.

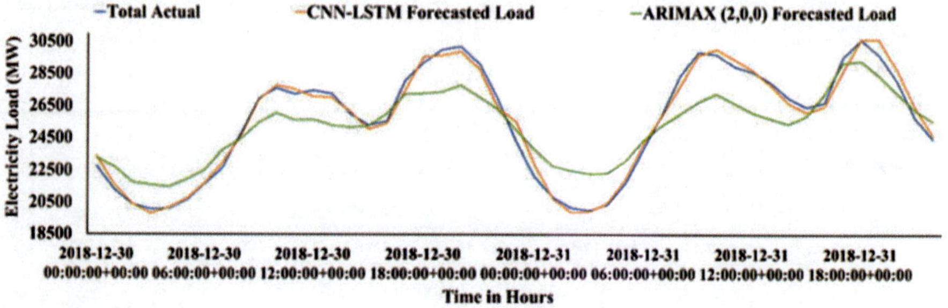

Fig. 7 Last 48 h forecasted load of the CNN-LSTM compared to the ARIMAX (2,0,0) along with actual load

6 Conclusion

In this paper, three deep learning models LSTM, stacked LSTM, and CNN-LSTM are acquainted for hourly electricity load forecasting using multivariate energy data consisting of weather data of the five cities of Spain. Three statistical models ARIMAX (2,0,0), ARIMAX (7,0,0), and ARIMAX (25,0,0) are also implemented in this work to concede comparative appraisement of the forecasting efficiency with the deep learning models. The achievement of the models is appraised by using five statistical metrics. The performance results concluded that CNN-LSTM model achieves the best forecast accuracy with RRMSE 1.528% in comparison with the other developed models.

Despite the best performance of the CNN-LSTM model, by integrating the model with different hyperparameters the forecasting accuracy can be increased. Also in this study, the Granger causality test is utilized for feature selection. From a large dataset, an effective selection of relevant features can enhance the model performance and outstandingly lessen the model training time. Consequently, for a better perception of the proposed models more feature selection methods like filter and wrapper-based techniques can be employed. In addition, along with the population of the cities, other factors, i.e., factories and industries, can be utilized for the weighted term and feature generation techniques can be applied to generate new features in future works.

References

1. Son H, Kim C (2017) Short-term forecasting of electricity demand for the residential sector using weather and social variables. Resour Conserv Recycl 123:200–207. https://doi.org/10.1016/J.RESCONREC.2016.01.016
2. Zhao GY, Liu ZY, He Y, Cao HJ, Guo YB (2017) Energy consumption in machining: classification, prediction, and reduction strategy. Energy 133:142–157. https://doi.org/10.1016/J.ENERGY.2017.05.110
3. Dehalwar V, Kalam A, Kolhe ML, Zayegh A (2017) Electricity load forecasting for urban area using weather forecast information. In: 2016 IEEE International Conference on Power and

Renewable Energy, ICPRE 2016, pp 355–359. https://doi.org/10.1109/ICPRE.2016.7871231
4. Bedi J, Toshniwal D (2018) Empirical mode decomposition based deep learning for electricity demand forecasting. IEEE Access 6:49144–49156. https://doi.org/10.1109/ACCESS.2018.2867681
5. Yukseltan E, Yucekaya A, Bilge AH (2020) Hourly electricity demand forecasting using Fourier analysis with feedback. Energ Strat Rev 31:100524. https://doi.org/10.1016/J.ESR.2020.100524
6. Biswas MAR, Robinson MD, Fumo N (2016) Prediction of residential building energy consumption: a neural network approach. Energy 117:84–92. https://doi.org/10.1016/J.ENERGY.2016.10.066
7. Zhao HX, Magoulès F (2012) A review on the prediction of building energy consumption. Renew Sustain Energy Rev 16:3586–3592. https://doi.org/10.1016/J.RSER.2012.02.049
8. Al-Musaylh MS, Deo RC, Adamowski JF, Li Y (2018) Short-term electricity demand forecasting with MARS, SVR and ARIMA models using aggregated demand data in Queensland, Australia. Adv Eng Informatics 35:1–16. https://doi.org/10.1016/J.AEI.2017.11.002
9. Hamzaçebi C (2016) Primary energy sources planning based on demand forecasting: the case of Turkey. J Energy Southern Afr 27:2–10. https://doi.org/10.17159/2413-3051/2016/V27I1A1560
10. Guo H, Chen Q, Xia Q, Kang C, Zhang X (2018) A monthly electricity consumption forecasting method based on vector error correction model and self-adaptive screening method. Int J Electr Power Energy Syst 95:427–439. https://doi.org/10.1016/J.IJEPES.2017.09.011
11. Li R, Jiang P, Yang H, Li C (2020) A novel hybrid forecasting scheme for electricity demand time series. Sustain Cities Soc 55:102036. https://doi.org/10.1016/J.SCS.2020.102036
12. Zhang G, Guo J (2020) A novel method for hourly electricity demand forecasting. IEEE Trans Power Syst 35:1351–1363. https://doi.org/10.1109/TPWRS.2019.2941277
13. Chandramitasari W, Kurniawan B, Fujimura S (2019) Building deep neural network model for short term electricity consumption forecasting. In: Proceeding—2018 international symposium on advanced intelligent informatics: revolutionize intelligent informatics spectrum for humanity, SAIN 2018, pp 43–48. https://doi.org/10.1109/SAIN.2018.8673340
14. Hourly energy demand generation and weather|Kaggle. https://www.kaggle.com/nicholasjhana/energy-consumption-generation-prices-and-weather. Last accessed 2021/08/28
15. Burt S (2007) The highest of the highs … extremes of atmospheric pressure in the British Isles, Part 2—the most intense anticyclones. Weather 62:31–41. https://doi.org/10.1002/wea.35
16. Fujita Scale—Tornado Damage Scale. factsjustforkids.com
17. Dickey DA, Fuller WA (2012) Distribution of the estimators for autoregressive time series with a unit root. https://doi.org/10.1080/01621459.1979.10482531. 74:427–431 (2012). https://doi.org/10.1080/01621459.1979.10482531
18. Kwiatkowski D, Phillips PCB, Schmidt P, Shin Y (1992) Testing the null hypothesis of stationarity against the alternative of a unit root: how sure are we that economic time series have a unit root? J Econometrics 54:159–178. https://doi.org/10.1016/0304-4076(92)90104-Y
19. Granger CWJ (1969) Investigating causal relations by econometric models and cross-spectral methods. Econometrica 37:424. https://doi.org/10.2307/1912791
20. Greff K, Srivastava RK, Koutnik J, Steunebrink BR, Schmidhuber J (2017) LSTM: a search space odyssey. IEEE Trans Neural Netw Learn Syst 28:2222–2232. https://doi.org/10.1109/TNNLS.2016.2582924
21. Hermans M, Schrauwen B (2013) Training and analysing deep recurrent neural networks. In: Advances in neural information processing systems, vol 26
22. Kim TY, Cho SB (2019) Predicting residential energy consumption using CNN-LSTM neural networks. Energy 182:72–81. https://doi.org/10.1016/J.ENERGY.2019.05.230
23. Fan J, Shan R, Cao X, Li P (2009) The analysis to tertiary-industry with ARIMAX model. J Math Res 1:156
24. Akaike H (1998) Information theory and an extension of the maximum likelihood principle, pp 199–213. https://doi.org/10.1007/978-1-4612-1694-0_15

25. Chai T, Draxler RR (2014) Root mean square error (RMSE) or mean absolute error (MAE)?—Arguments against avoiding RMSE in the literature. Geosci Model Dev 7:1247–1250. https://doi.org/10.5194/GMD-7-1247-2014
26. Prasad R, Deo RC, Li Y, Maraseni T (2017) Input selection and performance optimization of ANN-based streamflow forecasts in the drought-prone Murray Darling Basin region using IIS and MODWT algorithm. Atmos Res 197:42–63. https://doi.org/10.1016/J.ATMOSRES.2017.06.014
27. Mohammadi K, Shamshirband S, Anisi MH, Amjad Alam K, Petković D (2015) Support vector regression based prediction of global solar radiation on a horizontal surface. Energy Convers Manage 91:433–441. https://doi.org/10.1016/J.ENCONMAN.2014.12.015

Social Media Role in Pricing Value of Cryptocurrency

Anju Mishra, Vedansh Gupta, Sandeep Srivastava, Arvind K. Pandey, Lalan Kumar, and Tanupriya Choudhury

Abstract Cryptocurrency is a modern idea of decentralized digital currency. It is miles a shape of currency that exists handiest in digital form. To save you fraudulent transactions, a decentralized device is based on the block chain system to report and oversee transactions. For several years, cryptocurrency has been an ultra-modern subject matter, in particular inside the most youthful population. They invest their money for profit and for the fun additionally. Right here, social media plays a very critical function in manipulating the rate of cryptocurrency. As we recognize that social media is a completely effective device to mention anything to the general public. Now, the afternoon anyone is using social media and some bilinear humans use this approach and change the fee in crypto for his or her income and they advantage a totally massive profit. The impact of social media on the fee/price of cryptocurrencies is the focus of this study. And how this may manipulate the charge of cryptocurrencies. This has a look at additionally specializes in the prevailing situation and destiny potentialities of cryptocurrencies.

Keywords Cryptocurrency · Social media · Bilginer's · Doge coin · Elon Musk · Trend

1 Introduction

Cryptocurrencies are the latest digital exchange methods of the twenty-first century that use cryptography for security. It is also known as distributed digital currency.

A. Mishra (✉) · V. Gupta · S. Srivastava · L. Kumar
GLBITM, Greater Noida, India
e-mail: anju.pandey.12@gmail.com

A. K. Pandey
Modern College of Professional Studies, Mohan Nagar, Ghaziabad, India

T. Choudhury (✉)
School of Computer Science, University of Petroleum and Energy Studies (UPES), Dehradun, Uttarakhand 248007, India
e-mail: tanupriya@ddn.upes.ac.in

V. Skala et al. (eds.), *Machine Intelligence and Data Science Applications*, Lecture Notes on Data Engineering and Communications Technologies 132,
https://doi.org/10.1007/978-981-19-2347-0_63

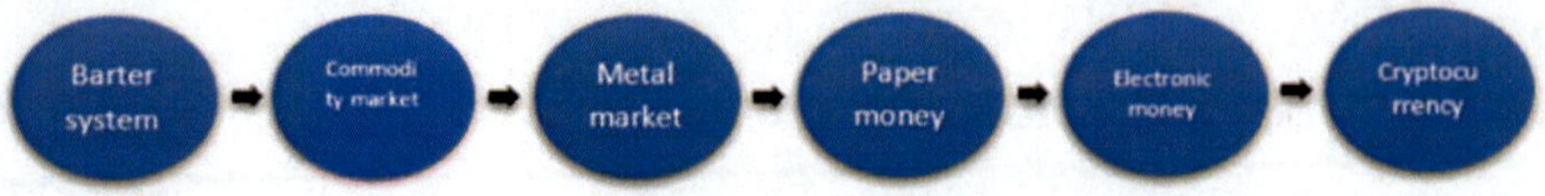

Fig. 1 Evolution of trade

The word 'crypto' comes from the ancient Greek word, 'Krypto's, which means encryption or encryption [1]. Digital currency was firstly introduced in 1990s in the form of store value of peer-to-peer (P2P) payments that did not require bank authorization because of the decentralized mode. Instead of a centralized authority, digital currency transactions are verified and records are kept by a decentralized system that uses cryptography (Fig. 1).

The mode of trade is changing time-to-time according to the need, security [2], technology advancement, people's requirements, rapid changing in resources, demand, and the presence of government policies. In these days, cryptocurrencies like Bitcoin, Ethereum, Ripple, Dogecoin, etc., are trending in the financial market. People are showing their interest in buying these cryptocurrencies. Bitcoin was the first decentralized created in the year 2009. Cryptocurrency does not available in the physical form as it is in the digital value in the web [3].

1.1 Types of Currencies

Its purpose is to act as a medium of trade. There are around 4000 digital currencies available on the internet in 2021, which really is a lot of cryptocurrencies [4]. New currency can be generated by anytime. Bitcoin is currently having the largest blockchain network [5], followed by Ripple, Ethereum and Litecoin. Here, some famous cryptocurrencies which are as follow [1, 3, 6–10].

1.1.1 Bitcoin (BTC)

Bitcoin is the most common and famous cryptocurrency [11–15]. On August18, 2008, the domain name bitcoin.org was registered. It operates without a central authority or banks, thanks to peer-to-peer electronic currency system technology and block chain technologies. Nowadays, we can do the transaction and exchange other currencies, products, and goods [8, 10].

Features of Bitcoin

Peer-to-peer transfers that are quick.

Payments from all across the world.

Processing fees are low.

1.1.2 Ethereum (ETH)

Ethereum is another type of cryptocurrencies which was proposed in late 2013 by VitalikButerin. Ethereum may be a decentralized, open-source blockchain that supports smart contracts and its own programming language, called Solidity. Since its inception in 2013, Ether's value has climbed dramatically, reaching slightly over $3000 per token in May, but it still behind Bitcoin's worth of about $40,000 per coin. [8]

1.1.3 Ripple (XRP)

Ripple is the native cryptocurrency developed by Ripple Lab. Ripple is a cryptocurrency and a digital payment network for financial transactions that was launched in 2012. XRP is another name for it. Ripple is used for payment settlement, asset exchange, and main use is international transaction and security transfer they use for this network of bank. The code of Ripple is written in the C++ programming language. Ripple transactions use less energy than Bitcoin, and its cost is very little cost. It is a global currency network that allows for quick, secure, and low-cost money transfers. We can trade any form of currencies using ripple [8].

1.1.4 Dogecoin (DOGE)

Dogecoin is a cryptocurrency created by two software engineers, Billy Marcus and Jackson Palmer. Launched in December 2013, Dogecoin features a Shiba Inu dog as its logo. It is an open-source cryptocurrency based on peer-to-peer. Dogon Coin became even more popular in 2021 because Elon Musk frequently mentioned or talked about Dogecoin [16] on his Twitter account. Also known as DOGE to the Moon. DOGE price rises very fast after many tweets by Elon Musk. The DOGE price is at Rs.0.34 on January 1, 2021 and then rises to Rs. 50.41 on May 8, 2021 in a few months. It is the most popular cryptocurrency in 2021, and the current cost on 31 Oct, 2021 is INR 22.41. Most people make huge profits from this coin. DOGE price is expected to rise further in the future [14, 17].

1.1.5 USD Coin

The USD Coin (USDC) promotes itself as the "world's digital dollar." The USDC is the outcome of the efforts of Goldman Sachs, Baidu, and IDG Capital investments. It was developed by the circle, a worldwide financial organization. USD Coin (USDC) is a significant advancement in the way we utilize money. Digital dollars function similarly to other digital content: they travel at the speed of the internet, can be transferred in the same manner that we share content, and are less expensive and more secure than current payment systems [8].

2 Review of Literature

H. S. Narman and A. D. Uulu (2020) write in their Impacts of Positive and Negative Comments of Social Media Users to Cryptocurrency research paper they analyze the perspective of user on social media's good and bad influence on cryptocurrencies. He also noticed that the percentage of people who are positive about investing in digital bitcoin is larger than the percentage of those who are negative about it. The raw data for the research study was gathered from Reddit.com. He helps the new investors firstly to learn the user's relation between price and mood change [18].

H. S. Narman, A. D. Uulu and J. Liu(2018) focus in their Profile Analysis for Cryptocurrency in Social Media research paper that they analyze the education level of that investors who are investing their money in the cryptocurrency. And they find the %age of the user's educations according to their standards. Like highest ratio of the education comes from middle school, i.e., 60% approx., then 50% from college, then 30% come from high school, and at the last 10% in the other level [19].

A. Mittal, V. Dhiman, A. Singh and C. Prakash (2019) focus on their Short-Term Bitcoin Price Fluctuation Prediction Using Social Media and Web Search Data research paper that the pricing of bitcoin is heavily influenced by social media platforms. They take the data of tweet volume, Google trend, tweet sentiment. They find that tweet sentiment is the biggest platform which effect the price most as compare to others. They apply some algorithm on the raw data, and they find that Bitcoin price is predicted by the 77.01% form tweet volume and 66.06% from Google trend [20].

A. Jain, S. Tripathi, H. D. Dwivedi and P. Saxena (2018) in his Forecasting Price of Cryptocurrencies Using Tweets Sentiment Analysis. Discuss about the prediction between the two digital currencies, i.e., about the basis of a user's tweet on cryptocurrencies, Bitcoin and Litecoin were created. He finds that Bitcoin is not much effective by the social media as compare to Litecoin. And they also said that social media is play a very big role in manipulating the price of the cryptocurrency. This is done by some of the popular users, user network [21].

3 Objective of the Study

Through this paper we learn the concept of cryptocurrency, and its type. To understand, how social media effect the changing in the price of cryptocurrency. This paper is based on secondary data previously done by others and referring to various sources such as research paper, articles, websites and statutory reports, news, and tweets. I have gone through with the previous research paper they all talk about the technology, blockchain, and price prediction. I have analyzed the reason that social media effect the price of the cryptocurrency by doing some comparative analysis research. Through this search paper I want to share my research. In this research paper, I have focus on that coins which are recently affected by the social media. I have talk about that coins whose price change very fast and go to their highest price ever.

4 Social Media Effect the Price of Cryptocurrency by the Following Reasons

We all know that social [17, 22, 23] media may be a very powerful tool. Now each day, everyone uses social media and it's a neighborhood of life. We will do anything with social media. For now, it goes without saying that social media [24, 25] plays a significant part in our everyday lives. It is how we communicate with our loved ones, receive news, or just wait. We all know that social media's influence is growing, and it's is reasonable to say that it might now have an impact on cryptocurrency pricing. On our favorite social networking applications, we now receive daily news. With popular cryptocurrencies like Bitcoin, Ethereum, and Dogecoin hitting new all-time highs on a regular basis, it is easy to assume that these digital currencies are on the rise [1]. Here are some details that social media to affect the pricing of cryptocurrencies which are as follow.

4.1 The Power of the Trend

In the current era, we all follow the latest trends. If any big celebrity says something about something, then this become a new trend. Like this cryptocurrency also be a part of this (Fig. 2).

Because of a TikTok trend, the price in DOGE will be affected in July 2020. One of the oldest Dogecoin-pumping tick chat videos was published by user Jamezg97. Since then the challenge has garnered hundreds of thousands of views on the app and started trending on Twitter. The competition-based coin stood at over $ 878 million

Fig. 2 TikTok video news [15]

in 24-h volume as of Wednesday afternoon, up 683% from the average daily volume of the previous year [1, 26].

4.2 *Billionaires and Big Companies*

Billionaires are the people that can easily manipulate everything by just saying a few words about that and then public follow them. Recently, the same happened by a billionaire the most recent real Elon Musk, the founder of Tesla, is an example. Tesla, an electric car and sustainable energy company, revealed in early February 2021 that it had purchased 1.5 billion dollars' worth of Bitcoin. And they also said that they will accept the payment of their product by the BTC [1] (Fig. 3).

Following this, BTC reached an all-time high and broke beyond the 44,000 USD barrier for the first time. And he also tweets about the DOGE coin. He said DOGE coin DOGE TO THE MOON. He pumps the DOGE at their highest price, but after some time other big companies crashed the market and the value of DOGE coin goes down. And then he also tweets that Tesla is not accepting the Bitcoin for car purchase after that price bitcoin fall back [1, 14, 17] (Figs. 4 and 5).

Fig. 3 Elon Musk tweet [27]

Fig. 4 Elon Musk [28]

Fig. 5 Tweet snap [29]

4.3 *YouTubers*

Nowadays, YouTube is a very helpful tool by a Google for getting the daily tasks easy. Every age population uses YouTube. Here is a solution to any problem, just searching that problem. Like this there are many YouTubers who create videos and tell the people to buy this and this currency to gain profit.

4.4 *Social Channels like (Facebook, Twitter, and Especially Reddit)*

Facebook, Twitter, and Reddit are the biggest social news channels in the world. They are the one who can connect us with the real-world people's thought's, their point of interview and, etc. They paid amount to the members posting the content. There are multiple communities who are there to share their views and also they invest their money in the crypto so they give advice to the public to also do invest in the same crypto because if people continuous buy a coin, then the value (price) of the coin will increase and if they sell the coins then the value (price) will decrease [9].

Fig. 6 Doge coin price graph [26]

5 Recently the Coins Which Affect the Most

In the blow, I have shown the graph of several cryptocurrencies which show the flow of price value with respect to year. I have mention three crypto coin named Dogecoin, Bitcoin, and Ethereum.

5.1 *Doge Coin*

In the Fig. 6, it represents the flow of Doge Coin in respect to price and year. This data is based on 2019–2021. Development rates surpassing following remarks from Tesla CEO Elon Musk on January 28 and 29, Doge Coin's value increased by about 216% to 0.023535 US dollars.

5.2 *Bitcoin*

In the Fig. 7, it represents the flow of Bitcoin in respect to price and year. This data is based on year Oct 2013 to Jul 24,2021. The price of the Bitcoin change when Tesla tweet that they will accept the bitcoin as a purchase of Tesla Car. Later on, they decline about that and then the price of Bitcoin falls.

5.3 *Ethereum*

In the Fig. 8, it represents the flow of Ethereum in respect to price and year. This data is based on year Aug 2015 to Jul 26, 2021. The price of Ethereum is keep growing

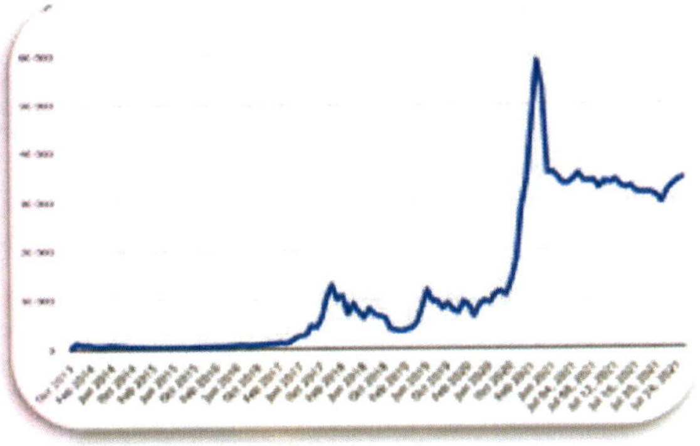

Fig. 7 Bitcoin price graph [26]

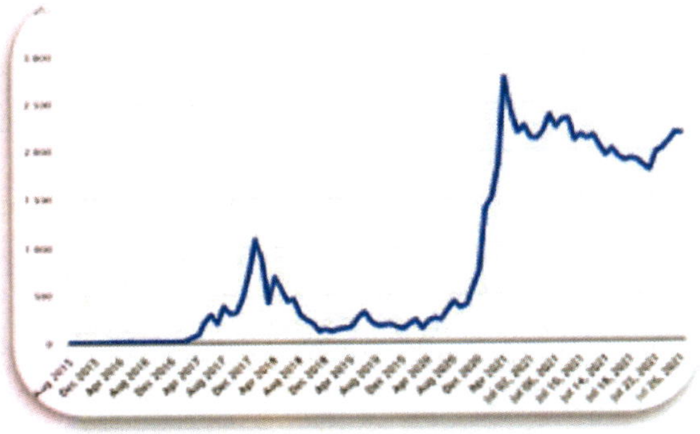

Fig. 8 Ethereum coin price graph [26]

form the year 2021. It touches the highest point, i.e., 2500 U.S. dollars. After the Bitcoin Ethereum price also went up in 2021.

6 Conclusion and Future Scope

The main focus of this paper is that, unlike previous work on cryptocurrencies, we will focus on consumers and their views on cryptocurrencies rather than on their technical and technical aspects. I analyzed the data and discovered that social media plays a significant impact in fluctuating bitcoin prices. There are different reasons like some ballooners are there they start the trend of something and people follow them. YouTubers are creating many videos on YouTube, and they manipulate the mind of the investors. Some influencers also do the same things. And We all know that the cryptocurrency is a modern innovation which has become a global phenomenon. Cryptocurrency is an attractive model of payment methods that are effective and secure. They give us effective and attractive mode of payment. Though cryptocurrency we can boost our business and companies. By continuously following their activity on social media related to that particular currency. By follow them, we have to be very careful and have patience. We have only invested that amount of money, which we can bear the loss because in cryptocurrency, and there is a lot of loss risk while investing in the cryptocurrency. In the current scenario, the market of the cryptocurrencies is low as compare to starting of the year. If you want to take entry in the world of cryptocurrency, then you can start with the small amount with your extra saving. Never invest your all saving in the cryptocurrency because there will anything happen to the next when you invest the money.

As a future scope, I will enhance my research paper by adding more reasons of social media that affect the price of the cryptocurrency. There may be some more reasons that I left in this paper so in the future I will try to give some more factors.

References

1. Inamdar A, Bhagtani A, Bhatt S, Shetty PM (2019) Predicting cryptocurrency value using sentiment analysis. Int Conf Intelli Comput Control Syst (ICCS) 2019:932–934. https://doi.org/10.1109/ICCS45141.2019.9065838
2. Tomar R, Patni JC, Dumka A, Anand A (2015) Blind watermarking technique for grey scale image using block level discrete cosine transform (DCT). In: Emerging ICT for bridging the future—proceedings of the 49th annual convention of the Computer Society of India CSI 2, pp 81–89
3. Here is the list of the cryptocurrency. https://en.wikipedia.org/wiki/List_of_cryptocurrencies
4. Yiying W, Yeze Z (2019) Cryptocurrency price analysis with artificial intelligence. In: 2019 5th International Conference on Information Management (ICIM), pp 97–101. https://doi.org/10.1109/INFOMAN.2019.8714700
5. Vujičić D, Jagodić D, Ranđić S (2018) Blockchain technology, bitcoin, and Ethereum: a brief overview. In: 2018 17th international symposium INFOTEH-JAHORINA (INFOTEH), pp 1–6. https://doi.org/10.1109/INFOTEH.2018.8345547
6. Introduction to the cryptocurrency. https://en.wikipedia.org/wiki/Cryptocurrency
7. Nagamani K, Pruthu R, Sai Teja V (2021) Applications of blockchain in cryptocurrency: bitcoin and dogecoin. Int J Res Eng Sci Manage 4(7)
8. Vargas I (2021) The relationship between social media and crypto prices, 03/10/2021. https://paxful.com/blog/social-media-and-crypto-price/
9. Shehhi AA, Oudah M, Aung Z (2014) Investigating factors behind choosing a cryptocurrency. IEEE Int Conf Industrial Eng Eng Manage 2014:1443–1447. https://doi.org/10.1109/IEEM.2014.7058877
10. https://www.investopedia.com/tech/most-important-cryptocurrencies-other-than-bitcoin/
11. Agrawal S, Shah HJ, Maheshwari R, Sai Shivakumar P (2020) The sanctioning of cryptocurrency: positive and negative ramifications in India, vol 10, issue 10, Oct 2020, Print ISSN No. 2249-555X. https://doi.org/10.36106/ijar
12. Thelwall M (2018) Can social news websites pay for content and curation? The SteemIt cryptocurrency model. J Inf Sci 44(6):736–751. https://doi.org/10.1177/016555151774829
13. Liang J, Li L, Chen W, Zeng DD (2019) Towards an understanding of cryptocurrency: a comparative analysis of cryptocurrency, foreign exchange, and stock, May 2019. https://doi.org/10.1109/ISI.2019.8823373 IEEE ISI2019
14. Osborne C (2014, May 14) The dirge of Dogecoin: cryptocurrency doomed to failure (Online). Available: http://www.zdnet.com/the-dirge-of-dogecoincryptocurrency-doomed-to-failure-7000029384

15. Kumar S, Gaur MS, Sagar Sharma P, Munjal D (2021) A novel approach of symmetric key cryptography. In: 2021 2nd International Conference on Intelligent Engineering and Management (ICIEM), pp 593–598. https://doi.org/10.1109/ICIEM51511.2021.9445343
16. https://indianexpress.com/article/business/world-market/dubaicoin-dbix-surges-in-24-hours-of-launch-government-calls-website-promoting-cryptocurrency-elaborate-phishing-campaign-7334068/
17. Kumar S, Karnani G, Gaur MS, Mishra A (2021) Cloud security using hybrid cryptography algorithms. In: 2021 2nd International Conference on Intelligent Engineering and Management (ICIEM), pp 599–604. https://doi.org/10.1109/Iciem51511.2021.9445377
18. Narman HS, Uulu AD (2020) Impacts of positive and negative comments of social media users to cryptocurrency. In: 2020 International Conference on Computing, Networking and Communications (ICNC), pp 187–192. https://doi.org/10.1109/ICNC47757.2020.9049693
19. Narman HS, Uulu AD, Liu J (2018) Profile analysis for cryptocurrency in social media. IEEE Int Symp Signal Process Inf Technol (ISSPIT) 2018:229–234. https://doi.org/10.1109/ISSPIT.2018.8642634
20. Mittal A, Dhiman V, Singh A, Prakash C (2019) Short-term bitcoin price fluctuation prediction using social media and web search data. In: 2019 twelfth international conference on contemporary computing (IC3), pp 1–6. https://doi.org/10.1109/IC3.2019.8844899
21. Jain A, Tripathi S, Dwivedi HD, Saxena P (2018) Forecasting price of cryptocurrencies using tweets sentiment analysis. In: 2018 eleventh international conference on contemporary computing (IC3), pp 1–7. https://doi.org/10.1109/IC3.2018.8530659
22. Anil Kumar VV, Swathy P (2019) A study on opportunities and challenges of cryptocurrency in India with special reference to bitcoin, vol 6, issue 1, Jan–March 2019, E Issn 2348-1269, Print Issn 2349-5138
23. Tomar R, Tiwari R, Sarishma (2019) Information delivery system for early forest fire detection using Internet of Things. In: Singh M, Gupta P, Tyagi V, Flusser J, Ören T, Kashyap R (eds) Advances in computing and data sciences, ICACDS 2019. Communications in Computer and Information Science, vol 1045. Springer, Singapore. https://doi.org/10.1007/978-981-13-9939-8_42
24. Gupta SS, Thakral A, Choudhury T (2018) Social media security analysis of threats and security measures. Int Conf Adv Comput Commun Eng (ICACCE) 2018:115–120. https://doi.org/10.1109/ICACCE.2018.8441710
25. Singh N, Sharma T, Thakral A, Choudhury T (2018) Detection of fake profile in online social networks using machine learning. Int Conf Adv Comput Commun Eng (ICACCE) 2018:231–234. https://doi.org/10.1109/ICACCE.2018.8441713
26. Graph of the Cryptocurrency. https://www.statista.com/statistics/806453/price-of-ethereum/
27. https://coinwizer.com/wp-content/uploads/2021/03/elon-musk.jpg
28. https://img.republicworld.com/republicprod/stories/promolarge/xhdpi/yazul6mtkflwdqmc_1618486091.jpeg
29. https://encryptedtbn0.gstatic.com/images?q=tbn:ANd9GcTvmW63kL9GqC7vOtcsD9QgW7CBkkdmTlQp4Q&usqp=CAU

Twitter Spatio-temporal Topic Dynamics and Sentiment Analysis During the First COVID-19 Lockdown in India

Arunkumar Dhandapani, Anandkumar Balasubramaniam, Thirunavukarasu Balasubramaniam, and Anand Paul

Abstract The advent of COVID-19 has tremendously affected the global economy. More people have suffered, and some of them even lost their lives. As the worst part, many countries are seeing second and third waves of COVID-19 cases despite the vaccination. India is one such country that is worse affected by the COVID-19. Understanding what people of India think and how they express their thoughts through social media platforms like Twitter has a vast significance. Therefore, in this project, we aim to use COVID-19-related tweets during the first COVID-19 lockdown in India and apply the nonnegative tensor factorization (NTF) algorithm to elicit spatio-temporal topic dynamics. While knowing the sentiments of people is important, identifying sentiments for each tweet is time-consuming and hard to interpret. Therefore, in this paper, we propose to apply sentiment analysis on the topics identified using NTF.

Keywords Nonnegative tensor factorization · COVID-19 · Topic modelling · Sentiment analysis · India

1 Introduction

COVID-19 was first detected in China, and it spreads all over the world, which has brought a drastic change to mankind. Millions of people got affected and lost their lives to this Virus. People are still struggling to return to their normal and

A. Dhandapani (✉)
Department of Computer Science and Engineering, VSB College of Engineering Technical Campus, Coimbatore, Tamil Nadu, India
e-mail: arunkumardhandapani2000@gmail.com

A. Balasubramaniam · A. Paul
School of Computer Science and Engineering, Kyungpook National University, Daegu, South Korea

T. Balasubramaniam
School of Computer Science, Queensland University of Technology, Brisbane, Australia
e-mail: thirunavukarasu.balas@qut.edu.au

V. Skala et al. (eds.), *Machine Intelligence and Data Science Applications*, Lecture Notes on Data Engineering and Communications Technologies 132,
https://doi.org/10.1007/978-981-19-2347-0_64

peaceful life as this dreadful virus keeps spreading. During the pandemic of COVID-19, people all over the world have been affected, ranging from ordinary people to organizations, geopolitical entities, and entire countries. People have been expressing their sentiments about the pandemic on Twitter which is an essential outlet for people to discuss their feelings and opinions [1, 2].

India is not an exception to this intimidating virus. The entire nation got shattered, and there was a tremendous loss in the economy as people lost their jobs, shops were closed, people were restricted from roaming outside that leads to discontinue their business. Corporate companies implemented work-from-home system and education institutions moved to online class. All these changes made people use mobiles/PCs and social media intensively. People started to express their thoughts as posts on social media platforms such as Twitter. As a result, an enormous amount of data gets generated, and the analysis of this data will benefit mankind.

With the technical advancements, the tweets on Twitter are associated with spatio-temporal information. For example, each tweet records the time when the tweet is posted and where it is posted from. Since billions of tweets are generated by millions of Twitter users, understanding what people generally talk about will become very difficult. Therefore, topic modelling approaches such as latent Dirichlet allocation [3], nonnegative matrix factorization [4], and nonnegative tensor factorization [5] have been used to elicit the general topics discussed by a huge number of users [6]. The identification of such topics will hugely benefit governments and other related agencies to take necessary steps for the well-being of people.

NTF-based methods are predominantly used when the dataset is multifaceted [7]. In this research, the dataset we have can be represented in multidimensional space as we have text (a dimension) associated with two other dimensions (time and location). Therefore, the usage of NTF will not only help to extract topics but also extract the spatial(location) and temporal(time) patterns. This is because NTF can be directly used to elicit patterns. We redirect the readers to [7] for more details on NTF and their different usages.

The contributions of this paper are twofold:

1. Firstly, we apply NTF, a machine learning algorithm to elicit spatio-temporal topic dynamics from tweets posted during the first COVID-19 lockdown in India. This will help understand how people react to the first-ever national lockdown the current generation has seen due to a pandemic.
2. Secondly, we apply sentiment analysis on the topics identified using NTF to capture the emotion and how people feel during the lockdown. Moreover, this proposed approach will generalize the sentiments for the huge data without identifying sentiments for each tweet.

The rest of the paper is organized as follows. In the Sect. 2, we discuss some of the recent related works. In Sect. 3, we discuss the proposed approach. In Sect. 4, we present the results and finally conclude in Sect. 5.

2 Related Work

Topic modelling on COVID-19-related data is conducted widely. Recently, COVID-19 vaccine-related discussion on Twitter has been conducted and identified various topics related to vaccination for COVID-19 [8]. Sentiment analysis and emotions for each topic were also identified but the spatio-temporal association is not specified. In another work, country-wise Twitter sentiment analysis during the COVID-19 outbreak is conducted [9]. Sentiments and emotions from citizens from 12 different countries were analysed with topics. But these existing works did not focus on finding the spatio-temporal patterns of the topics. Recently, [7] used NTF to identify the topics related to COVID-19 and learned the spatio-temporal patterns. However, the work is conducted on the Australian Twitter dataset, and the sentiment analysis is not conducted. On the other hand, in this paper, we use NTF to find the topics and their spatio-temporal patterns as well as the sentimental analysis on the topics.

In another recent work, sentiment analysis on COVID-19 Twitter India data has been conducted [10]. The sentiments of tweets across different states in India in various months were identified. However, this work does not consider the elicitation of topics, and the spatio-temporal association is not preserved. In [9] the sentiment analysis on COVID-19 during the first lockdown has been conducted and emotions of the people were identified. However, this work does not focus on learning the topics and their spatial-temporal association. Similarly, there exists many works [11] where the sentimental analysis has been conducted, but these works have not considered topic modelling or spatio-temporal dynamics of sentiments.

To overcome the above mentioned challenges, we in this research use NTF and sentimental analysis to elicit Twitter spatio-temporal topic dynamics and their sentimental analysis. More specifically, we focus on the COVID-19 dataset from Twitter that has been generated during the first lockdown in India. Figure. 1 represents the overall flowchart of the proposed approach.

3 Twitter Spatio-temporal Topic Dynamics and Sentiment Analysis

3.1 Dataset

In this research, we used the COVID-19 dataset collected during the first lockdown (20-Mar-2020 till 31-May-2020) [12]. This dataset consists of tweets posted by users in India, and each tweet is associated with location and time. From the original collection of 648957 tweets, we randomly sampled 2% (12980) tweets for our analysis. This will allow capturing the generalized topics.

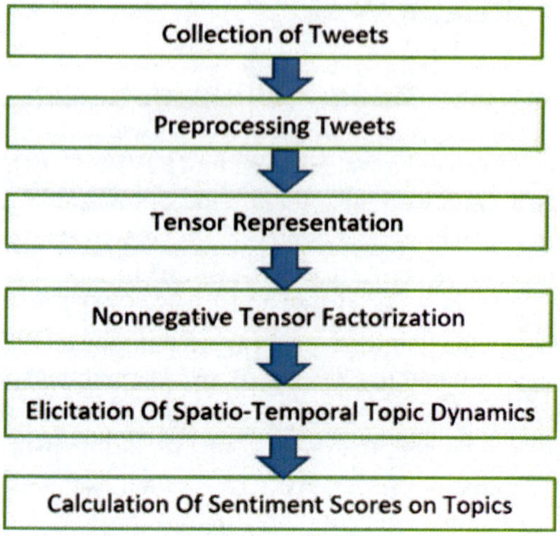

Fig. 1 Overall flowchart

3.2 *Data Representation*

Let us consider the twitter dataset where $\mathcal{D} = \{d_1, d_2, \ldots, d_S\}$ be the set of S tweets from $\mathcal{U} = \{u_1, u_2, \ldots, u_Q\}$ set of Q users. As each tweet is a collection of terms, let $\mathcal{E} = \{e_1, e_2, \ldots, e_M\}$ be the set of M unique terms for the entire collection of tweets. For each tweet, we also have associated locations and time. Let us consider $\mathcal{L} = \{l_1, l_2, \ldots, l_N\}$ be the set of N locations and $\mathcal{T} = \{t_1, t_2, \ldots, t_O\}$ be the set of O time slots. Time slots can be defined based on what type of interpretation is required. For example, it can be 24 h or 7 days. In this paper, we use 24 h as time slots to understand how the topic dynamics are in 24 h. The association between terms, locations, and their time slots can be represented as a tensor, $\mathcal{X} \in \mathbb{R}^{M \times N \times O}$. For the dataset described in Sect. 3.1, there are 3861 unique terms, 1468 unique locations, and 24 hours as time slots. This will generate a tensor of size (3861 × 1468 × 24).

3.3 *NTF*

For the tensor $\mathcal{X}$ represented in a three-dimensional as shown in Fig. 2, the objective is to decompose/factorize the tensor $\mathcal{X}$ into three factor matrices $\mathbf{E} \in \mathbb{R}^{M \times R}$, $\mathbf{L} \in \mathbb{R}^{N \times R}$, and $\mathbf{T} \in \mathbb{R}^{O \times R}$ representing each mode/dimension terms, location, and time slots, respectively. The objective function of NTF as per CP factorization [13] is given as,

$$\min_{\mathbf{E} \geq 0, \mathbf{L} \geq 0, \mathbf{T} \geq 0} f(\mathbf{E}, \mathbf{L}, \mathbf{T}) = \|\mathcal{X} - [\![\mathbf{E}, \mathbf{L}, \mathbf{T}]\!]\|^2 \tag{1}$$

This objective function is solved using factorization algorithms such as coordinate descent. In this paper, we used an advanced factorization algorithm, saturating

Fig. 2 Tensor representation

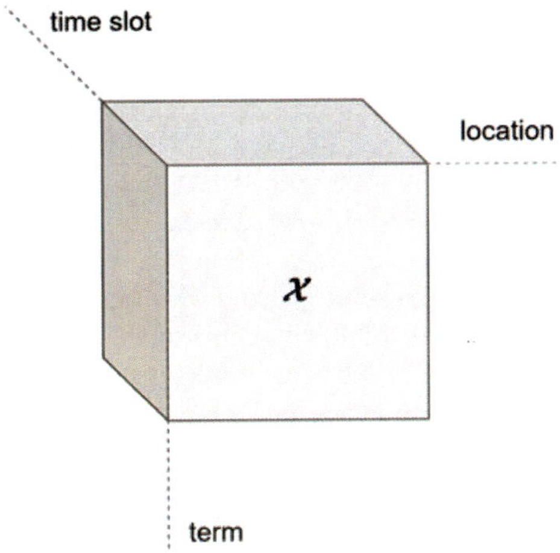

coordinate descent to solve the objective function and to learn the factor matrices [6, 14]. Each column in the factor matrix indicates a pattern. The columns from terms factor matrix **E** indicates topics, and each columns from location, and time factor matrices **L**, and **T** indicate the spatio-temporal patterns. For more details, we redirect the readers to [7]. The R in the factor matrix indicates the rank of the tensor, which determines the number of patterns we elicit from the tensor. In this paper, we set the rank R to 10 so that 10 topics/spatio-temporal patterns can be identified.

3.4 Sentiment Analysis on the Topics from NTF

Once the topics are identified using NTF, we then apply sentiment analysis on the topics. For each topic (rank R), we rank the terms based on the magnitude of the vector. Top-M terms are then selected as the representative of the topic, and the sentiment analysis is conducted on the topic to find the positive/negative sentiment.

Let the vector $\mathbf{e}_i$ be the ith column of the term factor matrix **E**. This vector (topic) will holds value for each term and the magnitude of the value determines the representative terms of this topic. For each column $\mathbf{e}_i$ in **E**, we find the top-M terms. We then concatenate the top-M terms to form a sentence. Therefore, if we have R topics, we form R sentences, $\mathcal{V} = \{v_1, v_2, ..., v_R\}$ that generalize the tweets in terms of topics.

For each sentence, we calculate the sentiment score using the TextBlob library in python [15]. It uses a Naive Bayes analyzer [16] that is pre-trained on a huge movie review dataset to classify the topics as positive or negative. As a result, we get a sentiment score for each topic ranging from 0 to 1, where 0 indicates negative sentiment and 1 indicates positive sentiment.

4 Results

Figures 3, 4, 5, 6, 7, 8, 9, 10, 11 and 12 present 10 topics and their spatio-temporal dynamics. Since the dataset used is during the first lockdown in India, the topics mostly reflect the spread of COVID-19 and COVID-19 test results. At the same time, a few people started to express their displeasure with other countries by asking everyone to buy local products. Please refer to the figure captions for a detailed summary of each topic and their saptio-temporal dynamics in India. One large point at the middle of all spatial patterns indicates India in general as the exact location is unavailable in the dataset. As shown in Table 1, except topic 6 (Fig. 8c) and topic 10 (Fig. 12c), all other topics show negative sentiments. This reflects that people want to help each other during a tough time like this (refer to Fig. 12c) and fight back COVID-19 (refer to Fig. 8c). The results are focused on the Delhi and Mumbai as the majority of tweets were posted from these locations. However, the methodology is more generic which can work for wider distribution.

5 Conclusion

The COVID-19 pandemic has changed everyone's life in some way or another. While we started to fight COVID-19, understanding more about what impact it has created in the people is important to effectively deal with it. In this paper, we propose to apply the NTF algorithm to learn the spatio-temporal topic dynamics of COVID-19 during the first lockdown in India. We also apply sentiment analysis on the topics to better understand the generalized sentiments of people. This paper demonstrates the usage of advanced machine learning algorithms to leverage huge text data generated in social media for the betterment of people.

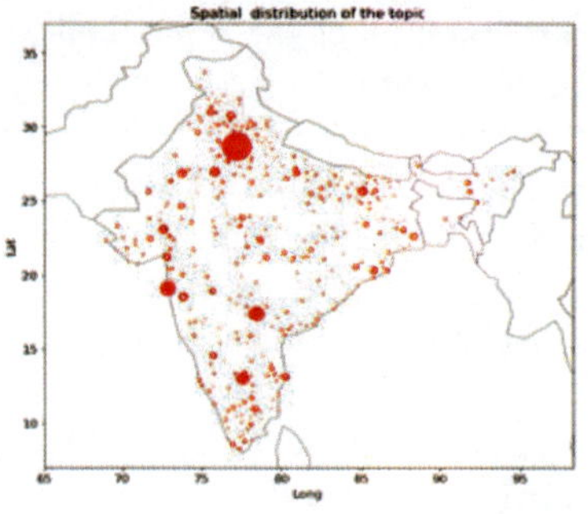

(a) Spatial distribution of the topic.

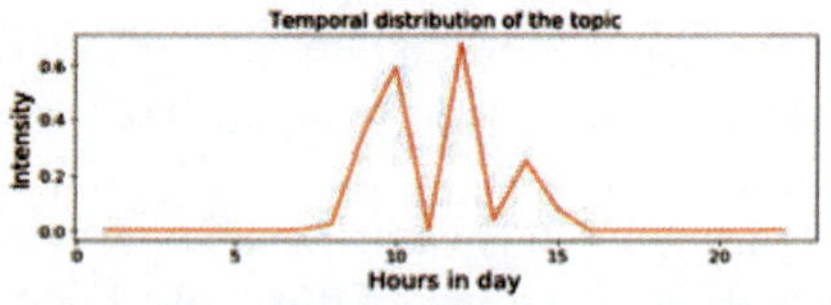

(b) Temporal distribution of the topic.

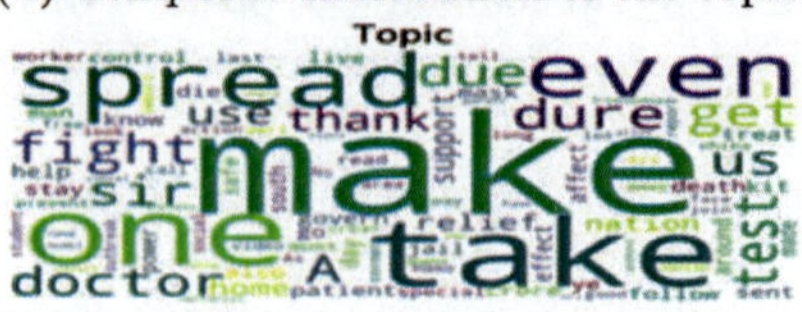

(c) Topic

Fig. 3 Topic 1. Summary: Spread of COVID-19 in India. Most of the tweets originated from Delhi, and most of the tweets are posted between 8 a.m. in morning and 1 p.m. in the afternoon

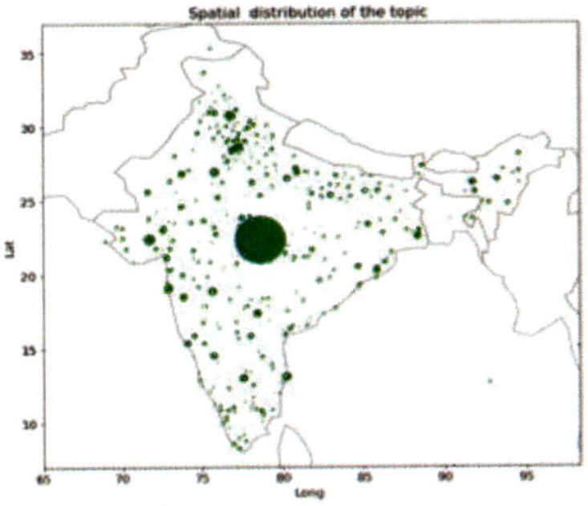

(a) Spatial distribution of the topic.

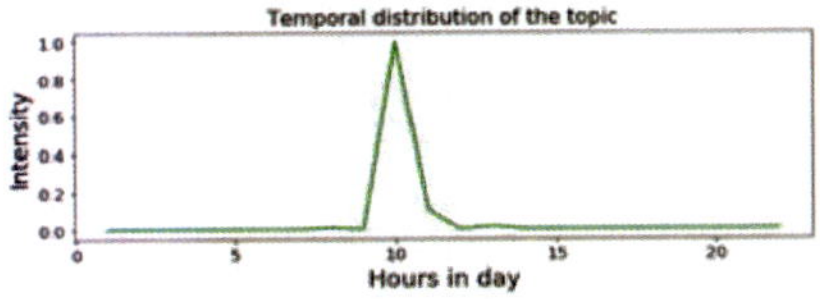

(b) Temporal distribution of the topic.

(c) Topic.

Fig. 4 Topic 2. Summary: Reports on tests and positive COVID-19 cases. People from Delhi tend to talk more about this than other states. They talk about this mostly between 9 a.m. in Morning and 12 p.m. in the afternoon

Table 1 Sentiment score of the topics elicted using NTF

Topic	Sentiment score
1 (Fig. 3c)	0.0057
2 (Fig. 4c)	0.0246
3 (Fig. 5c)	0.0454
4 (Fig. 6c)	0.0341
5 (Fig. 7c)	0.0994
6 (Fig. 8c)	**0.3477**
7 (Fig. 9c)	0.0373
8 (Fig. 10c)	0.0149
9 (Fig. 11c)	0.0989
10 (Fig. 12c)	**0.1301**

Closer to 1 indicates positive sentiment, and closer to 0 indicates negative sentiment

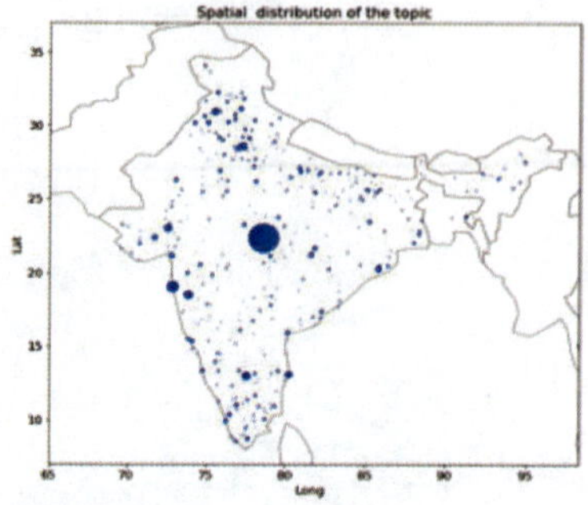

(a) Spatial distribution of the topic.

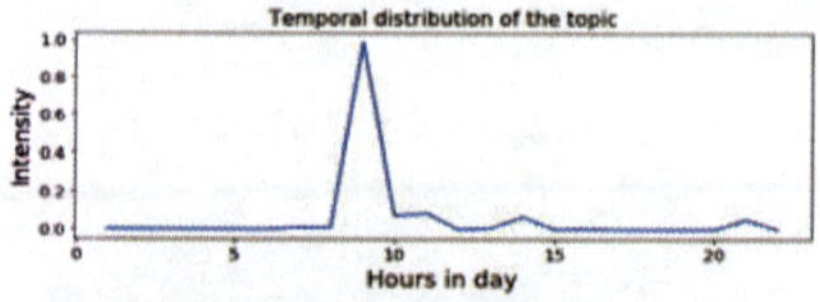

(b) Temporal distribution of the topic.

(c) Topic.

Fig. 5 Topic 3. Summary: Reports on positive cases and deaths associated with COVID-19. Mostly, the tweets are posted between 8 a.m. and 10 a.m. in the morning. While the topic 3 is similar to topic 2, and the temporal patterns of topic 3 and topic 2 are also looks same, the spatial pattern is different

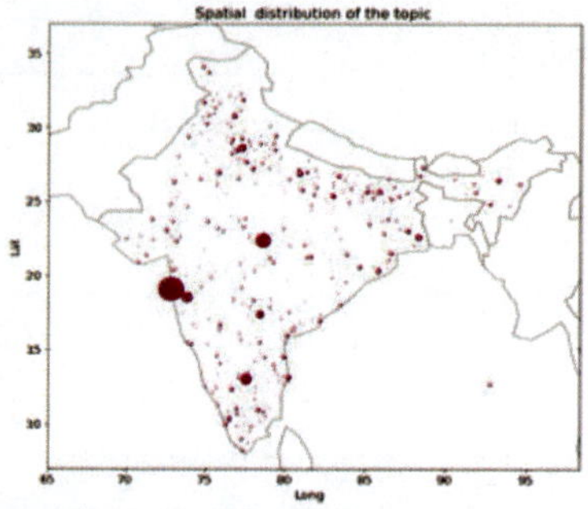

(a) Spatial distribution of the topic.

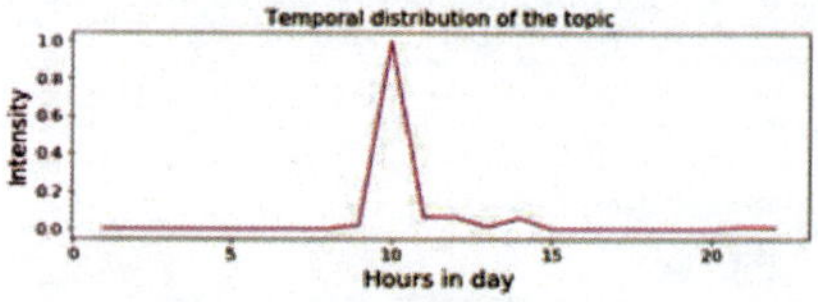

(b) Temporal distribution of the topic.

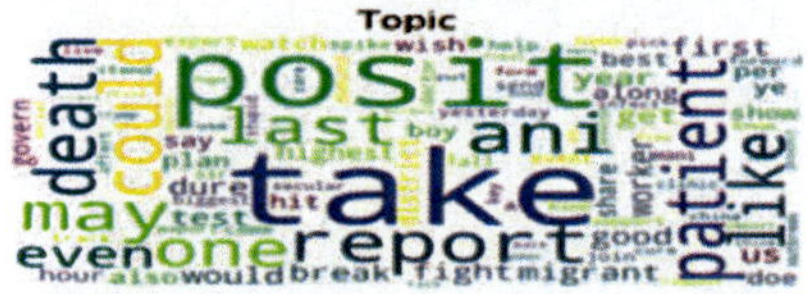

(c) Topic.

Fig. 6 Topic 4. Summary: Reports on positive cases and deaths associated with COVID-19. Mostly the tweets are originated from Mumbai posted between 9 a.m. and 11 a.m. in the morning

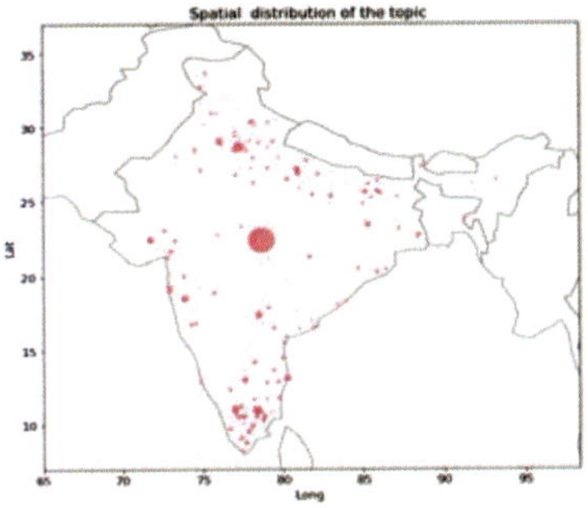

(a) Spatial distribution of the topic.

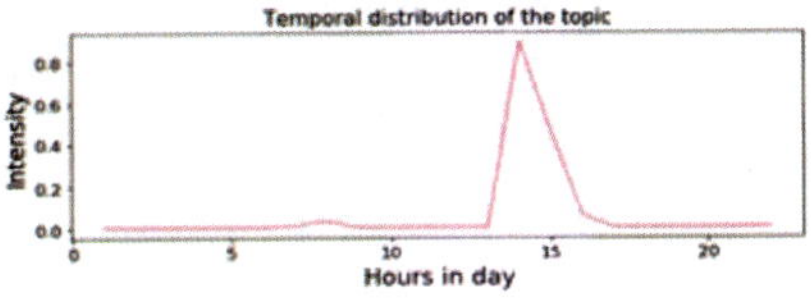

(b) Temporal distribution of the topic.

(c) Topic.

Fig. 7 Topic 5. Summary: Buy and spend local. This topic is wide spoken and most of the tweets are between 2 p.m. in afternoon and 4 p.m. in evening

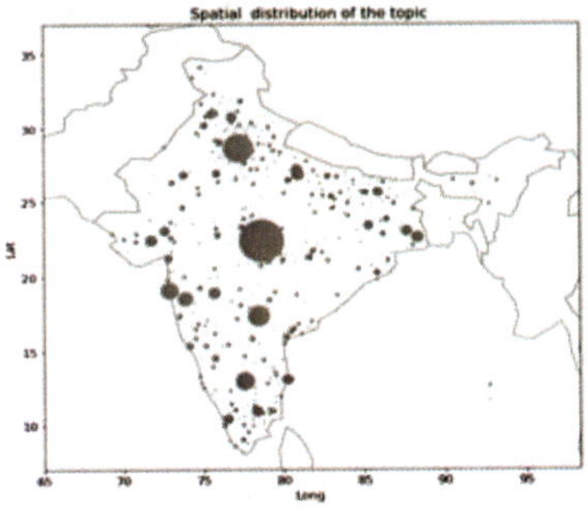

(a) Spatial distribution of the topic.

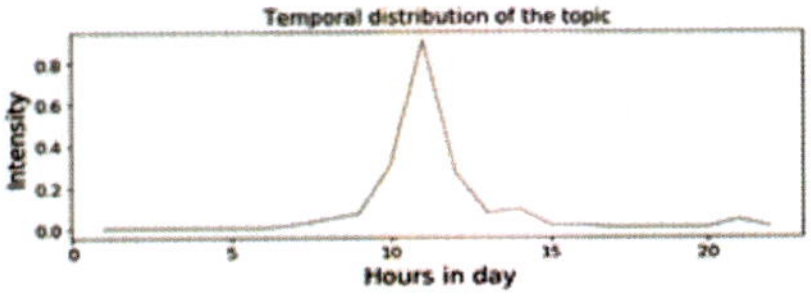

(b) Temporal distribution of the topic.

(c) Topic.

Fig. 8 Topic 6. Summary: Fight against COVID-19. This topic is wide spread across the nation but people from Delhi tend to talk more and most of the tweets are posted between 9 a.m. in Morning and 1 pm in the afternoon

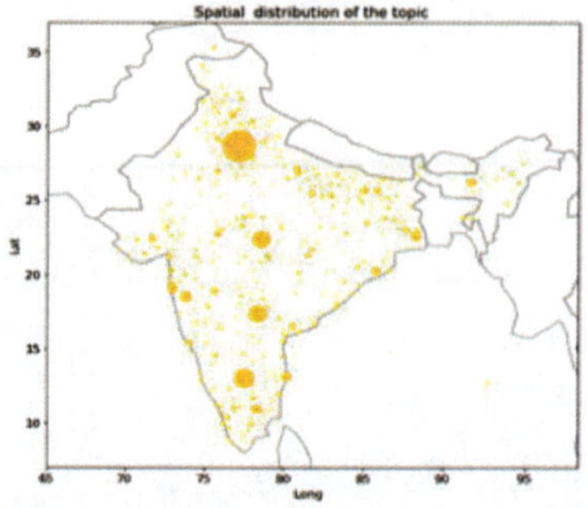

(a) Spatial distribution of the topic.

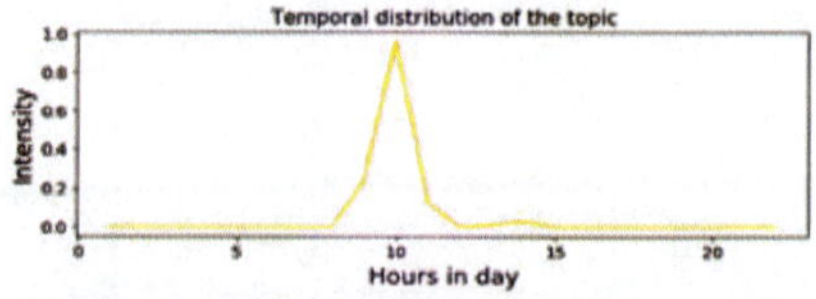

(b) Temporal distribution of the topic.

(c) Topic.

Fig. 9 Topic 7. Summary: Report on tests, positive cases and deaths associated with COVID-19. This topic is widespread across the nation but most of the tweets are from Delhi posted between 8 a.m. in the morning and 12 p.m. in the afternoon. Unlike Topic 6 which reflects same spatio-temporal patterns, the topics are different

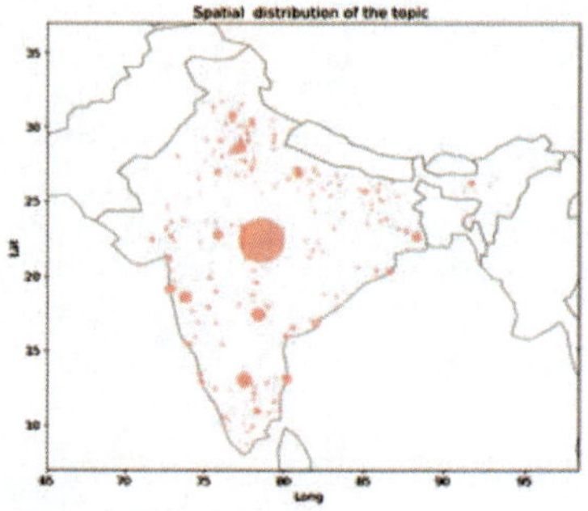

(a) Spatial distribution of the topic.

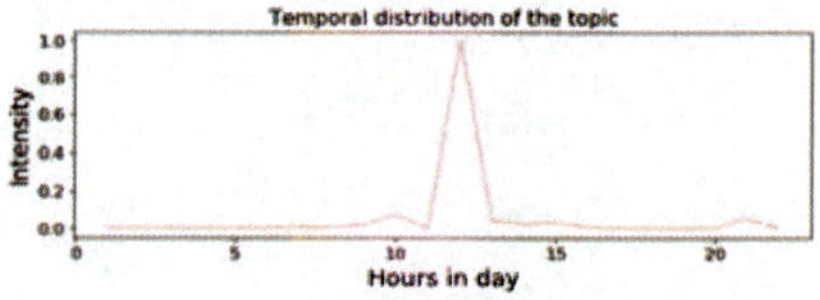

(b) Temporal distribution of the topic.

(c) Topic.

Fig. 10 Topic 8. Summary: This topic is similar to that of topic 6, but here the temporal range differs from that of Topic 6 as most of the tweets are between 11 a.m. and 1 p.m. in the afternoon

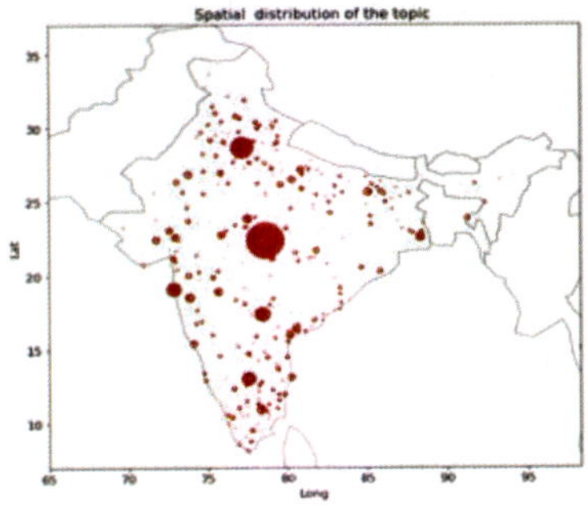

(a) Spatial distribution of the topic.

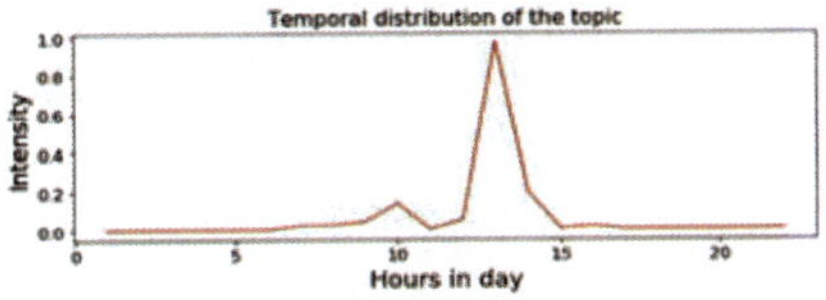

(b) Temporal distribution of the topic.

(c) Topic.

Fig. 11 Topic 9. Summary: This topic is similar to that of topic 6 but shows a different temporal pattern. The tweets originate between 12 p.m. and 3 p.m. in the afternoon

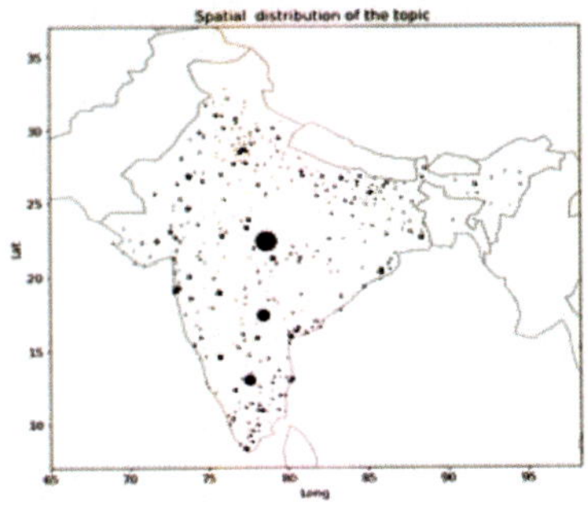

(a) Spatial distribution of the topic.

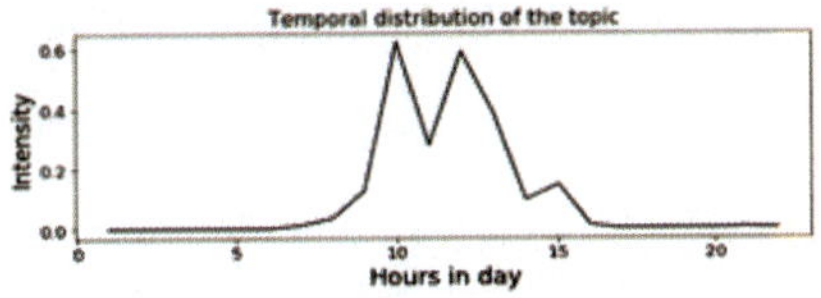

(b) Temporal distribution of the topic.

(c) Topic.

Fig. 12 Topic 10. Summary: Helping the people and donating funds. The tweets from this topic are wide spread throughout the nation, and most of the tweets are between 8 a.m. in the morning and 3 p.m. in afternoon. This is one of the long-range pattern of all the temporal patterns

Acknowledgements This research was supported by the National Research Foundation of Korea (Grant No. 2020R1A2C1012196), and in part by the School of Computer Science and Engineering, Ministry of Education, Kyungpook National University, South Korea, through the BK21 Four Project, AI-Driven Convergence Software Education Research Program, under Grant 4199990 214394.

References

1. Bashar MA, Nayak R, Balasubramaniam T (2020) Topic, sentiment and impact analysis: covid19 information seeking on social media. arXiv preprint arXiv:2008.12435
2. Barkur G, Vibha GBK (2020) Sentiment analysis of nationwide lockdown due to covid 19 outbreak: evidence from India. Asian J Psychiatry 51:102089
3. Blei DM, Ng AY, Jordan MI (2003) Latent dirichlet allocation. J Mach Learn Res 3:993–1022
4. Wang YX, Zhang YJ (2012) Nonnegative matrix factorization: a comprehensive review. IEEE Trans Knowledge Data Eng 25(6):1336–1353
5. Kolda TG, Bader BW (2009) Tensor decompositions and applications. SIAM Rev 51(3):455–500
6. Balasubramaniam T, Nayak R, Luong K, Bashar MA (2021) Identifying covid-19 misinformation tweets and learning their spatio-temporal topic dynamics using nonnegative coupled matrix tensor factorization. Soc Netw Anal Mining 11(1):1–19
7. Balasubramaniam T, Nayak R, Bashar MA (2020) Understanding the spatio-temporal topic dynamics of covid-19 using nonnegative tensor factorization: a case study. In: 2020 IEEE symposium series on computational intelligence (SSCI). IEEE, pp 1218–1225
8. Xue J, Chen J, Hu R, Chen C, Zheng C, Su Y, Zhu T (2020) Twitter discussions and emotions about the covid-19 pandemic: machine learning approach. J Med Internet Res 22(11):e20550
9. Dubey AD (2020) Twitter sentiment analysis during covid-19 outbreak. Available at SSRN 3572023
10. Vijay T, Chawla A, Dhanka B, Karmakar P (2020) Sentiment analysis on covid-19 twitter data. In: 2020 5th IEEE international conference on recent advances and innovations in engineering (ICRAIE), pp 1–7
11. AlAgha I (2021) Topic modeling and sentiment analysis of twitter discussions on covid-19 from spatial and temporal perspectives. J Inf Sci Theo Pract 9(1):35–53
12. Lamsal R (2020) Coronavirus (covid-19) tweets dataset (2020). https://dx.doi.org/10.21227/781w-ef42
13. Carroll JD, Chang JJ (1970) Analysis of individual differences in multidimensional scaling via an n-way generalization of "eckart-young" decomposition. Psychometrika 35(3):283–319
14. Balasubramaniam T, Nayak R, Yuen C (2020) Efficient nonnegative tensor factorization via saturating coordinate descent. ACM Trans Knowl Disc Data (TKDD) 14(4):1–28
15. Kaur C, Sharma A (2020) Twitter sentiment analysis on coronavirus using textblob. Tech Rep EasyChair
16. Parveen H, Pandey S (2016) Sentiment analysis on twitter data-set using naive bayes algorithm. In: 2016 2nd International conference on applied and theoretical computing and communication technology (iCATccT), IEEE, pp 416–419

Natural Language Interface in Dogri to Database

Shubhnandan S. Jamwal and Vijay Singh Sen

Abstract Database is one of the most common terms used in today's digital era in almost every domain. Digital applications use database for convenient storage and retrieval of data in an organized manner. Database management system is used to perform data operations while ensuring consistency, security, sharing, and privacy of data besides backup and recovery of the data. However, application of database requires sound technical skills since a particular language in a specific format is required by the system to obtain desired data. Dogri language is mainly spoken in Jammu and Kashmir and in some of its neighboring border district of states Himachal Pradesh and Punjab. It is added in the 8th schedule of Indian constitution and is one of the 22 official languages of India. In this paper, a novel attempt has been made to transform the Dogri language query to its equivalent SQL query format. Emphasis has been made on the fundamental and mostly used select queries that the proposed system accepts as input which is then transformed in SQL-formatted query. Since user is independent of following any predefined format for querying the database, some preprocessing operations on data are inevitable. Once data preprocessing like tokenization and stopword removal was done, Dogri language query was partitioned into clauses. These clauses were mapped to the most likely clause in SQL format. Clauses were finally sorted in a manner that a well-formatted SQL query is obtained. Natural languages are highly ambiguous, about five versions of each query were taken, and around 150 queries were tested. The proposed model obtained an accuracy of 88% in transforming appropriately Dogri language query to their equivalent SQL-formatted query.

Keywords Machine translation · Dogri · NLP · SQL · GUI · Query translation · Natural language

S. S. Jamwal (✉) · V. S. Sen
PGDCSIT, University of Jammu, Jammu, Jammu and Kashmir, India
e-mail: jamwalsnj@gmail.com

V. S. Sen
e-mail: senvjju@gmail.com

V. Skala et al. (eds.), *Machine Intelligence and Data Science Applications*, Lecture Notes on Data Engineering and Communications Technologies 132,
https://doi.org/10.1007/978-981-19-2347-0_65

1 Introduction

Query to the database in natural languages could benefit millions of user who otherwise stay aloof of using database to their convenience. Nowadays, millions of people on social media platforms are connected and they interact in their native languages to each other. Technology giants like Google, Microsoft, Amazon, Apple, etc. are assisting their users by providing applications in their native languages. Consequently, users prefer regional languages across the globe. India is a diverse country where more than 19,500 languages or dialects are spoken as their mother tongue but around 96% population has one of the 22 scheduled languages as their mother tongue according to recent Indian census. In India, Dogri language is spoken by people residing in Jammu and Kashmir and in some parts of Punjab and Himachal Pradesh. Dogri is one of the 22 official languages scheduled in constitution of India. The major source of information in the digital world of today is the database, and it plays very important role in data management. It may also be observed that almost all IT applications are using databases for storing and retrieving the information. But a person having no knowledge of structured query language (SQL) is handicapped in managing the database. In order to simplify the complexity of SQL and to manage the query in regional language by common people, many researchers are proposing solutions in natural language instead of SQL, which has resulted in the new type of NLP application in the form of NLIDB. The researchers have analyzed the abilities of NLP to develop products like NLIDB to interact with the database in regional languages. The NLIDB uses the syntactic knowledge and the knowledge about the specific database. The NLIDB translates the whole query in NLP into the SQL for extracting the relevant information from the table, and it is pertinent to mention that these NLP products have created a revolution in extracting information from databases in the regional language and has opened the domains of the research. The objective of this paper is to input query in Dogri language and get the result from the database and because of which the users have no need to learn SQL queries in order to access the database.

2 Literature Review

Natural language interfaces to databases appeared in early seventies [1], and since then, a number of NLIDB systems have been developed. LUNAR was the first system that managed to handle 90% of requests without any error [2]. TEAM was designed to be easily configurable by database administrators with no knowledge of NLIDBs [3, 4], and ASK [5, 6] developed in 1983, allowed end users to teach the system new words during the interaction. Generic interactive natural language interface to database [7] was designed by using UML and implemented in visual basic .NET 2005. The need of NLIDB is felt by number of researchers and has proposed number of architectures. Shah et al. [8] proposed NLKBIDB, an architecture based on keyword that provides solution for syntactically correct and incorrect natural language input query. The experimentation results were based on the survey database on agriculture and solve 53% of syntactically incorrect query. Wudaru et al. [9] and others proposed NLIDB system using intermediate query approach which was implemented on Movie domain chatbot, and it was proposed that it can be extended to different domains.

Das and Balabantaray [10] presented MyNLIDB, which uses a schema graph created from the underlying database, Stanford part-of-speech parser, and dependency parser to convert NL query to SQL using pipeline processing and observed that it has good performance with respect to keyword mapping. In DBPal [11], based on deep learning techniques, is introduced in order to improve the performance of models for natural language to SQL translation.

Most of the previous NLIDB systems were a kind of question-answering system that generated a SQL query corresponding to a natural language query regardless of dialog contexts [12, 13]. NLIDB techniques allow users to access information stored in databases by requests in natural language [14] and generate natural language sentences, tables, and graphical representation [15]. In order to reduce the complexities of SQL query generation, Kim [16] proposed a dialog-based NLIDB system. This dialog-based system classifies users' intentions into domain actions using a maximum entropy model. Based on the user intentions and classifications, the entropy model system fills up predefined SQL templates [17]. Further, these templates generate proper SQL queries for natural language interface. Many NLIDB systems have been developed since 1960s [18] and have given a detailed review of the classification of NLIDBs and the techniques used for their evaluation [19]. Thorough analysis of NLP techniques, architecture, and insight into the subcomponents that constitute NLIDB systems has been made. Computational linguistics has advantages and complexities involved. An introduction to NLIDB systems, their comparison, and various challenges involved in the development and efficiency improvement have been discussed. NLIDB has been developed that supported multiple databases [20]. A neural network classifier that detected classifier specific to data and an adversarial mechanism identified these components in the natural language query. Author claimed to achieve results better than previous WikiSQL [21] dataset, and it could be used with other datasets without any model retraining. Dogri is a low-resourced language that lacks the basic tools for Dogri text handling. Small amount of work done in Dogri language includes identification of verb and stemming of words to their root form [22, 23].

3 Dogri-NLIDB

This section provides a comprehensive portrayal of the architecture and functioning of the proposed Dogri-NLIDB with suitable example. Since more and more number of users are accessing the databases through their mobile apps and Web sites, the need of NLIDB system has increased and there are number of interfaces available to retrieve data such as form based, natural language, and keyword based. The mechanism of data retrieval requires knowledge of SQL and database. The NLIDB [17] is

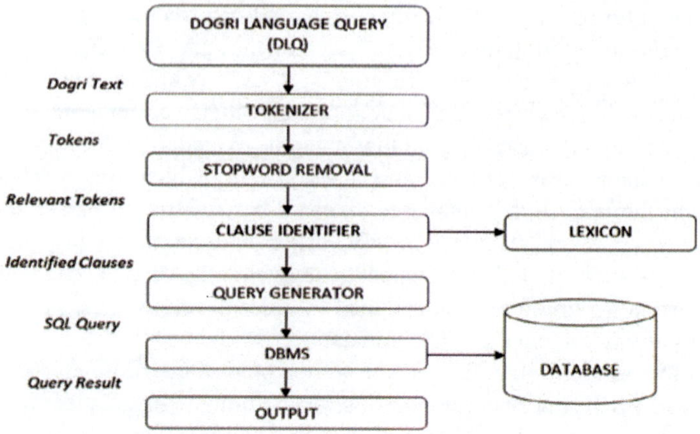

Fig. 1 Architecture of DLIDBMS

100% accurate when we answer "semantically tractable" questions which are only a small set of all the queries. In this research paper, we implemented the Dogri-NLIDB which converts a Dogri sentence into equivalent SQL query in order to make the data retrieval easy for a non-technical person. The goal of our system is to answer SQL queries written in Dogri language, and the objective is to answer the wider class of queries. Architecture of the Dogri-NLIDB consists of six components: Sect. 3.1 Dogri language query (DLQ), Sect. 3.2 tokenizer, Sect. 3.3 stopword removal, Sect.3.4 clause identifier, Sect. 3.5 query generator, and Sect. 3.6 database management system (DBMS). These phases function in sequence, and the output obtained in one acts as an input to the next phase. Architecture of Dogri language interface to database (DLIDB) is given in Fig. 1.

Lexicon in the architecture of the proposed model is a set of Dogri language word that helps to identify a particular clause category in the query. Clause identifier is explained in Sect. 3.4 with suitable example.

3.1 Dogri Language Query (DLQ)

This is the very first phase of the architecture that takes input. The input here is the query that needs to be translated to its corresponding equivalent SQL query so that the desired data can be retrieved from the database. For instance, the Dogri language query to the database is given in Fig. 2.

ओ बच्चेयां दे बारे दस्सो जुने जावा ओपन कोर्स रखेदा (DLQ)
Show those students who have chosen java as their open course subject

Fig. 2 Dogri language query

ओ	बच्चेयां	दे	बारे	दस्सो	जुने	जावा	ओपन	कोर्स	रखेदा
0	Bacheyan	de	Baare	dasso	juney	java	open	course	rakheda
Display the students data who have chosen for java as their open course									

Fig. 3 Tokenization of DLQ

3.2 *Tokenizer*

Tokenizer is the basic component that is used in text processing in variety of natural language tasks. Tokenizer simply tokenizes the text input, which is a Dogri language query in our case. The query here is split into tokens that are stored in an array data structure. The tokens here are the words of Dogri language, and the array of tokens obtained from the query is given in Fig. 3.

Query contains ten tokens, and each token need not be helpful in retrieving data but since the user here may not be technically sound enough, some unnecessary tokens are more likely to exist in the input query.

3.3 *Stopword Removal*

A token can possibly be a table name, an attribute of the table, a numeric value, a command name, or any other word that may not be required for processing but is used to make the query linguistically correct like prepositions etc. Since all the tokens present in token array may not be relevant to the query processing and data retrieval, it becomes necessary to remove all the irrelevant tokens so that equivalent SQL query is framed as intended by the user. This stopword removal phase does the job of removing all these unwanted tokens that are present in the token array. The output of this phase is the array of relevant tokens only that act as an input to the clause identifier. Stopwords/irrelevant tokens in the above query are given in Fig. 4.

Fig. 4 DLQ after stopword elimination

ओ	दे	बारे	दस्सो	रखेदा
O	de	baare	dasso	rakheda

Fig. 5 Clause identification in DLQ

बच्चेंयां	जुने	जावा	ओपन	कोर्स
Bacheyan	juney	java	open	course
Students	Who	Java	Open	Course

3.4 Clause Identifier

Once the relevant tokens are obtained after removing all the stopwords (unwanted tokens), clause identification is done. Query is divided into various clauses such as select clause and conditional clause. The identification of tokens maps to its appropriate corresponding clause in SQL format. We have broadly divided the query into three clauses, i.e., pre-conditional clause, conditional clause, and post-conditional clause (Fig. 5).

Now, each of the relevant tokens indicates for a particular clause like बच्चें यां(Students—pre-condition clause), जुने(those/who—Where clause) and जावा,ओपन कोर्स are the attributes of Student table and denote post-condition clause.

3.5 Query Generator

The clause identification is a crucial phase in generating the SQL equivalent of the Dogri query. Pre-conditional clause is mapped to the select clause that may contain functions like MAX(), AVG(), SUM(), COUNT(), etc., conditional clause to where clause, and post-conditional clause to the clause that can be a sub-conditional clause containing relational operators or this clause can be a query in itself. Finally, the above query is formatted in SQL format as: "Select * from Student where OpenCourse = Java".

3.6 Database Management System

This final phase takes SQL query as an input, executes the query, and returns the result that is displayed to the user. The output of the query taken for demonstration is shown in Fig. 6.

4 Algorithm of DLIDBMS

Description:

DLQ: Dogri language query

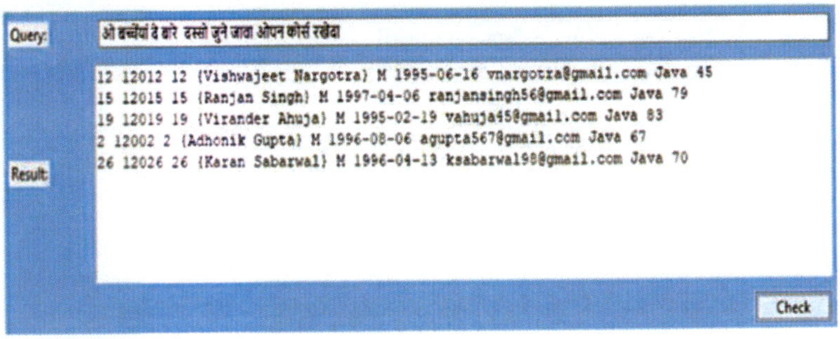

Fig. 6 Graphical user interface of DLIDBMS model

TKZR: Tokenizer
SWR: Stopword removal
CLID: Clause identifier
PreCon[], Con[], and PostCon[]: Pre-condition, condition, and post-condition lists
SQuery: SQL query
PLC: Preference list of clauses.

1. Read the Dogri language query (DLQ).
2. Break the DLQ into tokens using space as delimiter.
3. Stopword removal eliminates irrelevant tokens.
4. Clause identifier (CLID) identifies and categorizes DLQ tokens into one of the three categories, pre-conditional clause (PreCon[]), conditional clause (Con[]), and post-conditional clause (PostCon[]) list.
5. Mapper maps token Ti to equivalent SQL keyword Ki.
6. Generator formulates SQuery by arranging tokens in PreCon[], Con[], and PostCon[] lists using preference list of clauses (PLC).
7. SQuery is executed, and result is obtained from database.

5 Experimentation and Results

The graphical user interface of the proposed system takes the Dogri language query, displays the equivalent SQL query of the input query, and displays the output of the query as shown in Fig. 2: Since natural languages are highly ambiguous, about five versions of each query were taken, around 100 queries were tested, 95% of the queries were transformed appropriately to their equivalent SQL format, and the desired result is obtained by the proposed system. System is dictionary-based which implies that the tokens of the query in Dogri language are mapped to the keywords that are used in SQL queries, and a code snippet is shown in Fig. 7.

Dictionary contains a list of about 160 key–value pairs, and these are the Dogri language words that are most probable to be used by the user and words that are used in their equivalent SQL query. Since each user has his/her own choice of words that

```
{  "जावा":"'Java'", "कोर्स":  "subject", "जुने":  "WHERE", "बारे" :
"SELECT","दा":"SELECT","बच्चेंयां":"student", "नई":  "!=",
"ना":"name","बच्चे":"student","रोल_नंबर":"roll number","जिसदा":"WHERE",
"ज़्यादा":"max","नंबर":"marks","औसतन":"avg","घट":"<"
}
```

Fig. 7 Dictionary snippet

can be used to query the database, we have formulated each query from five different persons so that dictionary is enhanced to handle the queries and translation to the SQL query is appropriately accomplished. This consequently enables the user who may not be comfortable enough in handling with the database via SQL queries, to have the desired result. Some of the queries in Dogri language and their equivalent SQL query are tabulated in Fig. 8.

The experimentation is done on Windows operating system using Python programming language and SQL Database. Around 150 queries were tested, about 88% of the queries were appropriately translated to the equivalent SQL queries, and intended results were displayed by the system. System has been tested against queries

Dogri Language Query	SQL Query
ओ बच्चेंयां दे बारे दस्सो जुने जावा ओपन कोर्स रखेदा Display the students who have opted Java as open course	Select * from students where OpenCourse=Java
ओस बच्चे दा ना दस्सो जिसदा रोल नंबर 12021 ऐ Display the student data whose roll number is 2021	Select * from students where roll_number=12021
ओ बच्चेंयां दे बारे दस्सो जुने जावा ओपन कोर्स च नई रखेदा Display the students who have not opted Java as open course	Select * from students where OpenCourse!=Java
बच्चेंयां दे बारे दस्सो मत्ते नम्बरां दे साबा Display the students data in decreasing order of their marks	SELECT * FROM Students ORDER BY marks DESC
उन बच्चेंयां दे बारे दस्सो जुन्दा दा रोल नंबर 15 थमा बद्द ऐ Display the students data whose roll number is greater than 15	Select * from students where roll_number>15

Fig. 8 Snapshot of DLQs and equivalent SQL queries

that may retrieve data of all the attributes, selected attributes, based on basic functions (SUM, AVG, MAX, MIN, COUNT), and condition-based data retrieval using relational operators (=, <, <=, >, >=, <>). The usage of dictionary proves to be quiet beneficial in achieving the promising results. However, the complexity of the model further increase on user's part as user has freedom of using Dogri language words. According to the functioning of the system, it is mandatory to keep enhancing the dictionary of the system; in order to achieve this, the graphical GUI of the system allows user to raise the concern for undesired reporting of the database results. By raising concern, the user query will be stored by the system, which would later on be used by the system administrator to make the necessary changes in the system dictionary. The increase in the usage of system will result in enhancement of the system dictionary and less changes to be made in the dictionary over a period.

6 Conclusion and Future Scope

The proposed system DLIDB (Dogri language interface to database) accepts queries in Dogri language that is transformed into SQL query format which in turn is given to the database management system to get the desired result. Native people prefer regional languages, and the proposed system is a novel attempt in this direction. This system achieves significant level of accuracy in processing Dogri language query and returning the result. In this paper, fundamental queries have been covered that contain either "select clause" only or select clause along with "where clause." The tested queries covered the basic functions like SUM(), COUNT(), AVG(), MAX(), and MIN(); logical operators; and relational operators. Natural language processing is highly ambiguous task as there is no specific format for querying the database; it makes the natural language query processing a challenging task. Although different versions of the same query have been considered, it can be increased, so that more variations can be tested and system accuracy is enhanced. In future, we will be extending DLIDB to implementation compound and more complex queries that will retrieve data from multiple tables using join operations. The system can also have an interface that will process speech data so that user is exempted from typing in Dogri language.

References

1. Androutsopoulos I, Ritchie G, Thanisch P (1995) Natural language interfaces to databases—an introduction. Nat Lang Eng 1(1):29–81. https://doi.org/10.1017/S135132490000005X
2. Woods W, Kaplan R, Webber B (1972) The lunar sciences natural language information system. Bolt Beranek and Newman Inc., Cambridge, Massachusetts final report. B. B. N. report no 2378
3. Grosz BJ (1983) TEAM: a transportable natural-language interface system. In: Proceedings of the 1st conference on applied natural language processing, Santa Monica, California, pp 39–45

4. Grosz BJ, Appelt DE, Martin PA, Pereira FCN (1987) TEAM: an experiment in the design of transportable natural language interfaces. Artif Intell 32:173–243
5. Thompson BH, Thompson FB (1983) Introducing ASK, a simple knowledgeable system. In: Proceedings of the 1st conference on applied natural language processing, Santa Monica, California, pp 17–24
6. Thompson BH, Thompson FB (1985) ASK is transportable in half a dozen ways. ACM Trans Once Inf Syst 185–203
7. El-Mouadib FA, Zubi ZS, Almagrous AA, Irdess SE (2009) Generic interactive natural language interface to databases (GINLIDB). Int J Comput 3(3)
8. Shah A, Pareek J, Patel H, Panchal N (2013) NLKBIDB—natural language and keyword based interface to database. In: 2013 International conference on advances in computing, communications and informatics (ICACCI), pp 1569–1576
9. Wudaru V, Koditala N, Reddy A, Mamidi R (2019) Question answering on structured data using NLIDB approach. In: 2019 5th International conference on advanced computing and communication systems (ICACCS), pp 1–4
10. Das A, Balabantaray RC (2019) MyNLIDB: a natural language interface to database. In: 2019 International conference on information technology (ICIT), pp 234–238
11. (2020) Proceedings of the 2020 ACM SIGMOD international conference on management of data. Association for Computing Machinery, New York, NY, USA
12. Kim H, Seon C, Seo J (2005) A dialogue-based information retrieval assistant using shallow NLP techniques in online sales domains. IEICE Inf Syst 5:801–808
13. Warren D, Pereira F (1982) An efficient easily adaptable system for interpreting natural language queries. Comput Linguist 8
14. Senturk C (1997) Natural language interfaces to databases. In: The course of digital libraries, E6998-003
15. Jung H, Lee G (2002) Multi-lingual question answering with high portability on relational databases. In: Proceedings of COLING workshop on multilingual summarization and question answering
16. Kim H (2007) A dialogue-based NLIDB system in a schedule management domain. In: van Leeuwen J, Italiano GF, van der Hoek W, Meinel C, Sack H, Plášil F (eds) SOFSEM 2007: theory and practice of computer science, SOFSEM 2007. Lecture notes in computer science, vol 4362. Springer, Berlin, Heidelberg
17. Popescu AM, Etzioni O, Kautz H (2003) Towards a theory of natural language interfaces to databases. In: 8th International conference on intelligent user interfaces, Miami
18. Pazos RRA, González BJJ, Aguirre LMA, Martínez FJA, Fraire HHJ (2013) Natural language interfaces to databases: an analysis of the state of the art. In: Castillo O, Melin P, Kacprzyk J (eds) Recent advances on hybrid intelligent systems. Studies in computational intelligence, vol 451. Springer, Berlin, Heidelberg
19. Kumar A, Vaisla K (2013) Natural language interface to databases: development techniques. Elixir Comput Sci Eng 58:14724–14727
20. Wang W, Tian Y, Wang H, Ku W (2020) A natural language interface for database: achieving transfer-learnability using adversarial method for question understanding. In: 2020 IEEE 36th International conference on data engineering (ICDE), pp 97–108. https://doi.org/10.1109/ICDE48307.2020.00016
21. Zhong V, Xiong C, Socher R (2017) Seq2sql: generating structured queries from natural language using reinforcement learning
22. Jamwal S, Gupta P, Sen V (2021) Hybrid model for generation of verbs of Dogri language. In: 2020 Proceedings of MIDAS data driven approach towards disruptive technologies, pp 497–508
23. Gupta P, Jamwal S (2021) Designing and development of stemmer of Dogri using unsupervised learning. In: Soft computing for intelligent systems, pp 147–156

The Mutation Study of Social Media in Student's Life

Anju Mishra, Nikhil Kumar Varun, Sandeep Srivastava, Arvind K. Pandey, Lalan Kumar, and Tanupriya Choudhury

Abstract This paper is studying distinct social media platforms and their ups and downs. This paper calculates the different questionnaire on different ways about social media is destroying today's brilliant minds. Networking sites like Facebook, Twitter and so on put off students from their studies. Even though there is a loss of security and well-being, students tend to invest more time on social media. Social networking opens doors for interacting with friends, classmates and individuals with a common intrigue. Today, the principle point of the student ought to be education and their career. But they tend to lean towards social media more. That implies a decreased focus on learning and holding information. Technology is booming at a rapid rate, and the younger generations are the ones trapped in this rapid reform. For research methodology, questionnaires were distributed through email groups, to find out the various findings and to check whether student's well-being and academic performance are impacted by social media or not. The study also focuses on the frequency of audience on various social media platforms.

Keywords Communication · Cybersecurity · Digital well-being · Social media · Social networking

1 Introduction

The initial practice of communicating over long distances used composed messages conveyed by hand from one being to another, in other words, letters. The initial

A. Mishra (✉) · N. Kumar Varun · S. Srivastava · L. Kumar
GLBITM, Greater Noida, India
e-mail: anju.pandey.12@gmail.com

A. K. Pandey
Modern College of Professional Studies, Mohan Nagar, Ghaziabad, India

T. Choudhury (✉)
School of Computer Science, University of Petroleum and Energy Studies (UPES), Dehradun, Uttarakhand 248007, India
e-mail: tanupriya@ddn.upes.ac.in

V. Skala et al. (eds.), *Machine Intelligence and Data Science Applications*, Lecture Notes on Data Engineering and Communications Technologies 132,
https://doi.org/10.1007/978-981-19-2347-0_66

form of postal accommodation dates back to 550 B.C., and this ancient distribution system would become more global and efficient in future centuries. The telegraph came into the picture in 1792. This permitted messages conveyed over a long distance, faster than the traditional method. Although telegraph messages were brief, it was a progressive way to communicate news and intelligence. Two vital discoveries happened in the late nineteenth century: the phone in 1890 and the radio in 1891 [1].

Social media is a flip side for today's society. This world is drastically shifting towards social media. There are almost 65 major social media platforms now.

It was the year 1997 when Six Degrees came into the picture. It empowered users to upload a profile and make friends with others. Myspace and LinkedIn gained repute in the early twenty-first century. YouTube came out in 2005, creating a better approach for individuals to communicate and share across extraordinary distances. By 2006, Twitter and Facebook both got to be accessible to users throughout the globe. These platforms stay as a few of the foremost well-known social networks in the virtual world. Other sites like Pinterest, Spotify, Foursquare and Tumblr started appearing to fill particular social networking slots. Nowadays, there is a huge variety of social networking platforms. This makes an environment where clients can reach a large number of individuals without relinquishing the closeness of person-to-person (PTP) communication [2, 3].

Intercontinental digital population as of January 2020 is as follows [4]:

1. Operational Web users—4.54 billion
2. Sole mobile Web users—4.18 billion
3. Operational social media users—3.8 billion
4. Operational mobile social media users—3.75 billion.

China is the leading country with a total of 854 million Internet users as of the year 2019. As of the first quarter of 2020, the total Internet users in China is 903.59 million. Though China is a developing country, this results in the fast-growing Internet audience in China [5].

2 Businesses Evolved Due to Social Media

No matter what industry your line of work is in, social media offers the opportunity to set up your brand as a groundbreaker—the go-to source for data on subjects related to your specialty. The influence [7] of social media in business is developing at twist speed. With more and more individuals joining the platform and utilizing them regularly/efficiently, the social media industry is bound to become greater within the coming a long time. It is booming like never, sometime and recently [6, 18] (Fig. 1).

'Numbers don't lie', and social media statistics certainly appear that it proceeds to alter the approach one does business.

In Fig. 2, the number of audiences on various platforms and Facebook is dominating with the highest numbers.

Inc. 500 companies are present on various social media platforms from the year 2017 to 2019, in which LinkedIn is dominating of all with an average of 85.6% usage,

Fig. 1 Mostly used social media platforms

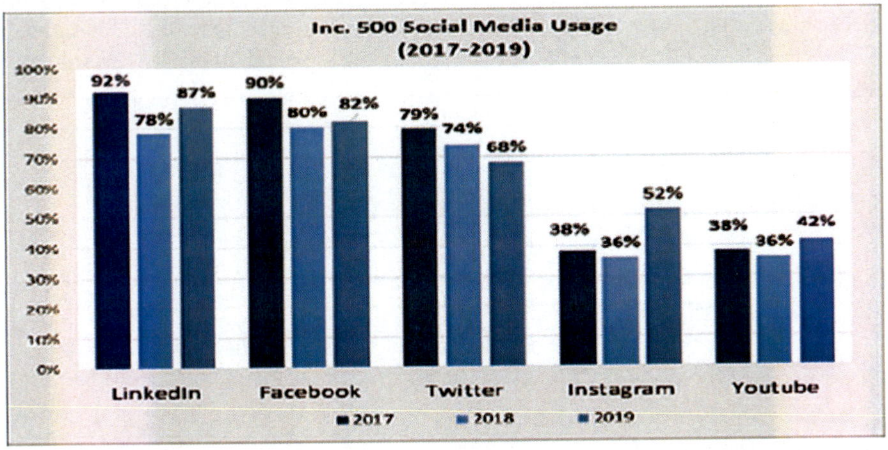

Fig. 2 Usage of social media in Inc. 500 companies: Centre for Marketing Research UMass Dartmouth

Facebook with an average of 84.0% usage, Twitter with an average of 73.66% usage, Instagram with an average of 42% usage and YouTube with an average of 38.66% usage.

LinkedIn is job-searching platform, and Facebook hosting the huge number of audiences is holding the second position because Facebook has an audience from every domain and not some specific type. Twitter is the micro-blogging site and is holding the third position in promoting business. Instagram is holding the fourth position because it comprises of millennials audience. YouTube is the video-promoting platform holding the fifth position because it holds the audience from all domains and not a specific one [8]. The rise in numbers of audiences on Instagram and YouTube is depicted [9].

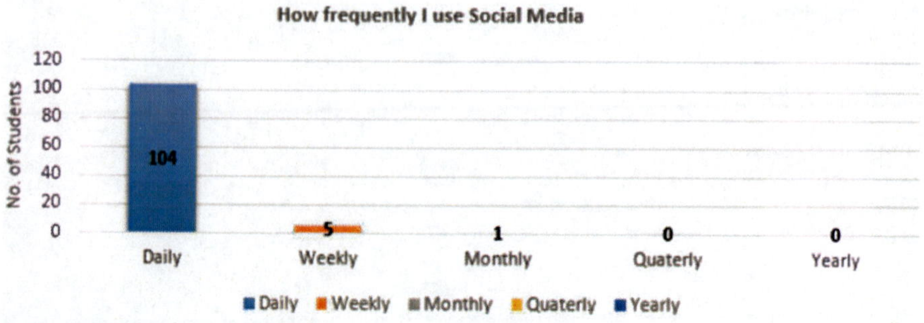

Fig. 3 Chart depicting how frequently social media is used

3 Research Methodology

The primary data for this research is collected through a survey. The data of the survey was collected from 110 respondents. In expansion, this collected data leads this research to consider the exploration of the effect of social media [15] on student's educational and psychological performance.

3.1 How Frequently I Use Social Media

In Fig. 3, from the data collected from 110 respondents, most of the users are daily user of social media.

From the data collected, it is concluded that none of the users uses social media platform quarterly or yearly. Social media acts as a basic need.

3.2 Time Spent on Social Media

In Fig. 4, the collected data can depict the usage of social media that most of the users spend 1–3 h a day on social media.

3.3 Influence of Social Media on Studies

In Fig. 5, the data collected can certainly depict the influence of social media on students' studies, which means that 63 out of 110 people think that using social media has a positive impact and sometimes harms studies.

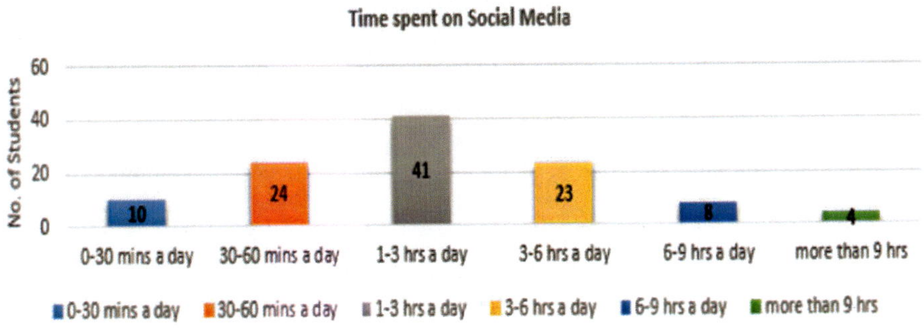

Fig. 4 Chart depicting time spent on social media

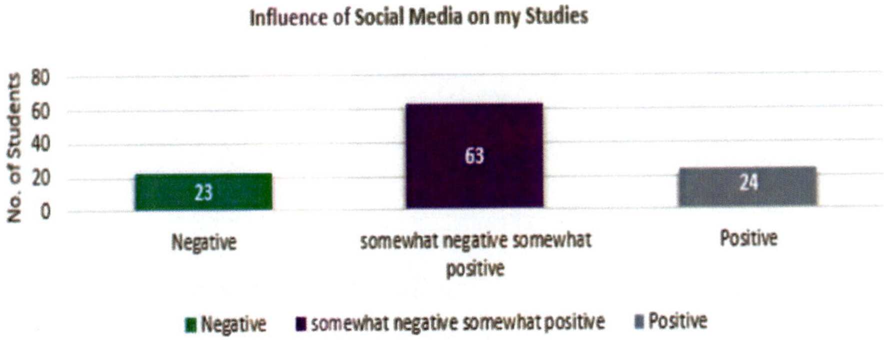

Fig. 5 Chart depicting the influence of social media on students' studies

There is a minute difference between the positive and negative effect of social media on studies. Referencing Table 1, the frequency for negative and positive is 23 and 24, respectively, out of 110 respondents.

Table 1 Table describing the influence of social media on studies

Effects	Frequency
Negative	23
Somewhat negative and somewhat positive	63
Positive	24
Difference between negative and positive	$\lvert 23 - 24 \rvert = 1$

Fig. 6 Chart depicting the influence of social media in time management

Table 2 Table describing the influence of social media on time management

Effects	Frequency
Complete waste of time	40
Somewhat wastes time and somewhat saves time	49
Very useful (saves time)	21
Difference between the waste of time and saves time	$\|40 - 21\| = 19$

3.4 *Influence of Social Media in Time Management*

In Fig. 6, the collected data can depict the influence of social media on time management. There is a huge difference between the two opposite aspects. Here, the table is given.

Referencing Table 2, the frequency for a complete waste of time and very useful (saves time) is 40 and 21, respectively.

The huge difference between the two contrasting effects of social media on time management is 19. And, for somewhat wastes time and somewhat saves time, the frequency is 49, which means 49 out of 110 respondents think that sometimes it wastes time and sometimes it saves time.

3.5 *Influence of Social Media on My Psychology*

In Fig. 7, the collected data can depict the influence of social media on one's psychology. Here, the table is given.

Referencing Table 3, the frequency for depressing and stress buster is 20 and 39, respectively. The huge difference between the two contrasting effects of social media on psychology is 19.

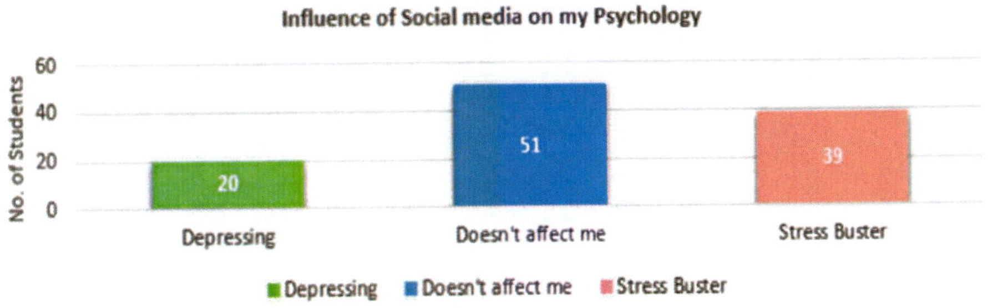

Fig. 7 Chart depicting impact of social media on psychology

Table 3 Table describing the impact of social media on student's psychology

Effects	Frequency
Depressing	20
Does not affect me	51
Stress buster	39
Difference between depressing and stress buster	$\|20 - 39\| = 19$

3.6 *Social Media Made My Life*

In Fig. 8, the collected data can depict the influence of social media on student's life. There is a huge difference between the two contrasting aspects. Here, the table is given.

Referencing Table 4, the frequency for difficult and easy is 19 and 39, respectively. The huge difference between the two contrasting effects of social media on one's life is 20. And, for somewhat difficult and somewhat easy, the frequency is 52. Considering the huge difference between two contrasting aspects can conclude that the use of social media plays both positive and negative impact on students' life depending on one's state of mind.

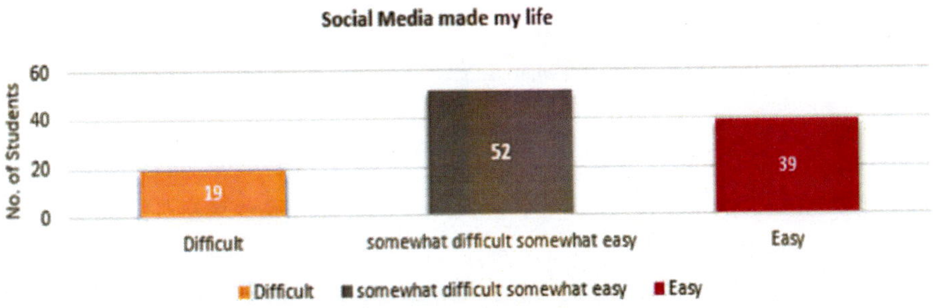

Fig. 8 Chart depicting the effect of social media on one's life

Table 4 Table depicting the impact of social media on one's life

Effects	Frequency
Difficult	19
Somewhat difficult and somewhat easy	52
Easy	39
Difference between difficult and easy	$\|19 - 39\| = 20$

3.7 *How Useful I Find Social Media*

In Fig. 9, the data collected can depict the usefulness of social media in a student's life. There is a difference between the two contrasting aspects. Here, the table is given.

Referencing Table 5, the frequency for garbage and useful is 30 and 46, respectively, out of 110 respondents. The difference between garbage and usefulness is 16. And, for somewhat garbage and somewhat useful, the frequency is 34.

The frequency of garbage aspect cannot be ignored, i.e. 30; that is a huge number which is $27.\overline{27}\%$ of total respondents.

Fig. 9 Chart depicting whether social media is useful or not

Table 5 Table depicting how useful is social media

Effects	Frequency
Garbage	30
Somewhat garbage and somewhat useful	34
Useful	46
Difference between garbage and useful	$\|30 - 46\| = 16$

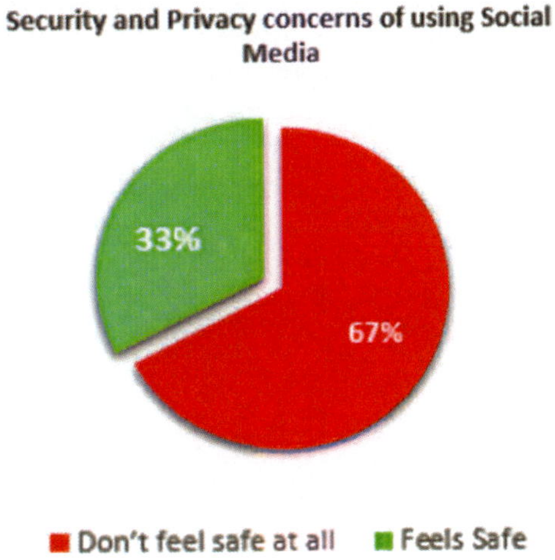

Fig. 10 Chart depicting the security and privacy concerns of using social media

Table 6 Table for the security and privacy concerns of using social media

Effects	Frequency	Percentage (%)
Do not feel safe at all	74	67.27–67
Feels safe	36	32.72–33
Difference between two contrasting aspects	$\|72 - 36\| = 38$	

3.8 *Security and Privacy Concerns on Social Media*

In Fig. 10, the collected data can depict the security [17, 18] and privacy [12] concerns on using social media. Here, the table is given.

Referencing Table 6, the frequency for two contrasting aspects is 74 and 36, i.e. for negative aspect (do not feel safe at all), it is 74, and for positive aspect (feels safe), it is 36. For negative, it is 67.27% of total respondents, and for positive, it is 32.27% of total respondents.

Difference between the frequencies for both aspects is 38. It is concerning that the weightage for negative aspect is more, i.e. large number of the population do not feel safe on social media because of security and privacy reasons [10, 11, 13, 14, 16].

3.9 *Overall, My Experience of Using Social Media Is*

In Fig. 11, the collected data can certainly depict the overall experience of social media in a student's life. Referencing Table 7, the frequency for negative aspects is 32 and for positive it is 78. For negative, it is 29% of total respondents, and for positive, it is 71% of total respondents.

Fig. 11 Chart depicting the overall experience of using social media in one's life

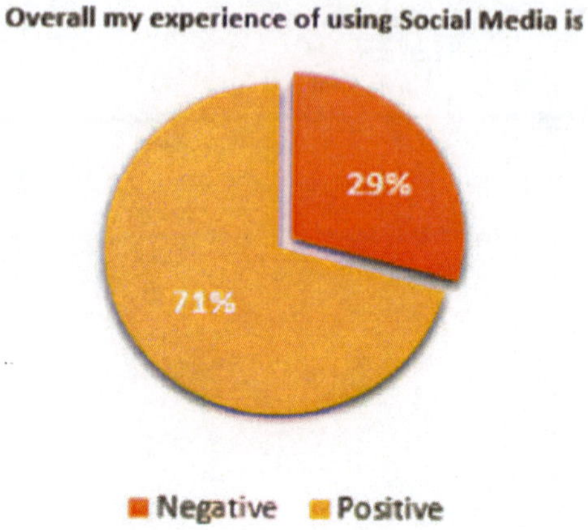

Table 7 Table for the overall experience of using social media

Effects	Frequency	Percentage (%)
Negative	32	29
Positive	78	71
Difference between negative and positive	$\|32 - 78\| = 46$	

4 Conclusion and Discussion

The aim was to study and understand the aspects of social media on a student's life academically and personally, i.e. in what different ways, one's life might get affected by social media. The research methodology focuses on nine different important areas related to social media users. The research methodology focuses on areas like:

- Frequent use
- Time spent
- Influence on studies
- Time management
- Psychology
- Nature of ease
- Usefulness
- Security and privacy
- Overall experience.

The study results demonstrate the mutation (transformation) of social media in student's life. The study shows how frequently social media is used by students. It shows that social media is being used by students quiet frequently. Most of the students use social media 1–3 h a day, which results in killing their time.

Talking about time management, most of the respondents think that it is a complete waste of time. And, about security and privacy concerns, it results that students do not feel safe on social media. They have privacy issues. Their overall experience is positive as they get to know about their friends and stay updated from day-to-day news.

References

1. Edosomwan S, Prakasan SK, Kouame D, Watson J, Seymour T (2011) The history of social media and its impact on business. J Appl Manag Entrepreneurship **16**:79–91
2. Gerpott TJ, Thomas S (2014) Empirical research on mobile internet usage: a meta-analysis of the literature. Telecommun Policy **38**(3):291–310. ISSN s0308-5961. https://doi.org/10.1016/j.telpol.2013.10.003. http://www.sciencedirect.com/science/article/pii/S0308596113001754
3. Dewan S, Ramaprasad J (2014) Social media, traditional media, and music sales. MIS Q 38(1):101–121
4. https://www.statista.com/statistics/617136/digital-population-worldwide/
5. https://www.statista.com/topics/1145/internet-usage-worldwide/
6. Grizane T, Jurgelane-Kaldava I (2017) Social media impact on business evaluation. Procedia Comput Sci **104**:190–196. https://doi.org/10.1016/j.procs.2017.01.103
7. Bansal P, Aggarwal B, Tomar R (2019) Low-voltage multi-input high trans-conductance amplifier using flipped voltage follower and its application in high pass filter. In: 2019 International conference on automation, computational and technology management (ICACTM), pp 525–529. https://doi.org/10.1109/ICACTM.2019.8776789
8. Pourkhani A, Abdipour K, Baher B, Moslehpour M (2019) The impact of social media in business growth and performance: a scientometrics analysis. Int J Data Network Sci 223–244. https://doi.org/10.5267/j.ijdns.2019.2.003
9. https://www.umassd.edu/cmr/research/2019-inc-500.html
10. Das R, Patel M (2017) Cyber security for social networking sites: issues, challenges and solutions. Int J Res Appl Sci Eng Technol V:833–838. https://doi.org/10.22214/ijraset.2017.4153
11. Kumar S, Gaur MS (2020) Call admission control in mobile multimedia network using grey wolf optimization. In: Solanki V, Hoang M, Lu Z, Pattnaik P (eds) Intelligent computing in engineering. Advances in intelligent systems and computing, vol 1125. Springer, Singapore. https://doi.org/10.1007/978-981-15-2780-7_27
12. Anuj Kumar Y, Tomar R, Kumar D, Gupta H (2012) Security and privacy concerns in cloud computing. Int J Adv Res Comput Sci Softw Eng 2(5)
13. Kumar S, Gaur MS (2019) Handoff prioritization to manage call admission control in mobile multimedia networks for healthcare. In: 2019 10th International conference on computing, communication and networking technologies (ICCCNT), Kanpur, India, pp 1–7. https://doi.org/10.1109/ICCCNT45670.2019.8944618
14. Kumar S, Kumar K (2019) Neuro-fuzzy based call admission control for next generation mobile multimedia networks. Int J Eng Adv Technol (IJEAT) 8(6). ISSN: 2249-8958
15. Narang S, Nalwa T, Choudhury T, Kashyap N (2018) An efficient method for security measurement in internet of things. In: 2018 International conference on communication, computing and internet of things (IC3IoT), pp 319–323. https://doi.org/10.1109/IC3IoT.2018.8668159
16. Gerlach J, Widjaja T, Buxmann P (2015) Handle with care: how online social network providers' privacy policies impact users' information sharing behaviour. J Strateg Inf Syst 24(1):33–43
17. Dhamija J, Choudhury T, Kumar P, Rathore YS (2017) An advancement towards efficient face recognition using live video feed: "for the future". In: 2017 3rd International conference on computational intelligence and networks (CINE), pp 53–56. https://doi.org/10.1109/CINE.2017.21

18. Al Alwan A, Rana NP, Dwivedi YK, Algharabat R (2017) Social media in marketing: a review and analysis of the existing literature. Telematics Inform. Available at http://www.sciencedirect.com/science/article/pii/S0736585317301077

GPS-Based Route Choice Model for Smart Transportation System: Bringing Intelligence into Vehicular Cloud

Sirisha Potluri, Sachi Nandan Mohanty, Katta Subba Rao, and Tanupriya Choudhury

Abstract An IoT-based intelligent performance evaluation strategy is anticipated to rationalize the communication among the vehicles, manage vehicle traffic, support vehicle drivers with privacy and safety, and provision significant applications for cloud users. GPS-based route choice model is substantial for handling the transportation complications of a big city, traffic engineering, and remote communication among vehicles. With this scenario in mind, we recommend a strategy to manage and evaluate the performance of the vehicular cloud. Different types of cloud services, storage mechanisms, resources, and information management are provided to the cloud users, commercial vehicles, emergency services, and disaster services through the vehicular cloud computing model. The performance of the vehicular cloud can be raised to meet the information requirements of the cloud users by using the proposed route choice model. The primary objective of the proposed model is to provide the finest traffic light control method using GPS inputs and vehicular area networks. An accurate traffic flow approximation method in vehicular area networks is useful to expect the amount of obtainable computational resources within a roadway subdivision given the unplanned arrivals and departures for smart transportation systems by bringing intelligence into the dynamic vehicular cloud.

S. Potluri (✉)
Department of CSE, Faculty of Science and Technology- IcfaiTech, The ICFAI Foundation for Higher Education, Shankarpally, Donthanpally, Hyderabad, Telangana 501203, India
e-mail: sirisha.vegunta@gmail.com

S. N. Mohanty
Department of Computer Science and Engineering, Vardhaman College of Engineering (Autonomous), Hyderabad 501218, India

K. S. Rao
Department of Computer Science and Engineering, B V Raju Institute of Technology, Narsapur, Medak, Telangana 502313, India
e-mail: subbarao.k@bvrit.ac.in

T. Choudhury (✉)
School of Computer Science, University of Petroleum and Energy Studies (UPES), Dehradun, Uttarakhand 248007, India
e-mail: tanupriya@ddn.upes.ac.in

V. Skala et al. (eds.), *Machine Intelligence and Data Science Applications*, Lecture Notes on Data Engineering and Communications Technologies 132,
https://doi.org/10.1007/978-981-19-2347-0_67

Keywords IoT · Cloud computing · Vehicular cloud · Wireless sensor network · Vehicular ad-hoc network · Mobile computing · Intelligent transportation system

1 Introduction—Vehicular Cloud Computing

Vehicular Cloud Computing (VCC) is a novel technological approach, which perceives and observes various advantages of cloud computing to assist the vehicle drivers of the VANETs through a pay-as-you-go model. The main objective of the VCC is to deliver the computational services at a low price to the drivers; henceforth, traffic congestion, overcrowding, accidents, driver's energy, travel time, and carbon footprint can be reduced. VCC provides better road safety using scalable and flexible wireless sensor network—WSN, advanced communication protocols for intelligent transport systems—ITS, and data readiness at all times using mobile cloud computing—MCC. To utilize various computing resources of vehicles, VCC is providing a scientifically viable and economically feasible solution by pooling up them together for better availability and accessibility. Mobile vehicular cloud is providing several prominent applications such as safe vehicle driving, inner-city traffic sensing, transport content distribution, portable advertising, and smart transportation. Vehicles gets information related to traffic congestion, road pavement navigations, surrounding vehicles mapping, environment details, information sharing, etc., through sensors. Vehicle users can explore the data with mapping and indexing. All these services are delivered to cloud users based on the computation of urban localities congestion depiction, representation of the urban pollution plan, collective representation of audio and video of accidents, traffic and transport violations, etc. [1–6].

2 Vehicular Cloud Computing Architecture

VCC architecture contains four layers, namely on-board components, communication, cloud services, and applications [7, 8]. Vehicular cloud computing architecture with various layers is presented in Fig. 1.

2.1 *On-Board Components*

The on-board components module is responsible for observing the well-being and mood of the vehicle driver. It gathers the inside information of the vehicle such as pressure and temperature by using various sensors of body, environment, smartphone, and internal and inertial navigation. On-board components use various intelligent

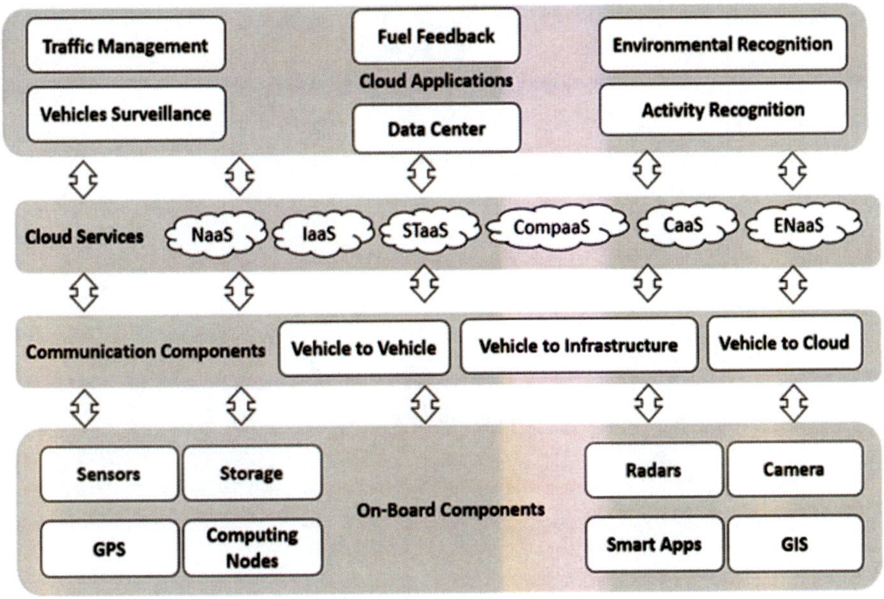

Fig. 1 The architecture of vehicular cloud computing

approaches to recognize vehicle driver behavior and predict the driver's responses and intents.

2.2 Communication

The communication component contains four modules, namely vehicle to pedestrian, vehicle to vehicle, vehicle to infrastructure, and vehicle to cloud. This layer guarantees active and effective communications through different protocols. All communication systems are denoted in Fig. 2.

- **Vehicle to Pedestrian**—V2P is a direct and shortest communication among vehicles and pedestrian/pedestrians within the nearby vicinity. Communication happens with exposed road users such as cyclists. V2P is accompanied directly or over and done with the support of network infrastructure.
- **Vehicle to Vehicle**—V2V is a communication capability to exchange information in a wireless environment about the speed and location of neighboring vehicles. It helps to avoid accidents, crashes, traffic congestion, and environmental pollution. A cooperative and flexible travel control mechanism is adapted through wireless communication by using on-board sensors of smart vehicles on the road to manage the traffic.

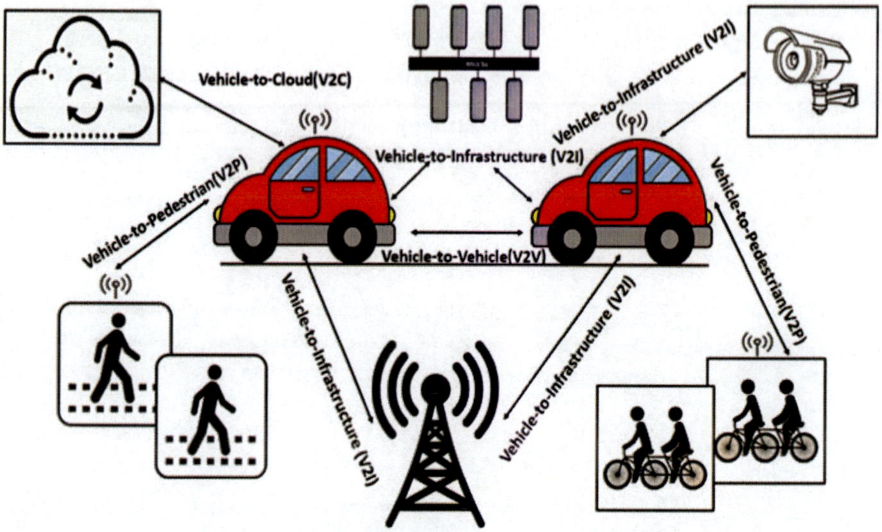

Fig. 2 Vehicle to everything communication

- **Vehicle to Infrastructure**—V2I is a communication capability to exchange information among vehicles and the infrastructure. Infrastructure modules include road symbols, lane/path markings, traffic lights, and other smart road setup linked via the wireless connection.
- **Vehicles to the Cloud**—V2C is a communication capability to exchange information among vehicles and cloud systems. V2C permits the vehicle to use information from other vehicles by using IoT through the cloud system.

2.3 Cloud Services

In the current market, there are various types of services which are offered by cloud computing, namely Network as a service (NaaS), Entertainment as a Service (ENaaS), Infrastructure as a service (IaaS), Storage as a Service (STaaS), Software as a service (SaaS), Information as a Service (INaaS), Platform as a Service (PaaS), and Cooperation as a Service (CaaS).

2.4 Cloud Applications

Cloud applications are programs run on the Internet besides processing and storage are observed in the cloud system. The cloud user works together with the cloud application through a web browser or an application, and the data processing is accomplished by a combination of the confined device and the cloud system. Various examples of cloud applications are traffic management, vehicles surveillance, fuel feedback, datacenter, environmental recognition, and vehicle driver activity recognition.

3 VCC Applications and Services

Vehicular Cloud Networking (VCN) offers an extensive collection of applications such as video surveillance, bandwidth management, real-time navigation, remote traffic management, data center, emergency management, and Internet of Vehicles. Vehicular networking is a noteworthy research area by its great features such as standardization, competent traffic management, road safety, and demonstration.

Various solutions were proposed by experts and researchers to resolve the challenges and problems of vehicular cloud networks. The vehicular cloud computing paradigm is one of the efficient solutions. VCC is an innovative hybrid computing technology that created significant impact and substantial resolution on traffic administration and management and road safety by promptly using vehicular cloud resources/services, such as computing power, storage capacity, and network facilities for decision-making.

The vehicular cloud computing model is a technologically and scientifically practical and economically feasible dynamic shifting paradigm for meeting intelligent vehicular networks concerning self-directed traffic, smart vehicle control, and flexible perception systems. Detailed elucidation of vehicular cloud applications is shown in Fig. 3. A comprehensive explanation of vehicular cloud services is given in Table 1.

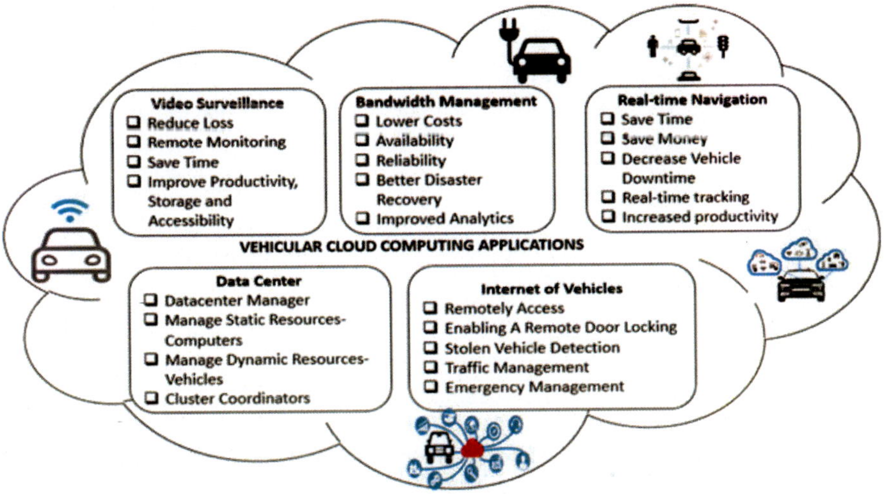

Fig. 3 Application of vehicular cloud computing

Table 1 Vehicular cloud computing services

Author	Cloud service	Description
Sonia Shahzadi et al.	IaaS—Infrastructure as a Service	Provides infrastructure as a service to the cloud user [9]
Priyanshu Srivastava et al.	PaaS—Platform as a Service	Provides platform as a service to the cloud user [10]
Sarbojit Banerjee et al.	SaaS—Software as a Service	Provides software as a service to the cloud user [11]
Jack F. Bravo-Torres et al.	NaaS—Network as a Service	It is a cloud service provisioning model using which vehicle users lease networking facilities from cloud service providers. NaaS permits cloud users to activate their networks without keeping their networking infrastructure [12]
Gurudatt Anil Kulkarni et al.	STaaS—Storage as a Service	STaaS is a coped service in which the cloud service provider provisions the cloud user with access to a data storage system. The cloud service can be brought on premises from infrastructure that is committed to a particular cloud user, or it can be transported from the public cloud as a pooled service that is obtained by cloud contribution and is payable conferring to one or more cloud service usage measures
Alisson Barbosa de Souza et al.	CaaS—Computation as a Service	CaaS is offered as a service when computing resources are delivered on request through virtual or physical resources as a service. Billing is based on a consumption-based model
Reinhard Bernsteiner et al.	ENaaS—Enterprise as a Service	ENaaS allows cloud users to access virtualized IT resources from public, hybrid, or private cloud services providers on a pay-as-you-use basis. These resources are computation servers, processing units, storage capacity, virtualization abilities, and networking infrastructure [13, 14]

4 Performance Evaluation in Vehicular Cloud

An intelligent performance evaluation strategy allows the vehicular cloud to identify the methods that utilize the under-used, accessible, and physical resources of the vehicles effectively. Dynamically fluctuating the number of available resources describes the vehicular cloud as a new difficult and complex model than the traditional cloud model. An efficient vehicular cloud strategy defines and analyzes the estimated number of vehicles in a roadway segment. These connected vehicles can aid as the primary component for the vehicular cloud, and they allow an extremely large set of applications that benefit the whole cloud-based automated traffic system. For this, a macroscopic traffic model can be used to examine the free-flow traffic, and a queuing theory to detect the queuing-up traffic in the roadway segment. Performance evaluation helps to improve the communications among vehicles, cloud, infrastructure, and pedestrians in a VANET. Our research is to design a system that can be applied to help vehicle drivers in altering their driving for better well-being and safety to save time and fuel costs [15–20].

5 Proposed GPS-Based Route Choice Model Architecture for Smart Transportation System

The proposed model architecture is in Fig. 4, consisting of users, GPS satellite, traffic management system, Internet, and sensor arrangement on roads to manage the vehicle. The architecture is demonstrating information management of sensors, traffic

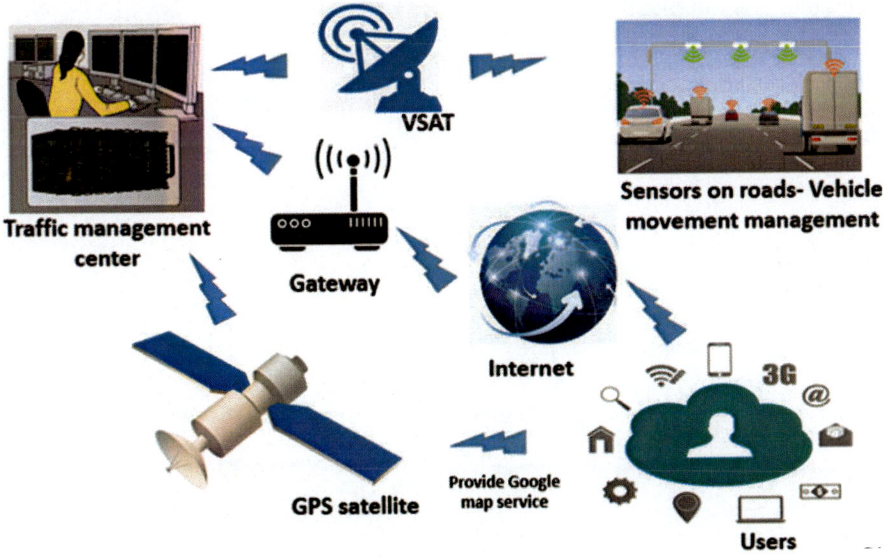

Fig. 4 Intelligent traffic control system

management through mobile computing, and free-flow traffic with dynamic routing mechanisms. The traffic management center gets information from the vehicle users by using the Internet and gateways. Sensors that are present on the roads continuously send the data to the traffic management center and manage the vehicle movements. Using a very small aperture terminal (VSAT) transmit/receive data over a satellite communication network. GPS satellite gets vehicle movement and guides them by providing the best routing through maps and images [21–30].

6 Mathematical Model for the Proposed Intelligent Performance Evaluation in Vehicular Cloud

6.1 *Packets—Inter-Arrival Rate*

Vehicles are treated as packets with different destinations and dimensions. For example, a unidirectional path of length Len, having sub-paths $P = \{p_1, p_2, \ldots, p_n\}$. The vehicles which are arriving follow a Poisson distribution with a mean value of 1/λ per unit time. Each subpath P_i has a continuous arrival rate λ_i where $i = 1, 2, 3, \ldots, N$. If S is a random speed of the vehicle, then Gaussian distribution is:

$$f(S) = \frac{1}{\sigma.\sqrt{2\pi}} \mathrm{e}^{-\frac{(v-\mu)^2}{2\sigma^2}} \tag{1}$$

where

μ Mean
σ Standard deviation
v Velocity.

The parameter A represents the intensity of vehicles and reflects the velocity of the inter vehicle arrangement.

$$A = \int_{v_{\min}}^{v_{\max}} \frac{f'(S)}{S} d(S) \tag{2}$$

$$f'(S) = \frac{f(S)}{\int_{v_{\min}}^{v_{\max}} f(S) d(S)} \tag{3}$$

Service times Se(t) of a vehicle that will go through the intersection is:

$$\mathrm{Se}(t) = \begin{cases} \frac{2t-\sqrt{4t^2-\frac{8l}{a_{\mathrm{avg}}}}}{2} \\ \sqrt{\frac{2l}{a_{\mathrm{avg}}}} \le t \le \frac{2v_{\max}-\sqrt{v_{\max}^2-a_{\mathrm{avg}}l}}{a_{\mathrm{avg}}} \\ \frac{l}{v_{\max}}, \frac{2v_{\max}-\sqrt{v_{\max}^2-a_{\mathrm{avg}}l}}{a_{\mathrm{avg}}} < t \le T_{\mathrm{ON}} \\ T - T_{\mathrm{ON}}, t < \sqrt{\frac{2l}{a_{\mathrm{avg}}}} \text{ or } T_{\mathrm{ON}} < t \le T \end{cases} \tag{4}$$

where

T_{ON} green light on
T_{OFF} green light off
$v_{\max}$ maximum speed
a_{avg} average acceleration rate.

6.2 *Free-Flow Model*

In the free-flow traffic module, the traffic is not congested by any complications such as traffic lights, traffic signals, traffic congestion, road diversion, and road maintenance. Vehicle driver can drive at any speed, on condition that vehicle remains within road controls. Hence, the traffic density in such cases is more probably low or medium. The Poisson process is an efficient and random process for representing vehicles' arrival.

Let $\{V_1, V_2, \ldots, V_N\}$ are the vehicles on the highway, the Poisson distribution function of relative velocities is:

$$f(v_{\mathrm{relv}}) = \frac{1}{\sqrt{\sigma_i^2 + \sigma_j^2 * 2\pi}} e^{\frac{(v-(v_i-v_j))^2}{2(\sigma_i^2+\sigma_j^2)}} \tag{5}$$

where

σ_i Deviation of vehicle *i*
σ_j Deviation of vehicle *j*.

Then, vehicle spatial density ρ on the road segment Sg is:

$$\rho = \frac{(t_1 - t_0)\lambda}{L_{\mathrm{Sg}}} \tag{6}$$

where

$[t_0 - t_1]$ The time interval
λ Vehicles arriving rate from t_0 to t_1.

6.3 *Queuing-Up Model*

Everywhere in cities, the roads are further expected to be crowded and jammed, due to a fast growing amount of vehicles and the traffic abilities of the roads. The bottleneck of the normal traffic queue is an intersection, which is the site or place of an accident. In common, the queue is certain to arise when the leaving rate is less than the coming rate.

The total delay time D_i of a random vehicle i in a roadway section is the summation of its all waiting times.

$$D_i = w_i + \tau_i \tag{7}$$

where

τ_i leaving time of vehicle i.

7 Results and Analysis

Figures 5, 6, 7, and 8 validate the stochastic features of a Poisson arrival process. It presents the probability of vehicular arrivals at various intervals, for example, 15-min intervals.

When traffic arrival rate is $\mu = 0.2$ vehicles per minute, two to three vehicles are anticipated to show up at the arrival location point during 15 min, with a probability rate of 23%. Similarly, when $\mu = 0.5$ vehicles per minute, more vehicles are anticipated to show up at the arrival location point. Significance of queuing theory greatly depends on assumption of arrival rate and departure rate and number of departure channels.

To make certain the significance, probability of vehicle arrivals is computed and represented in Figs. 5 and 6. Our proposed model is able to predict the best occurring sequences with various time intervals and shown better results with best probability

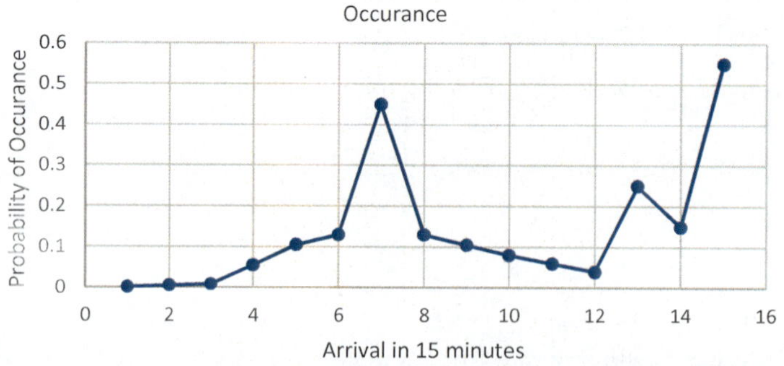

Fig. 5 Probability of vehicular arrivals at various intervals when mean = 0.5 vehicles/min

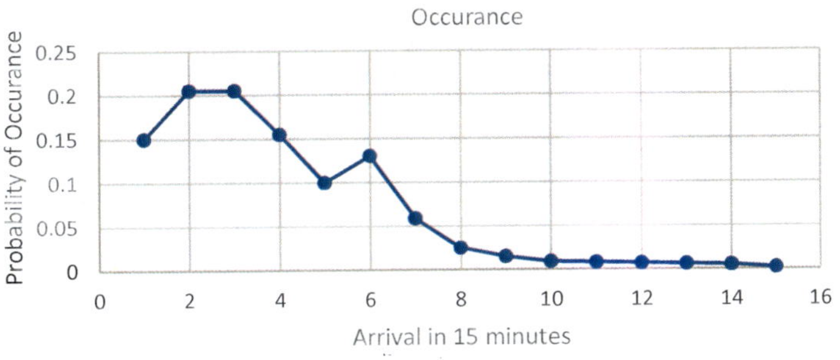

Fig. 6 Probability of vehicular arrivals at various intervals when mean = 0.2 vehicles/min

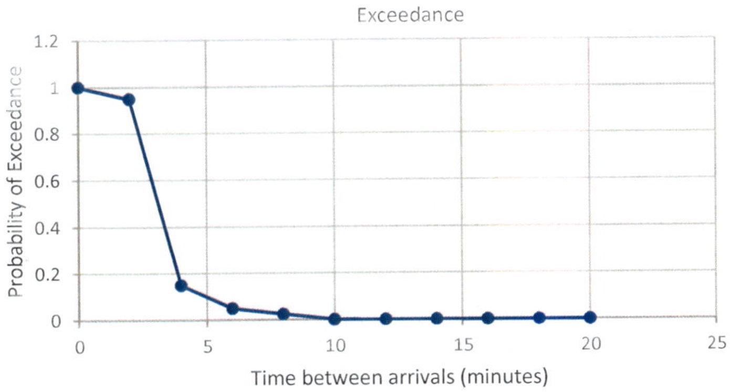

Fig. 7 Probability of exceedance at various intervals when mean = 0.5 vehicles/min

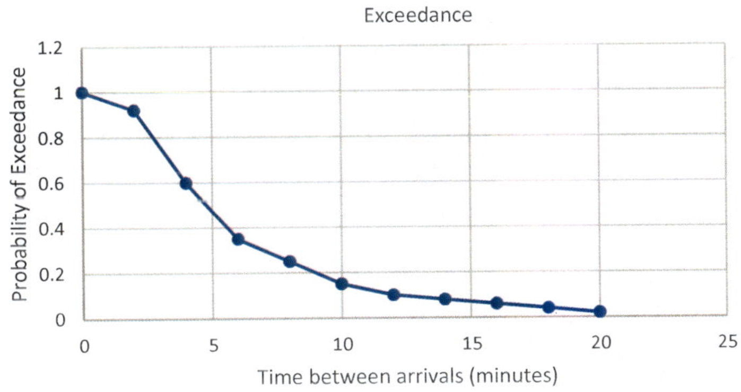

Fig. 8 Probability of exceedance at various intervals when mean = 0.2 vehicles/min

estimation. To make certain the significance, probability of exceedances is computed and represented in Figs. 7 and 8. Our proposed model is able to predict the best exceeding sequences with various time intervals and shown better results with best probability estimation.

8 Conclusion and Future Work

In this paper, we have proposed an efficient intelligent IoT-based computation resource provision scheme for a vehicular cloud computing system. The finest decision-making system is achieved through packet representation, queueing up the model and free-flow model. Limitations of the proposed model is that it is not investigating and examining the possessions of constraint tolerance to the optimal model in the vehicular cloud system, which may lead to developing a dynamic and more robust scheme. In the future, we plan to implement the same to improve the performance of the proposed model.

References

1. Prasad MR, Naik RL, Bapuji V (2013) Cloud computing: research issues and implications. Int J Cloud Comput Serv Sci 2:134–140. https://doi.org/10.11591/closer.v2i2.1963
2. Nagasubramanian G, Sankayya M, Al-Turjman F, Tsaramirsis G (2020) Parkinson data analysis and prediction system using multi-variant stacked auto encoder. IEEE Access 8:127004–127013. https://doi.org/10.1109/ACCESS.2020.3007140
3. Tsaramirsis G, Karamitsos I, Apostolopoulos C (2016) Smart parking: an IoT application for smart city. In: 2016 3rd International conference on computing for sustainable global development (INDIACom), pp 1412–1416
4. Indrani I, Revanth Svb, Durga SA, Potluri S, Mohanty SN (2021) Internet of cloud: secure and privacy preserving cloud model with IoT enabled service. In: Potluri S, Rao KS, Mohanty SN (eds) Cloud security. De Gruyter, Berlin, Boston, pp 83–104. https://doi.org/10.1515/9783110732573-005
5. Giannopoulos IK, Leros AK, Leros AP, Tsaramirsis G, Alassafi MO (2020) Real-time adaptive stochastic control of smart grid data traffic for security purposes. Sustain Cities Soc 63, 102473. ISSN 2210-6707. https://doi.org/10.1016/j.scs.2020.102473
6. Whaiduzzaman Md, Sookhak M, Gani A, Buyya R (2014) A survey on vehicular cloud computing. J Network Comput Appl 40:325–344. ISSN 1084-8045. https://doi.org/10.1016/j.jnca.2013.08.004
7. Boukerche A, De Grande R (2018) Vehicular cloud computing: architectures, applications, and mobility. Comput Networks 135. https://doi.org/10.1016/j.comnet.2018.01.004
8. Ahmad F, Kazim M, Adnane A, Awad A (2015) Vehicular cloud networks: architecture, applications and security issues. In: 2015 IEEE/ACM 8th International conference on utility and cloud computing (UCC), pp 571–576. https://doi.org/10.1109/UCC.2015.101
9. Shahzadi S, Iqbal M, Qayyum Z, Dagiuklas T (2017) Infrastructure as a service (IaaS): a comparative performance analysis of open-source cloud platforms. https://doi.org/10.1109/CAMAD.2017.8031522
10. Srivastava P, Khan R (2018) A review paper on cloud computing. Int J Adv Res Comput Sci Softw Eng 8:17. https://doi.org/10.23956/ijarcsse.v8i6.711

11. Kulkarni G, Sutar R, Gambhir J (2012) Cloud computing-storage as service. Int J Eng Res Appl 2:945–950
12. de Souza B, Leal Rego PA, de Souza JN (2019) Exploring computation offloading in vehicular clouds. In: 2019 IEEE 8th International conference on cloud networking (CloudNet), pp 1–4. https://doi.org/10.1109/CloudNet47604.2019.9064133
13. Bernsteiner R, Kilian D, Ebersberger B (2016) Mobile cloud computing for enterprise systems: a conceptual framework for research. Int J Interact Mobile Technol (iJIM) 10:72. https://doi.org/10.3991/ijim.v10i2.5511
14. Meneguette R (2016) A vehicular cloud-based framework for the intelligent transport management of big cities. Int J Distrib Sens Netw 2016:1–9. https://doi.org/10.1155/2016/8198597
15. Maroukian K, Apostolopoulos C, Tsaramirsis G (2017) Extending model driven engineering aspects to business engineering domain: a model driven business engineering approach. Int J Inf Technol 9:49–57. https://doi.org/10.1007/s41870-017-0009-8
16. Apostolopoulos C, Halikias G, Maroukian K, Tsaramirsis G (2016) Facilitating organisational decision making: a change risk assessment model case study. J Model Manag 11(2):694–721. https://doi.org/10.1108/JM2-05-2014-0035
17. Arkian HR, Atani RE, Diyanat A et al (2015) A cluster-based vehicular cloud architecture with learning-based resource management. J Supercomput 71:1401–1426. https://doi.org/10.1007/s11227-014-1370-z
18. Tsaramirsis G, Papoutsidakis M, Derbali M, Khan FQ, Michailidis F (2020) Towards smart gaming olfactory displays. Sensors 20(4):1002. https://doi.org/10.3390/s20041002
19. Alahmadi A, Musa MOI, El-Gorashi TEH, Elmirghani JMH (2019) Energy efficient resource allocation in vehicular cloud based architecture. In: 2019 21st International conference on transparent optical networks (ICTON), pp 1–6. https://doi.org/10.1109/ICTON.2019.8840547
20. Al-Rashed E, Al-Rousan M, Al-Ibrahim N (2017) Performance evaluation of wide-spread assignment schemes in a vehicular cloud. Veh Commun 9:144–153. ISSN 2214-2096. https://doi.org/10.1016/j.vehcom.2017.05.005
21. Liang H, Zhang X, Zhang J, Li Q, Zhou S, Zhao L (2019) A novel adaptive resource allocation model based on SMDP and reinforcement learning algorithm in vehicular cloud system. IEEE Trans Veh Technol 68(10):10018–10029. https://doi.org/10.1109/TVT.2019.2937842
22. Sun P, Samaan N. A novel VANET-assisted traffic control for supporting vehicular cloud computing. In: IEEE Transactions on intelligent transportation systems. https://doi.org/10.1109/TITS.2020.2994280
23. Midya S, Roy A, Majumder K, Phadikar S (2018) Multi-objective optimization technique for resource allocation and task scheduling in vehicular cloud architecture: a hybrid adaptive nature inspired approach. J Network Comput Appl 103:58–84. ISSN 1084-8045. https://doi.org/10.1016/j.jnca.2017.11.016
24. Hou L, Zheng K, Chatzimisios P, Feng Y (2018) A continuous-time Markov decision process-based resource allocation scheme in vehicular cloud for mobile video services. Comput Commun 118:140–147. ISSN 0140-3664. https://doi.org/10.1016/j.comcom.2017.10.011
25. Vijayarangam S, Megalai J, Krishnan S et al (2018) Vehicular cloud for smart driving using internet of things. J Med Syst 42:240. https://doi.org/10.1007/s10916-018-1105-4
26. Ahmed B, Malik AW, Hafeez T et al (2019) Services and simulation frameworks for vehicular cloud computing: a contemporary survey. J Wirel Commun Network 2019:4. https://doi.org/10.1186/s13638-018-1315-y
27. Chang BJ, Tsai YL, Liang YH (2017) Platoon-based cooperative adaptive cruise control for achieving active safe driving through mobile vehicular cloud computing. Wirel Pers Commun 97:5455–5481. https://doi.org/10.1007/s11277-017-4789-8
28. Jafari Kaleibar F, Abbaspour M (2021) SLA-based service provisioning approach in vehicular cloud network. Cluster Comput. https://doi.org/10.1007/s10586-021-03357-6

29. Sookhtsaraei R, Iraji M, Artin J et al (2021) Increasing the quality of services and resource utilization in vehicular cloud computing using best host selection methods. Cluster Comput 24:819–835. https://doi.org/10.1007/s10586-020-03159-2
30. Khadir AA, Seno SAH (2021) SDN-based offloading policy to reduce the delay in fog-vehicular networks. Peer Peer Netw Appl 14:1261–1275. https://doi.org/10.1007/s12083-020-01066-2

Social Listening on Budget—A Study of Sentimental Analysis and Prediction of Sentiments Using Text Analytics & Predictive Algorithms

A. Mansurali, P. Mary Jayanthi, R. Swamynathan, and Tanupriya Choudhury

Abstract Text analytics is a mining technique process applied on the text extracted from various platforms to have an understanding about the context and opinion. And this process of social listening is gaining its importance as it is the direct opinion and feedback from the social class or a consumer pool. Sentimental analysis is process of attaching sentiment to the processed text which is of people's opinions, attitudes, and judgments on certain context. This study is attempted to understand the sentiment and predict the sentiment on the most debated democratic element of the country, i.e., budget. The data for this study have been collected from Twitter using the APIs on the day of budget. With these data, various machine learning algorithms are applied for the classification problem. For the Twitter sentiment analysis, user sentiments about—#Budget2020 are found out by various sentiment classification techniques. Later, various machine learning models such as SVM, Naïve Bayes, random forest, and XGBoost were modeled. The best model comparison has been made based on evaluation metrics such as accuracy.

Keywords Budget · Sentiment · Twitter · Classification · Text data

---

A. Mansurali (✉)
PSG Institute of Management, Coimbatore, India
e-mail: mansurali@psgim.ac.in

P. Mary Jayanthi
Jaipuria Institute of Management, Jaipur, India

R. Swamynathan
Amity University, Dubai, India
e-mail: sramakrishnan@amityuniversity.ae

T. Choudhury (✉)
School of Computer Science, University of Petroleum and Energy Studies (UPES), Dehradun, Uttarakhand 248007, India
e-mail: tanupriya@ddn.upes.ac.in

V. Skala et al. (eds.), *Machine Intelligence and Data Science Applications*, Lecture Notes on Data Engineering and Communications Technologies 132,
https://doi.org/10.1007/978-981-19-2347-0_68

1 Introduction

Sentiment analysis [1–4] is process of analyzing the texts and classifying it into different sentiments such as positive, negative, and neutral. In Twitter [5, 6], there are more than 321 million active users and send about an average of 500 million tweets every day. The evolution of microblogging sites allows people to express their views and emotions on different platforms such as Facebook, Twitter, Reddit, and Tumblr. Companies are able to get various information from these opinions and emotions regarding their products or services from these microblogging sites. Companies study these customer reactions and respond to those customers on the microblogging sites itself. By the natural language processing (NLP) techniques and machine learning, the subjective information of a document can be extracted and can be classified into positive, negative, or neutral based on their polarity. This is a very useful technique for predicting the overall opinion about a product or a service by the company and also stock market for a company, like if majority of the people think positive about the company, probably the stock price for that particular company can go up and so on. To know more about human feelings and opinions, sentiment analysis from the extracted tweets is a powerful tool to be employed and understood [2]. In this study, sentiment analysis has been performed on Twitter tweets about the people's attitudes, opinions, thoughts, and judgments on certain entities. Twitter is one the most popular microblogging sites which have been used by people all around the globe. Nowadays, leaders of the world such as president, prime minister, government departments, and so on are sending out their messages to the public through their messages in these microblogging sites. Twitter has a wide impact as a tweet can lead to the rise/fall in stock market and impact the economic growth of the nation. The business entities and governments can leverage the information about the opinions and emotions of the user from the information available in Twitter, and they can make appropriate decisions. Sentiment analysis is one of the most important factors in decision-making process [7]. Twitter sentiment analysis can be done for various reasons such as the Twitter users have various intentions behind each tweet; Twitter users are available all around the globe; Twitter has varieties of audience ranging from common man to celebrities, politicians, and so on; the number of monthly active users is about 330 million in 2019.

2 Problem Statement

The tweets are short microblogs which range from 1 to 280 characters, which include hashtags too. Hashtags, which are suffixed with "—" "#" symbol on a set of words or phrases, are used by the users to have a discussion about a common point of interest. In this study, the "#Budget2020" had been considered for analysis, in a way to find out the opinions and sentiments of people about the 2020 Budget. Text mining techniques can be performed on the tweets to study the emotions behind the tweets.

With the text mining techniques, sentiment analysis on the Twitter data can be done to find out the sentiments and opinions based on the sentiment score assigned to each emotion. The sentiment of the tweets can be positive, negative, or neutral. In order to find the sentiments, various methods are used such as Bing, Afinn, Nrc, Syuzhet, bag of words. Twitter sentiment analysis can be done by various techniques such as lexicon-based approach and machine learning-based approach. The different machine learning algorithms [8] are implemented to predict the sentiments and to find the accuracy of the model. They are support vector machine (SVM), Naïve Bayes, random forest, XGBoost. From the above machine learning methods, the best model can be chosen and based on the accuracy which will be used for prediction of sentiments of people.

3 Scope of Twitter Sentiment Analysis

This project can be used by companies, politicians, and common public. This project is carried out with 5000 tweets extracted from Twitter, within two days from the release of Budget 2020. The tweets are limited to 140 characters, and Twitter contains only recent 7 days of tweet. The polarity of the tweets can be analyzed and used by companies to promote their products, obtain reviews, and various marketing activities can be carried out by analyzing the sentiments. Politicians can get the review of their recent programs carried out or any new releases. Sentiment analysis performed on Twitter data by Mishra [9] has given an understanding toward the opinion mining as a framework with the steps such as preprocessing, feature extraction, and polarity classification. By analyzing the tweets, we can understand how much positive or negative or neutral are people about a situation. There were scenarios like automation of sentiment analysis for the tweets done in two phases such as classifying as objective and subjective followed by positive and negative Barbosa and Feng [1].

4 Objective

1. To find out the sentiment of—Budget2020|| from the tweets.
2. To build different machine learning models to predict the sentiments.
3. Compare the models to find the best sentiment analysis model.

5 Literature Review

The literature review part discusses the research works related to methods used for sentiment analysis. Rezapour et al. [10] studied and evaluated an enhanced model that incorporated the hashtags into a lexicon for improving the accuracy of sentiment analysis model. In this study, the author has checked the effect of adding hand-annotated, corpus-based hashtags to a lexicon of sentiment. Then, they used an improved model to classify and rate the 2016 New York Republican and Democratic Primary Election candidates by collecting the tweets and identified the polarity. In the result, the findings in combination with the negation detection increased the accuracy of prediction by 7%. Judith and Shobha (2012) focused [6] on strategies for twittering the analysis of data to gain public opinion. Using the machine learning algorithm, a feature vector is created with the emotion representing tweet words and is fed to the classifier that classifies the feeling or opinion. The different techniques of Twitter data analysis are dictionary based and use the approaches to machine learning. Dictionary-based approach is used for classifying the sentiments, and the accuracy was found using the machine learning methods. Generally, classification algorithms are largely utilized in the area of sentiment classification and resulting in highest accuracy when random forest, decision tree, and ensemble methods are used.

Tang et al. (2014) in his research used the supervising algorithms for representing the words in the purpose of achieving sentiment analysis on the twitter tweets [11]. Through incorporating the sentiment information into three neural networks loss functions, they learn word embedding. Sentiment-specific embedding of words surpasses current neural models by huge margins. The drawback of this model used by this author is that it takes longer processing time. Pak et al. (2010) proposed a model to classify the tweets [12] as objective, positive, and negative. The training set they used was less efficient since it contains only tweets having emoticons. Turney et al. (2012) used bag-of-words method for sentiment analysis [13] in which the relationships between words were not at all considered and a document is represented as just a collection of words. To determine the sentiment for the whole document, sentiments of every word were determined, and those values are united with some aggregation functions. Kamps et al. (2004) used the lexical database [14] WordNet to determine the emotional content of a word along different dimensions. They developed a distance metric on WordNet and determined semantic polarity of adjectives. Kharde and Sonawane (2016) have given [15] an overview and a similar examination of existing methods for assessment mining like AI and dictionary-based methodologies, together with assessment measurements. The feature space used included retweets, hashtags, link, punctuation and exclamation marks in conjunction with features like prior polarity of words, and POS. Sentiment analysis has been made easier by various machine learning algorithms and resulting in higher accuracies. Bhumika (2017) in her research [5] insisted that the use of ensemble algorithms such as XGBoost ended up in highest accuracy while classifying the sentiments.

6 Methodology

5000 tweets of #Budget2020 are extracted in the form of CSV file with the help of api key, api secret, access token, and access token secret. The extracted data have been

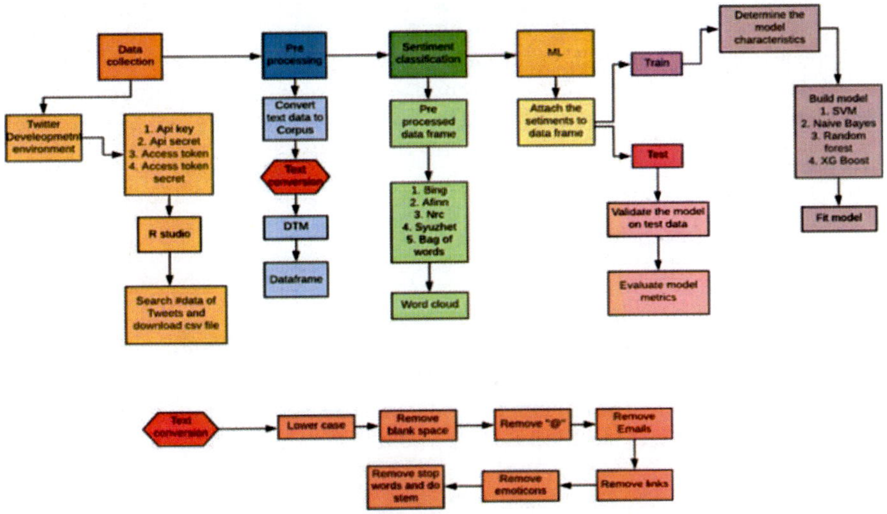

Fig. 1 Methodology of Twitter sentiment analysis

preprocessed, and the desired format of document term matrix has been prepared using packages and functions for finding the sentiments using NLP method with the help of "tm" package. The sentiments are classified using five different techniques such as Bing, Afinn, Nrc, Syuzhet, and bag of words based on their polarity, into positive, negative, or neutral. Sentiment [9, 16] problems become a classification problem when we require to predict the sentiments of a text data. SVM, Naïve Bayes, random forest, and XGBoost algorithms [17, 18] are used for model building with the training data, and the accuracy of the model has been obtained using the test data. The process followed in the research had been depicted detail in Fig. 1.

7 Analysis

The data used for sentiment analysis are the tweets about "#Budget2020." 5000 tweets are used for analysis. A sample of the data is shown below (Fig. 2).

	text	id	statusSource	screenName	retweetCount	isRetweet
1	RT @capt_amarinder: Sharing my article in @Ecor	1.22494E+18	<a href="http://twitter.com/do	bhawnas79921830	81	TRUE
2	RT @PriyaSanjay: There is not a lot for the #EV	1.22494E+18	<a href="https://mobile.twitter.	IndiaESA	1	TRUE
3	The last word on the budget #pettairaibudget #Bu	1.22494E+18	<a href="https://mobile.twitter.	Startupscout88	0	FALSE
4	RT @capt_amarinder: Sharing my article in @Ecor	1.22494E+18	<a href="http://twitter.com/do	gagndeepsingh11	81	TRUE
5	RT @OppressedHindu: Is Indian economy headed	1.22494E+18	<a href="http://twitter.com/do	rohittiwari22	51	TRUE
6	When the budget 2020 had an impact on the cosn	1.22494E+18	<a href="https://mobile.twitter.	Webvillee_media	0	FALSE
7	RT @capt_amarinder: Sharing my article in @Ecor	1.22494E+18	<a href="http://twitter.com/do	mann_deepp	81	TRUE
8	@EY_India This is a very comprehensive #incomet	1.22494E+18	<a href="https://mobile.twitter.	FinancialXpress	0	FALSE
9	RT @Taxsutra: Tax World reacts to Union Budget	1.22494E+18	<a href="https://mobile.twitter.	Nangiaadvisors	2	TRUE
10	RT @ABVPVoice: Welcome the reforms proposed	1.22494E+18	<a href="http://twitter.com/do	abvp4phagli	54	TRUE

Fig. 2 Sample of Twitter data

The extracted data consist of information about text, favorites, favorite's count, data created date and time, ID, user name, retweet status, re tweets count, latitude, longitude, and so on. For performing sentiment analysis, the text column has been extracted as it had the required information about the tweets.

7.1 Data Preprocessing

Preprocessing the data is an important process for the data mining algorithms. Twitter [19] data are an unstructured dataset which are a compilation of information from people who have entered their regional languages written in English, multiple languages, creative, emoticons, short writing spelling and misspelling, etc. Preprocessing in the NLP method is done with the help of tm package in R.

7.2 Sentiment Analysis

Algorithms such as Affin, Nrc, Bing, Syzhet algorithms are used for sentiment analysis. Different methods classified the words into positive, negative, and neutral words, and specifics of classification by each algorithm had been consolidated in Table 1.

Sentiment classification by Bing, Afinn, Nrc, Syuzhet, and bag-of-words algorithm is represented visually in Table 2.

7.3 Sentiment Classification

Machine leaning algorithms are very often helpful to classify and predict whether a document represents positive or negative sentiment. The four machine learning algorithms used for predicting the accuracy of sentiment analysis are, SVM, Naïve Bayes, random forest, XGBoost. Naïve Bayes algorithm was effective in classifying in an attempt to classify the movie dataset consisted of reviews [18]. Tyagi and Tripathi [19] had performed sentiment analysis on the Twitter data using technique of tokenization and employed the algorithms such as KNN, Naïve Bayes, and SVM for the sentiment classification (Fig. 8).

8 Findings

From the sentiment score, the polarity of the tweets is classified as positive, negative, or neutral. Among the 5000 tweets, different methods show varying polarity. Figure 9

Table 1 Methods of sentiment classification

Sentiment classification methods	Description
Bing	Minqing Hu and Bing Liu developed the Bing lexicon as the opinion lexicon. This consists of 6789 words, and the negative words are high compared to positive words. There are around 2006 positive words and 4783 negative words. Score assignment ranges from −1 to +1
Afinn	Finn Aruo Nielsen developed the Afinn lexicon as Afinn word database. This consists of Internet slang and obscene words. This lexicon consists of 2477 words. Negative words are high compared to positive words. There are around 1598 negative words and 878 positive word and a single neutral word
Nrc	The nrc lexicon is different from the above three lexicons. In this type of lexicon, along with the polarity of positive or negative words, a sentiment type is attached to each text based on the ten categories. This lexicon consists of 13,899 words in total, which is distributed in different categories as anger, anticipation, disgust, fear, joy, sadness, surprise, trust, positive, negative
Syuzhet	This package was developed in the Nebraska literary lab by Matthew L. Jockers. It consists of 10,748 words and a sentiment value associated with each word. Among that negative words are being dominated compared to positive words. There are 7161 negative words and 3587 positive words. The lexicons can be obtained by the function get_sentiments_dictionary()
Bag of words	This model focuses entirely on the terms, or sometimes a series of words, but generally doesn't pay attention to the so-to-speak "meaning." The word bag model normally contains a large collection, probably better thought of as a kind of dictionary which is known to be words that carry sentiments

shows the sentiments obtained from the five sentiment classification techniques such as Bing, Afinn, Nrc, Syuzhet, and bag of words.

It is seen that majority of the sentiments are neutral in case of Bing, Afinn, Nrc, and bag-of-words method. In Syuzhet method, the maximum sentiments are obtained as positive. Overall, the sentiments of the Twitter users regarding the #Budget2020 is neutral. The best model among SVM, Naïve Bayes, random forest, and XGBoost is obtained by comparing various parameters of the confusion matrix. In a process of achieving highest accuracy in the sentiment classification, a study by Gaye and Wulamu [4] has mentioned that the highest accuracy has been yielded by applying XGBoost algorithm on the data related to monitoring and communication systems. Table 3 shows the accuracy of various models for the five sentiment classification techniques. Being the dataset is just 5000 and with improvisation in tuning parameters—XGBoost learnt the data and over fitted. The overfitting can be avoided by

Table 2 Comparison of sentiments of all five methods

Figure 3	A score of 0 is considered as neutral. Scores >1 is considered as positive, and <1 is considered as negative sentiments. From the Bing method of sentiment classification, it is seen that majority of the sentiments in the tweets are classified as neutral. A part of the tweets is classified as positive
Figure 4	Afinn classifies the tweets between a sentiment score of −5 and +5. A score of 0 is considered as neutral. Scores >1 is considered as positive, and <1 is considered as negative sentiments. From the chart, it is observed that this method classified majority of the tweets as neutral
Figure 5	From the syuzhet method of sentiment classification, it is seen that majority of the tweets are classified as neutral based on the sentiment score. A score of 0 is considered as neutral. Scores >1 is considered as positive, and <1 is considered as negative sentiments
Figure 6	NRC method of sentiment analysis classifies the sentiments between the sentiment score of −5 and +5. A score of 0 is considered as neutral. Scores >1 is considered as positive, and <1 is considered as negative sentiments. From the chart, it is observed that majority of the tweets are classified as neutral, followed by positive. Negative tweets are comparatively less
Figure 7	In this method, the tweets are compared to a set of predefined bags of positive and negative words, and the sentiment scores are assigned to each tweet. From the chart, it is observed that majority of the tweets are classified as neutral. A score of 0 is considered as neutral. Scores >1 is considered as positive, and <1 is considered as negative sentiments

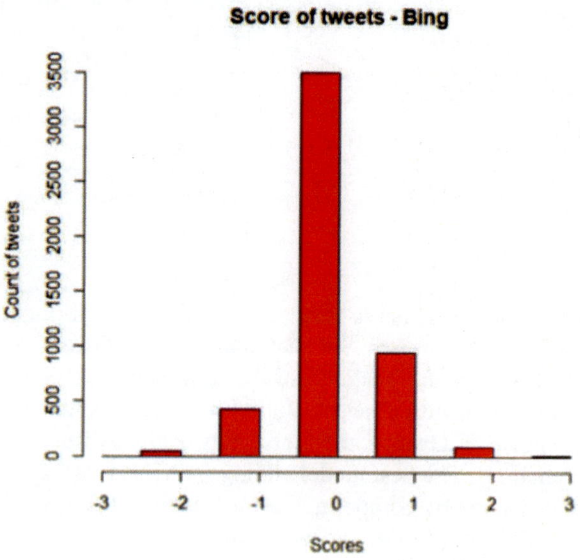

Fig. 3 Sentiment score of Bing

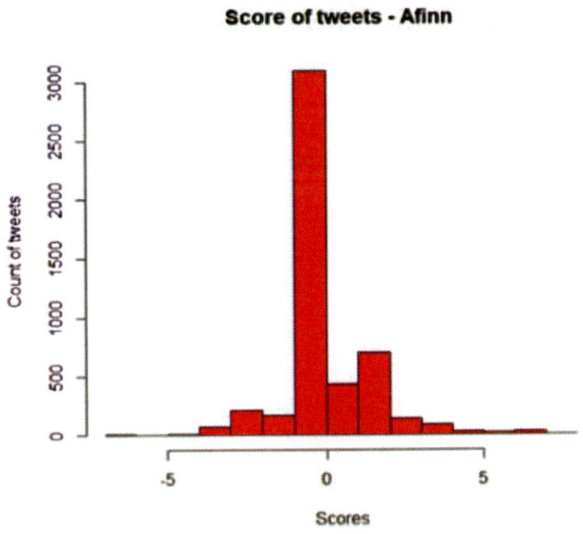

Fig. 4 Sentiment score of Afinn

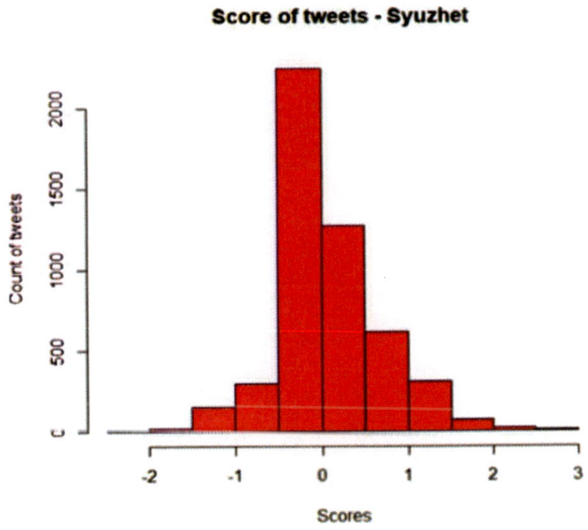

Fig. 5 Sentiment score of Syuzhet

supplying more data and also make the machines learn more about the people opinion could be future scope of the study.

Word cloud has done using the package called word cloud using R. Most frequent used words of the tweet from the document term matrix had been used to create the figure (Fig. 10).

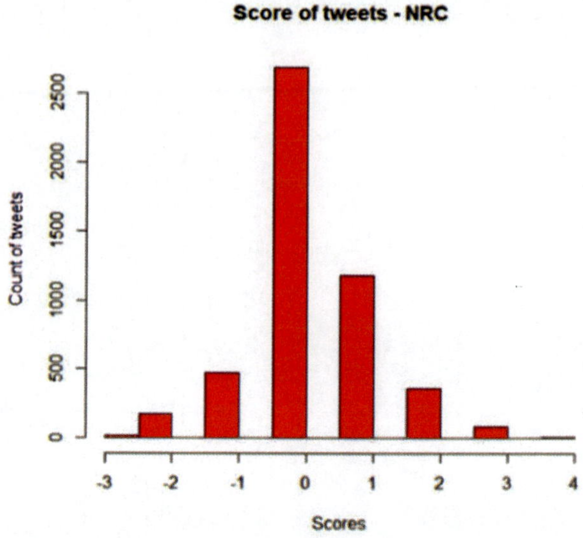

Fig. 6 Sentiment score of Nrc

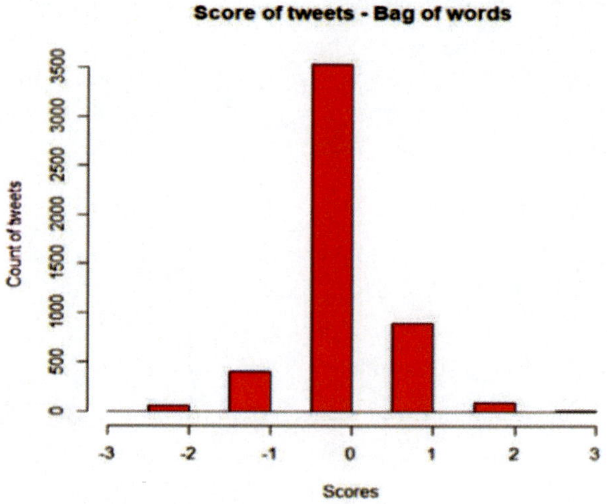

Fig. 7 Sentiment score of bag of words

9 Conclusion

Twitter sentiment analysis has become so popular among the business entities that majority of them are using it to know about the consumer sentiments. With the development of natural language processing and text mining capabilities, it has become easy for the business entities to mine the text data and process it for obtaining the emotions of the customers to extract business insights. The tweets of Indian #Budget2020 in Twitter have been extracted with the help of R platform. Preprocessing is the major step in sentiment analysis since the texts consists of punctuation,

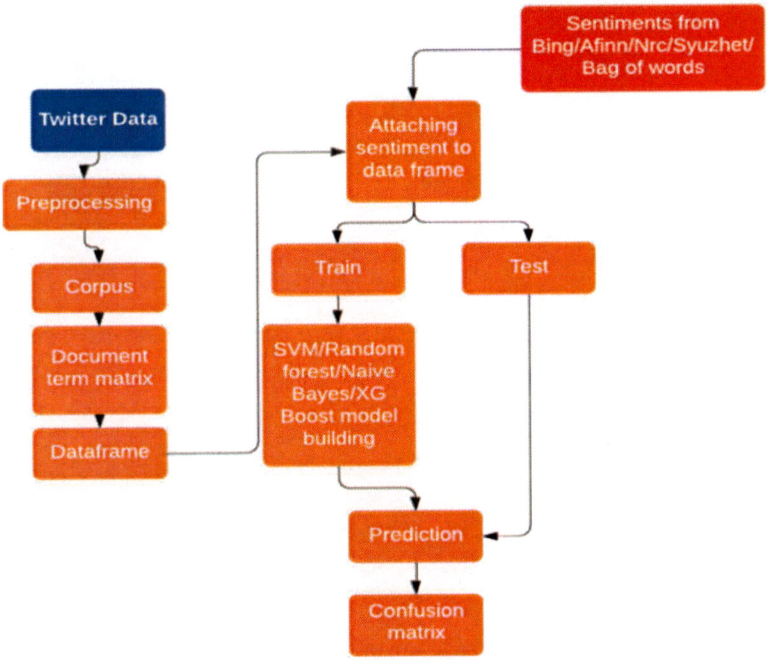

Fig. 8 Process of sentiment analysis by machine learning algorithm

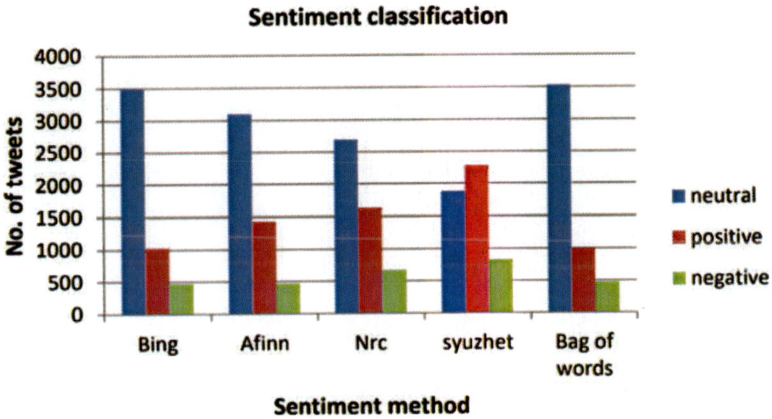

Fig. 9 Comparison of sentiment classification

emoticons, regional languages, and so on, which will affect in obtaining the exact opinions. Packages like tm and NLP are used for this. Sentiments have been classified based on their polarity into positive, negative, or neutral using various different techniques like Bing, Afinn, Nrc, Syuzhet, and bag of words. It was observed that the majority of the sentiments are classified as neutral by these techniques. Machine

Table 3 Comparison of accuracy

	SVM	Naïve Bayes	Random Forest	XGBoost
Bing	93.5	9.5	92.87	100
Affin	94.93	9.34	95.47	100
Nrc	91.33	13.47	88.93	100
Syuzhet	91.33	16.27	91.20	100
Bag of words	94.4	9.47	91.30	100

Fig. 10 Word cloud

learning algorithms such as SVM, Naïve Bayes, random forest, and XGBoost are modeled, and the sentiments are predicted. It was found that among all, XGBoost had the highest performance, and Naïve Bayes had lowest performance. Once the model has been built with high accuracy, they can be used for classifying the sentiments which will take less time. From the sentiment profiling and scoring, it is evident that the hashtag has positive sentiment associated which means people used the hashtag in a positive way to spread messages to the world. The limitation of the paper is that the extracted data are constrained to only 5000 tweets. In future, the size of data being extracted has to be increased to get highly accurate results on hashtag-level sentiment, and the automation of sentiment classification using more packages and algorithms would close the loop.

References

1. Barbosa L, Feng J (2010) Robust sentiment detection on twitter from biased and noisy data. In: Coling 2010: posters, pp 36–44
2. Devika MD, Sunitha C, Ganesh A (2016) Sentiment analysis: a comparative study on different approaches. Procedia Comput Sci 44–49
3. Dey L, Chakraborty S, Biswas A, Bose B, Tiwari S (2016) Sentiment analysis of review datasets using Naive Bayes and KNN classifier. Int J Inf Eng Electron Bus 54–62
4. Gaye B, Wulamu A (2019) Sentimental analysis for online reviews using machine learning algorithms. Int Res J Eng Technol (IRJET), 2395–0056
5. Gupta B, Negi M, Vishwakarma K, Rawat G, Badhani P, Tech B (2017) Study of Twitter sentiment analysis using machine learning algorithms on Python. Int J Comput Appl 29–34
6. Judith STS, Shobha MS (2015) A survey on Twitter data analysis. Int J Adv Res Comput Sci Softw Eng 536–540
7. Tiwari D, Singh N (2019) Ensemble approach for twitter sentiment analysis. IJ Information Technology and Computer Science, pp 20–26
8. Singh N, Sharma T, Thakral A, Choudhury T (2018) Detection of fake profile in online social networks using machine learning. Int Conf Adv Comput Commun Eng (ICACCE) 2018:231–234. https://doi.org/10.1109/ICACCE.2018.8441713
9. Mishra P, Rajnish R, Kumar P (2016) Sentiment analysis of Twitter data: case study on digital India. In: 2016 International conference on information technology (InCITe)—the next generation IT Summit on the theme-internet of things: connect your worlds, pp 148–153
10. Rezapour R, Wang L, Abdar O, Diesner J (2017) Identifying the overlap between election result and candidates' ranking based on hashtag-enhanced, lexicon-based sentiment analysis. In: 11th international conference on semantic computing (ICSC), pp 93–96
11. Tang D, Wei F, Yang N, Zhou M, Liu T, Qin B (2014) Learning sentiment specific word embedding for twitter sentiment classification. In: Proceedings of the 52nd annual meeting of the association for computational linguistics (Volume 1: Long Papers), pp 1555–1565
12. Pak A, Paroubek P (2010) Twitter as a Corpus for sentiment analysis and opinion mining. In: Proceedings of the seventh conference on international language resources and evaluation, pp 1320–1326
13. Turney PD (2012) Thumbs up or thumbs down?: semantic orientation applied to unsupervised classification of reviews. In: Proceedings of the 40th annual meeting on association for computational linguistics, pp 417–424, Association for Computational Linguistics
14. Kamps J, Marx M, Mokken RJ, Rijke MD (2004) Using wordnet to measure semantic orientations of adjectives. In: LREC, pp 1115–118
15. Kharde V, Sonawane S (2016) Sentiment analysis of Twitter data: a survey of techniques. Int J Comput Appl 139:11
16. Sarishma TR, Kumar S, Awasthi MK (2021) To Beacon or not?: speed based probabilistic adaptive Beaconing approach for vehicular ad-hoc networks. In: Paiva S, Lopes SI, Zitouni R, Gupta N, Lopes SF, Yonezawa T (eds) Science and technologies for smart cities. SmartCity360° 2020. Lecture notes of the institute for computer sciences, social informatics and telecommunications engineering, vol 372. Springer, Cham. https://doi.org/10.1007/978-3-030-76063-2_12
17. Gupta AS, Choudhury T (2018) Continuous and integrated software development using DevOps. In: 2018 international conference on advances in computing and communication engineering (ICACCE), pp 290–293. https://doi.org/10.1109/ICACCE.2018.8458052

18. Kumar S, Tomar R (2018) The role of artificial intelligence in space exploration. In: 2018 international conference on communication, computing and internet of things (IC3IoT), pp 499–503. https://doi.org/10.1109/IC3IoT.2018.8668161
19. Tyagi P, Tripathi RC (2019) A review towards the sentiment analysis techniques for the analysis of twitter data. In: Proceedings of 2nd international conference on advanced computing and software engineering (ICACSE)

An Application of Internet of Things for Cybersecurity and Control: Emerging Needs and Challenges

Sukanta Kumar Baral, Ramesh Chandra Rath, Richa Goel, and Tilottama Singh

Abstract With smartphones and other wireless portable devices, our post-PC world is becoming more participative, adaptable, and informative. To improve interactions between things, the new ecosystem, dubbed the Internet of Everything (IoE), brings together network wireless sensors, cloud computing, analytical data, interactive technologies, and intelligent gadgets. The Internet of Things (IoT) is evolving and providing intelligent solutions. We have e-solutions for everything from e-transportation to e-health. To overcome numerous obstacles, an organization needs a planned, well-thought-out cybersecurity strategy to secure its key infrastructure and information systems. The increasing trend in cyber-assaults on system infrastructure combined with the inherent weaknesses of the system gives cause for worry not just to suppliers but also to the customer. These safety issues should be solved so that user confidence is ensured in order to encourage widespread adoption and harvest the potential of IoT. This article, on the other hand, looks at the cybersecurity issues that in the foreseeable future are expected to fuel IoT research.

Keywords Internet of Things · Cyber-threats · System vulnerabilities

S. K. Baral (✉)
Department of Commerce, Faculty of Commerce and Management, Indira Gandhi National Tribal University (A Central University), Amarkantak, Madhya Pradesh, India
e-mail: sukanta.baral@igntu.ac.in

R. C. Rath
Guru Gobind Singh Educational Society's Technical Campus, Chas, Bokaro, approved by AICTE Government of India, New Delhi, and affiliated to Jharkhand University of Technology, Ranchi, Jharkhand, India

R. Goel · T. Singh
International Business Department, Amity University, Noida, Uttar Pradesh, India
e-mail: rgoel@amity.edu

T. Singh
e-mail: tsingh6@amity.edu

V. Skala et al. (eds.), *Machine Intelligence and Data Science Applications*, Lecture Notes on Data Engineering and Communications Technologies 132,
https://doi.org/10.1007/978-981-19-2347-0_69

1 Introduction

IBM (the technological company known as International Business Machines Corporation) predicts that by 2020, the number of linked devices will eventually outnumber the number of humans on the planet [1]. The Internet of Things (IoT) is an interconnected network of gadgets. A platform for sensors and actuators to communicate with the environment in order to generate a common operating picture [2]. A worldwide identity is the foundation of an IoT system of a single object. An item's access may only get static data from RFID tags in this case. IoT objects have a unique identification and may be accessed (interactively). IoT is the next breakthrough technology to completely integrate the Internet of Things (of things). Examples include wireless sensor networks (WSNs), RFID tags, and actuation nodes. Expanding communication networks to encompass physical objects, according to Karimi and Atkinson, would enhance the number of devices that can be linked as well as the quantity of data that can be transferred [3]. For a broad variety of devices, services and applications IoT offers omnipresent connection, intelligent computers, cellphones, workplace equipment, wireless cars, lighting, HVAC (heating, ventilation, and air conditioning), home items, and so on. To be IoT-enabled, a device ('thing') must be connected to a communication node. On a variety of platforms, networking technologies such as 3G, LTE, Wi-Fi, Bluetooth, ZigBee, Z-wave, Sigfox, and others provide IoT deployment connectivity services. As IoT progresses, the backbone of global information dissemination, data analysis (or calculation), and storage is anticipated to be cloud computing. In the near future, cloud-based solutions like physical devices with computing and communication capabilities will be able to engage with a wide range of applications, services, and technology thanks to Microsoft Azure and Amazon Web Services. R. Caceres and A. Friday focused on the cloud and the Internet of Things to examine the innovations, possibilities, and difficulties of an ubicomp ecosystem. Cloud computing is frequently viewed as a data receiver from omnipresent sensors as well as a platform for monitoring, analyzing, and comprehending IoT [4]. Cloud-based solutions for user engagement and connectivity are also available with appropriate Web interface applications. Gubbi et al. [2] say that for IoT to really emerge as a link between things and intelligence, traditional computer, and Internet connectivity platforms would need to extend beyond the usual mobile communication that connects humans. This fundamental foundation enables smart connectivity and context-aware computing. The IoT ecosystem now has billions of devices connected to it massive volumes of data will be generated that must be efficiently stored, processed, and communicated. Cloud-centric computing is required to enable IoT services as a virtual infrastructure [2] and monitoring, storage, analysis, visualization, and client service [2].

Gubbi et al. claimed that the IoT is "Internet" and "Thing" oriented. Internet as a service is embroiled in the Internet-centered design, whereas objects provide data [2]. With IoT services and applications, smart items take center stage in things-centered architecture [5]. A major problem is IoT-cloud integration since cloud models are experiencing numerous security challenges from infrastructure to service sectors.

These difficulties will most probably arise when the two systems combine and raise some basic research issues that need to be addressed.

2 Literature Survey

We, the researchers, have gone over a lot of literature in this part. As the extent and amplitude of ecological change increases, applying these principles at scales relevant to management becomes increasingly critical. Because complex systems are layered and hierarchical, planning for ecological resilience in the context of IoT needs a multiscale strategy. The administration and analysis of data provided by the devices used for application purposes become more important now-a-days. Gartner (2015) verified the massive scale of IoT, forecasting that 5.5 million new objects would be connected every day in 2016, and 6.4 billion connected devices would be in use worldwide, up 30% from 2015. The entire installed base of IoT-connected devices is expected to reach 30.9 billion units by 2025, up from 13.8 billion units predicted in 2021.

We, the researchers, conducted a wide survey both offline and online, asking respondents from various firms and sectors a series of questions linked to the aforementioned study subject "An Application of Internet of Things for Cyber Security and Control: Emerging Needs and Challenges." In this regard, we used both methods of data gathering and gathered a database of roughly 250 respondents from diverse sources like: Accordingly, an International Data Corporation (IDC) research says companies understand IoTs economic promise but are profoundly wary of the system's inherent security challenges [6]. The report also states that most corporate executives have acknowledged that they have limited awareness or underestimation of IoT's security risks. KPMG reported in separate research that infringement of security in consumption data, and recent assaults on cyber-infrastructure systems globally, caused IoT consumers to lose trust and shun suppliers of solution who do not take the necessary steps to secure their systems [7]. The problem of cybersecurity issues on IoT systems are an important worldwide issue that needs a comprehensive evaluation from both academics and industry. This study examines how security concerns may affect IoT adoption and customer confidence.

3 Research Objectives

The researchers have chosen the following research objectives to justify the aforementioned research endeavor:

- To understand the application process and the influence of the Internet of Things (IoT) on cybersecurity and control.
- To study the application and service domain of IoT in various social media.

- To study the advance models of IoT on supply chain and logistic systems for product design, production, and its control.

4 Hypothesis

In this section, we have taken two types of hypotheses such as null hypothesis (H_o) and alternative hypothesis (H_e) in order to justify the aforesaid research objectives.

4.1 *Null Hypothesis (H_o)*

This hypothesis refers about the application of IoT has no impact on cybersecurity and control when network and smart grid are weak or failure.

4.2 *Alternative Hypotheses (H_e)*

This hypothesis refers about the application of IoT on cybersecurity; it has a great impact and influence the function monitor and control all the security control and threats successfully.

5 The IoT Concept

Internet technology has transformed people's connectivity in unprecedented ways during the last decade. A "smart environment" is projected to emerge from the next revolution, which will connect various items. We need more data on demand utilising sophisticated and easy queries as we go from www (static web pages) to web (social networking web) [2]. The post-PC age is here, and cellphones and other devices are affecting the environment and the way people interact (including people). The new environment is more engaging and educational. As defined by Mark Weiser (creator of ubiquitous computing), "Sensors, actuators, displays, and computer components are seamlessly incorporated into everyday objects to create a smart physical environment" [8]. Indeed, Gubbi et al. [2] argued that cloud computing and IoT are shaping ubiquitous computing. In a contribution, Kevin Ashton claims that "By incorporating RFID and other sensors into common objects, we can build an internet of things, laying the foundation for a new era of machine perception" [9]. In the scientific and industrial eco-systems, the concept has emerged. They say that an IoT device is a heterogeneous object that has an Internet-based locatable, addressable, and readable computer that can communicate with any other entity and give and receive services anytime, anywhere. As a result, most "things"—including people and dogs—will be on the Internet in some form or another (IoE).

6 IoT Infrastructure

WSN, RFID tags, actuators, reading devices, cameras, controllers, GPS, sensors (magnetometers, wasp mote, ultrasonic, and infrared) have all been developed, as well as equipment processors, terminals, and other sensor gateways. In wireless networks or nodes, for shrinking, silicon-integrated circuits and nano-electronics are used in the majority of IoT devices, low cost, and better functionality. IoT hardware resources such as RFID tags and WSN hardware are still relevant. RFID allows for data transmission across small distances. An active or passive RF tag is attached to the item, as explained in Fig. 1.

RFID technology allows data to be transferred over a short distance. An RF reader/emitter and an active or passive radio frequency (RF) tag affixed to a monitored item make up the system. The power for a passive RFID tag comes from the reader, but the power for an active RFID tag comes from the embedded device.

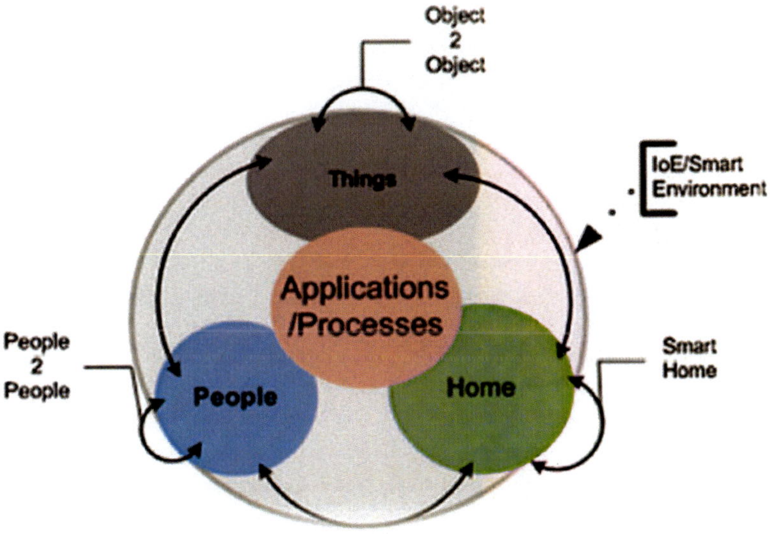

Fig. 1 Futuristic ecosystem model

7 Characteristics of IoT Devices

Some of the broad and important features found throughout the investigation, for example, IoT intelligence is a combination of computation and algorithms, software, and hardware that allows a system to be intelligent:

- Connectivity.
- Nature dynamics.
- Huge scale.
- Sensing.
- Heterogeneity.
- Security.

8 Definitions of IERC

The Internet of Things (IoT) is a self-configuring global network architecture based on standard and interoperable communication protocols that employ smart interfaces where physical and virtual "things" have identities, physical properties, and virtual personalities that can be tracked, as well as an RF reader/emitter, according to the IERC definition.

8.1 IERC Models of IoT

The following figures are referring about the IERC models of IoT focuses on a dynamic global network architecture, self-configuration capabilities, and standard interoperable communication processes and protocol which are connected to physical and virtual world. Again, these models also identify physical attribute and virtual personality such as use of intelligent, interfaces, seamless integration, and information network connected with digital world by which it enhances three components of IoT society knowledge integration (data integration, things integration, semantic integration in between real physical world and virtual cyber-word) as mentioned in Figs. 2 and 3.

An active RFID tag's built-in device powers it. A passive tag needs its reader's energy. WSN hardware consists of sensor interfaces, processors, power supplies, and transceivers. IoT software is supported by a variety of platforms, ranging from firmware operating systems to mobile application software (iOS/Android/Blackberry) and cloud-based applications. A network and device operating system are based on microkernels (e.g., Tiny OS, Tiny DB, Nano-RK, Lite OS, and VMs). HTML, JavaScript, Ajax, PHP, and Ruby are examples of other software. Aside from fundamental software services, IoT software integration provides data visualization, system integration, and remote access control (API).

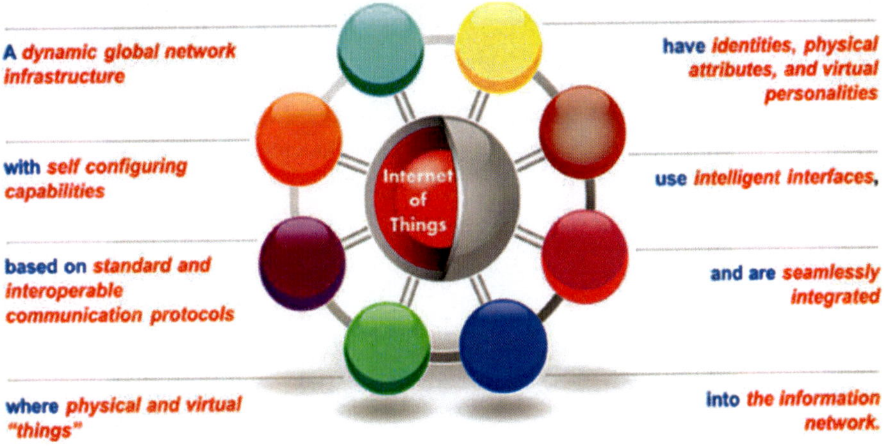

Fig. 2 Internet of Things (IoT)

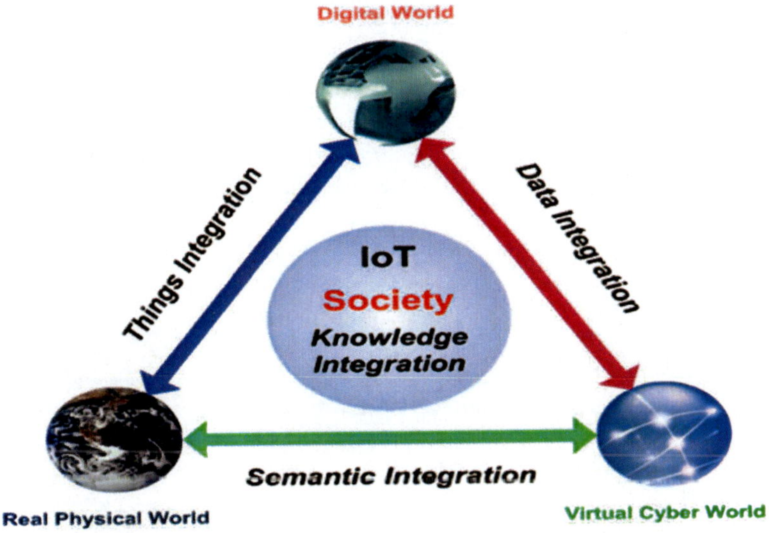

Fig. 3 IoT and society knowledge integration

According to Roman et al. [10], ensuring safety is critical to bringing IoT to life. Confidentiality, integrity, and data availability are among the security objectives (SOs) of traditional information systems (ISs) [11]. There are also unique IoT security issues. Examples of cloud-based vulnerabilities include those in the top ten of OWASP [12]. IoT systems and services are expected to be impacted by cloud-based vulnerabilities. Another key risk vector is substandard IoT goods and services [12]. Cloud-based vulnerabilities are anticipated to impact IoT systems and services. Substandard IoT products and services are another major risk vector. These may endanger IoT services. Touhill and Touhill [13] claim that unexpected equipment and system failures throughout the world, poor or inadequate maintenance, and erroneous and incorrect advice from untrained staff cost firms many hours of money and mission losses. The writers also claim the contract workforce underperforms. Incorrect data from corporate partners and IoT devices may be accepted in critical business decisions [13]. Furthermore, the infrastructure is vulnerable to different IoT cybersecurity threats due to the system's inherent flaws. Data, network interfaces, and ports are possible sources. The system's two-way connections expose it to network attacks and protocol failures.

9 IoT Application and Service Domains

Taking into account current developments in IoT platforms, many IoT applications, taking account of ongoing innovation in technology, services, and continual industrial requirements, are almost difficult to imagine. Some of the current uses include intelligent energy (intelligent grid and measurement), intelligent transportation and mobility, health care, retail and logistics, environmental monitoring, and intelligent manufacturing. All of these areas are linked to security yet each has unique concerns. Some are extremely clear and widely debated as security violations relating to the openness of heath care for abuse of highly personal and retail information that may be available for monetary/financial use but also for rapidly changing information.

9.1 *Smart Grid*

With IoT, a city can be made "smart." The topic involves sensing and monitoring activities in an intelligent environment. It addresses a broad variety of infrastructural and technical requirements. The energy domain/intelligent energy has a lot of IoT potential. Energy packets, like data packets, are handled by routers and gateways, which can choose the optimal path to their destination with the highest degree of integrity on their own [13]. The "IoEn" is a network architecture that provides real-time balancing of energy generation, storage, distribution, and demand by using common transceivers, gateways, and protocols. The smart grid is defined by the IEEE 2030 smart energy standard as having three primary components: WAN, FAN/NAN, and HAN (HAN).

9.2 *Smart Metering*

Vermesan and Friess claim IoEn will revolutionize electricity distribution, storage, and grid monitoring. IoT allows platform-based energy delivery where and when needed. So, power usage is traced from local (home) devices to national and international distribution hubs [13].

9.3 The Advanced Product Supply Chains Are Complex and Weakness

Increasing customer demands for shorter lead times, expanded products and services, and personalized experiences have all contributed to supply chain complexity, as explained in Fig. 4.

- In a mutually authenticated TLS session, a remote server establishes a secure channel.
- This secure channel can now be used to update/modify sections of the device's firmware/RTOS/apps/services and change the device's settings.

10 Data Table

After a careful survey, the following information are recorded in the data table for the proper conduction of research work analysis of result. The following information are given in Table 1.

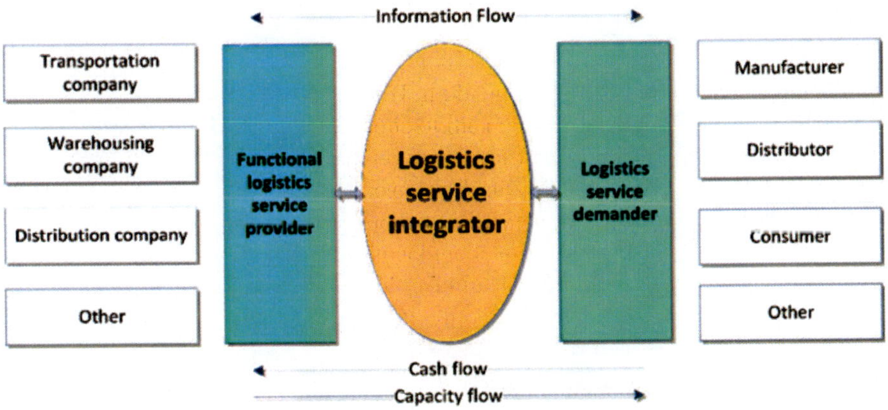

Fig. 4 Product supply chain model of complex and weak in attacks and service

Table 1 Application of Internet of Things (IoT) on cybersecurity and control

Companies/offices	No. of respondent	Use of models	Positive response	Percentage (%)
GOOGLE CLOUD	30	FESM	21(30)	70
AMAZON	30	SMS	24(30)	80
FLIP CARD	30	FESM	26(30)	86.66
SECURE THING Z	30	FESM	24(30)	80
AVINET	30	FESM	21(30)	70
MACRO CHIEF	30	FESM	24(30)	80
IAR ARM	30	FESM	28(30)	93.33
NP REMESAS	30	FESM	24(30)	80
SILICON	30	PSMC	23(30)	76.66
WOW AVNET	30	FESM	25(30)	83.33
Total	300		240(300)	80.00

10.1 Mean (PR)-24.0 Mean (NR) = 60/10 = 6.00

From the above table, we may understand mean value = sum total of respondents/number of users 300/10 = 30 respondents from each company. Out of 300 respondents, 240(300) is response positive, and the mean value is 24.0, and the negative respondent is 80/10 = 8.00.

11 Final Results

In this section, we have obtained a very good data from the respondents, out of 300 respondents 240 responded positively, and 60 numbers of respondents were not in favor of use of IoT in cybersecurity, and its control has a great impact and useful, given in Table 2.

Table 2 Final results

Users	10
Respondents	300
Mean value	30.00
Positive response (PR)	24.00
Negative response (NR)	8.00
Mean difference	16.00
Percentage of response (%)	80.00

12 Hypothesis Testing

We discovered in this section that the null hypothesis is false due to a lack of response support from the final result table and data table (against the Statement of Hypothesis) and most of respondent support the application of IoT for cybersecurity and its control when use of IoT in different constrains and platforms. Thus, we have rejected the null hypotheses and accept the alternative hypothec due to its significance in both the levels of alpha 0.1 and 0.5 points. Here, the semantic response models have been explained pie and graph chart in Figs. 5 and 6.

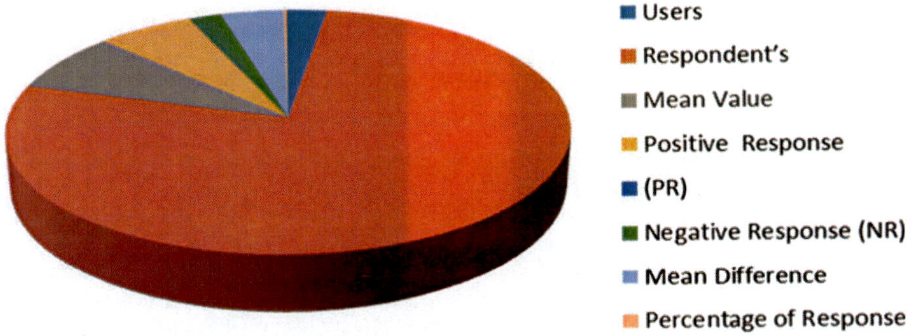

Fig. 5 Semantic response models in pie chart

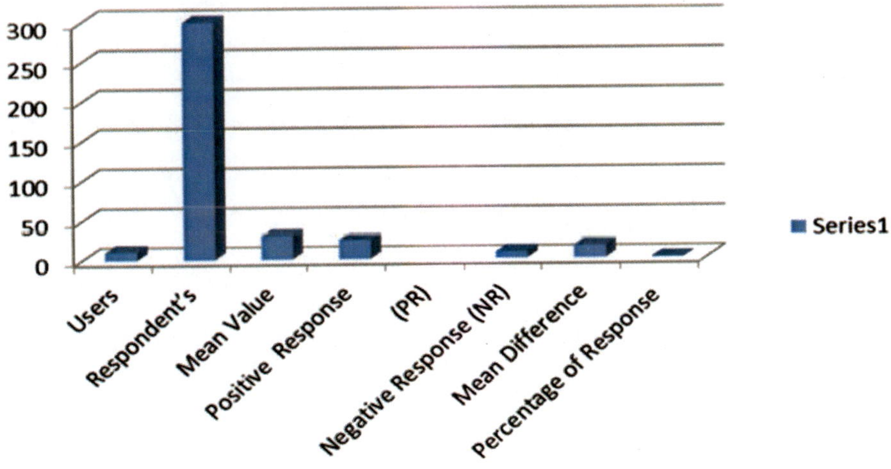

Fig. 6 Semantic response models in graph chart

13 Conclusion

With IoT infrastructure and apps becoming more vulnerable to cyber-attacks, this paper attempts to classify the underlying system vulnerabilities. Our presentation will highlight various IoT and service domain vulnerabilities. Installation of smart metering communication infrastructure was subjected to two different assault scenarios. Our results show that various threat actors can exploit vulnerable IoT devices (apps, hardware, software, or firmware) through created vectors. To conclude, it is critical to keep the dialogue going while also encouraging device manufacturers and component suppliers to create and implement solutions to combat cyber-threats.

References

1. Internet of Things (IoT) and non-IoT active device connections worldwide from 2010 to 2025. Retrieved from https://www.statista.com/statistics/1101442/iot-number-of-connected-devices-worldwide/
2. Gubbi J, Buyya R, Marusic S, Palaniswami M (2013) A vision, architectural components, and future prospects for the Internet of Things (IoT). 1645–1660 in Future Generation Computer Systems, vol 29(7)
3. Karimi K, Atkinson G (2013) What the internet of things (IoT) need to succeed. Free Scale, White Paper, and ARM
4. Caceres R, Friday A (2012) Progress, possibilities, and difficulties in ubicomp systems at 20. IEEE Pervasive Comput 11(1):14–21
5. López TS, Ranasinghe DC, Harrison M, McFarlane D (2012) Adding sense to the Internet of Things. Pers Ubiquitous Comput 16(3):291–308
6. Lund D, Turner V, MacGillivray C, Morales M (2014) From 2014 through 2020, a virtuous loop of proved value and demand is predicted for the Internet of Things (IoT) on a global and regional scale. IDC
7. Matuszak G, Bell G, Le D (2015) KPMG, December 2015, 132631–G, Security and the IoT ecosystem
8. Darianian M, Michael MP (2008) Smart home mobile RFID-based Internet-of-Things (IOT) systems and services. In: International conference on advanced computer theory and engineering, ICACTE'08, pp 116–120
9. Ashton K (2009) That 'Internet of Things' thing. RFID J 22:97–114
10. Roman R, Zhou J, Lopez J (2013) On the features and challenges of security and privacy in distributed internet of things. Comput Netw 57(10):2266–2279
11. Evans DL, Bond PJ, Bement Jr AL (2004) Security classification standards for federal information and information systems. U.S. Department of Commerce, Gaithersburg
12. Pandya D, Patel NJ (2016) Top ten vulnerability analyses in government websites, according to OWASP. Int J Enterp Comput Bus Syst 6(1)
13. Touhill GJ, Touhill JC (2014) Cyber security for executives: a practical guide. Wiley, Hoboken

Comparative Performance Evaluation of Supervised Classification Models on Large Static Malware Dataset

Kaniz Tasmim, Tamanna Akter, Nayan Banik, and Partha Chakraborty

Abstract Data security is one of the most difficult concerns that businesses and associations face. In recent years, the regularity and severity of cybercrime have increased, with new methods for stealing, altering, and destroying information, as well as deactivating information systems, appearing daily. This paper evaluates the performance of some classification techniques on a malware dataset. Machine learning is a type of data analytics that allows computers to perform specific tasks without being given explicit instructions. Algorithms of ML can 'reason' the properties of before unseen samples. In malware detection, an unseen before the sample could be a new file. As of late, ML abilities have been utilized to plan both static and dynamic malware identification. The process has mainly two stages: firstly, features are extracted from the feature vector, and then, a suitable classification method is then applied for the result. Our paper aims to discuss the far more accurate feature abstraction and supervised classification techniques available. Precisely, random forest, decision trees, support vector machines, gradient boosting, and k-nearest neighbors classifiers were estimated. We proceeded with the experiment in two ways. Both results show that the methodologies can gain more than 95% accuracy.

Keywords Cybercrime · Malware detection · Machine learning

1 Introduction

Malware refers to a variety of malicious software, such as viruses, spyware, and ransomware. Malware is software written by cybercriminals with the intent of causing widespread failure to information and services or gaining unauthorized entrance to a network. Malware is typically delivered via email as a link or file, with the user

K. Tasmim (✉) · T. Akter · N. Banik · P. Chakraborty
Department of Computer Science and Engineering, Comilla University, Cumilla 3506, Bangladesh
e-mail: tasmimtithe158@gmail.com

P. Chakraborty
e-mail: partha.chak@cou.ac.bd

V. Skala et al. (eds.), *Machine Intelligence and Data Science Applications*, Lecture Notes on Data Engineering and Communications Technologies 132,
https://doi.org/10.1007/978-981-19-2347-0_70

clicking on the link or opening the file to execute the malware. Every year, 0.35 million different malware samples get discovered. In the year, nearly, 700 million malicious assaults have been recorded. Still, there are over 0.98 billion malicious applications on the Internet [1]. Four businesses are attacked with ransomware per minute. Windows defender is a built-in security system in the windows operating system. Many antivirus programs are also available in both free and paid versions. However, the world's resources are insufficient to combat all malicious software. This is because malicious software's behaviors and characteristics change so quickly that an instructed program can't keep up. Malware is one of the dangers facing today's online users, including company owners, corporations, and institutions. Since then, now is the time to improve anti-malware techniques. A user never executes malware intensely on his system. Malware gets into a PC for unconscious browsing on the Internet, less security knowledge of the user [2]. Malware files can be executable code, scripts, and active content. To activate malware, a user must execute it. A computer can be slowed down or even crashed for the harmful effect of malware. It also steals confidential information such as credit card information, password, and provides the crackers. Ransomware is one of them. It steals confidential information and then threatens the owner of the information to pay ransoms or bitcoin to get them back. Sometimes strong ransomware can get full control of a system. Many malware detection techniques are created by engineers. Anti-malware tools [3] detect threats using various detection procedures. Many anti-malware programs support biometric recognition. On the other hand, malware creators are constantly coming up with new ways to get beyond detection. In practice, consequently, traditional approaches like signature-based detection analysis [4] to detect malware are undesirable when unknown computer viruses arise. Deep learning methodologies, which learn behaviors from malware datasets, can help to overcome these issues. In this case, ensemble methods may be useful. Random forests have already improved virus identification in IoT applications by facilitating the output of machine learning models [5]. As a result, one of the really essential cybersecurity requirements including both individuals and companies is malware prevention from system networks. A simple successful assault can jeopardize data and cause severe costs. As existing static or dynamic approaches cannot give adequate monitoring, particularly, when combined with zero assaults, machine learning-based solutions can be utilized [6]. The main points and concerns of machine learning-based malware detection are discussed in this paper, in addition to the most effective feature interpretations and classification techniques.

This paper has the following structure: Previous work in malware categorization tasks is mentioned in Sect. 2; methods utilized these paper are explained in Sect. 3, and Sect. 4 details the experiment's implementation as well as the outcomes, and the paper comes to an end in Sect. 5.

2 Related Works

In this part, we look at several ideas, awareness, and procedures for identifying and categorizing malware in publications, journals, and conference proceedings. In paper [7], they propose a robust and efficient approach for malware classification and detection by examining both signature-based and anomaly-based features using data mining and machine learning classification techniques. In paper [8], which uses the decision tree, Naive Bayes, and random forest algorithms for malware detection help in the implementation of centralized security in enterprise networks. The accuracy of the random forest algorithm was 97%, and the false-positive rate was 0.03. This paper proposes an advanced antivirus engine that can scan and detect files as strong viruses. This is accomplished by extracting system API calls. The accuracy of the proposed method was 98%. A novel malware analysis [9] they introduced a model for detecting and classifying distinct files (e.g., doc, pdf, exe, etc.) as safe or harmful using two different classifiers: macro and micro such as an Adware, Trojan, and Spyware. Using sub-signatures and machine learning technique, [10] discusses a malware detection system which is static. *N*-gram feature extraction algorithms provide enormous feature search spaces, which can be reduced by using known malware sub-signatures. To increase malware detection accuracy, an examination of several feature selection approaches is performed, as well as an analysis of multiple ML classifiers. When 4-g features are utilized for most of the evaluated ML classifiers, the result indicated that applying machine learning to evaluate n-gram with Snort sub-signature characteristics gives good malware detection accuracy of more than 99.78% and zero FPR. The survey [11] is aimed at reviewing and systematizing existing literature where machine learning is used to support malware analysis of Windows executables, i.e., portable executables, 64 recent papers were chosen based on their bibliographic significance, reviewed, and systematized using a taxonomy with three basic dimensions: (i) the specific goal of the analysis, (ii) the types of features extracted from PEs they consider, and (iii) the machine learning algorithms they use. Machine learning for static and dynamic malware analysis [12] is to look at executable binaries. Malware is analyzed using both static and dynamic features. More than 2300 features are extracted from dynamic analysis, and 92 features are extracted statically from binary file using PEFILE. The dynamic analysis has some limitations due to controlled network behavior, and it cannot be analyzed due to limited access of network. Gradient classifier accuracy for static features is 99.36%. Because malware is so dense, static analysis has some limitations. A malware detection method is based on API-Pair graphs [13] that is real time and adaptive. In concept, they propose a hybrid model that could detect malware in real time and adaptively learn new characteristics. A new API-Pair data format was used, and the resulting data were trained using the maximum entropy model, which met the weighting and adaptive learning goals. The model achieved the highest accuracy of 99.07% in general testing and maintained accuracies above 97% with the evolution of malware, according to the experimental results. The author of the paper [14] used the long short-term memory (LSTM) algorithm to achieve a high accuracy metric result. The

93.60% accuracy was achieved using 5000 samples, which proved the capability of the LSTM algorithm. A comparison study of the LSTM and the back-propagation (BP) algorithm has also been performed to illustrate the algorithm's capacity. The accuracy of the BP approach was 82%, which was lower than the LSTM algorithm. As a result, the LSTM algorithm was performed better than the BP algorithm.

3 Methodology

3.1 Dataset

In the paper, we have evaluated performance on a large malware dataset. The features of the dataset describe the static properties of malware. It has 56 features in total and 138,047 instances. It is a Drebin dataset and consists of samples of 179 malware groups [15]. The dataset contains string, floating-point, and integer-type data. No negative value and NaN value are included in the dataset. Features of a PEfile format file can be extracted using PEfile library of Python. All the features in the dataset indicate the properties which contain in a PEfile format file after downloading and before executing it on the desired operating system [16] 'legitimate' is the target of the dataset. It has two classes: 0 means malware, and 1 means ligate. First, 41,323 legitimate features others are malware features. It is a supervised learning classification-based problem.

3.2 Data Preprocessing

Preparing data are a necessary step in evaluating results. It has been demonstrated in numerous research papers that data prepossessing has a significant impact on results. Prepossessing is also important for removing data over-fitting and under-fitting in a model. In the dataset, first, 41,323 is legitimate features among 138,047 features. So, other 96,724 features are malware that is double then legitimate feature in the dataset. This can cause model biasing toward malware features. So, we removed the impurity of the dataset by removing 55,401 malware features from the dataset and then shuffled it. We worked with the dataset of 82,646 instances. In the dataset, two string types' fields are included which are as follows: name and md5. Classification models don't work with string values; they only work with numerical values. Also, name and md5 have unique values for every instances. So that we removed the two features from the dataset. The data were scaled using standard scalar with a mean of 0 and a standard deviation of 1. Later, in this paper, we converted the numerical value of label to categorical value and performed hyperparameter tuning and grid search on the whole dataset for decision tree.

3.3 Dimensionality Reduction

In method-1, we used correlation to analyze the data. The term 'correlation' refers to the relationship between two or more features. For a better computing experience, highly correlated features are removed. The correlated feature selection threshold was set at 0.80. This disables a total of 13 features. Following that classification models are used, and the results are recorded.

In method-2, PCA is applied for dimensionality reduction as the dataset has now 54 features that will complicate the task of classification models also cause time inefficiency. PCA is applied for n components 2, 10, and 15. Then, for each PCA-modified dataset, classification models are applied and recorded their result. In hyperparameter and grid search methodology, the correlated feature selection threshold is set at 0.90. This disables a total of eight features. The whole dataset is shuffled.

3.4 Dataset Splitting

In method-1, the dataset is split into train and test datasets. 70%, or 57,852 instances, were used for training and 30%, or 24,794 instances, were used for testing. In method-2, we used cross-validation with a value of 10 for training and testing models.

3.5 Classification Models

Support Vector Machine SVM is supervised machine intelligence technique such as tasks involving predictive analysis. These transforms data by performing some complex computations to determine the best boundary between the various outputs [17]. When the data have a clear separation margin, it works well. It is also good for features with a lot of dimensions. SVM is a memory-saving model. When there is a nonlinear boundary between classes in the feature space, nonlinear SVM is used. Given training vectors $x_i \in R^p, i = 1, \ldots, n$, in two classes, and a vector $y \in \{1, -1\}^n$, our goal is to find $w \in R^p$ and $b \in R$ such that the prediction [18] given by $sign(w^T \phi(x) + b)$ is correct for most samples. SVC solves the following primal problem:

$$\min_{w,b,\zeta} \frac{1}{2} w^T w + C \sum_{i=1}^{n} \zeta_i$$
$$\text{subject to } y_i(w^T \phi(x_i) + b) \geq 1 - \zeta_i,$$
$$\zeta_i \geq 0, i = 1, \ldots, n \quad (1)$$

Intuitively, we are trying to maximize the margin (by minimizing $\|w\|^2 = w^T w$ while incurring a penalty when a sample is misclassified or within the margin boundary. Ideally, the value $y_i(w^T \phi(x_i) + b)$ would be ≥ 1 for all samples, which indicates a perfect prediction.

Random Forest Random forest [19] is a classification model consists of a large individual decision tree that works as an ensemble. Each tree is defined as a class prediction. Results belong to the class that has the maximum votes. It provides higher accuracy through cross-validation. The classifier can handle missing values and work well with high-dimension dataset. It also handles over-fitting. Assume that x_i is a training sample where $i = 1, \ldots n$, with outputs y_i where $i = 1, \ldots n$. Then, bagging M times repeatedly to pick a positive sample with a substation of the training sample and build trees to fit these samples. For $m = 1, \ldots, M$: 1. Let X_m, Y_m be alternative samples 2. Train a classification or regression tree f_m on X_m, Y_m.

Decision tree The decision tree is a simple but effective model. To determine which class to belong to, the algorithm creates a tree, as its name implies, from root to leaf. Intermediate nodes define the attributes. The most important field for prediction is determined by a decision tree. This model does not necessitate a lot of computation. Assume that a training set $x_i \in R^n$, where $i = 1, \ldots, 1$ with responses $y \in R^l$, a decision tree repeatedly splits the feature space into similar groups of observations together [20]. If we let the data at node m be represented by the set Q, for each candidate split $\theta = (j, t_m)$ consisting of a feature j and the threshold t_m, partition the data into $Q_m^{\text{left}}(\theta)$ and $Q_m^{\text{right}}(\theta)$ subsets, where :

$$
\begin{aligned}
Q_m^{\text{left}}(\theta) &= \left\{(x, y) | x_j <= t_m\right\} \\
Q_m^{\text{right}}(\theta) &= Q_m \backslash Q_m^{\text{left}}(\theta)
\end{aligned}
\tag{2}
$$

Gradient Boosting Gradient boosting is an AdaBoosting model that has been normalized. The boosting method transforms weak students into strong students. Gradient boosting is a technique for improving train models in a gradual, additive, and sequential way. The techniques for identifying weak learners in AdaBoosting and gradient boosting are not the same. Gradient boosting uses a loss function, whereas AdaBoosting uses high-weight data points.

K-nearest Neighbors KNN is a very old model, but the model is used for ease of interpretation, low calculation time, and good predictive power. The model does not make any underlying assumptions about the distribution of data. It divides coordinates into clusters based on an attribute. In Fig. 1, illustrate the methodology for the classification models.

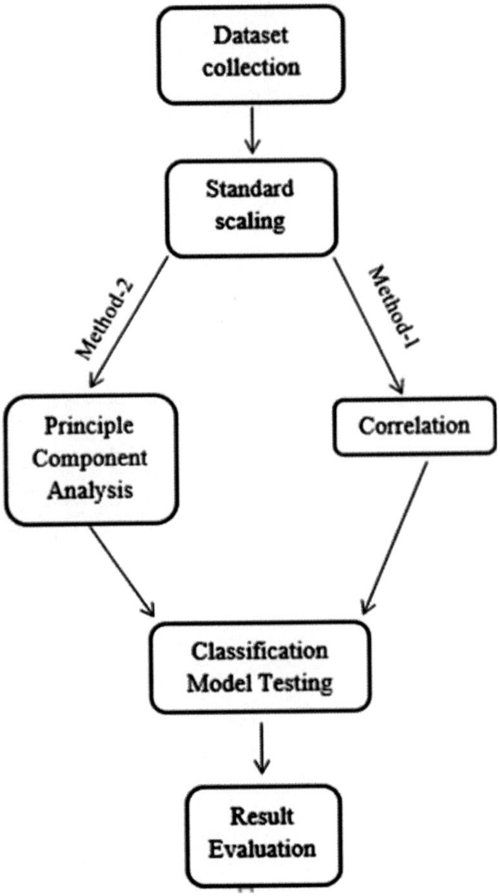

Fig. 1 Methodology

4 Result Evaluation

4.1 Setting of Experiments

DESKTOP-ARBG3D6 device is used for all experiments. Python's Scikit-learn is used for classifier training on the extracted features of the dataset.

4.2 Result of Method-1

The results of method-1 is shown in Table 1. K-nearest neighbors have 99% accuracy when neighbor = 1. An error rate vs neighbor's no. graph is plotted and shown in Fig. 2. It is shown that at neighbor = 1, error rate is minimum. Decision tree has

Table 1 Summary of results of supervised classification models for method-1

Performance measure	SVM linear	SVM poly	SVM rbf	SVM sigmoid	Random forest	Decision tree	Gradient boosting	KNN
Accuracy	0.98	0.98	0.98	0.94	0.99	0.86	0.96	0.99
Error	0.2	0.2	0.2	0.6	0.1	0.14	0.4	0.1
TPR	0.98	0.98	0.98	0.93	0.99	0.899	0.98	0.99
FPR	0.019	0.019	0.01	0.019	0.015	0.16	0.017	0.01
TNR	0.98	0.98	0.989	0.95	0.98	0.83	0.95	0.98
FNR	0.019	0.013	0.018	0.068	0.001	0.10	0.015	0.006
Precision-0	0.98	0.99	0.98	0.93	0.1	0.90	0.98	0.99
Precision-1	0.98	0.98	0.99	0.95	0.98	0.84	0.95	0.99
Recall-0	0.98	0.98	0.99	0.95	0.98	0.82	0.95	0.99
Recall-1	0.98	0.99	0.98	0.93	0.1	0.91	0.99	0.99
F1-score-0	0.98	0.98	0.99	0.94	0.99	0.86	0.97	0.99
F1-score-1	0.98	0.98	0.99	0.94	0.99	0.87	0.97	0.99

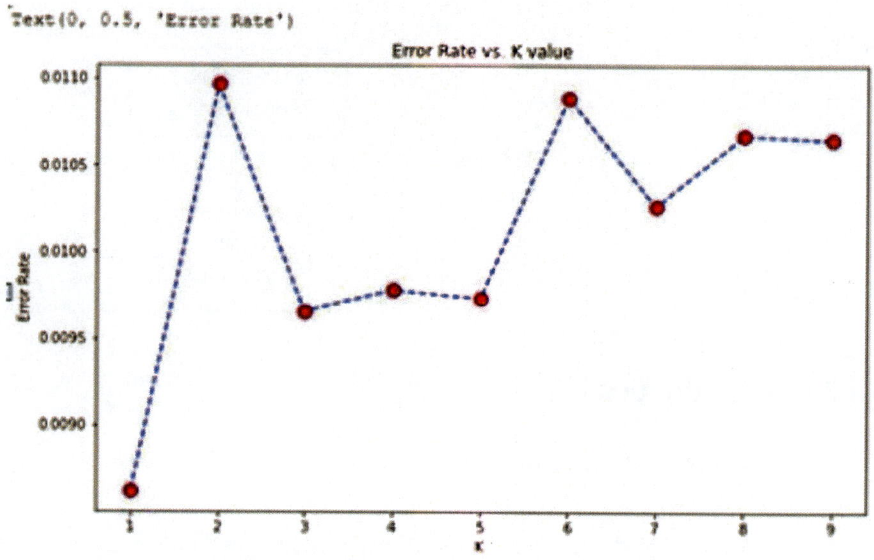

Fig. 2 Error rate for different *k* values

lowest number of accuracy. Confusion matrix for each model is shown in Fig. 3. The decision tree performed poorly, as shown in Table 1. It has an 86% accuracy rate and produces over 3000 incorrect results. When a dataset contains a feature that is critical in making a decision, DT performs well. The low accuracy of the DT model is also due to high variance. On the other hand, random forest has gained accuracy

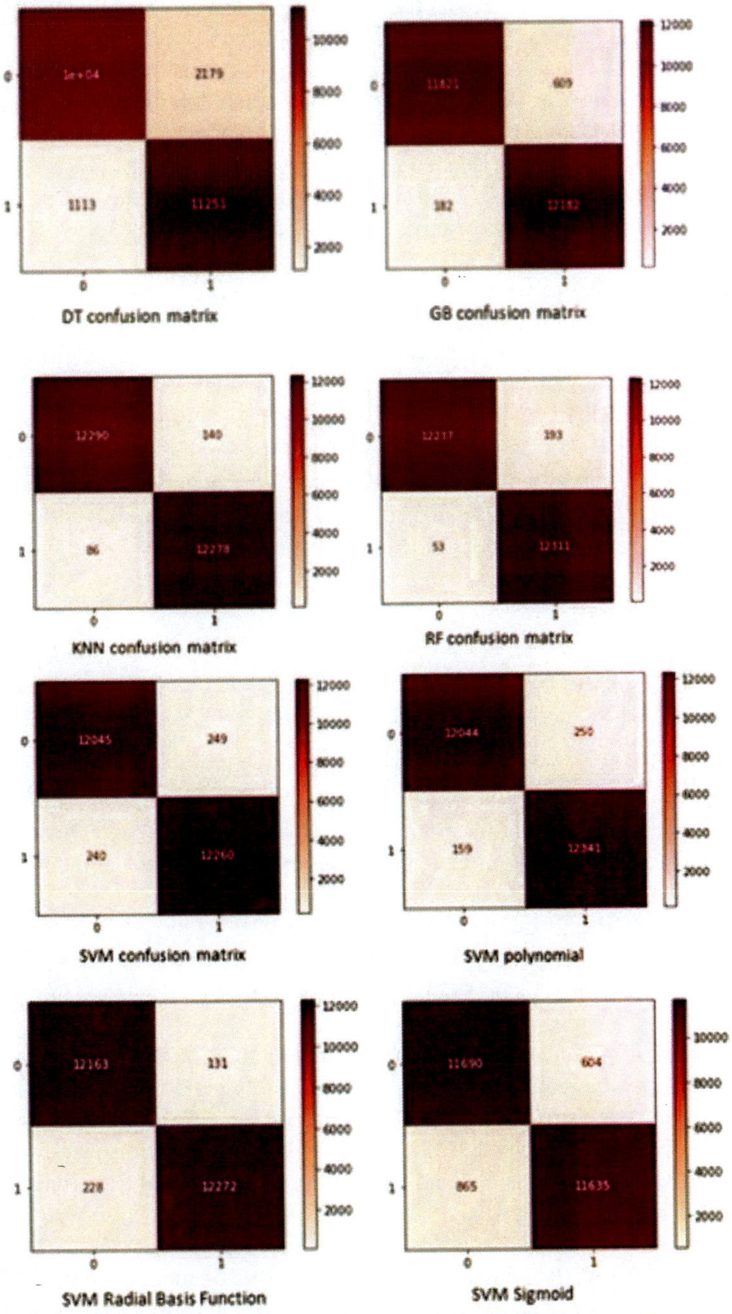

Fig. 3 Confusion matrixes for classification

of 99% and gives inaccurate results for less than 300 test samples. In reality, random forest reduces high variance by enameling multiple decision trees and selecting the class with the most votes. As a result, we get a lot of precision from it.

KNN model also gives 99% accuracy with less than 300 inaccurate test samples. Figure 2 shows that the error rate is lowest between 1 and 10 when the k value is 1. As the k value error rate rises, so does the overall error rate. KNN model tests samples according to the classification of its neighbor. When the dimension is less than 100 and there are many instances, KNN works well. It beats SVM; SVM is an upgraded version of KNN.

Multiple SVM kernels are used to train the dataset. Because the radial basis function kernel works best among them, classes are nonlinearly separable as well. In different kernels, different mathematical formulas are used to create hyperplanes between classes. SVM works best when there is a lot of sparsity among variables and a lot of instances. There is a lot of sparsity among the classes in the dataset. At last, gradient boosting is also an ensemble technique. It combines a modified version of decision trees sequentially as a base learner, or it follows boosting method. When n-estimator is 1 only one, DT is used to train the dataset. So, in this situation, this cannot handle the high variance problem of DT very well and resultant lower accuracy than random forest.

4.3 Results of Method-2

We used PCA to analyze the dataset, which had three components with values of 2, 10, and 15. Figure 4 depicts the two-components or two-dimensional feature space. Values are related to one another in a positive way. After that cross-validation is used to apply some classification models and ensemble methods to the dataset. Table 2 summarizes the results.

In PCA-2, SVM and gradient boosting have 92% accuracy; decision tree and KNN-1 acquired 96% accuracy, and random forest gained the best accuracy among all that is 97%. But, models on PCA-2 performed less well than models on PCA-10 and PCA-15.

Dimensionality reduction is beneficial. Intriguingly, The decision tree classification paradigm, which reduces dimensionality. DT obtained 96% accuracy on PCA-2 components and 99% accuracy on PCA-15 components. In feature space in method-1, DT had only 86% accuracy. SVM and gradient boosting perform well when there exist billion or million instances and hundred or thousand features. Random forest is an ensemble technique that found its best fit over here. KNN also well performed here for less dimensionality but more instances.

PCA-10 and PCA-15 are nearly identical in terms of model accuracy. Training models on PCA-15 have taken longer time then training model on PCA-10. As a result, we can use PCA-10 instead of PCA-15. Among others, random forest and KNN-1 provide the best accuracy.

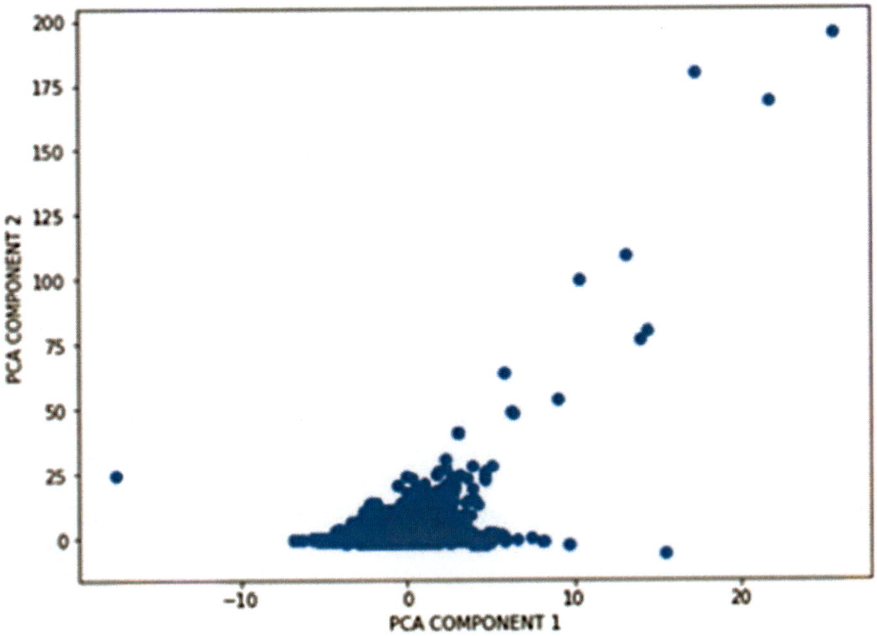

Fig. 4 PCA for two components

Table 2 Summary of results of supervised classification models for method-2

Performance measure	SVM rbf (%)	Random forest (%)	Decision tree (%)	Gradient boosting (%)	KNN (%)
PCA-2	92	97	96	92	96
PCA-10	98	99	98	97	99
PCA-15	98	99	99	97	99

At last, we have applied hyperparameter tuning on the whole dataset of 138,047 instances with 47 features for decision tree. Selected parameters are following Table 3.

Table 3 Selected hyperparameter

Parameters-type	Hyperparameter
Criterion	Gini, entropy
Max_depth	1,2,3,4,5,6,7

5 Conclusion

Data security or cybersecurity is the buzzword of the twenty-first century. The attributes of malware, spam, or other intruders change every microsecond to be more capable of not getting captured by cyber-defenders. The traditional methods would fail for these types of malware detection. The technique that can observe the pattern of behavior changes of intruders will be suitable for the detection of intruders. Machine learning can keep a great impact on the malware detection system. Machine learning provides the techniques which can take decisions by observing patterns. Almost, 99% accuracy is possible through the techniques. Comparative analysis helps us to determine better algorithmic techniques that ensure the reliability of detection. In this paper, we discuss two methodologies to analyze how the algorithms can relate them with the features. Our discussion is limited to supervised learning techniques. We did not acknowledge the performance of unsupervised learning techniques on these types of detection in the paper. Hyperparameter tuning is not performed for all the techniques. Improvement of machine learning techniques means more accurate observation and more accurate decision-making. Now, the time comes when AI should meet to cybersecurity.

References

1. Kumar A, Abhishek K, Shah K, Patel D, Jain Y, Chheda H, Nerurkar P (2020) Malware detection using machine learning. In: Iberoamerican knowledge graphs and semantic web conference. Springer, pp 61–71
2. Ham HS, Choi MJ (2013) Analysis of android malware detection performance using machine learning classifiers. In: 2013 international conference on ICT convergence (ICTC). IEEE, pp 490–495
3. Meng G, Xue Y, Mahinthan C, Narayanan A, Liu Y, Zhang J, Chen T (2016) Mystique: evolving android malware for auditing anti-malware tools. In: Proceedings of the 11th ACM on Asia conference on computer and communications security, pp 365–376
4. Sathyanarayan V, Kohli P, Bezawada B (2008) Signature generation and detection of malware families, vol 5107, pp 336–349
5. Chakraborty P, Yousuf MA, Rahman S (2021) Predicting level of visual focus of human's attention using machine learning approaches. In: Kaiser MS, Bandyopadhyay A, Mahmud M, Ray K (eds) Proceedings of International Conference on Trends in Computational and Cognitive Engineering. Adv Intell Syst Comput 1309. Springer, Singapore. https://doi.org/10.1007/978-981-33-4673-4_56
6. Sarker A, Chakraborty P, Sha SS, Khatun M, Hasan MR, Banerjee K (2020) Improvised technique for analysing data and detecting terrorist attack using machine learning approach based on twitter data. J Computer and Commun 8(7):50–62
7. Chakraborty P, Sultana S (2022) IoT-based smart home security and automation system. In Micro-Electronics and Telecommunication Engineering, (pp. 497–505).
8. Singhal P (2012) Malware detection module using machine learning algorithms to assist in centralized security in enterprise networks. Int J Netw Secur Appl 4(1):61–67. https://doi.org/10.5121/ijnsa.2012.4106

9. Sethi K, Chaudhary SK, Tripathy BK, Bera P (2018) A novel malware analysis framework for malware detection and classification using machine learning approach. In: Proceedings of the 19th international conference on distributed computing and networking, pp 1–4
10. Khammas B (2018) Malware detection using sub-signatures and machine learning technique. J Inform Secur Res 9(3):96–106
11. Ucci D, Aniello L, Baldoni R (2019) Survey of machine learning techniques for malware analysis. Comput Secur 81:123–147. https://doi.org/10.1016/j.cose.2018.11.001
12. Ijaz M, Durad MH, Ismail M (2019) Static and dynamic malware analysis using machine learning. In: 2019 16th international Bhurban conference on applied sciences and technology (IB-CAST), pp 687–691
13. Yang S, Li S, Chen W, Liu Y (2020) A real-time and adaptive-learning malware detection method based on api-pair graph. IEEE Access 8:208120–208135
14. Saleh A, Francis C (2019) A deep learning approach to malware detection in android platform. Int J Innov Technol Explor Eng 8(8):1043–1048
15. Arp D, Spreitzenbarth M, Hubner M, Gascon H, Rieck K, Siemens C (2014) Drebin: effective and explainable detection of android malware in your pocket. In: Ndss 14:23–26
16. Pe format—win32 apps—microsoft docs. https://docs.microsoft.com/en-us/windows/win32/debug/pe-format. Accessed on 08/29/2021
17. Wadkar M, Di Troia F, Stamp M (2020) Detecting malware evolution using support vector machines. Expert Syst Appl 143:113022
18. 4. support vector machines—scikit-learn 1.0.1 documentation. https://scikit-learn.org/stable/modules/svm.html. Accessed on 11/07/2021
19. Garcia FCC, Muga II FP (2016) Random forest for malware classification. arXiv preprint arXiv:1609.07770
20. 10. decision trees—scikit-learn 1.0.1 documentation. https://scikit-learn.org/stable/modules/tree.html. Accessed on 11/07/2021

Author Index

A
Abdul Wassay, Md, 481
Abhilash, S., 265
Abu Yousuf, Mohammad, 721
Adhikary, Sourav, 145
Agarwal, Ritu, 721
Agrawal, Arun Prakash, 587
Ahmed, Mazida A., 77
Akhand, Md. Nafis Tahmid, 363
Akter, Laboni, 513
Akter, Lubana, 639
Akter, Mshura, 639
Akter, Tamanna, 905
Alam, Kazi Saeed, 157
Alqahtani, Hamed, 609
AlSaud, Yousef, 481
Althar, Raghavendra Rao, 427
Anand, Abhineet, 481
Annapoorani, A., 265
Anurag, Kumar, 733
Anwar, Md. Musfique, 145
Arefin, Minhazul, 341
Atrey, Aryaman, 663

B
Balasubramaniam, Anandkumar, 831
Balasubramaniam, Thirunavukarasu, 831
Banerjee, Sheekar, 707
Banik, Nayan, 561, 905
Baral, Sukanta Kumar, 893
Bashir, Rumaan, 117
Bhandari, Neetika, 211
Bhardwaj, Prajwal, 621, 663
Bharti, Gaurav, 303
Bhowmick Das, Avijoy, 169
Bhushan, Bharat, 169
Bhuyan, Manash P., 77
Bisht, Diksha, 1
Biswas, Al Amin, 573
Biswas, Atri, 749
Biswas, Dipto, 11
Boruah, Parvez A., 77

C
Chadha, Harshita, 389
Chakraborty, Partha, 11, 49, 561, 573, 595, 721, 905
Chakraborty, Shourjya, 749
Chandra Das, Badhan, 573
Chandra Das, Shuvo, 595
Chandra Rath, Ramesh, 527
Chatterjee, Rajdeep, 749
Choudhary, Ankur, 25, 587, 733
Choudhury, Ridip D., 77
Choudhury, Tanupriya, 331, 469, 679, 721, 819, 853, 865, 879
Chowdhury, Al Akram, 125
Chowdhury, Mohammad Jabed Morshed, 145

D
Dalai, Asish Kumar, 377
Dangi, Sarishma, 1
Das, Sunanda, 363
Deb, Debarshi, 749
Deepa, J., 177
Dehraj, Pooja, 277

V. Skala et al. (eds.), *Machine Intelligence and Data Science Applications*, Lecture Notes on Data Engineering and Communications Technologies 132,
https://doi.org/10.1007/978-981-19-2347-0

Deka, Vaskar, 77
Dhandapani, Arunkumar, 831
Dharamsi, Sneha, 105
Dinesh, P., 265
Divya, V., 177
Dutta, Abhisek, 775

F
Fahad Hossain, Shah, 649
Fore, Vivudh, 621, 789

G
Gaurang Tiwari, Raj, 419
Giri, Kaiser J., 117
Goel, Richa, 527, 893
Gokula Krishnan, V., 177
Gupta, Vedansh, 819
Gupta, Vittesha, 399

H
Hasan, Mahmudul, 363
Hasan, Md. Jahid, 157
Hashim, Md., 277
Hoque, Mohammed Moshiul, 63
Hossen, Rakib, 341

I
Ishtiaque Ahmed, Nafiz, 35
Islam, Afroza, 721
Islam, Md. Ezharul, 707
Islam, Md. Tariqul, 695
Islam, Md. Toriqul, 227
Islam, Muhammad Nazrul, 63

J
Jahan, Hasnur, 649
Jain, Atishay, 355, 551
Jain, Ayush, 105
Jaiswal, Arunima, 399, 409
Jamwal, Shubhnandan S., 843
Janarthanan, R., 265
Jha, Girija, 239
Jhumur, Aminun Nahar, 707
Jindal, Vishesh, 313, 539
Jubier Ali, Md., 573

K
Kanti Podder, Suplab, 293
Karim, Ridoan, 227
Kaushik, Sunil, 679
Khosla, Dishant, 303
Khullar, Vikas, 419
Kibria, Kazi Atai, 125
Kovid, Raj K., 775
Kowshir Bitto, Abu, 649
Kumar Agarwal, Ambuj, 419
Kumar Baral, Sukanta, 527
Kumar Dey, Samrat, 639
Kumar Varun, Nikhil, 853
Kumar, Lalan, 819, 853
Kumar, Sandeep, 1, 551, 663

L
Lakshmi, Nalam, 789
Lenka, Arundhati, 195
Lohani, Kaustubh, 621, 663
Lone, Nawaz Ali, 117

M
Mahbubur Rahman, Md., 639
Maheshwari, Shikha, 481
Mahmud, Imran, 649
Malhi, Ashok S., 775
Mangalampalli, Sudheer, 457
Mangalampalli, Vamsi Krishna, 457
Mansurali, A., 879
Marfatia, Nimesh, 105
Mary Jayanthi, P., 879
Masuma, Fozilatunnesa, 561
Mehedi Hassan, Md., 513
Miah, Abu Saleh Musa, 125
Mishra, Anju, 819, 853
Misra, Alok, 419
Mithila, Sabia Khatun, 513
Mofizul Alam Mozumder, Md., 49
Mohanty, Sachi Nandan, 865
Moni Saha, Rakhi, 649
Moni, Mohammad Ali, 63

N
Nahid Hasan, S. K., 157
Naik, Archana, 95
Narayan Singh, Shailendra, 313, 539
Nasir Uddin, Mohammed, 341
Nasrin, Fatema, 35
Nayeem, Md., 805
Nor, Rozi Nor Haizan, 695

P
Pahwa, Payal, 211
Pandey, Arvind K., 819, 853
Panigrahy, Saroj Kumar, 377
Patgaonkar, Samarth, 105
Pathak, Nitish, 389
Paul, Anand, 831
Pavana, B. S., 95
Pervin, Shahanaz, 125
Pokkuluri, Kiran Sree, 457
Potluri, Sirisha, 865
Prakash Agrawal, Arun, 25, 733
Prakash, Chandra, 587

R
Raghavendra, C. K., 441
Rahman, Shagoto, 513
Raihan, M., 513
Raihan, Md. Mohsin Sarker, 513
Raj, Gaurav, 25, 587, 733
Rajendiran, P., 501
Rajput, Geetendra, 303
Raju, Pothuraju, 457
Ramnath, Maddula, 25
Rao, Katta Subba, 865
Rashid, Md. Mamunur, 125
Rath, Ramesh Chandra, 893
Ripon, Shamim, 761, 805
Rohini, A., 331, 469

S
Sabu, Kevin, 25
Saha, Chayti, 561
Saha, Dola, 639
Saha, Sourav, 595
Saha, Suman, 573
Saif Hasan, Md., 49
Saleh, Rakib Ahmed, 695
Samanta, Debabrata, 293, 427
Samantaray, Milan, 195
Samsuddoha, Md., 11
Sangeetha, J., 501
Sarkar, Atiquer Rahman, 125
Sarker, Aditi, 595
Sarker, Iqbal H., 63, 145, 609
Satapathy, Suneeta, 195
Sathyanaryanan, Vasanth, 789
Sen, Vijay Singh, 843
Seshamalini, M., 441
Sethi, Parminder Singh, 551, 789
Shahriar, Khandaker Tayef, 63, 609
Shahriar, Md. Ragib, 125
Shalem Raju, P. J. R., 457
Sharma, Ayush, 303, 587
Sharma, Manvinder, 303
Sharma, Neelam, 389
Shibly, Mir Moynuddin Ahmed, 761
Shin, Jungpil, 125
Shohel Arman, Md., 649
Shohug, Md. Mahfuj Hasan, 649
Sibi, Akhil, 733
Sijariya, Rajeev, 775
Sikdar, Ayan, 169
Sil, Riya, 169
Singh, Navtej, 621
Singh, Rahul, 733
Singh, Sarwan, 303
Singh, Tilottama, 893
Soni, Mahendra, 587
Sooda, Kavitha, 95
Srivastava, Sandeep, 819, 853
Sudalaimuthu, T., 331, 469
Sukthankar Sitaram, V., 527
Suvra Khan, Soumya, 313, 539
Swamynathan, R., 879
Swarnamalya, M., 441

T
Talukder, Kamrul Hasan, 513
Tani, Tanzina Akter, 761
Tasmim, Kaniz, 905
Tasnim, Nowshin, 609
Tatrapi, Rifah, 805
Thakur, Nikita, 409
Thomas, Benny, 293
Tomar, Ravi, 551, 621, 663, 789
Trivedi, Naresh Kumar, 481

U
Usha Devi N, S. S. S. N., 457

V
Vardhan Singh Rawat, Harsh, 1
Venkatesan, Hariraj, 501
Venkateswara Rao, Pinagadi, 177
Visalaxi, S., 331, 469
Vyas, Sonali, 227

X
Xavier, Sharon, 377

Y
Yadav, Akansha, 239

Z
Zaman, Shishir, 805